Relationship

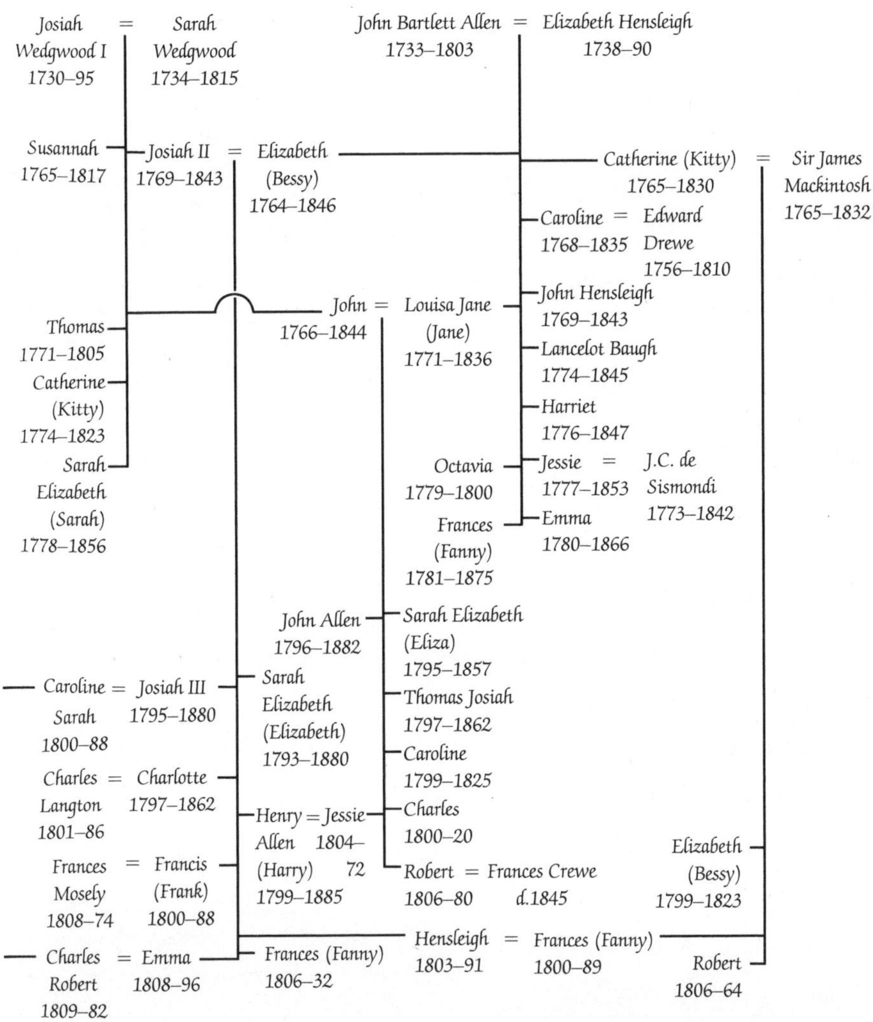

THE CORRESPONDENCE OF

CHARLES DARWIN

Editors

FREDERICK BURKHARDT† JAMES A. SECORD

SHEILA ANN DEAN SAMANTHA EVANS

SHELLEY INNES ALISON M. PEARN

PAUL WHITE

Associate Editors

ANNE SCHLABACH BURKHARDT

ROSEMARY CLARKSON

RUTH GOLDSTONE

SAM KUPER

MURIEL PALMER

ELIZABETH SMITH

ELLIS WEINBERGER

This edition of the Correspondence of Charles Darwin is sponsored by the American Council of Learned Societies. Its preparation is made possible by the co-operation of Cambridge University Library and the American Philosophical Society.

Support for editing has been received from the Alfred P. Sloan Foundation, the Andrew W. Mellon Foundation, the John Templeton Foundation, the National Endowment for the Humanities, the National Science Foundation, the Pew Charitable Trusts, the British Academy, the British Ecological Society, the Isaac Newton Trust, the Natural Environment Research Council, the Royal Society of London, the Stifterverband für die Deutsche Wissenschaft, and the Wellcome Trust. The National Endowment for the Humanities funding of the work was under grants nos. RE-23166-75-513, RE-27067-77-1359, RE-00082-80-1628, RE-20166-82, RE-20480-85, RE-20764-89, RE-20913-91, RE-21097-93, RE-21282-95, RZ-20018-97, RE-20393-99, and RZ-20849-02; the National Science Foundation funding of the work was under grants Nos. SOC-75-15840, SOC-76-82775, SES-7912492, SES-8517189, SBR-9020874, SBR-9616619, SES-0135528, and SES-0646230. Any opinions, findings, conclusions, or recommendations expressed in this publication are those of the editors and do not necessarily reflect the views of the grantors.

Oil portrait of Charles Darwin by Laura Russell, 1869.
By kind permission of the owner.

THE CORRESPONDENCE OF
CHARLES DARWIN

VOLUME 17 1869

CAMBRIDGE
UNIVERSITY PRESS

CAMBRIDGE UNIVERSITY PRESS
Cambridge, New York, Melbourne, Madrid, Cape Town, Singapore, São Paulo, Delhi

Cambridge University Press
The Edinburgh Building, Cambridge CB2 8RU, UK

Published in the United States of America by Cambridge University Press, New York

www.cambridge.org
Information on this title: www.cambridge.org/9780521190305

First published 2009

Citation:
Burkhardt, Frederick *et al.*, eds. 2009. *The correspondence of Charles Darwin.*
Vol 17. Cambridge: Cambridge University Press

Printed in the United Kingdom at the University Press, Cambridge

A catalogue record for this publication is available from the British Library

ISBN 978-0-521-19030-5 hardback

In Memoriam

WILLIAM T. GOLDEN

1909–2007

The Darwin Correspondence Project gratefully acknowledges the essential support of the following donors:

Patrons

Golden Family Foundation
James and Hilary Potter

Affiliates

Bern Dibner†
William T. Golden†
Kathleen Smith†

Friends

Jane Burkhardt
Pamela Davis
Florence Fearrington and James Needham
Gerald and Sue Friedman
John C. Greene
Daniel V. Grossman and Elizabeth Scott Andrews
Lawrence K. Grossman
Shirley Grossman, M. D.
Mary S. Hopkins
Robert McNeil
Michael Mathews
Victor Niederhoffer
Wendy L. Thompson
Daniel J. Wright

CONTENTS

ILLUSTRATIONS

CALENDAR LIST OF LETTERS

The following list is in the order of the entries in the *Calendar of the correspondence of Charles Darwin*. It includes all those letters that are listed in the *Calendar* for the year 1869, and those that have been redated into 1869. Alongside the *Calendar* numbers are the corrected dates of each letter. A date or comment printed in italic type indicates that the letter has been omitted from this volume.

Letters acquired after the publication of the first edition of the *Calendar*, in 1985, have been given numbers corresponding to the chronological ordering of the original *Calendar* listing with the addition of an alphabetical marker. Many of these letters are summarised in a 'Supplement' to a new edition of the *Calendar* (Cambridge University Press, 1994). The marker 'f' denotes letters acquired after the second edition of the *Calendar* went to press in 1994.

3426a. 31 Jan [1869]
4372. *[1865–77]*
5482. 3 Apr [1869]
5729. 14 [Jan] 1869
5753. *Cancel: part of 6680.*
5843. 6 Feb [1869]
5974. [3 Nov 1869]
6098a. 6 Apr [1869–71]
6218. *Cancel: part of 6657.*
6260. 28 June [1869]
6442. [before 20 Oct 1869]
6513. *26 Dec [1866]. To be published in next supplement.*
6519. [after 5 Aug 1869]
6520. *[before 10 Apr 1856]. To be published in next supplement.*
6521. [after 21 Apr 1869]
6522. [after 22 Feb 1869]
6523. *[1871]*
6524. *Date unknown. To be published in a future supplement.*
6525. *Cancel: draft of 5418f.*
6526. *[after June 1868?]*
6527. *[after 11 May 1868?]*
6528. *Cancel: part of 6421.*
6529. *[after 5 Oct 1868]*
6530. *[after 16 Oct 1868]*
6531. *[c. 23 Mar 1868]*
6532. *[after July 1868]*
6533. [1869]
6534. *[c. 10 Mar 1868]*
6535. [after 14 Oct 1869]
6536. *Cancel: part of 6488.*
6537. *[before 18 May 1868]*
6538. *[1856–61?] To be published in next supplement.*
6539. 26 [Nov 1869 or later]
6540. [before 21 Mar 1869]
6541. [Jan–Mar 1869]
6541a. [Jan 1869]
6542. 4 Jan 1869
6543. 4 Jan 1869
6544. 5 Jan 1869
6545. 5 Jan 1869
6546. *6 Jan [1868]*
6546f. [6–9? Jan 1869]
6547. 9 Jan 1869
6548. 11 Jan 1869
6549. 12 Jan 1869
6550. 13 Jan 1869
6551. 13 Jan 1869
6552. 13 Jan 1869
6553. 14 Jan 1869
6554. 15 Jan 1869
6555. 15 Jan 1869
6556. 15 Jan 1869
6557. 16 Jan [1869]
6558. 17 Jan 1869
6559. 17 Jan [1869]
6560. 18 Jan 1869
6561. 20 Jan 1869
6562. [before 21 Jan 1869]
6563. 21 Jan [1869]
6564. 21 Jan 1869
6565. 21 Jan [1869]
6566. 21 Jan 1869
6567. 22 Jan [1869]
6568. [22 Jan 1869]
6569. 22 Jan 1869
6570. 23 Jan [1869]
6571. 23 Jan 1869
6572. [before 23 Jan 1869]

INTRODUCTION

At the start of 1869, Darwin was hard at work making changes and additions for a fifth edition of *On the origin of species* (*Origin*). By the close of the previous year, he had taken stock of the criticisms that had surfaced since the fourth edition appeared at the end of 1866 and had told his cousin William Darwin Fox, 'My work will have to stop a bit for I must prepare a new edit. of that everlasting origin, & I am sick of correcting' (*Correspondence* vol. 16, letter to W. D. Fox, 12 December [1868]). He may have resented the interruption to his work on sexual selection and human evolution, but he spent forty-six days on the task (CD's 'Journal' (Appendix II)). As he remarked to his best friend, the botanist Joseph Dalton Hooker, 'If I lived 20 more years, & was able to work, how I sh^{d}. have to modify the "Origin", & how much the views on all points will have to be modified.— Well it is a beginning, & that is something' (letter to J. D. Hooker, [22 January 1869]). Much of the remainder of the year was spent researching and revising chapters for *The descent of man, and selection in relation to sex* (*Descent*), and gathering additional material on emotional expression. Yet the scope of Darwin's interests remained extremely broad, and many letters throughout the year touched on subjects such as South American geology, barnacle morphology, insectivorous plants, and earthworms, subjects that had exercised Darwin for decades, and that would continue to occupy him for many years to come.

Darwin's most substantial addition to *Origin* was a response to a critique of natural selection by Carl Wilhelm von Nägeli, a Swiss botanist and professor at Munich (Nägeli 1865). Darwin had considered Nägeli's paper important enough to have a translation made for his personal use. While not entirely dismissing natural selection, Nägeli had assigned it an auxilliary role, second to his own theory of perfectibility ('Vervollkommnung'), which he posited as the principal engine of change in the development of species. Darwin correctly assessed Nägeli's theory as a major challenge requiring a thorough and thoughtful response; indeed, over the next twenty years, Nägeli's critique inspired many to reassess their support for natural selection (see Cittadino 1990, pp. 122–3, 128–30). Nägeli had argued that natural selection could only operate on 'functional' characteristics, while his perfectibility principle could account for changes in most morphological features (Nägeli 1865, p. 29). Darwin sent a manuscript of his response (now missing) to Hooker, remarking: 'I should be extremely obliged if you would read it over, and see whether I have made any blunders, as is very likely to be the case' (letter to J. D. Hooker, 13 January 1869). Hooker went straight to a crucial point: 'I do not quite like the starting by shirking the question of what is a "Morphological character" … all characters i.e. all departures from a given structure are & must be morphological'. The comment highlights Darwin's apparent confusion about Nägeli's separation of features into functional and 'purely morphological'. The modern reader may well share Darwin's uncertainty, but Nägeli evidently believed the distinction was clear-cut. For example, he argued that while the function of leaves could be modified by natural selection, their arrangement, which he claimed had no value in terms of 'usefulness', could only be altered by his perfectibility principle (Nägeli 1865, pp. 28–9). In further letters, Hooker tried to provide Darwin with botanical examples he could use to challenge Nägeli; Darwin made full use of Hooker's assistance and many of his examples feature in the seven pages of the new edition that countered Nägeli's thesis (*Origin* 5th ed., pp. 151–7).

Another important criticism that Darwin sought to address in the fifth edition was that of the engineer Henry Charles Fleeming Jenkin. Darwin had been very impressed by Jenkin's 1867 review, which argued that any variation in an individual, however favourable, would not be preserved within a breeding population. Such variations, according to the prevailing theory of blending inheritance that Jenkin and Darwin both shared, would tend to be lost in intercrossing. Darwin tried to resolve this problem by emphasising variability within the breeding population as a whole; if a sufficient number of individuals possessed the favourable trait, then its perpetuation would be assured (*Origin* 5th ed., pp. 103–4). The terminology that Darwin and others employed in these matters ('individual differences', 'single variations') was sometimes ambiguous, however, and gave rise to misunderstandings. 'I must have expressed myself atrociously', Darwin wrote to Alfred Russel Wallace on 2 February, 'I meant to say exactly the reverse of what you have understood … I always thought individual differences [i.e., differences between individuals] more important, but I was blind & thought that single variations [i.e., variations that occurred in a single individual] might be preserved much oftener than I now see is possible or probable' (see also letter to A. R. Wallace, 22 January [1869], and letter from A. R. Wallace, 30 January 1869).

A further modification to the new edition of *Origin* was the result of correspondence between Darwin and the geologist James Croll. In the previous year, Croll had discussed topics such as the age of the earth and evidence of glaciation in North America, but it was his theory of alternate ice ages that piqued Darwin's interest the most. He wrote, 'this conclusion, which you have arrived at from physical considerations, explains so well whole classes of facts in distribution, that I must joyfully accept it; indeed I go so far as to think that your conclusion is strengthened by the facts in distribution' (letter to James Croll, 31 January [1869]). Darwin had argued (*Origin*, pp. 377–8) that plant species would migrate towards the equator during an ice age and that temperate species would survive at higher altitudes while tropical ones would inhabit lower levels. He had been forced to retrench that position following criticism from his friend Hooker, by admitting that the survival of tropical species was a difficulty for his theory (*Origin* 4th ed., pp. 450–1). Croll's theory, simply stated, proposed that ice ages alternated between hemispheres, so that a warmer, non-glaciated hemisphere where tropical species could survive would always exist.

In *Origin* 5th ed., pp. 450–61, Darwin accounted for the survival of tropical species using Croll's theory.

In the same letter to Croll, Darwin had expressed another worry: 'I am greatly troubled at the short duration of the world according to Sir W. Thompson, for I require for my theoretical views a very long period *before* the Cambrian formation' (letter to James Croll, 31 January [1869]). Croll could not supply Darwin with an estimate of the age of the earth much greater than that calculated by William Thomson, but he did point out, 'As regards determining the age of the earth's crust from the secular cooling of the globe I am not altogether satisfied with the plan. It would no doubt do if we had proper data to go by, but don't think we have got that yet' (letter from James Croll, 4 February 1869). Darwin did not directly challenge Thomson's estimate, but he added more on the process of subaerial denudation based on recent work of Croll, Andrew Crombie Ramsay, William Whitaker, and others (*Origin* 5th ed., pp. 352–4). Later in the year he was pleased to see Thomson's work challenged by both Thomas Henry Huxley and Wallace. He confided to Huxley, 'I find that the few sentences which I have sent to press in the Origin about the age of the world will do fairly well, though if I had read you first, perhaps I sh^{d} have been less deferential towards [Thomson]' (letter to T. H. Huxley, 19 March [1869]).

Once Darwin had completed revisions of the 'everlasting old Origin' (letter to Asa Gray, 1 June [1869]), he was able to return to work on *Descent.* He continued to receive material on sexual selection in various species from Britain and overseas. The dog-breeder George Cupples worked hard on Darwin's behalf, sending a steady stream of information on the proportion of the sexes in sheep, cattle, horses, and dogs, and circulating Darwin's queries to various contacts. As one Scottish sheep-farmer wrote: 'We all know very well who Mr Darwin is, although some of us do not think ourselves bound to agree with him in all his theories, if I could assist him however in his researches I would willingly do so' (letter from Robert Elliot to George Cupples, 21 June 1869). Details on mating behaviour and sexual characteristics continued to flood in, occasionally causing difficulties. The entomologist Frederick Smith, whom Darwin had asked to study the musical activities of insects, reported that one male field cricket had annoyed some London neighbours with his noisy courting of the female in the garden (letter from Frederick Smith, 8 October 1869). Albert Günther, assistant in the zoology department at the British Museum in London, supplied extensive data on differences between male and female fish, reptiles, and amphibians, while Roland Trimen in South Africa and John Jenner Weir in London sent more information on male and female butterflies, supplementing that received the previous year. Darwin also continued to receive assistance from his long-time correspondent, the pigeon and poultry fancier William Bernhard Tegetmeier, who sent him samples of juvenile plumage in chickens, as well as live chickens and eggs of various breeds. Darwin's cook evidently agreed to supervise the hatching of the eggs, though she doubted her ability to recognise the different varieties (letter to W. B. Tegetmeier, 25 February [1869]). The data contined to accumulate, and towards the end of the year Darwin complained: 'I am weary of everlasting males &

females, cocks & hens.—' (letter to J. D. Hooker, 13 November [1869]). Yet completion of the work was in sight. He was now polishing his chapters on sexual selection, and beginning to choose the artists to make engravings for *Descent*.

In 1869, Darwin still expected that *Descent* would include his extensive research on the expression of emotions in humans and other animals. He received more observations on infants, and on indigenous peoples in India and Africa. The American botanist Asa Gray and his wife, Jane Loring Gray, who spent part of the year in England and visited the Darwins in August, travelled to Egypt, and Jane furnished Darwin with several answers to his questionnaire: 'Passing slowly a common country cargo boat, the old man on board stood looking at us, with brow wrinkled & mouth compressed & upper lip raised— An expression, as I read it, of dislike & contempt— almost hatred—' (from Asa Gray and J. L. Gray, 8 and 9 May [1869]).

In addition to infants and non-Europeans, a group that particularly interested Darwin in his research were persons diagnosed as insane, for he believed that they expressed passions without reserve. Darwin initially contacted the physician Henry Maudsley, who had worked for some years in medical asylums. Maudsley forwarded Darwin's queries to James Crichton-Browne, the director of the West Riding Pauper Lunatic Asylum in Wakefield. Crichton-Browne sent a lengthy reply, adding: 'Should the sort of information which I have sent prove of any service to M^r. Darwin I can supply him with much more of the same description' (enclosure to letter from Henry Maudsley, 20 May 1869). Darwin had often complained of the difficulty of studying expression, and was always keen to find astute observers. Darwin wrote enthusiastically to Crichton-Browne on 22 May: 'I do not know how to thank you enough for your MS. observations on expression. They contain exactly and fully the information which I wanted; and besides being of the greatest use to me, are most interesting and so graphic as to be almost painful'. Thus began what would become Darwin's most detailed and lengthy correspondence on the subject of expression. Crichton-Browne had ambitions to make the asylum at Wakefield a centre of scientific research, and toward this end had begun to make pioneering use of photography to record the conditions of his patients, and to aid in diagnosis. Darwin was already familiar with the work of the French physiologist Guillaume Benjamin Amand Duchenne, who had used photography to capture the expression of persons whose facial muscles had been stimulated by an electrical current. Darwin sent Crichton-Browne his copy of Duchenne's photographic album, and began to receive, in turn, a large collection of photographs of the Wakefield patients, together with lengthy descriptions. One photograph of a woman with erect hair formed the basis of a woodcut that Darwin would later use in *Expression of the emotions in man and animals* (*Expression*), published in 1872, more than a year after *Descent*.

But even as Darwin's research on expression and sexual selection added strength to his views on the continuity between humans and the animal world, others were growing less certain. On 24 March, Wallace wrote to Darwin about a forthcoming article in the *Quarterly Review*: 'I venture for the *first time* on some limitations to the power of natural selection. [These] are the expression of a deep conviction founded on evidence which I have not alluded to in the article but which is to me absolutely unassailable'. In the article, Wallace claimed that certain human structures and higher capacities – a large brain, the delicate movements of the hand, sophisticated powers of language – could not have evolved through natural selection, because they conferred no advantage in the lower stages of human development. Savages, he asserted, had no need for them and made no use of them, and yet they were responsible for all of the higher achievements of human culture and civilization. Such features had only emerged, according to Wallace, through the agency of 'a Power which has guided the action of [natural] laws in definite directions and for special ends' ([Wallace] 1869b, pp. 393–4).

Darwin was astonished by Wallace's assertions: 'If you had not told me I sh^{d} have thought that they had been added by some one else. As you expected I differ grievously from you, & I am very sorry for it. I can see no necessity for calling in an additional & proximate cause in regard to Man' (letter to A. R. Wallace, 14 April 1869). More remarkable still were Wallace's reasons for believing in the providential origin and destiny of humans. Responding to Darwin's reply, Wallace wrote: 'My opinions on the subject have been modified solely by the consideration of a series of remarkable phenomena, physical & mental, which I have now had every opportunity of fully testing, & which demonstrate the existence of forces & influences not yet recognised by science. This will I know seem to you like some mental hallucination' (18 April 1869). Since his marriage to Annie Mitten in 1866, Wallace had become involved in the movement known as 'spiritualism', which had gained currency among the Victorian middle classes, and had attracted some interest from members of the scientific community as well. Wallace would later become an outspoken proponent of spiritualism, which he viewed as a wholly natural phenomenon, subject to scientific investigation and explanation. Despite their increasingly fundamental disagreements, Darwin and Wallace continued to display deep mutual respect and cordiality in their letters and publications. Darwin lavished praise on Wallace's long-awaited book, *The Malay Archipelago: the land of the orang-utan, and the bird of paradise* (Wallace 1869a; letter to A. R. Wallace, 22 March [1869]), and scolded him for again being too modest about his co-authorship of the theory of descent by natural selection: 'you are the only man I ever heard of who persistently does himself an injustice & never demands justice' (letter to A. R. Wallace, 14 April 1869).

While much of Darwin's attention in 1869 was focused on questions of human evolution and sexual selection, he continued to pursue his broad interests in geology, zoology, and botany. He revisited subjects that had occupied him early in his career, such as the geological structures of the South American cordillera (letter to Charles Lyell, 20 May 1869), and fossil discoveries in Patagonia and Wales (letter from T. H. Huxley, 7 May 1869, letter from W. B. Dawkins, 17 July 1869). He exchanged letters with Carl Friedrich Claus in Marburg, who was working on barnacle morphology and embryology and examining some of the same species that Darwin had investigated in depth (letter from C. F. Claus, 6 February 1869). In a letter to the *Gardeners' Chronicle*, he responded to the gardener and journalist David Taylor Fish in order to confirm the view expressed in his 1840 paper 'Formation of mould' of the astonishing role of earthworms in the formation of the soil (letter to *Gardeners' Chronicle*, 9 May [1869]).

In March, Darwin received specimens of the primitive and rare insectivorous plant *Drosophyllum lusitanicum* that had been painstakingly collected by William Chester Tait in Portugal. Darwin wanted to compare the insect-catching mechanism of *Drosophyllum* with that of *Drosera* (the sundew), a genus that he had studied in the early 1860s (letter to W. C. Tait, 12 and 16 March 1869). This research contributed to *Insectivorous plants*, published in 1875. Prompted by a request from a French student, Louis Rérolle, to translate *On the various contrivances by which British and foreign orchids are fertilised by insects* (*Orchids*), Darwin decided to add new material on the subject that he had been acquiring since its publication in 1862. Darwin asked his son William to examine the British orchid *Epipactis palustris* on the Isle of Wight in order to better ascertain its manner of pollination. William's contribution, and those of many others, were acknowledged in the separate publication that resulted from these researches, 'Notes on the fertilization of orchids'. Darwin was now easily able to observe many more orchids himself, as he had four greenhouses providing various degrees of temperature and humidity.

The year was marked by numerous requests for translations and foreign editions, showing both Darwin's growing international reputation and the continuing controversy over his work. The fifth edition of *Origin*, published in June, formed the basis for a new German edition (Bronn and Carus trans. 1870), prepared by Julius Victor Carus, who had worked on the previous German edition (Bronn and Carus trans. 1866), as well as on the German translation of *Variation* (Carus trans. 1868). The French translation proved more difficult, for Darwin learned that a third French edition of *Origin* had already been prepared without his consent and without regard for either the fourth or the fifth English editions (see letter from Victor Masson, 29 September 1869). The work had been undertaken, like the previous French editions, by Clémence Auguste Royer, and it had a new preface with Royer's own evolutionary views and critical commentary (Royer trans. 1870). Darwin complained to Hooker, 'Besides her enormously long & blasphemous preface to 1st Edit, she has added a 2^{d} Preface, abusing me like a pick-pocket for pangenesis, which of course has no relation to the Origin— Her motive being, I believe, because I did not employ [her] to translate "Domestic Animals"' (letter to J. D. Hooker, 19 November [1869]). Angered by these proceedings, Darwin arranged for another publisher, Charles Reinwald, and another translator, Jean Jacques Moulinié, to bring out a new French edition, incorporating his latest revisions (Moulinié trans. 1873). Reinwald and Moulinié had been engaged to produce the French edition of *Variation* (Moulinié trans. 1868), and CD now extended his permission for them to bring out his next book (*Descent*). As European and North American publishers sought to secure the rights to Darwin's books, both past and future, Darwin began to consider his own rights as an author in a period when international copyright laws were only just emerging. 'I am as yet quite ignorant on the subject,' he wrote to Moulinié on 23 October, 'yet it strikes me that I ought to receive some payment from the Publisher in Paris for the right of Translation; & I intend to make enquiries whether this is ever done.'

But the translation that was dearest to Darwin's heart was probably the one he commissioned and paid for himself: William Sweetland Dallas's edition of Fritz Müller's *Für Darwin* (Dallas trans. 1869). The book, an explication of Darwinian evolution through the natural history of crustaceans, had greatly impressed Darwin when it first appeared in 1864, and the ensuing correspondence with Müller was one of the most important of Darwin's entire career. When Darwin read the translation, he wrote to Müller in Brazil on 18 March: 'I can truly say that I look at the publication of y^r^ essay as one of the greatest honours ever conferred on me. . . . What an admirable illustration it affords of my whole doctrines! A man must indeed be a bigot in favour of separate acts of creation, if he is not completely staggered after reading y^r^ essay'. The work received a number of reviews, including one in the *Athenaeum* that quipped: 'It is to be feared that Dr. Müller's moral code is rather Crustacean than Christian'. Müller confessed to his brother Hermann that he thought this a rather good joke, although to Darwin he was more apologetic: 'I am sorry to learn … that some expressions of mine have shocked the religious feelings of one of the reviewers. I should indeed have suppressed or modified some passages, before offering my book to English readers.—' (letter from Fritz Müller, 15 June 1869).

More religious controversy over Darwinism broke out at the annual meeting of the British Association for the Advancement of Science, held in Exeter in August 1869. The controversy was directed in part at Darwin's most outspoken British supporter, Thomas Huxley, whose address 'The physical basis of life', published earlier in the year, was attacked in a series of papers by Anglican clergymen in the biology section of the meeting. Hooker described the session with some enthusiasm: 'The "punctum saliens" of the whole meeting was decidedly Huxley's answer to D^r^ M^c^Cann. He literally poured boiling oil over the bumptious man' (letter from J. D. Hooker, 7 September 1869). Huxley playfully groused that as usual Darwin's 'abominable heresies' got him into all sorts of trouble: 'Three parsons set upon you—and if you were the most malicious of men you could not have wished them to have made greater fools of themselves than they did' (letter from T. H. Huxley, 28 September 1869).

Problems of design and purposefulness in nature had been alluded to in a more thoughtful criticism of Darwin's work, an article by the Italian botanist Federico Delpino. The article addressed specifically Darwin's hypothesis of heredity, 'pangenesis', as presented in the second volume of *Variation.* Delpino raised many technical objections to Darwin's hypothesis, but also described it as 'monistic' and therefore as denying 'liberty' to the organism. Delpino's criticism was noted in a letter from Thomas Henry Farrer, who had been reading some of Delpino's botanical works for his views on dichogamy: 'In one of the books he assails you as a fatalist, and wanders into regions of free will & necessity, into which … I do not care to follow him' (letter from T. H. Farrer, 9 October 1869). Farrer ventured to summarise Delpino's position in the form of a poem, titled 'The biological teleologist', composed he said 'during a sleepless hour': 'The animal can comprehend/ A purpose, and attain an end:/ Whilst in its turn the humbler plant/ Can feel and satisfy a want—/ And thus we find throughout the line/ Freedom and Will and High Design—' (letter from T. H. Farrer, 13 October 1869). Darwin was sufficiently impressed by Delpino's criticism of pangensis, however, to have it translated and published in a British journal. Detailed discussion of pangenesis had been scarce in scientific literature, and the appearance of Delpino's review in *Scientific Opinion* allowed Darwin to publicly defend his provisional hypothesis in the form of a letter to the same journal: 'I can speak from recent experience, that he who has to consider complex cases of inheritance, as limited either separately or conjointly by sex, age, and season, with the inherited characters themselves and the form of inheritance liable to change from crossing and variability, will be able to disentangle the phenomena much more clearly, if he admits for the time our hypothesis with all its imperfections' (letter to *Scientific Opinion*, [before 20 October 1869]).

Scientific Opinion, launched towards the end of 1868, was one of several periodicals begun in these years, expanding and broadening the forums in which Darwinism was discussed. John Murray brought out the first issue of the *Academy* in September 1869. He described its aims to Darwin on 18 September: 'to convey early & trustworthy intelligence to English Readers on Literature Science & Art & to establish, if possible, a higher tone of Criticism than that now prevailing'. Here Murray was alluding particularly to the *Athenaeum*, the leading journal of its type for some time, which had carried highly unfavourable reviews of Darwin's works in recent years. 'I heartily wish your Periodical all success', Darwin wrote to Murray, 'I wish it had been weekly, as then perhaps it would have killed the Athenæum by a lingering death, & that to me would have been a pleasing sight' (letter to John Murray, [after 18 September 1869]). Much more influential in the long term, however, was *Nature*, the first issue of which appeared in November. Hooker was initially disappointed with it, but added, 'there is plenty of room for a *good* journal of the kind', and made a number of concrete suggestions to its publisher, Macmillan (letter from J. D. Hooker, 14 November 1869). Darwin was uniformly pleased. 'I like all scientific periodicals', he wrote to Hooker, regretting only that *Nature* did not review more foreign articles (letter to J. D. Hooker, 19 November [1869]).

Yet despite his enthusiasm for scientific journals, Darwin's most important medium of communication remained correspondence. His network of contacts and collaborators continued to grow and diversify in 1869. A gardener to the king of Prussia, Adolf Reuter, sent beautiful sketches of some curious examples of botanical reversion: 'as a small stone to [Darwin's] great building' (letter from Adolf Reuter, 23 September 1869). The physiologist William Thierry Preyer enclosed a paper on the struggle for existence recently given in a successful series of popular lectures on Darwinism in Bonn. Darwin's photograph was requested by the director of the national museum in Zagreb, Spiridion Brusina, to adorn diplomas of a new natural history society in Croatia (letter from Spiridion Brusina, 29 April 1869). The German zoologist Anton Dohrn solicited Darwin's support for a new zoological station on the Mediterranean coast; the enterprise would result in the establishment of the marine station at Naples, which remains an important institution to this day (letter from Anton Dohrn, 30 December 1869).

In his reply to Dohrn, Darwin remarked: 'You ask about my Health.— I cannot say much in its praise; but as long as I live the life of a hermit I am able to work some hours daily' (letter to Anton Dohrn, 4 January 1870). Darwin's health was generally much the same as it had been since his last period of prolonged illness in 1864 and 1865, although a particularly low spell in May prompted him to accompany his family to Barmouth in Wales, for an extended holiday. As often happened with Darwin, the enforced leisure left him feeling more tired, and he groaned to Hooker: 'it seems as soon as the stimulus of mental work stops, my whole strength gives way' (letter to J. D. Hooker, 22 June [1869]). Darwin also found it discouraging to reside in the vicinity where, in earlier years, he had energetically collected insects and studied geology: 'I have been much disappointed at gaining no strength & failing to climb even a hill, & I had longed once again to set foot on summit of a mountain.—' (letter to T. H. Huxley, 9 July [1869]). Earlier in the year, Darwin had suffered a more serious setback when he was thrown by his horse. Having been advised in 1866 by the doctor Henry Bence Jones to go riding for his health, Darwin had faithfully followed the prescription. Henrietta Emma Darwin wrote to her brother George on 10 April (DAR 245: 291) about the incident:

> We've had a very unpleasant event this week. The immaculate Tommy has thrown Father. they were cantering over Keston Common when Tommy tripped & fell bang down—so completely head over heels that his ears & the pommel of the saddle were the two parts muddied. Father of course imitated Tommy's movements & w^{d} not have been hurt at all if tommy had not hit him a fearful blow in the back. This numbed his back & prevented his getting up, but as it was a frequented place someone soon came to his help & he was taken into a house & lay down on a sofa for a bit— after $\frac{1}{2}$ an hour Tommy was caught & as the fly Father ordered was very long in coming he got on Tommy & was led home. he came in m. exhausted & in considerable pain.

Despite the seriousness of the accident, Darwin suffered no broken bones, and though in considerable pain, was soon 'able to hobble about the room a little'. News of the incident spread gradually through Darwin's network, and letters of concern were received for months afterwards.

In the Christmas season, Darwin spent some of his evenings listening to Emma read aloud from a new book by Darwin's half-cousin, Francis Galton. The work, *Hereditary genius: an inquiry into its laws and consequences*, described the inheritance of mental ability through generations of particular families. The chapter on 'men of science' contained an entry on the Darwins. Galton traced the pedigree of achievement from Darwin's paternal grandfather, Erasmus, to two of Darwin's sons (George and Leonard), who had recently excelled in their examinations. Darwin himself was described briefly as 'the illustrious modern naturalist; author of the "Theory of

Natural Selection"'. Darwin was full of praise for the book: 'I have only read about 50 pages … but I must exhale myself, else something will go wrong in my inside'. He did express some reserve about Galton's arguments for inheritance as the deciding factor in success: 'You have made a convert of an opponent in one sense, for I have always maintained that, excepting fools, men did not differ much in intellect, only in zeal & hard work; & I still think there is an *eminently* important difference' (letter to Francis Galton, 23 December [1869]). Only a few weeks earlier, Galton had enlisted Darwin's assistance in investigating the laws of inheritance through experiments on rabbits (letter from Francis Galton, 11 December 1869). This was the beginning of a series of systematic attempts to test Darwin's theory of pangenesis. Galton's research on the transmission of characters would form the basis of extensive correspondence in the next few years, and would lead in turn to the development of his own distinct theory of heredity, which diverged substantially from Darwin's own. Darwin advised and encouraged the younger man in his scientific work, much as he had done with his own sons. Reflecting on Darwin's pivotal role as mentor, and on the inspirational example of Darwin's own work, Galton wrote, 'I always think of you in the same way as converts from barbarism think of the teacher who first relieved them from the intollerable burden of their superstition … the appearance of your "Origin of Species" formed a real crisis in my life … and was the first to give me freedom of thought' (letter from Francis Galton, 24 December 1869).

ACKNOWLEDGMENTS

The editors are grateful to the late George Pember Darwin and to William Huxley Darwin for permission to publish the Darwin letters and manuscripts. They also thank the Syndics of Cambridge University Library and other owners of manuscript letters who have generously made them available.

The work for this edition has been supported by grants from the National Endowment for the Humanities (NEH), the National Science Foundation (NSF), and the Wellcome Trust. The Alfred P. Sloan Foundation, the Pew Charitable Trusts, and the Andrew W. Mellon Foundation provided grants to match NEH funding, and the Mellon Foundation awarded grants to Cambridge University that made it possible to put the entire Darwin correspondence into machine-readable form. Research and editorial work have also been supported by grants from the Royal Society of London, the British Academy, the British Ecological Society, the Isaac Newton Trust, the Jephcott Charitable Trust, the Natural Environment Research Council, the John Templeton Foundation, and the Wilkinson Charitable Foundation. The Stifterverband für die Deutsche Wissenschaft provided funds to translate and edit Darwin's correspondence with German naturalists.

Cambridge University Library, the American Philosophical Society (APS), and Cornell University have generously made working space and many services available to the editors.

Since the project began in 1975, the editors have been fortunate in benefiting from the interest, experience, and practical help of many people, and hope that they have adequately expressed their thanks to them individually as the work proceeded.

Without the expert help of John L. Dawson of the Literary and Linguistic Computing Centre of Cambridge University, the initial computerisation of the correspondence would not have been possible. Iain Burke, Peter Dunn, Ray Horne, Patricia Killiard, Adrian Miller, Merina Tuladhar, and Tomasz Waldoch have helped maintain and extend our access to the Library's electronic resources, and provided essential technical support. This volume has been typeset using LaTeX. We are grateful to Matthew Daws for his expert help in developing the current typesetting procedures, which built on those established by Simon Buck, and maintained and developed by Robin Fairbairns. Gratitude is expressed to the late C. A. Tripp for providing the editors with an invaluable tool for searching texts.

English Heritage has responded most generously to requests for information and for material from the collections at Down House, Downe. We are particularly grateful to Cathy Power, curator of Down House and Tori Reeve, former curator. Richard Darwin Keynes kindly made available Darwin family material in his possession.

Ursula Mommens has also provided letters and other materials that belonged to her grandfather, Francis Darwin.

Institutions and individuals all over the world have given indispensable help by making available photocopies of Darwin correspondence and other manuscripts in their collections. Those who furnished copies of letters for this volume can be found in the List of provenances. The editors are indebted to them, and to the many people who have provided information about the locations of particular letters.

The editors make daily use of the incomparable facilities of Cambridge University Library and have benefited greatly from its services and from the help and expertise of its staff, particularly the staff of the Manuscripts Department. We are especially grateful to the former University Librarian, Frederick W. Ratcliffe, to his successor, Peter K. Fox, and to the Keeper of Manuscripts and Archives, Patrick Zutshi, for their generous support. Other members of the library's staff who frequently respond to the editors' requests are: Marjolein Allen, Wendy Aylett, Frank Bowles, Kathleen Cann, Louise Clarke, Colin Clarkson, Jacqueline Cox, Les Goodey, John Hall, Tony Harper, Brian Jenkins, Morag Law, Stephen Lees, David Lowe, Peter Meadows, Anne Murray, William Noblett, Adam Perkins, John Reynolds, Jayne Ringrose, Mark Scudder, Clive Simmonds, Nicholas Smith, Anne Taylor, and Godfrey Waller. The fetchers in the Rare Books reading room have also patiently dealt with the editors' often complex requirements, as have the staff of the Map Room.

At the American Philosophical Society Library, a splendid collection of Darwiniana and works in the history of science has been available to the editors since the inception of the project. The editors have benefited from the co-operation of the late Whitfield J. Bell Jr, former secretary of the society, and the late Edward Carter II, Robert S. Cox, Roy C. Goodman, and Martin L. Leavitt, all of the APS Library.

Thanks are due to the faculty and staff of the Department of Science and Technology Studies, Cornell University, for accommodating one of the Project's editors, and to staff at Cornell University Libraries for help with research enquiries.

The editors would like to acknowledge the assistance of Rodney Dennis, Jennie Rathbun, and Susan Halpert of the Houghton Library, Constance Carter of the Science Division of the Library of Congress, and Judith Warnement and Jean Cargill of the Gray Herbarium of Harvard University, who have all been exceptionally helpful in providing material from the collections in their charge.

In Britain, the editors have received assistance from Lynda Brooks (librarian), Gina Douglas (former librarian), and Ben Sherwood of the Linnean Society of London; and from Michele Losse and Kiri Ross Jones of the Royal Botanic Gardens, Kew. We would also like to thank Anne Barrett, college archivist at the Imperial College of Science, Technology and Medicine; and Douglas Ferguson and Candace Guite of Christ's College, Cambridge.

Among the others who advise and assist the editors in their work are Nick Gill, Randal Keynes, Gene Kritsky, Gren Lucas, Carl F. Miller, Jim Moore, Garry J. Tee, Béatrice Willis, and Leonard Wilson. The editors are also pleased to acknowledge the invaluable support of the members of the Project's Advisory Committee.

For help with particular enquiries in volume 17, the editors would like to thank, besides those already mentioned, Barbara Beeker, Chiara Ceci, Peter Dance, Joy Harvey, Nick Jardine, Jenan Jobst of the LaSalle County Genealogy Guild, Pat Knight of Essex County Library, Canada, Diana McFarland, Fredrik Pleijel of Göteborg University, Alyson Price of the Harold Acton Library, British Institute of Florence, Anne Secord, Janet Wareham of the Washington County Historical Library, and Kathie Way of the Natural History Museum, London.

Thanks are also due to all former staff of the Darwin Correspondence Project, including: Sarah Benton, Charlotte Bowman, Heidi Bradshaw, Janet Browne, P. Thomas Carroll, Mario Di Gregorio, Rhonda Edwards, Kate Fletcher, Hedy Franks, Joy Harvey, Arne Hessenbruch, Thomas Junker, David Kohn, Jyothi Krishnan-Unni, Kathleen Lane, Sarah Lavelle, Anna K. Mayer, William Montgomery, Perry O'Donovan, Stephen V. Pocock, Duncan Porter, Marsha L. Richmond, the late Peter Saunders, Andrew Sclater, Anne Secord, Tracey Slotta, the late Sydney Smith, Emma Spary, Nora Carroll Stevenson, Jonathan R. Topham, Tyler Veak, and Sarah Wilmot.

The editors also gratefully acknowledge the invaluable assistance of Charissa Varma and Philip Hardie.

We are most grateful to Margot Levy for providing the index to the current volume, and look forward to continuing to work with Margot in the future.

The editors are grateful to the owner of Laura Russell's 1869 oil portrait of Darwin for supplying an image of it for use as the frontispiece to this volume. We also wish to thank Randal Keynes for writing an essay to accompany the portrait (see Appendix V).

This book is dedicated to the memory of William T. Golden, who admired Darwin, supported the Project's work, and served on the Advisory Committee until his death in 2007.

PROVENANCES

The following list gives the locations of the original versions of the letters printed in this volume. The editors are grateful to all the institutions and individuals listed for allowing access to the letters in their care. Access to material in DAR 261 and DAR 263, formerly at Down House, Downe, Kent, England, is courtesy of English Heritage.

American Philosophical Society, Philadelphia, Pennsylvania, USA
Annals and Magazine of Natural History
Archive of the Berlin-Brandenburgische Akademie der Wissenschaften, Berlin, Germany
Archives de la famille de Candolle, Geneva, Switzerland
Archives of the Croatian Academy of Sciences and Arts, Zagreb, Croatia
Archives, New York Botanical Garden, Bronx, New York, USA
Athenæum
L'Autographe S.A.
Bayerische Staatsbibliothek, Munich, Germany
Bibliothèque de Genève, Geneva, Switzerland
Bodleian Library, Oxford, England
Boston Society of Natural History (papers at Museum of Science, Boston, Massachusetts, USA)
British Library, London, England
Cambridge University Library, Manuscripts Department, Cambridge, England
Christie's London (dealers), London, England
Cleveland Health Sciences Library, Cleveland, Ohio, USA
Cobbold 1885
Cupples 1894
Edinburgh University Library, Edinburgh, Scotland
Ernst-Haeckel-Haus, Friedrich-Schiller-Universität, Jena, Germany
Field Museum of Natural History, Chicago, Illinois, USA
Gardeners' Chronicle and Agricultural Gazette
Gray Herbarium of Harvard University, Cambridge, Massachusetts, USA
History of Science Collections, University of Oklahoma Libraries, Norman, Oklahoma, USA
E. C. Hodgkin (private collection)
Houghton Library, Harvard University, Cambridge, Massachusetts, USA
The Huntingdon Library, San Marino, California, USA
Imperial College of Science, Technology, and Medicine, London, England

Dr Hildegard Jaeger (private collection)
John Hay Library, Brown University, Providence, Rhode Island, USA
King Edward VI High School, Stafford, England
C. C. Kohler (private collection)
Det Konegelige Bibliotek, Copenhagen, Denmark
Linnean Society of London, Piccadilly, London, England
K, M. Lyell ed. 1881
McGill University Libraries, Rare Books and Special Collections Division, Montréal, Québec, Canada
Mitchell Library, State Library of New South Wales, Sydney, N.S.W., Australia
National Library of Australia, Canberra, Australia
Trustees of the National Library of Scotland, Edinburgh, Scotland
Nature
Paget ed. 1901
Phillips, Son and Neale (dealers)
The Pierpont Morgan Library, New York, USA
Rendell's (dealers)
Rothamsted Research, Harpenden, Hertfordshire, England
Trustees of the Royal Botanic Gardens, Kew, Richmond, Surrey, England
Royal Horticultural Society, London, England
Schlesinger Library, Radcliffe Institute, Harvard University, Cambridge, Massachusetts, USA
Scientific Opinion
Shrewsbury School Library, Shrewsbury, England
Sotheby's, London (dealers)
Spruce 1908
Staatsbibliothek zu Berlin – Preus+sischer Kulturbesitz, Berlin, Germany
Stadsarchief Amsterdam, Amsterdam, The Netherlands
Clifford D. Stromberg (private collection)
Alan R. Tait (private collection)
UCL Library Services, Special Collections, London, England
University of Chicago Library, Chicago, Illinois, USA
University Library of Lund, Lund, Sweden
Württembergische Landesbibliothek Stuttgart, Germany

A NOTE ON EDITORIAL POLICY

The first and chief objective of this edition is to provide complete and authoritative texts of Darwin's correspondence. For every letter to or from Darwin, the text that is available to the editors is always given in full. The editors have occasionally included letters that are not to or from Darwin if they are relevant to the published correspondence. Volumes of the *Correspondence* are published in chronological order. Occasional supplements will be published containing letters that have come to light or have been redated since the relevant volumes of the *Correspondence* appeared. Letters that can only be given a wide date range, in some instances spanning several decades, are printed in the supplement following the volume containing letters at the end of their date range. The first such supplement was in volume 7 and included letters from 1828 to 1857. The second supplement was in volume 13, and included letters from 1822 to 1864.

Dating of letters and identification of correspondents

In so far as it is possible, the letters have been dated, arranged in chronological order, and the recipients or senders identified. Darwin seldom wrote the full date on his letters and, unless the addressee was well known to him, usually wrote only 'Dear Sir' or 'Dear Madam'. After the adoption of adhesive postage stamps in the 1840s, the separate covers that came into use with them were usually not preserved, and thus the dates and the names of many recipients of Darwin's letters have had to be derived from other evidence. The notes made by Francis Darwin on letters sent to him for his editions of his father's correspondence have been helpful, as have matching letters in the correspondence, but many dates and recipients have had to be deduced from the subject-matter or references in the letters themselves.

Transcription policy

Whenever possible, transcriptions have been made from manuscripts. If the manuscript was inaccessible but a photocopy or other facsimile version was available, that version has been used as the source. In many cases, the editors have had recourse to Francis Darwin's large collection of copies of letters, compiled in the 1880s. Other copies, published letters, or drafts have been transcribed when they provided texts that were otherwise unavailable.

The method of transcription employed in this edition is adapted from that described by Fredson Bowers in 'Transcription of manuscripts: the record of variants', *Studies in Bibliography* 29 (1976): 212–64. This system is based on accepted principles of modern textual editing and has been widely adopted in literary editions.

The case for using the principles and techniques of this form of textual editing for historical and non-literary documents, both in manuscript and print, has been forcefully argued by G. Thomas Tanselle in 'The editing of historical documents', *Studies in Bibliography* 31 (1978): 1–56. The editors of the *Correspondence* followed Dr Tanselle in his conclusion that a 'scholarly edition of letters or journals should not contain a text which has editorially been corrected, made consistent, or otherwise smoothed out' (p. 48), but they have not wholly subscribed to the statement made earlier in the article that: 'In the case of notebooks, diaries, letters and the like, whatever state they are in constitutes their finished form, and the question of whether the writer "intended" something else is irrelevant' (p. 47). The editors have preserved the spelling, punctuation, and grammar of the original, but they have found it impossible to set aside entirely the question of authorial intent. One obvious reason is that in reading Darwin's writing, there must necessarily be reliance upon both context and intent. Even when Darwin's general intent is clear, there are cases in which alternative readings are, or may be, possible, and therefore the transcription decided upon must to some extent be conjectural. Where the editors are uncertain of their transcription, the doubtful text has been enclosed in italic square brackets.

A major editorial decision was to adopt the so-called 'clear-text' method of transcription, which so far as possible keeps the text free of brackets recording deletions, insertions, and other alterations in the places at which they occur. Darwin's changes are, however, recorded in the back matter of the volume, under 'Manuscript alterations and comments', in notes keyed to the printed text by paragraph and line number. All lines above the first paragraph of the letter (that is, date, address, or salutation) are referred to as paragraph '0'. Separate paragraph numbers are used for subscriptions and postscripts. This practice enables the reader who wishes to do so to reconstruct the manuscript versions of Darwin's autograph letters, while furnishing printed versions that are uninterrupted by editorial interpolations. The Manuscript alterations and comments record all alterations made by Darwin in his letters and any editorial amendments made in transcription, and also where part of a letter has been written by an amanuensis; they do not record alterations made by amanuenses. No attempt has been made to record systematically all alterations to the text of copies of Darwin letters included in the correspondence, but ambiguous passages in copies are noted. The editors believe it would be impracticable to attempt to go further without reliable information about the texts of the original versions of the letters concerned. Letters to Darwin have been transcribed without recording any of the writers' alterations unless they reflect significant changes in substance or impede the sense; in such cases footnotes bring them to the reader's attention.

Misspellings have been preserved, even when it is clear that they were unintentional: for instance, 'lawer' for 'lawyer'. Such errors often indicate excitement or haste and may exhibit, over a series of letters, a habit of carelessness in writing to a particular correspondent or about a particular subject.

Capital letters have also been transcribed as they occur except in certain cases, such as 'm', 'k', and 'c', which are frequently written somewhat larger than others as initial letters of words. In these cases an attempt has been made to follow the normal practice of the writers.

In some instances that are not misspellings in a strict sense, editorial corrections have been made. In his early manuscripts and letters Darwin consistently wrote 'bl' so that it looks like 'lb' as in 'albe' for 'able', 'talbe' for 'table'. Because the form of the letters is so consistent in different words, the editors consider that this is most unlikely to be a misspelling but must be explained simply as a peculiarity of Darwin's handwriting. Consequently, the affected words have been transcribed as normally spelled and no record of any alteration is given in the textual apparatus. Elsewhere, though, there are misformed letters that the editors have recorded because they do, or could, affect the meaning of the word in which they appear. The main example is the occasional inadvertent crossing of 'l'. When the editors are satisfied that the intended letter was 'l' and not 't', as, for example, in 'stippers' or 'istand', then 'l' has been transcribed, but the actual form of the word in the manuscript has been given in the Manuscript alterations and comments.

If the only source for a letter is a copy, the editors have frequently retained corrections made to the text when it is clear that they were based upon comparison with the original. Francis Darwin's corrections of misreadings by copyists have usually been followed; corrections to the text that appear to be editorial alterations have not been retained.

Editorial interpolations in the text are in square brackets. Italic square brackets enclose conjectured readings and descriptions of illegible passages. To avoid confusion, in the few instances in which Darwin himself used square brackets, they have been altered by the editors to parentheses with the change recorded in the Manuscript alterations and comments. In letters to Darwin, square brackets have been changed to parentheses silently.

Material that is irrecoverable because the manuscript has been torn or damaged is indicated by angle brackets; any text supplied within them is obviously the responsibility of the editors. Occasionally, the editors are able to supply missing sections of text by using ultraviolet light (where text has been lost owing to damp) or by reference to transcripts or photocopies of manuscript material made before the damage occurred.

Words and passages that have been underlined for emphasis are printed in italics in accordance with conventional practice. Where the author of a letter has indicated greater emphasis by underlining a word or passage two or more times, the text is printed in bold type.

Paragraphs are often not clearly indicated in the letters. Darwin and others sometimes marked a change of subject by leaving a somewhat larger space than usual between sentences; sometimes Darwin employed a longer dash. In these cases, and when the subject is clearly changed in very long stretches of text, a new paragraph has been started by the editors without comment. The beginnings of letters, valedictions, and postscripts are also treated as new paragraphs regardless of whether they appear as new paragraphs in the manuscript. Special manuscript devices delimiting sections or paragraphs, for example, blank spaces left between sections of text and lines drawn across the page, are treated as normal paragraph indicators and are not specially marked or recorded unless their omission leaves the text unclear.

Occasionally punctuation marking the end of a clause or sentence is not present in the manuscript, but the author has made his or her intention clear by allowing, for example, extra space or a line break to function as punctuation. In such cases, the editors have inserted an extra space following the sentence or clause to set it off from the following text.

Additions to a letter that run over into the margins, or are continued at its head or foot, are transcribed at the point in the text at which the editors believe they were intended to be read. The placement of such an addition is only recorded in a footnote if it seems to the editors to have some significance or if the position at which it should be transcribed is unclear. Enclosures are transcribed following the letter.

The hand-drawn illustrations and diagrams that occur in some letters are reproduced as faithfully as possible and are usually positioned as they were in the original text. In some cases, however, it has been necessary to reduce the size of a diagram or enhance an outline for clarity; any such alterations are recorded in footnotes. The location of diagrams within a letter is sometimes changed for typesetting reasons. Tables have been reproduced as close to the original format as possible, given typesetting constraints.

Some Darwin letters and a few letters to Darwin are known only from entries in the catalogues of book and manuscript dealers or mentions in other published sources. Whatever information these sources provide about the content of such letters has been reproduced without substantial change. Any errors detected are included in footnotes.

Format of published letters

The format in which the transcriptions are printed in the *Correspondence* is as follows:

1. *Order of letters.* The letters are arranged in chronological sequence. A letter that can be dated only approximately is placed at the earliest date on which the editors believe it could have been written. The basis of a date supplied by the editors is given in a footnote unless it is derived from a postmark, watermark, or endorsement that is recorded in the physical description of the letter (see section 4, below). Letters with the same date, or with a range of dates commencing with that date, are printed in the alphabetical order of their senders or recipients unless their contents dictate a clear alternative order. Letters dated only to a year or a range of years precede letters that are dated to a particular month or range of months, and these, in turn, precede those that are dated to a particular day or range of dates commencing with a particular day.

2. *Headline.* This gives the name of the sender or recipient of the letter and its date. The date is given in a standard form, but those elements not taken directly

from the letter text are supplied in square brackets. The name of the sender or recipient is enclosed in square brackets only where the editors regard the attribution as doubtful.

3. *The letter text.* The transcribed text follows as closely as possible the layout of the source, although no attempt is made to produce a type-facsimile of the manuscript: word-spacing and line-division in the running text are not adhered to. Similarly, the typography of printed sources is not replicated. Dates and addresses given by authors are transcribed as they appear, except that if both the date and the address are at the head of the letter they are always printed on separate lines with the address first, regardless of the manuscript order. If no address is given on a letter by Darwin, the editors have supplied one, when able to do so, in square brackets at the head of the letter. Similarly, if Darwin was writing from an address different from the one given on the letter, his actual location is given in square brackets. Addresses on printed stationery are transcribed in italics. Addresses, dates, and valedictions have been run into single lines to save space, but the positions of line-breaks in the original are marked by vertical bars.

4. *Physical description.* All letters are complete and in the hand of the sender unless otherwise indicated. If a letter was written by an amanuensis, or exists only as a draft or a copy, or is incomplete, or is in some other way unusual, then the editors provide the information needed to complete the description. Postmarks, endorsements, and watermarks are recorded only when they are evidence for the date or address of the letter.

5. *Source.* The final line provides the provenance of the text. Some sources are given in abbreviated form (for example, DAR 140: 18) but are listed in full in the List of provenances unless the source is a published work. Letters in private collections are also indicated. References to published works are given in author–date or short-title form, with full titles and publication details supplied in the Bibliography at the end of the volume.

6. *Darwin's annotations.* Darwin frequently made notes in the margins of the letters he received, scored significant passages, and crossed through details that were of no further interest to him. These annotations are transcribed or described following the letter text. They are keyed to the letter text by paragraph and line numbers. Most notes are short, but occasionally they run from a paragraph to several pages, and sometimes they are written on separate sheets appended to the letter. Extended notes relating to a letter are transcribed whenever practicable following the annotations as 'CD notes'.

Quotations from Darwin manuscripts in footnotes and elsewhere, and the text of his annotations and notes on letters, are transcribed in 'descriptive' style. In this method the alterations in the text are recorded in brackets at the places where they occur. For example:

> 'See Daubeny ['vol. 1' *del*] for *descriptions of volcanoes in [*interl*] S.A.' *ink*

means that Darwin originally wrote in ink 'See Daubeny vol. 1 for S.A.' and then deleted 'vol. 1' and inserted 'descriptions of volcanoes in' after 'for'. The asterisk before 'descriptions' marks the beginning of the interlined phrase, which ends at the bracket. The asterisk is used when the alteration applies to more than the immediately preceding word. The final text can be read simply by skipping the material in brackets. Descriptive style is also used in the Manuscript alterations and comments.

Editorial matter

Each volume is self-contained, having its own index, bibliography, and biographical register. A chronology of Darwin's activities covering the period of each volume and translations of foreign-language letters are supplied, and additional appendixes give supplementary material where appropriate to assist the understanding of the correspondence. A cumulative index is planned once the edition is complete. References are supplied for all persons, publications, and subjects mentioned, even though some repetition of material in earlier volumes is involved.

If the name of a person mentioned in a letter is incomplete or incorrectly spelled, the full, correct form is given in a footnote. Brief biographies of persons mentioned in the letters, and dates of each correspondent's letters to and from Darwin in the current volume, are given in the Biographical register and index to correspondents. Where a personal name serves as a company name, it is listed according to the family name but retains its original order: for example, 'E. Schweizerbart'sche Verlagsbuchhandlung' is listed under 'S', not 'E'.

Short titles are used for references to Darwin's books and articles and to collections of his letters (e.g., *Descent*, 'Parallel roads of Glen Roy', *LL*). They are also used for some standard reference works and for works with no identifiable author (e.g., *Alum. Cantab.*, *Wellesley index*, *DNB*). For all other works, author–date references are used. References to the Bible are to the authorised King James version unless otherwise stated. Words not in *Chambers dictionary* are usually defined in the footnotes with a source supplied. The full titles and publication details of all books and papers referred to are given in the Bibliography. References to archival material, for instance that in the Darwin Archive at Cambridge University Library, are not necessarily exhaustive.

Darwin and his correspondents writing in English consistently used the term 'fertilisation' for the processes that are now distinguished as fertilisation (the fusion of female and male gametes) and pollination (the transfer of pollen from anther to stigma); the first usage known to the editors of a distinct term for pollination in English was in 1873 (letter from A. W. Bennett, 12 July 1873 (*Calendar* no. 8976)). 'Fertilisation' in Darwin's letters and publications often, but not always, can be regarded as referring to what is now termed pollination. In the footnotes, the editors, where possible, have used the modern terms where these can assist in explaining the details of experimental work. When Darwin or his correspondents are quoted directly, their original usage is never altered.

The editors use the abbreviation 'CD' for Charles Darwin throughout the footnotes. A list of all abbreviations used by the editors in this volume is given on p. xxxviii.

The Wedgwood and Darwin

Robert Waring Darwin = Susannah Wedgwood
1766–1848 1765–1817

- Henry Parker = Marianne
  1788–1856 1798–1858
  - Robert b.1825
  - Henry 1827–92
  - Francis 1829–71
  - Charles b.1831
  - Mary Susan 1836–93
- Caroline Sarah = Josiah III (Jos)
  1800–88 1795–1880
  - Sophy Marianne 1838–9
  - Katherine Elizabeth Sophy (Sophy) 1842–1911
  - Margaret Susan 1843–1937
  - Lucy Caroline 1846–1919
- Susan Elizabeth 1803–66
- Erasmus Alvey 1804–81
- Charles Robert = Emma 1808–96
  1809–82
  - William Erasmus 1839–1914
  - Anne Elizabeth 1841–51
  - Mary Eleanor Sept.–Oct. 1842
  - Henrietta Emma (Etty) 1843–1927
  - George Howard 1845–1912
  - Elizabeth 1847–1926
  - Francis (Frank) 1848–1925
  - Leonard 1850–1943
  - Horace 1851–1928
  - Charles Waring 1856–8
- Charles Langton = Emily Catherine (Catherine) 1810–66
  1801–86

Families up to 1869

Josiah Wedgwood II 1769–1843 = Elizabeth (Bessy) Allen 1764–1846

- Sarah Elizabeth (Elizabeth) 1793–1880
- Mary Anne 1796–8
- Charlotte 1797–1862 = Charles Langton 1801–86
  - Edmund 1841–75
- Henry Allen (Harry) 1799–1885 = Jessie Wedgwood 1804–72
  - Louise Frances 1834–1903
  - Caroline b.1836
  - John Darwin 1840–70
  - Anne Jane 1841–77
  - Arthur 1843–1900
  - Rowland 1847–1921
- Francis (Frank) 1800–88 = Frances Mosley 1808–74
  - Godfrey 1833–1905
  - Amy 1835–1910
  - Cicely Mary 1837–1917
  - Clement 1840–89
  - Lawrence 1844–1913
  - Constance Rose 1846–1903
  - Mabel Frances 1852–1930
- Hensleigh 1803–91 = Frances (Fanny) Mackintosh 1800–89
  - Frances Julia (Snow) 1833–1913
  - James Mackintosh (Bro) 1834–64
  - Ernest Hensleigh 1838–98
  - Katherine Euphemia (Effie) 1839–1931
  - Alfred Allen 1842–92
  - Hope Elizabeth 1844–1934
- Frances (Fanny) 1806–32

ABBREVIATIONS

AL	autograph letter
ALS	autograph letter signed
DS	document signed
LS	letter in hand of amanuensis, signed by sender
LS(A)	letter in hand of amanuensis with additions by sender
Mem	memorandum
(S)	signed with sender's name by amanuensis
CD	Charles Darwin
CUL	Cambridge University Library
DAR	Darwin Archive, Cambridge University Library
del	deleted
illeg	illegible
interl	interlined
underl	underlined

TRANSCRIPTION CONVENTIONS

[some text]	'some text' is an editorial insertion
/some text/	'some text' is the conjectured reading of an ambiguous word or passage
[some text]	'some text' is a description of a word or passage that cannot be transcribed, e.g., '*3 words illeg*'
⟨ ⟩	word(s) destroyed
⟨some text⟩	'some text' is a suggested reading for a destroyed word or passage
⟨*some text*⟩	'some text' is a description of a destroyed word or passage, e.g., '*3 lines excised*'

THE CORRESPONDENCE OF CHARLES DARWIN
1869

From James Paget [1869][1]

I enclose a note from Lord Fitzwilliam about his horse with zebra-marks.[2] The case seems as striking as I believed.

Incomplete
Paget ed. 1901, p. 408

[1] The year is given in the printed source.
[2] The note from William Wentworth Fitzwilliam has not been found. CD had discussed stripes in horses in *Variation* 1: 61–4 and 2: 41.

From William Erasmus Darwin [January 1869][1]

N. Staffordshire—[2]		48 . 15
Gr. Northern[3]		62 . 2 . 7
Great Eastern[4]		87 . 12 . 4
Midland[5]		8 . 16 . 3
Monmouth[6]		54 · 9 · 7
Shrewsbury Street mortgage }		11 . 14 .
Castle Morton tithes[7]		19 · 3 · 1
City Bonds		34 · 14 · 9
Penarth Harbour[8] to 31 Dec 1866 }		213 . 1 . 1
	7/10)	540 . 8 . 8
		378 . 6 .
Business Bonds		4 17 . 6
		£383— 3—6

PS. I invested the £160 you did not think should be divided in Russian Gov. Bonds.—

Your affect son | W. E. Darwin

Incomplete
DAR 262.11: 6 (EH 88206193)

CD ANNOTATION
1.1 N. . . . 3—6 1.15] 'Jan. 1869' *ink*

[1] The date is established by CD's annotation.
[2] North Staffordshire Railway Company.
[3] Great Northern Railway Company.
[4] Great Eastern Railway Company.
[5] Midland Railway Company.
[6] Monmouth Canal and Railway Company.
[7] CD's father, Robert Waring Darwin, had inherited a share of Castle Morton tithes from Josiah Wedgwood I (DAR 227.9: 52), and this share or some part of it had evidently been bequeathed to CD.
[8] Penarth Harbour, Dock & Railway Company.

From Vladimir Onofrievich Kovalevsky [January–March 1869][1]

My dear Sir

I did not answer your first letter immediately, and yesterday I received your second letter with questions concerning the Edition of your "Origin"[2] The publisher is not living here but at Moscow, I write to him to day and shall directly after receiving his answer send it to you. There are two russian editions of the "Origin", no remark is made to show from which Engl-edition the translation was made but I think it is from the Second Engl. edition; the second Russ. edition is merely a reprint from the first without any corrections or additions.[3] At all events if there is no hope of a new edition appearing in a short time I will with your permission translate all the additions of the Fifth Engl. edition of the "Origin" and print them aparte as a supplement.—[4]

I am very astonished that Messr. Petter & Cassel have complained to you of my not fulfilling an engagement towards them[5]—but as there never was an engagement I dont well understand what they are complaining for,—the matter is simply thus: One of my partners asked me to go, during my visit in England, at Messr Cassel and ask them if "in case there will be a Russian edition of an illustrated Bible they will sell coppercasts of their great illustrated Bible",—he gave me full power to treat about the price;—Mrs Cassel consented to sell coppercasts at 6d per sq. inch.— After my returning in Russia I knew that all endeavours of my partner to have a permission for printing an illustrated Bible were in vain, all the holy & Scriptural books being considered as litterary property of the Chief Church establishment (Synode)—and in this sense I wrote to Mrs Cassel, saying that our Government did not allow to print holy books to private persons and therefore there could be no russian Edition of the Illust. Bible, thanking them for their readiness to furnish *casts* I could make no use of them.— Not one single inch of casts was commanded, and not one made for me, so that I consider Mrs Cassel pretension as wholly unfounded, but I thank you very much you wrote me about it.

My brother is now in the University of Casan, he will be very pleased with the good opinion you formed on account of his Ascidiae— [6] The year beeing at Triest he made the full development of the "*Pyrosoma*" and came to entirely different conclusions with Mr Huxley, who wrote his memoir I think only studying preserved specimens.—[7]

Yours very truly | V. Kovalevsky

As soon as I receive an answer from Moscow, I shall communicate it to you.—

DAR 169: 54

[1] The date range is established by the relationship between this letter and the letter from V. O. Kovalevsky, 13 September 1869, and by the dates of CD's preparation of the fifth edition of *Origin* (see 'Journal' (Appendix II)).

[2] CD's two letters to Kovalevsky have not been found.

[3] The Russian editions of *Origin* (Rachinskii trans. 1864 and Rachinskii trans. 1865) were published by the firm A. I. Glazunov.

[4] *Origin* 5th ed. was published in 1869; Kovalevsky's proposed supplement was not printed.

[5] The reference is to the publishing firm, Cassell, Petter, & Galpin. No letter from the firm to CD registering a complaint about Kovalevsky has been found, but see *Correspondence* vol. 16, letter to Cassell, Petter & Galpin, [August–December 1868].

[6] Alexander Onufrievich Kovalevsky was a professor at Kazan from 1868 to 1869 (*DSB*). CD had received a copy of 'Entwickelungsgeschichte der einfachen Ascidien' (Developmental history of simple ascidians; A. O. Kovalevsky 1866) in 1867 (see *Correspondence* vol. 15, letter from V. O. Kovalevsky 15 March 1867 and n. 6). CD's heavily annotated copy of the paper is in the Darwin Pamphlet Collection–CUL. CD referred to Kovalevsky's discovery of the relation between larval ascidians and vertebrates in *Descent* 1: 205.

[7] *Pyrosoma* is a genus of colonial pelagic tunicates. Kovalevsky refers to Thomas Henry Huxley and his paper 'On the anatomy and development of *Pyrosoma*' (T. H. Huxley 1859). When A. O. Kovalevsky published his observations on development of *Pyrosoma* in 1875 (A. O. Kovalevsky 1875), he referred favourably to Huxley's work.

From George Cupples 4 January 1869

The Cottage | Guard Bridge | Fifeshire. N.B.
Jany 4th. 1869.

Dear Mr Darwin,

(If I may be permitted so to say,)

I seize on any pretext in order to have an opportunity of wishing you the best wishes of the season, with health improved from that reported in your last kind communication to me. I trust the seaside change did good, so as to refresh you for further work in the great cause.[1]

My said pretext is merely this—that I have lately got a fine *Pyrenean* dog, and find there is an singular peculiarity in the breed (the breed being an old one, the main foundation of others better-known)—viz. that they have *six* well-developed toes on the hind-foot, and I understand this is always the case when they are pure. In this country, when there are *five*, I think (as in "*collies*",) the dew-claw is considered an excrescence—and a dog of high-breed, as in deerhounds, is bound to have only *four* hinder-toes.[2]

I do not know that this is worth mentioning to you—but you will perhaps pardon the officiousness as before, on the ground of the deferential homage meant to be expressed. And it may stand for what it is worth.

With sincere regards | I am | Dear Sir, ever truly yours | George Cupples

Charles Darwin, Esq.
F.R.S. &c. &c. &c.

P.S. I have been and am receiving some interesting notes on *in-breeding* from Mr Wright (who wrote to you some time ago.)[3] I do not know if he made any such statements to you—but he has had large experience—and if ever there is anything of the sort you might want, I could get it for you.

G.C.

DAR 161: 287

[1] CD had been at Freshwater on the Isle of Wight from 17 July until 20 August 1868 (see *Correspondence* vol. 16, Appendix II).

[2] In *Variation* 2: 317–18, CD noted the development of the rudimentary fifth toe in some large breeds of dogs.

[3] John Wright had written to CD on 11 June 1868, sending information on female preference in mating and proportional numbers of sexes for horses and dogs (*Correspondence* vol. 16). He later sent information on the size of male and female deer-hounds (see *Correspondence* vol. 16, letter from George Cupples, 13 July 1868).

From James Orton 4 January 1869

Rochester, New York, U.S.
Jan. 4. 1869.

Charles Darwin, Esq
Dear Sir—

I take the liberty to make a request.

I have lately returned from a Scientific Expedition, under the auspices of the Smithsonian Institution, across equatorial America, via Quito, the Napo and the Amazons.[1] My party made collections in every department of Natural History, to which I added a continuous series of barometrical and other physical observations across the continent, an account of which was published in Silliman's Journal for last September.[2] I made a special study of the geology of the Ecuadorian volcanoes and the Amazonian valley. Of fossils, I found at Paita on the coast three forms additional to those described by D'Orbigny in your 'Observations',— Cinthium laeviuscula, Ostrea gallus and a new sp. of Ampullina.[3] In the high valley of Quito near Riobamba, I discovered a vast deposit of mammalian bones, chiefly of the mastodon, horse, tapir and llamas.[4]

In the valley of the Amazons at Pebas, 2200 miles above Pará, I discovered a bed of fossil shells in the peculiar clay formation which overspreads the whole valley. Those determined are Neritina pupa (a familiar West Indian sp.), new sp. of Turbonilla, Mesalia and Tellina, and representatives of a new genus related to Isocardia.[5] You will recollect that Bates, Wallace and Agassiz found no fossils in the Valley, and the glacial theory of the last rests somewhat on that "negative evidence".[6] You may be interested also, in the fact that on a little lake on the slope of Antisana, 13,300 ft. above the Pacific, I found a grebe-like bird with undeveloped wings and scarcely able to walk, but at home on the water.[7] Associated with it were Penguins (in appearance): they were so shy we failed to secure one.

I will also state that I met a peccari, a "cock of the rock" and fine reptilian sp. common to both sides of the Andes,—Esmeraldas and Napo.[8]

The collections are in the hands of eminent naturalists, and will be reported on at an early day.[9] I am myself preparing, and have nearly ready for the press, a Narrative of the Expedition, modeled after your charming Voyage of the Beagle,[10] weaving in the most important scientific results.

And now, my dear Sir, this is the burden of this letter,—I desire the high honor and privilege of dedicating this work on "The Andes and the Amazons" to one whose name is so pleasantly associated with our southern continent.[11]

With the very highest respect I am, dear Sir, | Your obedient Servant | James Orton.

DAR 173: 37

CD ANNOTATION

3.1 In the . . . valley. 3.2] *scored pencil*

1 Orton led an expedition to the Andes from 1867 to 1868; the expedition was sponsored by Williams College, Massachusetts, and the Smithsonian Institution, Washington DC (Elliott 1979). Quito, the capital city of Ecuador, is in Pichincha province in the north central part of the country. The Napo river flows through north-eastern Ecuador and Peru into the upper Amazon river.

2 Orton refers to his paper 'Physical observations on the Andes and the Amazons' (Orton 1868). The *American Journal of Science and Arts* was popularly known as 'Silliman's Journal' after its founder, Benjamin Silliman.

3 In *South America*, p. 130, CD referred to fossil gastropods from Paita (a province in north-western Peru) that had been described by Alcide Charles Victor Dessalines d'Orbigny. In Orton 1868, p. 116 n., Orton mentions *Cerithium laeviuscula*, *Ostrea gallus*, and *Ampullina ortoni.*

4 Riobamba is a city in Chimborazo province in central Ecuador. The deposit of bones is mentioned in Orton 1870, p. 154.

5 Pebas is a town in north-eastern Peru on the banks of the upper Amazon; Pará is a region (now state) in northern Brazil. The species *Neritina pupa*, *Turbonilla minuscula*, *Mesalia ortoni*, *Tellina amazonensis*, as well as two others are mentioned in Orton 1870, p. 283. The passage is scored in CD's copy (see n. 11, below).

6 Orton refers to Henry Walter Bates, Alfred Russel Wallace, and Louis Agassiz. In Orton 1870, p. 282 n., Orton refers to *A journey in Brazil* for Agassiz's view that the sandstone deposits in the area had resulted from glacial drift and showed no sign of marine origin (see J. L. R. Agassiz and Agassiz 1868, pp. 250, 411, and 424).

7 Antisana is a volcano in north central Ecuador near Quito. The grebe was later identified as *Podiceps occipitalis* (the silvery grebe; see Orton 1870, p. 146). CD had collected specimens of this bird (identified as *P. kalipareus* in *Zoology* pt. 3, p. 136) during the *Beagle* voyage.

8 Esmeraldas and Napo are provinces in northern Ecuador on the western and eastern sides respectively of the Andes.

9 For a list of the naturalists who identified and catalogued the specimens collected, see Orton 1870, pp. xii–xiii.

10 *Journal of researches.*

11 Orton refers to *The Andes and the Amazon* (Orton 1870). The dedication reads, 'To Charles Darwin, M.A., F.R.S., F.L.S., F.G.S., whose profound researches have thrown so much light upon every department of science, and whose charming "Voyage of the Beagle" has so pleasantly associated his name with our southern continent, these sketches of the Andes and the Amazon are, by permission most respectfully dedicated.' CD's annotated copy is in the Darwin Library–CUL (see *Marginalia* 1: 644–5).

To Julius Victor Carus 5 January 1869

Down. | *Bromley.* | *Kent. S.E.*
Jan 5. 1869.

My dear Sir

I am going to beg a favour of you which I shall be very much obliged if you will grant.

The females of the Merino breed of sheep do not possess horns, & I am particularly anxious to learn at what age the horns are developed in the young rams, in comparison with other breeds in which both sexes have horns. I am very anxious to know whether the horns in the young Merino rams appear *earlier* or *later*, or grow *quicker* or *slower* than in other breeds of sheep in which both sexes are horned in the same district. Merinos appear to be extinct in England, but as Saxony is their head quarters perhaps you c^d communicate with some agriculturist, & procure for me trustworthy information, & this w^d be a very great kindness.[1]

It caused me very great regret that owing to my absence from home I did not have the pleasure of making your acquaintance when you were in England.[2]

My dear Sir | yours very sincerely | Ch. Darwin

P.S. Do you know how your translation of my last book has sold?[3]

LS(A)
Staatsbibliothek zu Berlin – Preußischer Kulturbesitz, Slg. Darmstaedter Lc 1859 Darwin, Charles, Bl. 37/38

1 In the nineteenth century Saxony was a centre for the breeding of merino sheep; the breed had been popular in England earlier in the century but had declined dramatically (see Ponting 1980, pp. 16–21, 23–4).

2 Carus had been in England for the meeting of the British Association for the Advancement of Science at Norwich in August 1868, but CD was staying at Freshwater on the Isle of Wight from 17 July until 20 August 1868 (see *Correspondence* vol. 16, letter to J. V. Carus, 16 August [1868], and Appendix II).

3 CD refers to the German translation of *Variation* (Carus trans. 1868).

To Sven Nilsson 5 January 1869

Down. | *Bromley.* | *Kent. S.E.*
Jan 5 1869

Dear & Honoured Sir

I thank you sincerely for your great kindness in having taken so much trouble to oblige me.

The information which you have sent is exactly what I wanted, & will be of the greatest value to me.[1]

I am very proud to be able to add your photograph to my collection of eminent naturalists.[2]

With my best thanks & the most sincere respect | I remain my dear Sir | yours truly obliged | Charles Darwin

LS
University Library of Lund

1 In his letter of 31 December 1868 (*Correspondence* vol. 16), Nilsson had sent CD information on the age at which reindeer acquired antlers and on the possession of antlers by both sexes. In *Descent* 1: 288, CD cited Nilsson for the information.

2 Nilsson had enclosed his photograph with his letter of 31 December 1868, but it has not been found in the Darwin Archive–CUL.

To George Cupples [6–9? January 1869][1]

For example, and before I in the least knew what the result would be, I fixed on this as a crucial instance,—the reindeer *alone* (of *cervidæ*) has horns in both sexes, therefore, according to my rule, their horns should be developed very early in life;[2] and I now hear from Sweden that these appear within two or three weeks after birth:[3] whilst with all other deer, in which the horns are confined to the male, these do not, as far I have hitherto ascertained, appear till nearly a year after birth. So it is with the horns of antelopes. Now you will see that if a large part of the variation in stature occurs late in life among male deer-hounds, this variation will (on such a principle) tend to be transmitted to the males alone, and will not affect the females as any other ordinary variation would do.[4]

Incomplete
Cupples 1894, p. 165

1 The date is conjectured from the relationship between this letter and the letters from George Cupples, 4 January 1869, 11 January 1869, and 21 January 1869, and the letter to Sven Nilsson, 5 January 1869. The responses contained in Cupples's letter of 21 January 1869 indicate that he had passed on CD's query about the first appearance of horns in the sexes of different species of deer.

2 The reindeer is *Rangifer tarandus* (family Cervidae). For CD's theory of the emergence of characteristics at different developmental stages, see *Descent* 2: 244.

[3] CD had received the information from Sven Nilsson (see letter to Sven Nilsson, 5 January 1869 and n. 1).

[4] In *Descent* 2: 261–2, CD cited Cupples for information on the development of deer-hounds.

From George Maw 9 January 1869

Benthall Hall, | nr. Broseley.
Jany 9th. 69

Dear Sir

I am looking forward to visiting Gibraltar. Tangiers & Tetuan & if I find it tolerably safe shall try & get as far south as Mount Anna (20 miles S of Tetuan & 7219 feet high) for the purpose of collecting plants & examining the geology. I also hope to visit the district N of Malaga.[1]

I write this to say that if there is anything in the parts I am about to visit about which I can collect information that will be of use to you (as to local breeds of domesticated animals &c &c &c.) I shall be most happy to place myself at your service if you will name any points on which you require special observations (or specimens collected) I think I shall sail from Liverpool on the 3rd. of Feby & be absent Five or Six weeks.

I have taken the liberty of forwarding for the favor of yr acceptance a copy of a paper I have recently written on Ferruginous variegation.[2]

I am dr Sir | yrs very truly | Geo Maw

C Darwin Esqre

We have here a Tortoiseshell cat exhibiting rather curious characters Her fur is almost as long as that of the long haired Persian or French, but there is no long-haired cat in the neighbourhood the mother was a Tabby I think & all the other kittens of the same litter & succeeding litters were of the ordinary kind. If you would like to see her I shall be very happy to bring her up with me the next time I am in Town & can also ascertain all particulars as to her pedigree

DAR 171: 102

[1] Tangiers (also spelled Tanger) and Tetuan (or Tétouan) are cities in northern Morocco. Málaga is a port city in southern Spain.

[2] Maw refers to his paper 'On the disposition of iron in variegated strata' (Maw 1868). CD's copy has not been found.

From George Cupples 11 January 1869

The Cottage | Guard Bridge | Fifeshire. N.B.
Jany. 11/69.

Dear Mr Darwin,

Thanks for your kind letter.[1] I have no doubt I can get exact information on the point you refer to—and by this post write on *fallow-deer* to *Mr Wright* (whom I shall request to communicate direct with you.)[2] I also write to Mr Macneil of Colonsay as to red-deer and roe-deer—and about the same to my friend the head-forester of the great Breadalbane forest of Black Mount, who has been more than 40 years among deer.[3] I shall write to you with their replies as to all three species.

My brochure on deerhounds, I hope, may get out this year—but has been kept back by other literary matters.[4]

Yours very truly | George Cupples

Charles Darwin, Esqr.

DAR 161: 288

[1] Cupples probably refers to the letter to George Cupples, [6–9? January 1869].

[2] Cupples refers to John Wright. See letter from George Cupples, 4 January 1869 and n. 3.

[3] Cupples refers to Archibald McNeill and Peter Robertson. McNeill had sent information to Cupples on the size of male and female deer-hounds and Robertson had given information on the proportion of the sexes in the breed (see *Correspondence* vol. 16, letter from George Cupples, 13 July 1868).

[4] Cupples's only work on deer-hounds (Cupples 1894) was published posthumously.

From Fritz Müller 12 January 1869

Itajahy, S^a Catharina, Brazil
January 12. 1869.

My dear Sir

As the Eschscholzia-seed, which I formerly sent you, has not germinated, I send you some fresh seeds from my plants, which this summer again have proved self-impotent.[1]

The plants raised from your seeds have suffered so much from the extraordinary heat and the heavy thunderstorm, which in November followed after an extremely rainy winter and spring, that I could make but very few experiments.[2] Only four plants escaped and three of these flowered tolerably well. My own plants, which were introduced here about six years ago, suffered but little from the weather and none has perished.—

Here are the experiments, I tried on the Eschscholzia raised from your seeds:

First plant:

1.)	*Octbr. 23.*	A flower fertilized with its own pollen.
	Novbr. 15.	Germen, 12^{mm} long, begins to whither.
2)	*Novbr. 3.*	Fertilized two flowers, one (*a*) with pollen from a distinct flower of the plant, the other (*b*) with pollen from a distinct plant.
	Novbr. 5.	Stigmas of the flower (a) fresh, those of (b) withering
	Novbr. 9	Germen of (a) 12^{mm}, of (b) 26^{mm} long
	Novbr. 11	. . . . 19^{mm}, ... 47^{mm}—
	Novbr. 15	. . . . 30^{mm}, ... 56^{mm}—.

	Novbr. 30.	Fruits ripe; the pod (a) 32^{mm} long, with 10 seeds, 4 of which are very small; the pod (b) 58^{mm} long, with 59 good seeds.
3)	*Novbr. 9*	Repeated the same experiment.
	Novbr. 10	Stigmas of (a) fresh, those of (b) withering.
	Novbr. 15	Germen of (a) 11^{mm}, that of (b) 18^{mm} long
	Novbr. 18.	— — — — – 12^{mm}, — — — 49^{mm} long. The pod (a) fell off unripe; the pod (b), 53^{mm} long, yielded 45 seeds (Decbr. 4.)

Second plant.

1.)	*Novbr. 1.*	Fertilized two flowers with each other's pollen.
	Novbr. 2.	Stigmas withering, having remained in a horizontal position in one of the flowers, while the had become [exerted] in the other.
	Novbr. 9.	Germens 16^{mm} and 46^{mm} long
	Novbr. 11.	— — — 17^{mm} — 50^{mm} —
	Novbr. 15.	The smaller germen withered, the larger 53^{mm} long
	Novbr. 30.	Pod ripe, 56^{mm} long, with 24 apparently good seeds (a remarkably small number for so large a pod.)
2.)	*Novbr. 19,*	midday. Fertilized one flower (a) with pollen from a distinct flower of the plant; a second flower (b) with pollen from a distinct plant; a third flower (c) simply protected from insects.
	Novbr. 22.	Germens of (a) and (c) 10^{mm}, that of (b) 16^{mm} long
	Novbr. 27.	Germen of (a) 40^{mm}, of (b) 64^{mm}, of (c) 32^{mm} long. The extremity of the germen (c) putrifying.
	Novbr. 30.	Germen of (a) 40^{mm}, that of (b) lost by an accident, that of (c) withered.
	Decbr. 13.	The pod (b) ripe, 45^{mm} long, with 9 seeds.
3.)	*Novbr. 25.*	Fertilized two flowers (a, b) with pollen from distinct flowers of the plants, and one flower (c) with pollen from a distinct plant.
	Novbr. 30.	Germen of (a) not increased, (b) 10^{mm}, (c) 27^{mm} long.
	Decbr. 6.	Germen (a) and (b) withered; (c) 40^{mm} long.
	Decrb. 17.	Germen (c) putrified.

A third plant, all the germens of which withered or putrified, showed the same difference in the growth of the germens fertilized with the same plants and a distinct plant's pollen: The experiments ought to be repeated on a larger scale and on *healthy* plants.

I have lately met with a most curious grass, the *Streptochæta* Nees.[3] The rhachis of the simple ear is prolonged beyond the last floret and its extremity is club-shaped and densely covered by strangely curled short, stiff hairs. Each floret has an extremely long barb, which adheres to the club-shaped extremity of the rhachis and then contracts spirally in reversed directions, just as a tendril would do after having caught a support. This is effected long before the ear begins to peep out from the enclosing leaves. When the seeds are ripe, the florets hang down, by means of their tendril-like barbs, from the tip of the rhachis,—probably till they are carried away and disseminated by some passing-by animal. It is curious that a contrivance for dissemination sh^d be formed at so early a period, long before the flowers expand.— Under a morphological point of view, the Streptochæta is not less remarkable; *Endlicher*, indeed, calls it "gramen ad modum paradoxum".[4] There are not only six stamens and three stigmas (as in a few other grasses, Pharus, ec.), but also *six leaves in two whorls* surrounding the sexual organs. Thus *Streptochæta* forms a connecting link between common grasses and other monocotyledons, in the same way, in which some "paradoxal" animals (such as Lepidosiren, Ornithorhynchus ec.) connect classes now widely separated.[5] I think we may look at it as a very old form, much less modified in structure of its flowers, than any other living grass.—

Ear of Streptochæta Nees. (nat. size.)

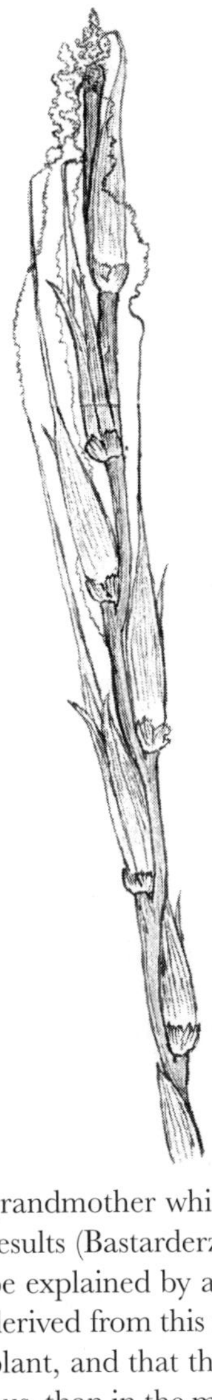

Our Brazilian cuckoos, according to Burmeister, agree in their habits with the North-american species;[6] but we have here another bird, not belonging to the cuckoo-family, which, like the European cuckoo, lays all her eggs, as far as I can make out, in other birds' nests. In the nests of the Tico-tico (Zonotrichia matutina Gray)[7] and other small birds a white egg, about 24^mm long and 18^mm in diameter, with the two ends equally rounded, is occasionally found. Sometimes even two of these eggs are found in the same nest. I have inquired several boys, who attend much to birds' nests, but none had ever seen a nest containing only these white eggs and which therefore might have been supposed to be the own nest of the parent-bird— I am told, that this bird is a *Trupialis*.[8] I wished to raise young birds; but one, which I found this summer was devoured by a snake or bird of prey, when three days old, and another was destroyed by a thunderstorm.—

On an ear of maize I found, last year, six dirty bluish-grey grains, the rest being pale-yellow. I sowed separately the bluish and the yellow grains; on 30 pale-yellow ears, from the yellow seeds. There were but 5 bluish grains and 4 were spotted with violet. On 3 pale-yellow ears, from the bluish seeds, there were 581 pale-yellow and white, 193 bluish grains and 78 of mixed colours (yellow with bluish ring, or spotted with violet, or dark bluish-grey with a white central spot ec.)— 9 ears from the yellow and one from the bluish seed had reverted to the colour of their grandmother which had dark brown grains. *Gaertner*, as you know, obtained similar results (Bastarderzeugung pg. 322):[9] On the hypothesis of Pangenesis the fact might be explained by assuming, that in the seeds of each individual flower the gemmulæ derived from this flower are more numerous than those from any other flower of the plant, and that therefore in the offspring flowers of the same kind are more numerous, than in the mother-plant.—[10] Do you think, that this will hold good as a general rule? I intend to try some experiments on plants which bear at the same time flowers with 4, with 5 and with 6 petals.—

With sincere respect | very faithfully yours | Fritz Müller.

DAR 76: B34–5

CD ANNOTATIONS

Top of letter: 'I find that the seed supplied by Cattell is Eschotzia crocea—[Probably] as we have experimented on distinct species' *ink*[11]

End of letter: '(*Precise* facts on inheritance from individual flowers— I think new— good for Pangenesis)' *blue crayon*[12]

1 Müller first discussed the self-sterility of *Eschscholzia californica* (the California poppy) in a letter of 2 August 1866 (*Correspondence* vol. 14). No earlier letter from Müller mentioning enclosed seeds of *E. californica* has been found.

2 CD had twice sent Müller seeds of *Eschscholzia californica* in 1868 (see *Correspondence* vol. 16, letters from Fritz Müller, 22 April 1868 and 17 June 1868).

3 The grass is now *Streptochaeta* Schrader *ex* Nees (Mabberley 1997).

4 Stephan Endlicher described the plant as 'Gramen brasiliense, admodum paradoxum' (Brazilian grass, very strange; Endlicher 1836–42, p. 103).

5 Most grasses have a reduced number of stamens and stigmas (three and two respectively is typical) compared with other monocotyledons, and lack the outer whorl (a complete flower has four whorls: pistil, stamens, corolla, and calyx). For more on the floral structure of *Streptochaeta* in its evolutionary context, see Rudall and Bateman 2004, pp. 32–4, and Rudall *et al.* 2005, p. 1437. Müller also refers to the genera *Lepidosiren* (lungfishes; at this time all lungfishes were placed in this genus, which is now restricted to the South American lungfish; see Pauly 2004, pp. 130–2) and *Ornithorhynchus* (the duck-billed platypus). In the nineteenth century, the former genus was often seen as a link between fish and reptiles, the latter as a link between birds and mammals.

6 For Karl Hermann Konrad Burmeister's description of the nesting habits of Brazilian cuckoos, see Burmeister 1854–6, 2: 251–2.

7 *Zonotrichia matutina* is now considered to be a subspecies of *Z. capensis* (the rufous-collared sparrow; tico-tico is the Portuguese common name).

8 *Trupialis* is now *Sturnella*; this genus does not include nest parasites, but like *Zonotrichia matutina* is itself parasitised by *Molothrus bonariensis* (the shiny cowbird).

9 Karl Friedrich von Gärtner reported his experimental results on *Zea mays* in *Versuche und Beobachtungen über die Bastarderzeugung im Pflanzenreich* (Experiments and observations on hybridisation in the plant kingdom; Gärtner 1849, pp. 322–7; for CD's annotations to his copy, see *Marginalia* 1: 256–98).

10 For CD's hypothesis of pangenesis, see *Variation* 2: 357–404. CD had theorised that a certain number of gemmules (heritable units) were required for the development of each cell or part and that the diffusion of gemmules was not homogeneous (see *Variation* 2: 375 n. 29, and 379–81).

11 CD refers to *Eschscholzia crocea*, now considered to be a synonym of *E. californica*, and to John Cattell, a nurseryman based in Westerham, Kent (*Post Office directory of the six home counties, 1859–66*).

12 CD refers to Müller's observations on maize (see above, n. 10).

From Lydia Ernestine Becker 13 January 1869

Manchester National Society for Women's Suffrage | 232 Brunswick St | Manchester
Jany 13. 1869

My dear Sir

I am just enjoying the long deferred pleasure of studying your book on animals and plants under domestication and I hope you will not think it wrong in me before

I come to the object of my note to say with what pleasure and wonder I endeavour to follow the facts and reasonings there unfolded. Especially the chapter on Pangenesis, which seems like a revelation.[1] The power of imagination which can thus realise the Unseen seems akin to inspiration— Though you only announce it as a provisional hypothesis to connect hitherto unexplained facts, the key fits the puzzle so beautifully that I cannot help supposing it to be the right one—and nothing but the presentation to my mind of a more probable theory could detach it from the firm hold it has taken of yours If I might venture a prophecy I would say that the hypothesis of Pangenesis is likely to have at least as great a result in giving value and coherence to scientific thought and investigation as the earlier one with which your name is connected.

But of course I should not have intruded on your attention to offer an opinion on your book which I fear you may think presumptuous— It is with reference to a passage p. 154—vol 2—on parrots under domestication that I write to you—[2] When you wrote it I think you could hardly have been aware of what you may probably have heard now—the existence of a colony of parrots naturalised at Northrepps Hall near Cromer—the residence of the Dowager Lady Buxton—[3] I dare say you know all I am going to relate if so, my apology must be the supposition that you might prefer the chance of hearing more than once of an interesting fact to the possibility of never hearing it at all.

At the last meeting of the British Association a party of the members were entertained at Northrepps Hall.[4] On entering the grounds, a beautiful sight presented itself— The garden was surrounded with lofty trees whose branches were all alive with great white cockatoos—green parrots and gray—flying about like pigeons. At the entrance was a large elm tree bearing a signboard, on which was inscribed "Parrot's nest and a ladder was placed for the convenience of any enterprising member of the B.A. who might be inclined to take a peep at the erection a large heap of twigs like an enormous rook's nest, but resting on a thick branch— We were informed that a young brood had just tenanted the nest and it had been hoped they would be obliging enough to remain till the scientific visitors came to see them but they had perversely flown the day before— I saw what I suppose to be this young brood nursing in a large cage—six baby-parrots all in a row with their little red tails sprouting nicely—

We were told that the birds had been domesticated there for thirty years, that they breed freely and were perfectly naturalised—but they have the shelter of a conservatory provided for winter— They are at perfect liberty, and sometimes fly miles away—but when they do this, they frequently get shot— Many have been destroyed in this way—

I saw an old green parrot who I was told was the original progenitor in the colony—and had the honour of carrying her about on my hand— The birds were tamer than ordinary pigeons— they would fly down for food—perch on one's hand, and allow themselves to be carried about without any symptoms of alarm—

Mr. Charles Buxton M.P—read a very interesting paper giving an account of their history and habits.[5] I did not take notes, so I cannot give you many particulars but doubtless he would be most willing to afford you full information respecting them—

He said that in the middle of the day the cockatoos retired to roost in the trees, and then the whole colony was invisible— he said it was very wonderful how such large white birds could contrive to hide so completely among the branches of English fruit trees, that not a feather could be seen.

They had come out for the evening by the time we got there and the sight was one of the most charming I have ever seen— But it makes one sigh to see these glorious tropical birds so thoroughly at home in our lands, and to reflect that but for the barbarous an inhuman practice of shooting every lovely feathered stranger that puts in an appearance among us we might hope to see our parks and woods decked with these gorgeous creatures as our gardens and conservatories are with tropical plants.—

We were told that the liberty in which these birds lived did not destroy their imitative faculties—many of them learned to talk—

Again apologising if I have troubled you with a needless communication

Believe me to be | Yours very truly | Lydia E. Becker

Charles Darwin Esq.

DAR 160: 105, 116

1 Becker refers to *Variation* and to the chapter on CD's hypothesis of pangenesis (*Variation* 2: 357–404).
2 In *Variation* 2: 155, CD had discussed the difficulty of breeding parrots in captivity.
3 Hannah Buxton.
4 In 1868, the meeting of the British Association for the Advancement of Science was held at Norwich from 19 to 26 August.
5 Buxton's paper, 'Acclimatization of parrots at Northrepps Hall, Norfolk', was published in *Annals and Magazine of Natural History* (Buxton 1868). In *Variation* 2d ed., 2: 138, CD cited Buxton on parrots breeding. He also referred to Buxton's paper in *Descent* 1: 76, 2: 109.

To Joseph Dalton Hooker 13 January 1869

Down. | *Bromley.* | *Kent.*
Jan 13. 69

My dear Hooker,

As I believe you will not much grudge the trouble I send by this post 13 pages of M.S, **well written**, in answer to Nägeli for the new Edit. of the Origin.[1] I should be extremely obliged if you would read it over, and see whether I have made any blunders, as is very likely to be the case. I have quoted Aug. St Hilaire from my own notes, & when I made them I certainly understood what he meant, but now I am not so sure that I do, and therefore am the more likely to blunder.[2] I quote Asa Gray on Mimulus and Dr Masters on Saponaria from the Abstracts given in the Gardiner's Chron: of papers read to the Linn. Soc. 1856 Ap. 15 & Nov. 18; and these papers I have not got; but the Abstract in the G. Chron. seems clear.[3]

Besides looking out for blunders I should of course be grateful for any criticism, but I do not want to be unreasonable, knowing well how busy you are.

Ever yours | Ch. Darwin

The style of my M.S. will no doubt require improvement.—

LS(A)
DAR 94: 110–11

[1] The manuscript has not been found, but see *Origin* 5th ed., pp. 151–7. Carl Wilhelm von Nägeli's monograph *Entstehung und Begriff der naturhistorischen Art* (The origin and concept of natural historical species; Nägeli 1865) was about the mechanisms and principles operative in the development and transmutation of species. CD's copy of Nägeli 1865 is heavily annotated and there is a partial manuscript translation, beginning on page 15 of the German text, along with a page of notes by CD, in the Darwin Pamphlet Collection–CUL. Nägeli had opposed what he termed the theory of 'usefulness' ('Nützlichkeitstheorie', that is, CD's transmutation theory) to his own theory of perfectibility ('Vervollkommnung'), which posited an inner tendency towards a more complex organisation (Nägeli 1865, p. 29). He claimed that CD's theory could only account for functional changes whereas his own theory explained development in purely morphological characteristics that had no functional advantage. For more on Nägeli's theory, see *Correspondence* vol. 16, letter to J. D. Hooker, 25 December [1868].

[2] In *Origin* 5th ed., pp. 155–6, CD referred to Auguste de Sainte-Hilaire's discussion of varying numbers of ovaria in *Zanthoxylon*, varying numbers of locules in *Helianthemum*, and different forms of *Gomphia oleaeformis* (probably a misspelling of *G. oleaefolia*). Sainte-Hilaire discussed *Gomphia* in Sainte-Hilaire 1841, pp. 507–8.

[3] For the reference to Gray on *Mimulus*, see *Origin* 5th ed., p. 155 (see also *Gardeners' Chronicle*, 19 April 1856, p. 263, and Gray 1856). In *Origin* 5th ed., p. 156, CD mentioned Maxwell Tylden Masters's observations on *Saponaria officinalis* (see also *Gardeners' Chronicle*, 6 December 1856, p. 806, and Masters 1856).

To George Maw 13 January 1869

Down. | Bromley. | Kent.
Jan. 13. 1869.

My dear Sir,

I am very much obliged for your kindness in offering to observe anything for me in your tour, which I hope will be in every way successful.[1]

There are two points, but it is a thousand to one against your being able to aid me.

The rams alone of Merino sheep have horns, and I want much to know how many months or weeks after birth the horns appear. The answer would be much more useful, if there exists in the same country a breed of sheep with both sexes horned, and if I could learn how many months after birth the horns in such breeds first appear.[2]

Secondly I have wished during many years for a living specimen of the Drosophyllum; I know that it grows near Gibraltar, but as it is rare my wish to compare it with Drosera is I fear quite hopeless.[3]

Many thanks for your Memoir which relates to a subject that formerly interested me greatly.

The illustrations are admirable and must have put you to much expense and the whole subject cost much labour.[4]

My dear Sir | Yours very faithfully | Charles Darwin

LS
Royal Horticultural Society, Lindley Library

1 See letter from George Maw, 9 January 1869.
2 For more on CD's interest in merino sheep, see the letter to J. V. Carus, 5 January 1869.
3 *Drosophyllum* is a genus of insectivorous plants; *D. lusitanicum* (the Portuguese dewy pine) is the only species. It is found in parts of Portugal, Spain, and northern Morocco. CD first began experiments on *Drosera rotundifolia* (the common sundew) in the summer of 1860, and worked on other species of that genus as well (see *Correspondence* vol. 8). He had advised Joseph Dalton Hooker to acquire a specimen of *Drosophyllum* as early as 1864, but Hooker was evidently unsuccessful (see *Correspondence* vol. 12, letter to J. D. Hooker, 26 November [1864] and n. 15).
4 See letter from George Maw, 9 January 1869 and n. 2. Maw 1868 contains five pages of coloured illustrations showing the disposition of iron in variegated strata.

From J. D. Hooker 14 [January] 1869[1]

Royal Gardens Kew
Dec 14/69

Dear Darwin

I have been 10 days in Country & find on my return that Oliver has your letter but overlooked your request about Rutaceous flowers. We fail to find the reference in S^t. Hilaire, but Payer points out that the variations displayed by the uppermost flower, are by no means confined to it.[2] Please let me know more precisely the point anent ovules you want illustrations of— is it

1 of ovules of different position in same cell (as Sabia)

2 of ovules differing in position in different cells of same ovary (as Linnæa & Diervilla).

3. of ovules differing in position in different flowers of same plant.

My wife was staying with some wealthy well educated (soi disant) people in Suffolk who never heard of Darwin or Darwinism.— she describes them as "etiolated." except one who took you for your grandfather.[3]

Babington is "very much surprized at Dr Hooker's advocacy of Darwinian views at Norwich, & observes that it has greatly disappointed many of Dr Hooker's friends & well-wishers".[4] I feel like the Parrot which was in the habit of saying in a tone of great contempt, after the family-prayers were over, "My God". Or like the Turk in Hogarth's picture, calmly smoking his pipe as he gazes in through the window of a Church when the congregation are in a state of religious excitement.—[5]

Ever yours affec | J D Hooker

I was at Hardwick & Barton & heard such loud praises of your boy.[6]

DAR 48: A78, DAR 103: 3

CD ANNOTATION

4.1 3. . . . plant.] *scored blue crayon*

Top of first page: 'Spanish chesnut | Vaucher [*illeg*] | Rue generally | ovules differing in position as in different [*illeg*] of same plant | Is agreement good for [*2 words illeg*]' *pencil, underl and del pencil*; 'Viola nana' *pencil*[7]

1 Hooker evidently wrote December in error. The month is established by the relationship between this letter, the letter to J. D. Hooker, 29 December 1868 (*Correspondence* vol. 16), and the letter to J. D. Hooker, 16 January [1869].

2 In his letter to Hooker of 25 December [1868] (*Correspondence* vol. 16), CD had asked about morphological differences in the flowers of rue and had commented 'Aug. St Hilaire insists on the locularity of the ovarium varying on the same plant in some of the rutaceæ.' For Auguste de Sainte-Hilaire's remarks on locules of the ovarium, see Sainte-Hilaire 1841, pp. 477, 481–7, and 550–8. No comment by Sainte-Hilaire on some members of the family Rutaceae having ovaria with varying numbers of locules on the same plant has been found. Jean Baptiste Payer discussed variation in flowers of rue in Payer 1857, pp. 73–6 (the comment cited is on p. 73).

3 Hooker refers to Frances Harriet Hooker and Erasmus Darwin.

4 The references are to Charles Cardale Babington and to the presidential address Hooker gave on 19 August 1868 at the meeting of British Association for the Advancement of Science in Norwich.

5 William Hogarth's 1762 engraving is entitled 'Credulity, superstition, and fanaticism'.

6 Charles James Fox Bunbury lived at Barton Hall, Great Barton, near Bury St Edmunds, Suffolk (*Post Office directory of Cambridge, Norfolk, and Suffolk* 1865). According to Bunbury's journal, on 7 January 1869, he and Hooker were guests at Hardwick House, near Bury St Edmunds (F. J. Bunbury ed. 1891–3, *Later life* 2: 4, 6). George Howard Darwin had recently been made a fellow of Trinity College, Cambridge (see *Correspondence* vol. 16, letter from E. A. Darwin, [11 October 1868] and n. 2).

7 CD's annotations are for his letter to J. D. Hooker, 16 January [1869].

From George Maw 14 January 1869

Benthall Hall, | nr. Broseley.
Jany 14. 69

Dear Sir

I am very pleased to find I can be of any use to you. I will make a note as to the age of horning of Merino Lambs & try & get some information[1]

With reference to the drosphyllum could you send me a dried specimen to look at which I will return you.[2] I am not acquainted with the plant nor is it mentioned by Kelaart in his Flora Calpensis or by Munby in the little work he published on the Flora of the N African coast but if I can see a dried specimen I will do my best to find it & send it by post as I shall be some weeks in the neighbourhood of Gibraltar & am generally lucky in finding plants I want to get.[3]

I am Dr Sir | Very truly | Geo Maw

C Darwin Esq

DAR 171: 103

1 See letter to George Maw, 13 January 1869.

2 Maw refers to the insectivorous plant *Drosophyllum* (see letter to George Maw, 13 January 1869 and n. 3).

3 Maw refers to Edward Frederick Kelaart's *Flora Calpensis* (Kelaart 1846) and Giles Munby's *Flore d'Algèrie* (Munby 1847).

From George Cupples 15 January 1869

Guard Bridge | Fifeshire. N.B.
Jany. 15/69

Dear Mr Darwin,

I enclose a letter just now received from Mr McNeill of Colonsay in reference to the deer's horns.[1] I trust it may be of some definite value, though (as usual with some people) it is discursive and rambling, apparently. I had furnished him with an exact quotation of your queries.[2]

I await reply from the Forester of Black Mount,[3] who, I have no doubt, can tell about all three species.

Mr Wright,[4] I presume, will communicate direct.

I venture to ask that in Mr M^c^Neill's case (as he wrote previously about the *relative sizes* of Deerhound dog and bitch) you would send him a line or two in reply.[5] Not that he expects this—indeed *no one* could wish to trespass on *your* time—but he is always ready to serve if he can. He is brother to Lord Colonsay,[6] and himself a scholarly Highland gentleman, as well as an Edinburgh lawyer of high standing—the author of the portion of *Scrope's* book referring to the "long-descended race of Deerhounds".[7]

At your leisure and convenience you might perhaps return Mr M^c^Neill's letter. Any others I may receive on the question shall not give you such trouble. I *do* hope they may furnish the facts required—and that the Spring may recruit your health.

Ever truly yours | George Cupples

To | Charles Darwin, Esq. | &c &c—

DAR 161: 289

1 Archibald McNeill's letter has not been found.
2 See letter to George Cupples, [6–9? January 1869].
3 Peter Robertson.
4 John Wright.
5 For the information sent by McNeill, see *Correspondence* vol. 16, letter from George Cupples, 13 January 1868.
6 Duncan McNeill.
7 Cupples refers to William Scrope and Scrope 1838, pp. 333–59. The chapter was entitled 'A description of the highland deerhound, with an account of a day's deer-coursing in the island of Jura. (Communicated by Archibald Macneill, Esq., of Colonsay)'.

From Thomas Henry Farrer 15 January 1869

3 Gloucester Terrace | R.P
15. Jan/69

My dear Mr Darwin

You always contrive to send something pleasant. It comes alas! in the midst of much Red Tape which leaves no time for Botany.[1] New Ministers laudably attempting economies—many of them, I fear, impossible, give much trouble: almost as much, though more satisfactorily, as expensive & jobbing Ministers.[2]

Would Mrs Darwin or Miss D. kindly give me one line to say whether I shall send the beautiful French Book your son so kindly lent me back to Cambridge I heard with much regret that an accident had kept him there.[3]

Believe me | Most truly yours | T H Farrer

C Darwin Esq FRS

DAR 164: 51

[1] CD's letter to Farrer has not been found. Farrer was the sole permanent secretary of the Board of Trade (*ODNB*).

[2] Parliamentary elections had taken place in November 1868; re-elected member John Bright became president of the Board of Trade on 11 December 1868 (*The Times*, 12 December 1868, p. 6).

[3] Farrer refers to Emma Darwin and Henrietta Emma Darwin. George Howard Darwin had sent Farrer *Traité générale de botanique descriptive et analytique* (Le Maout and Decaisne 1868) from Trinity College, Cambridge (*Correspondence* vol. 16, letter to T. H. Farrer, 26 November 1868 and n. 3).

From J. D. Hooker 15 January 1869

Royal Gardens Kew
Jany 15/69.

N°. 2

Dear Darwin

I do not quite like the starting by shirking the question of what is a "Morphological character"—[1] you *imply* that it is a term of indefinite meaning— you talk of what— "*he* calls M. characters" & of what "*I* presume likewise to be M. characters"— I think that Non-Scientific readers will at once say, "how little these men know of what they write so much about, when their fundamental terms have no definite meaning"— all characters i.e. all departures from a given structure are & must be morphological.— all originate in the fact, that every individual varies from it's parent; & this from being subject to "the direct & definite action of the conditions of life"—(an admirable definition, Weismann's is not intelligible to me—if sense at all)[2]

p. 3. at A. this is very mildly put— would it not better meet Nægeli's objection, which seems to point to *histological characters* (& to which & symmetry he probably confines his use of term "morphology.") to add "nor do we know the uses of all the special *tissues* of any one organ"[3]

p. 4. At B. Furthermore, though these arrangements of leaves are reducible to mathematical laws, &—might hence be presupposed to be the most constant of all the laws of vegetable growth, and to be absolute & irrefragable, they prove not to be so—shewing that even here is variation which no one could call *progressive*! capable of transmission & ready for the action of selection.[4]

p. 5. What is *Viola nana*?, we can find no such name in our books.[5]

p. 6. Better add which Chesnut— I have referred to Payer, he distinctly says it is not constantly the terminal flower[6]

7— In Adoxa calyx-lobes is better than *sepals*—but this is hypercriticism. I have added note on Himal. sp or form.[7]

8. Leave out Asa Gray— all the world knew it.— for seeds read *fruit* or *achene*.[8]

DC. divided Umbelliferae primarily by this character! it is *exploded* in Gen. Pl.[9]

10 A little confusion at top of page— I have set it right.

Ever yr affec | J D Hooker

DAR 103: 1–2

CD ANNOTATIONS
1.1 I do ... at all) 1.9] *crossed pencil*
1.1 I do ... character"— 1.2] *scored blue crayon*; 'I do **not** know— if not *[illeg]* variation but as I am *[illeg]* I *[illeg]*' *added pencil*
3.1 p. 4. ... selection. 3.5] *enclosed in square brackets blue crayon*
4.1 p. 5. ... right. 9.1] *crossed pencil*

1 CD had sent Hooker a thirteen-page manuscript containing arguments against Carl Wilhelm von Nägeli's theory of perfectibility (see letter to J. D. Hooker, 13 January 1869 and n. 1).
2 In the version that appeared in print in *Origin* 5th ed., pp. 151–7, CD referred simply to 'morphological characters'. For August Weismann's discussion of morphological variation, see Weismann 1868, pp. 27–9. CD's annotations to his copy of Weismann 1868 contain a discussion of Weismann's interpretation of a morphological character (see *Marginalia* 1: 860–1).
3 In the printed version (*Origin* 5th ed., p. 152), CD wrote: 'No one will maintain that we as yet know the uses of all the parts of any one plant, or the functions of each cell in any one organ.'
4 In the printed version (*Origin* 5th ed., p. 152), CD referred to the possible explanation of the angles at which leaves are placed relative to the stem, but did not mention mathematical laws.
5 In the printed version (*Origin* 5th ed., p. 153), CD referred to 'some species of Viola'. *Viola nana* is a synonym of *V. tricolor*; CD had received seeds of *V. nana* from John Scott (see *Correspondence* vol. 15, letter from John Scott, 24 September 1867 and n. 7).
6 In the printed version (*Origin* 5th ed., p. 154), CD referred to the varying angles of divergence in leaves of the Spanish chestnut, but did not discuss the flowers. The reference to chestnut flowers by Jean Baptiste Payer has not been found.
7 In the printed version (*Origin* 5th ed., p. 154), CD referred to variations in the calyx-lobes of *Adoxa*. A Himalayan form is not mentioned.
8 In the printed version (*Origin* 5th ed., p. 154), CD referred to the 'achenes or seeds' of central and circumferential flowers of many Compositae and Umbelliferae differing in character.
9 Augustin Pyramus de Candolle had divided the Umbelliferae (now Apiaceae) into three suborders, Orthospermae, Campylospermae, and Coelospermae, based on the character of the seeds (A. P. de Candolle 1829, p. 6 *et seq.*). For the classification of Umbelliferae in *Genera plantarum*, see Bentham and Hooker 1862–83, 1: 859–931. In *Origin* 5th ed., p. 154, CD noted that in certain Umbelliferae, exterior seeds were orthospermous but the central one was coelospermous.

To J. D. Hooker 16 January [1869][1]

Down. | Bromley. | Kent. S.E.
Jan 16th.

My Dear Hooker

Your two notes and remarks are of the utmost value, and I am greatly obliged to you.[2] Your criticism on the term "morphological" seems quite just, but I do not see how I can avoid using it.[3] I found after writing to you in Vaucher about the rue but from what you say I will speak more cautiously.[4] It is the Spanish Chesnut that varies in divergence.[5] Seeds named "Viola nana" were sent me from Calcutta by Scott; I must refer to the plants as an "Indian species", for though they have produced hundreds of Closed flowers, they have not boarn one perfect flower[6] You ask whether I want illustrations "*of ovules differing in position in different flowers on the same plant*". If you know of such cases, I should certainly much like to hear them. Again you speak of the angle of leaf-divergeance varying, and the variations being *transmitted*; was the latter point put in in a hurry to round the sentence; or do you really know of cases?[7]

Whilst looking for notes on the variability in the divisions of the ovarium, position of the ovules, æstivation, &c, &c, I found remarks, written 15 or 20 ago, showing that I then supposed that characters which were nearly uniform throughout whole groups must be of high vital importance to the plants themselves; consequently I was greatly puzzled how with organisms having very different habits of life this uniformity could have been acquired through Natural Selection.[8] Now I am much inclined to believe, in accordance with the view—given towards the close of my MS., that the near approach to uniformity in such structures depends on their not being of vital importance and therefore not being acted on by natural selection.[9] If you have reflected on this point, what do you think of it? I hope that you approved of the argument deduced from the modifications in the small closed flowers.

Hearty thanks for all you assistance and I will give you no more trouble on this subject.

Yours affectionately | Ch. Darwin

It is rich about Babington, but I daresay many others have said the same thing.[10] It is only about 2 years since last Edit. of Origin, & I am fairly disgusted to find how much I have to modify & how much I ought to add; but I have determined not to add much.—[11] Fleming Jenkins has given me much trouble, but has been of more real use to me, than any other Essay or Review.[12]

LS(A)
DAR 94: 112–13

1 The year is established by the relationship between this letter and the letter from J. D. Hooker, 15 January 1869.

2 CD refers to the letters from J. D. Hooker, 14 [January] 1869 and 15 January 1869.

3 See letter from J. D. Hooker, 15 January 1869 and nn. 1 and 2.

4 See letter from J. D. Hooker, 14 [January] 1869 and n. 2. CD refers to Jean-Pierre-Etienne Vaucher and Vaucher 1841, 1: 557–70. Vaucher's comments on numbers of locules of the ovarium refer to the genus *Peganum* (African rue, now in the family Zygophyllaceae; see Vaucher 1841, 1: 562). In *Origin* 5th ed., p. 155, CD referred to differences in the number of ovaries in some species of the genus *Zanthoxylon* (a synonym of *Zanthoxylum*, prickly ash, family Rutaceae).

5 See letter from J. D. Hooker, 15 January 1869 and n. 6.

6 See letter from J. D. Hooker, 15 January 1869 and n. 5. CD refers to John Scott.

7 See letter from J. D. Hooker, 15 January 1869.

8 The notes CD refers to have not been identified.

9 See *Origin* 5th ed., pp. 156–7.

10 CD refers to Charles Cardale Babington. See letter from J. D. Hooker, 14 [January] 1869 and n. 4.

11 The fourth edition of *Origin* was published in the autumn of 1866 (on the delay in releasing the book, see *Correspondence* vol. 14, letter from John Murray, 18 July [1866]).

12 Henry Charles Fleeming Jenkin had written an article in the *North British Review* criticising some aspects of CD's theory ([Jenkin] 1867). For more on the review and responses to it, see *Correspondence* vol 15. In *Origin* 5th ed., pp. 104–5, CD responded to Jenkin's argument that a variation occurring in a single individual, no matter how advantageous, would not be perpetuated (see [Jenkin] 1867, pp. 288–92; see also Gayon 1998, pp. 97–102).

To George Maw 17 January [1869]

Down. | *Bromley.* | *Kent. S.E.*
Jan 17th

My dear Sir

I have never seen the Drosophyllum.— It is described as somewhat shrubby, with narrow elongate leaves, studded with glandulose hairs, as in Drosera; it was indeed called a Drosera by Linnæus.—[1] It grows in sandy places in Portugal, Andalusia & Teneriffe; & as I said has lately been found near Gibraltar.— My object, if I cd get a living specimen, is to compare its habits & structure with Drosera, on which I have made very numerous observations still unpublished.[2]

With very many thanks | Yours very faithfully | Ch. Darwin

Postmark: 'JA 18 69'
Royal Horticultural Society, Lindley Library

1 Carl von Linné (Carolus Linnaeus).
2 See letter to George Maw, 13 January 1869 and n. 3.

From William Winwood Reade 17 January 1869

Cre. Hon Charles Heddle. | Sierra Leone
Jany 17. '69.

My dear Sir

I have postponed writing to you before in the hope that I might be able to send you some information of value before I started for the interior, which I do in two or three days to search for the source of the Niger. I have unfortunately very little to say— As regards expression I can at present answer but one query.[1] The head is shaken laterally in negative & nodded in affirmative among the Gold Coast tribes.[2] I find it exceedingly difficult to seize expressions if not prepared for them (as one wd. be in case of a surgical operation). Expression is sudden & momentary— In smiling by the bye I observed two crescent wrinkles at the corners of the mouth, concave side nearest mouth— However I shall try to get something more in that way for you

There is one breed of sheep only on the Gold Coast (where I have spent most of my time since coming out). The rams have horns & manes: the females neither. When castrated not only horns but hair is affected.[3] I have frequently asked about the age at wh. the horns sprout but cd. get no answer. Europeans out here do not observe such things, & the blacks have no idea of time. However I have set three persons to watch, & hope to get a reply on my return from the interior, wh. will be in a few months. There is a breed with both sexes horned in Senegambia (with dewlaps) and another breed of enormous size in the Niger Country in wh. the female has rudimentary horns.[4]

Women in Africa at all events among the more intelligent pagan tribes have no difficulty about getting the husbands they want, although it is considered unwomanly to *ask* a man to marry them, & that on the Gold Coast I am informed they never do— They are quite capable of falling in love, and even of forming tender passionate

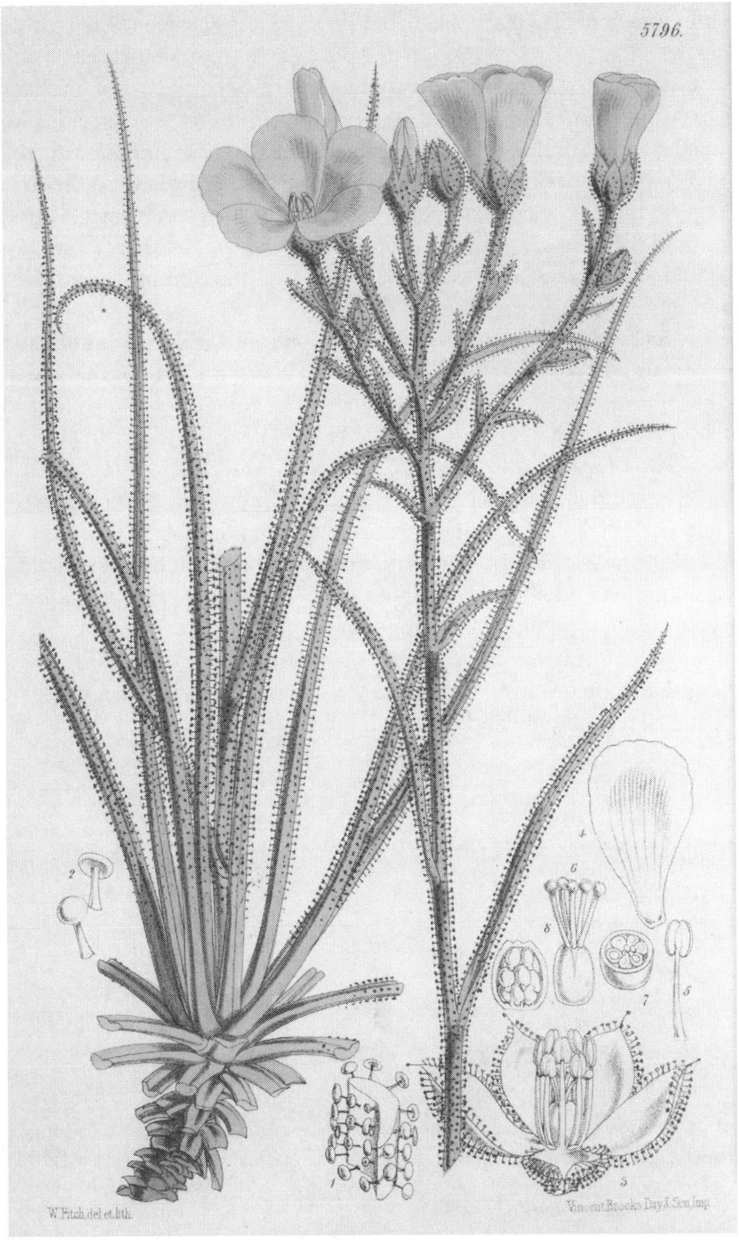

Drosophyllum lusitanicum.
Curtis's Botanical Magazine 3d ser. 25 (1869): plate 5796.
By permission of the Syndics of Cambridge University Library.

and faithful attachments.[5] As a rule tribes do not intermarry but I have little doubt that there are many exceptions. Women do not like to marry strangers visiting their country because they do not like to leave their own families; on the other hand when a Foula or Mandingo comes among the Soosovo or Timmanies (pagan coast tribes in this neighbourhood) they try to make him settle & marry amongst them.[6] They like to breed from superior men that the offspring may be of use to the town a *fact* rather opposed to the vulgar ideas about African want of foresight. Girls are not forced to marry the choice of the family in certain tribes: in others (of a lower kind) she is The intermarriage-between-tribes question is an important one. I shall pay attention to it.

I am not likely to see the gorilla up here. At present it is only found a little above Cape St. John and Loango.[7] But so little is known about the interior that the question of its habitat, must be left open. Chimpanzees are found in the bush here. If I see one I will note down whether there is more hair on back or front.

My plans on returning from the Kong Mountains[8] (if I get there) are not settled. I want much to study the Upper Niger & Haussa Country[9] before I leave Africa but cannot tell whether I shall have the opportunity. I am not likely however to return home just yet. Any queries you may send to me shall be faithfully attended to, as far as lies in my power. The worst of it out here is that one gets no assistance; things which every resident ought to know one has to find out for oneself often in a few days—if one can.

Trusting that your health is permitting you to continue labouring in science, and so benefiting all who inquire and observe | I remain | Yours truly | Winwood Reade

DAR 83: 165–6

CD ANNOTATIONS
1.1 I have … tribes. 1.5] *crossed pencil*
1.4 The head] *opening square bracket pencil*
1.5 I find … you— 1.9] *crossed ink*
2.1 The rams … affected 2.3] *scored pencil*; *triple scored blue crayon*; 'Just [male **mane**]' *added blue crayon*
3.1 Women … they want, 3.2] *double scored pencil*
3.1 Women … to it. 3.14] *crossed ink and pencil*
4.1 I am … observe 6.2] *crossed ink*
Top of letter: 'Expression & Sexual Selection & Man | Azara on ♀ choice[10] | All marry' *blue crayon*; 'MacClellan'[11] *pencil*; 'Rams manes & Horns lost when **castrated**' *ink*

1 The last extant letter from Reade is that of 23 May 1868 (*Correspondence* vol. 16), in which Reade mentioned observations about expression that he recalled from an earlier stay in Africa. CD had enclosed a copy of his queries on expression in his letter to Reade of 21 May [1868] (*ibid.*). For a copy of CD's queries on expression, see Appendix VI.

2 The Gold Coast was the name of the former British colony that now makes up the larger part of the country of Ghana (*Columbia gazetteer of the world* s.v. Ghana). See query number 17 of CD's queries on expression (Appendix VI).

3 Reade refers to the West African dwarf or Djallonke sheep (see I. L. Mason 1951, p. 32). CD had asked Reade to make observations on sheep in his letter of 21 May [1868] (*Correspondence* vol. 16). In *Descent* 2: 247; 284–5, CD cited Reade for information on the horns and the mane of this breed.

[4] The breeds referred to have not been identified. Senegambia was the former name of the region of west Africa that now makes up the countries of Senegal, Gambia, and Mali (*Columbia gazetteer of the world*).

[5] In his letter of 21 May [1868] (*Correspondence* vol. 16), CD had asked Reade to observe whether and to what extent women of 'barbarous tribes' were able to choose husbands. In *Descent* 2: 374, CD quoted this section of Reade's letter.

[6] The people Reade refers to as 'Mandingo' are now known as Mandinka; 'Timmanie' are now called Temne. Reade probably also refers to the Susu or Sousou people (see Reade 1873, 2: 374).

[7] Loango was a former African state in the basin of the Kouilou and Niari rivers, now largely in south-western Republic of the Congo (*EB*; see also Reade's map of the area he called gorilla country in Reade 1873, 1: facing p. 1). Cape San Juan is the northern boundary of Corisco Bay in Equatorial Guinea and Gabon.

[8] On the mythical Kong mountains, see Bassett and Porter 1991.

[9] The territory of the Haussa or Hausa people was in present-day northern Nigeria with additional territory in Mali and Cameroon (*Columbia gazetteer of the world*).

[10] CD refers to Félix d'Azara and probably to Azara 1809.

[11] CD may refer to McLennan 1865, a work on primitive marriage.

From J. D. Hooker 18 January 1869

Royal Gardens Kew
Jany 18/69.

My dear Darwin

I do not see either how you can avoid using the term "Morphological", but can you not use it, without leaving the reader to suppose that it has no definite sense: a very slight modification of what you say when alluding to Nægelis limitation of it would effect this I think.[1]

I am writing to Scott who has writ a clever & good paper on fern cultivation in Calcutta about *Viola nana*.[2]

I should not have implied that variations in leaf divergence *were* transmitted, but that they might be inherited (like any other variation)—that they are there *to be* acted upon by N.S. (ie to be propagated but that if such a variation occurs, there is no reason why it should not be transmitted, & if transmitted why N.S. should not determine its prevalence & subsequent constancy as a specific mark.—[3] If you have kept my letter please look & let me know if I have implied more than this.— I should extremely like to graft a Chesnut branch in such a variation from the normal leaf divergence occurred, & sow the seed a similar branch produced.

I know no case of ovules differing in position in the different flowers of one plant, except perhaps in monsters— I think Henslow gave me a Primrose in which the ovules were basal (as normally they should be) in most flowers, & they were parietal in others.—[4] it was otherwise monstrous.

I was much struck with your conclusion that the near approach to uniformity in an organ throughout a group, implied it's functional inutility— it is no doubt true.[5] I had a sort of gleam of this truth when considering the fact you once pointed out to me, that the calli of Oncidium though so essential to the plant for physiological purpose—are still very variable—[6] it then suggested the converse, which you have so well evolved. But what an apparent contradiction it involves—or paradox at least—that classification & system is founded on the least useful modifications:—& this explains a very common observation, that physiology i.e. the operations of active plant life, does not much help the systematist. And yet there is something uncomfortable in the idea that system is based on modifications the active exigencies of which are no longer in play. It seems frightfully paradoxical to say that the quinary arrangement of Dicotyledons is a matter of no moment to the Dicotyledons as such;—& yet that this is true is proved by the fact that such Dicots: as are ternary or quaternary are as good Dicots: as their quinary brethren. It is a tremendous upset to Owen's doctrines, or rather his *writings*, for these in no way rise to the dignity of *doctrines*. The "law of necessary correlation" is—nowheres.—[7]

Have you seen Herbert' Spencers "Appendix"— it appears to me to be astonishingly able, & the wriggle out of Materialism to be worthy of — — (i.e you or me)—[8] But how dreadfully difficult he is to read mark *or* digest, (for no human creature ever did or ever will do all three)— The subtlety of his reasoning is to me wonderful—

I have got tremendously pitched into for quoting him in my address as I expected: & for declaring the power above us to be inscrutable.[9] My last flagellation is from Pritchard the Astronomer who blames me for not being—complimentary enough to the Almighty— I have answered him that I think the concluding 3 Verses of Palgrave's poem is enough for the occasion—[10] He says I have quoted you Erroneously about Lythrum pollen— I have no time to look— I hope not—& it is too late if I have.[11]

Ever yrs affec | J D Hooker

DAR 103: 4–7

CD ANNOTATIONS

1.1 can you … I think. 1.4] *scored blue crayon*; '(Origin)' *added pencil, square brackets in original*
2.1 I am … monstrous. 4.4] *crossed pencil*
3.1 I should … produced. 3.8] *scored pencil*
3.2 that they … propagated 3.3] *crossed ink*
5.7 this explains … systematist. 5.9] *double scored pencil*
5.9 And yet … brethren. 5.14] *scored pencil*
5.14 It is … nowheres.— 5.16] *scored pencil*
6.1 Have you … Almighty— 7.4] *crossed pencil*
7.5 He says … I have. 7.7] *scored pencil*
Top of letter: 'Keep for Classification' *pencil*

1 See letter from J. D. Hooker, 15 January 1869 and nn. 1 and 2, and letter to J. D. Hooker, 16 January [1869]. Hooker refers to Carl Wilhelm von Nägeli and Nägeli's concept of a morphological character as given in Nägeli 1865 (see especially pp. 27–8).

2 John Scott's 'List of the higher cryptogams cultivated in the Royal Botanic Gardens, Calcutta' (Scott 1869) was published in the *Journal of the Agricultural and Horticultural Society of India*. In his letter of 16 January [1869], CD mentioned that he had received seeds of *Viola nana* from Scott (see also letter from J. D. Hooker, 15 January 1869 and n. 5).

3 In his letter of 16 January [1869], CD had asked Hooker whether he knew of cases where a variation in the angle of leaf divergence had been transmitted. See also letter from J. D. Hooker, 15 January 1869 and n. 4.

[4] See letter to J. D. Hooker, 16 January [1869] and n. 8. Hooker refers to John Stevens Henslow. For more on Henslow's work on primroses, see Walters and Stow 2001, pp. 163–6.

[5] See letter to J. D. Hooker, 16 January [1869] and n. 9.

[6] In orchid terminology, 'callus' refers to the crest or fleshy outgrowth of the lip or 'labellum' as CD called it (see Dressler 1981, p. 307). CD never used the term 'callus' in publication. In *Orchids*, p. 283, CD had noted that the labellum of *Oncidium* was characterised by 'all sorts of singular protuberances', and speculated that insects might be attracted to food within the labellum. In *Orchids* 2d ed., p. 271, CD added Fritz Müller's observation that prominences on the labellum of *Oncidium* often appeared 'gnawed'.

[7] Hooker refers to Richard Owen's support for the concept of 'correlation of parts' first developed by Georges Cuvier. For more on Owen's understanding and defence of the concept, see Rupke 1994, pp. 136–7.

[8] Hooker refers to a letter written by Herbert Spencer to the editor of the *North American Review*, 5 December 1868, and appended to the first volume of *Principles of biology* (Spencer 1864–7, 1: 479–92) with the title, 'On alleged "spontaneous generation," and on the hypothesis of physiological units'. CD's annotated copy is in the Darwin Pamphlet Collection–CUL. Most bound copies of Spencer 1864–7 lack this appendix, which evidently was sent only to subscribers such as CD and Hooker, who had received the book in instalments. The appendix appears in later editions at the end of the first volume. For Spencer's rejection of the allegation that his system was 'materialistic', see pp. 490–2.

[9] In his presidential address to the meeting of the British Association for the Advancement of Science held at Norwich in August 1868 (J. D. Hooker 1868, p. lxxiv), Hooker quoted from Spencer's *First principles* (Spencer 1860–2), 'If religion and science are to be reconciled, the basis of the reconciliation must be this deepest, widest, and most certain of facts, that the power which the universe manifests to us is utterly inscrutable.'

[10] Hooker refers to Charles Pritchard. Hooker had ended his address with a long quotation from a poem 'The reign of law' by Francis Turner Palgrave (see J. D. Hooker 1868, p. lxxv).

[11] For Hooker's description of CD's paper 'Three forms of *Lythrum salicaria*', see J. D. Hooker 1868, p. lxvii. Hooker reported correctly that CD had identified six forms of pollen, but his statement that 'five at least are essential to complete fertility' may be compared with CD's claim that 'regularly five kinds of pollen are elaborated by this one species of *Lythrum*' ('Three forms of *Lythrum salicaria*', p. 189; *Collected papers* 2: 123).

From Alfred Russel Wallace 20 January 1869

9, St. Mark's Crescent | NW.
Jany. 20th. 1869

Dear Darwin

It will give me very great pleasure if you will allow me to dedicate my little book of Malayan Travels to you, although it will be far too small and unpretending a work to be worthy of that honour.[1] Still, I have done what I can to make it a vehicle for communicating a taste for the higher branches of Natural History, and I know that you will judge it only too favourably.

We are in the middle of the 2nd. Vol. and if the printers will get on, shall be out next month.[2]

Have you seen in the last Number of the "Quarterly Journal of Science", the excellent remarks on Fraser's article on Nat. Selection failing as to Man?[3] In one page it gets to the heart of the question & I have written to the Editor to ask who the Author is.[4]

My friend Spruce's paper on Palms is to be read tomorrow evening at the "Linnæan".[5] He tells me it contains a discovery which he calls "alternation of function." He found a clump of *Geonema* all of which *were females*, and the next year the *same clump* were *all males*! He has found other facts analogous to this, & I have no doubt the subject is one that will interest you.[6]

Hoping you are pretty well, and are getting on steadily with your next volumes,—and with kind regards to Mrs. Darwin and all your circle,[7]

Believe me | Dear Darwin | Yours very faithfully | Alfred R. Wallace

P.S. Have you seen the admirable article in "The Guardian"! on Lyell's *Principles*? It is most excellent & liberal. It is written by Rev^d. Geo. Buckle, of Tiverton Vicarage, Bath, who I met at Norwich and found a thoroughly scientific & liberal parson.[8] Perhaps you have heard that I have undertaken to write an article for the Quarterly! on the same subject, to make up for that on "Modern Geology" last year not mentioning Sir C Lyell.[9] Really what with the Tories passing radical reform bills & the Church periodicals advocating Darwinianism, the Millenium must be at hand.[10]

A R W.

DAR 106: B73–4

1 Wallace refers to *The Malay Archipelago* (Wallace 1869a). Wallace had travelled throughout the Malay Archipelago from 1854 until 1862, and had started 'a small book' about his travels by January 1864 (*Correspondence* vol. 12, letter from A. R. Wallace, 2 January 1864 and n. 10). CD enquired about the book's progress in his letters to Wallace of 22 September [1865] (*Correspondence* vol. 13) and 22 January 1866 (*Correspondence* vol. 14). The dedication reads, 'To Charles Darwin, author of "The Origin of Species," I dedicate this book, not only as a token of personal esteem and friendship but also to express my deep admiration for his genius and his works.'

2 Wallace 1869a was published in March 1869 (*Publishers' Circular*, 15 March 1869, pp. 162, 171).

3 Wallace refers to an anonymous article, 'The alleged failure of natural selection in the case of man', that appeared in the January 1869 issue of *Quarterly Review of Science* (Anon. 1869a). The writer challenged the conclusions of an article that appeared in *Fraser's Magazine*, 'On the failure of "natural selection" in the case of man' ([Greg] 1868). The article in *Fraser's* was unsigned, but the author was later identified as William Rathbone Greg (see *Descent* 1: 167).

4 James Samuelson was the editor of *Quarterly Review of Science*. The author of the article (Anon. 1869a) has not been identified.

5 Richard Spruce's paper, '*Palmae Amazonicae*, sive enumeratio palmarum in itinere suo per regiones Americae aequatoriales lectarum' (Spruce 1869) was read on 21 January 1869.

6 Spruce discussed 'alternation of function' in two species of *Geonoma* in Spruce 1869, pp. 95–6, and hypothesised that the alternation of male and female flowers at different times on the same plant was an intermediate step towards complete 'dioicity' (that is, dioecism). CD had long been interested in transitional states between hermaphroditism and unisexuality in plants and animals (see *Notebooks*, Notebook C, 167; Notebook D, 157–9). For CD's interest in the possible relationship between heterostyly and dioecism, see *Correspondence* vol. 14. CD later devoted a chapter to the subject in *Forms of flowers* (pp. 278–309).

7 Wallace refers to *Descent* and to Emma Darwin.

8 George Buckle was a regular contributor of articles to the *Guardian* (a weekly church newspaper published from 1846 until 1951; North 1997, 3: 2122–3), and an active member of the British Association for the Advancement of Science (*The Times*, 4 January 1900, p. 8). His unsigned review of the tenth edition of Charles Lyell's *Principles of geology* (C. Lyell 1867–8) appeared in the *Guardian*, 30 December 1868, pp. 1463–4.

9 Wallace refers to a review of the fourth edition of Roderick Impey Murchison's *Siluria* (Murchison 1867), 'Sir Roderick Murchison and modern schools of geology' ([Geikie] 1868). Archibald Geikie did not mention Lyell by name but wrote, 'Uniformitarian writers have argued fallaciously against the hypothesis of violent action in the past' (*ibid.*, p. 203). Wallace's review of C. Lyell 1867–8 appeared in the April 1869 issue of the *Quarterly Review* ([Wallace] 1869b).

10 The Conservative party under Benjamin Disraeli had passed the Reform Act in 1867 (Mitchell *et al.* 1988). See n. 8, above.

From William Kencely Bridgman to John Gwyn Jeffreys [before 21 January 1869][1]

Norwich

⟨ ⟩

My Dear Sir

Thank you very much indeed for the Autograph— it is curious how often with great thinkers the fingers reveal their inability to keep pace with their "Copy" it is just the style of writing I should have expected.

With respect to the minor tetraodons[2] my Son made one expedition after them some time back but very few were obtainable— the waters were high & the current ⟨ ⟩ that at present it was useless— When the water is low in the summer is the time when they may be had in abundance & I will endeavour then D V not to forget them

Yours faithfully | W K Bridgman

J G Jeffreys Esq

Mr Darwin has told us lots about fertilization by insects but I think I have just made a curious discovery which I don't remember ever to have seen noticed— Gardening is one of my recreations & last year I procured some Aucubas ⟨ ⟩ from Standish but not succeeding with it I procured ⟨ ⟩ male plants to save my own this year—[3] *[well]* ⟨ ⟩ ⟨t⟩hese is now in flower & I have it in my office to ⟨ ⟩ ⟨un⟩der close inspection— I have cut over 100 flowers ⟨I obser⟩ved that *after the anthers began to burst* a drop ⟨ ⟩ made its appearance on the disc enclosed by ⟨ ⟩ ⟨s⟩tamens— I at first thought it was from over ⟨ ⟩or excess of moisture in the plant but a moment's reflection showed me that it was honey from the nectaries & the object

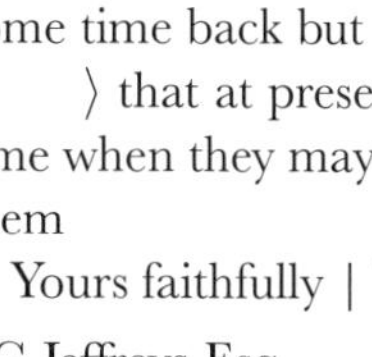

became at once apparent— the Aucuba is *Diœcious* & consequently the pollen has to be specially carried to the female flowers which are very inconspicuous— *the honey appears* ***only*** *when the pol⟨len⟩ is ready*— the bee cannot get it without rubbing its head ⟨&⟩ proboscis over the pollen— it is curious that *no honey appears until the anthers begin to burst*— apparently lest it should be carried off before the pollen is ripe— now does this occur with the fuchsia &c? & does the Bee know of honey being there by ⟨th⟩e state of the anthers?— is this new? | WKB

DAR 160: 306

CD ANNOTATIONS
1.1 Thank . . . forget them 2.5] *crossed ink*
Top of letter: 'On coincide⟨ ⟩ & secretion ⟨ ⟩' *ink*; 'Dichogamy' *red crayon*

1 The date is established by the relationship between this letter and the letter to W. K. Bridgman, 21 January [1869].
2 Bridgman refers to the genus *Tetraodon* of the family Tetraodontidae, puffer fish.
3 John Standish's nursery had received plants collected by Robert Fortune in Japan in 1860 (Shephard 2003, p. 145). Fortune had introduced *Aucuba japonica* (Japanese laurel) to England a few years earlier (*ODNB*). Bridgman alludes to CD's book, *Orchids*.

To W. K. Bridgman 21 January [1869][1]

Down. | Bromley | Kent S.E.
Jan 21.

Dear Sir.

Mr Jeffreys has forwarded your letter which has interested me, as I remember speculating some few years ago, whether Aucuba was fertilised by the wind or insects.[2]

I do not remember whether the coincidence of the secretion of nectar & the opening of the Anthers has ever been observed and it would take a long search to discover— But it is a pretty fact which ought on some fitting occasion to be published I rather doubt whether it will prove a general rule— So many flowers open their anthers before the Corolla even expands— The following fact has some bearing on your observation:—I have observed (& I think recorded dates) that during many days the flowers of the common Polygala—Linaria cymballaria and some of the smaller clovers may all be fully expanded and not a bee will visit them, and then suddenly swarms of bees will visit the flowers, which marks I believe the first secretion of nectar.—[3]

Again the stipules of the common Vetch secrete nectar (of no service for fertilisation) whilst the sun shines, & crowds of Hive Bees suck the stipules; the sun is obscured & in a very short time not a bee can be seen; again there is sunshine and again bees.—[4] I observed this several times on the same day & wondered how the news spread so rapidly through the Hive, but I suppose that all the bees knew the fact, & did not need to communicate it.—

I wish I could have answered your query more precisely & I beg leave to remain— | Dear Sir. | Yours very faithfully. | Ch. Darwin.

Copy
DAR 143: 146

1 The year is established by the relationship between this letter and the letter from W. K. Bridgman, 23 January 1869.
2 See letter from W. K. Bridgman to J. G. Jeffreys, [before 21 January 1869].
3 In *Cross and self fertilisation*, p. 422, CD reported on the sudden increase in the activity of bees visiting certain flowers, including *Linaria cymbalaria* (now *Cymbalaria muralis*, ivy-leaved toadflax) and *Polygala vulgaris* (common milkwort), and hypothesised that the secretion of nectar might be accompanied by 'some odour' alerting bees to its presence.

[4] In *Cross and self fertilisation*, pp. 403 and 422, CD discussed the excretion of fluid on the stipules of *Vicia sativa* (vetch) and its relation to the amount of sunshine present.

From George Cupples 21 January 1869

Guard Bridge | Fifeshire. N.B.
Jany. 21/69

Dear Mr Darwin,

I have just now received the enclosed reply from the Breadalbane Head-Forester—and send the *original* for various reasons, principally because I fear it is not by any means clear & definite enough, and rambles off from the points, which I gave in your own words exactly, and in full.[1]

About *roe-deer*, and on any other point—I would write to him again, as well as to others—if you have time just to suggest what are the deficiencies, or what precise details you would wish enlarged upon.

I have not heard about *fallow-deer* from Mr Wright, and hope he has written to you.[2] If not, I can easily stir up both him and others. Indeed there might be more little facts I could obtain, when they occur to you.

In haste yours most truly | George Cupples

Charles Darwin, Esqr.

P.S. Have you any good Red deer Antlers? If not, I think I could get them for you.

[Enclosure][3]

Forest Lodge
16 Jany

Mr Cupples
Dear Sir

I have Received your Letters and regarding the Red deers Horns, they don't appear untill 10 or 11 Months after Birth and the Roe Buck about 9 months but it depends a good deal upon the Pasture and the climate, in low grounds or in Parks the horns appear much earlier than in high ground where the climate is colder,[4] a stag born in a park where there is a rich Pasture the Head of Antlers are as good at 4 year old as the wild deer on the mountains are at 6 years a stag is not at his best untill 10 years, the Hind carrys her calve between 8 and 9 months but I may mention that the young males when they are castrated will never have horns, or if a Stag is casterated when the horns are at full growth he will never cast the horns, they will remain as they are when cut with the Velvet or without the Velvet, those deer that we castrated when young are in Season when the Stags are out of Season we call them Heaviers and are very fine venison when they are 8 or 9 years old; I cannot give you any information about Fallow deer as we have none in this quarter but I think the horns will appear much earlier than on the wild deer being on richer Pasture and more Sheltered than on the high mountains

I was glade to Receive your likeness for to put in my album I am Sorry that I cannot Send you mine at present although I have a number of my pictures in large Plate in different ways by Sir Edwin Landseer[5] and other artists I have no carte de visits but when I will be in Town I will be glade to get one for you I do not know of any fine deerhounds except those I mentioned to you before in Glengarry I believe they are handsome pair of deerhounds

I remain | Dear Sir | Yours very Respectfully
Peter Robertson

DAR 161: 290; DAR 86: A81–2

1 See letter to George Cupples, [6–9? January 1869], and letter from George Cupples, 11 January 1869 and n. 3. Cupples had evidently quoted from CD's letter in requesting information from Peter Robertson.
2 John Wright had earlier sent information on deer-hounds via Cupples (see *Correspondence* vol. 16, letter from George Cupples, 13 July 1868).
3 Cupples added the following annotation to the top of Robertson's letter: 'from Mr (Peter) Robertson | above forty years Head-Forester to the Marquis of Breadalbane | (Black Mount)'.
4 In *Descent* 1: 288, CD cited Robertson and Cupples for information on the roebuck and red deer.
5 Edwin Henry Landseer.

From Maxwell Tylden Masters 21 January 1869

Gardeners' Chronicle | & Agricultural Gazette Office, | 41, Wellington Street, Strand, W.C.
Jan 21 *1869*

My dear Sir/

Did you ever get the piebald potato?[1] I greatly fear you did not— I have another w^h. I should like to send you though the colors have faded—and also a spray of Holly for a full account of w^h. see the Chronicle of Saturday (tomorrow) (under Science Committee Hort. Soc.) and also a statement of Rev^d. F W Radclyffe as to grafting potatoes in Home Corresp— I have written to him for further information—[2]

Mr Fish who sends the Holly also sends some fasciated branches of Sycamore for *your* inspection but they are in no wise different from the ordinary fasciated branches.

What is the best means of sending small parcels to you

faithfully y^rs | Maxwell. T. Masters

Ch^as. Darwin Esq

DAR 171: 79

1 After reading CD's account of Friedrich Hildebrand's success in grafting a red variety of potato with a white (*Variation* 1: 396, 2d printing), Masters had arranged to have a potato of mixed purple and white colour sent to CD in November 1868, but expressed doubt about the occurrence of graft-hybridisation in potatoes (see *Correspondence* vol. 16, letter from M. T. Masters, 2 November 1868).
2 Masters refers to an account and specimens sent by David Taylor Fish of a yellow-berried holly that had produced a shoot with orange-red berries (*Gardeners' Chronicle*, 23 January 1869, p. 83) and to a letter from William Frederick Radclyffe on hybrid potatoes (*ibid.*, p. 81).

To Mary Somerville 21 January [1869][1]

Down. | Bromley. | Kent. S.E.
Jan 21st.

Dear Madam

I beg leave to thank you very sincerely for your great kindness in sending me a copy of your beautiful work, received yesterday, on Molecular & Microscopic Science, and for the honour you have thus conferred on me.[2]

I have not yet had time to do more than turn over the pages, & to read the part about Orchids, in which you give an excellent summary of the subject.

With my best thanks & the most sincere respect, | I beg permission to remain | dear Madam, | yours truly obliged | Charles Darwin

LS
The Bodleian Library (Somerville)

1 The year is established by the publication date of Mary Somerville's book (see n. 2, below).
2 Somerville's book, *On molecular and microscopic science* (Somerville 1869) was published at the beginning of January 1869 (*Publishers' Circular*, 16 January 1869, p. 10). CD's copy is in the Darwin Library–Down. Somerville, having received CD's permission, used fourteen illustrations from *Orchids* in the book (Somerville 1869, 1: 389–403; see also *Correspondence* vol 14, letter from Mary Somerville, 30 October 1866).

From J. V. Carus 22 January 1869

Leipzig,
Jan. 22d. | 1869.

My dear Sir,

As soon as your letter reached me I tried to get the directions of some trustworthy keepers of Merino-herds.[1] I found only two. Both of them have answered my questions, viz. yours, but one of them, Mr Steiger,[2] has not yet sent me his reply, as he intended to give it in a more elaborate form. At the same time he asked me, if he might be allowed to ask you directly about some points on which he should like to have your opinion Of course I told him he may do so. The other man was for a long time "administrator" of the oldest flock of Merinos we have in Saxony, that of Lohmen, it is Mr Sison.[3] He sent me with his answer a number of the "Agronomische Zeitung" containing his description of the Lohmen-herd. This I send you by the same post.[4]

His answer is as follows (I should send you his letter, but his handwriting is the old Saxon one, and I am afraid, you would not be able to make it out):

"The flock has been kept by intercrossing since 1765. The formation of the horns goes on in the same rate as the development of the rest of the body. When the body is developed early and vigorously, the horns do so also. The lamb has already when it is born in most cases rudiments of the horns; four to six weeks later the points of the horns are formed. After one year the young ram, when it is well fed, presents itself with its noble form of the head crowned by the horns, and now the mark is burnt

Merino ram from Lohmen. Sison 1867, facing p. 216.
By permission of the Syndics of Cambridge University Library.

on the horn. Rarely only and exceptionally it is necessary to repeat this marking by burning. At 1½ year the ram is moderately able for propagating.

"The formation of the horns keeps the same rate as the formation of (the body and) the teeth. When the animal has got its complete dentition, the growth of the horns seems to have come to an end. The exterior of the horns is vigorous, I should like to say shining; they are fatty in touching them; the colour is dark. Then slowly the colour gets lighter, the surface more opaque, the horn seems dry, rough, "chalky"; finally it gets brittle; and if the ram is a good fighter, his horns are at the age of 8 to 12 years scarcely imblamable.

"There occur also exceptionally horns in ewes, their bodies are always stouter, the wool vigorous. A family of horned ewes is no rarity, but all horns are smaller.

"In the muttons (the castrated rams) the horns are weaker.

"There occur also rams without horns; but these are not liked, especially by the shepherds. Their progeny is partly horned, partly hornless; on the whole they inherit well.

"In comparing the Merinos with other breeds within the reach of my neighbourhood, I could not observe any special differences. Also with them the formation of the horns is in relation to the development of the body. Yet the form of the horns of the Merinos is finer, nobler than in the other cross breeds of wool and flesh sheep.

"Special notices" on the formation of the horns are never made in our registers. Yet all the matters concerning them are well known especially to the shepherds"

Thus far Mr Sison, who seems to be a goodnatured downright herdsman. Since August 1868 he has his flock no more, as Lohmen, a crown-estate, was given to a new tenant by way of auction, so that it seems as if our Government would give it up to compete with foreign wool. This Mr Sison complains eagerly.

I am sorry I cannot give you more to-day. For even Mr Sison's notices don't keep strictly to the points, which I copied him from your letter.

As soon as Mr Steiger, the other Merino-man, sends me his Memoire, I shall try if I can send it to you entirely or if I translate the most important passages.

Meanwhile let me tell you how glad I was to be of any service to you

If my translation has sold even the publisher cannot tell before Easter-fair, when all the booksellers settle their accounts[5]

Believe me | My dear Sir | Yours very sincerely, | Prof. J. Victor Carus

DAR 86: A43–4

CD ANNOTATIONS

1.1 As soon … post. 1.10] *crossed blue crayon*
3.1 The formation … formed. 3.5] *scored pencil*
5.1 "There occur … in ewes,] *underl pencil*
5.2 A family … smaller.] *underl pencil*
6.1 the horns … weaker.] *underl pencil*
8.1 "In comparing … sheep. 8.4] *double scored pencil*
9.1 "Special … accounts 14.2] *crossed blue crayon*
Bottom of third page: "Sheep Horns—' *red crayon*

1 See letter to J. V. Carus, 5 January 1869.

2 Steiger has not been further identified.

3 Robert Sison. The oldest flock of merinos in Saxony was founded in 1765 from the Spanish Escurial flock and was initially sent to Stolpen (eighteen miles east of Dresden). From here, rams were sent to state farms, including that at nearby Lohmen. Initially, the rams were crossed with a local hornless Saxon sheep (Wood and Orel 2001).

4 CD's copy of Sison's article, 'Lohmen. Königliche Merino Stammschäferei Sachsens v. 1765' (Sison 1867) is in the Darwin Pamphlet Collection–CUL.

5 Carus refers to the German edition of *Variation* (Carus trans. 1868). Although the first volume had been printed in December 1867, the second did not appear until July 1868. The publisher, Eduard Koch, had informed CD that sales of the book would improve when both volumes were available (see *Correspondence* vol. 16, letter from Eduard Koch, 21 July 1868).

To J. D. Hooker [22 January 1869]

Down. | Bromley. | Kent. S.E.

My dear Hooker

Your letter is quite splenditious.[1] I am greatly tempted, but shall I hope refrain, from using some of your remarks. (quinary & Dicots &c.) in my Chapt. on Classification.—[2]

It is very true what you say about unimportant characters being so important systematically; yet it is hardly paradoxical, bearing in mind that the natural system is genetic, & that we have to discover the genealogies anyhow.[3] Hence, such parts as organs of generation are so useful for classification, though not concerned with the manner of life— Hence use for same purpose of of rudimentary organs &c &c.— You cannot think what a relief it is that you do not object to this view; for it removes *partly* a heavy burden from my shoulders.— If I lived 20 more years, & was able to work, how I sh$^{d.}$ have to modify the "Origin", & how much the views on all points will have to be modified.— Well it is a beginning, & that is something.

About the Leaves you say ". . . . even here is variation, capable of transmission"; I perhaps wrongly inferred that you might know of actual cases of transmission.—

I hope you will have a lateral branch of Spanish Chesnut grafted, to see, if when upright it will retain same divergence; but surely it w^{d} not be inherited, as both tendencies *must* be transmitted in every seed.—[4]

I did notice a *slight* error about Lythrum;[5] but your Address is so deeply embedded in other, *as yet*, unclassed pamphlets, I cannot find it just now.— How odd old P.[6] sh$^{d.}$ have detected it. I sh^{d} like to see any published flagellation of your Heterodoxy.—

I have received H. Spencer's appendix on spontaneous generation & on his confused physiological units; but I do not remember any discussion on materialism;[7] I read, however, with care only the part on spont. gen. & was struck with vivid admiration at its ability: He is a wonderful man.—

My dear old fellow | Yours affec^{ly} | C. Darwin

I hope you take in, to encourage, "Scientific Opinion": some articles have interested me a good deal; but the Nat. Hist part seems not well done, or not much attended to.—[8]

I have just received Mrs Somerville's Book:[9] it seems a most strange Hotchpotch.—

Endorsement: 'Ja^{y} 22/69.'
DAR 94: 114—15

1 CD used the word 'splenditious' in two other letters to Hooker (see letters to J. D. Hooker, 1 June [1865] (*Correspondence* vol. 13) and 7 August [1869]). Henrietta Emma Darwin used a variant, 'splendicious', in a letter to William Erasmus Darwin dated [18 November 1862] (DAR 219.1: 67), which suggests that it may have been a family word.

2 For Hooker's remarks on the quinary arrangement of dicotyledons, see letter from J. D. Hooker, 18 January 1869.

3 See letter from J. D. Hooker, 18 January 1869.

4 See letters from J. D. Hooker, 15 January 1869 and n. 6, and 18 January 1869. Hooker had suggested the grafting experiment to determine whether the variation in angle of divergence in the leaves would be transmitted in seeds produced by such a grafted branch.

5 See letter from J. D. Hooker, 18 January 1869 and n. 11.

6 Charles Pritchard.

7 See letter from J. D. Hooker, 18 January 1869 and n. 8.

8 *Scientific Opinion* was a weekly popular-science newspaper, published from 1868 to 1870, after which it merged with the *English Mechanic and Mirror of Science* (North 1989, 6: 4271).

9 Somerville 1869. See letter to Mary Somerville, 21 January [1869].

To A. R. Wallace 22 January [1869][1]

Down. | Bromley. | Kent. S.E.
Jan 22nd.

My dear Wallace

Your intended dedication pleases me much & I look at it as a *great* honour & this is nothing more than the truth.[2] I am glad to hear for Lyell's sake & on general grounds that you are going to write in the Quarterly.[3] Some little time ago I was actually wishing that you wrote in the Quarterly, as I knew that you occasionally contributed to periodicals, & I thought that your articles would thus be more widely read.

Thank you for telling me about the Guardian which I will borrow from Lyell.[4] I did note the article in the Q. Journal of Science & put it aside to read again with the articles in Frazer & the Spectator.[5]

I have been interrupted in my regular work in preparing a new edit of the Origin, which has cost me much labour & which I hope I have considerably improved in two or three important points.[6] I always thought individual differences more important than single variations, but now I have come to the conclusion that they are of paramount importance, & in this I believe I agree with you. Fleming Jenkyn's arguments have convinced me.[7]

I heartily congratulate you on your new book being so nearly finished be

Believe me, | My dear Wallace | yours very sincerely | Ch. Darwin

LS(A)
British Library Board (Add 46434, f. 165–6)

1 The year is established by the relationship between this letter and the letter from A. R. Wallace, 20 January 1869.

2 CD refers to the dedication of *The Malay Archipelago* (Wallace 1869a). See letter from A. R. Wallace, 20 January 1869 and n. 1.

3 See letter from A. R. Wallace, 20 January 1869 and n. 9. CD refers to Charles Lyell.

4 See letter from A. R. Wallace, 20 January 1869 and n. 8.

5 See letter from A. R. Wallace, 20 January 1869 and nn. 3 and 4. CD also alludes to an article in the *Spectator* (Anon. 1868), written in response to the article in *Fraser's* ([Greg] 1868).

6 CD had started work on the fifth edition of *Origin* on 26 December 1868 (*Correspondence* vol. 16, Appendix II). He finished corrections on 10 February 1869 (see 'Journal', Appendix II).

7 The *North British Review* had published an unsigned article, with several criticisms of CD's theory, by Henry Charles Fleeming Jenkin in June 1867 ([Jenkin] 1867). Wallace responded to these criticisms in his article 'Creation by law' in the *Quarterly Journal of Science* (Wallace 1867; see *Correspondence* vol. 15). For CD's discussion of 'individual differences' and his response to Jenkin's argument on how rarely variations in a single individual could be perpetuated, see *Origin* 5th ed., pp. 104–5.

From Thomas Campbell Eyton [before 23 January 1869][1]

Eyton, | Wellington, | Salop.

Dear Darwin

I am writing or rather intending to write on a very curious subject Why do salmon rise or why do not they always rise at the fly the first short paper or rather letter to

Buckland will be published in one of the next no[s] of Land and Water[2] Negretti & Zambra[3] are preparing some instruments for me viz

Electrometer
Hygrometer
Barometer
Thermometer

and compass to be fitted in a box of a portable form to go with me fishing by means of which I shall always be able to ascertain the state of the atmosphere at any time and tabulating the observations I hope to come to some results.

Have in your travels made any observations of the sort or observed anything that will help conclusions if you have or can give me any hints I shall be much obliged to you. Hoping that you are in better health than when you wrote last and that all your party are well also

I remain yours truly | Tho C Eyton

I shall a supplement out shortly to Osteologia[4] Have you seen Alphonse Milne Edwards book on the fossil birds of France.[5] I was yesterday trying to make from the old book of what country the rabbit was originally a native but was not very successful.[6]

DAR 163: 42

1 The date is established by the date of publication of Eyton's letter to *Land and Water*; see n. 2, below.

2 The letter was published in *Land and Water*, 23 January 1869, p. 58. Eyton refers to Francis Trevelyan Buckland.

3 Negretti & Zambra were a firm of scientific-instrument makers and photographers whose main premises were at 59 Cornhill, London. The firm had been founded in 1850 by Italian immigrants (Clifton 1995).

4 Eyton refers to the first supplement to *Osteologia avium* (Eyton 1867–75), published in 1869. CD's copy of the book is in the Darwin Library–Down, but the supplements have not been found.

5 The first volume and atlas of Milne-Edwards 1867–71 were published in 1867 and 1868.

6 The book Eyton refers to has not been identified.

From W. K. Bridgman 23 January 1869

Norwich
Jan 23/69

Dear Sir

I hope you will allow me to thank you which I do very much indeed for your kind & very interesting letter—[1] The observation respecting the apparent simultaneous appearance of nectar with the bursting of the anthers was only casually made with a common hand lens when cutting off the flowers to obtain the pollen—[2]

Since then, however I have arranged a Binocular with $1\frac{1}{2}$ object glass. 2@. —eye pieces & a Beck's side reflector in such a manner as to receive a continuous view of any individual flower from its first opening to the shedding of the pollen—[3] by this means I am enabled to obtain a distinct & sufficiently enlarged view of the disc as to be able to detect the slightest exudation immediately it appears— I have several male plants in different stages of forwardness & by examining a succession of flowers & noting the times of the anthers beginning to burst, & the first appearance of the nectar together with the time of day it occurs the presence or absence of sunlight &c. &c. something perhaps may be learnt on the subject— That fertilisation is designed to be carried on here by organic agency, I think, scarcely admits of a doubt, but whether by *night* feeders or by day feeders I am inclined to believe is not quite so clear—

The flower is singularly favorable in structure for examination, the clumping filament with its anthers & the nectary being in the *[field]* at the same time if you would like to have a few expanded & unexpanded flowers I shall have very great pleasure in sending some in all their stages

I have plenty of female plants but none yet in flower—

With respect to the Polygala, Linaria &c it will be a very interesting exp^{t} to try & discover whether the absence of nectar be synchronous with the absence of the bees or perhaps it may be more owing to the state of the weather to which we know them to be remarkably sensitive although it is just possible that there may be something which serves them as a "Beggars mark" with which we are yet unacquainted—[4]

Again thanking you allow me | to remain Dear Sir | Yours faithfully | W K Bridgman

C Darwin Esq FRS

DAR 160: 307

1 See letter to W. K. Bridgman, 21 January [1869].
2 See letter from W. K. Bridgman to J. G. Jeffreys, [before 21 January 1869].
3 Bridgman refers to a binocular microscope with a parabolic side reflector attachment, manufactured by Richard & Joseph Beck, 31 Cornhill, London (*Post Office London directory* 1868).
4 See letter to W. K. Bridgman, 21 January [1869] and n. 3. 'Beggars mark' may be a reference to the secret sign-language used by tramps (Rose 1988, p. 33).

To James Orton 23 January [1869][1]

Down. | Bromley. | Kent. S.E.
Jan 23rd

Dear Sir

Owing to your letter of January the 4th being addressed to London I received it only yesterday. I assure you that I am greatly pleased at the honour which you intend to confer on me by dedicating your new work to me.[2] You seem to have had a most interesting expedition, & I am very glad that you intend to publish a general account of the results.[3]

Your discovery of marine shells high up the Amazons possesses *extreme* interest, not only in itself, but as one more most striking instance how rash it is to assert that any deposit is not a marine formation because it does not contain fossils. Agassiz will not at all approve of this discovery; & if I am not mistaken marine shells have also been found in the large island at the mouth of the river. As for myself I never believed for a moment in Agassiz's idea of the origin of the Amazonian formation.[4]

I shall be very curious to read a full account of the wingless grebe-like bird which you have discovered, & of the great deposit of bones; but I fear of the latter you will not have been able to bring away many specimens.[5]

With my very sincere thanks for the intended honour & for your kind letter, I remain | Dear Sir, | yours very faithfully | Charles Darwin

LS
History of Science Collections, University of Oklahoma Libraries

[1] The year is established by the relationship between this letter and the letter from James Orton, 4 January 1869.
[2] See letter from James Orton, 4 January 1869 and n. 11.
[3] See letter from James Orton, 4 January 1869 and n. 1.
[4] See letter from James Orton, 4 January 1869 and n. 6. CD refers to Louis Agassiz.
[5] See letter from James Orton, 4 January 1869 and nn. 4 and 7.

From C. F. Claus[1] 24 January 1869

Marburg
24/1 1869

Hochgeehrter Herr!

Indem ich mir erlaube, Ihnen beifolgende kleine Arbeit über die Fortpflanzungsweise von *Leptodera*[2] ehrerbietigst zu übersenden, darf ich wohl gleichzeitig mir die Freiheit nehmen, einige Zeilen an Sie zu richten und einem schon längst von mir gehegten Wunsche zu entsprechen, meine innige Hochachtung und Verehrung Ihnen auszudrücken. Sie werden, hochverehrter Herr, wohl wissen, dass unter den deutschen Zoologen zahlreiche Bewunderer und Anhänger Ihrer Selectionstheorie bemüht sind, nach Thatsachen zu forschen, die noch weitere Stützen für die Transmutation der Arten liefern, ohne die uns der Zusammenhang der organischen Lebenserscheinungen ein Räthsel bleibt. Ich selbst zähle mich auch zu jenen, ohne freilich wie *Haeckel*, dessen Werke Sie sicher trotz zahlreicher trefflicher und anregender Ideen mit mir als einen *nicht reifen*, zu enthusiastischen Versuch zu einer Naturphilosophie bezeichnen werden, die Nothwendigkeit überzeugender Beweismittel aus dem Auge zu verlieren.[3] Gerade in diesem Winter beschäftigt mich der Gegenstand besonders, und ich bin durch eine öffentliche Vorlesung, die ich vor einem ansehnlichen Auditorium von Studirenden halte ("*Darlegung und Kritik der Darwinschen Lehre*") gezwungen, das Material möglichst sorgfaltig durchzuarbeiten und auf Beweiskraft zu prüfen. Dabei ist es mir leider zur Klarheit geworden, dass H. nicht mit der nothigen Schärfe scheidet und überall seinen Glaubensausdruck für Beweis hält. Dem Wagnerschen Migrationsgesetz, das ja in Ihren Werken bereits vollkommen enthalten ist, kann ich unmöglich den ausschliessenden Werth, den der Verfasser beansprucht, zusprechen,[4] offenbar hat er Ihre Lehre gar nicht ausreichend verstanden und den Einfluss der *mittelbaren* potentiellen Anpassung bei der natürl-Züchtung nicht entsprechend gewürdigt. Es ist möglich, dass ich den

Inhalt meiner Vorlesungen als Versuch zu einer Prüfung der Beweismittel Ihrer Lehre in einer kleinen Broschüre veröffentliche. Doch will ich hiermit vorsichtig sein, den Gegenstand noch mehrmals zum Vortrag bringen und vorläufig lieber Anderen überlassen, die Häckelschen Selbsttäuschungen zu zurückzuweisen.[5] Vorläufig bin ich mit einer Publication eines anderen Gegenstandes beschäftigt, von dem ich denken sollte, dass Sie vielleicht meinen Beobachtungen darüber einiges Interesse schenken, "Die Entwicklung der Lepaden".[6] Auch hier ist es ja Ihr Werk, welches die Grundlage bildet, wie es aber keinen Gegenstand gibt, der nicht nach Jahren von Neuem erfasst, eine weitere Fortführung—selbst nach der ausgezeichnetsten früheren Bearbeitung—ermöglichte, so auch hier. Zwar ist bereits von *Pagenstecher* ein Versuch gemacht, aber leider nicht mit der ausreichenden Sorgfalt und Gründlichkeit, die um Ihre Angaben zu erweitern nothwendig ist.[7]

Nun möchte ich mit der kleinen Untersuchung nicht eher abschliessen, als ich die Gewissheit habe, das mir zu Gebote stehende Material nicht vergrössern zu können. Sollten Sie verehrter Herr im Stande und Willens sein, mir einiges Material von Puppen[8] und jungen in den Puppen steckenden Lepaden oder aus *Balaniden* recht bald mitzutheilen, so würde ich Ihnen zu grossem Danke verpflichtet sein. Insbesondere läge mir viel daran, die Puppen von *Proteolepas*, *Ibla Cryptophialus Alcippe* wie überhaupt diese sonderbaren Formen und deren Complemental males untersuchen zu können.[9]

Indem ich Sie bitte, meinen Wunsch nicht für zu unbescheiden halten zu wollen sondern nach Massgabe des Bestrebens eines naturae curiosi[10] gütigst zu entschuldigen, zeichne ich als Ihr | hochachtungsvoll ergebener | C Claus | Professor der Zoologie

DAR 161: 176

1 For a translation of this letter, see Appendix I.

2 Claus enclosed an offprint of an article on the organisation and reproduction of the nematode *Leptodera appendiculata* (Claus 1868a). CD's copy is in the Darwin Pamphlet Collection–CUL.

3 Claus refers to Ernst Haeckel. For more on Claus's objections to Haeckel's version of Darwinism, see Nyhart 1995, pp. 186–93.

4 Claus refers to Moritz Wagner and the 'law of migration' elucidated in Wagner 1868a and 1868b. In a letter to August Weismann of 22 October 1868 (*Correspondence* vol. 16), CD had remarked that Wagner greatly overrated the necessity for emigration and isolation. In *Origin* 5th ed., p. 120, CD referred to Wagner's essay, noting that while geographical isolation was important for preventing crosses between new varieties, it was not necessary for the formation of new species.

5 Claus never published the lectures on Darwinian theory; but in later work he was highly critical of many aspects of Haeckel's work, most notably the latter's gastraea theory (see Claus 1874 and Claus 1876; see also Di Gregorio 2005, pp. 297–301).

6 Claus enclosed an offprint of an article on the cypris-like larvae of cirripedes and their transformation into sedentary adults (Claus 1868b). The offprint is in the Darwin Pamphlet Collection–CUL, and is lightly annotated.

7 Claus refers to Heinrich Alexander Pagenstecher and Pagenstecher 1863.

8 Claus followed CD's usage in referring to the last larval stage in barnacles as 'pupal' (see *Living Cirripedia* (1854), p. 13); this stage is now more commonly referred to as the cypris stage (see *Living Cirripedia* (1851), p. 14).

9 The genus *Alcippe* is now *Trypetesa*. *Proteolepas* is no longer classed as a cirripede but instead is classed as a parasitic isopod of the family Cabiropidae (see *Living Cirripedia* (1854), p. 588). Claus refers to CD's

discovery of 'complemental males' in several species of Cirripedia. CD had observed minute males attached to the bodies of hermaphrodites, and differing greatly from them in size and structure (see *Living Cirripedia* (1851), pp. 231–2 and 281–93, *Living Cirripedia* (1854), pp. 23–30, and Newman 1993, pp. 377–81). In his letter to J. D. Hooker, 10 May 1848 (*Correspondence* vol. 4), CD remarked that the discovery was suggestive of the development of separate sexes, and thus of significance for his theory of species (see also *Correspondence* vol. 4, Appendix II, pp. 399–400). CD later suggested that complemental males might have developed in cirripedes to enable intercrossing between two hermaphrodites (see 'On the males and complemental males of certain cirripedes, and on rudimentary structures', *Nature* 8 (1873): 431–2; *Collected papers* 2: 177–82).

10 *Naturae curiosi*: inquirer into nature (Latin).

To T. C. Eyton 24 January 1869

Down. | Bromley. | Kent. S.E.
Jan 24 1869

Dear Eyton

I am sorry to say that I cannot give you the least information about the subject of your enquiry.[1]

It is a very interesting one, but I sh^{d} fear that it would prove very difficult.

I am glad to hear that you are continuing to work on your great subject of the osteology of birds.[2] I have not seen Alph. M. Edwards' book, but I hear that it is excellent.[3]

Can you give me information on one point? viz. how many months after birth do the horns first appear in the young males of the fallow deer?[4]

If you can tell me with any degree of accuracy I should be *particularly* obliged.

If I do not hear I shall understand that you cannot do so.

Believe me dear Eyton | yours very sincerely | Charles Darwin

LS
American Philosophical Society (359)

1 See letter from T. C. Eyton, [before 23 January 1869] and n. 2.

2 See letter from T. C. Eyton, [before 23 January 1869] and n. 4.

3 CD refers to Alphonse Milne Edwards. See letter from T. C. Eyton, [before 23 January 1869] and n. 5.

4 For more on CD's interest in the development of horns, see letter to George Cupples, [6–9? January 1869] and n. 2.

From Archibald McNeill 24 January 1869

73 G^{t}. King Street | Edin^{h}.
24th. Jany. 1869

Dear Sir

I am much pleased to learn from yours of the 18th. that the information sent by me to M^{r}. Cupples has been interesting to you—[1]

As to the question on w^{h}. you now wish information vizt. whether the branches on the horns of deer are intended for use or ornament I feel some diff^{y}., but am of opinion that they are probably intended for both purposes— For while I know that some of the branches are useful in attack and defence, there are others that I never saw bro^{t}. into action and I am inclined to think your remark correct that a single point at the top of the horn w^{d}. be more effective in combat than three points and in fact stags frequently have only one point—but the brow antlers w^{h}. incline downwards are a great protection to the forehead, & as the their points are turned outwards they are used effectually in attack—[2]

I have very seldom had a dog killed by a deer for my dogs were *deer* hounds & not *Stag* hounds and were used only for coursing, and as a Stag will not turn to bay till he is exhausted a good deer hound will run up to him & pull him down before he is blowen— on this subject you may perhaps find in M^{r}. Scropes Book on deer stalking some of the information you desire—[3]

In regard to the other branches I am not aware that they are much used in their fights with each other but I have seen stags in a half tame state use them when fighting with Bulls by bringing the antlers into contact with the Bulls forehead & in every case with success—

I may here mention what is not Gen^{y}. known that I have frequently seen Stags in a half tame serve Cows, but of course there was no proginy—

The long hair on the Stags neck may be intended for ornamint but it is also a great protecton to him when hunted by dogs who generally seize the throat—[4]

I am sorry that I cannot give more specific information on a Subject in w^{h}. I take great interest having had for many years an opportunity of studing the habits of diff^{t}. sorts of animals

Believe me to be | Yours faith^{y}. | Arch^{d}. M^{c}.Neill

DAR 83: 175–6

CD ANNOTATIONS
1.1 I am … dog killed 3.1] *crossed ink*
2.2 I feel … purposes— 2.3] *double scored ink*
3.1 by a deer … desire— 3.5] *crossed blue crayon*
4.1 In … success— 4.4] *crossed ink*
5.1 I may … cannot 7.1] *crossed blue crayon*
6.1 The long … throat— 6.2] *double scored ink*
7.1 give … animals 7.3] *crossed ink*

1 CD's letter has not been found, but see the letter from George Cupples, 15 January 1869.
2 In *Descent* 2: 253, CD cited McNeill for this information.
3 McNeill refers to William Scrope and Scrope 1838.
4 CD cited McNeill for this information in *Descent* 2: 268.

From J. D. Hooker [25 January 1869][1]

Royal Gardens Kew
Monday

Dear Darwin

Just one last thought anent genetic characters of no value to the plant–: is not the fact, that characters of primary value in system are so often of no use, an argument in favour of your conclusion, that such characters as are of no use if not in any way detrimental, are not necessarily eliminated, but may be retained *ad infinitum*—[2]

On the other hand, is it not an argument against the theory of characters acquired by the **individual** being hereditary—thus—if hereditary modifications that never come into play, do not die out; is it likely that non hereditary modifications brought into play by the *individual* (for its own special use) should be transmitted?—

I dare say you have put all this in the Origin,—& I have forgot it.

We dined at the Lyell's the other day & I was put in Elysium between that lovely Mrs Norton who I met at your house, & took down, whose face is nearest an Angel's (barring the Nose) of any I ever saw, in a dream.—& Miss Symonds, dau[r] of the Pendock man, such a nice girl—with the lovely name of "Hyacinthe".[3] Poor old Murchison was there, looking rather "got up"— he is a good old soul. & now that he wears a stick in the house, he looks quite picturesque, with his star & high cravat.[4] The Huxleys & Lady Eastlake[5] & Leonard Lyell were there, no one else; & it was a delightful little party.— Lyell complained of a cold, but looked remarkably well— I wish he would not babble to me about the Copley medal. I do *hate* it. & it is such bad taste. & all the more as, *on my soul*, I would rather not have the Copley than have it with the draw-back of an after-dinner speech at the R.S. anniversary.[6]

We go to a tremendous wedding at Bury S[t] Edmunds this week, of our old friend the Rev C. Babington, of Cambridge, now Rector of Cockfield in Suffolk. it is to be a wonderful affair I hear. We go to Cockfield on Wednesday, & I return on Thursday night to Kew.[7]

Ever yrs affec | J D Hooker

Have you seen a curious notice of the Cannibal tribes of S. Africa, very horrid— printed in S. Africa.[8]

See p 490 of Appendix D[r] Spencer on Materialism.[9]

DAR 103: 8–9

[1] The date is established by the reference to the wedding of Churchill Babington (see n. 7, below). In 1869, the Monday preceding 28 January was 25 January.

[2] See letters to J. D. Hooker, 13 January 1869 and 16 January [1869], and letters from J. D. Hooker, 14 [January] 1869 and 15 January 1869.

[3] Hooker refers to Charles and Mary Elizabeth Lyell, Susan Ridley Norton, Hyacinth Symonds, and William Samuel Symonds, rector of Pendock, Worcestershire.

[4] Roderick Impey Murchison was made KCB (Order of the Bath) in 1863 and was therefore entitled to wear the Knight Commander star (*ODNB*).

[5] Hooker refers to Thomas Henry and Henrietta Anne Huxley, and Elizabeth Eastlake.

[6] Hooker received the Copley Medal of the Royal Society in 1887 (*ODNB*). His name was not put forward as a candidate for the medal in 1868 or 1869 (see *Minutes of Council of the Royal Society* 3 (1870): 415, 473).

[7] Churchill Babington was married on Thursday 28 January 1869 at St James Church, Bury Saint Edmunds, Suffolk (General Register Office, marriage certificate no. 407).

[8] The notice has not been identified, but on 23 February 1869 a paper 'The cave cannibals of South Africa' by Mr Layland was read and later appeared in *Journal of the Ethnological Society of London* n.s. 2 (1869–70): 76–80. Layland has not been further identified.

[9] The reference is to Herbert Spencer. See letter from J. D. Hooker, 18 January 1869 and n. 8.

From T. C. Eyton 26 January [1869][1]

Eyton, | Wellington, | Salop.

Dear Darwin

The fawns have no horns for the first 12 months after they are droped they then grow short ones and are called prickets on feeling the head of a dead fawn about Xmas a slight knob is to be felt[2]

yours truly | Tho C Eyton

Jan 26

DAR 86: A51

[1] The year is established by the relationship between this letter and the letter to T. C. Eyton, 24 January 1869.

[2] See letter to T. C. Eyton, 24 January 1869 and n. 4. In *Descent* 1: 288 n. 24, CD cited Eyton for information on the appearance of horns in fallow deer.

From William Chester Tait 26 January 1869

Oporto

Jan[ry]. 26th. 1869

Charles Darwin Esq[r]. | Down | Bromley | Kent

Sir,

I hope that you will excuse the liberty I have taken in addressing you though not having the pleasure of being personally acquainted.

Perhaps the subject if not the matter may prove a sufficient extenuation. Your "Origin of Species"[1] I read almost as soon as it was published was very much struck with it and felt convinced that you had *solved* a great problem in the world's history and given a new direction to thought. I am now finishing the first volume of your Variation of animals and plants under domestication[2] and every now and then many circumstances have been brought to my recollection in connection with the subject I have therefore made notes of them and should you wish it would forward any that may be likely to interest and prove useful to you in the consolidation of your theory, on the principle that small stones may fill up crevices.

At the end of your "Origin of Species" you make an appeal to young naturalists,[3] perhaps you meant to include also mere observers and I venture to put myself in this latter class, being 24 years of age and having always been fond of natural history especially ornithology and thinking with Gilbert White that labour should be divided and every place should have its own monographer I have made occasional ornithological notes having in view more particularly the making of a list of birds seen in this neighbourhood, then the rest of Portugal and those migrating through.[4] My occupation (being in business as a commission merchant) will not allow me to apportion more than a limited time to my favourite study but as I am fond of shooting curiosities in natural history sometimes come before my notice This country having been hitherto isolated from the rest of Europe in a great measure has never been properly explored and I have little doubt that much might be learnt here—

What I wish to bring before your notice is a curious fact with regard to the pointer dog which is occasionally born here without more than a stump of a tail varying in size— I know of one very decided case which first made me enquire about them and I am told by several people that it is by no means a very uncommon occurrence and have since heard of no less than 3 other cases 2 of them said to be now living in this city this coupled with the fact that it is a general custom here to cut the pointer's tail off leaving only a stump and that the greater part of them are thus mutilated, leads me to suppose that the oft repeated cutting of the tail has had a strong influence on the birth of tail-less dogs and dogs born with only *half a tail* I do not know that an analogous case has ever been mentioned and after strict verification and enquiry it may be found to bear on your hypothesis of Pangenesis & your general theory.[5] I think that I have observed a similar case with regard to the comb of game cocks. Though no doubt selection in early times before cutting the comb was thought of, may have favoured those birds with smallest combs still man's interference in dubbing them has for a long time rendered selection on that point unnecessary.

I intend to try to make the tail-less feature constant by breeding tail-less dogs interse which would be the more curious if it were proved that it is an inheritance entirely produced by man's interference.

I should expect a more marked analogous case in the vegetable Kingdom—but unfortunately I know little or nothing of that branch. If you think this case of any interest I should very much like to sustain a correspondence on this and similar observations the more so as I am almost isolated here from those of similar tastes.

Hoping you will excuse my boldness | I remain | Sir, | Yours faithfully
William C. Tait.

Via France— | Mr. William C Tait
Oporto | Portugal

Will find me—

DAR 178: 43

1 *Origin.*

2 *Variation.*

3 See *Origin*, p. 482.

4 Tait alludes to a statement in White's *Natural history of Selborne* (letter VII to Daines Barrington), 'Men that undertake only one district are much more likely to advance natural knowledge than those that grasp at more than they can possibly be acquainted with: every kingdom, every province, should have its own monographer.' See White 1825, 1: 222–3.

5 Tait refers to CD's provisional hypothesis of pangenesis (*Variation* 2: 357–404). On variation caused by the direct action of changed conditions, see *Variation* 2: 394–7.

To John Phillips 27 January 1869

Down. | Bromley. | Kent. S.E.
Jan 27 1869

My dear Phillips

Your book on Vesuvius has not arrived, but has perhaps been sent to my London address, or may come by Railway. I must however send you my very sincere thanks for this present, & even more so your very kind note[1]

yours very sincerely | Ch. Darwin

LS
American Philosophical Society (360)

[1] Phillips 1869 had just been published (*Publishers' Circular*, 1 February 1869, pp. 67, 73). CD's copy is in the Darwin Library–Down. CD refers to the London address (6 Queen Anne Street) of his brother, Erasmus Alvey Darwin. Phillips's letter to CD has not been found.

To C. F. Claus 28 January 1869

Down. | Bromley. | Kent. S.E.
Jan 28 1869

Dear & esteemed Sir

Your letter has given me great pleasure, for permit me to say that I have long honoured your name as that of one of the best naturalists in Europe. I am very much obliged for your two Memoirs, which I am sure will interest me, but I am a poor German scholar & I have another book which I must finish reading first.[1]

I earnestly hope that you will publish your lecture, as the judgment of so experienced a naturalist must be of value to every one; & for myself I much wish to read your criticisms & facts. Of course I think very highly of Häckel's abilities & his works; I should indeed be very insensible to praise if I did not do so; but I dare say he is too enthusiastic & too bold in drawing conclusions.[2] I have been particularly glad to hear what you say about M. Wagner's interesting pamphlet. Perhaps I have underrated the importance of isolation, but I am now correcting a new edit. of my "Origin of species", & I do not find that I have much or any thing to alter on this head.[3]

I cannot tell you how glad I am to hear that you intend to study the metamorphoses of the Lepads.[4] As you are a much better dissector than I am, & have so much knowledge of the lower Crustaceans, I have no doubt that you will find *much* that I have overlooked or misunderstood.

I still cannot avoid believing that the lateral horns of the larvæ in the first stage include the antennæ of the second stage.[5] I wish you could be induced to examine the curious organ which I suppose I falsely called the acoustic sack.[6] But beyond every thing I hope that you will be able to examine the Complementary males of Scalpellum, which no one else has ever seen.[7] I shall never forget my astonishment when I first made out their nature. I sh[d] think you w[d] be able to find some where specimens in spirits. I grieve to say I can give you very little help. When I finished my work I returned almost all the specimens to their owners. The few which I kept have been entirely forgotten, & now I find to my *great* disappointment that some which w[d] have been useful to you are quite dried up. I have however the following specimens not quite spoilt.

(1) Xenobalanus, that curious form resembling Lepas but really allied to Coronula—in pretty good state[8]

(2) A single small Scalpellum; & I am nearly sure that with only a hand-lens, I see a complementary male on each side, but I do not like to cut off the valve to examine carefully.

(3) The shell of Concholepas, after being treated with acid, & which formerly *abounded* with Cryptophialus but whether they are in a good state I will not say.[9]

(4) Some fragments of shell with Alcippe,[10] but again I do not know in what state.—

Lastly 17 slides with various fragments & minute specimens; I can see one male Alcippe in a pretty good state, & some larvæ of the first stage of Scalpellum; these larvæ are of remarkable size.—

Now w^d these specimens be worth your acceptance? They need not be returned, & I can send a list of the slides, which however are in a poor condition. If you think these specimens worth sending you must tell me the best means of sending them. They will all go in a very small box, but much too heavy for the post.

Lastly you must let me have the pleasure of sending you in 2 or 3 months time the New edit. of the Origin some parts of which I hope I have improved.[11]

With the most sincere esteem & respect, I remain Dear Sir Yours very faithfully | Ch. Darwin

P.S. Would you have the great kindness to send me your Photograph, as I have a collection of the best naturalists.—

LS(A)
Staatsbibliothek zu Berlin – Preußischer Kulturbesitz, Slg. Darmstaedter Lc 1859 Darwin, Charles, Bl. 205/206

1 See letter from C. F. Claus, 24 January 1869 and nn. 2 and 6. CD refers to Claus 1868a and 1868b.

2 For Claus's criticism of Ernst Haeckel, see letter from C. F. Claus, 24 January 1869 and n. 5.

3 See letter from C. F. Claus, 24 January 1869 and n. 4. CD began working on the fifth edition of *Origin* in December 1868 (*Correspondence* vol. 16, Appendix II). For CD's response to Moritz Wagner, see *Origin* 5th ed., p. 120.

4 In his letter of 24 January 1869, Claus informed CD that he was working on the metamorphosis of pedunculated barnacles (family Lepadidae) and included a paper on the subject (Claus 1868b).

5 For CD's understanding of the homologies of the appendages in cirripedes, see *Correspondence* vol. 15, letter to Anton Dohrn, 26 November [1867] and n. 5.

6 For CD's interpretation of the 'curious organ' as an auditory sac and subsequent criticism of his view, see *Correspondence* vol. 13, letter to Fritz Müller, 10 August [1865] and n. 8.

7 See letter from C. F. Claus, 24 January 1869 and n. 9. For the criticism of CD's discovery of 'complemental' males, see *Correspondence* vol. 13, letter to Fritz Müller, 10 August [1865] and n. 9.

8 *Xenobalanus* and *Coronula* are genera of sessile barnacles in the family Coronulidae (whale barnacles). See *Living Cirripedia* (1854), p. 651 and plate 17. In *Xenobalanus*, the body is everted into a stalked structure that superficially resembles the peduncle of *Lepas* (for more on the attachment structures and morphology of whale barnacles, see Seilacher 2005).

9 *Concholepas* is a gastropod genus (family Muricidae). *Cryptophialus* is a genus of burrowing barnacles, first discovered by CD on the *Beagle* voyage; this genus inspired CD's later work on Cirripedia (see *Living Cirripedia* (1851), p. v; see also *Autobiography*, p. 117).

10 The genus *Alcippe* is now *Trypetesa*.

11 The fifth edition of *Origin* was published in May 1869 (*Publishers' Circular*, 1 July 1869, p. 386).

From David Forbes 30 January 1869

11, York Place, | Portman Square, | London. W.
30 Jany. 1869.

My dear Mr Darwin

I have to thank you for Jenzsch's book which I had previously read (after a trial of patience) and put down as consummate rubbish.[1] Ideal wanderings of the brain and as I today have received a letter from Professor Zirkel of Kiel the most eminent of the Continental petrologists and observers (of the microscope in reference to rocks)—& in which he also mentions Jenzschs book I enclose you herewith a copy of his remarks in the original—[2] It certainly would be a rather severe reflection on the microscopic powers of Sorby[3] Zirkel and myself who have so many years worked on the subject if such a Flora and Fauna could exist—rocks which we have examined—and I know that Zirkel has carefully examined the Metaphysics alluded to by Jensch[4]—who undoubtedly is *mad* or something like it—

I am always thinking about writing a monograph on the Aymara Indian race of Bolivia, for the Ethnological but since Xmas I have been quite out of sorts with a very severe attack of Bronchitis which has also affected my head and made me very confused—[5] Might I ask you when you purpose to bring out your book on man?—[6] I really must endeavour to find some leisure for getting up some data for you—but I have been so thoroughly overwhelmed with work that I have not had an instants rest to sit down and collect my notes or thoughts on the subject.[7]

Wishing you a very happy new year and many of them believe me | My dear Mr Darwin | Yours most sincerely | David Forbes

[Enclosure][8]

Extract from a letter from Professor Zirkel of Kiel to DF.

Was sagt man denn in England zu der neuen Entdeckung von Jenzsch: eine mikroscopische fossile Flora und Fauna in den gemengtheilen von Eruptivgesteinen? Alle, mit denen ich mündlich oder brieflich darüber verhandelt habe, halten die Sache für Humbug— Das ist, bis ich eines bessern belehrt werde auch meine Ansicht. Ein neues heiteres Landschaftsbild: gegend Quarz im Melaphyr[9] im Vorgrunde ein idyllisch weidendes 'Infusorium', welches mit ausgestrecktem Rüssel gemüthlich grüne Algenzellschichten benagt, daneben ein Räderthierchen in geburtswehen begriffen, in Hintergrund einige Fluidcavities mit beweglichen Bläschen. Ein Bild, würdig der Pinsels von Sir Edwin Landseer![10] Da *uns* eine solche Entdeckung entschlüpfen muste! Ich habe an Jenzsch geschrieben und ihn gebeten, mir für einen Tag einige Praeparate zu schicken und mir einige Melaphyr scherben zu schenken, damit ich mich selbst Praeparate machen könne; ich erhielt zur Antwort, er brauche seine Praeparate *selbst* und seine Melaphyrhandstücke seien so fest verpackt, das er selbst nicht mehr dazu kommen könne. "Sapienti sat."[11]

Wenn Sie diese famose Schrift die natürlich allen Ultrametamorphiken und Dilettanten fabelhaft imponieren etwa noch nicht haben, so kann ich Sie Ihnen sc⟨hic⟩ken, ich besitze sie dreimal.

DAR 164: 143–143/2

1 Forbes refers to Gustav Jenzsch and Jenzsch 1868, a copy of which CD evidently sent to Forbes.
2 Forbes refers to Ferdinand Zirkel. See enclosure.
3 Henry Clifton Sorby.
4 Forbes may refer to Jenzsch's explanation of how volcanic rock was capable of supporting the microscopic flora and fauna he claimed to have discovered (see Jenzsch 1868, pp. 17–21).
5 Forbes's paper 'On the Aymara Indians of Bolivia and Peru' was published in 1870 in the *Journal of the Ethnological Society of London* (D. Forbes 1870).
6 Forbes refers to *Descent.*
7 For more on Forbes's work on the Aymara, see *Correspondence* vol. 16, letter from David Forbes, 26 March 1868.
8 For a translation of the enclosure, see Appendix I.
9 Melaphyre: 'any of various dark-coloured porphyritic (esp. basaltic) rocks. Later: (specifically) an altered basalt, esp. of Carboniferous or Permian age' (*OED*).
10 Edwin Henry Landseer.
11 *Sapienti sat.*: abbreviation of '*Verbum sapienti satis est*', a word to the wise is sufficient. (Latin.)

From A. R. Wallace 30 January 1869

9, St Mark's Crescent | N.W.
Jany. 30th. 1869

Dear Darwin

Will you tell me *where* are Fleming Jenkyn's arguments on importance of single variations.[1] Because I at present hold most strongly the contrary opinion, that it is the individual differences or *general variability* of species that enables them to become modified and adapted to new conditions.

Variations or "sports" may be important in modifying an animal in one direction, as in *colour* for instance, but how it can possibly work in changes requiring co-ordination of many parts, as in *Orchids* for example, I cannot conceive. And as all the more important structural modifications of animals & plants imply much coordination, it appears to me that the chances are millions to one against *individual variations* ever coinciding so as to render the required modification possible.

However let me read first what has convinced you.

You may tell Mrs. Darwin that I have now a daughter.[2]

Give my kind regards to her & all your family.

Very truly yours | Alfred R. Wallace—

DAR 106: B75–6

1 See letter to A. R. Wallace, 22 January [1869] and n. 7. Wallace refers to Henry Charles Fleeming Jenkin.
2 Wallace refers to Emma Darwin and to Violet Wallace, who was born on 27 January 1869 (Raby 2001, p. 199).

To James Croll 31 January [1869][1]

Down. | *Bromley.* | *Kent. S.E.*
Jany 31

My dear Sir

Tomorrow I will return registered your book, which I have kept so long.

I am most sincerely obliged for its loan, & especially for the MS. without which I sh^{d} have been afraid of making mistakes.[2]

If you require it the M.S shall be returned. Your results have been of more use to me than I think any other set of papers which I can remember.

Sir C. Lyell who is staying here is very unwilling to admit the greater warmth of the S. hemisphere during the glacial period in the N;[3] but, as I have told him, this conclusion, which you have arrived at from physical considerations, explains so well whole classes of facts in distribution, that I must joyfully accept it; indeed I go so far as to think that your conclusion is strengthened by the facts in distribution.[4] Your discussion on the flowing of the great ice-cap southward is most interesting.[5]

I suppose that you have read Mr Moseley's recent discussion on the force of gravity being quite insufficient to account for the downward movement of glaciers:[6] if he is right, do you not think that the unknown force may make more intelligible the extension of the great northern ice cap. Notwithstanding your excellent remarks on the work which can be effected within a million years, I am greatly troubled at the short duration of the world according to Sir W. Thompson, for I require for my theoretical views a very long period *before* the Cambrian formation.[7] If it w^{d} not trouble you I sh^{d} like to hear what you think of Lyell's remarks on the magnetic force which comes from the sun to the earth; might not this penetrate the crust of the earth & then be converted into heat.[8] This w^{d} give a somewhat longer time during which the crust might have been solid; & this is the argument on which Sir W. Thompson seems chiefly to rest. You seem to argue chiefly on the expenditure of energy of all kinds by the sun, & in this respect Lyell's remark wd have no bearing.

My new edition of the "origin" will be published, I suppose in about two months, & for the chance of y^{r} liking to have a copy, I will send one.[9]

With my very sincere thanks for all your kind assistance | I remain | yours very faithfully | Charles Darwin

I wish that you would turn your astronomical knowledge to the consideration whether the form of the globe does not become periodically slightly changed, so as to account for the many repeated ups & downs of the surface in all parts of the world.— I have always thought that some cosmical cause would some day be discovered

LS(A)
American Philosophical Society (361)

[1] The year is established by the relationship between this letter and the letter from James Croll, 2 December 1868 (*Correspondence* vol. 16).

[2] Croll had sent CD a manuscript summarising his views on climate change, together with a book containing all of his published papers up to that time and possibly some unpublished work. Croll had

evidently paginated the contents of the book and referred to the material by this system of pagination. Likely sources of some, but not all, of the material have been identified (see *Correspondence* vol. 16, letters from James Croll, 2 December 1868 and [2 December 1868]). CD had informed Croll that he wanted to keep the book until Christmas so that George Howard Darwin could read some of the papers (*Correspondence* vol. 16, letter to James Croll, 4 December [1868]).

3 According to Emma Darwin's diary (DAR 242), Charles Lyell arrived at Down on 29 January 1869. See C. Lyell 1867–8, 1: 268–73.

4 CD refers to Croll's theory of alternate ice ages in northern and southern hemispheres. See *Correspondence* vol. 16, letter from James Croll, [2 December 1868]. In *Origin*, pp. 377–8, CD argued that temperate plants could have crossed through the tropical regions near the equator during a glacial period by means of mountain ranges, which would have remained cooler, while tropical plants were preserved in the greater heat of the low-lying areas. This would account for the existence of similar species in both the northern and southern temperate zones. Joseph Dalton Hooker had challenged the explanation, arguing it would require 'so very cool a greenhouse' for temperate plants to cross the equator that tropical species would not survive, and suggested the distribution could be better explained by land-bridges and continental extension (see *Correspondence* vol. 11, letter to J. D. Hooker, 5 March [1863] and n. 17). In the fourth edition of *Origin*, CD admitted that the survival of tropical species was a difficulty for his theory (*Origin* 4th ed., pp. 450–1). CD here alludes to the fact that Croll's theory of alternate ice ages would solve the difficulty by allowing for a warmer non-glaciated hemisphere where tropical species could survive. In *Origin* 5th ed., pp. 450–61, CD accounted for the survival of tropical species using Croll's theory.

5 In his letter of [2 December 1868] (*Correspondence* vol. 16), Croll referred CD to his book of papers for explanations of the flow of ice over North America. The references have not been identified (see n. 2, above).

6 At the 7 January 1869 meeting of the Royal Society of London, Henry Moseley read a paper, 'On the mechanical possibility of the descent of glaciers by their weight only' (Moseley 1869a). An abbreviated version of the paper appeared in the 13 January 1869 issue of *Scientific Opinion* (Moseley 1869b).

7 The physicist William Thomson (later Lord Kelvin) had calculated the age of the earth based on its cooling from a molten ball. He concluded that the earth could not possibly have existed with its crust in a similar state as at present for the length of time proposed by geologists (see W. Thomson 1865). In *Origin* 5th ed., p. 379, CD noted that based on Thomson's calculation of the age of the earth and Croll's estimation that 60 million years had passed since the Cambrian period, the time left for the pre-Cambrian period (140 million years) was 'hardly sufficient for the development of the varied forms of life which certainly existed towards the close of the Cambrian period'.

8 In C. Lyell 1867–8, 2: 232, Lyell had suggested that the magnetic force of the sun was one of the principal means by which heat lost by conduction into space might be restored to the planet.

9 The fifth edition of *Origin* was published in May 1869 (*Publishers' Circular*, 1 July 1869, p. 386).

From Charles Phipps Haussoullier 31 January 1869

52 Grand rue de Batignolles, Paris.
31st. January 1869.

Sir,

I come to beg of you the authorization of translating into french your Naturalist's Voyage round the World.[1]

Without overrating my abilities, I am led to believe that my knowledge of the English language, & my general ⟨sc⟩ientific education, will allow me to do justice to your work, the more so that I have alrea⟨dy⟩ made many translations.

Yours most respectfully | Ch. Phipps Haussoullier.

Ch. Darwin Esq[re].

DAR 171: 362

CD ANNOTATION
Top of letter: '(Thierry Editor)' *pencil*

[1] The reference is to *Journal of researches*, part of which had been translated into French with the title *Voyages d'un naturaliste* (Montgolfier and Belloc trans. 1860). A translation of the full text was not published until 1875 (Barbier trans. 1875).

To Richard Kippist 31 January [1869?][1]

Down. | *Bromley.* | *Kent. S.E.*
Jan 31

My dear Sir

You are most perfectly welcome to the Fragmenta,[2] & I shall be delighted if they are of the slightest use to you.—

Yours sincerely | C. Darwin

Linnean Society of London (pasted in Mueller 1858–82, vol. 1)

[1] The year is conjectured from the date range of the accession of volume 5 of Mueller 1858–82 recorded by the library of the Linnean Society of London (see n. 2, below).

[2] The reference is to Mueller 1858–82. The work consists of twelve volumes, two of which (vols. 1 and 5) were dedicated to CD by the author. CD's copy of volume 5 had been sent by Ferdinand von Mueller in care of Joseph Dalton Hooker (see *Correspondence* vol. 16, letter from J. D. Hooker, 25 July 1868 and n. 3). CD's gift of vol. 5 to the Linnean Society was recorded in *Proceedings of the Linnean Society of London*, 'Additions to the library of the Linnean Society', 4 June 1868 – 15 June 1869, p. xix. CD probably sent the library his copy of vol. 1 at the same time, but the library had already received that volume in parts between 1858/9 and 1862/3. It is likely that CD wrote this letter in reply to a request from Kippist to keep CD's copy of vol. 1 for his personal collection. The letter was found at the Linnean Society library, tipped into CD's copy of vol. 1, which was stamped 'Linnean Society of London 29 Sept. 1943'.

From D. Appleton & Co. to Asa Gray 1 February 1869

Statement of Sales of Darwin's Origin of Species *to 1st* Feby *1869*

by D. APPLETON & CO. for account of Asa Gray[1]

On hand last account	350	*On hand this date*	70
Printed since	250	*Given away*	
		Sold to date	530
	600		600

Sold 530 *Copies.* 2^{50} *Rate* 5% — *$* 66.25
Feby 1/68 By Balance — 153.11
$219.36

DAR 159: A83

[1] This statement gives sales figures for the US edition of *Origin*. Gray had negotiated the original terms of publication of the book with Appleton & Co. (see *Correspondence* vol. 8, Appendix IV).

From Charles Spence Bate 1 February 1869

Plymouth
Feby 1— 69

My dear Sir

One word to tell you that a little time since—My Boys saw two crabs, which evidently are Portunus puber and Carcinas Mænas fighting on the Shore, the Portunus which they call "the devil crab" seized the other, threw it on its back, & then tore off every limb of the *Carcinus* successively & left the poor limbless wretch to perish—[1] I have occasionally seen evidence of this by occasionally finding bodies of living carcini without any legs for which I could not account until now— I thought that you might like to know this fact, but it must not trouble you to reply—

DAR 82: 71

CD ANNOTATION
Top of letter: '**Perhaps** worth quoting to show power of fighting.—' *ink* '(Fighting) | Crustacea' *blue crayon*

[1] *Portunus puber* is now *Necora puber* (the velvet swimming crab). *Carcinus maenas* is the common shore crab. In *Descent* 1: 332, CD cited Bate for this information.

From S. A. Merrell 1 February 1869

Council Bluffs Iowa
February 1/1869

Mr Darwin
Dear Sir

I have lately procured a work of yours—"(The Origin of Species)" which I am now reading from time to time— I had read the "Vestiges" many years before and my own reflections on the variation of Animals and plants under domestic culture had fully prepared me to welcome your book long before I had the pleasure of seeing it[1]

I have also the work of the elder Darwin (Erasmus) entitled "Zoonomia or the laws of Animal Life", which in the last century anticipated the theory promulged in the Vestiges.[2]

My present object in writing to you is to correct a statement of Dr Brewer's in your Book to the purport that the American Cuckoo never uses the nests of other birds for depositing its eggs and rearing its young.[3]

I remember well when a boy living at home in Illinois, to have seen a case of the kind in my father's orchard.

My attention was called to it by noticing a great commotion among the birds, and seeing a cuckoo slip away from the nest which was built in the top of an apple tree, I lost no time in climbing up to it.

Guess my surprise on reaching it, at seeing a young Cuckoo and a young Jay bird crouched in the nest side by side.

It was a nest of the Blue Jay—Garrulus cristatus. The young cuckoo was about twice the size of the Jay, and looked as though he had chiefly monopolized the food from both parents.[4]

It was nearly full feathered when I discovered it, and seemed to shrink from me as though he felt conscious of the essential meanness of his position. To my young mind, there was an expression of guilt in its eyes as it crouched itself back into the nest, and I was greatly tempted to throw it out of the nest at once. However, the novelty of the thing prevailed over my sense of justice and I left them both in the nest together until they were gone away—

This is the only instance of the kind I have ever seen, and indeed is the only time I ever saw the young of the Cuckoo at any time

But there could be not the slightest chance of mistake as to the identity of the birds for both the young birds were nearly full feathered especially the young cuckoo—

If to know this fact will be of any service to you I shall feel amply paid for the trouble of writing to you

I am, Sir | very truly yours | S A Merrell

DAR 171: 157, 157/1

[1] Merrell refers to *Origin* and to *Vestiges of the natural history of creation* ([Chambers] 1844). For CD's assessment of [Chambers] 1844, see *Correspondence* vol. 3, letter to J. D. Hooker, [7 January 1845] and n. 5.

[2] Merrell refers to E. Darwin 1794–6. CD had appended a note to his discussion of Jean Baptiste de Lamarck's views on species change in the historical sketch for the authorised American edition of *Origin*: 'It is curious how completely my grandfather, Dr. Erasmus Darwin, anticipated these erroneous views in his Zoonomia (vol. i. pp. 500–510), published in 1794.' (*Origin* US ed., p. vi n.). The note appeared in all subsequent editions of *Origin*. CD also discussed *Vestiges of the natural history of creation* ([Chambers] 1844) in the historical sketch, but did not specifically link Erasmus Darwin's views to those of Robert Chambers.

[3] Thomas Mayo Brewer had been cited on this point in *Origin*, p. 217.

[4] CD added Merrell's information to *Origin* 5th ed., p. 266.

To John Murray [1–3 February 1869][1]

Down. | Bromley. | Kent. S.E.

My dear Sir

M. Ch. Phipps Haussoullier of 52 Grand Rue des Batignolles, Paris, has applied to me for authorization for translating into French my Naturalist's Voyage.[2] I should be *very* glad of this. As you have copyright, will be so kind as to grant permission, & if so, please to send me the proper authorization that I may forward it to him— I hope that you will not object, & believe me yours very sincerely | Ch. Darwin

P.S. How has this Book sold since you printed an Edit. uniform with Origin?[3]

National Library of Scotland (John Murray Archive) (Ms. 42152 ff. 192–3)

1 The date is established by the relationship between this letter, the letter from C. P. Haussoullier, 31 January 1869, and the letter from John Murray, 3 February [1869].
2 CD refers to Charles Phipps Haussoullier and to *Journal of researches* (see letter from C. P. Haussoullier, 31 January 1869 and n. 1).
3 CD refers to *Journal of researches* (1860). For more on the publishing history of the book, including changes to the title and appearance of different editions, see Freeman 1977.

To John Warburton Robinson 1 February 1869

Due from the Rev^{d}. J. W. Robinson on account of School[1]
27"12"5
Ch. Darwin

Feb. 1. 1869.—[2]

ADS
American Philosophical Society (Getz 6587)

1 CD had acted as treasurer for the Down National School accounts, but had passed on the responsibility to Robinson in 1868. In December 1868, Robinson went to Ireland for three months; CD had expressed concern about school subscriptions in Robinson's absence (see *Correspondence* vol. 16, letter to J. B. Innes, 1 December 1868) and evidently took over the treasureship again in 1869 (see letter from John Lubbock, 8 February 1869, and Bromley Public Library, Downe National School Annual Report 1869 (P123/5/7)).
2 The document is countersigned 'J. Robinson'.

From J. V. Carus 2 February 1869

39, Elsterstrasse, Leipzig
Febr 2 1869.

My dear Sir,

I send you hereby the translation of Mr Steiger's answer, the latter part in a somewhat condensed form.[1] You see he has had just as little opportunity of comparing different races as Mr Sison; and even his observations on the growth of the horn are just as vague as Sison's.[2] So I regret, not to be able to give you more facts to the purpose. But on the whole both concur in the fact that the horns do not appear earlier in the Merinos than in other breeds.

In the case you should wish for more information I am afraid I shall not be able to get it here in Saxony. I wrote to Dresden to ask our Secretary to the Government for Agriculture for directions of breeders, and I got Mr Steiger's name.[3] I then applied to our Agricultural school, now in the way of being connected with the University, and I received Mr Sison's address[4] Now I confess I don't know where to apply for more.

I hope you are pretty well. We all look forward for good news from you.

I remain | My dear Sir | Yours very sincerely | J. Victor Carus

[Enclosure]

Mr. *Steiger's* answer.

On reading again Mr. Darwin's book on the origin of species, I see that I am too little of a scientific observer for being able to enter into a thorough discussion of my views. If not in the whole extent yet on the whole I have found Darwin's principles to be true. But that is just, what strikes me, that I don't see, why a thing is proved by so many words, what I found true by a thirty years working in practical breeding. I think it ridiculous with Mr. Darwin to call a great many varieties "races", which are nothing but little variations of one species and cannot yet claim that title. I lay more stress than Darwin on other causes, not only on the selection.

Nearly all Saxon races are Merinos, in modern times there have some English Southdowns[5] come true My observations have been made therefore only on Merinos and I am not able to compare the growth of the horns in Merinos with other breeds. Our Merino-males are all horned, a small part also of the females. They are the descendants of those Spanish Merinos which had been brought to Saxony in the last half of the 18th. century.[6] My flock is descended, also on the part of the mother, from this *[illeg]* Spanish immigration. In the Rambouillet and Negretti breeds it was said, that in the pure breed the ewes are not horned.[7] But it is proved that even in Spain the ewes bear sometimes horns. With us about 3–5 percent of the ewes are horned, while the males are all so, except some few exceptions, which occur now very seldom. These not-horned rams (the so-called Kalben-böke)[8] are not liked, properly speaking without reason.

The horns of the Merino rams are immediately after birth mostly present as small projections on the skull; they go on growing nearly equally as the whole animal. They are fully developed when the animal has attained its full growth, in the third year of its age. In the horned ewes the horns are also visible after birth but they remain small, about the 6th. or 8th. part of the weight of a male horn, the growth is nearly equal, in the same rate as the animal grows.

I tried to find if the formation of horns wants a larger amount of food; but all I found is, that the males want on the whole much more food than the ewes.

The castrated rams want less; but in these the horns are also often much smaller, so that I am inclined to think that the secretion or formation of the semen wants most of the food. In castrated animals all the other parts grow much quicker. Animals, on which the not-fleshy parts are covered with wool, are very difficult to breed and to keep so, and more difficult the male animals than the female and the castrated.

One observation is perfectly sure: that the horn of the male, as well as that of the female animal corresponds exactly to the nature of the whole animal, just as any other part. And this is the point which I should have liked to discuss with Mr Darwin.

While almost all breeders believe to be able to breed single parts or forms of single parts, I think the differences rest on illusion. The head of the English flesh-cattle is not absolutely smaller, but only so in relation to the greater amount of flesh on the fore part of the body, etc. etc.

Just the same holds good with the texture of the parts. Flesh, wool, skeleton, all that stands in close harmony.

DAR 86: A45; DAR 161: 69

CD ANNOTATIONS
Enclosure:
2.1 On reading . . . reason 3.12] *crossed blue crayon*
3.9 3–5 percent] '$\frac{3-5}{100}$ Horned' *blue crayon*
4.1 The horns . . . animal. 4.2] *double scored blue crayon*
4.4 In the . . . grows. 4.6] *double scored blue crayon*
7.1 One . . . harmony. 9.2] *crossed blue crayon*
End of enclosure: 'All parts change in correlation.—' *blue crayon*; '(I think I must say that at least no great difference in development, as Horns appear so very early in Merinos.)'[9] *ink; square brackets in MS*

1 See letter from J. V. Carus, 22 January 1869. Steiger has not been further identified.
2 Carus refers to Robert Sison and the observations on merino sheep in Sison 1867 (see letter from J. V. Carus, 22 January 1869 and nn. 3 and 4).
3 Dresden was the capital of the Prussian province of Saxony (Sachsen). The government official has not been identified.
4 Carus refers to the Landwirtschaftliche Institut (agricultural institute) of the University of Leipzig; the institute was founded in 1869 (Kirchner 1904, p. 3).
5 The Southdown is a native English breed, the name of which derived from the South Down hills around Lewes in East Sussex. For more on the development of the breed, see Wood and Orel 2001, pp. 77–8.
6 See letter from J. V. Carus, 22 January 1869 and n. 3.
7 The Rambouillet merino, developed from several Spanish flocks, was named after the royal experimental farm near Versailles in France (see Wood and Orel 2001, pp. 137–8). The Negretti breed was developed in Austria-Hungary from Spanish Negretti and Infantado flocks (*ibid.*, pp. 179–80).
8 Kalben-böcke: calf-rams (German).
9 See letter to J. V. Carus, 5 February 1869 and n. 2.

From Andrew Crombie Ramsay 2 February 1869

Jermyn S^t^
2 Feby 1869

My Dear Sir

Lyell has just told me that you are remodelling in the Origin of Species what you said abt the Weald.[1] Have you got a copy of the *2nd Edition* of my Physical Geology & Geography of G^t^ Britain?[2]

In it you will find 1st stated that the theory of Subaerial denudation applied to the Weald & the Origin of the Wealden escarpments, & I would be obliged if you would look at it. If I did not send you a copy please let me know & I will send one. The book is some years old; & very different from the 1st Edition which was very imperfect & little more than a shorthand writers Report.[3]

Yours very sincerely | A C Ramsay

DAR 176: 14

[1] CD was preparing the fifth edition of *Origin* (see 'Journal' (Appendix II)). Ramsay refers to Charles Lyell. In *Origin*, pp. 285–7, CD had calculated that the time required for the denudation of the Weald in Kent was nearly three hundred million years. This estimate was challenged by a reviewer of *Origin* in the *Saturday Review*, 24 December 1859, pp. 775–6. CD had accepted this criticism as justified and the section on the Weald was altered or omitted in subsequent editions (see *Correspondence* vol. 8 and Peckham ed. 1959, pp. 483–4). CD had also received information from Ramsay on the depth of deposits of the Weald (see *Correspondence* vol. 7, letter from A. C. Ramsay, 29 December 1858).

[2] This edition (Ramsay 1864) has not been found in the Darwin Libraries at CUL or Down, but CD possessed other editions (see *Marginalia* 1: 695–6).

[3] CD's copy of the first edition of Ramsay's book (Ramsay 1863) is in the Darwin Library–CUL.

To W. C. Tait 2 February [1869][1]

Down. | *Bromley.* | *Kent. S.E.*
Feb. 2

Dear Sir

I thank you for your very obliging letter, & am pleased to hear that you have been interested in my works.[2] As you have been so kind as to offer me some notes, I sh^{d} be glad to receive any facts about which you feel positive; but I should mention that owing to having so many other subjects in hand I do not intend to follow up any further the subject of variation under domestication. With respect to the tailless dogs, there would be I fear much difficulty in determining how far the unknown causes, which occasionally lead in other countries dogs to be born without tails, have acted more energetically in Lisbon; & how far the result has followed from the cutting off of the tail; but if you c^{d} render your case highly probable it would be very interesting.[3] Since the publication of my book I have heard of several other apparent cases of inherited mutilations.[4] I would beg a favour of you, if you c^{d} get acquainted with any good botanist, viz to send me a living young plant of the **rare** Drosophyllum Lusitanicum, which grows in sandy places in Portugal.[5] I have long wished to try a series of experiments on this plant.

With your taste for natural history, you must feel very isolated, & I can fully sympathize with you.

I am sorry to say that my health is weak, so that I have little spare strength or time to do more than to answer very briefly the numerous letters which I receive from various quarters. With every good wish pray believe me dear Sir | yours very faithfully | Charles Darwin

P.S. At present I am engaged on the secondary sexual differences in animals of all kinds, & sh^{d} be grateful for any information.—

LS(A)
Alan R. Tait (private collection)

[1] The year is established by the relationship between this letter and the letter from W. C. Tait, 26 January 1869.

[2] See letter from W. C. Tait, 26 January 1869.

[3] See letter from W. C. Tait, 26 January 1869 and n. 5.

[4] See, for example, *Correspondence* vol. 16, letter from J. J. Weir, 5 April 1868, in which John Jenner Weir related how his son had been born with a mark on his knee similar to a scar that Weir had on his own knee as the result of a cut.

[5] CD had already asked George Maw to try to obtain a specimen of *Drosophyllum lusitanicum* (see letter to George Maw, 13 January 1869 and n. 3).

To A. R. Wallace 2 February [1869][1]

Down. | Bromley. | Kent. S.E.
Feb 2[d]

My dear Wallace

I must have expressed myself atrociously; I meant to say exactly the reverse of what you have understood.[2] F. Jenkins argued in *N. Brit. R.* against single variations ever being perpetuated & has convinced me, though not in quite so broad a manner as here put.—[3] I always thought individual differences more important, but I was blind & thought that single variations might be preserved much oftener than I now see is possible or probable.— I mentioned this in my former note merely because I believed that you had come to similar conclusion, & I like much to be in accord with you.— I believe I was mainly deceived by single variations offering such simple illustrations, as when man selects.—

We heartily congratulate you on the birth of your little daughter.—[4] Yours very sincerely | C. Darwin

British Library Board (Add 46434, f. 168–9)

[1] The year is established by the relationship between this letter and the letter from A. R. Wallace, 30 January 1869.

[2] See letter from A. R. Wallace, 30 January 1869 and n. 1.

[3] CD refers to Henry Charles Fleeming Jenkin's article in the *North British Review* ([Jenkin] 1867).

[4] Violet Wallace. See letter from A. R. Wallace, 30 January 1869 and n. 2.

From John Murray 3 February [1869][1]

50[A], Albemarle S[t]. | W.
Feb[y] 3

My Dear Sir

You have my hearty consent to the Translation to French of your Naturalists Journal—not that either your or mine is now needed. The book has been out so long that the term of Protection has expired—[2]

What number of copies am I to print of Facts Argument for Darwin?[3] I am at a loss to advise not knowing what sort of demand to expect. I think however you w[d] be safe with 750.

Clowes is prepared to receive copy of New Edition of Variation of Species whenever you are ready to begin[4]

My Dear Sir | yours very faithfully | John Murray

Chas Darwin Esq

DAR 171: 363

1 The year is established by the relationship between this letter and the letter from C. P. Haussoullier, 31 January 1869.

2 Murray refers to *Journal of researches*, which was first published in 1839 (see letter to John Murray, [1–3 February 1869]). Murray also refers to the term of protection under copyright, which was twenty-eight years under the 1814 Copyright Act, and was extended to the author's life plus seven years under the 1842 Copyright Act (Seville 2006, p. 9). Evidently, the terms of the earlier act applied here.

3 Murray refers to *Facts and arguments for Darwin* (Dallas trans. 1869), which was a translation of Fritz Müller's *Für Darwin* (F. Müller 1864a).

4 William Clowes & Sons were the printers employed by Murray. The reference is to the fifth edition of *Origin* (Murray evidently wrote 'Variation' in error). CD finished corrections for the edition on 10 February 1869 (see 'Journal' (Appendix II)).

To A. C. Ramsay 3 February [1869][1]

Down. | *Bromley.* | *Kent. S.E.*
Feb. 3[d]

Dear Ramsay

I am very much obliged for your note.— You *did* send me a copy, & I read it with greatest interest, & I marked the very discussion to which you refer.[2] I had already put you, of course, at the head of the subaerial denudationists, but I am ashamed to say that from having read M[r] Whittaker's excellent article on escapement subsequently to yours, I confounded the facts together & thought that M[r] Whittaker had first brought this most important & to my mind quite conclusive evidence to bear.[3] Now I shall be able to rectify this.— I struck out in 2[d] or 3[d] Edit. all the rubbish which I wrote about the weald, & now I am adding a paragraph on the importance of sub-aerial denudation.—[4]

Believe me | Dear Ramsay | Yours very sincerely | Ch. Darwin

DAR 261.9: 4 (EH 88205901)

1 The year is established by the relationship between this letter and the letter from A. C. Ramsay, 2 February 1869.

2 CD refers to the second edition of *The physical geology and geography of Great Britain* (Ramsay 1864). CD's copy of this edition has not been found, but he refers to pp. 63–6 (see letter from A. C. Ramsay, 2 February 1869 and n. 2).

3 CD refers to William Whitaker's article 'On subaërial denudation, and on cliffs and escarpments of the chalk and lower tertiary beds' (Whitaker 1867). Two offprints of the article, one annotated, are in the Darwin Pamphlet Collection–CUL.

4 See letter from A. C. Ramsay, 2 February 1869 and n. 1. In *Origin* 5th ed., pp. 349–52, CD discussed the evidence for subaerial denudation and cited both Ramsay and Whitaker for observations on the escarpment in the Wealden district.

From James Croll 4 February 1869

Edinburgh
Feby 4th 1869.

Dear Sir,

Your favour with book came duly to hand, and I am glad to hear that some of the papers have been of a little use to you.[1] I am very much pleased to learn that you consider the facts in distribution favourable to some of the views expressed in my paper on climate[2]

I have not as yet been able to overtake that part of the question relating to the condition of the hemisphere whose winters occur in perihelion. I have no doubts that when this part of the subject has been fully discussed Sir Charles Lyell will agree with me;[3] the facts in favour of a warm climate are so numerous and strong

It is a pity that Sir Charles should have made those remarks on the "secular loss of heat in the solar system" vol. II p 213.[4] He must have done it without due consideration of that point

If there is one thing more than another in physics, regarding which we have absolute certainty, it is that the Solar system is losing its store of energy. We not only know this fact, but we have a means of determining the actual rate at which it is losing its power.

3,869,000 foot-pounds of energy in the form of heat is radiated off every square foot of the sun's surface per second. In other words the quantity of energy thrown off into space by the sun is equal to a *7000 horse* power engine working on every square foot of its surface. And when we reflect that all this prodigious expenditure has been going on during countless geological ages we may well ask the question what is the secret of the sun's great strength? Gravitation only affords up to the present time 20,000,000 years heat

There must be some other source in addition to that of gravitation. It is strange that that other possible sources did not suggest itself to Sir William Thomson and other physicians when working at this question.[5] It is perfectly obvious that the sun or rather the matter which composes the sun might have been in possession of heat prior to condensation

In this case it is difficult to say how old the sun may be for we do not know what this original store may have amounted to. In my paper I assumed a certain relation between the amount of original heat and that produced by gravitation viz 234 to 95 (Phil. Mag May 1868)[6] but as I stated, I may be wrong It may be more than this, or it may be less. This proportion some 70 000 000 years

The introduction of this new element, changes the entirely the conditions of the problem and I have no doubt that the whole matter will have to be re-considered. And it is quite possible that we may yet be able to get considerably more than one hundred million of years although very much beyond this we are brought to a limit by other considerations.

As regards determining the age of the earth's crust from the secular cooling of the globe I am not altogether satisfied with the plan.[7] It would no doubt do if we had proper data to go by, but I don't think we have got that yet.

I think that you may quite fairly assume a very long period before the Cambrian formation, even according to Sir William Thomson's theory for supposing the earth to have orginally been in a molten condition, a solid crust would very rapidly form, and if this crust would not break up and sink, the globe at the surface would be cool and suitable for life although a short way down below the surface the heat was intense. This results from the slow rate at which the crust is able to conduct the heat from within

It is some years since I read Sir William's paper on the Secular cooling of the globe,[8] but I think he states the above as his opinion, at all events I heard him once say in a lecture on the subject that supposing the earth to be in a molten state, in a few thousand years you could walk on its surface and hardly be sensible of the heat from within

Electricity and Magnetism used to be my favourite study, but for the past four years I have been paying little attention to what was going on in that department.

A relation between the spots of the sun and the manifestation of electric phenomena on the earth does not *necessarily imply any transmission of* electric forces from the one body to the other

One thing is certain that it is but an infinitesimal quantity of the forces of nature that ever assume the electric or magnetic form. Electrical phenomenon is very imposing and this is the reason why so much is attributed to it. A thunder storm is something very striking but Faraday has shown that more electricity is evolved in the silent decomposition of a few *grains* of water in the cell of a battery than would be required to produce the most violent flash of lightning[9]

It is owing to *high tension* that electricity makes such a display in passing from the statical to the dynamical state But when you estimate the amount of energy thus displayed in foot-pounds it is often very little

The quantity of energy in the form of electricity coming from the sun (if there be any at all) is certainly trifling compared with what comes in the form of heat.[10] I believe that no physicist will call this in question.

Your suggestion as to to the possibility of a cosmical cause for the ups and downs of the crust never occurred to my mind. I can see no possible way at present how the thing can be; but I shall ponder certainly over it.

I have never heard of Mr Moseley's paper.[11] My curiosity is very much excited and perhaps you will be so kind as to let me know when the paper appeared.

Edinburgh with all its books and learning is miserably behind in scientific literature. Since I came here I hardly know what is going on in the scientific world around. One can get plenty of good solid books on science, but the current news and literature of the subject is not to be found anywhere.

Edinburgh I fear is falling behind.

I need hardly say that a present of a copy the Origin of Species from its Author will be esteemed worth a dozen of copies out of a shop.

I trust you will make out to read this rather long affair, written hurriedly to catch the post.

I am, | yours very truly, | James Croll

Charles Darwin - Esqr. FRS

Keep the M.S. since it is of no use to me.

DAR 161: 263

1 See letter to James Croll, 31 January [1869] and n. 2.
2 See letter to James Croll, 31 January [1869] and n. 4.

[3] See letter to James Croll, 31 January [1869].

[4] See letter to James Croll, 31 January [1869] and n. 8. In C. Lyell 1867–8, 2: 213, Lyell had questioned whether the theory that the sun, as well as the planets, was constantly losing heat should be accepted, and suggested the possibility that some form of 'regenerating force' might be discovered.

[5] For Thomson's calculation of the approximate amount of heat produced by the sun's gravitation, see Thomson 1862a.

[6] Croll refers to calculations in his article 'On geological time, and the probable date of the glacial and the upper miocene period' (Croll 1868, p. 371).

[7] See letter to James Croll, 31 January [1869] and n. 7.

[8] Croll refers to Thomson's paper 'On the secular cooling of the earth' (Thomson 1862b).

[9] Croll refers to Michael Faraday (for the calculation, see Faraday 1834, p. 119).

[10] See letter to James Croll, 31 January [1869] and nn. 7 and 8.

[11] Croll refers to Henry Moseley. See letter to James Croll, 31 January [1869] and n. 6.

To J. V. Carus 5 February 1869

Down. | Bromley. | Kent. S.E.
Feb 5 1869

My dear Sir

Your kindness has been very great & I thank you cordially. Although I felt much curiosity respecting the horns of the merinos I did not intend to cause you so unreasonable an amount of trouble.[1]

It seems evident from your two correspondents that if there is any difference at all in the rate of development of the horns in the two cases, it must be extremely slight. There is so great a difference in this respect between the reindeer & other deer, & between antelopes in which one or both sexes have horns, that I fully expected something of the kind w^{d} be observed in sheep.[2] If you have any opportunity, I beg that you will return my sincere thanks to Messrs Sison & Steiger.[3]

I received the newspaper safely with excellent wood cut.[4]

During the last 6 weeks I have been at work in preparing a new Ed. of the Origin, & have found a good deal to alter & I hope I have in some degree improved it.[5] When published, in about 2 month's time, I will send you a copy.

Pray again accept my very sincere thanks for your kindness & believe me | yours very faithfully | Charles Darwin

LS

Staatsbibliothek zu Berlin – Preußischer Kulturbesiz, Slg. Darmstaedter Lc 1859, Darwin, Charles, Bl. 39/40

[1] See letters from J. V. Carus, 22 January 1869 and 2 February 1869.

[2] For more on CD's interest in the development of horns, see the letter to George Cupples, [6–9? January 1869] and n. 2. For CD's earlier correspondence on the age of development of antlers or horns in deer and antelope, see *Correspondence* vol. 16. In *Descent* 1: 286–90, CD summarised the results of his inquiries on the development of horns and noted that the results regarding merinos were anomalous (*ibid.*, pp. 289–90).

[3] CD refers to Robert Sison and a breeder named Steiger, who supplied Carus with information on the development of horns in merinos. Steiger has not been further identified.

[4] Carus had sent an article by Sison (Sison 1867) that contained a woodcut of one of the rams of the royal flock at Lohmen, Saxony. The article is in the Darwin Pamphlet Collection–CUL. See plate on p. 34.

[5] CD had started work on the fifth edition of *Origin* on 26 December 1868 (*Correspondence* vol. 16, Appendix II). He finished corrections on 10 February 1869 (see 'Journal' (Appendix II)). It was published in June 1869 (letter from R. F. Cooke, 22 June 1869).

From Frederik Anthony von Hartsen[1] 5 February 1869

Cannes
5 Febr—1869

Herrn Charles Darwin Hochwohlgeboren.
Hochgeehrter Herr

Ich würde mir nicht unterstehen, Eu Hw. wieder zu schreiben, wäre es nicht aus herzlichem Interesse zu Ihrer Forschungen, und zur Wissenschaft überhaupt.[2]

Es ist gegen Ihre Theorien Manches eingewandt worden. Nun scheint es mir dass die meisten Gegner dieser Theorien derselben viel zu schwere Anforderungen stellen. Darum ist m E. zur Apologie der Theorien nöthig, von vornherein dieses zuzugestehen,: es ist in den meisten Fällen ganz unmöglich, *zu beweisen* dass die sämmtlichen Individuen einer Gruppe einen gemeinschaftlichen Urheber gehabt haben. M. E kann man in den meisten Fallen nicht weiter gehen als, zu beweisen dass solches nicht *ungereimt* sein würde.

Ich sage, es sei selten möglich zu beweisen dass die Individuen einer natürlichen Gruppe einen gemeinschaftlichen Urahn gehabt haben. Was ich meine ist Folgendes. Ist es möglich dass ein Naturvorgang durch *natural selection* aus einen einfachen Organismus eine gewisse Gruppe höherer Organismen bildet, so lässt sich nicht einsehen warum dieses nur auf *einem* Puncte der Erde geshehen soll. Ich kann mir, zum Beispiel, vorstellen dass in Ostindien aus einer sehr einfachen Pflanze allmählig *Palmen* enstanden sind, und dass in Süd-Amerika, ganz unabhängig davon, *ebenfalls* aus einer sehr einfachen Pflanze Palmen entstanden sind. Kurz, ich kann mir denken dass auf 2, 3 ja mehrere Puncte der Erde, *dasselbe* Genus, ja vielleicht *dieselbe* Species, aus dem Anorganischen Reiche hervorgegangen ist.

Nach Ihrer Theorie also lässt sich m E sehr wohl annehmen dass diese oder jene Gruppe von Organisme nicht *Eins*, sondern *mehrere* Centra des Ursprungs gehabt hat. Es würde also die Theorie ihren Werth behalten, wäre es sogar dass die Individuen einer Gruppe, aus geographischen Gründen, unmöglich einen gemeinschaftlichen Ursprung gehabt haben könnten. Ueberhaupt kann ich mir denken dass die anorganische Masse auf *mehreren* Puncten der Erde lebendige Wesen hervorgebracht hat und dass diese Wesen sich auf mehreren Puncten *auf ähnliche Weise* weitergebildet haben, so dass z B. die Schnecken Englands einen anderen Ursprung wie die Schnecken Amerikas haben.

Diese Bemerkung schien mir vielleicht zur Entkräftung geographischer Einwände gegen Ihre Lehre nicht ungeeignet zu sein.[3] Daher dass ich so frei bin, sie Ihnen mit zu theilen.—

"Carl Vogt" hat in Holland Vorlesungen gehalten und darin Ihre Theorien mit seinem Materialismus verwebt.[4] Manche Leute sind dadurch veranlasst zur Ansicht dass Materialismus und Atheismus das letzte Wort des Darwinismus ist. Ich habe nun eine populäre Darstellung Ihrer Lehren verfasst, und dabei gezeigt dass dieselben über die Hauptfragen der Religion gar nichts entscheiden.[5] Es ist aber nicht warscheinlich dass Eu Hw Holländisch lesen, daher Holländische Schriften für Sie wenig Interesse haben dürften.—

Prof. Fechner schreibt mir dass ein D^r. Pfaff Ihre Ansichten critisirt hat.[6] Ich hoffe die Schrift von Pfaff zu lesen, lege aber auf Einwände von Deutscher Seite wenig werth, weil dieselben meistens aus Vorurtheil stammen.[7] Nur bedauere ich dass wir nicht wissen wie das Organische Leben auf *anderen Sternen* entsteht. Eine vergleichende "Organismogenie" wäre sehr schätzbar! Vielleicht kommen wir noch einmal dazu!

Herr Traherne Moggridge geht glücklich besser.[8] Er ist ein sehr enthousiastischer Discipel von Ihnen!

Und hiermit danke ich ergebenst für die freundliche Zuschrift mit welcher Sie, nach Veranlassung meines Essay's mich beehrt haben.[9] Hatte ich gewusst dass Sie mir eine so ausführliche Antwort zugedacht hatten, ich würde vielleicht nicht den Muth gehabt haben, Ihnen meine Schrift zu übersenden. Mit herzlichen Wünschen für Ihr Wohlsein erlebe ich | Hochgeehrter Herr | Ihr Hochw. ergebenster | F. A. v. Hartsen

DAR 166: 114

1 For a translation of this letter, see Appendix I.

2 See *Correspondence* vol. 16, letter from F. A. von Hartsen, 5 January 1868. The abbreviation 'Eu Hw.' is not standard, but Hartsen evidently intended 'Eure Hochwohlgeboren'.

3 In *Origin*, p. 351, CD had argued that species of the same genus, although widely separated geographically, must originally have proceeded from the same source, as they had descended from the same progenitor. He then discussed geographical distribution and the means of dispersal of species to new locations (see *Origin*, pp. 356–410). In subsequent editions of *Origin*, CD added material to counter objections to his theory based on difficulties associated with dispersal.

4 For more on Vogt's popular lectures of 1868 and 1869, see Kockerbeck ed. 1999, pp. 23–4.

5 Hartsen refers to his book, *Darwin en de Godsdienst* (Darwin and religion; Hartsen 1869b).

6 Hartsen refers to Gustav Theodor Fechner and Friedrich Pfaff.

7 In his book, *Die neuesten Forschungen und Theorien auf dem Gebiete der Schöpfungsgeschichte* (Latest research and theories on the history of creation; Pfaff 1868), Pfaff argued that CD's theory was untenable in the light of geological findings.

8 Hartsen refers to John Traherne Moggridge. In his letter of 5 January 1868 (*Correspondence* vol. 16), Hartsen had mentioned that Moggridge's health was bad.

9 CD's reply to Hartsen's letter of 5 January 1868 (*Correspondence* vol. 16) has not been found. Hartsen had sent CD a copy of his *Grundlegung von Aesthetik, Moral und Erziehung* (Foundation of aesthetics, morals and education; Hartsen 1869a).

To William Bernhard Tegetmeier 5 February [1869][1]

Down. | Bromley. | Kent. S.E.
Feb 5

My dear Sir

Have you quite thrown me overboard as too troublesome? I have not heard from you for an age.— I wrote some two months ago asking you to send as soon as you could any extracts & facts, which you told me you had collected, about number of sexes— Also any account of even one or two breeds of the fowl,—as colour of plumage of hen & chickens of Pile Game or Golden Hamburghs.— Will you not aid me so far?—[2] I hope that are well & that all goes well with you.—[3]

My regular work has been interrupted for 6 weeks & more by preparing a new Edit. of the "Origin," which is wearisome work; but I shall now finish in 2 or 3 days.[4]

Yours sincerely | Ch. Darwin

Archives of the New York Botanical Garden (Charles Finney Cox collection) (ser. 14)

1 The year is established by CD's reference to finishing work on a new edition of *Origin* (see n. 4, below).

2 CD had asked for information on the proportion of sexes in domestic animals in a letter published in the *Field*, 22 February 1868 (see *Correspondence* vol. 16, letter to W. B. Tegetmeier, 11 February [1868] and n. 2). Tegetmeier had promised to tabulate results and forward them (*ibid.*, letter from W. B. Tegetmeier, 9 March 1868). In his letter of 8 September [1868] (*ibid.*), CD had asked Tegetmeier to forward whatever replies he had received to date, and also asked for information on the plumage of a few breeds of fowl, mentioning the pile game breed as an example. The last extant letter from Tegetmeier is that of 25 May 1868 (*ibid.*).

3 Tegetmeier had evidently mentioned that his health had been poor in September 1868 (see *Correspondence* vol. 16, letter to W. B. Tegetmeier, 8 September [1868]).

4 CD finished corrections for the fifth edition of *Origin* on 10 February 1869 (see 'Journal' (Appendix II)).

From C. F. Claus[1] 6 February 1869

Marburg
d 6/II 1869

Hochgeehrter Herr!

Sie haben mir durch Ihre gütigen Zeilen eine ebenso grosse Freude als Ehre erwiesen, und weiss ich in der That kaum, wie ich mich für das in Aussicht gestellte Untersuchungsmaterial dankbar beweisen kann.[2] Ich hoffe durch eine möglichst sorgfältige und gewissenhafte Prüfung jener so interessanten und wichtigen Objecte mich Ihrer grossen Güte und Ihres Vertrauens einigermassen würdig zu zeigen. Ich werde die bezeichneten Gegenstände mit dem grössten Danke acceptiren. Da ich bislang nur die Larven und Puppen von Lepadiden untersuchen könnte, so würde ich Ihnen auch besonders verpflichtet sein, wenn Sie mir vielleicht die Gelegenheit verschaffen könnten, die Puppen von Balanus zu examiniren; ich sollte denken, die Post würde ein solch' kleines Kästchen, welches ich Sie bitten würde, an das Königliche zoologische Institut zu adressiren, leicht annehmen und beförderen—

In einer kleinen Notiz, die ich neben mehreren Arbeiten mir erlaubt habe an Sie in Kreuzband abzusenden, habe ich schon 1862 den Versuch gemacht, den *Cirripedienleib* auf den der *Copepoden* morphologisch zurückzuführen.[3] Meine neueren Studien der Larven, Puppen und jungen in Puppen eingeschlossenen Cirripedien, haben hoffe ich meine damalige Zurückführung ziemlich sicher gestellt.[4] Die Organe in den Hörnern der Larve sind jedenfalls Sinnesorgane und mit Fäden ausgestattet den

Sinnesfäden (Leydigschen Organen) niederer Crustaceen vergleichbar, sie sind offenbar dasselbe, was Sie als Gehörorgane in der Puppe dargelegt haben, deren *Stirnecken* den Stirnhörnern der Larve entsprechen.[5] Die feinere Bestimmung dieser Organe ist leider nur an frischen Puppen möglich, ich konnte leider nur Spiritusexemplare untersuchen.

Bezüglich der Schrift von Wagner[6] dürfte wohl auch hervorzuheben sein, dass die allmähligen tellurischen und klimatischen Veränderungen als Ursache der Veränderungen in den Lebensbedingungen und Wechselbeziehungen der Organismen viel zu gering geschätzt worden sind. Die Migration mag besonders in Betracht kommen zur Erklärung der räumlich neben einander enstehenden geographischen Varietäten, die letztern dagegen zur Erklärung der *zeitlich* aufeinanderfolgenden Abänderungen. Im letzteren Falle ist die Migration secundär und Folge nicht wie Wagner glaubt primäre Ursache. Auch haben Sie, sollte ich denken, durch die Darlegung der 2 erdachten Beispiele, insbesondere der langbeinigen schnellern und kurzer beinigen Wolfrasse sehr wahrscheinlich gemacht, dass auch ohne so strenge Isolierung, wie sie Hochgebirge, Ströme und Meere bedingen, Varietäten räumlich neben einander hervorgehn können. Das hat Wagner nicht widerlegt, sondern nur seine Behauptung einfach entgegengehalten—.

Es gereicht mir zur besondern Ehre, Ihnen, hochverehrter Herr, meine Photographie beilegen zu dürfen.[7] Ich würde mich glücklich schätzen, die Ihrige mit Ihrer eigenhändigen Namensunterschrift zu besitzen und in meinem zoologischen Album an die Spitze der Naturforscher stellen zu können, die durch ihre Arbeiten in die Entwicklung der Wissenschaft gewaltig eingreifend, umgestaltend und aufbauend gewirkt haben.

Dass *R Leuckart*[8] einen Ruf nach Leipzig erhalten und angenommen hat, wird Ihnen wohl noch nicht bekannt sein.

Mit ausgezeichter Hochachtung und Verehrung unterzeichne ich als Ihr ergebenster | C Claus

P.S. In der kleinen Schrift "über die Copepodenfauna Nizza's im Sinne Darwin's habe ich die Bemerkungen Fr. Müller's über die zwei Männchen von Tanaïs d. (Riecher und Packer) in ihrem Werthe in Zweifel gezogen.[9] Ich darf mir wohl erlauben, Ihnen uber die *Naupliuslarven* der Garneelen meine Zweifel nicht vorzuenthalten. Fr. Müller hat nicht bewiesen, weder dass die sog. *Naupliuslarven* aus den Eiern von *Peneus* hervorgingen noch dass die Zoëastadien nach Abstreifung der Haut aus den Naupliuslarven sich entwickelten. Das neben einander Vorkommen der Naupliusformen und Zoestadien berechtigt nicht zu dem Schlusse, dass diese die jüngeren Stadien der letzteren sein, *was noch zu beweisen ist.*[10]

Es mag vollkommen wahr sein, aber die *Sicherheit* fehlt. Meine Zweifel aber sind nunmehr geradezu angeregt durch die *Grösse* der Eier von einer nicht näher bestimmten Peneus art. Ein so grosses Ei lässt auf die vollkommenere Ausbildung des Embryos und der ausschlüpfenden Larve schliessen—

In etwa *vier* Wochen fangen die Ferien an und ich habe dann *sechs* Wochen freie Zeit, um mich ausschlieslich der Cirripedienarbeit widmen zu können. Sollten Sie indessen, hochverehrter Herr, bis dahin nicht wohl Zeit haben, mir das Material zuzusenden, so würde ich auch bis in den Sommer warten können, ich möchte um keinen Preis Ihre kostbare Zeit im unrechten Augenblick in Anspruch nehmen. Ihr | C Claus

DAR 161: 177

1 For a translation of this letter, see Appendix I.

2 CD had offered to send some specimens of barnacles. See letter to C. F. Claus, 28 January 1869.

3 Claus refers to his paper on the morphological relations between copepods and other crustacean groups, including cirripedes (Claus 1862; for the section on cirripedes, see pp. 161–2). CD's copy of the paper has not been found.

4 With his letter of 24 January 1869, Claus had enclosed an offprint of his article on the cypris-like larvae of cirripedes and their transformation into sedentary adults (Claus 1868b).

5 Claus refers to CD's description of acoustic organs in the final pre-adult stage (cypris) of *Lepas australis* (*Living Cirripedia* (1851), p. 15). For more on CD's view of the homologies of the naupliar, cyprid, and adult horns and antennae, see Newman 1993, pp. 372–4. In modern terminology, the 'Leydig organ' is the organ found near the oesophagus of sharks and other members of the subclass Elasmobranchii, and is associated with the immune system. Claus probably refers to an organ described by Franz von Leydig in a monograph on *Daphnia* as an analogue of the green gland in decapods (Leydig 1860, pp. 25–6).

6 Claus refers to Moritz Wagner and Wagner 1868a or 1868b. See letter from C. F. Claus, 24 January 1869 and n. 4.

7 The photograph has not been found.

8 Rudolf Leuckart.

9 Claus refers to his study, *Die Copepoden-Fauna von Nizza. Ein Beitrag zur Charakteristik der Formen und deren Abänderungen 'im Sinne Darwin's'* (The copepod fauna of Nice. A contribution on the diagnosis of its form and its modification 'in the Darwinian sense'; Claus 1866). In *Für Darwin* (F. Müller 1864a, pp. 13–16; Dallas trans. 1869, pp. 20–4), Fritz Müller had described two forms of male in *Tanais dubius* (now *Leptochelia dubia*). Claus questioned whether one might be the sexually mature form of the other, as was the case with the amphipod *Phronima* (see Claus 1866, pp. 1 and 2). In *Origin* 4th ed., p. 50, CD had noted Müller's discovery of the two forms of male; the passage remained unchanged in subsequent editions. In *Descent* 1: 328–9, CD cited Müller at greater length on *Tanais.*

10 Müller had suggested that *Penaeus* emerged from the egg in its nauplius state in a paper, 'Die Verwandlung der Garneelen' (F. Müller 1863, p. 8), that appeared in English as 'On the metamorphoses of the prawns' (F. Müller 1864b, p. 104). Müller's observation was only confirmed much later (Kishinouye 1900), but Claus evidently accepted Müller's account when he made his own observations on development in the genus (see Claus 1876, pp. 10–13).

To James Croll 6 February [1869][1]

Down. | Bromley. | Kent. S.E.
Feb 6

My dear Sir

I am very much obliged for your long &, to me, extremely interesting letter. It is consolatory to me that you are enclined to give a little more age to the world.[2]

I read Mr Moseley's article in "Scientific Opinion" about 3 or 4 weeks ago; I have had the house searched but cannot find the copy. The article was given as extracted from the Proc. Royal Soc., but I have looked in the two last numbers which I have received, & it is not in them. Hence I suppose the author or Sec^y. sent an abstract beforehand, & I suppose it will appear in the next number of the proceedings.[3] The article interested me, though I could not follow all the reasonings, as I hear he is a sound man.

I was reminded of my crude notion that the cause of elevations, volcanic phenomena &c was cosmical, by my son telling me about Cap[t] Clark's paper in Phil. Trans. (which you probably know) on the globe being a little flattened at the equator, & that this stands in relation to relative position of continents & oceans.[4]

It w[d] be a great gain if someone could shew a cause of the many changes of level in the crust of the earth.

With very sincere thanks | believe me | yours very faithfully | Charles Darwin

LS(A)
American Philosophical Society (B/D25.284)

1 The year is established by the relationship between this letter and the letter from James Croll, 4 February 1869.

2 Croll had argued that the earth would have formed a solid crust very rapidly, so one could assume 'a very long period before the Cambrian formation' (letter from James Croll, 4 February 1869).

3 CD refers to Henry Moseley and Moseley 1869a . See letter to James Croll, 31 January [1869] and n. 6. A fuller version of the paper appeared in *Proceedings of the Royal Society of London* (Moseley 1869b).

4 In the conclusion to his paper 'Volcanic phenomena and the formation of mountain chains', p. 631 (*Collected papers* 1: 82), CD had noted that 'even the attraction of the planetary bodies on a sphere not solid throughout' had been suggested as one of the forces modifying the earth's surface. CD also refers to Alexander Ross Clarke and Clarke 1866, and probably to George Howard Darwin.

To George Howard Darwin 6 February [1869][1]

Feb 6

My dear George

Perhaps you w[d]. like to read enclosed letter from Croll. & you can some time return it.— What he means by Gravitation & Heat I do not understand.[2]

Sir J. Lubbock has been calling here, & he seemed to regret that you had not accepted the Eton proposal, as a grand opening for making a great school scientific; but he agreed that without you felt a decided taste for the work, it w[d]. not answer.—[3] He thought scientific masters w[d] soon occupy as high & as profitable a place as the classical masters.

Yours affect | C. Darwin

DAR 210.1: 4

1 The year is established by the relationship between this letter and the letter from James Croll, 4 February 1869.

2 CD evidently enclosed the letter from James Croll, 4 February 1869. Gravitational heat is energy released when masses aggregate, converting their original potential energy into heat energy (see Thomson 1862a). See letter from James Croll, 4 February 1869 and n. 5.

3 George had recently visited Eton and evidently had been considering a teaching position there (see letter from G. H. Darwin, 6 February 1869). John Lubbock had been a pupil at Eton from 1845 to 1848; he was later critical of the emphasis on classics to the exclusion of other subjects (Hutchinson 1914, 1: 16–17). At Eton, physical science was introduced for the Fifth Form in 1869 (Lyte 1911, p. 529).

From G. H. Darwin 6 February 1869

Amateur Dramatic Club [Cambridge]
Saturday

My dear Father,

I have looked thro' every number of the Philosophical Mag. for the last year & cannot find any translation of a paper by Delaunay; And Strutt who takes it in has not seen any thing and so I can do no more unless you can get another reference.[1] Strutt has been here since I have been up—but has gone away today as the Fellows' meeting is over. I asked him about the Magnetic Storm business & he seemed to consider that any effect that cd be produced by the Sun's energy in retarding the cooling of the earth must be utterly insignificant.[2]

My visit to Eton was a great success; I got down there are 11.30 & went & saw Rawlins, who shewed me all the lions.[3] I don't wonder at Etonians liking the the place, it is so divine. At 2 we went to luncheon (or dinner) & there I met 3 other masters 2 of whom I knew a little thro' Cambridge, & the 4th. was a nephew of Browning's & of the same name.[4] He is one of their shining lights & was an extremely pleasant man— he actually offered to give me a bed when I came there again.[5] I had heard of him before thro' Colvin.—[6]

After lunch I went & saw Browning's house & a brother of a Cambridge man in it. And then R. & I took a loaf round Windsor Castle & I came back to London dined at Uncle Ras'[7] & came on here.

I suppose Frank has written to you & told of his sprained ankle. He is just beginning to hobble about, & he is reading v. hard. Pryor seems quite indefatigable in coaching him.[8]

Last night Jim turned up at 8.30 & came & tea'd with me & Strutt dropt in & had tea After tea L. & I talked such mathematics that it drove Jim away to Frank's.[9] Jim went off this mg. at 1. after loafing about all the A.M.

That letter that was forwarded to me was from Mr. Hamilton telling me that the Solicitor Mr. Hollams was willing to take me for 6 mos.[10]

Bristowe[11] told me that I had better not go till after 6 mos. with a conveyancer. If you don't think it giving too much bother I shd be v. much obliged if you wd write to Grove[12] (at once) & ask him whether he thinks it wd be good for me to go to a Solicitors **taking into consideration** the fact that I am going to Equity & that in all proby. the Solicitor would treat me just like any other clerk.

I saw the bust & am just a little disappointed[13]

The college has paid me a scholar's dividend. Pollock a co-fellow with me told me that he had'nt got one & so I believed him[14]

Yrs | G H Darwin

Feb. 6. 69

DAR 210.2: 6

[1] George refers to the *Philosophical Magazine* and to Charles Eugène Delaunay. CD was probably looking for the translation of an article by Delaunay, 'Sur l'hypothèse de la fluidité intéieure du globe terrestre'

(Delaunay 1868a). The article had been translated in *Geological Magazine* as 'On the hypothesis of the internal fluidity of the terrestrial globe' (Delaunay 1868b). CD may have become aware of it from an article by David Forbes, 'The nature of the interior of the earth', that appeared in the April 1869 issue of *Popular Science Review*; Forbes referred to Delaunay but did not give a citation (D. Forbes 1869, p. 127). For more on Delaunay's hypothesis of the earth's fluid core, see Brush 1979, pp. 239–40. George also refers to John William Strutt.

2 Charles Lyell had suggested that the magnetic force of the sun might contribute to the earth's heat, but James Croll had argued the amount of electrical energy would be 'trifling' (see letter to James Croll, 31 January [1869] and n. 8, and letter from James Croll, 4 February 1869).

3 See letter to G. H. Darwin, 6 February [1869] and n. 3. George refers to Francis Hay Rawlins, who was a pupil at Eton at the time.

4 Oscar Browning, an assistant master at Eton, was not, in fact, related to the poet Robert Browning (Browning 1910, p. 6).

5 Browning ran a boarding-house for Eton scholars from 1862 (*ODNB*).

6 Walter Mytton Colvin.

7 Erasmus Alvey Darwin.

8 George refers to Francis Darwin and Marlborough Robert Pryor.

9 Jim and 'L.' have not been identified.

10 George refers to John Hollams and possibly to Alfred Douglas Hamilton; he was considering studying law.

11 Probably Henry Fox Bristowe.

12 William Robert Grove.

13 George may refer to the bust of CD by Thomas Woolner. CD sat for the artist in 1868 (see *Correspondence* vol. 16, letter to J. D. Hooker, 26 November [1868]).

14 Dividends made up of fines, rents, and other dues were paid by colleges to fellows (Brooke ed. 1988–2004, 3: 136–7). George refers to Frederick Pollock.

To David Forbes 7 February [1869][1]

Down. | Bromley. | Kent. S.E.
Feb 7th.

My dear Mr Forbes

I am much obliged for your note, and am sorry that I troubled you with the pamplet, as you already possessed it. I dare say the man is mad. He wrote to me boasting that he should revolutionize geology.[2]

With respect to my own book, the subject grows so, that I really cannot say when I shall go to press; but I know that in about two months, I shall have to go again over that part of my MS., where I should like to insert a short abstract of your most interesting observations on the Aymara Indiens, and I should be most grateful if you could let me have it by that time.[3] I am glad to hear that you are thinking of publishing a paper on the Indians in extenso.[4]

I hope that you will be soon well,[5] and pray believe me | yours very sincerely | Ch. Darwin

P.S. Owing to your letter being addressed to Q. Anne St, I did not receive until yesterday.[6]

LS(A)
American Philosophical Society (B/D25.300)

1 The year is established by the relationship between this letter and the letter from David Forbes, 30 January 1869.
2 See letter from David Forbes, 30 January 1869. CD refers to Gustav Jenzsch and Jenzsch 1868. Jenzsch's letter to CD has not been found.
3 *Descent* was published in 1871, by which time CD was able to refer to Forbes's published work on the Aymara people (D. Forbes 1870).
4 See letter from David Forbes, 30 January 1869 and n. 5.
5 In his letter of 30 January 1869, Forbes mentioned suffering a bad attack of bronchitis.
6 Forbes evidently sent his letter to the address of CD's brother, Erasmus Alvey Darwin. In 1869, CD stayed with his brother from 16 to 24 February (see 'Journal' (Appendix II)).

From John Lubbock 8 February 1869

15, Lombard Street. E.C.
Feb 8/69

My dear Mr Darwin,

I am very much obliged to you for the trouble you have taken about the school accounts, which I have asked Mrs Lubbock to return today.[1]

I much regret to find the funds at such low water, but I really think we might get more subscribers who only wait to be asked. For instance all my brothers are parishioners as well[2] Mrs Lubbock & myself & the clerk[3] were the only communicants, & the present state of things must be very demoralizing to the parishioners

DAR 170: 71

1 Lubbock refers to the accounts for the Down National School for boys and to Ellen Frances Lubbock. In a letter to J. B. Innes of 1 December 1868 (*Correspondence* vol. 16), CD wrote that he had handed over the school accounts to John Warburton Robinson, who had recently become curate of Down, but that Robinson had gone to Ireland for an extended stay and that parishioners were afraid in any case to pay subscriptions to Robinson. CD evidently had taken charge of the accounts in Robinson's absence (see letter to J. W. Robinson, 1 February 1869 and n. 1). This incident was the latest in a series of problems with the curacy (see *Correspondence* vol. 16).
2 Lubbock had seven brothers (*Burke's peerage* 1999).
3 The clerk has not been identified.

From Georg Recht[1] 9 February 1869

München Schommergasse N.2/3
den 9*ten* Februar 1869

Euer Wohlgeboren!

Indem ich Ihnen meine Schrift, "die Weltgesetze" übersende,[2] möchte ich beweisen, dass auch ich Sie als den Meister der Naturwissenschaften verehre. Es würde mir eine unendliche Ehre sein, wenn Sie dieselbe der wissenschaftlichen Society in London mittheilen und mir Ihre Anschauung über den Werth oder den Unwerth derselben, ja vielleicht auch die Ansicht der erwähnten Society mittheilen würden, da es meine Verhaeltnisse kaum zulassen, dass ich ein besonderes Exemplar mit einem Schreiben an dieselbe absende.[3] Ich habe mit meiner Familie nur taeglich Einen Gulden zu verzehren und leide den grösten Mangel am Allernöthigsten.

Dieses ist auch der Grund warum ich die genannte Schrift durch die Gefaelligkeit eines Freundes in Auszug mittheilen konnte, und die weitere Ausführung, so wie die Verglichung meiner Anschauung mit den bisherigen Lehren ausser Acht lassen musste.

Ich wäre glücklich, wenn die Verhaeltnisse es mir gestatten würden, weiter an der Aufgabe die mir Viel zu versprechen scheint fortzuarbeiten, doch das liegt in höhern Haenden.

Mit tiefster Verehrung | Euer Wohlgeboren | ergebenster Dr. Recht

DAR 176: 74

1 For a translation of this letter, see Appendix I.

2 Recht refers to his monograph, *Die Entwicklung der Weltgesetze* (Development of the laws of the universe; Recht 1869).

3 CD evidently sent the copy of Recht 1869 to the Royal Society of London; it is still in their collection (Royal Society Library, Tract 525/5). An inscription at the front of the book reads, 'Herrn Charles Darwin | Esq. M. A | der Verfasser' (to Charles Darwin Esq. M.A., from the author).

From Ferdinand Maria Malven 12 February [1869][1]

Vienna (Austria)
february 12.

Most venerated Sir.

The undersigned begs leave to acquaint You, that in a letter which a few days ago he adressed to the Editors of one of our most renowned and influential Newspapers, the "New free Press", he *suggested* the celebration of the to-days 60th. anniversary of your birthday.—[2] He expressed therein the wish, that our learned Societies might You give a token of their acknowledgment of your immeasurable and immortal scientifical discoveries and achievements.— Believe me Sir, your name is here and everywhere in Germany as worshiped as that of our ever-lamented immortal *Humboldt* whose decease by a noteworthy coincidence happened in the same year where your mighty pioneering work appeared!—[3] May you excuse my bad English, I never having been in England nor being in the habit to write or to speak English.

May it please you Sir, to agree the tribute of profound respect and veneration of your most devoted and adherent servant.| ferd. Maria Malven, | librarian to the Imp. and Royal | Austrian board of trade, | companion of the royal order of the saviour of Greece.—

DAR 171: 35

1 The year is established by the reference to CD's sixtieth birthday.

2 Malven refers to the *Neue Freie Presse*, a liberal Viennese daily paper founded in 1864 (Wandruszka 1958, p. 64).

3 Alexander von Humboldt died in 1859, the year *Origin* was published.

From Edward Wilson 12 February 1869

Hayes | Bromley, Kent.
12 Feby 1869

My dear Mr Darwin

For long past I have been promising myself a leisure day in which I could drive over & return you the wonderful book of the beavers, which you were so kind as to lend me.[1] But one day it is "fund Committee of Society of Arts", another "Council for Colonial Society" another "Emigration Association" & so on till one really never gets a day to oneself.[2] "No rest but the grave" &c as the song says![3] I trust that I have not inconvenienced you or hindered the great work by retaining it? It is very interesting & the man himself quite an enthusiast in the cause of his most intelligent friends. The particulars of their dam-building are certainly most astonishing.

With kindest regards to M[rs] Darwin & the young ladies[4] I am Dear Sir | Yours very truly | Edw Wilson

DAR 181: 126

1 Wilson refers to *The American beaver and his works* (Morgan 1868). CD's annotated copy is in the Darwin Library–CUL.

2 Wilson was known for his philanthropic activities. He was a founder of the Colonial Society in 1868 and served as a member of the council (*ODNB*).

3 The final line of the song 'Pilgrim of love' by Henry Rowley Bishop was 'No rest but the grave for the pilgrim of love' (Bishop [1864]).

4 Wilson refers to Emma, Henrietta Emma, and Elizabeth Darwin.

To John Lubbock [before 13 February 1869][1]

My dear L.

Some friends of Miss M. with Hooker amongst them, are getting up a petition to Government to put her name on the civil Pension list.[2] I am going to sign & she has asked me to get a few other signatures.— As she has done nothing for science, you may, perhaps object to signing; but I sh[d] be very much obliged for a line to say yes or no.— I hardly know when the Petition will be sent on, but if you do agree to sign, I will forward it to you in due time.— Miss M. wrote the life of my grandfather matern J. W. & it is really a magnificent work;[3] she spent additional money on illustrations, & so I believe did not profit by the publication— Some of the Reviews speak of her history of the development of the art of Pottery, as valuable.[4] She has supported herself by literary & *[educational]* work, but is now growing old & has lost an aunt, who aided in her support. She seems to me on the whole to have a very fair claim to be put on the list of pensioners.—

Forgive me troubling you & believe me yours very sincerely | C. Darwin—

ADraftS
DAR 96: 70

[1] The date is established by the relationship between this letter and the letter from John Lubbock, 13 February [1869].
[2] CD refers to Eliza Meteyard and Joseph Dalton Hooker. Meteyard was awarded a Civil List pension of £60 on 5 April 1869 (Meteyard 1970, 1: introduction [25]).
[3] CD refers to his maternal grandfather, Josiah Wedgwood I, and to Meteyard 1865–6. CD had provided Meteyard with material for the book (see *Correspondence* vol. 13).
[4] For more on the reception of Meteyard 1865–6, see Meteyard 1970, 1: introduction [20–3].

From John Lubbock 13 February [1869][1]

13 Feb.

My dear Mr. Darwin

If you & Hooker say it is all right I shall feel quite justified in signing Miss Meteyards petition.[2]

I am really anxious about the Down school & think if the Parish was properly Canvassed the necessary funds would be forthcoming without pressing unduly on a few.[3]

It seems that none of my brothers (except Henry)[4] had even been spoken to on the subject.

I have been very nearly promising to stand for Westminster if Smith is turned out, but my senior partner Mr Ellice advised me so strongly against doing so that I have declined.[5]

Ever | Yours very sincerely | John Lubbock

DAR 170: 72

[1] The year is established by the relationship between this letter and the letter from Eliza Meteyard, 19 February 1869.
[2] Lubbock refers to Joseph Dalton Hooker and Eliza Meteyard. See letter to John Lubbock, [before 13 February 1869] and n. 2.
[3] See letter from John Lubbock, 8 February 1869 and n. 1.
[4] Henry James Lubbock.
[5] Lubbock stood for Parliament in the general election of November 1868, but lost by a narrow margin (see *Correspondence* vol. 16, letter from John Lubbock, 26 November [1868]). William Henry Smith was elected in the district of Westminster. The large amount of money he spent in the campaign led to suspicion of bribery and a petition was raised to remove him, but Smith retained his seat (*ODNB*). Russell Ellice was a partner in the banking firm Robarts, Lubbock & Co. (*Post Office London directory* 1868).

From G. H. Darwin 14 February 1869

Trinity Coll.
Feb 14. 69.

My dear Father,

Grove's letter is satisfactory in one way that it settles that I go to a Solicitor sometime; but his opinion about the when is so diametrically opposite to that of Bristowe & Godfrey Lushington that I do'nt at all know what to do.—[1] Mr. Hamilton wrote to me to propose that I shd come & be introduced to Mr. Hollams sometime, & so I have proposed to come up this week,—or if not to see him just before I start for Paris next month.[2] It is wonderfully kind of Mr. H. to take so much bother about me.

Crolls letter interested me. The thing wh. you do'nt understand about gravitation & heat, I believe is as follows— If the Solar system were gaseous & then were to begin to condense towards a nucleus (the Sun) owing to the mutual attractions of its particles, (i.e gravitation) an immense quantity of heat wd be given out & thus wd suffice to keep the Sun hot for 20,000000 yrs. (—I forget the exact number).— This I think is called the heat due to gravitation.[3]

I must begin to settle my plans about Paris soon; my general idea is to go to a Pension for some time & then perhaps go to a family for a while. I know a man here who was in a family & found it appallingly dull. I have covered 35 sheets of paper the size of this when open (on one side) with French exercises written in a small hand— there's business for you. My law & mathematics are getting on too—but the quaternions are so hard they take a deal of reading.

Frank spends all his time on cutting up toads with Pryor—[4] I think he ought to have a chance of the scholarship, but I'm afraid he does'nt do himself justice in an exam.

We are going to meet Boyd Dawkins at Newton's tonight[5] I do'nt know who he is, but I suppose you do; however I shall have to pretend I do. Clifford, I believe, is going to be one of the ladies' lecturers at S. Kensington along with Milman &.c.—[6]

Fitzmaurice[7] has been here for yesterday & today, he breakfasted with me this m^{g}. He says he is coming to Paris at Easter, but I don't suppose I shall see much of him as I fancy he has got relations there. I have not got Croll's letters with me, but I will send them in a day or two.[8]

I dined in Kings the other night with a man called Leigh—who is the tutor & I was introduced to the University Librarian & he said to me—I believe I met you out walking at Xmas & there were two dogs that fought—& I found that we had met in a turnip field near Reeve's cottage—& Bob & his dog had fought.[9] I did'nt remember him a bit— but I was awfully surprized.—

I wonder whether you are in town.—[10]

Your affectionate Son | G H Darwin

DAR 210.2: 7

1 George refers to William Robert Grove and Henry Fox Bristowe. See letter from G. H. Darwin, 6 February 1869.

2 George refers to John Hollams and probably to Alfred Douglas Hamilton. See letter from G. H. Darwin, 6 February 1869. George left for Paris on 5 March 1869 (see letter from G. H. Darwin, [23 February 1869] and n. 7).

3 See letter from James Croll, 4 February 1869, and letter to G. H. Darwin, 6 February [1869] and n. 2.

4 George refers to Francis Darwin and Marlborough Robert Pryor.

5 George refers to William Boyd Dawkins and Alfred Newton.

6 William Kingdon Clifford delivered ten lectures on geometry to ladies at South Kensington (Clifford 1882, pp. xxii, 628–37). George may also refer to Archibald John Scott Milman.

7 Edmond George Petty-Fitzmaurice.

8 George refers to letters from James Croll, [2 December 1868] (*Correspondence* vol. 16) and 4 February 1869. See also letter to James Croll, 31 January [1869] and n. 2.

9 George refers to Augustus Austen Leigh and Henry Bradshaw. William Reeves (b. 1814/15) was a local gamekeeper and Bob was CD's dog (Freeman 1978).

[10] CD was in London from 16 to 24 February 1869 (see CD's 'Journal' (Appendix II)).

From James Croll 15 February 1869

Edinburgh.
Feby 15th 1869.

Dear Sir,

I am much obliged to you for the copy of Scientific Opinion you sent me containing an abstract of Canon Moseley's admirable paper on the motion of glaciers[1]

I have a curious incident to tell you, regarding the matter. The reading of the abstract thoroughly convinced me that the generally received theory of glacier motion, which I believe is that of Tyndall's,[2] must be given up and that some other explanation must be sought. Strange to say, this conviction had hardly got time to settle down, if I may so express myself, before, what I fully believe to be the true cause suggested itself. The cause is so simple, so beautiful, and obvious that I wonder that it should have escaped observation so long.

I have drawn up a short account of the matter and have sent it to the Phil. Mag I have requested Dr Francis to send me two copies of proof and when they come to hand I shall send you one[3]

I am | Yours very truly | James Croll

Charles Darwin Esq MA FRS

DAR 161: 264

[1] Croll refers to Henry Moseley. See letter to James Croll, 31 January [1869] and n. 6.

[2] For John Tyndall's account of his own as well as earlier theories of glacier motion, see Tyndall 1860.

[3] Croll's paper 'On the physical cause of the motion of glaciers' (Croll 1869) appeared in the March 1869 issue of *Philosophical Magazine.* CD's copy has not been found. William Francis was the editor of *Philosophical Magazine.*

From Jean Jacques Moulinié 15 February 1869

Geneva
15th February 1869

Dear sir,

In perusing a french periodical intitled Journal d'Agriculture pratique, published weekly in Paris, I fall on an article published by a Mr. Carrière,[1] concerning a very curious case of variation observed on the wild radish (*Raphanus raphanistru⟨m⟩*); and describing the remarkable results obtained in a very short time, by selection and a rational cultivation of that wild plant. Thinking that the case, owing to the paper in which it is published, and which is perhaps not much known in England, might not come to your knowledge,—and may be of some interest for you,—I inclose hereby,—with a rough outline of the woodcuts accompanying the notice, and showing the wonderful differences obtained after four generations only, between the wild radish roots and those of the derived varieties improved by selection,—an abstract of the author's paper, describing the experiment, its results, and the very simple process followed by him.[2]

Hoping that my little abstract, from your being still unacquainted with its object, may have some interest for you, and still more that this letter may find you in good health, pray, dear Sir, present my respectful compliments to Mrs. Darwin, remember me kindly to all the members of your amiable and hospitable family,[3] and believe me | your's ever and truly devoted | J. J. Moulinié

[Enclosure]

Journal d'Agriculture pratique. T. I, No. 5, 1869. 33e Année; p. 159. Paris.— Rédacteur en chef, E. Lecouteux.[4]

The wild radish improved, an insight on the origin of cultivated plants, in reference to *Raphanus raphanistrum.* | by Mr. E. A. Carrière.

The *Raphanus raphanistrum* is a wild plant well known by agricultors as a very inconvenient weed, for its tendancy to invade with much tenacity cultivated fields. It belongs to the Cruciferae, and has some likeness with the *Sinapis arvense*,[5] with which it is often mistaken. Its flowers are sometimes of a pale yellow, sometimes white or streaked with lilac. Having remarked some analogy between the fruit of the garden radish and that of the wild *R. raphanistrum*, the author gathered some of this last's seed, sowed it in good soil, during five years, each time selecting with care the seeds borne by the plants modified in the most advantageous manner. The experiment was pursued in two different conditions; in Paris, in the dry and light soil of the Museum's nursery gardens; and in the country, in a heavier, calcareous and clayish ground. The results of the two trials were similar but not identical. In Paris the elongated napiform shape was predominant in the roots; in the country it was the reverse. In Paris the roots were all white or pink, while in those raised in the country, besides those two colours, a great many were very dark lilac or brown, some nearly black, and others of all shapes and colours representing all the various known forms of radishes & turnips, and much alike them, though not identical. One particular root (fig. 3) quite similar to the large China-radish, had a particular savour, between that of radish and turnip. Another (fig. 5) had a fine lilac skin, with flesh of the same colour and was consequently amongst vegetables, analogous to the red fleshed beetroot, the lilac potato; and certain flesh-coloured peeches & pears amongst fruit.

Table of characters of *R. raphanistrum* wild, and of its cultivated varieties, compared.

R. raphanistrum.	Varieties *derived.*
Flowers, pale yellow or white, sometimes slightly streaked with lilac.	*Flowers*, white, yellow or lilac-pink, sometimes of one colour, oftener streaked.
Pods, small, inclined, hard.	*Pods*, variable in size and shape, inclined, sometimes erect, sometimes as large and long as the pods of Madagascar radish, and then fleshy and good for eating.

Roots filiform, dry, fibrous, uniform, always white, hard, subligneous, uneatable.

Roots large, sometimes enormous, fleshy, varying considerably in shape and colour; flesh, white, sometimes pinkish or yellowish, sometimes lilac, juicy, soft, and eatable.

Figure 1,	represents half-size, the roots of the wild plant at full growth, weighing 22 grammes, white, hard, and quite uneatable.
Fig. 2.	root white, top lilac; total length 0.45 centimeter;—largest diameter 0.06 cent.:— weight 345 grammes.
Fig. 3.	Colour fine dark reddish pink, with top nearly lilac, hardly to be distinguished from the China-radish.— Total length, 0.40 cent.—diameter, 0.09 ct.—weight, 445 gr.
Fig. 4.	White, with a lilac hue at the summit.— Length 0.25 ct.—diameter, 0.07 ct.—weight 201 gr.
Fig. 5.	Dark black-lilac, veined; flesh lilac, with darker shades and streaks to a depth of 0.01 ct.;—the rest of the pulp being of a purplish white.— Length 22 ct.—diameter 0.07 ct.—weight 145 gr.
Fig. 6	Dark black chesnut skin; flesh milky white, very soft. Length, 27 ct.,—diameter, 0.06ct.;—weight 87 gr.
Fig. 7	Skin delicate and thin, of a fine pink colour; flesh of a very juicy and melting quality. This root spreads near the surface like certain breeds of turnips, and owing to its wide and flattened shape, much ressembles the fine radishes that gardeners select for seed. Length 0.12 ct.— Diameter, 0.06 ct.—Weight, 68 gr.
Fig. 8	This root has a pinkish brick colour: skin wrinkled, as embroidered; flesh pink, with red streaks or veins to 0.01 centimeter's depth, the rest being of a light flesh-coloured white. Length, 0.26 ct;—diameter 0.13 ct:—weight 625 gr.
Fig. 9.	White, smooth and glossy skin; ressembling much a good kind of turnip. Length, 0.32 ct.; diameter, 0.10 ct.—weight 651 gr.
Fig. 10	Fruit branch of the wild radish, *R. raphanistrum*,—natural size.
Fig. 11.	Pod of the same, improved by cultivation. Natural size.

All these roots though so different in shape and colour, were juicy, with a plain and marked flavour of radish, in some much like that of black radish. Tasted with care, a certain sweet flavour can be detected in several, reminding somewhat that of turnips, very slight in the raw state, but strong when cooked, the astringency of the radish disappearing completely. In the roots exposed to the air, and undergoing decomposition, the same smell of turnips was sensible. Their flesh was not quite similar to the radish, but more consistent, hardly sweet, and somewhat flowery, consequently more nutritious.

Here is a plant that can be ranked neither amongst radishes nor turnips, but partakes of the two, of radish when raw, of turnips when cooked, though generally the radish flavour is preponderant. This vegetable has been found excellent by all who have tasted it; the roots of all size are never hollow, and can be kept with all their qualities, for months in the cellar, and be used during the whole winter.

After some remarks on the effects of exterior conditions, and particularly the influence of the soil in modifying plants, shown by the different results obtained in the country soil, where the short roots were preponderant; while in the light and deep soil of Paris, they were exclusively long, white or slightly purplish (fig. 2), though all were derived from the same seed, the author concludes his paper in giving the best process to be used, and how selection must be practised in order to obtain in the quickest manner the desired modifications.

The object being the greatest growth of the underground part of the plant, he starts from seeds taken on the wild individuals which present already the largest roots, and continues in the same way with the offspring. The period of sowing the seed is not indifferent and must vary according to the part of the plant that is to be improved, so that it may come to its full growth.

To get large pods or improve any other aerial part, the seed must be sown in April or May, in a rich soil, in the best conditions of growth possible, and the best seed of the best plants selected for the next generations till the desired end be obtained.

To improve and increase the size of the roots, the seed must be sown in the beginning of September, so that the plants should not grow to seed the same year. When the cold season approaches, they are to be pulled out; and kept during the winter, sheltered from the frost, in a cool place, all their leaves, excepting those of the central part of the collar, being taken away. Next spring as soon as frost will be over, the finest of these wintered plants will be replanted in the best conditions, and their mature seed sown the same year, and the same process will have to be continued till the breed be fixed. Four generations and five years have been sufficient to give the results figured 2–9.

To improve rapidly and transform the *R. raphanistrum*, gather seed on the wild plant, sow it the next spring in a good soil, in favouring its growth by all possible manner. On the most vigourous plants choose again the finest seed, to be sown the same autumn, and continue in the same manner till the desired results be obtained.

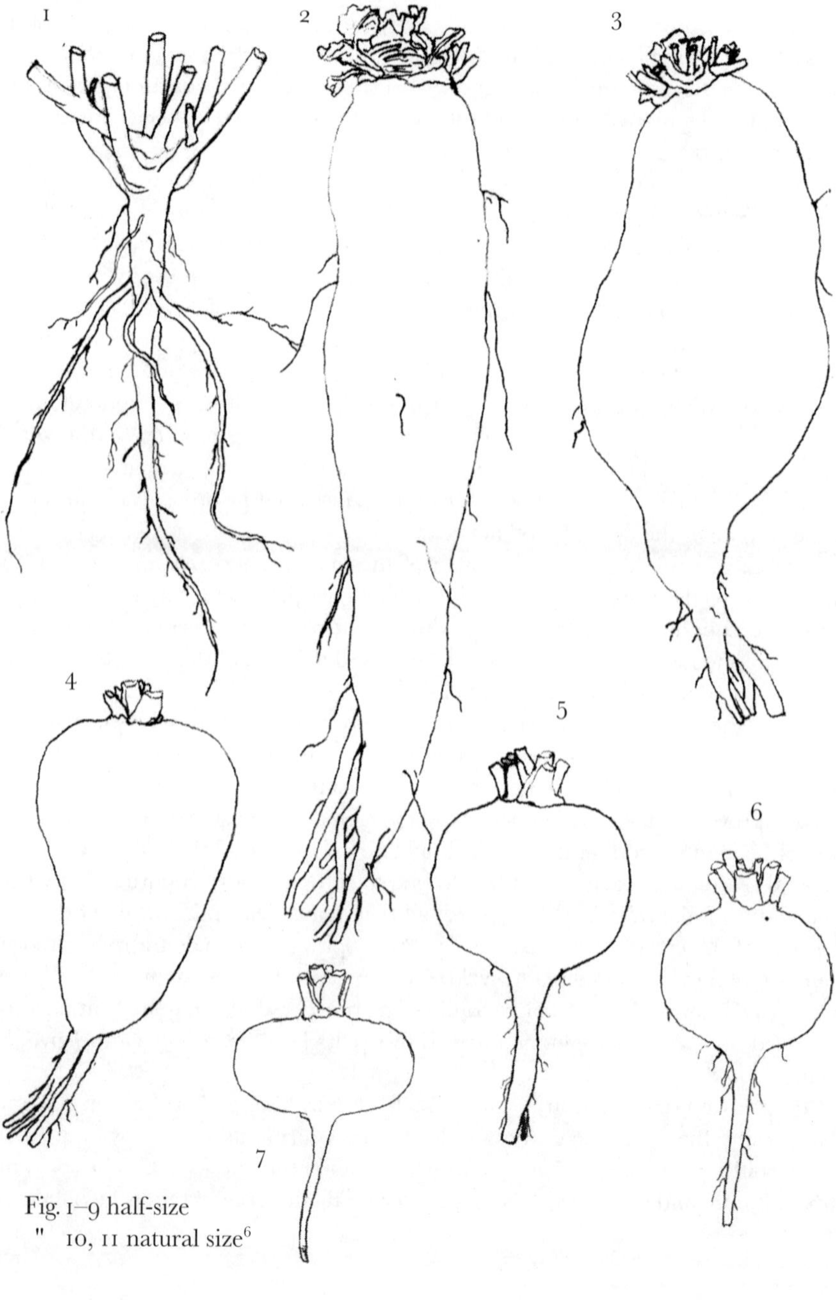

Fig. 1–9 half-size
" 10, 11 natural size[6]

8

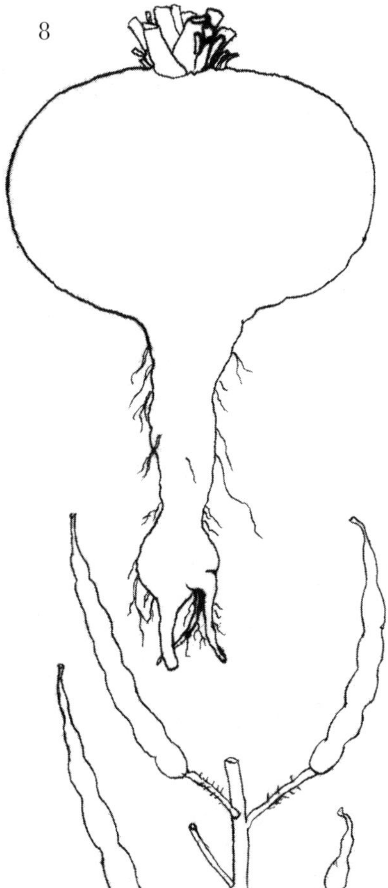

9

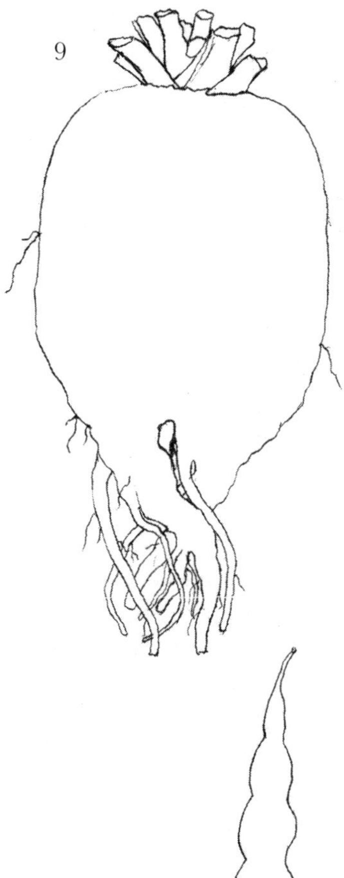

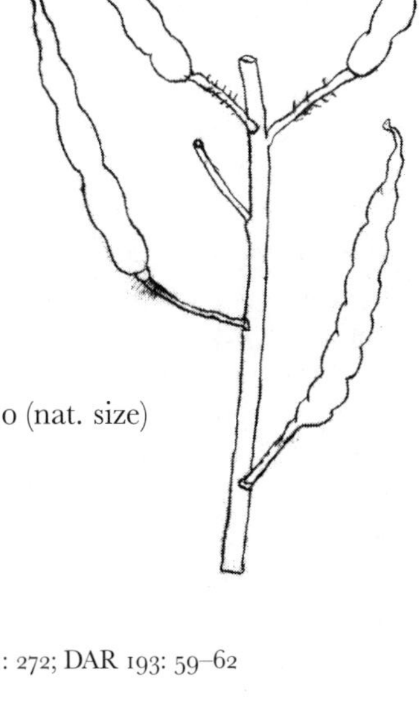

10 (nat. size)

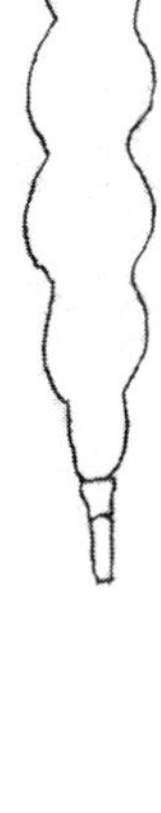

11 (nat. size)

DAR 171: 272; DAR 193: 59–62

CD ANNOTATION
Top of enclosure: 'Effects of conditions & Selection' *pencil*

1 Elie Abel Carrière.
2 CD later received a copy of a revised version of the article, *Origine des plantes domestiques démontrée par la culture du radis sauvage* (Origin of domestic plants shown by the cultivation of wild radish; Carrière 1869) from the author; this copy is in the Darwin Pamphlet Collection–CUL. CD cited Carrière's paper and gave information on the radish in *Variation* 2d ed., 1: 345.
3 Moulinié had visited CD on 31 August 1868 (Emma Darwin's diary (DAR 242)). See also *Correspondence* vol. 16, letter from J. J. Moulinié, 7 September 1868.
4 Edouard Lecouteux (editor-in-chief).
5 Moulinié refers to *Sinapis arvensis* (wild mustard).
6 Diagrams are reproduced at eighty per cent of original size.

From Philip de Malpas Grey-Egerton 18 February 1869

Athenæum Club
Feb 18 1869

Dear Darwin

The habits of the Fallow Deer as regards the breeding season are precisely the same as those of the Red Deer. The males herd together and keep aloof from the females during the growth of the horns and until the velvet is stripped and the horns hard. Soon after the horns are clean, the throat begins to enlarge simultaneously, I believe, with the enlargement of the prostate and Cowpers glands; this is always hastened by cold weather and frosty nights, and retarded by warm weather. They then begin to roar, at first only occasionally, and moderately and go in search of the herds of females, breaking up themselves into smaller herds. The large *Stags* have generally one or two sentinels with them who keep watch; but I have never seen this occur with Fallow Deer. When the males join the herds of females, the fighting begins for mastery. The battle is always prefaced by roaring loud and long, the Stags walking round and round each other Suddenly they come together with a clash, and getting the horns fixed, a desperate struggle ensues, each trying to force the other to turn.[1] When this occurs the victor dashes into the other and endeavours to plunge his *brow antlers* into him as he runs away. The upper portion of the horns is used only for fencing, the brow antler being the weapon of offence. So forcible is the blow when well directed that I have frequently had bucks killed, generally by the antler traversing the muscles of the haunch and breaking the femur. The roar of the Stag is very like that of a bull. The volume of sound is not so great, and yet I think it is heard at a greater distance. The roar of the Fallow Buck is harsher and more monotonous, something between a bark and a roar. Very shortly after they begin to roar, they lose their condition. All the internal fat dissolves, they have a very rank smell and loose their appetite. There is little food found in the stomach and that not properly digested and all the bloodvessels of the interior are gorged. The males are always endeavouring to keep together as large a seraglio as they can but the ladies seem to have no predilections. It generally happens that while two large Stags are fighting for the mastery, the smaller Stags, which always hover around the herd at a respectfull distance, take the opportunity of engaging the females.[2] I may mention a circumstance which I do not find alluded to in any book of natural history I have read, which is this, the deciduous horned are the only ruminants in which the males have a season as well as the females. In all the other ruminants, as also in most of the mammalia the male is—*nunquam non paratus*[3]—if the female be in season. I shall be most happy to answer any further questions you may send me addressed to Oulton Park Tarperley as I return there tomorrow

Believe me | Yours very truly | P. M Grey Egerton

DAR 83: 179–81

CD ANNOTATIONS
1.11 The battle … femur. 1.18] *crossed ink*
1.18 haunch] 'haunch' *above ink*
1.26 It generally … tomorrow 1.34] *crossed ink*

1 In *Descent* 2: 253, CD cited Egerton on fighting technique in red and fallow deer.
2 In *Descent* 2: 269, CD cited Egerton on females mating with males not engaged in fighting.
3 *Nunquam non paratus*: never unprepared (Latin).

From Ernest Faivre[1] 18 February 1869

Lyon
18 Fevrier 1869

Monsieur,

Un de Mes éleves a traduit en Francais la plus grande partie de votre Interessant ouvrage sur la *Fertilisation des Orchidées*;[2] Avant de Pousser plus loin son Entreprise, il sollicite de votre Bienveillance, quelques Indications que je me permets de Vous demander.

Votre livre sur la fécondation des Orchidées est-il traduit en Francais? Seriez vous disposé a en autoriser la traduction, et pourrait-elle être faite En prenant pour Guide le Volume publié en 1862 a Londres chez John Murray?

Depuis ce temps la Science a fait des progrès et peut être préparez-vous en ce moment une Édition nouvelle? Dans le cas contraire il vous conviendrait sans doute de revoir l'Édition parue et d'y ajouter des notes; Vous serait-il agréable que la traduction vous fut préablement et successivement soumise?

Je prends la liberté Monsieur, de vous demander votre avis sur des questions que préoccupent fort mon jeune disciple, et j'ose Esperer que vous voudrez bien nous faire connaitre Votre sentiment.

Peut être trouverez vous a l'exécution de l'oeuvre des difficultés que nous n'avons pas prévues? Je vous prie de vouloir bien ne pas nous les dissimuler.

En tous cas, Monsieur, soyez sûr que rien ne sera fait sans votre Consentement Exprès.

J'ai vivement regretté, de ne pouvoir m'entretenir avec vous l'orsque je me suis rendu a Londres au mois d'Aout dernier, a l'Occasion du Congrès de Norwich, auquel j'étais Invité.[3]

Sachant votre état de santé, j'ai craint d'être Importun et de troubler un repos qui vous est bien nécessaire.

Permettez que je vous remercie de l'Envoi de votre second Volume sur la Variation des animaux et des Plantes, j'en fais une lecture, je devrais dire, une étude attentive.[4]

Je n'ose abuser plus longtemps de vos précieux moments et je vous prie, d'agréer, | Monsieur, l'Assurance de mes sentiments d'admiration et de profond Respect.

Ernest Faivre | Professeur à la faculté des sciences de Lyon | Conservateur du Jardin Botanique etc.

DAR 164: 4

1 For a translation of this letter, see Appendix I.
2 Faivre refers to Louis Rérolle and to *Orchids*.
3 Faivre refers to the meeting of the British Association for the Advancement of Science held in Norwich from 19 to 26 August 1868 (*Report of the 38th meeting of the British Association for the Advancement of Science*, p. lvii).
4 Faivre refers to the second volume of the French translation of *Variation* (Moulinié trans. 1868).

To Eliza Meteyard [18 February 1869][1]

6 Queen Anne St | W
Thursday

Dear Madam

I have signed the petition myself & have procured the signature of Sir Charles Lyell & have then forwarded the momorial to Sir John Lubbock asking him to return it to you & I hope before this you have received it.[2]

Yours very faithfully | Ch. Darwin

LS
American Philosophical Society (Getz 6618)

1 The date is established by the relationship between this letter and the letter from Eliza Meteyard, 19 February 1869. The Thursday before 19 February 1869 was 18 February.
2 See letter to John Lubbock, [before 13 February 1869], and letter from John Lubbock, 13 February [1869].

From Hugo Thiel 18 February 1869

Bonn
Febr. 18th. 1869.

Charles Darwin Esqu. Dr. M. M.A. etc etc.
Dear Sir!

You will kindly excuse my taking the liberty of sending you a little pamphlet of mine.[1] Only the introduction of this little work can be of any interest to you and I should be very glad if you appreciate this attempt to succeed in better understanding the social questions by adopting your ideas of all organical development.

At this opportunity I am obliged to express how much especially agriculture is indebted to you for the sciential foundation you laid to our practical endeavours of breeding animals and plants. You can be sure that in spite of a few opponents the number of your disciples in Germany is daily increasing, at the Royal agricultural Academy at Poppelsdorf f.i. all our experiments are founded upon your doctrines.

Believe me your most obedient | D^r. H. Thiel | Member of the Bonn University and the Royal agricultural Academy at Poppelsdorf.

DAR 178: 89

1 CD's annotated copy of Thiel's article, 'Ueber einige Formen der landwirtschaftlichen Genossenschaften' (Thiel [1868?]) is in the Darwin Pamphlet Collection–CUL.

From Eliza Meteyard 19 February 1869

wildwood. | Hampstead. N.W
Feb. 19. 1869.

Dear Sir,

I received the memorial from Sir John Lubbock—quite safely—the evening before last.[1] It contains his valuable signature as also your own—and that of Sir Charles Lyell. When you see or write to these gentlemen, will you offer to them my very earnest thanks for their true goodness. And will you please receive the same yourself—with the addition of my deep respect. Such names are of the highest value to me—not only for the intrinsic weight they bear—in relation to the matter in question—but to myself as a silent guarantee of opinion. Believe me truly and respectfully grateful to you.

The memorial is now with Mr Winter Jones[2] at the British Museum—for his and others' signatures. But I am somewhat in a commotion—Mr Smiles having written urgently last night—for the memorial to be sent in to Mr Gladstone.[3] There will therefore be only time to add the names of my kind friends—Mr Duffus Hardy and Mr Bennett Woodcroft.[4] There are many names of great value—awaiting the memorial—as those of Mr Warren De La Rue—Mr Millais[5]—and others—but there is now no time for application.

Believe me—with great respect, | your's obliged & faithfully | Eliza Meteyard

Charles Darwin Esq.

DAR 171: 162

1 Meteyard refers to a petition to put her name on the Civil Pension list (see letter to John Lubbock, [before 13 February 1869] and n. 2).

2 John Winter Jones.

3 Samuel Smiles and William Ewart Gladstone. For more on the pension scheme, see Meteyard 1970, 1: introduction [25].

4 Thomas Duffus Hardy and Bennet Woodcroft.

[5] John Everett Millais.

From W. C. Tait 19 February 1869

Oporto
Febry 19th 1869.

Charles Darwin Esqr. | Down | Bromley | Kent
Dear Sir

I received your kind letter of the 2nd. and delayed writing until I could give you the result of my endeavours to obtain for you the "Drosophyllum Lusitanicum"[1] I applied to my friend Mr. Alexander Grant who I am told is a good botanist and he told me he knows the plant very well and has found it at a place about 8 miles from here called "San Pedro da Cova".[2] He gave me a description of the plant and I went to look for it at the place indicated but although I looked in the fields all round the village & asked many people about, I was not able to find it.

I have again conversed with Mr. Grant about it and intend to have another search for it and as Mr Grant has given me the address of a villager who has brought specimens of the plant to him I hope to be more successful on my second expedition. Another locality for it is the Penedo da Meditaçao near Coimbra and if I cannot find it at S Pedra da Cova I shall go to Coimbra for it by rail.[3]

Mr. Grant's ill health did not permit him to accompany me otherwise he said he would have great pleasure in assisting your researches in any way— He informs me that a rare Polygola a milkwort is found at S. Pedro. And the fern "Cheilanthes fragrans" at a place called Vallongo a short distance from S Pedro— That part of the country is interesting from the fact that a change takes place in the geological formation from that of granite about Oporto to Slate at Vallongo and on the Vallongo hills there runs a large vein of Quartz—supposed to have been worked in ancient times for gold. Antimony is also found there— I have picked up there traces of copper, antimony, and iron and near the Village of Vallongo there is a slate quarry which is worked by an english company— The hills are perfectly honey combed by ancient excavations and the works must have been conducted on an immense scale— Mr. Edward Allen of the Public Library[4] and Who is interested in archeological remains tells me that the workings are pre-historical and he intends to make a survey of them.

I intend to look in the botanical authority Brotero with respect to the "Drosophyllum Lusitanicum" which however he calls "Spergula Droseroides"[5] and you depend upon getting the plant (many specimens) from me sooner or later would you kindly inform me to whom I am to address the box when shipping it by steamer to London.

With regard to tail less dogs I have obtained a pointer bitch which is said to have been born without a tail and have crossed her with a pointer dog which is known to have been born without a tail and no doubt the result will appear in due course.[6] There are certain peculiarities in the tails of these dogs they are generally about four inches in length (*about the length usually left when cut*) and the skin of the tail encloses it perfectly and loosely like a bag whereas in a terrier whose tail was cut when young the end is hard and the skin appears to have dried on the bone. The hills about here and in the Douro district are covered with prickly furze and any dog with an entire tail which enters it on the scent of the red legged Partridge is liable to come home with a bleeding and lacerated tail and in the same way but in a lesser degree when hunting for quail in the indian corn fields. I am told that in the Alto Douro district the tail is cut on this account but from my own observations down here the practise seems to arise from or more correctly perhaps followed from peculiar ideas of *beauty*!

I have another pointer which has two dew claws on each of her hind feet and there is a popular superstition that dogs born with this peculiarity are safe from the effects of the bite of a mad dog these superstitions another cause perhaps of the perpetuation of varieties of domestic animals.

I am told by one sportsman that he possessed a dog and bitch born without tails and that nearly all the pups were also born without tails— Another one tells me that he also has a dog and bitch of the same kind and he does not remember that any of the many pups were born with a tail—and two people have told me that they never knew a pointer dog born without a tail which did not prove a good dog for game but at the same time there did not appear to have been any systematic breeding from such a strain the process of "curtailing" being so general and simple—

If I remember rightly I have seen in some book I think "Whites Natural History of Selbourne" that at the period of migration there is a perceptible increase in the size of the pectoral wing muscle of a certain bird[7] Now it strikes me that if such be the case may there not be an abnormal condition of the brain too, urging it to take flight after previously congregating perhaps the organ of inhabitiveness or its opposite Then again imitation may have a great effect on the young birds— But this is such an obscure subject that it would be very difficult to come to any secure conclusion—

Thanks to the spirit of your works I have been made alive to the necessity of strict and cautious enquiry but speculation is also a great aid if used in moderation—

Has it not struck you that the orange colour is very variable?

In my next letter I hope to be able to tell you that the "Drosophyllum Lusitanicum" is ready to be sent and to mention several circumstances which lead me to suppose that certain variations are eliminated in nature if to the disadvantage of the species— If you will point out any particular subject of observation I would most willingly devote special attention to it and remain | Dear Sir, | Yours faithfully | William C. Tait.

DAR 178: 44

1 See letter to W. C. Tait, 2 February [1869].

2 Tait lived in Oporto (now Porto), about 16 km west of São Pedro da Cova. Alexander Grant has not been further identified.

3 Coimbra is about 100 km south-east of Oporto.

4 Edward Allen has not been further identified.

5 For Félix de Avellar Brotero's description of *Drosophyllum lusitanicum* under the name *Spergula droseroides*, see Brotero 1804, 2: 215–16. See also letter from W. C. Tait, 10 May 1869 and enclosure.

6 For Tait's earlier remarks on tailless dogs, see the letter from W. C. Tait, 26 January 1869.

7 Tait refers to Gilbert White's *Natural history and antiquities of Selborne* (White 1789). The passage referred to has not been found.

From Archibald McNeill 20 February 1869

73 G^{t}. King Street | Edin^{r}.
20th. Feby. 1869

Dear Sir

I am in receipt of yours of the 15th. & have much pleasure in furnishing you with any information of w^{h}. I am possessed—[1]

I am of opinion that the females of Red Deer are not gathered to the males by the call but that the males seek out the females—

My observation is this, that a small herd of females has generally one large or master Stag with them who is constantly bellowing & chasing off the younger Stags from the herd This Stag occasionally leaves the females & goes to one of the holes he has made on the margin of some small pool near to the place where the females resort & there continues bellowing, tearing the ground with his horns & rolling in the mud— He seems to be in a state of furor but why he rolls in wet mud till he is covered all over with it I do not understand as he might easily cool himself in the water of any stream or loch—

Stags usually (in a wild state) feed during the night & lie down in ferns or under wood at the approach of day, but during the rutting season they are so restless that they constantly keep roaming about & roaring tho' no doubt they roar most & loudest during the night—

I have never heard a young stag under 3 years old roar,[2] nor am I aware that Stags roar when fighting tho' their fights almost invariably take place during the rutting season—

The Stags emit no strong odour during the rutting Season but their flesh at that season is strong tasted—

I may mention that during the rutting season the stags eat little & are so much reduced during the months of oct^{r}. & nov^{r}. that many of them die during the winter—

I do not know whether or not there is any enlarg^{t}. of the larynx during the rutting season but I can easily conceive that it sh^{d}. be so, as the neck swells & the whole frame seems to be in such a state of irritation that the Stags are constantly passing urine—

Believe me to be | Dear Sir | Yours truly | Arch^{d}. M^{c}.Neill

DAR 83: 177–8

CD ANNOTATIONS

1.1 I am … possessed— 1.2] *crossed pencil*
2.1 I am … females— 2.2] *crossed ink; double scored pencil*
3.1 My … the night— 4.4] *crossed pencil*
5.1 I have … season— 5.3] *crossed ink; double scored pencil*
6.1 The … tasted— 6.2] *scored pencil*
7.1 I may … winter— 7.2] *crossed pencil*

8.1 I . . . urine— 8.3] *crossed ink*
8.2 easily . . . swells] *scored pencil*
Top of letter: 'Voice & odour' *pencil*

1 CD's letter has not been found, but see the letter from George Cupples, 15 January 1869 and n. 5, and the letter from Archibald McNeill, 24 January 1869.
2 In *Descent* 2: 274–5, CD cited McNeill for this information.

To Robert Francis Cooke [22 February 1869][1]

6 Queen Anne St
Monday,

My dear Sir

I have just been applied to for permission for a French translation of my Orchis book.[2] As Mr Murray has never objected, I have assumed that he will not in this case. I write now to ask whether you will supply, *if required*, stereotypes of the woodcuts at prime cost, or only at a slight increase;[3] for if there is any large charge, I feel nearly sure that the work will not appear, which I should regret. If in your power would you let me have an answer by tomorrow evening as we return home early on Wednesday morning.[4]

Dear Sir | yours very faithfully | Ch. Darwin

LS
National Library of Scotland (John Murray Archive) (Ms. 42152 f. 200)

1 The date is established by the relationship between this letter and the letter from Ernest Faivre, 18 February 1869. In 1869, the Monday following 18 February was 22 February.
2 See letter from Ernest Faivre, 18 February 1869.
3 John Murray had supplied woodcuts for the German translation of *Orchids* (Bronn trans. 1862; see *Correspondence* vol. 10, letter to H. G. Bronn 25 April [1862]). CD had also given approval for Murray to supply the woodcuts from *Orchids* to Mary Somerville for Somerville 1869 (see *Correspondence* vol. 14, letter to M. E. Lyell, [19 October 1866] and nn. 2 and 3).
4 CD was in London from 16 to 24 February 1869 (see CD's 'Journal' (Appendix II)).

From R. F. Cooke 22 February 1869

50A, Albemarle Street, London, W.
Feby. 22 *1869*

My dear Sir

There will be no objection to the French Transn, or difficulty about the woodcuts to the Orchids.[1]

They shall be supplied at or near cost price.

I have not yet received the necessary information about "Fritz Muller" but I will not fail to let you know the moment I do.[2]

Yours faithfully | Robt. Cooke

Charles Darwin Esq

DAR 171: 364

CD ANNOTATION

Top of letter: '*copious notes*— will not let me buy the paper— Postage is Franca | Thin paper— | *Exact estimate* | ['*Postage*' *del*]' *pencil*[3]

1 See letter to R. F. Cooke, [22 February 1869] and n. 2.

2 Cooke refers to the English translation of Fritz Müller's *Für Darwin* (F. Müller 1864a; Dallas trans. 1869). See letter from R. F. Cooke, 23 February 1869.

3 CD's annotations appear to be notes relating to expenses for Dallas trans. 1869 (see letter from R. F. Cooke, 23 February 1869).

From George Henslow 22 February 1869

St. Johns Parsonage | S. Johns Wood | N.W.
Feb 22[d]/69

Dear Mr Darwin

I have got an idea which I am endeavouring to support by collecting materials of various kinds, & I take the liberty of writing to you to ask if you can assist me. It is this.— that mimetic forms of animals have arisen through the *direct cause* of prenatal impressions received through the *sense of sight*; & aided by *natural selection.*

The grounds for this idea are these

1) the human fœtus is frequently affected by the sight of objects which produce shocks upon the mother.— this is pooh-poohed by many but *I think*, I have succeeded in establishing it, nevertheless.[1]

2[d]) Cases amongst animals a cow A pregnant by Bull B is, however, kept in a field with C (Bull or Cow or Bullock) The result is that the calf may resemble C more than B.

Virgil states somewhere, but have not to hand to refer that if a mare is to bring forth a particular color place such a colored animal before her while being covered by the stallion.—[2]

compare Jacob's rod.—[3] &c &c

Now I want to collect cases from Breeders of cattle, if possible; & knowing from yr writings that you had many communications from them on other points,—I was anxious to know if they stated anything to the above effect. & if so whether you would kindly let me hear it.

No case where Pangenesis could apply is of any use. For I want to try & substantiate my idea, where there has been no *connection* at all: no hereditary descent &c. &c all of which cases would be explained by Pangenesis.[4]

If you have no instances to refer to, could you oblige me with the names & addresses of those breeders with whom you have had correspondence? You will see that my idea is that in consequence of animals constantly living in the same localities & surrounded by the same objects generation after generation; the fœtus has by the slow accumulation of mental force—gradually become assimilated in form or appearance to the surrounding objects.—

Yrs very truly | Geo Henslow

DAR 166: 168

[1] The ancient belief in the effect of maternal imagination on the development of the foetus was well established by the early eighteenth century, when it was used to account for a wide variety of malformations (D. Todd 1995, pp. 47–52). For more on popular belief in the effect of maternal imagination, see Huet 1993, D. Wilson 1993, and D. Todd 1995. For CD's scepticism on the topic, see *Correspondence* vol. 16, letter to J. J. Weir, 18 April [1868].

[2] Virgil discusses horse breeding in book 3 of the *Georgics*, but the reference has not been identified.

[3] Henslow refers to the biblical account of Jacob's placing speckled rods before sheep so that the offspring would be speckled (see Gen. 30: 37–9).

[4] CD published his provisional hypothesis of pangenesis in *Variation* 2: 357–404. CD suggested that minute particles (gemmules) circulated in the bodily fluids and were capable of generating new cells, remaining dormant until required. He thought his hypothesis could explain both sexual and asexual reproduction, as well as reversion and the regrowth of body parts. For CD's discussion of pangenesis with correspondents, see, for example, *Correspondence* vol. 13, letter to T. H. Huxley, 27 May [1865], and *Correspondence* vol. 14, letter to J. D. Hooker, 4 April [1866].

To J. J. Moulinié 22 February [1869][1]

Down Bromley Kent [6 Queen Anne Street]
Feb. 22^d

My dear Sir

I am truly obliged to you for your extraordinary kindness in having taken such great trouble in translating the long paper & copying the diagrams. The paper has interested me much, not only in showing so much variation & the effects of selection within a short space of time; but more especially from the variations having been different under different conditions. For of such cases I have very few instances.— I sh^d never have heard of this paper.—[2]

My health continues the same; I am always what is called "a poor Devil"; but am able to work for two or three hours daily.—

M^rs. Darwin begs to be kindly remembered to you,[3] & with my best thanks pray believe me | My dear Sir | Yours very faithfully | Charles Darwin.

Bibliothèque de Genève, Ms. suppl. 66, ff. 9–10

[1] The year is established by the relationship between this letter and the letter from J. J. Moulinié, 15 February 1869.

[2] See letter from J. J. Moulinié, 15 February 1869 and nn. 1 and 2. Moulinié had summarised in English an article by Elie Abel Carrière (Carrière 1869a), published in a French weekly agricultural journal, on selective breeding of wild radishes. CD later received a copy of a revised version of the article (Carrière 1869b) from the author.

[3] Moulinié had visited CD on 31 August 1868 (Emma Darwin's diary (DAR 242)).

From George Henslow [after 22 February 1869][1]

S. Johns Parsonage | S. Johns wood | N.W.

Dear Mr Darwin,

I send you a "case" which I think might have some little interest for you.—[2] as follows:

A Ewe of the Kentish breed, belonging to Mr Beard of Norton nr Canterbury,[3] gave birth to twins, *one* of which was a pure Kentish breed, like its mother *and father*; the other a *pure Southdown*[4] & continued as such through life. I put the follg questions to him, & the replies are Mr Beard's answers.[5]

1. In what particulars did the twin lamb resemble the Southdown breed?

1. as to Head? *Face dark.*

2. as to body? *wool short.*

3. as to limbs? *dark.*

The Resemblance in all points was like a down.

Had the parent ewe *ever* been covered by a Southdown ram? *not known.*

Had any of the ancestry of the ewe ever been crossed by a Southdown ram? *not known.*

Had the ewe any opportunity of **seeing** a flock, or *even one constantly*, while *pregnant*, or before that time?

Do not think it had.

Do you think Pangenesis would be the probable explanation of the phenomenon? by some *early* crossing not known, if so, it seems curious that it should reappear when *twins* were born & in only *one* of them!

Yrs very truly | G Henslow

DAR 166: 167

[1] The date is established by the relationship between this letter and the letter from George Henslow, 22 February 1869.
[2] See letter from George Henslow, 22 February 1869 and n. 1.
[3] Mr Beard has not been further identified.
[4] Henslow probably refers to the Romney, an ancient longwool breed, named after Romney Marsh in south Kent, where it originated. For Southdown sheep, see the letter from J. V. Carus, 2 February 1869 and n. 5.
[5] The underlined answers, from '*Face dark.*' to '*Do not think it had.*' are not in Henslow's hand.

From R. F. Cooke 23 February 1869

50A, Albemarle Street, London, W.
Feby 23 *1869*

My dear Sir

I think the expense of Fritz Muller, will be about £86 for 750 Copies—or less than £10 extra for 1000.[1] I will order a stoutish paper & as the book will be 160 pages we might put 5/ or 6/– on it & then there would be a profit.[2]

Yours faithfully | Robt Cooke

C. Darwin Esq

DAR 171: 365

CD ANNOTATION

Verso of letter: '100 copies at 5s = 500s = 25 £ | 500 copies 125 £' *ink*

[1] See letter from R. F. Cooke, 22 February 1869. Cooke refers to the English translation of Fritz Müller's *Für Darwin* (F. Müller 1864a; Dallas trans. 1869).

[2] *Facts and arguments for Darwin* (Dallas trans. 1869) was published in March 1869; the retail price was 6*s*. (*Publishers' Circular*, 1 April 1869, p. 201).

To R. F. Cooke 23 [February 1869][1]

6. Queen Anne St
Tuesday 23d. night.

My dear Sir

After much hesitation I will fix for 1000 copies, though I fear that it is foolish.—[2]

With respect to price I will leave it your kind decision what is best: I have paid for woodcuts & translation 20£.— Nevertheless I shd think that 6s ought to be the utmost price.—[3]

I enclose list of (7) copies for Reviews, & (12) for presentation to friends, most of which I will forward as they have to go abroad.— Kindly inform me by letter when the 8 copies are sent to Down.—

Dear Sir | Yours very faithfully | Ch. Darwin

Fritz Muller for Reviews.
(1) Scientific Opinion
(2) Athenæum
(3) Quarterly Journal of Science
(4) Popular Science Review
(5) Annals & Mag. of Natural History
(6) Intellectual observer
(7) Geological Magazine
Fritz Muller

8 copies to be sent to C. Darwin care of Down Postman, Bromley, Kent *Per. Rail* (Kindly inform me when sent by Rail.)

3 copies to W. D. Dallas Esq.[4] Geological Soc. Somerset House

1 Copy to "C. Spence Bate Esqre.[5] 8, Mulgrave Terrace Plymouth." marking "From Author.— " (Making 12 copies)

National Library of Scotland (John Murray Archive) (Ms. 42152 ff. 194–7)

[1] The date is established by the relationship between this letter and the letter from R. F. Cooke, 23 February 1869.

[2] According to John Murray's ledger book (John Murray Archive, National Library of Scotland), 1000 copies of *Facts and arguments for Darwin* (Dallas trans. 1869) were printed. Murray received a ten per cent commission for sales and warehousing. After one year, a loss of £21 2*s*. 3*d*. was recorded, but over the next ten years small profits were recorded each year.

[3] See letter from R. F. Cooke, 23 February 1869 and n. 2.

[4] William Sweetland Dallas.

[5] Charles Spence Bate.

From G. H. Darwin [23 February 1869][1]

Trin. Coll.
Tuesday.

My dear Father,

I enclose Croll's letter.[2] Altho' Miller told me that Delaunay's paper was translated in the Geolog. magazine, yet when I went to the Phil. Soc. rooms, I cd find no such Mag. & therefore I am up a tree again.[3]

I want to know whether you will have the 70£, wh. wd naturally be put to my acc.t at the bank on the 1st of Ap., put to my acc.t now, as I shall want to take 50 or 60£ to Paris. I am going to write to S.t Barbe to find out the best way to take it.[4] Frank[5] has gone up to town to be teethed today, but I suppose is coming back this evening. I have received a letter from M.r Hamilton asking me to dinner on the 4th. to meet Hollams.[6]

I shall come home on Tuesday & probably start for Paris on Friday[7]

Your affectionate Son | G H Darwin

DAR 210.2: 8

CD ANNOTATIONS

Top of letter: 'Papa has written to Bank to put the 70£ to your account & 12£.10 for Thrale'[8] *pencil, del pencil*; 'Emma— Had you not better off Ruck'[9] *pencil*

1 The date is established by the relationship between this letter and the letter from G. H. Darwin, 14 February 1869, by the reference to a dinner engagement with Alfred Douglas Hamilton, and by an entry in CD's bank account for 25 February 1869 (see nn. 4, 6, and 7, below). In 1869, the Tuesday before 25 February was 23 February.

2 George refers to James Croll. See letter from G. H. Darwin, 14 February 1869 and n. 8.

3 George was trying to find a translation of a paper by Charles Eugène Delaunay (see letter from G. H. Darwin, 6 February 1869 and n. 1). *Geological Magazine* was first published in 1864. George also refers to the Cambridge Philosophical Society.

4 CD recorded a payment of £70 to George on 25 February 1869 (CD's Account books–banking account (Down House MS)). Mr St Barbe has not been further identified.

5 Francis Darwin.

6 In his letter of 14 February 1869, George had mentioned that Hamilton would introduce him to John Hollams either that week or just before his departure for Paris.

7 In 1869, 4 March was a Thursday, so George must have returned to Down on 2 March and left for Paris on 5 March.

8 In addition to the £70 paid to George, CD recorded a payment of £12 10*s* for Thrale 'per George' (CD's Account books–banking account (Down House MS)). Thrale was evidently a horse-dealer (DAR 210.2: 4, DAR 239.23: 1.58), but has not been further identified.

9 The reference is probably to Richard Matthews Ruck, who was a student at the Royal Military Academy, Woolwich, where Leonard Darwin was also a student.

To James Croll 24 February [1869][1]

Down. | *Bromley.* | *Kent. S.E.*
Feb. 24th

My dear Sir

I am very much obliged for the Proofs, which have interested me greatly.[2] I cannot for a moment pretend to form any judgment, but your view seems to me very ingenious; if accepted it will be a most satisfactory & great step in our knowledge of glacier-movements

In Haste | Your's very sincerely | Ch. Darwin

American Philosophical Society (271)

[1] The year is established by the relationship between this letter and the letter from James Croll, 15 February 1869.

[2] CD refers to proof-sheets of 'On the physical cause of the motion of glaciers' (Croll 1869; see letter from James Croll, 15 February 1869 and n. 3).

From Henry Holland to Erasmus Alvey Darwin 24 February [1869][1]

Brook Street
Feb 24th

My Dear Erasmus

If Charles has not yet gone will you put this note into his hands,—otherwise forward it to him.

Looking at some old notes, I find that Quetelet's work—"Sur la Theories des Probabilités &c" was published in 1846, & reviewed by Sir J Herschel in one of the Edinburgh Reviews of the year 1850[2]

Lacroix, as well as Laplace, has written a work sur "le Calcul des Probabilités".[3] I have read the latter only. It is a very interesting volume like every thing that comes from the hands of Laplace

As far as I recollect Herschel's Review of Quetelet, it will give Charles most of the facts that would interest him for his particular objects. But I may be mistaken in this

My old notes furnish me with some other facts

Out of *3,630,000* registered births in England & Wales the excess of Males is very little more than $2\frac{1}{2}$ per cent. It results from this, as matter of exact calculation, that a deviation of $2\frac{1}{2}$ per cent from the law so determined, would involve the same improbability as an ace being thrown *643 times* in succession from a single die.

I see that in England, France & Belgium the rates of illegitimate children to others is as 1. to 13

Either in Quetelet's work, or elsewhere, I have seen some curious facts, as to the mean type of stature in different countries or districts.[4] I recollect that the French mean stature is consid^y below both Belgian & English.—the highest at that time the *adult agricultural type* in Lancashire.

I shall be glad if anything I have thus hastily written is worth having

Ever yours very sincerely | H. Holland

DAR 86: A79–80

CD ANNOTATIONS

1.1 If ... 1850 2.3] *crossed pencil*
2.1 "Sur ... 1850 2.3] *scored blue crayon*; 'Read' *in margin blue crayon*
3.1 Lacroix ... facts 4.1] *crossed blue crayon*
7.1 I see ... having 9.1] *crossed blue crayon and pencil*
Verso of letter: 'Babbage Edinburgh Journal of Science I. 1829. p85–104 on Proportionate number of sexes under different circumstances.—'[5] *ink, crossed pencil*

[1] The year is established from the date and subject matter of the letter. CD worked on the proportion of the sexes in various animals between 1868 and 1870 as part of his research for *Descent*, which was published on 24 February 1871 (CD's 'Journal' (DAR 158)). The only year during this period when he visited London around 24 February was 1869, when he was in London from 16 to 24 February (CD's 'Journal' (Appendix II)).

[2] Holland refers to Lambert Adolphe Jacques Quetelet and to Quetelet's study on the theory of probability applied to moral and political science (Quetelet 1846). He also refers to John Frederick William Herschel and probably to an anonymous essay review of Quetelet 1846 published in the *Edinburgh Review* for July 1850 ([Herschel] 1850), which Holland evidently knew had been written by Herschel. CD may have already read either the translation of Quetelet 1846 (Quetelet 1849) or the review (see *Correspondence* vol. 4, letter to W. D. Fox, 4 September [1850]).

[3] Holland refers to Sylvestre François Lacroix and Lacroix 1816, and to Pierre Simon Marquis de Laplace and Laplace 1814.

[4] Holland may refer to Quetelet 1835.

[5] CD refers to Charles Babbage and Babbage 1829. CD cited this work in *Descent* 1: 302.

To W. C. Tait 24 February 1869

Down. | *Bromley.* | *Kent.*
Feb. 24th /1869

Dear Sir

Your kindness is extraordinarily great about the Drosophyllum.[1] If the Box (which if possible had better be covered only with bars?) must be addressed to some one in London, please address it to my brother's House

"6 Queen Anne St.,
Cavendish Sqre"

Otherwise address it as at head of this note "per Rail".— In either case please mark outside "*Living Plants, to be forwarded immediately*".—

You seem to be collecting numerous statements about the Dogs' tails;[2] I wd suggest that you shd keep a list of the cases, observing whether the dogs in the several cases are known *not* to be related; for the number of cases alone wd be evidence of the intervention of some new cause.—

Did I mention in my former letter that I am very anxious to learn about the rate of development of the Horns in breeds of sheep in which the Rams alone are horned (viz merinos) & in common breeds in which both sexes are horned,—especially if inhabiting same country & fed in nearly the same manner.[3] What ought to be observed, is, whether the horns are sensibly *larger* or *smaller*, at one or at two or at 3 months age, in the one breed than in the other. I hear from Saxony that in Merinos rams the horns can just be felt at birth.[4] I shall much like to hear your facts on the elimination under nature of disadvantageous variations.—

Dear Sir | Yours truly obliged | Ch. Darwin

If you find the Dros. I shd much wish to know whether it grows in sandy, peaty or clay soil—whether damp or dry—whether sunny or shady.—

Alan R. Tait (private collection)

[1] See letter from W. C. Tait, 19 February 1869.

[2] See letters from W. C. Tait, 26 January 1869 and 19 February 1869.

[3] In his letter of 2 February [1869], CD had mentioned that he was studying secondary sexual differences in various animals, but did not specifically mention his interest in the rate of development of horns in different breeds of sheep.

[4] CD had received this information from Robert Sison (see letter from J. V. Carus, 22 January 1869 and n. 3).

From Philip Gidley King 25 February 1869

Sydney
Feb. 25/69.

My dear Mr. Darwin.

On reading the enclosed I could not help thinking of you. I suppose the variety is a sport—but who can say what might not be made of it by crossing & judicious selection—.[1]

Did you ever get any answers to your queries about the habits and manners of our black fellows.[2] I tried to make some answers but I found myself unable to distinguish the aboriginal manner from the acquired habit. All blacks I have associated with have been more or less civilized.—

I often think of you & read yr works. I obtained the first copy of yr Domesticated Animals &c that arrived in the Colony. & noticed with much pleasure yr mention of my name—.[3]

I hope you enj⟨oy⟩ better health than f⟨or⟩merly— you & I will soon ⟨be the⟩ last relics of the ⟨old⟩ Beagle.[4] Don't trouble to reply to this—tho I am always pleased to get a line from you. I may take a trip home one of these days & will hunt you up.

Ever yrs sincerely | Philip Gidley King

DAR 169: 28

[1] The enclosure has not been found.

[2] King refers to CD's queries on expression (see *Correspondence* vol. 15, Appendix IV). For the replies CD received on Australian Aboriginal peoples, see *Correspondence* vols. 15 and 16. No other correspondence between CD and King on this subject has been found.

[3] In *Variation* 1: 21, 28, CD referred to information received from King on dingoes.

[4] King was a midshipman on HMS *Beagle* between 1831 and 1836 (*Aust. dict. biog.*).

To W. B. Tegetmeier 25 February [1869]

Down. | Bromley. | Kent. S.E.
Feb. 25th

My dear Sir

With respect to Fowls, I shall be truly obliged if you can assist me on the following points.

Does the plumage of young birds when about 7 or 8 weeks old closely resemble that of the adults in

Cuckoo Dorkings (or other Cuckoo sub-breeds)

& *Sebright Bantams*

& *Silver Laced Polish*: with respect to these latter I want specially to know what is the character in colour of the topknot at 7 or 8 weeks old.—[1]

In *Buff Cochins*, the adult sexes are nearly alike or not very different, whilst in *Partridge Cochins*, the adults in your Poultry Book differ considerably; therefore I sh[d] like to learn the character of the young. Could you aid me about Cochins? My cook[2] doubts rather whether she could manage to remember & keep distinct so many as 5 lots of fowls; but if you could procure me eggs of *Buff & Partridge* Cochins, perhaps this would be the best plan. Also eggs of *Pile Game, Golden Pencilled Hamburghs* & *Golden Spangled Polish*: so that there will be 3 or 5 lots of eggs.— Golden Spangled Polish will be very interesting, in comparison with silver Polish, because I see in former that the topknot differs in colour in adult males & females; therefore I want much to know what is character of top-knot in chickens 7–8 weeks old.—

My Cook says she shall have hens ready in under a fortnight.— If eggs are sent by Rail, address them to "C. Darwin care of Down Postman, Bromley Kent"—or I could send carrier on any Thursday morning to Field Office.—[3]

You said you w[d]. ask M[r]. Teebay what breed of sheep his estimate of sexes referred to. Will you further ask him what is colour of Black Hamburgh chickens about 7–8 old; he is such a splendid observer, he may perhaps remember.— His statement about the chickens of all 4 sub-breeds[4]

When shall you be able to send me the m.s. about sexes of cattle, dogs &c fowls, ducks &c.; not that I am in a hurry, for my subject has grown very large in my endeavour to trace sexual differences througout the animal kingdom.[5]

Forgive me for earnestly begging you to keep some memorandum of the time consumed in correspondence, personal enquiries &c &c &c for me; for it would greatly distress me to think that I had prevented your writing Reviews or Books, which would have been more profitable & more interesting to you. You will be able by keeping a memorandum to put some value on the time which you so very kindly give to me, & I shall then feel at ease & thoroughily grateful to you. In simple truth I shall otherwise feel great perplexity & discomfort.—[6]

With sincere thanks | Believe me, my dear Sir | Yours very sincerely | Ch. Darwin

Endorsement: '/69'
Archives of the New York Botanical Garden (Charles Finney Cox collection) (ser. 14)

1 In *Descent* 1: 294–5, CD noted that young birds of breeds of fowl in which both sexes are similarly coloured resemble the adult from an early stage. He mentioned Cuckoo sub-breeds and the Sebright bantam and discussed the early appearance of the crest in Polish breeds. Tegetmeier was not cited for the information.

2 Margaret Evans.

3 A carrier service to and from London was operated every Thursday by George Snow, the village coal-dealer (*Post Office directory of the six home counties* 1866, Freeman 1978).

4 Richard Teebay's article on different breeds of spangled Hamburgh chickens was published in *The poultry book* (Tegetmeier 1867, pp. 154–8). CD's annotated copy is in the Darwin Library–CUL (see *Marginalia* 1: 800–3). CD cited the article in *Descent* 1: 281. Teebay's information on sexes of sheep has not been found.

5 See letter to W. B. Tegetmeier, 5 February [1869] and n. 2.

6 CD had earlier asked Tegetmeier to keep an account of the time spent working on the results of his questions on proportion of sexes (*Correspondence* vol. 16, letter to W. B. Tegetmeier, 15 February [1868] and n. 3).

To Hugo Thiel 25 February 1869

Feb. 25. 1869.

Dear Sir.

On my return home after a short absence I found your very courteous note & pamplet and I hasten to thank you for both and for the very honourable mention which you make of my name—[1] You will readily believe how much interested I am in observing that you apply to moral and social questions analogous views to those which I have used in regard to the modification of species— It did not occur to me formerly that my views could be extended to such widely different, and most important subjects. With much respect I beg leave to remain

Dear Sir. | Yours faithfully & obliged. | Charles Darwin.

Copy
DAR 148: 71

1 See letter from Hugo Thiel, 18 February 1869 and n. 1. At the beginning of his essay, Thiel cites CD for having established with convincing clarity the struggle for existence as the basic principle of all organic development (Thiel [1868?], p. 132). Thiel goes on to analyse aspects of human social development from a Darwinian standpoint. CD was in London from 16 to 24 February 1869 (CD's 'Journal' (Appendix II)).

From C. S. Bate 1 March 1869

My dear Sir

I have by a friend been asked if there be any known case of Hybridism between the Fox & the Dog: I said yes but only could recall a far back recollection of having been informed that a specimen that I saw was the case & on turning to your *[recent]* work I find that you state vol 1 p 31 that Female dogs will take the male Fox, & in page 151 vol II you state that there is no case on record of a fox breeding in confinement—[1]

The Question which has caused some litle discussion here is Whether there be any *undoubtful* case of a young animal the result of a cross between a *fox and a dog*: It is here among Hunters generally denied— Am I asking too great a favor, in requesting you to answer this question for me—

an acquaintance of mine here has a white cat with blue Eyes I believe that it is not deaf— would particulars be of service to you[2]

I am dear Sir | Yours very faithfully | C. Spence Bate

March 1/69 | Plymouth

DAR 160: 54

1 Bate refers to *Variation* 1: 31 and 2: 151.

2 In *Variation* 2: 329, CD referred to deafness in blue-eyed white cats in his discussion of correlated variability.

From Edward Blyth 1 March 1869

7 Princess Terrace, | Regents Pk, N.W.,
March 1 /69.

My Dear Sir,

In the common Indian antelope (*Antilope cervicapra* and *Cervicapra bezoartica*) the females are usually hornless, & the males have straight and tensely spirated horns, approaching sometimes to 2 ft. in length, and of which the twist is more or less intense in different specimens. In the castrated buck, of which I have seen several, (& there is a fine one mounted as a stuffed specimen in the Calcutta museum) the horns are not spirated, but arch and gyre round to the front, with a considerable spread apart, and the colour of the coat remains permanently chestnut-brown as in the doe. In exceedingly rare instances (of which three have come to my knowledge) the doe bears horns similar to those of the castrated buck but smaller and more slender.[1] Such a doe is figured as a supposed new species in one of the Indian sporting periodicals, if I mistake not in one of the latest numbers of the *Bengal Sporting Magazine*, New Series—

In the nilgai, also, the female is hornless, & the castrated buck develops small and slender horns, while in this instance likewise the coat remains of the same rufous-brown colour as that of the female & young, instead of assuming the colour which is styled *nil* (pronounced *neel*) or blue in the mature male.[2]

In the axis deer (*Cervus axis*)[3] the mature male has a considerably darker hue than the female, which also is not assumed by the castrated buck or "*[*beaver*]*". The African *Antelope* (*Hippotragus*) *leucophaeus* was long considered to be a lost species; but I believe it to be merely the mature male of what is known as the *A.* (*H.*) *equinus.*[4] An allied species, the *A. (H.) niger* would appear to resemble in the difference of colouring of the sexes the *Bos sondaicus*, & the castrated animal would doubtless retain the colouring of the female.[5] Ditto with the *A. (Kobus) maria* of the banks of the White Nile.[6] The Indian antelope has this peculiarity (so far as I am aware) that the black colour of the buck appears & disappears according to the condition of sexual excitement, by changes of hue in the actual coat, without any renovation of the vesture—[7]

Yours Sincerely, E. Blyth

P.S. Only half of my article on the life of animals in wild nature appears in this week's **Land & Water.**[8] On Saturday I went to Tottenham in Essex to inspect a lot of fossils & subfossils just exhumed in the course of excavations for a reservoir. The usual species, but two rather interesting human skulls. I have secured all for the Museum of the College of Surgeons.[9]

DAR 83: 152–3

CD ANNOTATIONS

1.1 In the] 'Keep' *added pencil*
2.1 In the ... vesture— 3.10] *crossed ink*
3.1 the mature male ... "*[*beaver*]*". 3.2] *double scored blue crayon*
3.2 The African ... Nile. 3.8] *crossed blue crayon*
3.7 Ditto ... Nile. 3.8] *double scored blue crayon*

Top of letter: 'Desmareste p. 455 ♀ Antelope with Horns like other species.'[10] *ink*; 'The Horns of ♀ are modified' *blue crayon*; 'permanently' *pencil*; 'Keep for early part of Chapt' *blue crayon*

[1] *Cervicapra bezoartica* is now considered a synonym of *Antilope cervicapra* (the blackbuck). CD referred to the species by another synonym, *Antilope bezoartica*, in *Descent* 2: 246, where Blyth is cited for this information.

[2] In *Descent* 2: 287–8, CD discussed sexual differences in colour in the nilgai (*Boselaphus tragocamelus*), which he referred to as the nilghau, *Portax picta*. He cited Blyth for some of this information.

[3] *Cervus axis* is now *Axis axis*.

[4] *Hippotragus leucophaeus* (the blue antelope or bluebuck) probably became extinct around 1800 (see Lichtenstein 1811–12, 1: 265). Blyth evidently believed it to be a colour variant of *Hippotragus equinus* (the roan antelope). Blyth had made similar claims in earlier letters (see *Correspondence* vol. 15, letter from Edward Blyth, 24 February 1867, and *Correspondence* vol. 16, letter from Edward Blyth, [before 25 March 1868]).

[5] Blyth refers to *Hippotragus niger* (the sable antelope, referred to by CD as *Antilope niger*) and *Bos sondaicus* (now *B. javanicus*, the banteng). In *Descent* 2: 289, CD noted the colour differences in the sexes of these species.

[6] *Kobus maria* is now *K. megaceros* (the Nile lechwe).

[7] In *Descent* 2: 288–9, CD cited Blyth on changes in the colour of the male *Antilope bezoartica*.

[8] An article on the musk sheep of Arctic America appeared in *Land and Water*, 6 March 1869, pp. 151–2.

[9] Blyth refers to the Hunterian Museum at the Royal College of Surgeons in London.

[10] Desmarest 1820, p. 455, contains a description of *Antilope euchore* (now *Antidorcas marsupialis*, the springbok). CD's annotated copy is in the Darwin Library–CUL (see *Marginalia* 1: 192–4).

From Georg Recht[1] 1 March 1869

München Schommergasse N. 2/III
den 1sten März 1869

Euer Wohlgeboren!
Edler Gönner & Wohlthaeter!

Gestern erhielt ich Ihre Anweisung ausbezahlt, und das erste Geld davon will ich verwenden, Ihnen meinen Dank zu sagen für diese Stütze, die mich auf ein Paar Monate aller Sorgen überhebt, und es mir moeglich macht, meine Schrift weiter zu versenden.[2] Meinen herzlichen Dank für diese unerwartete Unterstützung. Als kleine Erkenntlichkeit kann ich Ihnen das Ihnen bestimmte Exemplar übersenden.[3]

Ich kann mich vor der Hand nicht weiter über den Inhalt desselben aussprechen, nur wollen Sie mir erlauben, zu bemerken, dass das Princip d'Alembert die Hauptstütze desselben bildet. Dieses Princip heisst: Im Zustande einer jeden Bewegung stehen Kraft & Hindernisse im Gleichgewichte.[4]

In dieser einfachen form erscheint das erwaehnte Princip als selbstverstaendlich und von keiner weitern Bedeutung, als dass es die nöthigen Gleichungen gibt. Allein bringt man damit in Verbindung, dass die Hindernisse bei einer Bewegung keine rückwirkende, sondern nur eine hemmende Kraft sind, so entsteht dadurch die gleichmässige Bewegung, und ihre Geschwindigkeit als das Maass derselben, während eine rückwirkende Kraft den Zustand der Ruhe erzeugen würde. Bei der gleichmässigen oder gleichförmigen Bewegung ist das Moment der Hindernisse gleich dem Momente der Kraft oder dem Momente der entbundenen Waerme.

Überall in der Natur erblicken wir die gleichmaessige Bewegung unter Einwirkung zweier Kraefte oder zweier Geschwindigkeiten und zwar von den Grundelementen des Weltalls—Raum & Zeit—bis herab zu den Organen der geringsten Thiere. Der Raum ist das ruhige gegebene oder hemmende Element, die Zeit das bewegliche, und so besteht jedes Organ aus zwei solchen Elementen. Jedem beweglichen Knochen z. B. steht ein fester oder hemmender gegenüber. Die Anwendung des d'Alembertischen Principes auf die Natur insbesondere auf die Chemie verspricht meines Erachtens ungeheuere Resultate und macht diese Wiesenschaft dem Calcul zugaenglich. Schon jetzt wollte ich ein grosses Buch schreiben aus lauter Folgerungen beshehend, deren Wahrheit mehr oder weniger durch die Erfahrung bestaetigt sind, und die Wahrheit dieser Folgerungen laesst mich schliessen, dass ich in meiner Schrift die richtigen Principien niedergelegt habe.[5]

Ich bitte um Entschuldigung wegen der belehrend scheinenden Form meines Schreibens, indem ich auf eine andere Gelegenheit warte, um Ihnen meine weitern Resultate mitzutheilen. Nur einen Satz erlauben Sie mir noch; naemlich: Die Haut des Menschen mit ihren spiralfoermig gewundenen Drüsen zerfaellt in innere & aeussere. Die aeussere ernährt den Menschen mit Licht durch die Augen, mit Electricitaet durch die Ohren & mit Waerme durch das Gefühl; die innere durch Geruch & Luft, durch Geschmack und durch Verdauung.

Damit mochte ich aussprechen, dass ich sechs anstatt fünf Sinne am Menschen unterscheide, und dass jeder Sinn ein Ernaehrungsorgan ist.

Mit der herzlichsten Bitte, dass beifolgende Exemplar als ein Zeichen meiner Achthung & Verehrung aufzunehmen, sage ich Ihnen wiederholt meinen Dank.[6]

Euer Wohlgeboren | ergebenster | D Recht

DAR 176: 75

1 For a translation of this letter, see Appendix I.

2 See letter from Georg Recht, 9 February 1869. A payment of £3 to Recht on 13 February 1869 is recorded in CD's Account Books–banking account (Down House MS).

3 See letter from Georg Recht, 9 February 1869 and nn. 2 and 3.

4 Recht refers to Jean le Rond d'Alembert and to the principle of classical mechanics first elucidated in his *Traité de dynamique* (Alembert 1743).

5 Recht's proposed larger work was apparently never completed. Recht also refers to Recht 1869 (see n. 6, below).

6 Recht enclosed a copy of his work on the development of world laws (Recht 1869; see also n. 3, above).

From W. C. Tait 2 March 1869

Oporto
March 2nd 1869

Charles Darwin Esqr., | Down | Bromley | Kent

Dear Sir,

Since writing on the 19th ult*mo* I have again been to S Pedro da Cova and have this time found the plant there in great abundance and enclose in this a specimen (the only one of those I found which had a flower)[1] It grows on the dry hill sides with the distance of a few feet between each plant. I am afraid that you will find some difficulty in keeping this species alive as it seems to like the sun and a stony soil and the most extraordinary circumstance is that it seems to live partly on insects! A great number of flies were adhering to the gum glands of the plants and some of them appeared to be under the process of absorption.

The villagers call it "Caça-moscas" or "Fly-catcher" also "a Herva das impigems" or "the skin disease plant" but Brotero's Portuguese name is "Herva pinheira orvalhada" or the "Dew pine grass"

The village boys told me that if hung in the cottage it caught flies and is used for that purpose and for skin diseases it is bruised in a pestel and mortar before application.[2]

M^r. Grant[3] tells me he has tried to grow it in his garden but without success and thinks the reason of failure was the difference in quality of soil—

A plant which shrivelled up on the way home has revived in a tumbler of water and the glandular secretion has appeared again. A villager promised to bring me some live plants of this species in flower pots and I expect him on Thursday afternoon (4th.) I shall send them if possible by the "Beta" S.S. leaving on Friday or Saturday and if you have not forwarded an address in London by that date I shall ask my correspondent to take charge of them until you send for them. I should recommend great care with the soil of which I hope to send you a fair quantity. I have found a fern near our house and which M^r. Grant says is new to him and he thinks it has never been described as a portuguese fern. I enclose a small leaf hoping you will be so kind as to inform me of the name &c.[4] I have collected the ferns in this neighbourhood and cultivated in our garden the Ivy leaves Fern "Asplenium palmatum"[5] Some leaves have varied very much Some are split up in all directions the lines of spores converging from the sides of the clefts (an instance of correlation perhaps) The enclosed fern varies in the number of branclets from the two lower branches—from being actually counting to three in number.[6] The "Asplenium Scolopendium" Harts Tongue I have often seen split at the end—[7]

With regard to tail-less dogs until your letter reached me I had never read or heard of such a thing as inherited mutilations but am now much interested in the instances mentioned in your work and therefore the supposed case which I have observed seems all the more remarkable.[8] Do you not think that mutilations of bone would be more likely to be inherited than flesh mutilations?

I remember my father many years ago telling me that the Chinese dwarfed trees by pruning the roots and that he thought the small mandarin Orange had been produced in this manner. Would not this also come under the catalogue of inherited mutilations? I met by chance a former owner of my tail-less pointer and he tells me that its mother was tail-less and that he is certain that my dog was born without a tail (or short tail) and that she has already had some pups without tails also—

I think that sterility of crosses between different species and also in a great measure inheritance has been fixed by natural selection to the advantage of the species.[9] There would be fine confusion if all species were prolific when crossed but most of them would probably be eliminated in a very short time if not protected in an unnatural manner (such as by Domestication) from close competition with their more specially adapted neighbours. Selection (or the law of the greater being more worthy than the less) seems to have had such an eternity of time for action that I should not hesitate to ascribe to it the production of the sympathetic element in animals. It is well known that union is strength and we see instances of one society banded together under the social impulses against other societies or individuals. The only difficulty against your theory which seems to me unexplained is that mentioned by the Duke of Argyle in his "reign of law" that of beauty such as humming birds and I should like to see a satisfactory explanation given.[10] I think that though selected for the benefit of the species it has had collateral effects and that beauty depends as much on our own selected senses as on the object For instance in a book called "Helionde or Adventures in the Sun" a fanciful publication but yet very interesting it mentions that our appreciation of the curve or line of beauty arises from the formation of the muscles of the eye[11] it also depends upon our own consciousnesses which vary in different individuals hence "de gustibus non disputandum"[12]

Dr. Bree in his work on the "Birds of Europe" says that colour of birds is dependant on the *form* of the minute particles of the feathers—[13] It has been suspected that taste is the appreciation of this form of minute particles of the object tasted, perhaps smell is dependant on the same process

I have seen some curious instances of varying habits with birds in a wild state with regard to nidification. There is a blackbird which has built several times in our garden and I have always found pieces of newspaper in the nest— A robin also in our garden built twice on the ground and the second time the nest was made in a minature small brick house made by a small brother of mine it hatched the eggs and when nearly fledged the young birds fell a prey to our cat who most effectually eliminated that brood— When at school in England I met with many instances of birds making their nests on the ground contrary to the custom of the species I have also met with small and imperfect eggs of wild birds. Heredity tends of course to make these instances rare—

I enclose some feathers of a Laughing Dove which died under my care this bird had a deformed bill and I think that from an inability to preen its feathers properly it met its death.[14] The sheath of the feather has not been taken off. This might have occurred in nature very well and been eliminated in the same manner— I have noted several similar cases of variations in tame and wild animals. I think that the laughing dove of which I possess the fawn and also the white kinds are descended from some other than the Columba Livia possibly from the turtle dove?[15] The markings of the tail feathers are similar the voice (as you have noticed) and the general shape and appearance— The portuguese call the Turtle dove and the laughing dove by the same name "Rôla" (from the voice) whereas they call the pidgeon "poruba" If it could be proved that the laughing dove was descended from the turtle dove and that the laughing dove and common pidgeon would interbreed and produce fertile offspring would it not be an interesting case of elimination of sterility of crosses between different species? White seems to be the simplest color in nature under the "least action" law otherwise why should eggs laid by wild birds in holes of trees, rocks &c. be in general white—and why should white varieties (as noticed by you) be the most constant.[16]

I was much interested in that portion of your "Origin of species" in which you speak of the Geological succession of life.[17] Do you not think it probable that there has been variation in the situation of the axis of the earth? and that there has been a slow and regular change in the formations of the land above water and the sea the land being alternately above and below the water the bulging out of the equator from the centrifugal force would give the varying tropical districts and would connect the New World with the Old Hemisphere— A great many small collateral circumstances make me think this but principally an explanation of the chalk formation of England in one of the Magazines also that geological deposits shew that climate of changed *very much*, that the sea seems to be leaving this continent and encroaching on Greenland. This is a very interesting speculation to me and I should *very much* like your opinion on it. Your very large experience and collected observations would render you very able to give a correct judgement

I remain | Dear Sir, | Yours very truly, | William C. Tait

DAR 58.1: 18a–f

CD ANNOTATIONS
Top of letter: 'On Drosophyllum L William C. Tait of Oporto.—' *ink*

1 See letter from W. C. Tait, 19 February 1869. Tait refers to *Drosophyllum lusitanicum*; the flower specimen has not been found.

2 Tait refers to Félix de Avellar Brotero and Brotero 1804, 2: 216. Tait enclosed an extract from Brotero 1804, 2: 215–17, in his letter of 10 May 1869. CD cited Tait for parts of this information that he used in *Insectivorous plants*, p. 332.

3 Alexander Grant has not been further identified. See letter from W. C. Tait, 19 February 1869 and n. 2.

4 The leaf specimen has not been found.

5 *Asplenium palmatum* is a synonym of *A. hemionitis* ('Feto-de-folha-de-hera' is the Portuguese common name).

6 The fern specimen has not been found.

7 *Asplenium scolopendrium* is the hart's-tongue fern.

8 See letter to W. C. Tait, 2 February [1869]. Tait refers to *Variation* 2: 22–4.

9 In *Variation* 2: 185–9, CD discussed whether the sterility of hybrids could be produced through natural selection, and concluded that, in general, it could not. For CD's changing views on the subject of the sterility of hybrids as a selected quality, see *Correspondence* vol. 13, letter to M. E. Wichura, 3 February [1865] and n. 9. In 1868, CD and Alfred Russel Wallace had discussed this question extensively (see *Correspondence* vol. 16).

10 Tait refers to George Douglas Campbell, eighth duke of Argyll, and to Campbell 1867. Campbell suggested that specialised beaks did not give the greatest possible advantage to humming-birds, which had equal access to a wide range of flora, and that the rule governing the proliferation of humming-bird species had as its object rather 'the mere multiplying of Life, and the fitting of new Forms for new spheres of enjoyment' (Campbell 1867, pp. 241–2). For CD's response to Campbell's argument, see *Correspondence* vol. 15, letter to Charles Kingsley, 10 June [1867].

11 See Whiting 1855, pp. 98–9 n. 1. The concept of the 'line of beauty' was initially discussed by William Hogarth in *The analysis of beauty* (Hogarth 1753).

12 *De gustibus non disputandum*: there's no accounting for taste (Latin).

13 Tait refers to Charles Robert Bree and Bree 1859–63.

14 The laughing dove is *Streptopelia senegalensis*. The enclosed feathers have not been found.

15 Turtle dove: *Streptopelia turtur*; *Columba livia*: the rock-pigeon. In *Variation* 1: 131–224, CD discussed characteristics of different domestic varieties of pigeon and concluded that they were all descended from *C. livia*.

16 For CD's discussion of the sterility of first crosses between species, see *Origin*, pp. 254–67. In *Variation* 2: 20, CD noted that white varieties of flowers were generally the most constant.

17 Tait refers to chapter ten of *Origin*, pp. 312–45.

To Thomas Roscoe Rede Stebbing 3 March 1869

Down. | Bromley. | Kent.
March. 3rd. 1869

Dear Sir,

I am very much obliged to you for your kindness in sending me your spirited & interesting lecture.[1] If a layman had delivered the same address, he would have done good service in spreading what, as I hope and believe, is to a large extent the truth; but a clergyman in delivering such an address does, as it appears to me, much more good by his power to shake ignorant prejudices, & by setting, if I may be permitted to say so, an admirable example of liberality.

With sincere respect I beg leave to remain— | Dear Sir, | Yours faithfully & obliged | Charles Darwin

LS(A)
American Philosophical Society (362)

1 The lecture was probably one on 'Darwinism' given in February 1869 before the Torquay Natural History Society and published in Stebbing's book *Essays on Darwinism* (Stebbing 1871). CD's copy has not been found.

From T. C. Eyton 4 March 1869

Eyton, | Wellington, | Salop.

Dear Darwin

I yesterday evening forwarded to you the head of a fawn which died yesterday skin and all it was droped last june so you will see exactly the state of the[1] now I have a series of horns afterwards up to eight years old, Hoping you are better than you were when last you wrote[2] I remain | yours truly | Tho C Eyton

March 4 /69

DAR 86: 50

CD ANNOTATION
End of letter: '(I presume the Fawn born in June now March 3d has no Horns.) Fallow-deer' *ink*

1 Eyton apparently refers to the state of the animal's horns. See CD's annotation.

2 CD's last extant letter to Eyton was that of 24 January 1869, but in that letter he did not refer to his health.

From T. R. R. Stebbing 5 March 1869

Tor Crest Hall, | Torquay.
March 5th. 1869.

My dear Sir,

It may be interesting to you to know that I began the perusal of your works with the common prejudices strong upon me, and a very impertinent expectation that I should find a damaging flaw somewhere or another in your argument.

It was a vast surprize to me and soon became a great pleasure to find what was the real character of these insidious writings, which even men of considerable scientific attainments seemed anxious to taboo. Of course I had to undergo the useful pang of giving up some opinions on which I had formerly been very positive; but in return there was laid before me a view of the world's history, so simple, so harmonious, so unlike the chaotic shreds and patches that most histories are made up of, that I should have felt genuine admiration, even had I not been also convinced.

Incidentally, I may add, it was no slight pleasure to "an orthodox clergyman" to find in his Author's calm and temperate style that Science might make gigantic strides without offering such collateral opinions as, if true, would certainly dispense with clergymen altogether, whether orthodox or otherwise.

The kind letter in which you approve my slight attempt to get people to study your works for themselves has afforded me one more gratification.[1]

Believe me | Yours most faithfully | Thomas R. R. Stebbing.

Charles Darwin Esq

DAR 177: 248

[1] See letter to T. R. R. Stebbing, 3 March 1869.

From W. C. Tait 5 March 1869

Oporto
March 5th 1869.

Charles Darwin Esqr. | Down | Bromley | Kent

Dear Sir,

I this morning received your letter of the 24th. ultmo. and am glad to say that the plants are now in my possession and I am sending 12 by the "*Beta*" S.S leaving tomorrow morning.[1] I send them in a shallow wooden box covered over with bars and the Captain has promised to take care of them during the voyage he has also given me the address of a man in London who will forward them on to your address thus obviating troubling your brother who lives at a considerable distance from the wharf.[2] I have therefore asked

Mr. T. Allen[3]
British & Foreign Wharf
Lower East Smithfield
London E.C.

to receive and forward the plants without delay immediately they arrive.

I hope you will receive them in good condition and if you should wish me to forward some more I shall be most happy to do so.

Please notice the manner in which the flies are held on the leaves and the strong resinous smell of the plant. The man who brought me the plant made a mistake I think in putting too rich soil over the natural soil which is at the bottom of the flower pots and from which you will be able to judge more or less what is suitable for them.

The situation in which I found them was dry and sunny but from my short experience with them the plants seem to like plenty of water, my resuscitated plant looking very well under the hydropathic treatment. I do not know how they will get on without their animal diet.

I shall make notes on all the peculiar circumstances relating to tail-less dogs and also as to systematic selection[4] I am also making enquiries about what I suppose is an interesting case of correlation of growth in the cultivation of the common orange.

Every enquiry shall be made about sheep also but unfortunately this northern part of Portugal is too mountainous for merino sheep which breed (or a similar one) is kept on the plains of the Alemtejo near Lisbon.[5]

I am in the habit of buying wool for shipment to England but do not see any thing of the sheep, the Wool district being up the river round the Wine country I shall try to pick up as much reliable information as possible. I have noticed the effect of selection (artificial) on the quality of Wool from various districts but the pasturage has an immense influence and the quality of the Wool may be foretold by knowing the locality from which it comes.

On the Gerez Mountains[6] near the Borders of Spain almost due NW from this city there is a curious wild goat I am told it is of a very rare species—[7] This serra is very interesting on account of its great wildness. I saw the old trees growing there covered with moss and falling around from sheer old age or too much crowding Several rare plants are to be found there and large birds of prey are common Wild boar and roebuck may be shot there. And the scenery is very fine–

Have you ever heard that if a fig shoot is planted upside down the tree from it will be pendulous? I have been told so and am trying whether such is the case— I shall be most happy to try to obtain any plants or birds you may wish for and remain | Dear Sir, | Yours very truly, | William C. Tait.

DAR 178: 45

CD ANNOTATIONS
1.1 I this … wharf. 1.6] *crossed blue crayon*
3.1 Please … plant. 3.2] *double scored blue crayon*
4.1 The … diet. 4.4] *double scored blue crayon*; 'Tap-root' *added blue crayon*
5.1 I shall … crowding 8.4] *crossed blue crayon*

1 See letter to W. C. Tait, 24 February 1869.

2 CD had suggested the plants could be sent to Erasmus Alvey Darwin's address in London (see letter to W. C. Tait, 24 February 1869).

3 T. Allen has not been further identified.

4 See letter to W. C. Tait, 24 February 1869 and n. 2.

[5] See letter to W. C. Tait, 24 February 1869 and n. 4. Alemtejo (now Alentejo) was a province in south-east Portugal (*Columbia gazetteer of the world*). For more on the breeding of merino sheep in Portugal, see Wood and Orel 2001, pp. 23–4.

[6] Tait refers to the Serra Do Gerês, a short mountain chain that extends from north-west Portugal into Spanish Galicia (*EB*).

[7] Tait probably refers to the Geres goat (*Capra pyrenaica lusitanica*, a subspecies of the Spanish ibex), which became extinct around the end of the nineteenth century (Moço *et al.* 2006, p. 351).

To A. R. Wallace 5 March [1869][1]

Down. | Bromley. | Kent. S.E.
March 5th

My dear Wallace

I was delighted at receiving your Book this morning. Its whole appearance & the illustrations, with which it so profusely ornamented, are quite beautiful. Blessings on you & your publisher for having the pages cut & gilded.—[2] As for the Dedication, putting quite aside how far I deserve what you say, it seems to me decidedly the best expressed dedication, which I have ever met.[3]

The reading will probably last me a month, for I dare not have it read aloud, as I know that it will set me thinking. I see that many points will interest me greatly.— When I have finished, if I have anything particular to say, I will write again. Accept my cordial thanks. The dedication is a thing for my children's children to be proud of.—

Yours most sincerely | Ch. Darwin

British Library Board (Add 46434, f. 170–71)

[1] The year is established by the reference to Wallace's book (see nn. 2 and 3, below).

[2] CD refers to *The Malay Archipelago* (Wallace 1869a), published at the beginning of March 1869 (*Publishers' Circular*, 15 March 1869, p. 171). CD's annotated copy is in the Darwin Library–CUL (see *Marginalia* 1: 840–2). CD was a strong advocate of producing books with the pages already cut (see *Correspondence* vol. 15, letter to *Athenæum*, 1 January 1867).

[3] See letter from A. R. Wallace, 20 January 1869 and n. 1.

To T. C. Eyton 6 March [1869][1]

Down. | Bromley. | Kent. S.E.
March 6th

Dear Eyton

I received the Fawn's head, safely, & I am much obliged to you for so kindly sending it; though your previous information sufficed for my purpose.—[2]

I am pleased to hear from my son at Trinity that he is a friend of your son.[3] It is a resurrection of good old times. Yours very sincerely | Ch. Darwin

American Philosophical Society (366)

[1] The year is established by the relationship between this letter and the letter from T. C. Eyton, 4 March 1869.

[2] See letter from T. C. Eyton, 4 March 1869. For the earlier information, see the letter from T. C. Eyton, 26 January [1869].

[3] Francis Darwin and William Campbell Eyton were students at Trinity College, Cambridge.

To J. D. Hooker 8 March [1869][1]

Mar 8—

Dear Hooker

I think that part of the enclosed letter is well worth your reading, viz about the grass of which the awns like a tendril catch its own rachis. The first part about Escholtzia, which is self-sterile in a much greater degree in S. Brazil than here, is not worth your reading. At the end of the note however there is a good & precise case of the transmission of character from the individual flowers on the same spike.[2]

Please return the letter.

I am rather glad to have the excuse of sending it, as I want to hear a little news of you. About myself I have no news, as I am going on in my old routine. The subject of sexual selection grows bigger & bigger as I progress but I suppose I shall some day end it.[3] We are a very small party here at present for Lizzy is gone to Germany & Henrietta is in a very poor state & has been confined to her bed room for the last fortnight.[4]

As you always puff me up so I must tell you that I have just been applied to to permit a French trans. of my Orchis book. It bothers me a good deal to know how much to add, for I have an immense amount of new matter; want of time however will compel me I think to make but few additions.[5]

yours affectionately | Ch Darwin

Though I am so despised by the great guns of the Institute, I presume I am rising in estimation amongst the mob, for another man has applied to translate my Journal of Travels.—[6] Here is a boasting note.

L(S)(A)
DAR 94: 116-17

[1] The year is established by the relationship between this letter and the letter from Fritz Müller, 12 January 1869.

[2] CD evidently enclosed the letter from Fritz Müller, 12 January 1869. CD refers to Müller's experiments with *Eschscholzia californica* and maize.

[3] CD refers to his research for *Descent.*

[4] According to Emma Darwin's diary (DAR 242), Elizabeth Darwin left on 5 March, and Henrietta Emma Darwin was very ill from 23 February until 9 March 1869.

[5] The French translation of *Orchids* by Louis Rérolle was published in 1870 (Rérolle trans. 1870). CD's additions to the French edition were in the form of footnotes ending 'C. D., mai 1869'. CD published an English version of these notes in *Annals and Magazine of Natural History* in July 1869 ('Fertilization of orchids'). Much of this material was later incorporated into the text of the second edition of *Orchids.*

[6] CD probably alludes to the Académie des Sciences and to the negative reception of Darwinian views among the French scientific élite (see Browne 2002, pp. 260–1). Charles Phipps Haussoullier had requested permission to translate CD's *Journal of researches* into French (see letter from C. P. Haussoullier, 31 January 1869 and n. 1).

To R. F. Cooke 9 March [1869][1]

Down Bromley | Kent.
Mar 9.

Dear Sir
I think

For
Darwin
Müller

is the best form of lettering for the back.[2] Please to hurry on the binding & have the goodness to send me the **nine** copies as soon as you can addressed differently to what I before told you viz.[3]

C. Darwin Esq
Orpington Station
Kent. | per South East Railway.

Be so kind as to inform me when the parcel is sent off that I may know when to send for them.

Dear Sir | yours faithfully | Ch Darwin

Please attend to copy for Mr Spence Bate & to 3 for Mr Dallas & for Reviews[4]

LS(A)
National Library of Scotland (John Murray Archive) (Ms. 42152 ff. 198–9)

1 The year is established by the relationship between this letter and the letter from R. F. Cooke, 10 March 1869.
2 CD refers to the lettering on the spine of Dallas trans. 1869, the translation of Fritz Müller's *Für Darwin* (F. Müller 1864a).
3 See letter to R. F. Cooke, 23 [February 1869].
4 See letter to R. F. Cooke, 23 [February 1869]. CD refers to Charles Spence Bate and William Sweetland Dallas.

From R. F. Cooke 10 March 1869

50^A, Albemarle S^t. | W.
March 10. 1869

My dear Sir

On the otherside I send you a statement of the actual expenses of Fritz Muller & what the Vol will produce, if sold at 6/—[1]

I shall wait to hear, if you agree to 6/— before I publish.

By this post I also send you a copy, so that you will see the thickness. The lettering in all other copies will be as you wished.[2]

I hope to send off your other copies tomorrow evening.

Yrs faithfully | Robt Cooke

C. Darwin Esq

DAR 171: 366

1 The statement referred to has not been found. Cooke refers to Dallas trans. 1869.
2 Cooke refers to CD's request regarding the lettering on the spine of the book (see letter to R. F. Cooke, 9 March [1869]). The lettering in the end read, 'Facts | For | Darwin | Muller'.

To Thomas Henry Huxley 10 March 1869

Down. | Bromley. | Kent. S.E.
March 10/69/

My dear Huxley

I fear that you will hate me, but I could not bear to refuse the request in the enclosed note, coming from a man whom I much like & respect.— Accordingly I send the long letter by this day's post, 21 pages, but in excellent hand-writing.[1]

Possibly, but not probably, it may be worth your while to read, before you answer all your critics, what a red-hot Comtist, lawyer & able man has to say for his prophet.—[2] If you can let me have a line, saying that you will consider Vernon Lushingtons remarks & that will satisfy him.

There is no peace for the wicked or the good in this world.

Ever yours | C. Darwin

Imperial College of Science, Technology, and Medicine Archives (Huxley 5: 262)

[1] The letter, which was from Vernon Lushington, has not been found.

[2] Lushington was a supporter of the positivist philosophy of Auguste Comte. In an article that appeared in the *Fortnightly Review* of 1 February 1869, Huxley had criticised positivism, commenting, 'M. Comte's philosophy in practice might be compendiously described as Catholicism *minus* Christianity' (T. H. Huxley 1869a, p. 141). Lushington's letter probably contained a response to Huxley's critique.

From Henry Landor 10 March 1869

Amherstburg | Ontario | Canada
March 10th 1869

Dear Sir

I wish to call your attention to a mode of transportation of boulders and with boulders dirt and animal & vegetable life, not noticed in your wonderful book on Origin of Species, but which must have very considerable influence in distributing species over large areas in this part of the world, and doubtless in every other country in similar circumstances.

The theories on the formation of ground ice in running streams, are none of them without objections, which cannot at present as far as I can see be explained, but I will not allude to them further than is necessary to explain facts of observation.

When the temperature falls to Zero or 10 degrees above it, a curious phenomenon occurs in rivers, after two or even one nights frost. In those parts where the current is too rapid for the formation of surface ice, the water freezes at the bottom, adhering to the ground or stones on the bottom. As day after day of sharp frost continues, the ground ice thickens; at first it is more or less tubular, the tubes in the direction of the current. It would seem that there are tubular currents of small diameter and the water freezes between them, allowing pipes as it were for the passage of the swifter water. The diameter of these is small, and the cakes of ice often break off the stones or bottom of the river, and fill the surface of the stream with this porous ice. In those masses of ice which continue to adhere to the bottom thickening continues daily, and in time by its power of flotation, it raises itself and the stone or dirt to which it was anchored to the surface. The block of ice and dirt or boulder as the case may be, then floats away to the nearest part of the river where the surface is frozen over. Then it becomes frozen in with the mass of surface ice, with its blocks of earth or stone enclosed, until the general thaw in the spring. The ice on the surface in one night loses its tubular character, if it has even retained it, until flotation, and any one seeing the boulder or dirt attached to it, would find it impossible to say whether it came from the bank or bottom of the river, unless he had watched it to its position.[1]

In the spring these blocks are carried away, some dropping in the river itself; many carried into the great lakes,[2] and far into them, before they thaw sufficiently to drop their burden of earth or boulders. Those streams that flow into the Mississippi carry their contents into that river which may carry them through degrees of latitude before they thaw & deposit.

I am not a man of science, only a common sense observer, but I am sure that here we have a mode of distribution that ought not to be overlooked. Every year this goes on in every river of Canada, and in all the States in similar latitude & condition of climate, and has gone for nameless centuries. The accumulated results of centuries of this work must tell, and this also shews that boulders are brought from considerable distances, and in lapse of time in considerable quantities, and are distributed over large areas. Ice is drifted by every changing wind across the largest of these great lakes again and again in every winter. Even this year a man has been drifted on ice across Lake Huron. Therefore these boulders are dropped all over the bottom of these lakes. It may be said that the result

Incomplete
DAR 205.2 (Letters): 244

CD ANNOTATIONS
3.12 The block ... enclosed, 3.15] *underl pencil*
Top of letter: 'Only Probable [*interl*] Distribution by Ground-ice in Canada— | from Dr. Landor' *ink*

1 CD mentioned icebergs as a means of transport in *Origin*, pp. 362, 381–2, 393–4, and 399.
2 The Great Lakes between the United States and Canada include Lakes Ontario, Erie, Huron, Michigan, and Superior; Amherstburg lay at the east end of Lake Erie.

From A. R. Wallace 10 March 1869

9, St. Mark's Crescent | NW.
March 10th. 1869

Dear Darwin

Thanks for your kind note. I could not persuade Mr. Macmillan to cut more than 25 copies for my own friends, & he even seemed to think this a sign of most strange and barbarous taste.[1]

Mr. Weir's paper on the kinds of larvæ &c. eaten or rejected by insectivorous birds, was read at the last meeting of the Entomological Society & was most interesting &

satisfactory.[2] His observations & experiments, so far as they have yet gone, confirm in *every instance* my hypothetical explanation of the colours of catterpillars. He finds that all nocturnal feeding obscure coloured catterpillars, all *green* & *brown* and *mimicking* catterpillars, are greedily eaten by almost every insectivorous bird. On the other hand every gaily coloured spotted or banded species, which never conceal themselves, & all spiny & hairy kinds are *invariably rejected*, either without or after trial.[3]

He has also come to the curious & rather unexpected conclusion, that hairy & spiny catterpillars are not protected by their hairs,—but by their nauseous taste, the hairs being merely an external mark of their uneatableness, like the gay colours of others— He deduces this from two kinds of facts—1st. That very young catterpillars before the hairs are developed are equally rejected,—and 2nd—that in many cases the smooth *pupæ* & even the *perfect insect* of the same species, are equally rejected.

His facts it is true are at present not very numerous, but they all *point one way*. They seem to me to lend an immense support to my view of the great importance of protection in determining colour,—for it has not only prevented the eatable species from ever acquiring bright colour spots or markings injurious to them, but it has also conferred on all the nauseous species distinguishing marks to render their uneatableness more protective to them than it would otherwise be.

When you have read my book I shall be glad of any hints for corrections if it comes to another edition— I was horrified myself by coming accidentally on several verbal inelegancies after all my trouble in correcting, & I have no doubt there are many more important errors.

Believe me Dear Darwin | Yours very truly | Alfred R. Wallace

DAR 106: B77–8

CD ANNOTATIONS

End of letter: 'Sondaicus | Volcanic outburst'[4] *pencil*

1 See letter to A. R. Wallace, 5 March [1869] and n. 2. Wallace refers to his publisher, Alexander Macmillan, and to Wallace 1869a.

2 John Jenner Weir's paper 'On insects and insectivorous birds' (Weir 1869) was read on 1 March 1869. CD cited it in *Descent* 1: 417.

3 Wallace had been asked by CD to suggest a solution to the problem of bright colours in caterpillars as well as butterflies. CD had maintained that bright colours in butterflies were due to sexual selection (see *Correspondence* vol. 15, letter to A. R. Wallace, 23 February 1867). Wallace suggested that Weir might conduct experiments in his aviary to test Wallace's hypothesis that bright colours served as protection (see *Correspondence* vol. 15, letter from A. R. Wallace, 24 February [1867]). CD and Wallace continued to discuss the problem in 1868 (see *Correspondence* vol. 16).

4 CD's annotations are notes for his reply to Wallace. See letter to A. R. Wallace, 22 March [1869].

From George Cupples 11 March 1869

The Cottage | Guard Bridge | Fifeshire. N.B
(Thursday) March 11/69

Dear Mr Darwin,

I was going to write further to you at any rate, and to ask whether additional observations this season, by myself and others, might not be still of use. Of this I shall speak after answering your question.[1]

I *do not* think that at 5 weeks old (or even so soon as 5 *months* old) the difference in stature between puppies of the opposite sexes is equal or appears equal to the difference when they are adult. This, I think, is the question. I must answer it more fully, however. 1st I am confident that, although there is probably then a difference, invariably in favour of the male in all deerhound dogs—yet it does not show itself on the average so decidedly at that early age, in stature, length, strength, or weight—I take these all together. Hence, so far as this conviction is traversed by the record of weights I gave, I should say those weights are probably to be viewed as exceptional, accidental, and insufficiently-extended data. The record was the first of the kind I ever made— it is not easy to weigh puppies, because one is afraid to hurt, expose, or risk their falling, when inexperienced in the process, and not provided with all necessary apparatus for weighing. The *mother* & the *nurse-bitch* are to be considered. The *mother* may be worth laying stress upon— She was herself the smallest bitch in a litter of seven (4 dogs & 3 bitches)— she always "throws" large dogs and small bitches very obviously (her sisters do so not quite so obviously, in one case when the same dog was father.) She (the mother of the pups weighed), I may mention, is one of the bitches whose case I gave you as strongly marking sexual *preference*—notwithstanding the fact about the Russian interloper, she evinces a marked preference in every case for the large Deerhound (*in his prime*) to whom she was first put, and who has all along been concerned in her various litters, in fact never *failing* even when admitted secondarily (it seems.)[2]

2d. The adult difference of stature &c—may be pretty correctly but roughly stated from all experience and testimony, thus—

(in modern times)
Height of male ranging from 27 inches (very low) to 33 or 34 inches (rare)
— female — from 23 in (*not* abnormal) to 27½ or 28 uncommon
one female only in all the pedigrees I know of, stood 30 inches
Length— in proportion.
Girth of chest &c. in more than proportion favourable to the male, I should say, from exercise
Weight of male from 80 lbs (low) up to 120 lbs or considerably *more*
— female from 50 lbs to 70 or 80 at farthest.

These figures might be brought to *accuracy*, if it were required, with *time* given me.

3. The male grows in stature till from 12mos. to 18mos. in length similarly— In weight &c.—from 18 to 24mos. The female grows in stature till from 9mos to 14 or 15mos in length similarly. In weight & general size from till 12mos. to 15mos.

This, I think, is as correct as any amount of opinion & judgment (apart from scientific statistics) would give it. And for *general* purposes Nos. 2 & 3 may be depended upon.

The *table* meanwhile should perhaps yield to these.

As to further observations proposed or obtainable:—

In regard to *weights of puppies, male & female, at same age*— I am confident that *Mr Wright*[3] has a series of records. I shall therefore endeavour to get these from him at once, if your time allows them yet to come in to consideration.

I could obtain records of the same kind, I believe, from a very careful observer (in Gloucestershire, a friend of Mr Wright's and mine)[4]—in regard to Cross-bred puppies, however. They are specially bred with a view to reproducing a large Wolfhound strain (like the old Irish)—on the stock of the Scotch deerhound.

I could myself, during the whole course of this year, take careful comparative weights from the litters of no less than *five* Deerhound bitches of my own—to be put to *three* or four different dogs (Deerhound) of my own. If this is not too long a time to talk of, the result would be perfectly reliable, now that I have begun.

At present I have two *dog*-puppies alone, between 2 & 3 weeks old, from the bitch whose former puppies were weighed. The others will come successively, from summer to autumn. (The rest of the present litter died when newly-born—females in a minority of 3 to 4 dogs this time, as before.)

I have just now two pure Newfoundland puppies, dog & bitch, of one litter—3 to 4 months old. I could weigh and measure them weekly. I could also command measurements on the part of two or three gentlemen who have got bitches from me.

I may be able to find out yet about Mr Barr's[5] statistics of male & female numbers—and I might obtain other statistics in this direction. Would it be of any value to have cattle statistics here—and those of sheep, pigs &c.? Such could perhaps be had. Where *prize* poultry are bred, would *their* statistics be of any service?

I feel strongly how puerile from a scientific point of view the various statements are, which the desire to serve has led me to offer. Such as they are, however, they are free from bias—as I have of course no notion of your drift or analogical anticipation. Should it at any time occur to you that a statement of your expectation or *question* might surely & usefully be proposed. I have only to say that I think I should now be able to look-out through it with the most religious integrity, and to give candid replies. Sometimes this seems to help the inexpert. But I am aware you cannot wait till observers have grown up round you.

Pray believe me ever truly yours | George Cupples

Charles Darwin, Esq! | &c &—

[Enclosure]

Note of comparative weights of 7 Deerhound puppies (4 Dogs & 3 bitches) as taken week by week from the age of 2 weeks to that of 7 weeks. With additional note of weights—and supplementary statement &c.

The mother of these puppies is a Scotch Deerhound bitch, rather small—about 24 inches high at top of shoulder-blade. The *paternity* requires remark— it was an undoubted case of there being *two* fathers to the litter—the one being a Scotch Deerhound (1st. prize breed) above 30 inches high at shoulder—the other a young Russian deerhound or wolfhound (understood to be virtually the same race of dogs) his height then $27\frac{1}{2}$ inches—since grown to 28 inches. From the Scotch Deerhound father there were unquestionably 5 puppies (3 dogs & 1 bitch) from the Russian 2 puppies (dog & bitch) marked as such in the table. The four dogs were left to be suckled by the mother—the three bitches were transferred to a foster-mother, a colley in good condition, whose own puppies were removed, so that the bitches may be supposed to have been on at least equal terms with the dogs.

Born May 12th. 1868

Weight on May 26.	Dogs	No. 1. —	5lbs	14oz.	1
when 14 days old.		No. 2 —	5.	8.	
		No 3 —	5.	4	
		No 4 —	5.	4* from the Russian Sire.	
	Bitches	No 1 —	5lbs	4oz	
		No 2 —	4.	14.	
		No 3 —	2 –	6* from the Russian sire.	
Weight on morning of June 3d when 3 weeks old	Dogs	No. 1 —	7 –	4	2
		2 —	6 –	14	
		3 —	6 –	$13\frac{1}{2}$	
		4 —	6 –	6*	
	Bitches	No 1 —	5 –	14	
		No. 2 —	5 –	13	
		No 3 —	2 –	12*	
Weight on evening of June 10th 4 weeks old	Dogs	No 1. —	9 –	6oz	3
		No 2 —	9 –	4	
		3 —	9 –	—	
		4 —	8 –	10*	
	Bitches	1 —	7 –	4	
		2 —	7 –	4	
		3 —	3 –	8*	
Weight June 17th when 5 weeks old	Dogs	No. 1 —	9lbs	6oz	4
		– 2 —	9	$5\frac{1}{2}$	
		– 3 —	9	2	
		– 4 —	9	2*	
	Bitches	No. 1 —	5	14	
		2 —	6	14	
		3 —	3	9*	
Weight June 24th. 6 weeks old	Dogs	No 1 —	10	$6\frac{3}{4}$	5
		2 —	10	$6\frac{1}{2}$	
		3 —	10	$6\frac{1}{4}$	
		4 —	10	6*	

	Bitches	Nº 1 —	7	½ oz	
		2 —	7	¾	
		3 —	3	12*	
Weight July 2d. when fully 7 weeks old	Dogs	Nº. 1 —	11lbs	2oz	6
		2 —	11	12	
		3 —	10	10	
		4 —	10	6*	
	Bitches	Nº. 1 —	8	8	
		2 —	7	12	
		3 —	4	2*	

At the same time with these were being weighed 3 bitch puppies of another litter, (the whole litter consisting of 4 bitches) a fortnight older. (A)

when a fortnight old they weighed as follows

all Bitches	Nº. 1. —	4lbs	10¾oz	compare with 1st. weighing of *Dogs* above
	— 2 —	3	11	
	— 3 —	3	8	

About same time Mr Wright[5] (of Yeldersley) had a litter of 3 bitches, which he weighed at a fortnight old—as follows–

Nº. 1 —	5½lbs —	These were the largest he ever bred
2 —	5lbs	(bitches) and have grown to be large,
3 —	5lbs	but not more than 27½ inches high or 28.

weight of my litter of bitches, marked (A) on July 2d. when fully 9 weeks old, weighed along with the dog-puppies 7 weeks old

Nº 1 —	13lbs		(B)
2 —	12lbs	10oz	
3 —	10	12	

weight comparative of Nº. 1 bitch in (A). Nº 1 bitch weighed 57 lbs. height 25½ inches (of the litter A)

weighed Decr. 22/68—with height
as against

(C)

Dog-pup Nº 1 weighed same day–with height

Nº. 1 dog weighed 72 lbs. height 26½ inches.

and

Dog-pup Nº 4 weighed same day with height

Nº 4 dog (*Russian) weighed 64 lbs height 26½ inches

Additional remarks.

Having unfortunately taken no previous notes of the sort, I cannot say more myself about the matter of the relative size & growth of the sexes. Mr Macneil's[6] statements may be depended on, so far as they go. Mr Wright (John Wright, Esqr., Yeldersley House, near Derby) is of all men I know of, the likeliest to have made careful & continued notes on these points—indeed I believe he has various such memoranda.

Cross-bred families of dogs seem to differ less in size sexually than pure races do. But Mastiffs, Newfoundlands, &c—do not differ as deerhounds do in this respect.

The difference in deerhounds seems to be similar to that in the human race, however. *[Quere]* In the South Sea Islands, where there are very large men, do we find that the women are large? I think not. So among some of the finest & largest races of Indians (North American)—see the discovery of a Northwest Passage by Land, published of late years.[7]

Mr Mcneil speaks of the deerhound bitch not being used much in the field, except by way of experiment—viz—to judge of her quality & spirit. N.B. All Mr Macneil's remarks refer to *Coursing deer unwounded*—not to modern deer-stalking.

The comparative prominence of the Deerhound dog may be traced far back in a legendary way. The celebrated dogs of Ossian &c were *male*— Bran. Luath, Phorp. &c. but the bitch *Sceolan* is mentioned, as mother of Bran. + + In the mythical accounts of the first coming of the Scoti or Milesians to Ireland from the East, the "greyhound" (bitch) *Partholan* is specially associated with the event. + + In one of the old Irish stories about Ossian's return to earth, or revival from trance, after the fall of the Fenian heroes—is a curious illustration of the fact that the old hunters must have been rigorous in their *selection* of puppies for the chase. Ossian is represented as taking up successively a litter of "greyhound" puppies which he sees lying, and he dashes each against a rock, to try which will bear it best, or survive the test, so as to become a good dog on the hill if reared to maturity.—[8] So much for the slight traces backward as to Deerhound peculiarity.

Richardson, a well informed writer on dogs, says of crosses between greyhound & foxhound.[9] "I have seen smooth deerhounds in Scotland— & merely a cross between ordinary greyhound and foxhound. In such case it is better that the greyhound should be father, as more likely to obtain size & power, combined with swiftness. This is more particularly to be attended to when it is the rough greyhound to which you resort, for amongst all the rough greyhounds, and more especially those of Ireland and Scotland (?) there exists a greater disparity of size between male and female than in the case of any other member of the canine family. E.g., of a litter of pups—a dog shall grow to the height of, say, 30 inches at the shoulder—and not a female of the same litter shall exceed 24 inches."

The preceding statement of the disparity is not exaggerated—it is, I think, a literally correct one. But I should suppose that it extends through all the real (old) greyhound family (from which at all Civilized periods men seem to have bred a factitious *smooth* greyhound by crossing and acclimatization.) I have reason to see that it does in the Russian or Siberian greyhound (wolfhound or deerhound)—e.g. the puppies N°. 4 of the dogs, and N°. 3 of the bitches, same litter. The dog is now twice the size of the bitch—and at 14 days old, he appears to have been twice the weight.

Of the numerical proportions of the sexes born.

My own limited experience cannot go for anything here, I should say. On the whole, my *impression* has been that I have had more females born.

N.B. Most of the bitches are drowned—they are almost invariably the devoted victims to necessity and Malthusian principle. This, in one way, furnishes a kind of guide to the memory. All races of dogs are subject to the same rule in this respect.

Mr Barr's statement—out of 117 litters, the females were in excess.[10] Mr Barr was quite unbiased in this statement, so far as I am concerned— since I gave him no idea of my own impression. He is a careful man, zealous about dogs—still I think it unlikely that he had made any detailed record—and the difficulty is, that he has left Manchester, owing to illness, to live abroad, as I understood. I have not his present address, if he is still alive—but shall try to find out.

Mr Wright is, I think, the likeliest here also. He takes the most rigorous view of the extent of statistics necessary for any conclusion on such a point. I will write to him about it—but if Mr Darwin himself were to do so direct, in the first place, it would be the surer way.

Greyhound (Turf) records would be probably the most capable of affording correct light on the question.

An Advertisement, *repeated*, in the Field, with a special notice from the Editor—which could be had, I think—might have a useful effect, and bring in statistics from quarters otherwise inaccessible.[11]

If *Deerhounds* are *particularly* wanted in this view, I *might* find out something, and shall take means to do so.

DAR 83: 133–8, DAR 161: 282

CD ANNOTATIONS
1.1 I . . . question. 1.3] *crossed blue crayon*
2.1 I . . . puppies 2.2] *scored blue crayon*
3.1 but . . . uncommon 3.5] *double scored blue crayon*; 'correct' *in margin blue crayon*
8.2 *cross in margin blue pencil*
10.1 'If you felt interest yourself' *beginning of paragraph pencil*
10.1 *cross in margin blue crayon*
11.1 At . . . me. 12.3] *crossed pencil*
13.1 *two crosses in margin blue crayon*
14.1 *cross in margin blue crayon*
Enclosure: Additional remarks
1.1 Having . . . years. 3.5] *crossed blue crayon*
1.2 Mr Macneil's . . . go. 1.3] *scored blue crayon*
5.1 *three crosses in margin blue and red crayon*
5.2 The celebrated . . . the event. 5.5] *double scored blue crayon*
5.8 Ossian] *after opening square bracket blue crayon*
5.8 Ossian is . . . peculiarity. 5.12] *double scored blue crayon*
6.1 *Richardson* . . . weight. 7.8] *crossed blue crayon*
Enclosure: Of the numerical proportions of the sexes born
1.1 My . . . so. 7.2] *crossed blue crayon*
5.2 *cross at end of paragraph blue crayon*

1 CD's letter has not been found, but an envelope addressed to Cupples with a postmark of 9 March 1869

and annotated by Cupples 'rec'd March 11 | much abt dogs' is in the collection of the American Philosophical Society.

2 See *Correspondence* vol. 16, letter from George Cupples, 11–13 May 1868 and n. 6.

3 John Wright.

4 Cupples's friend from Gloucestershire was George Augustus Graham (see letter from George Cupples, 20 June 1869).

5 J. G. R. Barr.

6 Archibald McNeill.

7 See Milton and Cheadle [1865], p. 335.

8 Cupples refers to dogs mentioned in Celtic mythology. The Milesians were the fifth and last Celtic people to invade and settle in Ireland, purported to be descendants of Míl Espáine and Scota. Ossian or Osín was a warrior and poet, the narrator of the 'Fenian cycle' of Irish mythology (Briggs [1976], MacKillop 1998). In *Descent* 2: 262, CD cited Cupples on ancient references to dogs.

9 See H. D. Richardson 1847, p. 42. Cupples's transcription of the passage contains an omission and some minor errors.

10 See *Correspondence* vol. 16, letter from J. G. R. Barr to George Cupples, [after 11 May 1868?].

11 CD began gathering information on sex ratios in February 1868 and asked the editor of the *Field*, William Bernhard Tegetmeier, to insert a request for information on the proportion of the sexes in domestic animals (see *Correspondence* vol. 16, letter to W. B. Tegetmeier, 11 February [1868]). The request appeared in the 22 February 1868 issue of the *Field*, p. 144. Tegetmeier stated that the request had come from CD, and that it was made 'in the interests of scientific research'.

From J. D. Hooker 11 March 1869

Royal Gardens Kew
March 11 /69.

Dear Darwin

That is a capital letter of Fritz Mueller's, many thanks for it.—[1]

We were very glad to hear of you,—but this is bad news of Henrietta— give her our love, & best wishes for her recovery.[2] I hope you are not anxious about her. I am delighted to hear of the Orchis-book, if that does not goad them into electing you to the Academy I do not know what will— I would not make many additions, it is very full as it is—[3]

As for me I am working like an "Injine" but not to much scientific purpose— Firstly I am threatened with being sent to S<u>t</u>. Petersburgh by Govt. to represent British Botanists & Horticulturalists (God help them) at the approaching Congress which the Emperor has taken up.—[4] I hate the sort of thing—but shall have to go.—about 12th May.— Consequently I am mugging up French as hard as I can, with a French *Baron*! for 2 hours from London who gives leçons on pronunciation,—with French novels with my wife, & French talk with the "spirituelle" Miss Symonds, who has been 10 days here & has captivated us entirely—[5] Then I am working away at this British Flora,[6] & lastly getting 3 months ahead with my current duties, with the view of extending my travels from St. Petersburgh for 2 months to the S.E. perhaps—& picking myself up somewhere.

Have you read Huxley upon Protoplasm?.[7] & what do you think of it— I am on the look out for his G. Soc. Address, which Lyell says is "something of extraordinary" (you see I am getting on in French!)[8]

We are all well— no news of Willy yet— Charlie is launched at Marlborough & takes a good place, which he probably will not keep—[9]

We fertilized a large *Aucuba*[10] last year with the pollen of **various** varieties (or species) & there appear on the bush here & there, besides the common form of fruit

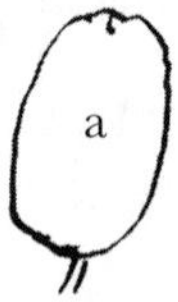

, a few of another form

whose pollen was amongst that used There are none of intermediate form & *b* is still green, though all *a* is ripe I will try to have definite experiments made.

Ever yrs affec | J D Hooker

DAR 103: 10–11

1 In his letter to Hooker of 8 March [1869], CD enclosed the letter from Fritz Müller, 12 January 1869.

2 Henrietta Emma Darwin had been ill for a couple of weeks (see letter to J. D. Hooker, 8 March [1869] and n. 4).

3 Hooker refers to the proposed translation into French of *Orchids* and to the Académie des Sciences (see letter to J. D. Hooker, 8 March [1869] and nn. 5 and 6).

4 The International Horticultural Exhibition at St Petersburg opened on 16 May 1869. For reports on the exhibition, including reports on the attendance of Emperor Alexander II, see *Gardeners' Chronicle* (1869): 553–4, 582, 586–8, 609–10, 614. Hooker did not go as a government representative because the government was not willing to pay expenses. For more on Hooker's trip, see L. Huxley ed. 1918, 2: 85–9, and the letter from J. D. Hooker to James Hector, 23 April 1869 (Yaldwyn and Hobbs eds. 1998, pp. 109–10).

5 Hooker refers to Frances Harriet Hooker and Hyacinth Symonds.

6 In his letter of [28 November 1868] (*Correspondence* vol. 16), Hooker had informed CD that he was going to write a 'British Flora' adapted to students' purposes. *The student's flora of the British Islands* was published in 1870 (J. D. Hooker 1870).

7 Thomas Henry Huxley's article, 'On the physical basis of life' was published in the *Fortnightly Review*, 1 February 1869 (T. H. Huxley 1869a). Huxley equated 'protoplasm' with the physical basis of life (T. H. Huxley 1869a, p. 129).

8 Hooker refers to the anniversary address given by Huxley in his capacity as president of the Geological Society of London on 19 February 1869 (T. H. Huxley 1869c). In the address, Huxley challenged the accuracy of William Thomson's calculations of the cooling of the earth. Hooker also refers to Charles Lyell.

9 William Henslow Hooker had been sent to New Zealand for his health (see *Correspondence* vol. 16, letter from J. D. Hooker, 30 August 1868, and letter to J. D. Hooker, 26 November [1868] and n. 8). He left England on 11 November 1868 on the *Matoaka*, which arrived in New Zealand on 8 February 1869 (Yaldwyn and Hobbs eds. 1998, pp. 102, 108 n. 7). Hooker also refers to Charles Paget Hooker and Marlborough College, Wiltshire.

10 *Aucuba* is a genus of dioecious shrubs in the family Cornaceae (Mabberley 1997).

From T. H. Huxley 11 March 1869

Jermyn St
March 11th 1869

My dear Darwin

I "know quite enough about Mr Vernon Lushington to have paid every attention to what he has to say", even if you had not been his ambassador[1]

I glanced over his letter when I returned home last night very tired with my two nights chairmanship at the Ethnological & the Geological Societies—[2]

Most of it is fair enough, though, I must say not helping me to any novel considerations— Two paragraphs, however, contained opinions which Mr Lushington is at perfectly liberty to entertain; but not, I think, to express to me—

The one is, that I shaped what I had to say at Edinburgh with a view of stirring up the prejudices of the Scotch presbyterians (imagine how many presbyterians I had in my audience!) against Comte[3]

The other, is the concluding paragraph, in which Mr Lushington recommends me to "*read Comte*"—clearly implying, that I have criticized Comte without reading him—

You will know how far I am likely to have committed either of the immoralities thus laid to my charge—

At any rate, I do not think I care to enter into some direct relations with any one who so heedlessly & unjustifiably assumes me to be guilty of them— Therefore I shall content myself with acknowledging the receipt of Mr Lushington's letter through you

Yours very faithfully | T. H. Huxley

DAR 166: 317

1 See letter to T. H. Huxley, 10 March 1869 and n. 2. Lushington's letter has not been found.

2 The Ethnological Society of London met on 9 March 1869 (*Journal of the Ethnological Society of London* n.s. 1 (1868–9): ix). The Geological Society of London met on 10 March 1869 (*Quarterly Journal of the Geological Society of London* 25 (1869): 235). Huxley was president of both societies in 1869.

3 The paper that appeared in the *Fortnightly Review* for 1 February 1869 (T. H. Huxley 1869a) was based on a talk that Huxley had given in Edinburgh on 8 November 1868 in which he criticised the philosophy of Auguste Comte (see letter to T. H. Huxley, 10 March 1869 and n. 2).

From R. F. Cooke 12 March 1869

50A, Albemarle Street, London, W.
March 12 *1869*

My dear Sir

The binder played me false yesterday & we have only received a supply of copies this afternoon & we have now sent you to *Orpington Station*, 9 copies—3 to Mr Dallas— & 1 to Spence Bate.[1]

The copies to Reviews go out also this day.[2] I have sent a few more, than the 7 you named. It is too late I fear to alter the lettering, but I wrote to the binder to do so, if in time.[3]

Yours faithfully | Robt Cooke

C. Darwin Esq

DAR 171: 367

1 Cooke refers to William Sweetland Dallas, Charles Spence Bate, and copies of Dallas trans. 1869 (see letter to R. F. Cooke, 9 March [1869]).

2 For the list of journals that received review copies of Dallas trans. 1869, see the letter to R. F. Cooke, 23 [February 1869].
3 Cooke refers to the lettering of the spine of the book (see letter to R. F. Cooke, 9 March [1869], and letter from R. F. Cooke, 10 March 1869 and n. 2).

To T. H. Huxley 12 March [1869][1]

Down. | *Bromley.* | *Kent. S.E.*
Mar 12.

My dear Huxley

I read Lushington's letter in a rather hurried manner, as if it had been a printed essay & not addressed to you; & I suppose in consequence of this the offensive passages did not strike me.[2] Now that they are pointed out I fully agree that he had no right to address them to you; & this I have now told him.[3] Assuredly if I had noticed these passages I sh^d^ have refused to be the channel of communication. I suspect when Mr Lushington urges you to *read* Comte[4] he meant to express to study & digest his works. Red-hot disciples do not weigh their words when their master is attacked, & on this plea I hope you will make some allowance. I feel sure that I need not tell you that you are about the last man in England towards whom I w^d^ consciously offer myself as a channel of offence.

This note requires no answer.

ever yours most truly | Charles Darwin

LS
Imperial College of Science, Technology, and Medicine Archives (Huxley 5: 264)

1 The year is established by the relationship between this letter and the letter from T. H. Huxley, 11 March 1869.
2 CD refers to Vernon Lushington. See letter from T. H. Huxley, 11 March 1869 and n. 1.
3 See letter to Vernon Lushington, [12 March 1869].
4 Auguste Comte.

To Vernon Lushington [12 March 1869][1]

My dear M^r^. Lushington.

I have heard from Prof. Huxley who *[3 words illeg]* desires me to acknowledge for him the receipt of your letter.[2] He thinks that although you have at full liberty to hold any opinion in regard to him, you have no right, & now that it is pointed out, I agree with him, to imply in a letter to him firstly imply that he has criticised Comte without having read him, & secondly that he had stated what he had to say with a view to stir up the prejudices of the S. presb^s^.[3] I have said in answer that I presumed you meant by reading Comte, carefully studying & digesting his works.— — I feel sure that you did not wish to offend Huxley, as this w^d^ have prevented your influencing his opinions.—

I was glad to read your letter, as I wished to see a considered statement of his claims. No doubt the law of progress from the theological to the positive point of view, is an important one, if true on which I cannot judge, & I sh[d] think the attempt to reduce the social system to a science state seems important.— On one point I have for many years vehemently /protested/ in my own /mind/, viz against the schemes of suggesting subject for me to follow, I cannot conceive any scheme better adapted for stopping originality & great discoveries— I remember in consequence doubting, whether Conte however able could have the mind of a discoverer. This will appear as a dreadful piece of heresy in your eyes.—

Pray believe me | Yours very sincerely | C. D.

Draft
DAR 96: 74

1 The date is established by the relationship between this letter and the letter to T. H. Huxley, 12 March [1869].
2 CD refers to Thomas Henry Huxley. See letter from T. H. Huxley, 11 March 1869.
3 CD refers to Auguste Comte and 'Scotch Presbyterians' (see letter from T. H. Huxley, 11 March 1869 and n. 3). For more on the reception of Comte in Britain, see Kent 1978.

To W. C. Tait 12 and 16 March 1869

Down, Bromley, Kent. S.E.
March 12, 1869

Dear Sir

I have received your two letters of Mar. 2nd and 5th; and I really do not know how to thank you enough for your extraordinary kindness and energy.[1] I am glad to hear that the inhabitants notice the power of the Drosophyllum to catch flies, for this is the subject of my studies. I have observed during several years the manner in which this is effected, and the results produced in several species of Drosera, and in the wonderful American Dionæa, the leaves of which catch insects just like a steel rat-trap.[2] Hence I was most anxious to learn how the Drosophyllum would act, so that the Director of the Royal Gardens at Kew wrote some years ago to Portugal to obtain specimens for me, but quite failed.[3] So you see what a favour you have conferred on me. With Drosera it is nothing less than marvellous how minute a fraction of a grain of any nitrogenised matter the plant can detect; and how differently it behaves when matter not containing Nitrogen of the same consistence, whether fluid or solid, is applied to the glands. It is also exquisitely sensitive to a weight of even the $\frac{1}{70,000}$ of a grain. From what I can see of the glands on *Drosophyllum* I suspect that I shall find only the Commencement, or nascent state of the wonderful capacities of the *Drosera*, and this will be eminently interesting to me. My MS on this subject has been nearly ready for publication during some years, but when I shall have strength and time to publish I know not.[4] And now to turn to other points in your letter: I am quite ignorant of ferns and cannot name your specimen.[5] The variability of ferns passes all bounds. With respect to your Laugher Pigeon, if the same with the two sub-breeds which I kept, I feel sure from the structure of the skeleton, &c. that it is a descendant of

C. Livia.[6] In regard to Beauty I do not feel the difficulty which you and some others experience. In the last Edit. of my *Origin* I have discussed the question, but necessarily very briefly.[7] A new and I hope amended edit. of the *Origin* is now passing thro' the press, and will be published in a month or two, and it will give me great pleasure to send you a copy.[8] Is there any place in London where parcels are received for you, or shall I send it by post? With reference to dogs' tails no doubt you are aware that a rudimentary stump is regularly inherited by certain breeds of sheep dogs, and by manx cats.[9] You speak of a change in the position of the axis of the earth: this is a subject quite beyond me, but I believe the Astronomers reject the idea. Nevertheless I have long suspected that some periodical astronomical or cosmical cause must be the agent of the incessant oscillations of level in the earth's crust. About a month ago I suggested this to a man well capable of judging but he could not conceive any such agency; he promised however to keep it in mind.[10] I wish I had time and strength to write to you more fully. I had intended to send this letter off at once, but on reflection, will keep it till I receive the plants.

With the most sincere thanks for your very great kindness I remain dear Sir | yours very faithfully | Charles Darwin

P.S. 2nd. Did you ever see my little Book on the Fertilisation of Orchids by Insect Agency: the facts are really curious. If you have it not, shall I send you a copy? Do you think you would care about the Nat. Hist. and wonderful spontaneous movement of Climbing Plants. I wrote a little pamphlet on this subject— should you care for it?[11]

March 16th

My dear Sir

I have been anxiously expecting the plants for several days, but I suppose the ship was delayed by northern winds. At last they have just arrived; but most unfortunately the weather for the last week has been very cold, and the soil in the pots had become as dry as dust on a road, and the poor plants have suffered greatly. Some, however, look so green in the centre that I fully expect that they will recover; if only 2 or 3 leaves will grow, and the glands secrete the viscid fluid, I shall be able to make my observations, to which I look forward with great interest, for the whole structure is very different from Drosera.— I have had long consultations with my gardener,[12] & we mean to treat them in different ways, & carefully observe what suits best.— I have hopes of deep pot kept dry in upper part, but standing in shallow saucer with water. I have put them in greenhouse, & will then try a warmer house on a few, & so on, for I have 4 Houses of different temperatures. I do hope that some will revive: if 2 or 3 revive I will give one to the Royal Bot. Garden at Kew, where it will be treasured. How wonderfully kind you have been, & what trouble you have taken in packing them. Not one pot broken!! This dreadful cold frost has been most unfortunate.—

Yours *most truly* obliged | C. Darwin

Copy, LS(A)[13]
DAR 147: 541; Sotheby's London Sale (19 July 1990)

1 See letters from W. C. Tait, 2 March 1869 and 5 March 1869.

2 CD had carried out a series of experiments on *Drosera* and *Dionaea* between 1860 and 1862 (see *Correspondence* vols. 8–10); however, he decided to postpone this line of research until after *Variation* was completed (see *Correspondence* vol. 10, letter to Edward Cresy, 15 September [1862]). The genus *Dionaea* has only one species, *Dionaea muscipula*, the Venus fly-trap.

3 For CD's request to Joseph Dalton Hooker, see *Correspondence* vol. 12, letter to J. D. Hooker, 26 November [1864].

4 There is an undated rough draft in the Darwin Archive–CUL (DAR 61), with the title 'On the movement of the leaves of the Drosera; & on their power of detecting nitrogenous compounds'.

5 See letter from W. C. Tait, 2 March 1869.

6 See letter from W. C. Tait, 2 March 1869 and nn. 14 and 15. CD confused the laugher pigeon (a variety of *Columba livia*) with the bird Tait had mentioned, the laughing dove (*Streptopelia senegalensis*).

7 See letter from W. C. Tait, 2 March 1869 and n. 10. CD had considered the problem of beauty in *Origin* 4th ed., pp. 238–40.

8 The fifth edition of *Origin* was published in May 1869 (*Publishers' Circular*, 1 July 1869, p. 386).

9 See letter from W. C. Tait, 2 March 1869.

10 See letter from W. C. Tait, 2 March 1869. CD had written to James Croll, 'I wish that you would turn your astronomical knowledge to the consideration whether the form of the globe does not become periodically slightly changed, so as to account for the many repeated ups & downs of the surface in all parts of the world' (see letter to James Croll, 31 January [1869]).

11 CD refers to *Orchids* and *Climbing plants.*

12 CD probably refers to Henry Lettington.

13 Half of the last paragraph of the letter, from 'interest' (line 7), and the valediction, have been transcribed from a facsimile in a sale catalogue; the rest has been transcribed from a copy made for Francis Darwin (DAR 147: 541).

To Fritz Müller 14 March 1869

Down. | Bromley. | Kent. S.E.
March 14th 1869

My dear Sir

I fear that your patience will have been completely exhausted; but at last the Translation is published.[1] The delay has been caused by the Translator; but he is not to blame for he had to move a large family several hundred miles & entered on his new office at the busiest time of the year, so that it was really impossible for him to do anything.[2] I have this day sent off 3 copies to you by Book Post.

I do hope that you will be contented with the appearance & with the Translation. I have not read it yet, but Dallas generally translates well & was greatly interested with your work.— I have sent copies "From Author" to C. Spence Bate, Dana in U. States, your Brother H. Müller, Max Schutz of Bonn & Oscar Schmidt at Gratz.—[3] I am ashamed to say that I cannot find your instructions about the presentation copies, though they are somewhere safe, so I wrote to your brother & he suggested the two last names.— If you will write I will send copies to anyone else & of course more to yourself. I have sent copies to 7 of such reviews, as generally treat of Scientific works.—[4]

My publisher[5] can form no idea whether the book will sell, but fears it is too purely scientific for England. There will be 1000 copies printed.

I received some time ago a very interesting letter from you with many facts about Oxalis, & about the non-seeding & spreading of one species.—[6] I may mention that our common *O. acetosella* varies much in length of pistils & stamens, so that I at first though it was certainly dimorphic, but proved it by experiment not to be so.—[7] Borreria has after all seeded well with me when crossed by opposite form, but very sparingly when self-fertilised.[8] Your case of Faramea astonishes me.— Are you sure there is no mistake— the difference in size of flower & wonderful difference in size & structure of pollen-grains naturally makes one rather sceptical.[9] I never fail to admire & to be surprised at the number of points to which you attend.

I go on slowly at my next book & though I never am idle, I make but slow progress, for I am often interrupted by being unwell & my subject of sexual selection has grown into a very large one. I have, also, had to correct a new Edit. of my "Origin", & this has taken me six week, for science progresses at Rail-road speed.[10] I cannot tell you how rejoiced I am that your Book is at last out; for whether it sells largely or not, I am certain it will produce a great effect on all capable judges, though these are few in number.

Believe me | Yours sincerely & cordially | Ch. Darwin

P.S. I have just received your letter of Jan. 12th.— I am greatly interested by what you say on Escholzia— I wish your plant had succeeded better. It seems pretty clear that the species is much more self-sterile under the climate of Brazil than here, & this seems to me important result.—[11] I have no spare seeds at present, but will send for some from nurseryman, which though not so good for our purpose will be worth trying. I can send some of my own in autumn You could simply cover up *separately* 2 or 3 single plants, & see if they will seed without aid,—mine did abundantly.—

Very many thanks for seeds of Oxalis;[12] how I wish I had more strength & time to carry on these experiments, but when I write in morning; I have hardly heart to do anything in the afternoon. Your grass is **most** wonderful.— You ought to send account to Bot. Zeitung.[13] Cd. you not ascertain whether the bracts are sensitive, & how soon they become apical in the bud?— —

Your Bird is, I have no doubt the Molothrus, mentioned in my Journal of Travels p. 52, as representing a N. American species, but with Cuckoo-like Habits.—[14]

I knew that seeds from same spike transmitted to a certain extent their proper qualities; but as far as I know no one has hitherto shown how far this holds good; & the fact is very interesting. The experiment wd be well worth trying with flowers bearing different numbers of petals.— Your explanation agrees beautifully with the hypothesis of Pangenesis, & delights me—[15] If you try other cases, do draw up a paper on subject of inheritance of separate flowers for Bot. Zeitung or some Journal.— Most men, as far as my experience goes, are too ready to publish; but you seem to enjoy making most interesting observations & discoveries & are sadly too slow in publishing.— Your case of the grass seems so wonderful, that I am sure Dr. Hooker will much like to read it & I will send it off at once.—[16]

With sincere admiration | Yours very truly | C. Darwin

P.S. I have sent one copy to Mr Wallace,[17] as I am sure he wd enjoy the work & cannot afford to buy many Books

P.S. I have just received following statement from M[r] Murray

	£ s d
Printing	32 . 0 . 0
Paper	16 .10 .
Translation & Stereotypes	21
Advertising	21 .10
Binding	34 ·
	125 . 0 . 0

Produce of 1000 copies at 6[d] each

	192
Expences	125
	67
Commission for Mr. Murray	20
Profit	47

(less, copies for Presentation, Reviews & Public Libraries as by Law)

(I am vexed to see that on Title my name is more conspicuous than yours, which I especially objected to & cautioned Printers, after seeing one Proof.—)

I see that the seed that the nurseryman has sent me is that of Escholzia crocea.— Perhaps my observations & yours have been made on distinct species!!!![18]

British Library (Loan 10:27) (by kind permission of English Heritage)

1 CD refers to *Facts and arguments for Darwin* (Dallas trans. 1869), the translation of *Für Darwin* (F. Müller 1864a).

2 William Sweetland Dallas had been curator of the museum of the Yorkshire Philosophical Society before his appointment as assistant secretary to the Geological Society of London in November 1868 (see *Correspondence* vol. 16, letter from W. S. Dallas, 27 November 1868).

3 CD refers to Charles Spence Bate, James Dwight Dana, Hermann Müller, Max Johann Sigismund Schultze, and Oskar Schmidt. CD also refers to Graz, Austria.

4 For the list of journals, see the letter to R. F. Cooke, 23 [February 1869].

5 John Murray.

6 See *Correspondence* vol. 16, letter from Fritz Müller, 31 October 1868 and n. 8.

7 CD's experiments on heterostyly in *Oxalis acetosella* (wood sorrel) were made in 1862; his observations were later published in *Forms of flowers*, p. 182.

8 Müller had sent CD a specimen of *Borreria* (a genus now subsumed within *Spermacoce*; see Mabberley 1997) in 1867; CD reported in August that seeds of *Borreria* had come up (*Correspondence* vol. 15).

9 See *Correspondence* vol. 16, letter from Fritz Müller, 31 October 1868. In *Forms of flowers*, p. 129, CD included an illustration by Müller of the different forms of flowers and pollen-grains in *Faramea*. What Müller described as possibly a species of *Faramea* was subsequently identified as *Rudgea jasminoides* from specimens Müller sent to the Royal Botanic Gardens, Kew (see Baker 1965, pp. 27–9).

10 CD refers to *Descent* and *Origin* 5th ed.

11 See letter from Fritz Müller, 12 January 1869.

12 Müller had probably sent the seeds of *Oxalis* with his letter of 31 October 1868 (*Correspondence* vol. 16) or with his letter of 12 January 1869.

13 See letter from Fritz Müller, 12 January 1869 and nn. 3–5. Müller did not publish his observations on *Streptochaeta* in *Botanische Zeitung*, but did publish some observations on *Streptochaeta* in 'Einige Nachträge zu Hildebrand's Buch: die Verbreitungsmittel der Pflanzen' (F. Müller 1885, pp. 440–2).

14 See letter from Fritz Müller, 12 January 1869 and nn. 6–8. For CD's description of *Molothrus niger* (now *M. bonariensis*) see *Journal of researches* 2d ed., p. 52.

15 See letter from Fritz Müller, 12 January 1869 and n. 10.

16 CD enclosed Müller's letter of 12 January 1869 with his letter to Joseph Dalton Hooker of 8 March [1869].

17 Alfred Russel Wallace.

18 Müller and CD had been studying the degree of self-sterility in *Eschscholzia californica* (the California poppy) and had exchanged seeds of this species (see letter from Fritz Müller, 12 January 1869 and nn. 1 and 2). *E. crocea* is now considered to be a variety of *E. californica.*

From Fritz Müller 14 March 1869

Itajahy, S[a] Catharina, Brazil
March 14, 1869

My dear Sir

I have received your kind letter of Novbr. 28 and am much obliged for it.[1] In my last letter I told you of a bird with cuckoo-like habits;[2] afterwards I remembered of the account, which you give in your 'Journal of Researches' on the *Molothrus niger* and I think, our bird will belong to the same or an allied species; at least it has also the habit of standing on the backs of cattle.[3]

I am now experimenting on a species of *Abutilon*, the seeds of which I had brought last year from the Rio Capivary,[4] and which I have found to be quite self-impotent. Only three plants have as yet flowered; all the flowers of the first two plants, when crossed, have produced or are producing fine pods, but two flowers of the first plant, fertilized with pollen of the third, have fallen off. I am very curious to know from further experiments, whether these two plants (1st and 3[d]) will continue to prove mutually sterile. In this case, the sterility would probably be owing to their being raised (as are all my plants) from seeds of the same pod.

In your book on Variation under Domestication you mention a remarkable plant of *Begonia frigida*, producing hermaphrodite flowers with inferiour perianth. I have lately found an analogous wild plant of another *Begonia*, which is here a common weed. In this plant all the male flowers show a strong tendency to become hermaphrodite, one, two or three of the central stamens being transformed more or less completely into pistils. No two of these male flowers appear to be exactly alike and almost every day affords a new and surprising modification.— Here are some cases:[5]

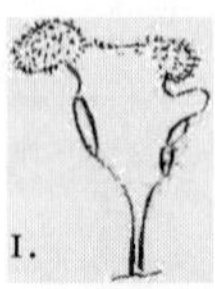

1. A single stamen modified; connectivum[6] much dilated; on either margin a short anthera with good pollen; at the end, well developed stigmatic papillæ.

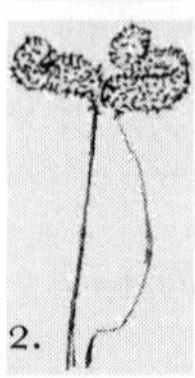

2. A single stamen modified; a well-developed stigma; neither anthers nor ovula.

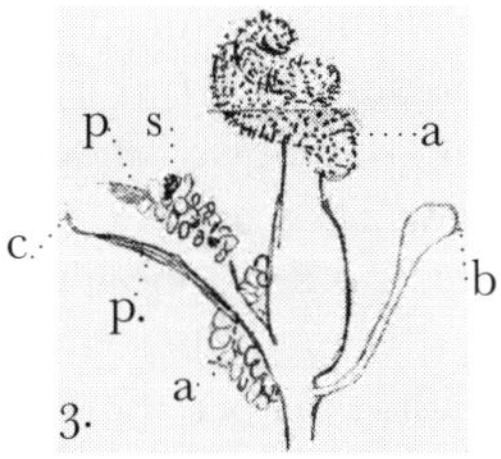

3. Three modified stamens, united at the base.

a. well developed stigma; no pollen; numerous ovula, differing in nothing from those of the normal ♀ flowers.

b. club-shaped, without pollen, ovula, and stigmatic papillæ.

c. pollen on both margins of the connectivum; ovula on the convex margin; apex of the connectivum smooth, without stigmatic papillæ, *but* one of the ovula transformed into a stigma!—

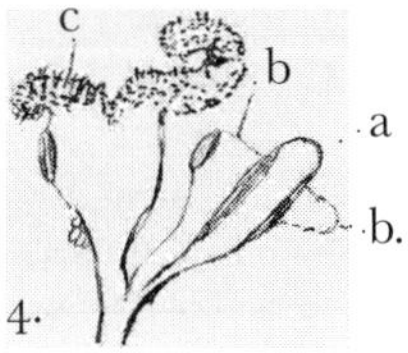

4. Three stamens united. *a.* not modified.

b. connectivum much dilated; pollen on either margin; neither ovula, nor stigmatic papillæ.

c. well developed stigmatic papillæ; pollen (a small quantity) on one margin alone of the much dilated connectivum a few ovula.

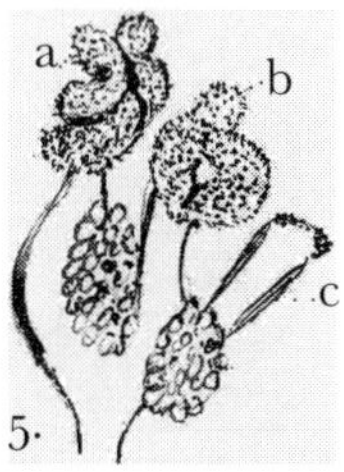

5. Three stamens, modified and united:

a and *b* without pollen, with large stigmas and numerous ovula.

c nearly normal, only the tip of the connectivum being somewhat enlarged and provided with small stigmatic papillæ.

Once I saw (fig. 3, 5), in the midst of the white ovula, a dark yellow body of a club-shaped form, having nearly the size of an ovulum, covered by club-shaped papillæ exactly ressembling in shape and colour those of the stigma, so that, in this case, an ovulum appeared to have been transformed into a stigma!—

Since I have found this plant, I have been looking out for others and yesterday I have, at length, met with a second specimen (growing within 2 yards distance from the first), which promises to offer still more curious modifications. Some of the male flowers of this second plant have been transformed completely into female ones with superior perianth, but distinguished from the normal ♀ flowers by the perianth having (as in the male flowers) two large broad outer and two small narrow inner segments (whilst the female flowers have 5 segments, 1 being smaller), and by their having 4–5 stigmas and as many alæ on the ovarium (the female flowers have 3.).— In one of these abnormal female flowers there were some naked ovula between the stigmas beside those included in the ovarium.— In the first plant all the ovula of the male flowers are naked.— There are some unripe pods on the second plant, all of which are produced by normal ♀ flowers; as soon as they are ripe, I shall send you seeds of this second plant also.

There is now flowering in my garden a *Salvia*, almost all the flowers of which are visited by large humble-bees, which in order to reach the honey make a hole at the base of the corolla. [*Specimen*][7] As some authors look at instinct as a force, which compels all the individuals of a species to act in the same way under the same circumstances, I was much interested by observing how differently the several humble-bees behaved. One humble-bee invariably alighted on the upper-lip and thence crawled down on the back of the corolla to the calyx, where it turned to the right hand; another always flew directly to the spot, where it had to open the tube of the corolla; a third behaved in different ways on different flowers.

Believe me, my dear Sir, | Yours very faithfully | Fritz Müller

Linnean Society of London (SP 859)

1 See *Correspondence* vol. 16, letter to Fritz Müller, 28 November 1868.
2 See letter from Fritz Müller, 12 January 1869 and n. 8.
3 See *Journal of researches* 2d ed., pp. 52–3, and letter to Fritz Müller, 14 March 1869 and n. 15.
4 Rio Capivary (now Capivari) is a small river, a tributary of the Rio Grande, in Bahia, Brazil.
5 See *Variation* 1: 365–6. Müller later published a fuller description of his observations in *Botanische Zeitung* (F. Müller 1870b).
6 The connectivum or connective is the portion of stamen distinct from filament that connects the lobes of the anther (Jackson 1900).
7 Müller pasted specimens of the flower next to his description.

From Ernest Faivre[1] 15 March 1869

le 15 Mars 1869

Monsieur,

Je Veux faire une courte Réponse a votre lettre du 21 Fevrier dernier, en laissant a mon élève, Mr Rérolle, le soin de la completer et d'Entrer ainsi en rapports directs avec vous[2]

La traduction de votre important travail sur les orchidées est décídée en Principe: elle aura certainement de l'importance, grâce aux notes que vous voulez bien promettre de faire Parvenir.

La question est maintenant de trouver un Éditeur, et j'espère qu'elle sera résolue vers le Quinze du Mois Prochain.

Aussitôt le traité fait et conclu, nous aurons l'honneur de vous en informer immediatement; et il est a partir de ce moment qu'il conviendra de se mettre en rapport avec votre Éditeur de Londres au sujet du Stereotypes et de vous parler des notes que vous avez bien voulu promettre.[3]

Je me réjouis d'avance, de pouvoir Contribuer, même d'une manière indirecte, a faire connaître en France une de vos productions les plus originales et les plus dignes d'Interêt, et je vous prie d'agréer, Monsieur, l'expression de mes sentiments de Respecteuse admiration. | Ernest Faivre

DAR 176: 129

1 For a translation of this letter, see Appendix I.

[2] CD's letter has not been found, but see the letter to R. F. Cooke, [22 February 1869]. Faivre had asked whether CD would agree to a translation of *Orchids* by Louis Rérolle (see letter from Ernest Faivre, 18 February 1869). See also the letter from Louis Rérolle, 17 March 1869, which was written on the second leaf of the same folio as this letter.

[3] CD had already written to his publisher, John Murray, about providing stereotypes (see letter to R. F. Cooke, [22 February 1869]). For more on CD's notes to the French translation of *Orchids*, see letter to J. D. Hooker, 8 March [1869] and n. 5.

To J. D. Hooker 17 March [1869][1]

Down. | *Bromley.* | *Kent. S.E.*
March 17th

My dear Hooker

This is a wonderful piece of news about St. Petersburgh, & I sh^d. think you would be sure to find it very interesting. I thought you were a perfect master of the French language, & respected & envied you accordingly.— But you make my blood run cold at the cool way you talk of going for a couple of months S.E. from St. Petersburgh. & turning up God knows where.[2] I will have you tried by a court martial of Botanists & have you shot. Perhaps the grapes are sour, which makes me so savage & virtuous at your deserting your work.—

Henrietta has been very unwell, & though baddish today, she is now better & comes down stairs;[3] but she has a fearfully poor constitution, & can digest hardly anything; & all this has followed from a mere cold!—

Thanks for telling me about the Aucuba— now for Heaven sake, try *yourself* some careful experiments & *publish yourself* details with figures, & observe range of variation in form when the vars'. own pollen has been used. The case really has highest physiological importance, not to mention pangenesis.—[4]

Yesterday I received a splendid cargo from Oporto of 12 plants of Drosophyllum Lusitanicum; but the Box was almost open & the frost has injured them cruelly; yet I hope & think some will recover. The structure seems very different from Drosera, & I think the plant will prove the single basis or starting point for Drosera & Dionæa.[5] The plants are called by the Portuguese Boys fly-catchers.[6] If some spare plants recover, I suppose you w^d like some for Kew— But I hear that a merchant at Oporto c^d. not cultivate them in his garden, so it seems that they are ticklish things. It is odd that they grow in very dry soil. Huxley told me that he had been attacking Sir W. Thompson. & I fancy mathematicians in general, so he seems to have followed your audacious lead.[7] You must write me one other note before you start for St. Petersburgh.

Yours affect^y. | C. Darwin

P.S A good paper from a man like you on the direct action of the pollen in Aucuba—w^d settle the question for ever.—

DAR 94: 118–20

[1] The year is established by the relationship between this letter and the letter from J. D. Hooker, 11 March 1869.

[2] See letter from J. D. Hooker, 11 March 1869 and n. 4.

[3] CD refers to Henrietta Emma Darwin. See letter to J. D. Hooker, 8 March [1869] and n. 4.

[4] See letter from J. D. Hooker, 11 March 1869. CD was interested in the direct action of the male element upon the female as a phenomenon to be explained by his hypothesis of pangenesis (*Variation* 2: 365–6).

[5] See letter to W. C. Tait, 12 and 16 March 1869 and n. 2. The Portuguese city of Oporto is now known as Porto. In *Insectivorous plants*, p. 358, CD inferred that the 'common parent form' of the Droseraceae 'partook of the characters of Drosophyllum, Roridula, and Byblis'.

[6] See letter from W. C. Tait, 2 March 1869.

[7] CD alludes to Thomas Henry Huxley's address to the Geological Society of London on 19 February 1869 (T. H. Huxley 1869c), in which he challenged the accuracy of William Thomson's calculations on the cooling of the earth. In his presidential address to the meeting of the British Association for the Advancement of Science at Norwich in August 1868, Hooker had also challenged astronomers' objections to geologists' theories on the age of the earth (see J.D. Hooker 1868, pp. lxxi–lxxii).

From T. H. Huxley 17 March 1869

Geological Survey of England & Wales

My dear Darwin

After I had sent my letter to you the other day I thought how stupid I had been not to put in a slip of paper to say it was meant for Lushington's edification.[1]

I made sure you would understand that I wished it to be sent on & wrote it (standing on the points of my toes & with my tail up very stiff with that end in view[2]

I am getting so weary of people writing to propose controversy to me upon one point or other, that I begin to wish the article had never been written The fighting in itself is not particularly objectionable but its the wasting time

I begin to understand your sufferings over the 'Origin'—[3] A good book is comparable to a piece of meat & fools are as flies, who swarm to it, each for the purpose of depositing & hatching his own particular maggot of an idea

Ever yours | T. H. Huxley

March 17. 1869

DAR 166: 318

[1] Huxley refers to Vernon Lushington. See letter from T. H. Huxley, 11 March 1869.

[2] CD wrote his own letter to Lushington rather than simply passing on Huxley's letter of 11 March 1869 (see letter to Vernon Lushington, [12 March 1869], and letter to T. H. Huxley, 12 March [1869]).

[3] *Origin*.

From Louis Rérolle 17 March 1869

Sir

Allow me to offer you many thanks, concerning the letter you wrote to Mr Faivre, in which you authorise the translation of your Orchis book.[1] I will be happy to present to the French naturalists your remarkable and interesting experiments.

The importance of this undertaking required on my part, a few day's reflection, before coming to a decision. I pray you to excuse the delay of the present letter. I have made up my mind to publish the translation of your work, making all the rectifications you think fit. I have not yet taken any engagement with a publisher; but I will soon apply to one, at Paris.

As soon as the business is settled with him, I will receive with pleasure and gratitude your precious foot-notes; they will certainly add much value to the publication.[2] I will then inquire of your publisher the terms on which the stereotypes of the wood-cuts can be supplied.

I have now translated the whole book, save only the twenty last pages; I shall have to review and copy it, but previously I will persue the practical study of the Orchid-flowers mentioned in your volume. Mr Faivre is so kind as to leave at my disposal the very rich collection of the botanical garden of Lyons;[3] I will carefully examen under his direction every flower you have described, in order to exclude any chance of error. Though I cannot exactly know at what time the translation will be published, yet I hope it will appear at the end of the present year, or at least, in the course of the two first months of the following.

I thank you much, Sir, for your kindness in offering to send me your work on climbing plants; I have heard of it, and will read it with the greatest interest.

Allow me, Sir, to assure you of the deep respect with which I have the honour to be | Your very obedient servant | L. Rérolle

17 march 69.
4, place d'Ainay. Lyon.

DAR 176: 129

1 See letter from Ernest Faivre, 15 March 1869 and n. 2. Rérolle refers to *Orchids*.

2 For more on CD's notes to the French edition of *Orchids*, see the letter to J. D. Hooker, 8 March [1869] and n. 5.

3 Ernest Faivre was the director of the botanic gardens at Lyons (*DBF*).

To Fritz Müller 18 March 1869

Down. | Bromley. | Kent. S.E.
Mar 18 1869

My dear Sir

Since I wrote a few days ago & sent off three copies of y^r book, I have read the English translation, & cannot deny myself the pleasure of once again expressing to you my warm admiration.[1] I might, but will not, repeat my thanks for the very honourable manner in which you often mention my name; but I can truly say that I look at the publication of y^r essay as one of the greatest honours ever conferred on me. Nothing can be more profound & striking than y^r observations on development & classification. I am very glad that you have added y^r justification in regard to the metamophoses of insects, for y^r conclusion now seems in the highest degree probable.[2] I have re-read many parts; especially that on cirripedes with the livliest interest. I had almost forgotten y^r discussion on the retrograde development of the Rhizocephala.[3] What an admirable illustration it affords of my whole doctrines! A man must indeed be a bigot in favour of separate acts of creation, if he is not completely staggered after reading y^r essay; but I fear that it is too deep for English readers, except for a select few.

With cordial gratitude & respect believe me my dear Sir | yours truly | Charles Darwin

LS
British Library (Loan 10:28) (by kind permission of English Heritage)

[1] See letter to Fritz Müller, 14 March 1869 and n. 1. CD's annotated copy of Dallas trans. 1869 is in the Rare Books Room–CUL (see *Marginalia* 1: 609).

[2] For the text of Müller's addition on insect metamorphosis, see Dallas trans. 1869, pp. 119–21. Müller had argued that metamorphosis in insects was not inherited from a primitive ancestor but had been acquired at a later period.

[3] Müller had hypothesised that Rhizocephala probably developed from a conventional cirripede, with its cement ducts converted into nutritive filaments and without any protective covering, limbs, or mouth (see Dallas trans. 1869, pp. 137–9).

From C. F. Claus[1] 19 March 1869

Marburg
d 19. III. 69.

Hochgeehrter Herr!

In Ihrem so freundlichen Briefe vom 10. Febr. d. J. theilten Sie mir mit, dass Sie die Herrn *Williams* und *Norgate* beauftragt hätten, auf dem Wege des Buchhandels das für mich bestimmte Cirripedien-Material an mich gelangen zu lassen.[2] Da inzwischen 5 Wochen verflossen sind und das Paquetchen noch nicht eingetroffen ist, auch eine durch die Elwertsche Buchhandlung[3] nach Leipzig gerichtete Anfrage zu keinem Resultate führte, so fange ich an Besorgniss zu hegen, dass möglicherweise eine Verwechselung das Paquetchen an eine andere Adresse oder an einen anderen Ort geführt hat, oder sonst irgend ein anderes Malheur passirt ist und erlaube mir Sie davon in Kenntniss zu setzen mit der Bitte, die Herren Williams und Norgate zu unterrichten, da auf diese Weise ein möglicher Irrthum am besten beseitigt wird. Es thut mir sehr leid, Ihre Güte mit dieser vielleicht etwas unbescheidenen aber doch wohl begründeten Bitte missbrauchen zu müssen, ich hoffe aber Sie werden meine Freimuth gütigst entschuldigen.

Mit ausgezeichneter Hochachtung | Ihr Sie verehrender | C Claus

DAR 161: 178

[1] For a translation of this letter, see Appendix I.
[2] CD's letter of 10 February 1868 has not been found, but see the letter from C. F. Claus, 6 February 1869.
[3] Claus refers to the Marburg publisher and bookseller, N. G. Elwertsche Verlagsbuchhandlung (Würffel 2000).

To T. H. Huxley 19 March [1869][1]

Down. | Bromley. | Kent. S.E.
March 19th

My dear Huxley

Thanks for your Address.[2] People complain of the unequal distribution of wealth, but it a much greater shame & injustice that any one man shd. have the power to write so many brilliant essays as you have lately done. There is no one who writes like you.— I have hugely enjoyed your attack on Thompson; but if I were in your shoes I should tremble for my life.— I agree with all you say, except that I must think that you draw too great a distinction between your evolutionists & the uniformitarians.[3]

I find that the few sentences which I have sent to press in the Origin about the age of the world will do fairly well, though if I had read you first, perhaps I shd have been less deferential towards Thompson.[4]

Many thanks for your note received a day or two ago with yourself represented as a bristly little Terrier.—[5]

Ever yours | C. Darwin

Imperial College of Science, Technology, and Medicine Archives (Huxley 5: 266)

[1] The year is established by the relationship between this letter and the letter from T. H. Huxley, 17 March 1869.
[2] CD refers to Huxley's presidential address to the Geological Society of London on 19 February 1869 (T. H. Huxley 1869c).
[3] Huxley's address was a response to William Thomson's statements concerning the conflicting calculations of astronomers and geologists in regard to the age of the earth (see T. H. Huxley 1869c, p. xxxviii). For Huxley's comparison of catastrophism, uniformitarianism, and evolutionism, see T. H. Huxley 1869c, pp. xlvi–xlvii.
[4] In *Origin* 5th ed., pp. 354, 379, CD discussed Thomson's estimates of the age of the earth, but did not dispute the evidence for his findings.
[5] See letter from T. H. Huxley, 17 March 1869.

To John Dean Caton 20 March 1869

Down Bromley Kent. S.E.
March 20 1869.

Dear Sir

Your former kindness leads me to think that you will excuse me asking you two or three questions—[1]

In books of Nat Hist the male of the Elk or Wapiti is said to be darker colored than the female; if this is the case is he darker in his winter or breeding dress, or during the Summer?[2]

Is it possible that you could enquire for me from any hunters who have had much experience with the American bison, whether the thick mane of the bull protects him in his battles with other bulls? This might possibly be known to witnesses of their contests or by the position of scars from old wounds. Lastly, though it is very unlikely that you should be able to answer me, Do the females of any of the deer which you have kept show a preference for one male over another?[3] I have lately felt much perplexed & have made enquiries from those who have closely observed deer in England what is the meaning of the terminal branches of the horns, for a single point would be a more effectual weapon?[4]

I have lately been making much use of, and & quoting yr most interesting paper—[5]

Pray believe me dear Sir | Yours faithfully & obliged | Charles Darwin.

Copy
DAR 143: 254

[1] Caton's offprint of his paper, 'American Cervus' (Caton 1868) had been forwarded to CD, and Caton later replied to CD's questions (see *Correspondence* vol. 16, letter to J. D. Caton, 18 September 1868, and letter from J. D. Caton, 2 [October] 1868). CD's annotated copy of Caton 1868 is in the Darwin Pamphlet Collection–CUL. The paper was also published in Caton 1880, pp. 146–75.

[2] Caton discussed the elk or wapiti (*Cervus canadensis*) in Caton 1868.

[3] Caton kept 'Virginia deer' (*Cervus virginianus*; now *Odocoileus virginianus*, white-tailed deer) on his property in Illinois; he included his observations of the deer in Caton 1868.

[4] No letter from CD requesting information on terminal branches of horns has been found, but see the letter from George Cupples, 15 January 1869, and the letter from Archibald McNeill, 24 January 1869.

[5] For CD's references to Caton 1868, see *Descent* 1: 288, 2: 245, 253, 282, 288, and 290.

To Julius Dub 20 March 1869

Dear Sir,

I sh[d] be pleased at the appearance of your work, as I naturally wish my views to be widely understood; but I am doubtful whether the publisher of the German editions might not feel injured, if you give many passages, literally from the Origin.[1] You must judge on this head for yourself, for I cannot authorise the appearance of any work, which could interfere with the right of the existing Translation.[2] For myself, I believe of that more the subject of my work is discussed, the more the original is likely to sell. D[r]. Rolle has published a popular exposition of my views, & I have not heard that the German publisher complained.[3] I presume that you do not wish that your book sh[d] appear under my authorisation, as I could not give this without receiving the distinct permission of the German publisher.— A new English Edition of the Origin is now passing through to press, & in a month or so all the sheets will be printed off. The Edt contains some important additions in relation to the views of Nageli, Mr Wagner— the glacial period &c.[4] I sh[d] strongly advise you *[2 words illeg]* given the time to read it carefully & if you agree to do so, I would gladly send you clean sheets before the work is published. The smaller corrections are numerous, partly in consequence of the evidence being stronger or weaker on ⟨each⟩ point & more especially as I have become convinced, that slight individual differences are cases of more importance than I firstly supposed & that single variations have largely led to ⟨every⟩ result.—

If on reflection you are convinced that you can ⟨conveniently⟩ publish your new work without injury to the G. ⟨Translation⟩, I shall be pleased, & in this case be so kind as to inform me & I will send you the sheets of the new Edition.

Believe me | Dear Sir | Yours faithfully | C. Darwin

Dr Julius Dub
March 20/1869/

AdraftS
DAR 96: 62, 64

1 CD refers to *Kurze Darstellung der Lehre Darwin's über die Entstehung der Arten der Organismen: mit erläuternden Bemerkungen* (Brief description of Darwin's theory on the origin of species, with explanatory remarks; Dub 1870), and to *Origin*.

2 CD refers to Bronn and Carus trans. 1867.

3 CD refers to Friedrich Rolle and to Rolle 1863, which he had received in parts in 1862 (see *Correspondence* vol. 10, letter to Friedrich Rolle, 17 October [1862]). CD's German publisher was E. Schweizerbart'sche Verlagsbuchhandlung.

4 Carl Wilhelm von Nägeli and Moritz Wagner had written on different aspects of CD's theory in Nägeli 1865 and Wagner 1868a and 1868b. CD responded to these works in *Origin* 5th ed., pp. 120, 151–7. He also added material based on James Croll's theory of alternate ice ages (*ibid.*, pp. 451–61).

From Samuel Owen Glenie 20 March 1869

Trincomale. [Ceylon]
20: March. 1869.

Dear Sir

I hope that the following brief account of some chickens of the black boned variety of the common fowl may interest you,—[1]

I procured a cock and hen of the pure breed and carefully isolated them. Both were perfectly white. The hen laid 11 eggs, and hatched 11 chickens about a month ago.— Of these 11 chickens 3 have died. Eight were living this morning. The first plumage of these eight (of all of them) is pure white. I do not mean the down, which was white also, but the feathers.— Six of them are decidedly of the black-boned variety, with black eyes, skins, and mouths, i,e, the roof the mouth.— The other two are in no way distinguishable from ordinary fowls with white plumage. The three that died were also of the black-boned kind.— As I am quite sure that the parent birds were isolated, and that no other eggs were introduced, I cannot account for the the difference between the *nine* and the *two* chickens— The natives tell me that for the hen of this variety to hatch chickens of both descriptions is an ordinary occurrence, but then, they never take any pains to keep the hens & cocks of this breed separate from other fowls.— If my account is correct, and I am pretty sure it is, of the isolation of the parent birds, and non introduction of any other eggs, it would seem that this blackboned variety is not a distinct species.—

I requested our mutual friend M^r. Thwaites to inform you that I had often seen, & had had on my table, black boned fowls with plumage not only not white but differing quite as much in colour as in ordinary fowls.—[2]

I left the breast bone of one of these blackboned fowls in a spot in my room where the common red ants could easily get at it, and in three or 4 days the apparently perfectly black bone was quite white— The ants had eaten the membrane which covered the bone, and thus proved that the black colour was not in the bone itself—

Believe me | Yours faithfully | S. O. Glenie

P.S. I had almost forgotten to send you an extract from a letter lately received from M^r. Holdsworth the naturalist employed by the Ceylon Govt to examine their Pearl Banks— Mr Holdsworth is well known, I believe to Messrs Gosse & Huxley.[3]

"There is a curious habit among both Singalese & Tamils not strictly belonging to M^r. Darwin's questions, but interesting, because undoubtedly natural and universal.[4] You may be familiar with it. It is, when beckoning to any one, the movement of the hand forwards and downward instead of upwards and backward, as is the practice among Western nations. I do not know how general in India the custom is, but it undoubtedly obtains among the Malabars.[5] Also in pointing to an object the hand is not directed definitely towards it, but the outstretched arm is raised quickly above the head through a perpendicular line cutting the object I was much bothered with this mode of pointing when I first went into the jungle with natives. I often found myself looking among the treetops for what was close to or on the ground Can you give me any explanation of these customs?—

As far as my observation extends both these modes of beckoning and pointing are the common modes of the people here.— I do not know, nor can I even guess, why they differ so much from Europeans in these matters, which, after all are only two amidst a very number of customs in which they differ from Occidentals.—

S. O. Glenie

DAR 165: 56

CD ANNOTATION

End of letter: 'Trincomale | Ceylon. March 20/1869'

1 For CD's original query on the plumage of black-boned fowl, see *Correspondence* vol. 16, letter to G. H. K. Thwaites, 19 May [1868].

2 Glenie refers to George Henry Kendrick Thwaites. Glenie had himself written to CD about the variations in the plumage of black-boned fowl in a letter of 27 November 1868 (*Correspondence* vol. 16).

3 The Pearl Banks, in the Gulf of Mannar, south of Mannar Island, Ceylon (Sri Lanka), were an important pearl fishery at the time. Glenie refers to Edmund William Hunt Holdsworth, Philip Henry Gosse, and Thomas Henry Huxley.

4 The reference is to the queries on expression that CD sent to correspondents all over the world as part of his research for *Expression*. For more on the queries and the responses, see *Correspondence* vols. 15 and 16.

5 The Malabar coast is in south-west India. In the nineteenth century, Europeans often referred to the different ethnic groups living in the region as Malabars (*OED*).

To Charles Lyell 20 March [1869][1]

Down. | *Bromley.* | *Kent. S.E.*
March 20th

My dear Lyell

I am obliged to you for sending me Wallace's letter, which I have been glad to read.— I did not pretend to form any opinion on Moseley's or Croll's papers; but the latter *seemed* to me very ingenious.[2] As Mosely has studied mechanics all his life, I cannot suppose that he has overlooked the considerations specified by Wallace. I sh^{d}. have though that the close, fine & parallel veins in the ice of glacier, along the planes of movement, & the regular & gentle curve, which a straight row of sticks soon & invariably assumes, were opposed to the belief that a glacier descended by a succession of abrupt-fractures first at one point & then at another. But as I have said the subject is quite beyond me.—

What a capital book Wallace has published; his discussions on Geograph. Distrib. are inimitably good— I have not yet finished the book.[3] I sh^{d}. think it w^{d} have a great success.— I am working very hard at many subjects.—

Ever yours | Most truly | Ch. Darwin

American Philosophical Society (367)

1 The year is established by the reference to the recent publication of Wallace 1869a (see n. 3, below).

2 Alfred Russel Wallace's letter has not been found. CD had first mentioned Henry Moseley's paper on glacial motion (Moseley 1869a) in a letter to James Croll of 31 January [1869]. Croll sent CD a pre-publication copy of his article on the subject (Croll 1870; see letter from James Croll, 15 February 1869 and n. 3).

3 CD refers to Wallace 1869a (see letter to A. R. Wallace, 5 March [1869] and n. 2).

To William Thierry Preyer [before 21 March 1869][1]

Charles Darwin

His father's father	D^{r}. Erasmus Darwin, author of Zoonomia
His mother's father	Josiah Wedgwood[2]
His father, born when? where? died when?	D^{r}. Robert Darwin,[3] Physician, died 1848 at Shrewsbury.
Brothers and sisters of M^{r}. Charles Darwin	Four sisters and one brother
Charles Darwin born where?	At Shrewsbury
" " " when? (day, month & year)	Feb. 12. 1809.
Education at home, where? how long?—	
At school where? how long?	At Shrewsbury School for 7 years.

Influence of parents	
" " tutors	
" " circumstances	
Holidays spent where? at home?	At home
Health, vigour	Good health
University. Which Universities? Studies begin when? Lectures heard.	Edinburgh 1825 for two years, thence to Cambridge where he took his degree in 1831.
Names of University (at Cambridge, Edinburgh &c) Professors of influence	The Rev^d. Professor Henslow;[4] the Professor of Botany had great influence in deciding my taste for natural science.
Favourite university studies	His education only really began when he went on board the Beagle.
University studies end when?	
Examinations; B.A when? where?	1831 at Cambridge
M.A. " "	I forget date (1837?)[5]
The Voyage round the World, Dec. 27. 1831–Oct. 2. 1836	Published 1839[6]
How he came to be naturalist of the expedition	Captain Fitzroy[7] offered to give up part of his cabin to any naturalist who would accompany H.M.S. Beagle in her surveying voyage and circum-navigation. M^r. D. volunteered his services without salary, but on condition that he[8]
	Wollaston Medal was awarded me by the Geological Soc.^y. of London.
(b) F.L.S. when?	
(c) F. Geol. S. when?	M.A.—F.G.S.—F.L.S.—F.R.S.E.—Acadd, Reg. Sci. Berol. et Holm.— Soc. Reg. Sci. Upsal.— Acad. Cœs. Nat. Cur. Dresd.— Soc. Sci. Nesc.— et Acad. Nat. Sci. Philad. Soc. Honor.— Also St. Petersburgh, Edinburgh and Dublin and I believe some others, but I have no list.[9]
Additional biographical notes	From my earliest youth I was fond of collecting all sorts of specimens and observing habits of wild animals. I was an ardent sportsman and this made me very idle. At Edinburgh though I had the opportunity I neglected from foolish disgust to study anatomy, and I have regretted this all my

life. I observed at Edinburgh the lower marine animals. At Cambridge I collected Coleoptera energetically and dabbled a little in Geology; but I studied no subject scientifically; only for amusement. I never worked till I joined the Beagle and then I worked with all my heart.

C.D.

At Edinburgh Dr. Grant,[10] now Professor at the London University, led me to collect and observe the lower marine animals.

Copy
DAR 147: 262–5

[1] The date is established by the relationship between this letter and the letter from W. T. Preyer, 21 March 1869, and by Preyer's statement that the questionnaire was completed by CD early in 1869 when he was 60 (Preyer 1896, p. 148). A German translation of this letter is reproduced in Preyer 1896, pp. 143–6. Preyer's sketch of CD's life was published in *Ausland*, 2 April 1870.

[2] Josiah Wedgwood I.

[3] Robert Waring Darwin.

[4] John Stevens Henslow.

[5] CD did receive his MA in 1837 (*Alum. Cantab.*).

[6] The reference is to *Journal of researches.*

[7] Robert FitzRoy. For more on the circumstances leading to FitzRoy's request for a companion on the *Beagle* voyage, see Browne 1995, pp. 145–9.

[8] The text breaks off at this point in the copy, but probably continued, 'retain complete freedom in the disposition of his collections'.

[9] For a complete list of CD's society memberships, see Freeman 1978, pp. 107–10.

[10] Robert Edmond Grant.

From W. T. Preyer 21 March 1869

Bonn, University,
March 21, 1869.

Dear Sir—

I take the liberty of sending you a popular lecture on the struggle for existence, wherein (before an auditory of more than 600 persons at Cologne) I have attempted to popularize one of the principles of your theory. I differ from your statements only in very few instances e.g. note 9 (p. 42) and note 23 (p. 45) and should be quite unhappy if I have misunderstood any one of your ideas.[1]

Perhaps it will interest you to hear that I held a course of lectures in the University here last Winter (Oct.–March) "Über die Darwinsche Theorie" constantly heard by 200–250 students. Never before has so numerous an assembly followed the lectures of a "Privatdocent" in this place, whatever the subject might have been. This single fact shows the enormous progress your doctrine is making in Germany.

You would very much oblige me and your scientific friends if you would kindly correct and complete the list of your works which I have compiled as well as it was possible for me.[2] I dare say the four titles a, b, c, d are in the main identical with N^r. 1). Without wishing in the least to trespass on your precious time I may beg you to send me back the corrected and completed list of your works before the 15th. of April.[3] I have been appointed professor of physiology at the University of Jena last week and am anxious to possess the correct list before I leave Bonn.

I hope in a future work to bring some new important facts in favour of your theory and concluding this letter venture to trouble you once more: Could you send me a copy of your's and M^r. Wallace's paper, printed in the proceedings of the Linnean Society 1858?[4] I nowhere can find the proceedings here.

I remain | Your's | Most truly | W^m. Preyer

DAR 174: 67

[1] Preyer refers to *Der Kampf um das Dasein* (The struggle for existence; Preyer 1869). CD's annotated copy is in the Darwin Pamphlet Collection–CUL. In note 9, Preyer contended that the struggle for existence was not one of individuals against each other, but a competition among all living beings (Preyer 1869, pp. 42–3). In note 23, Preyer argued that CD's conclusion that supernumerary mammae in women were due to reversion (*Variation* 2: 57) did not take into account the fact that mammae had been reported on the shoulder or back (Preyer 1869, p. 45; this section is scored in CD's copy). In *Descent* 1: 125 n. 38, CD cited Preyer for this information.

[2] The list has not been found, but see *Ausland*, 2 April 1870. For CD's publications, see Freeman 1977.

[3] The list of works was part of Preyer's sketch of CD's life, which he submitted to *Preußische Jahrbucher* in April 1869 (see also letter to W. T. Preyer, [before 21 March 1869] and n. 1); the article was not accepted for publication (Preyer 1896, pp. 148–9).

[4] Preyer refers to Darwin and Wallace 1858.

From George Cupples 22 March 1869

Guard Bridge | Fifeshire. N.B.
March 22/69.

Dear Mr Darwin,

Today I was about to write to you at any rate—when I got your note.[1]

In the first place I have to say that I shall during the whole of this season look out in every possible way for statistics as to comparative number in the sexes of any animals whose birth-records have been at all taken.[2]

Secondly I shall weigh every litter of puppies I may have—beginning at a week old—and I will secure aid in the same way from several other breeders of deerhounds and similar dogs, most of whom have been in the habit of weighing their puppies at any rate.

Lastly—to return to the trifling results of previous contributions on my part. These would appear to have gone against your own drift, so far—but on looking at the figures you bring out in comparison between male and female, if you mean that the adult dog & bitch measure as 90 to 70 lbs, I cannot agree that this proportion truly marks the difference. A bitch of 70 lbs would, if I can judge in the rough way, be a well-sized if not a large bitch—whereas a dog of 90 lbs (deerhound) would not be remarkable. I think the dogs go up much higher, and the bitches not so— then a bitch is fleshier, and the dog has generally been weighed in *running* condition—i.e. when lean. I lately measured a dog (Russian Wolfhound so-called, from the Emperor's Kennels, with nothing of the Deerhound breed about him, but still of the *hound* type). he stood 33 inches high at shoulder, or nearly so—and in low condition weighed 152 lbs. The Scotch Deerhound & the Irish Greyhound (Wolf-dog, same race as Deerhound) is universally testified of as standing, in former days when thoroughly used & bred, up to 34, 35, and even 36 inches. His weight must have gone up near 200 lbs. Quere—did the bitches share in the present degeneracy? N.B. Within the last 20 years, the effect of English breeding and Dogshows on Deerhounds has been to raise the *average* and extreme height of the Deerhound again, about an inch, or more—i.e. from an average of 28 inches to 29 &c., and this without the benefit of suitable active use in the Forest. I believe that the previous resort to *crossing*, in Scotland, obviously brought down the difference between male & female.

I merely throw out these hasty notes, trusting to clear the ground at least. It is for you to judge whether the Deerhound (as the present race of dogs for about 3000 years at any rate) is worth building anything upon. If so, then any trouble that can come near statistical accuracy nowadays is worth taking in the matter.

If Wolves could in any way be reached, it might be important—but I suppose they cannot. The weight of a large Russian wolf has been found in one case 101 lbs (or 110, I am not sure at this moment which) in *running* condition (as a wolf always is.) His height I suppose from from 28 up to 29 inches at the shoulder—his shoulder being disproportionately high, I believe. In some of the old manuals about Wolfhunting in France, I think there are minutiae about the difference between the sexes in some points.

As to the comparative size of the young puppies, my belief is that the *weighings* given you by me (however correct) are not to be depended on as representing the normal facts. 90 to 83 is what you find here, as I understand it, for the proportion. The proper way would be to weigh a number of litters, when the puppies are all left with their *own* mother—and where the total number does not exceed 4 or 6 puppies. No breeder does otherwise, as a rule. It has to be considered that sometimes 12 or even 14 puppies may be born—at other times 2, 3, 4 &c. Mr Wright's[3] bitch-puppies (3 born, I think) weighed near, or more than, my dog-puppies (of a 7-born litter.) At the same time I had 4 bitch-puppies (4 born, 1 dying in a day or two, 1 dying in a week)—and they weighed near Mr Wright's.

In your last note you say "On reflection I daresay the Bitch-puppies so soon exceeding in weight the Dog-puppies was caused by their foster-mother having fewer to nurse." I suppose there is a lapsus styli[4] here—you mean the Dogs exceeding the bitches so soon, I presume. The *bitches* were with the foster-mother. But I incline to agree that a difference may have been exaggerated by this. The mother, I suspect, was the more favourable nurse, ceteris paribus.[5]

In short, As your *drift* would be contradicted apparently by this very insufficient instance, I am glad to *repeat* what I said before I knew the drift—that I believe the difference does *not* mark itself to that degree at so early an age.

Most anxiously do I wish I could contribute even the slightest point of (pregnant) fact—and shall see what this year can do.

It has sometimes been a question with me whether—on the one hand the mother ever gives any early preference to the male puppy as bigger, and on the other hand whether the forwardness and precocity of the female puppy is ever so soon marked as to give itself an advantage, or perhaps to redress any balance in this respect. Such queries are, I suppose, mere ineptitude.— Again as to number of the sexes, I should certainly like to know when you hear anything from the Greyhound quarter—or from Stud-books. The *male* in man, I presume, is undoubted during the short period in which statistics have been taken.

Please to excuse my prolixity. My next steps will be to communicate with Mr Wright and two or three others—and I shall endeavour not to trouble you with mere trifles.

Believe me Most truly yours | George Cupples

Charles Darwin, Esq.

DAR 161: 291

CD ANNOTATIONS
1.1 Today ... far 4.2] *crossed blue pencil*
4.4 adult ... marks 4.5] *double scored blue pencil*
4.5 judge ... female. 4.20] *crossed blue pencil*
5.3 If ... matter. 5.4] *double scored blue pencil*
8.2 weight ... here— 8.3] *scored blue pencil*

1 CD's letter has not been found.
2 For more on CD's interest in proportion of sexes in domestic animals, see the letter to W. B. Tegetmeier, 5 February [1869] and n. 2.
3 John Wright.
4 'Lapsus styli': slip of the pen (Latin).
5 'Ceteris paribus': all other things being equal (Latin). CD hypothesised that characteristics acquired later in life tended to be transmitted to one sex alone (see *Descent* 2: 260–2). Under this hypothesis, CD expected the size of the male and female puppies to remain similar, only diverging at a later stage of development.

From Karl von Seidlitz[1] 22 March 1869

Livonia | Dorpat
^{10}/22 März 1869[2]

Verehrter Herr!

Aus Ihrem Briefe vom 31 Jan. 1869 sehe ich, dass mein Buchhändler Ihnen dasjenige Buch *nicht* überschickt hat, welches ich grade als Beweis davon verehren wollte, dass Ihre bahnbrechenden Ideen auch bei uns schon seit Jahren Anerkennung gefunden haben.[3]

Dass *Vererbung* ein mächtiger Factor in der Entwicklungs Geschichte der organischen Wesen sey, ist schon lange angenommen worden. Ich möchte aber noch weiter gehen, und die *Vererbungskraft* als eine immanente Eigenschaft aller irgendwie organisirten Materie—vom ersten formlosen Schleim bis zu den ausgebildetsten Lebewesen—angesehen wissen. Nur vermöge dieser Grundeigenschaft kann die Entwickelung aller organischen Wesen auf unserm Erdballe vorsich gegangen seyn und auch künftig immer weiter schreiten. Was die Gravitation in der anorganischen Welt ist,—das ist die Vererbungs-Kraft in der organischen. In diesem Sinne habe ich auf pag. 52–54 meine Ansicht ausgedrückt.

Da meine, unter Kreuzband beifolgende Broschüre wohl schwerlich in den Buchhandel gelangt ist, so würde vielleicht eine Anzeige in irgend einem Englischen Journale kenntlich machen, in welcher Weise Ihre Hypothese in dem sogenannten Lande der Eisbären aufgefasst worden ist.

Mit ausgezeichneter Hochachtung | Ihr ergebenster | Dr. C. Seidlitz

DAR 177: 132

1 For a translation of this letter, see Appendix I.

2 Dorpat: now Tartu, in Estonia. Seidlitz gives both the Julian (10 March) and Gregorian (22 March) calendar dates.

3 Seidlitz sent his monograph *Ueber die Vererbung der Lebensformen, Eigenschaften und Fähigkeiten organischer Wesen auf ihre Nachkommen in Bezug auf Physiologie und praktische Heilkunst* (On the transmission of life structures, properties and vitality of organic beings to their descendants in relation to physiology and practical medicine; Seidlitz 1865). CD's copy is in the Darwin Pamphlet Collection–CUL.

To A. R. Wallace 22 March [1869][1]

Down. | Bromley. | Kent. S.E.
March 22[d]

My dear Wallace

I have finished y[r] book; it seems to me excellent, & at the same time most pleasant to read.[2] That you ever returned alive is wonderful after all y[r] risks from illness & sea voyages, especially that most interesting one to Waigiou & back.[3] Of all the impressions which I have rec[d] from y[r] book the strongest is that y[r] perseverance in the cause of science was heroic. Your descriptions of catching the splendid butterflies have made me quite envious, & at the same time have made me feel almost young again, so vividly have they brought before my mind old days when I collected, tho' I never made such captures as yours. Certainly collecting is the best sport in the world. I shall be astonished if y[r] book has not a great success; & your splendid generalizations on Geog. Distrib., with which I am familiar from y[r] papers, will be new to most of your readers. I think I enjoyed most the Timor case, as it is best demonstrated; but perhaps Celebes is really the most valuable.[4] I sh[d] prefer looking at the whole Asiatic continent as having formerly been more African in its fauna, than admitting the former existence of a continent across the Indian ocean.[5]

Decaisne's paper on the flora of Timor in which he points out its close relation to that of the Mascarene I.s supports y^{r} view.[6] On the other hand I might advance the giraffes &c in the Sevalik deposits.[7] How I wish some one w^{d} collect the plants of Banca! The puzzle of Java, Sumatra & Borneo is like the 3 geese & foxes: I have a wish to extend Malacca thro' Banca to part of Java & thus make 3 parallel peninsulas, but I cannot get the geese & foxes across the river.[8]

Many parts of y^{r} book have interested me much: I always wished to hear an independent judgment about the Rajah Brooke, & now I have been delighted with y^{r} splendid eulogium on him.[9]

With respect to the fewness & inconspicuousness of the flowers in the Tropics, may it not be accounted for by the hosts of insects, so that there is no need for the flowers to be conspicuous. As according to Humboldt fewer plants are social in the tropical than in the temperate regions, the flowers in the former w^{d} not make so great a show.[10]

In y^{r} note you speak of observing some inelegancies of style, I notice none. All is as clear as daylight. I have detected 2 or 3 errata.

In vol 1. you write Sond*i*acus is this not an error?[11]

Vol 2. p 236 for *Western* side of Aru read *Eastern.*

p. 315 Do you not mean the horns of the moose? for the elk has not palmated horns.[12]

I have only one criticism of a general nature, & I am not sure that other geologists w^{d} agree with me: you repeatedly speak as if the pouring out of lava &c from Volcanoes actually caused the subsidence of an adjoining area; I quite agree that areas undergoing opposite movements are some how connected;[13] but volcanic out bursts must I think be looked at as mere accidents in the swelling up of a great dome or surface of *plutonic* rocks,—& there seems no more reason to conclude that such swelling or elevation in mass is the cause of the subsidence, than that the subsidence is the cause of the elevation; which latter view is indeed held by some geologists.

I have regretted to find so little about the habits of the many animals which you have seen.

In Vol 2. p. 399 I wish I c^{d} see the connection between variations having been first or long ago selected & their appearance at an earlier age in birds of Paradise than the variations which have subsequently arisen & been selected. In fact I do not understand y^{r} explanation of the curious order of development of the ornaments of these birds.[14]

Will you please to tell me whether you are sure that the female Casuarius (Vol. 2. 150 sits on her eggs as well as the male; for if I am not mistaken Bartlett told me that the male alone, who is less brightly coloured about the neck, sits on the eggs.[15]

In Vol. 2 p. 255 you speak of male savages ornamenting themselves more than the women, of which I have heard before; now have you any notion whether they do this to please themselves, or to excite the admiration of their fellow men, or to please the women, or, as is perhaps probable from all 3 motives.

Finally let me congratulate you heartily on having written so excellent a book, full of thought on all sorts of subjects. Once again let me thank you for the very great honour which you have done me by your dedication

Believe me | My dear Wallace | Yours very sincerely | Ch. Darwin

Vol. 2. p. 455 When in New Zealand I thought the inhabitants a mixed race, with the type of Tahiti preponderating over some darker race with more frizzled hair; & now that the stone instruments reveal the existence of ancient inhabitants is it not probable that these Is^{ds} were inhabited by true Papuans. Judging from descriptions the pure Tahitians must differ much from your Papuans.—

LS(A)
British Library Board (Add 46434, f. 173–7)

1 The year is established by the relationship between this letter and the letter from A. R. Wallace, 10 March 1869.

2 CD refers to *The Malay Archipelago* (Wallace 1869a; see letter from A. R. Wallace, 10 March 1869).

3 See Wallace 1869a, 2: 331–48.

4 See Wallace 1869a, 1: 316–30 and 424–47. Wallace concluded that the 'miscellaneous character' of the fauna of Timor supported CD's view that oceanic islands had never been connected to the mainland (*ibid.*, p. 329). He argued that the animal forms on Celebes suggested the possibility of a former continent in the Indian Ocean with a link to Africa (*ibid.*, p. 445).

5 See Wallace 1869a, 1: 445.

6 CD refers to Joseph Decaisne and Decaisne 1834, p. 341. The Mascarene Islands are in the Indian Ocean east of Madagascar.

7 CD refers to the vast beds of vertebrate fossils found in the Siwalik (Shiwalik) mountain range of Nepal and India. Fossils of several animals now found only in Africa have been found in these beds.

8 Banca (now Bangka) is an Indonesian island in the Java Sea south-east of Sumatra, separated by the narrow Bangka Strait (*Columbia gazetteer of the world*). CD may be alluding to the traditional game of fox and geese (*OED*). Wallace had noted that the flora of Java, Sumatra, and Borneo was imperfectly known and also suggested that Banca, although geographically close to Sumatra, was geologically similar to Malacca and Singapore (see Wallace 1869a, 1: 215, 225–6).

9 Wallace had lived in the house of James Brooke, raja of Sarawak, whenever he was in Sarawak (Wallace 1869a, 1: 54). For Wallace's assessment of Brooke's government and his character, see Wallace 1869a, 1: 144–7.

10 See Wallace 1869a, 1: 127–9. CD refers to Alexander von Humboldt; the reference to the lesser number of social plants in tropical regions has not been identified.

11 Wallace mistakenly referred to *Bos sondiacus* rather than *B. sondaicus* (now *B. javanicus*, the banteng; see Wallace 1869a, 1: 220).

12 CD refers to Wallace's discussion of the horned fly or deerfly, *Elaphomia alcicornis* (now *Phytalmia alcicornis*; Wallace 1869a, 2: 314–15), whose horns Wallace compared to the horns of an elk.

13 See, for example, Wallace 1869a, 1: 17.

14 Wallace had hypothesised that since colour was the most variable feature in birds-of-paradise, difference in colour would be the first sexual difference to appear in young birds (Wallace 1869a, 2: 399–400). For CD's annotations on this section in his copy of Wallace 1869a, see *Marginalia* 1: 842.

15 Abraham Dee Bartlett was superintendent of the Zoological Gardens, Regent's Park, London (*Modern English biography*). The Zoological Society's gardens first received a cassowary in 1857, and acquired a pair of Bennett's cassowaries (*Casuarius bennettii*) in 1858. No letter from Bartlett discussing cassowaries has been found, but in a note dated 15 February 1857 (DAR 84.2: 178), CD wrote, 'Casuarius galleatus male alone incubates & takes care of young, but ♀ in development, & bright colours of naked skin & appendages all about neck, decidedly most developed so that anyone w$^{d.}$ think she was male: Can their difference relate to safety in incubation?'

From Harrison William Weir 23 March 1869

Weirleigh | Brenchley Kent.
March 23rd. 1869

Dear Sir

Since I last wrote to you I have been collecting information as to the numbers of the different sexes in the families of birds and Animals[1]

As regards Pigs

Most of the breeders that I have consulted on the matter tell me that the average of males to females is about 7 to 6,[2] but my own observation *though limited* makes them nearly if not quite 8 males to 6 females and one litter that I met with of **13** (of which I purchased 2) there were no less than **12** males and only **1** female. Once at Halifax I was shewn one of *neither* sex a little of *each*.

As regards Pigeons

There are more cocks then hens, I have often bred two cocks in a nest, but seldom two hens.[3] The hen in the nest is generally the weaker bird of the two, and is much more liable to die in rearing. Supposing that you were breeding from two blues, the hen of the two young ones frequently (perhaps *not* very so) comes out a *silver* color, but I do not remember a visa versa case. In breeding from two reds the hen in nest is sometimes *yellow* while the cock bird is red.

With regards to the likes and dislikes

Some of the fanciers have informed me that their hens will often take a fancy to one particular cock and often leave their own mates this I have also known to be the case with my own birds but not frequent

Pheasants.

Mr. Baker[4] of Leadenhall market tells me that in breeding the above, he generally gets 4 to 5 Cocks to 1 hen. When out shooting (where we have killed *all rising* or rather shot at) *I* have noticed the *great* preponderance of cocks, when the bag has been laid out.

Pigeons

Mr. Ridpeth of Manchester keeping blue Rock pigeons has noticed that they *drive off* any *other color*, such as yellow, reds, and whites[5]

Tame Rabbits.

I have kept rabbits many years, and have noticed the far greater number of bucks than does. I kept an account last year 1868 on purpose for you and out of **three** litters **15** in all I only had *one* doe. I will keep an acct again *this* year.[6]

I have had my Rabbits some generations back, but curiously enough I bred a **long** hair last year It is **grey** and a buck. The hair is *very* long. I have kept it on purpose to see if the breed from it will be long haired. I have one litter now about 6 weeks old from it *all* smooth I shall try again as an experiment or if you would like to have it I shall be most happy to send it you. I found that the doe rabbits did not care so much for his attentions as the smooth buck *Rabbits* are different from any other any animals I know. They will receive the attention of the buck *though with young*, even up to the *day before* kindling and if put to different bucks seem not particular as to how many this is curious so *excuse* my mentioning it as I know of no other animal like it.

Breeding singing birds

I am informed that the best *singing* bird (cock) generally gets a mate first when they are bred in rooms.

A Poultry breeder told me a few days since (He keeps Bramah Poutras and Spanish) that *some years* his Spanish *run mostly* to cocks and the Bramahs the opposite.[7] Then the next year the Bramah are mostly cocks possibly, but he seems to have the idea that there is something in the *season* that helps the breeding in some way, this (if true) seems to carry out my notice of the sheep some seasons having more twins

Incomplete
DAR 86: C10, DAR 84.1: 118

CD ANNOTATIONS
1.1 Since … birds 1.2] *crossed blue crayon*
2.1 As … *each.* 3.5] *crossed pencil*
5.3 that … bird 5.6] *crossed blue crayon*
6.1 With … the case 7.3] *crossed blue crayon*
10.1 Pigeons … that 11.1] *crossed blue crayon*
11.1 they … whites 11.2] *crossed blue crayon*
12.1 Tame Rabbits. … *this* year. 13.3] *crossed pencil*
13.2 out of … *one* doe. 13.3] *tick added blue crayon*
14.1 I have one … *all* smooth 14.4] *tick added blue crayon*
14.1 I have … like it. 14.9] *crossed blue crayon*
14.4 I shall … send it you. 14.5] *scored blue crayon; tick added blue crayon*
15.1 Breeding … rooms. 16.2] 'Harrison Weir' *added pencil*
17.1 A Poultry … more twins 17.5] *crossed blue crayon*
Top of letter: 'Harrison Weir' *ink*
End of letter: 'than at others.' *ink*

1 The last extant letter from Weir was that of 28 March 1868 (*Correspondence* vol. 16).
2 In *Descent* 1: 305, CD cited Weir for this information.
3 In *Descent* 1: 306, CD cited Weir for this information.
4 Charles Newcomb Baker.
5 In *Descent* 2: 118, CD cited Weir for this information. Weir refers to Thomas Hardwick Ridpeth.
6 In *Descent* 1: 305, CD cited Weir for this information.
7 Weir refers to Brahmaputra (often shortened to 'Brahma') and Spanish varieties of domestic fowl.

From A. R. Wallace 24 March 1869

9, St. Mark's Crescent, N.W.
March 24th. 1869

Dear Darwin

Allow me first to thank you for sending me a copy of the translation of Fritz Muller's interesting little work, which I have read with great pleasure;[1] but whether it is owing to my extreme ignorance of aquatic animals, or to my having formed too high expectations of the book, it has given me the impression that the author has not made the most of his subject, & has not put either his facts or his arguments in the clear & powerful manner which you yourself would have done. If no one else makes a similar observation I must impute it to my own dulness in matters of minute anatomy & embryology.

Many thanks for your *corrections* in my book.[2] It only shows what absurd errors a person having once written may go over & over again & never see. As to *Elk* & *Moose*, I have always understood that they are the same animal,—the Americans *improperly* calling their Wapiti deer the Elk.[3]

I rather demur to the view that volcanoes are due to any general heaving up; and it seems to me that when a *vent* is once established the amount of pressure which would have caused a very slow upheaval of a wide district may be all expended in ejecting volcanic matter, and the downward pressure of adjacent districts *may* assist the ejecting process, & be rendered possible by the existence of the vent.[4]

I was sure that many naturalists would be disappointed at the scantiness of my notes on habits of animals &c. But the fact is my facts of this kind are very few, owing to my having devoted myself so completely and almost exclusively to the work of *collecting*,—and I have no inclination whatever to write page after page, as some people can, of descriptions of animals & their habits, founded on very scanty observations.

As to the acquirement of plumage of Birds of Paradise perhaps I have been too hasty in assuming there is the kind of connection I have intimated as probable; yet I cannot help believing that it is true.[5] My reasoning was somewhat as follows: in the process by which the male birds acquired their plumage,—*colour* would be first acquired, because variations in colour are most frequent & most extensive,—developments of special plumage in the *head* & *tail* next,—for the same reason,—& plumes of body, last. As all these modifications are very great in quantity there was probably a long period during which the *males* differed from the *females* only in colour,—after that for another long period the long tail cirrhi were developed,—& then the body plumage would begin to be selected to give the last touch of improvement. The development of these different kinds of ornaments in three different parts of the body can not go on *at once*, but *successively*. The brilliant head & breast feathers, appeared after the first moult when there was nothing else to appear,— when the long tail cirrhi became selected & developed they required another year's growth & nutriment, & were therefore obliged to *follow* after the colour of head &c. the growth of which was already *established* in the system,—so the side plumes, requiring another years exclusive nutriment, could only follow after the already established tail cirrhi.

The same rule seems to follow in many other cases. The peacock takes three years I am told to acquire his fully developed train, an *unusual ornament*,— the Waxwing does not get the red tips to the secondaries till the 2nd. year & to the tertiaries still later. The Ruby hummer[6] of N. America gets his gorget only the 2nd. year,—& Gosse says the *Long tailed* Jamaica hummer only gets the long feathers of the tail a year after he has got the otherwise perfect male plumage.[7] As in all cases it seems to be the *least common* kind of sexual ornament that appears last, it appears a fair conclusion that those appear last which were selected last, & because they were selected last. I fear no cases of domesticated varieties will help us with facts, because sexual characters have never been selected to any thing like the enormous degree in which they occur in nature. My statement about the *male* & *female* Cassowary sitting alternately was of course derived from the natives, as I never once even saw the adult birds myself.[8] I sh[d] doubt the male's being less brightly coloured,—as these birds have no enemies.

Male savages ornament themselves from personal vanity I sh[d] say, & partly no doubt with a view to admiration by the ladies.

You do not say how far you think my defence of the Dutch policy is successful.[9] That is a point on which I expect the most severe criticism, as it is a subject on which so much prejudice exists in this country. As I went out *with* those prejudices, I ought to be at all events an impartial judge.

In my forthcoming article in the "Quarterly", I venture for the *first time* on some limitations to the power of natural selection.[10] I am afraid that Huxley[11] & perhaps yourself will think them weak & unphilosophical. I merely wish you to know that they are in no way put in to please the Quarterly readers,—you will hardly suspect me of that,—but are the expression of a deep conviction founded on evidence which I have not alluded to in the article but which is to me absolutely unassailable.

With many thanks for your kind remarks on my book | Believe me Dear Darwin | Yours very faithfully | Alfred R. Wallace—

DAR 84.1: 112–15

CD ANNOTATIONS
1.1 Allow me … scanty observations. 4.6] *crossed blue crayon*
6.11 My statement … my book 10.1] *crossed ink*
Top of letter: 'Pheasant Crossoptilon'[12] *pencil*
End of letter: 'Case bears on character appearing late in life.' *pencil*

1 Wallace refers to Dallas trans. 1869 (see letter to Fritz Müller, 14 March 1869 and nn. 1 and 17).
2 See letter to A. R. Wallace, 22 March [1869].
3 See letter to A. R. Wallace, 22 March [1869] and n. 12. The wapiti or (American) elk is *Cervus canadensis*; the moose or (European) elk is *Alces alces*.
4 See letter to A. R. Wallace, 22 March [1869].
5 See letter to A. R. Wallace, 22 March [1869] and n. 14.
6 Wallace refers to the ruby-throated humming-bird (*Archilochus colubris*).
7 Philip Henry Gosse had commented on the development of tail feathers in the male long-tailed hummingbird (*Trochilus polytmus*, now known as the red-billed streamertail) in Gosse 1847, p. 101.
8 See letter to A. R. Wallace, 22 March [1869] and n. 15.
9 For Wallace's discussion of Dutch colonial practices, see Wallace 1869a, 1: 398–403.
10 Wallace refers to his review of Charles Lyell's *Principles of geology* and *Elements of geology* (Lyell 1865 and 1867–8) in the *Quarterly Review* ([Wallace] 1869b). Wallace suggested that natural selection could not explain the origin of the higher intellectual nature of humans or some of their physical characteristics, such as the absence of body hair (see [Wallace] 1869b, pp. 391–3).
11 Thomas Henry Huxley.
12 CD's annotations are notes for his reply (see letter to A. R. Wallace, 27 March [1869]).

From Martin Wilckens[1] 25 March 1869

Pogarth | *bei Prieborn in Schlesien* | Preussen
25 März 1869.

Herrn Charles Darwin | in | Down Bromley Kent S.E.
Hochgeehrter Herr

Ich erlaube mir Ihnen beifolgend zwei kleine Arbeiten von mir zu übersenden, eine frühere, welche sich mit Ihrer "on the origin of species &c" beschäftigt, sowie einen Jahresbericht über die wissenschaftlichen Fortschritte der Thierzucht, welche vom Standpunkte der Thierzucht Ihr von Carus übersetztes Werk einer Kritik unterzieht.[2] Ich bitte Sie mir den Empfang zu bestätigen u. hoffe dass Ihnen eine gegnerische Ansicht auf diesem Gebiete nicht unwillkommen sein wird.

In grösster Hochachtung zeichnet | Ihr ergebenster | Dr. M. Wilckens. | Gutsbesitzer.

DAR 96: 67v

1 For a translation of this letter, see Appendix I.

2 Wilckens enclosed 'Darwin's Theorie in Beziehung zur landwirtschaftl. Viehzucht' (Darwin's theory in relation to cattle breeding; Wilckens 1866) and 'Bericht über die wissenschaftlichen Fortschritte der Thierzucht im Jahre 1867/68' (Report on scientific advances in animal husbandry; Wilckens 1869). CD's annotated copies are in the Darwin Pamphlet Collection–CUL. Wilckens also refers to the German editions of *Origin* (Bronn trans. 1862) and *Variation* (Carus trans. 1868) and to Julius Victor Carus.

To Martin Wilckens [after 25 March 1869][1]

I am very much obliged to you for your kindness in sending me your 2 publications.—[2] I have within the last few days received several works, which I must read first, but as soon as these are finished, I shall have great pleasure, I am sure, in studying your remarks & observations—

with my best thanks Dear Sir—

ADraft
DAR 96: 67r

1 The date is established by the relationship between this letter and the letter from Martin Wilckens, 25 March 1869.

2 See letter from Martin Wilckens, 25 March 1869.

To A. R. Wallace 27 March [1869][1]

Down. | *Bromley.* | *Kent. S.E.*
March 27th

My dear Wallace

I must send a line to thank you, but this note will require no answer.— The very morning after writing I found that Elk was used for moose in Sweden, but I had been reading lately about Elk & moose in N. America.—[2]

As you put the case in your letter, which I think differs somewhat from your book, I am inclined to agree, & had thought that a feather c^d. hardly be increased in length until it had first grown to full length, & therefore it w^d be increased late in life & transmitted to corresponding age.— But the Crossoptilon pheasant & even common pheasant shows that the tail feathers can be developed very early.[3]

Thanks for other facts, which I will reflect on, when I go again over my M.S.

I read all that you said about the Dutch Government with much interest, but I do not feel I know enough to form any opinion against yours.[4]

I shall be intensely Curious to read the Quarterly: I hope you have not murdered too completely your own & my child.[5] I have lately i.e. in new Edit, of Origin been moderating my zeal, & attributing much more to mere useless variability.— I did think I w^d send you the sheet, but I daresay you w^d not care to see it, in which I discuss Nägelis essay on Nat. selection, not affecting characters of no functional importance, & which yet are of high classificatory importance. Hooker is pretty well satisfied with what I have said on this head.[6] It will be curious if we have hit on similar conclusions.— You are about the last man in England, who would deviate a hair's breadth from his conviction to please any Editor in the world.—

Yours very sincerely | Ch. Darwin

P.S. After all I have thought of one question, but if I receive no answer I shall understand that (as is probable) you have nothing to say:—I have seen it remarked that the men & women of certain tribes differ a little in shade or tint: but have you ever seen or heard of any difference in tint between the 2 sexes, which did not appear to follow from a difference in habits of life?—

British Library Board (Add 46434, f. 178–80)

1 The year is established by the relationship between this letter and the letter from A. R. Wallace, 24 March 1869.
2 See letter from A. R. Wallace, 24 March 1869 and n. 3.
3 *Crossoptilon* is the genus of eared pheasants. In *Descent* 2: 166, CD noted that the tails of both sexes were equal in length.
4 CD refers to Wallace's discussion of Dutch colonial practices in *The Malay Archipelago* (see Wallace 1869a, 1: 398–403).
5 CD refers to Wallace's article in the *Quarterly Review* (see letter from A. R. Wallace, 24 March 1869 and n. 10).
6 CD had discussed with Joseph Dalton Hooker revisions for the fifth edition of *Origin* to answer criticisms made by Carl Wilhelm von Nägeli (see letter to J. D. Hooker, 13 January 1869 and n. 1, and letters from J. D. Hooker, 15 January 1869, 18 January 1869, and [25 January 1869]).

From J. D. Hooker to Emma Darwin 29 March 1869

Royal Gardens Kew
March 29 /69.

My dear Mrs Darwin

I shall be delighted to come on the 17th & I hope by that time to have recovered my heart; which Miss Symonds took away with her on Saturday.[1]

Please tell Mr Darwin that I am arranging for the *Aucuba* experiments; with all precautions.[2] I send a letter of Mr. I. A. Henrys, with message for C.D. & a very interesting one of Sir H. Barklys, marking in both the passages in pencil—as I know he is bored beyond endurance with correspondence.[3]

Frances sends her love she is very busy, preparing for Russia with me—[4]

You do not mention Henrietta— we are most anxious to hear how she is.[5]

If Mr Darwin can write please let him tell me what he thinks of Huxley's address.[6] I think it *very* clever.

Ever Sincerely Yrs | Jos D Hooker

I should be extremely glad of Drosophyllum plants— We have had seeds & seedlings (raised ourselves) & live plants, which I have had tried in 3 different places & by as many persons, but all died.[7]

[Enclosure]

Redit, Mauritius
13th Feby 1869

My dear Dr Hooker,

I am in your debt for two letters dated 7th & 29th Decr. The former enclosed one from Mr Baker[8] about ferns, which I defer answering till next Mail, by which time I hope to hear again from him about our Bourbon & Madagascan discoveries.

He rather puzzles us, by finally referring Lady Barkly's Hymenophyllum H. emersum to H. tenellum (as representative of Jacquin's old Adiantum tenellum,) for this latter was identified by your Father in the Sp. Filicum with the H. ricciæfolium which figures in Bojer's Hortus Mauritianus, and is described as like H. Polyanthos a Fern for which Mr Baker gives Mauritius credit in the Synopsis.[9]

Query does the Kew Herbarium contain specimens of both H. tenellum & H. polyanthos from this Island?

What complicates the matter is that Bojer gives H. tenellum (of Don) as a synonym of our widely different H. gracile.[10]

I will not take up your time with more ferny talk, but I must mention that our friend Mrs Higginson, the daughter of the former Governor of that name,[11] and who has been the companion of many of our explorations and has added to the List of Mauritius Ferns Goniopteris pectifera, & Pellæa *[annluni]* &c, —goes home by this Mail, & will probably find her way in due course to Kew, very full of the subject. She takes with her Plants of the terrestrial Orchid which Dr Meller fancied might be 'Dryopeia cordata', but which you thought new.[12] It has flowered with her, and is clearly not the Plant described in Lindley[13] as it has 8 to 10 large greenish flowers on its erect stem.

I had no opportunity of examining the structure of the Flowers on the Pollen, but if it flowers with me I will do so. Mrs Higginson has a drawing of the Plant in flower, which will probably enable you to judge if it be new. I have not received any letter from Meller since he left this, but I heard that he has written to me by sailing Ship, and I heard from Dr V. Mueller[14] by the Mail that he had been at Melbourne & gone on to Sydney.

Horne[15] as you remark is very competent to replace him in the case of the Botanic Gardens, and he is superintending our Sugar Cane Experiments to my entire satisfaction.

With regard to his explanation of the Islets on the Coasts, rumour says that Horne was not bold enough to land at the chief of them—Isle Ronde, a very dangerous matter from the surf, even in calmer conditions, & that his companion young Duncan alone got on shore.[16]

The Island however has been lately visited by another Party in which was included the American Consul Colonel Pike, who like most Yankees knows something of everything, & who has written a very interesting Paper on its Geology, with a description of its Fauna & Flora, which is to be read at the next meeting of our Local Royal Society, & of which I will send you a copy.[17]

There can be no doubt that this little rock lying but a few Miles from Mauritius offers some of the strangest problems as to the distribution of Plants and Animals that can be met with.

It has not only its peculiar Pandanus—as you say—which by the bye has been christened at Rotterdam P. Vandermeerschii—after one of our Surveyors here who has sent many Plants to Holland,— but it has its peculiar Palm the Jubæa spectabilis of Duncans catalogue a name I presume derived from Kew.[18]

Another a Latania glaucophylla is found on others of these Islands & I believe in Bourbon, whilst it has also Areca alba in common with Mauritius.[19]

As regards its Zoology— Colonel Pike proved that the tradition of its having Snakes (which are unknown here as in most volcanic Islands) is true, by bringing back one in Spirits of the genus *[*Heteræum*]* about 2 foot long and 2 inches round, & he also captured several large lizards quite unknown here.[20]

Having at a little distance the appearance of other great Island Cones about a Mile in diameter at the base & 1000 feet in height to the apex, it has generally been deemed from the days of Bory[21] & downwards a Volcano, but Colonel Pike describes it as consisting of beds of friable Sandstone upheaved at angles of inclination more and more acute all round the axis of the Cone from the base to the summit.

He certainly has given me specimens of red Sandstone.

Is it not allowable to suppose that Round Island therefore is the remnant of a much more ancient Island or Continent which once existed where Mauritius is now situated and of which traces are to be found in the Casts of the stems of Submarine forests found by Dr Ayres in Gabriel Island & by Mr Charles in the Islands off Grand Port.[22] If so how many hints in the chain of animal & vegetable existence may not have perished in the Volcanic eruptions which produced the Mascarene groups! I have just had an opportunity of reading Dr Wrights letter to the Principal of his University about the Seychelles, & must own I am a good deal disappointed.[23] He was I see to publish the first part of his Flora of the groups in November, & I must order it to be sent to me.[24] I presume he is competent for the task, but I hope he will be fair to those who have gone before him. I see he makes great capital out of his discovery of Nepenthes Wardii; but surely its existence was well known to your father & Mr Newton certainly took it to Kew in flower before Dr Wright set foot at Mahé.[25] Thanks for your hint about the Tropical African Flora. I have applied to Mr Layard for copies.[26] Many thanks also for completing Lady Barkly's set of the Synopsis Filicum which she will highly value.[27] I hope you will be able to make your expedition to Russia and return in renovated health to your post.[28]

Very truly Yours
Henry Barkly

DAR 103: 12–13; Royal Botanic Gardens, Kew (DC 188: 141–2)

1 Hooker refers to Hyacinth Symonds, whom Hooker had met earlier in the year (see letter from J. D. Hooker, [25 January 1869]). According to Emma Darwin's diary (DAR 242), Hooker visited on 25 April 1869, but no visit is recorded for 17 April.

2 See letter to J. D. Hooker, 17 March [1869] and n. 4.

3 Hooker refers to Isaac Anderson-Henry and Henry Barkly. The letter from Anderson-Henry has not been found.

4 Frances Harriet Hooker was preparing for Hooker's trip to Russia for the International Horticultural Exhibition at St Petersburg (see letter from J. D. Hooker, 11 March 1869 and n. 4).

5 Henrietta Emma Darwin had been ill for a couple of weeks (see letter to J. D. Hooker, 8 March [1869] and n. 4).

6 Hooker refers to Thomas Henry Huxley's presidential address to the Geological Society of London on 19 February 1869 (T. H. Huxley 1869c).

7 CD had received plants of *Drosophyllum lusitanicum* from Portugal (see letter to J. D. Hooker, 17 March [1869] and nn. 5 and 6).

8 John Gilbert Baker.

9 The fern *Hymenophyllum tenellum* (*H. emersum*, *H. ricciaefolium*, *H. polyanthos*, and *Adiantum tenellum* are synonyms) is now *Mecodium tenellum*. Barkly also refers to Wenzel Bojer and Bojer 1837, p. 411, to William Jackson Hooker and W. J. Hooker 1846–64, 1: 107–8, and to W. J. Hooker and Baker 1868, p. 60.

10 See Bojer 1837, p. 411. *Hymenophyllum gracile* is now *H. inaequale.*

11 Barkly refers to Mary Higginson and James Macaulay Higginson.

12 The names *Goniopteris pectifera* and *Pellaea annuli* were evidently not published. The orchid described by Charles James Meller as *Dryopeia cordata* is *Disperis cordata* (*Dryopeia discolor* is a synonym).

13 Barkly refers to John Lindley and Lindley 1830–40.

14 Ferdinand von Mueller.

15 John Horne.

16 Barkly refers to Round Island and to J. W. Duncan.

17 Nicolas Pike's paper 'Geology, fauna and flora of Round Island, Mauritius' appeared in *Transactions of the Royal Society of Arts and Sciences of Mauritius* n.s. 4 (1870): 11–22.

18 Barkly refers to the screwpine species, *Pandanus vandermeeschii*; Vandermeesch has not been identified. *Jubaea spectabilis* is now *J. chilensis*, the Chilean wine palm. Barkly also refers to James Duncan's *Catalogue of plants in the Royal Botanic Garden, Mauritius* (Duncan 1863; see Barkly 1870, pp. 94–5 for his discussion of the misuse of the name *J. spectabilis* for a species of Round Island palm).

19 *Latania glaucophylla* is now *L. loddigesii*, the blue latan palm. *Areca alba* is now *Dictosperma album*, the hurricane palm.

20 Barkly may intend *Herpetodryas*, a snake genus he mentioned in his paper, 'Notes on the flora and fauna of Round Island' (Barkly 1870, p. 98). The two snakes found on Round Island are *Casarea dussumieri*, the Round Island keel-scaled boa, and *Bolyeria multicarinata*, the Round Island burrowing boa; they are now in their own family, Bolyeriidae (Mauritius snakes). Round Island has some endemic gecko and skink species, the largest of which are *Phelsuma guentheri*, Gunther's gecko, and *Leiolopisma telfairii*, Telfair's skink.

[21] Jean Baptiste Georges Marie Bory de Saint-Vincent.

[22] For some observations by Philip Burnard Ayres on the geology of Gabriel Island, see Ayres 1860. Mr Charles has not been identified. Grand Port is an area in the south-east part of the island of Mauritius near the city of Mahebourg.

[23] Barkly refers to Edward Perceval Wright, who was professor of botany at Trinity College, Dublin (*ODNB*).

[24] Wright's paper, 'Contributions towards a knowledge of the flora of the Seychelles' was read at the Royal Irish Academy in December 1868 (Wright 1868).

[25] *Nepenthes wardii* is now *N. pervillei*, the name it was originally given in 1852 (see *Museum Botanicum Lugduno* 2 (1852): 10). Edward Newton was colonial secretary of Mauritius. Mahé is the largest island of the Seychelles.

[26] The first volume of *Flora of tropical Africa* had been published in 1868 (Oliver 1868–77). Edgar Leopold Layard was the curator of the South African Museum.

[27] Barkly refers to Anne Maria Barkly and to *Synopsis filicum* (W. J. Hooker and Baker 1868) which updated the earlier five volume work, *Species filicum* (W. J. Hooker 1846–64).

[28] Hooker visited Russia in May 1869 to attend the International Horticultural Exhibition at St Petersburg (see letter from J. D. Hooker, 11 March 1869 and n. 4).

To W. T. Preyer 29 March 1869

March 29. 1869.

Dear Sir.

I thank you sincerely for your very obliging letter, and the present of your two Essays—It is highly gratifying to me to hear of the great success of your lectures (on Darwinianism) though I must attribute this to your powers as a lecturer— Unfortunately I am a very poor German Scholar and I have not yet had time (but assuredly will soon find time) to read your essay excepting the two notes— Had I known of the "Mammæ Erraticæ" I should not have hazarded the speculation on Reversion— About the term "Struggle for Existence" I have always felt some doubts but was unable to draw any distinct line between the two ideas therein included, I suspect that the German term Kampf &c. does not give quite the same idea.—[1] The words "Struggle for Existence" express I think exactly what concurrency does— It is correct to say in English that two men struggle for existence who may be hunting for the same food during a famine, and likewise when a single man is hunting for food; or again it may be said that a man struggles for existence against the waves of the sea when shipwrecked. I have never hitherto made a list of all my works, but your letter has led me to do so. I now send a copy, though it gives fuller details than you can possibly want.[2]

It is the custom in our scientific societies to publish abstracts before the papers appear in full; but I have not thought it worth while to hunt these up. Your N°. 12 is probably a translation of one such abstract, or perhaps of extracts of some of my letters. Your N°. 18 I presume is an extract from my Journal of Travels.[3]

Whilst searching for the paper which you require I found one or two other trifling articles, which I send to you by this post.

I congratulate you on your election to your Professorship,[4] and with very sincere thanks for your kind interest about me, | I remain, Dear Sir | Yours very faithfully | Charles Darwin.

Copy
DAR 147: 254–6

[1] See letter from W. T. Preyer, 21 March 1869 and n. 1.
[2] Preyer had drafted a list of CD's works that he sent to CD to be corrected and completed (see letter from W. T. Preyer, 21 March 1869 and n. 2). CD's list has not been found, but see Preyer 1896, pp. 72–99.
[3] The numbers were evidently from the list Preyer sent. The 'Journal of Travels' is *Journal of researches*.
[4] Preyer had just been made professor of physiology at Jena (see letter from W. T. Preyer, 21 March 1869).

From Armand de Quatrefages[1] 29 March 1869

Paris
29 Mars 69

Mon cher Confrere

Je voulais ne répondre à votre lettre que lorsque j'aurais eu le plaisir de voir M^r votre fils.[2] Mais il parait que le destin ne veut pas que j'aie ce plaisir. Il est venu deux fois chez moi, je suis allé deux fois chez lui, toujours sans que nous nous soyons rencontrés. Et pourtant je passe des trois mois entiers sans sortir si ce n'est pour aller à l'Institut au le soir.[3] J'espère que je finirai par être plus heureux et que j'aurai le plaisir de faire sa connaissance un de ces jours.

Je vous remercie de la manière dont vous avez apprécié mes études sur votre théorie. Il m'était devenu impossible de ne pas entreprendre cette discussion. Au nom du transformisme en général, au nom de vos idées en particulier, je voyais chaque jour attaquer et représenter comme des vieilleries usées, ce que je regarde précisément comme les fondements des Sciences naturelles. J'avais été souvent personnellement désigné comme un homme arriéré et qui fermait volontairement les yeux aux progrès de la Science.[4] J'ai donc été obligé de me défendre. A mon tour, j'ai du discuter les théories qu'on m'opposait. Vous êtes incontestablement le chef de toutes les théories transformistes, le seul qui ayez proposé une théorie fondée uniquement sur des considérations scientifiques et embrassant l'ensemble des problêmes posés par la nature vivante. C'est donc avec vous que j'avais surtout à lutter. J'ai fait mon possible pour montrer que la raison scientifique est de mon coté. Mais j'espère n'avoir jamais méconnu la grandeur de votre œuvre, et en terminant cette longue etude dans un dernier article qui paraitra le 1er Avril, j'ai été heureux d'exprimer publiquement toute l'estime que je porte en vous à l'homme et au savant— Dès que j'aurai reçu mes *tirés-à-pàrt* le premier exemplaire vous sera envoyé.[5]

Je profite de l'occasion pour vous exprimer mes regrets au sujet d'une inexactitude qui s'est glissée dans mon troisieme article.— En examinant ce que vous dites de la *mesange* et du *casse-noix* dans votre livre sur l'espèce, j'avais mis une note dans la quelle je faisais remarquer que vous ne parliez pas précisément d'une généalogie; que vous n'affirmiez rien positivement. Cette note a disparu à l'impression définitive avec beaucoup d'autres. Probablement des arrangements typographiques auront été cause de sa disparition. Elle sera rétablie dans l'édition en volume qui sera faite de mes articles Mais je tiens à vous dire dès à présent que je ne suis pas coupable du fait.[6]

J'espère bien, Monsieur et cher confrere, que des dissidences d'opinion n'altereront jamais nos relations. C'est dans cet espoir que je vous prie d'agréer l'expression de mes sentiments les plus cordialement dévoués | De Quatrefages

DAR 175: 5

1 For a translation of this letter, see Appendix I.

2 CD's letter has not been found. George Howard Darwin had probably travelled to Paris on 5 March 1869 (see letter from G. H. Darwin, [23 February 1869] and n. 7).

3 Quatrefages refers to the Institut de France. The institute was the home of five learned societies, including the Académie des Sciences, of which Quatrefages was a member (*DSB*).

4 For more on the French debates over transformism, see Conry 1974, pp. 359–92. See also *Correspondence* vol. 16, letter from Camille Dareste, 3 April 1868.

5 Quatrefages later sent CD an offprint of a series of five articles he had written for the *Revue des Deux Mondes* between December 1868 and April 1869, 'Origines des espèces animales et végétales' (Quatrefages 1868–9). CD's annotated copy of the offprint is in the Darwin Pamphlet Collection–CUL.

6 In Quatrefages 1868–9, p. 63, Quatrefages had questioned whether CD had demonstrated that the nuthatch was the 'grandson' of the titmouse. When Quatrefages later published a book based on these articles, a note was added giving a reference to the section of *Origin* where CD had commented on the titmouse and nuthatch (see *Origin* 4th ed., p. 281), but there was no mention in the note that CD had not, in fact, suggested a genealogical connection between these species (Quatrefages 1870, p. 156).

To J. D. Hooker 30 March [1869][1]

Down. | Bromley. | Kent. S.E.
March 30th

My dear Hooker

Many thanks for sight of the letters.[2] The account of the islet off Mauritius is most wonderful. I wish I c^d believe that the island was formed of upheaved sandstone, as it w^d. indicate former *non*-volcanic land.— I do not know whether Sir H. B is a geologist; but until some one, who is, sees the sandstone, I must believe that a steep conical islet, near a volcanic island, formed of *friable* sandstone, with the beds rising in inclination to the summit, consists of tuff.

You ought to urge Sir H. B. to have every plant, shell, insect & reptile collected. On any view, former continent, or waifs & strays (which I rather lean to, as snake & reptile different) the case is most interesting.—[3]

You are a real good man to come here on the 17th—[4]

I thought Huxleys address wonderfully brilliant,—some of his sayings most happy.— I wrote indeed to him to say that it was a shame that any one, man sh^d. have the power to write so many brilliant essays.—[5] I do not think it right to make a class of Evolutionists, as distinct from Uniformitarians, simply because they are ready to speculate on first origin of our planet.— I think he has here sacrificed reality to make a striking point.[6]

Yours affect[y] | C. Darwin

Henrietta is a little better, & comes down daily.—[7]

Judging from Report in Gard. Chron. how splendidly Bentham has come out on the Descent of Species: I shall be very much interested in reading it in larger type in our Journal.—[8]

DAR 94: 121–2

[1] The year is established by the relationship between this letter and the letter from J. D. Hooker to Emma Darwin, 29 March 1869.

[2] Hooker had enclosed letters from Henry Barkly and Isaac Anderson-Henry with his letter to Emma Darwin of 29 March 1869.

[3] See the letter from Barkly to Hooker enclosed in Hooker's letter to Emma Darwin of 29 March 1869.

[4] See letter from J. D. Hooker to Emma Darwin, 29 March 1869 and n. 1.

[5] CD refers to Thomas Henry Huxley's presidential address to the Geological Society of London on 19 February 1869 (T. H. Huxley 1869c). See letter to T. H. Huxley, 19 March [1869].

[6] See letter to T. H. Huxley, 19 March [1869] and n. 3.

[7] Henrietta Emma Darwin had been ill for a couple of weeks (see letter to J. D. Hooker, 8 March [1869] and n. 4).

[8] CD refers to the report in *Gardeners' Chronicle*, 27 March 1869, pp. 336–7, of the Linnean Society meeting of 4 March 1869 at which George Bentham read a paper, 'Revision of the genus *Cassia*'. Bentham was quoted as affirming 'the great principle that natural affinity results from community of descent'. Bentham's paper was later published in *Transactions of the Linnean Society of London* (Bentham 1869a).

To C. F. Claus 31 March 1869

Down. | Bromley. | Kent. S.E.
March. 31/69/

Dear Sir

On receiving your letter of the 19th, I wrote at once to M[essrs] Williams & Norgate, but owing to some accident received their answer only this morning.[1] They inform me that they despatched my small parcel on Feb. 23 to Leipzig to be forwarded to you: & they expect that you will have by this time received it.— They do not mention who their correspondents in the Book-trade are in Leipzig..— I fear my parcel is not worth so much trouble to you..—

Dear Sir | Your's very sincerely | Ch. Darwin

Staatsbibliothek zu Berlin – Preußischer Kulturbesitz, Slg. Darmstaedter Lc 1859 Darwin, Charles, Bl. 208/209

[1] See letter from C. F. Claus, 19 March 1869 and n. 2. CD had sent Claus specimens of barnacles.

From Richard Spruce [before 1 April 1869][1]

Welburn | Castle-Howard | York.

⟨*8 lines missing*⟩ plain of ⟨ ⟩— I never again saw a plant of that fam⟨ily⟩ ⟨Had⟩ I been able to travel I should have gone again in quest of them, not only to try to obtain the information you wanted, but to follow up some observations of my own touching the origin of curious bags on the leaves of some of the species inhabited by ants.[2]

A few years ago I drew up a memoir on certain modifications of the organs of plants which I consider entirely the work of insects, altho' some of them take rank as specific and even generic characters. I have kept the Mst. by me hoping at least to go again to Kew and reexamine my specimens of those curious structures, but as the state of my health renders that an impossibility I have written out the memoir afresh, and send it to you by this day's post, requesting to be kind enough to look through it carefully and then hand it over to the Linnean Society to be read at one of their meetings.[3] I shall be thankful for a line to say how far you consider my inferences justified by the facts.

⟨*8 lines excised*⟩

It is a great defect of my relation that it contains no definite account of the ants infesting the plants, & I have no materials for supplying it. I have somewhere a small packet of ants that dropped out of my specimens of Amaioua saccifera,[4] but I cannot just now lay my hand on it.

I have often wished to write to thank you for your admirable writings on philosophic biology, most of which I have read since I returned home invalided 4 years ago. The 'Origin of Species', especially, filled up a void of which I had long been sensible in my own speculations. When a very young botanist I came to some of the same conclusions as yourself, but very empirically, & not by any process of laborious and accurate observation such as you have gone through. When I read Linnæus's definition of a species in the 'Philosophia Botanica', based on the supposed fact that no hybrid between *true* species could be fertile,[5] I said to myself "We ought surely to *experiment* unceasingly on the interbreeding of so-called species, for we may go on *describing* animals & plants to the end of all time and we shall never attain to the correct delimitation of a single species". Afterwards, when I tried to deduce some "law of species" from the writings of eminent botanists, I could find only this, running through the descriptions of every author: "*My* species are *good* species—those of other authors probably not"—the same as in any other orthodoxy & heterodoxy. I could then only resolve to examine everything for myself—to take names and descriptions as (at best) attempts to fix transitory aspects of things—and to try to trace what variation of form & structure was tending to. Your book makes it most clear that *variation in plants and animals is not oscillation round fixed centres but progress in directions determined by surrounding influences.*

But I am taking up too much of your precious time, so, wishing you still a long lease of life and energy to continue your most valuable labours, I subscribe myself | Yours very faithfully | Rich. Spruce.

Charles Darwin Esq.

Incomplete
DAR 177: 241

1 The date is established by the relationship between this letter and the letter to Richard Spruce, 1 April 1869.

2 Spruce probably refers to the former family Melastomeae in the order Melastomaceae (now the family Melastomataceae in the order Myrtales). CD had sent Spruce queries about Melastomataceae at the end of 1863, when Spruce was ill and coming to the end of his stay in South America (see *Correspondence* vol. 12, letter from A. R. Wallace, 2 January 1864 and nn. 12 and 13); CD's letter to Spruce has not been found. CD was interested in a possible new form of dimorphism in the Melastomaceae (see *Correspondence* vol. 10, letter to Asa Gray, 22 January [1862] and n. 13).

3 Spruce's paper was read at the Linnean Society on 15 April 1869, but was not published in the society's proceedings. The paper is printed in full in Spruce 1908, 2: 384–412, together with an account by Alfred Russel Wallace of events associated with its reading. Spruce also refers to the Royal Botanic Gardens, Kew.

4 *Amaioua saccifera* is a synonym of *Duroia saccifera*, a member of the family Rubiaceae. It has 'ant-houses' on its leaves (Willis 1973). A specimen of the wood of *A. saccifera* is in the Richard Spruce Collection, Royal Botanic Gardens, Kew.

5 Spruce refers to Carl von Linné and Linnaeus 1751: see especially sections 132 and 259. CD wrote on hybrid fertility in *Origin*, pp. 246–50; see *Origin* 4th ed., pp. 310–26, and *Correspondence* vol. 10, Appendix VI.

To Richard Spruce 1 April 1869[1]

The facts which you state are extraordinary, and quite new to me.[2] If you can prove that the effects produced by ants are really inherited, it would be a most remarkable fact, and would open up quite a new field of inquiry. You ask for my opinion; if you had asked a year or two ago I should have said that I could not believe that the visits of the ants could produce an inherited effect; but I have lately come to believe rather more in inherited mutilations. I have advanced in opposition to such a belief, galls not being inherited. After reading your paper I admit, Firstly, from the presence of sacs in plants of so many families, and their absence in certain species, that they must be due to some extraneous cause acting in tropical South America. Secondly, I admit that the cause must be the ants, either acting mechanically or, as may perhaps be suspected from the order to which they belong, from some secretion. Thirdly, I admit, from the generality of the sacs in certain species, and from your not having observed ants in certain cases (though may not the ants have paid previous visits?), that the sacs are probably inherited. But I cannot feel satisfied on this head. Have any of these plants produced their sacs in European hot-houses? Or have you observed the commencement of the sacs in young and unfolded leaves which could not *possibly* have been visited by the ants? If you have any such evidence, I would venture strongly to advise you to produce it …

I may add that you are not quite correct (towards the close of your paper) in supposing that I believe that insects directly modify the structure of flowers. I only believe that spontaneous variations adapted to the structure of certain insects flourish and are preserved.[3]

Incomplete
Spruce 1908, 2: 385

[1] The date is provided by Alfred Russel Wallace, the editor of Spruce 1908, from which this transcription of part of CD's letter is taken.
[2] See letter from Richard Spruce, [before 1 April 1869] and n. 3.
[3] Spruce had written, 'The inimitable researches of Mr. Darwin have rendered it (to my mind) almost certain that many of the deviations from symmetry in the form and direction of the parts of a flower have been brought about by the direct mechanical agency of insects' (Spruce 1908, 2: 408).

To Benjamin Dann Walsh 3 April [1869][1]

Down. | Beckenham | *Kent. S.E.*
April 3^d.

My dear Sir

I have received your kind letter of March 12th & the 3 numbers of the Amer. Entomologist.—[2]

I am very glad that you have proved your case about the dimorphism of the Cynips.—[3] I will send by this Post the specimens to Fred. Smith of Brit. Museum,[4] either for his private or the public collection as he may think best.—

What an admirable Journal your new one is: I have read two numbers, but not as yet that on Wasps,[5] but I shall begin it this evening.— Nothing in my opinion can be better than the two articles on Galls & the Cicadas. That is a most perplexing case about the 2 forms of the Cicadas.—[6] The galls interest me much, not so much as galls, but as a proof what organic changes extraneous matter can cause.—[7] There are numbers of curious facts, for example the sterility of the Hateful Grasshopper.—[8]

I will not so much trespass on your kindness as to accept the rare back numbers; but if you publish anything which you think would interest me (& you can judge almost as well as I can myself) I sh^d. be grateful for an odd number.—[9]

My dear Sir | Yours sincerely | Ch. Darwin

P.S. I have just finished your Wasp paper—[10] it is capital very interesting & very amusing.—

Postmark: London | Ap 5 | 69
Field Museum of Natural History, Chicago (Walsh 17)

[1] The year is established by the publication date of the *American Entomologist* articles sent by Walsh (see n. 2, below).
[2] Walsh's letter has not been found. CD refers to issues 4, 6, and 7 of the first volume of the *American Entomologist*, dated December 1868, February 1869, and March 1869 respectively. The issues are in the Darwin Pamphlet Collection–CUL; four articles by Walsh and Charles Valentine Riley (B. D. Walsh and Riley 1868a–b and 1869a–b) are annotated (see nn.3, 5, 6, and 8, below).
[3] Experimental proof that the gall wasp *Cynips quercus aciculata*, hitherto regarded as a distinct species, was one of the two female forms of *C. q. spongifica* was presented in B. D. Walsh and Riley 1869a, p. 103; Walsh had explained the experiment in his letter to CD of 12 November 1865 (*Correspondence* vol. 13). For Walsh's initial speculation on the subject, see B. D. Walsh 1864, pp. 447–8, and *Correspondence* vol. 12.
[4] Frederick Smith was an entomologist in the British Museum.
[5] Walsh and Riley started the *American Entomologist* in 1868 (see *Correspondence* vol. 16, letter from B. D. Walsh, 29 August 1868). The article on wasps is B. D. Walsh and Riley 1869b.

[6] In 'The periodical cicada' (B. D. Walsh and Riley 1868a), the authors described *Cicada septemdecim* (now *Magicicada septendecim*), which emerged every seventeen years, and a cognate species, which emerged every thirteen years; see also *Correspondence* vol. 16, letter from B. D. Walsh, 29 August 1868 and n. 17, and letter to B. D. Walsh, 31 October 1868.

[7] CD had cited Walsh's earlier paper on galls (B. D. Walsh 1866) in *Variation* 2: 282–5.

[8] In 'The hateful or Colorado grasshopper', Walsh and Riley discussed the sterility of the grasshoppers born on the plains; the fertile grasshoppers migrating from mountain canyons were destructive but their young ate little and did not breed ([B. D. Walsh and Riley] 1868b). The species, *Melanopus spretus*, also known as the Rocky Mountain locust, is now extinct (Lockwood 2004).

[9] Though no further correspondence with Walsh has been found, issues of the *American Entomologist* for April, September, and October 1869 are in the Darwin Archive–CUL.

[10] See n. 5, above.

From George M'Ilvaine Ramsey 5 April 1869

New York
April 5th 1869

Charles Darwin M.A. F.R.S.
Dear Sir

You have fought your battle, & are reaping the fruits of victory.

I have just begun.

I have a work (Cosmology) ready for the press—but alass neither the press nor people ready for the work—which is destined to overturn nearly all that has hitherto been taught of Astronomy, Geology, & Theology.[1]

Save only the *Darwin System.* & you sir well know that no one truth ever clashes with another.

I have discovered the origin of *Diurnal motion.*[2] & the results, as seen logically flowing therefrom are most wondrous in scope, diversity & magnificense.

The Advocates of *Periodic creations*, you are aware, cite against you, the abrupt *termination* of some-so called-species, & the *introduction* of others in the same locality—but in different geological periods.

My Work demonstrates to the *Scientific* World the *origin of Periods.* & thereby explains this *apparent* extinction, of some species.[3]

I wish first to explain to you that *Diurnal Motion*, is the result of *Sunshine*[4]

AL incomplete
Ursula Mommens Deposit, 'Amy Ruck letters' (DAR 271.6 (3))

[1] *Cosmology* (Ramsey 1870) was published by W. White and Company in Boston. Ramsey gave a lecture expounding the same ideas on 15 June 1876; this was published as Ramsey 1877. See also Ramsey 1897.

[2] Diurnal motion is the apparent movement of the sun, moon, and planets around earth resulting from the earth's rotation (*EB*).

[3] Ramsey argued that the combined forces of atmospheric pressure, the earth's rotation, and gravity resulted in periodic and dramatic shifts in the earth's axis: 'at long but undefined periods the earth is virtually capsized, causing geological and glacial periods and changes in the geographical position of the polar centres' (Ramsey 1870, quoted in Crumrine 1882, p. 959). See also Ramsey 1877.

[4] Ramsey wrote: 'Atmospheric motion is simply the result of sunshine on vapor … Diurnal rotation of the Earth, [is] the result of atmospheric pressure and motion' (Ramsey 1877, pp. 3, 10).

To ? 6 April [1869–71][1]

Beckenham, Kent
April 6th

[...] My experiment was intended solely to show that colour reappeared, and I choose kinds which breed turn to colour, as is certainly the case with spots and those which I tried[2] [...]

[...] I have recorded an undoubted case of wild rock Pigeons caught in Scotland having bred, in confinement[3] [...]

[et termine sa lettre en rappelant qu'il est accoutumé à être démenti et habitué au peu de confiance qu'accordent à la véracité de ses écrits ceux][4] [...] who do not take the trouble to learn what I have written [...]

Incomplete[5]
L'Autographe S.A. Catalogue 21

1 The date range is established by the address. According to the sale catalogue, the name of the town Bromley is corrected by hand to Beckenham. This form of address was used by CD between April 1869 and May 1871.

2 Spots are a breed of pigeon; see *Variation* 1: 156. The word transcribed in the catalogue as 'turn' was probably 'true'.

3 See *Variation* 1: 184–5, and *Correspondence* vol. 6, letter to Laurence Edmonston, 19 April [1857].

4 'He finishes his letter by recalling that he is used to being contradicted and accustomed to the little trust that is placed in the truth of his works by those ...'

5 The original letter is complete and is described in the sale catalogue as being one and a half pages long.

To W. C. Tait 7 April [1869][1]

Down. | Beckenham | *Kent. S.E.*
Ap. 7.

Dear Sir

I am delighted to say that your plants of Drosophyllum are recovering; the old leaves have rotted off & nice healthy new ones are forming. They succeed best in a deep pot with very light soil, standing in a saucer of water, & not watered from above, & they like a rather high temperature.[2]

The young leaves have caught a good many minute flies. The viscid fluid which is apparently attractive to the insects, is acid as in Drosera & Dionæa; & as in these plants softens the insects. I wished for negative results, but these have proved rather too negative for my taste; for I can detect no trace of movement from the application of nitrogenous matter nor from mechanical irritation. Although I have not the least doubt from analogy that the glands absorb the animalized fluid, I cannot prove this; but I shall have another trial on a more extensive scale with a very weak solution of carbonate of ammonia.[3]

I will send a couple of plants to Kew, where I hear they will be eagerly received. I hope my last letter came safe to hand.[4]

With very sincere thanks for your great kindness believe me dear Sir | yours faithfully | Ch. Darwin

LS
Postmark: Beckenham Ap. 8 | 69
C. C. Kohler (dealer)

1 The year is established by the relationship between this letter and the letter to W. C. Tait, 12 and 16 March 1869.

2 Tait had sent specimens of *Drosophyllum lusitanicum* from Portugal. See letter to W. C. Tait, 12 and 16 March 1869, and letter to J. D. Hooker, 17 March [1869].

3 CD described his observations of *Drosophyllum lusitanicum* in *Insectivorous plants*, pp. 332 *et seq.* CD experimented on *Drosera* and *Dionaea* in 1860 and 1861 (see *Correspondence* vols. 8 and 9; see also *Insectivorous plants*).

4 The 'Plants inwards' book at the Royal Botanic Gardens, Kew, records the receipt of two plants of *Drosophyllum lusitanicum* from CD on 29 September 1869. CD also refers to the letter to W. C. Tait, 12 and 16 March 1869.

From Alfred Newton 9 April 1869

Magd: Coll: Cambridge
9 April 1869.

My dear Sir,

Mr J. W. Clark who lately met your son Mr. George Darwin in Paris tells me from him you had some kind intention of helping us in our troubles, I therefore take the liberty of sending you a copy of the appeal to the Members of the Senate which we have circulated.[1]

It was with very great regret that I heard that your son Frank was not successful in the Examination for the Trinity Scholarship—but it affords me much satisfaction to learn, as I have done today, that he acquitted himself very creditably—especially in the *vivâ voce* part of the examination.[2] I believe that my regret is fully shared by nearly all the members of the University who take any interest in Natural History and I sincerely trust he may try his luck next year to find that he is a favoured being in this struggle for existence—

Believe me, my dear Sir, | Yours very truly | Alfred Newton

Charles Darwin, Esqre. M.A., F.R.S.

[Enclosure]

CAMBRIDGE
February 15, 1869.

GENTLEMEN,

We wish to draw your attention to the condition of the Museum of Zoology.

Want of funds has hitherto prevented the Museum Syndicate from fitting up this portion of the building: nor does it appear probable that they will, for many years at least, have sufficient money at their disposal to enable them to purchase the requisite cases and cabinets. The various collections, especially the Swainson collection, have suffered considerably from want of proper protection: the Strickland collection, which was presented to the University last year on the express understanding that cabinets should be provided for its reception with as little delay as possible, has not yet been unpacked: the Woodward collection of shells is very inconveniently located, and not displayed at all; and the Philosophical Society's collections remain in the same state as when they were presented to the University five years ago.[3] It is manifest that it would be highly prejudicial to the interests of Natural Science, and would appear discreditable to the University, to allow the Museum of Zoology to remain thus neglected any longer.

Under these circumstances, we appealed last year to the Colleges, with the approval of the Museum Syndicate, to subscribe out of their corporate funds a sum of £500, which we then thought would be sufficient to provide cabinets for the Swainson and Strickland collections. They have contributed only £385, and at the same time we find that we understated the sum required: as we were unable, at the time we made that statement, to do more than obtain a very rough estimate of the probable cost of the fittings and of the collateral expenses.

We now find that we shall require *at least* £800 to do the work thoroughly.

We are therefore compelled to appeal to the liberality of individual members of the University, who have already, on more than one occasion, generously subscribed towards supplying the wants of the Museum, in the hope that they will come to our assistance, and enable us to place this portion of the collections under our charge in a condition worthy of the University.

ALFRED NEWTON, | *Professor of Zoology and Comparative Anatomy.*
J. W. CLARK, | *Superintendent of the Museum of Zoology and Comparative Anatomy.*

The following subscriptions have been already received:—

	£	*s.*	*d.*
The Chancellor[4]	20	0	0
The Vice-Chancellor[5]	10	0	0
The Master of Trinity[6]	10	10	0
The Master of Jesus[7]	5	0	0
Professor Lightfoot[8]	10	0	0
Professor Kingsley[9]	5	0	0
Professor Birkbeck[10]	2	2	0
A. A. VanSittart, Esq., Trinity College[11]	10	0	0
Rev. R. Burn, Trinity College[12]	10	0	0
Rev. T. G. Bonney, S. John's College[13] .	5	0	0
L. Ewbank, Esq., Clare College[14]	5	0	0

Subscriptions may be paid to *Mr* J. W. Clark, *Trinity College, or to his account with Messrs* Mortlock.[15]

DAR 172: 47

[1] Newton refers to John Willis Clark and George Howard Darwin. The Senate was the governing body of the University of Cambridge.
[2] Francis Darwin was an undergraduate at Trinity College, Cambridge.
[3] Newton refers to Samuel Pickworth Woodward's mollusc collection and William Swainson's collection of bird skins. Hugh Edwin Strickland's collection of bird skins was originally donated to Oxford University, but moved to the University of Cambridge in 1867 (*ODNB*). The Cambridge Philosophical Society donated its natural history collections to the university in 1865 (A. R. Hall 1969, p. 28).
[4] William Cavendish.
[5] Edward Atkinson.
[6] William Hepworth Thompson.
[7] George Elwes Corrie.
[8] Joseph Barber Lightfoot.
[9] Charles Kingsley.
[10] William Lloyd Birkbeck.
[11] Augustus Arthur VanSittart.
[12] Robert Burn.
[13] Thomas George Bonney.
[14] Lucas Ewbank.
[15] Mortlock's, bankers to the University of Cambridge, was currently headed by Edmund John Mortlock.

From W. C. Tait 13 April 1869

Oporto
April 13th 1869

Charles Darwin Esqr.
Dear Sir,

I received your kind letter of the 12th. ultmo. and am sorry to learn that the plants did not arrive in good condition[1] but I do not despair of their recovery as out of four supernumerary and weakly plants brought to me at the same time 2 have so far recovered that they appear to be on the point of flowering and one has begun to catch flies again—

If those which I sent to you do not recovery I shall send you another lot of them and shall proceed in an improved method—

They shall be planted in larger and deeper pots and shall be kept in my garden until they have successfully passed over the transplanting stage and are ready to bear a change of climate.

I am very much obliged to you for your kind offer of copies of your origin of species, your Fertilization of Orchids by Insect agency and your pamphlet on Climbing Plants[2] I have seen only the first and shall value them very much— If you were to send them to

William C Tait | Oporto
care of Messrs. Rossari & Co.[3]
2 Crosby Square | *London.*

they would reach me safely.

I am very much interested in all subjects of Natural History preferring those however which lead to or bear on some general theory and I must confess that I am rather speculative in my thoughts and have little regard for preconceived opinions if I think they are false and am ready to embrace those which reason tells me are true.

You will find I think that the Drosophyllum catches insects by the viscid glands, but perhaps you will also find minute hooks or some contractive movement assisting to secure the insect & which would easily be developed into a fly trap such as the American Dionæa you mention.[4] I have looked at the plants through a microscope of insignificant power little better than the naked eye otherwise I feel sure some very curious processes might be discovered.

I would suggest experimentalizing as to the stupifying or poisoning power of the fluid on the insects caught. In my next I hope to send you Brotero's[5] description in Latin of this plant it may prove suggestive to you

I feel very much interested to know how your plants are progressing. If I go to England this summer which is very uncertain I shall take a large number of plants of this species with me that you may have a large number to try experiments on and be enabled to supply your friends with specimens—

Since writing the first portion of this letter I have just received your letter of the 7th. and am glad to learn good news of the plants.[6] You will find I think that they like plenty of sun. If you do not discover any contraction in the plant I suppose it is an earlier stage of development than you anticipated and would tend to shew how from one adaptation the plant rose to a higher one from a simple secretion to a complicated mechanism.

I have a plant in my garden which has a very curious structure The flower looks and smells like a piece of putrid meat and I am told that it attracts flies—[7] I shall examine it attentively this summer and try to find out why it does so. Perhaps it may aid the fertilisation as I do not remember seeing any signs of flycatching.

Respecting *sheep.* I have learnt that the merino breed is kept near Oporto but the Wool is *very inferior* to the specimens I saw in the Paris Exhibition of last year sent from Saxony and Silesia— I am told that the rams are horned and the ewes without horns but that sometimes the rams are born *without* horns and that sometimes the ewes are born *with* horns, and that the rams which are born without horns are more easily fattened than those which are born with horns—[8] Here then may be another instance of correlation— I hope to obtain for you full particulars as to the growth of the horns in the young but have not yet had an opportunity of speaking to the right man.

Dogs. One of my pointers the one I mentioned as having 2 dew claws on each of the hind feet had six pups a short time ago—[9] three of them have one dew claw on each hind foot, one has two dew claws on each hind foot, one has none, and one was born with one on each hind foot one of which has dropped off and the other is slenderly attached. Having been in the Isle of Man I have seen the Manx cats and cannot avoid thinking that the scarcity of trees on that island has had some effect on the cats there whose shape seems to approach that of the rabbit to a certain extent.

I have walked along the sea-shore here and far above highest tide marks the ground seems to shew signs of a receding sea This is a great country for boulder stones—immense numbers are scattered all over the north. When in England

I heard from a friend of mine that in a certain lake in Italy there were fresh water oysters and that it was supposed that at one time the lake was connected with the sea.[10]

Under your theory the gradual change in their constitution would be accounted for and perhaps the same cause which formed the Sahara desert is still drawing away the waters from the Mediterranean

If there be any other plants &c. in this country which you would like to have please tell me and I will try to obtain them for you. I should much like to be a member of the Zoological Society Regents Park and I have been told that by sending foreign animals birds &c. they make people corresponding and honorary members and as when in London I stay near Regents Park it would be very convenient to me. I dare say there may be birds, reptiles &c. here of which they would be glad to possess specimens.

I am sorry to hear that your health is not very good but hope it may improve soon Believe me to remain, | Dear Sir, | Yours very truly, | William C. Tait.

Your letter of the 12th. March was delayed many days on account of addressing it "Viâ Southampton" the Brazil packet leaves on the *9th*.[11]

DAR 178: 46

[1] See letter to W. C. Tait, 12 and 16 March 1869. Tait had sent plants of *Drosophyllum lusitanicum* from Portugal.

[2] Tait refers to *Origin* 5th ed., *Orchids*, and *Climbing plants*. See letter to W. C. Tait, 12 and 16 March 1869.

[3] Alexander Rossari was a silk merchant (*Post Office London directory* 1869). He has not been further identified.

[4] See letter to W. C. Tait, 12 and 16 March 1869 and n. 2.

[5] Félix de Avellar Brotero.

[6] See letter to W. C. Tait, 7 April [1869].

[7] The plant may have been a *Stapelia*.

[8] See letter to W. C. Tait, 24 February 1869, and letter from W. C. Tait, 5 March 1869.

[9] See letter from W. C. Tait, 19 February 1869.

[10] The friend and the lake have not been identified.

[11] Post was forward daily to Portugal via France, but it was cheaper to send it by the monthly packet to Brazil (*Post Office London directory* 1869, p. 2277).

To A. R. Wallace 14 April 1869

Down. | *Bromley.* | *Kent. S.E.*
Ap. 14. 1869

My dear Wallace

I have been wonderfully interested by your article, & I shd think Lyell will be much gratified by it.[1] I declare if I had been editor & had the power of directing you I shd have selected for discussion the very points which you have chosen. I have often said to younger geologists (for I began in the year 1830) that they did not know what a revolution Lyell had effected; nevertheless yr extracts from Cuvier have quite astonished me.[2] Though not able really to judge, I am inclined to put more confidence in Croll than you seem to do; but I have been much struck by many of y^{r} remarks on degradation.[3]

Thompson's views of the recent age of the world have been for some time one of my sorest troubles, & so I have been glad to read what you say.[4] Your exposition of Nat. selection seems to me inimitably good; there never lived a better expounder than you. I was also much pleased at y^{r} discussing the difference between our views & Lamarck's.[5] One sometimes sees the odious expression "Justice to myself compels me to say &c"; but you are the only man I ever heard of who persistently does himself an injustice & never demands justice. Indeed you ought in the review to have alluded to y^{r} paper in Linn. Journal, & I feel sure all our friends will agree in this.[6] But you cannot "Burke" yourself, however much you may try, as may be seen in half the articles which appear. I was asked but the other day by a German Prof. for y^{r} paper which I sent him.[7] Altogether I look at y^{r} article as appearing in the Q-ly as an immense triumph for our cause. I presume that y^{r} remarks on Man are those to which you alluded in y^{r} note.[8]

If you had not told me I sh^{d} have thought that they had been added by some one else. As you expected I differ grievously from you, & I am very sorry for it. I can see no necessity for calling in an additional & proximate cause in regard to Man.[9] But the subject is too long for a letter. I have been particularly glad to read y^{r} discussion because I am now writing & thinking much about man.

I hope that y^{r} Malay book sells well: I was extremely pleased with the Art. in the Q. J. of science, inasmuch as it is thoroughly appreciative of y^{r} work: Alas! you will probably agree with what the writer says about *the uses of the bamboo.*[10]

I hear that there is also a good article in the Sat. Rev., but have heard nothing more about it.[11]

Believe me my dear Wallace | yours ever sincerely | Ch. Darwin

P.S. I have had a baddish fall. My horse partly rolling over me, but I am getting rapidly well—[12]

LS
British Library Board (Add 46434, f. 181–3)

1 Wallace had published a review of the latest editions of Charles Lyell's *Principles of geology* and *Elements of geology* (C. Lyell 1865 and 1867–8) in the *Quarterly Review* ([Wallace] 1869b). See also letter from A. R. Wallace, 24 March 1869.

2 CD went on a geological tour of North Wales in 1831 (see *Correspondence* vol. 1, Lucas 2002, and Herbert 2005). Lyell was a proponent of the 'uniformitarian' school of geology, according to which geological phenomena had been created by the action of causes still in operation at the present day; Georges Cuvier had been a proponent of the opposing 'catastrophist' school, which maintained that geological phenomena had been formed by catastrophic events in the past, including, according to some, the biblical flood. In his article, Wallace quoted from Cuvier's *Essay on the theory of the earth* (Cuvier 1827).

3 CD refers to James Croll and his calculations of the date of the glacial epoch; see [Wallace] 1869b, pp. 376–9. Wallace argued that the rate of subaerial denudation of some present-day rivers was not a conclusive means of determining the age of the earth.

4 William Thomson held that the crust of the earth had solidified only 100 million years ago, and had criticised CD's estimate in the first edition of *Origin* of 300 million years for the denudation of the

Weald (see W. Thomson 1862a, pp. 391–2). See also *Correspondence* vol. 16, letter to G. H. Darwin, [9 December 1868].

5 In [Wallace] 1869b, pp. 381–2, Wallace criticised Jean Baptiste Lamarck's supposition of a 'tendency to progressive development'.

6 Charles Lyell and Joseph Dalton Hooker had communicated CD and Wallace's papers on the theory of natural selection to the Linnean Society in 1858 under the title 'On the tendency of species to form varieties; and on the perpetuation of varieties and species by natural means of selection' (Darwin and Wallace 1858; see *Correspondence* vol. 7).

7 CD refers to William Thierry Preyer. See letter from W. T. Preyer, 21 March 1869.

8 See letter from A. R. Wallace, 24 March 1869.

9 See letter from A. R. Wallace, 24 March 1869 and n. 10.

10 *The Malay Archipelago* (Wallace 1869a) was anonymously reviewed with another work in the *Quarterly Journal of Science* 6 (1869): 165–79. Of bamboo, the author wrote (*ibid.*, p. 172): 'if we wish for a mass of evidence in favour of design, before which Paley pales, we need only read the author's account of the Bamboo and its uses. . . . He shows that it is indispensable to the natives. Looking at their mental condition, they could not have existed without it, or some similar boon of Providence.' The reference is to William Paley.

11 See *Saturday Review* 27: 427–8, 456–7 (issues of 27 March and 3 April 1869).

12 On 8 or 9 April, CD was thrown by his horse, Tommy, and suffered serious bruising to his back. There is an account of the accident in the letter from H. E. Darwin to G. H. Darwin, 10 April [1869] (DAR 245: 291). Henrietta wrote that the accident happened on Thursday (8 April); Emma Darwin recorded in her diary (DAR 242) that it happened on Friday 9 April.

From Richard Spruce 15 April 1869

Welburn. | York.
15. April. 1869.

Dear Sir

I was much obliged for your kind ⟨le⟩tter & for the suggestions it contained.[1] I sh$^{d.}$ ⟨ha⟩ve acknowledged it earlier but that I have been little able to write. I have sought out your letter written many years ago to me about Melastomes. I find it asks for information chiefly about any dimorphism of the sexual organs, such as you have observed in the allied order of Lythraceæ.[2]

I recollect looking through my notes in vain for any record of such structure. The baccate[3] Melastomaceæ, such as I chiefly gathered, have stamens remarkably equable in length, although usually in a double row. The filaments and anthers, white in some, yellow in others, are apt to turn rose and even purple after shedding their pollen, which (in these species) is ⟨n⟩ever green.

The only observation I have preserved about their fertilisation you will find in my 'Report on Expedition to procure Red Bark Tree';[4] but as it is unlikely you should have a copy of that Report I will copy it here.

"The first plant which took my attention at Limon,[5] ⟨af⟩ter the Cinchona, was a beautiful epiphytal *Blakea*,[6] ⟨gro⟩wing from 12 to 18ft high with large coriaceous ⟨lea⟩ves & large rose-coloured flowers. At the base of ⟨the⟩ flower is a turgid involucre, of four large orbicular, widely & closely imbricated leaves, within which is secreted a limpid fluid. When the corolla falls away, the involucral leaves close firm⟨ly⟩ over the calyx, and do not open out, nor does the contained fluid dry up, until the globose ro⟨s⟩eate be⟨ ⟩ the size of a pea, is quite ripe. Another singu⟨lar⟩ character is the *syngenesious* anthers, with a minu⟨te⟩ pore at the apex of each cell, through which not a g⟨rain⟩ of pollen ever escapes, as I satisfied myself by repeated observation; fertilisation being effected through the agency of minute beetles, which abound in the flowers, and eat away the inner edge of the anther-cells—probably part of the pollen also."

This is what I wrote on the spot, but I have no doubt now that the pollen, liberated by the beetles, falls on the honeyed disk, whence it is carried off by winged insects & deposited on the stigmas of other flowers; for the stigma hangs out far beyond both petals & stamens, therefore out of reach of the pollen as it falls into the centre of the flower.

It is noteworthy that in this most geometrical of of all orders of flowering-plants—for the strongl⟨y⟩ ribbed leaves, often purple beneath, closely wired with the numerous parallel veins between the ribs, have reminded me of some delicate fabric of copper or cast-iron—the only departure from symmetry is in the style going out between the filaments on one side, so that the stigma becomes quite excentric to the flower Sometimes the stamens are slightly declinate in the same direction as the style.* I have observed t⟨hat⟩ flies alighting on the flower use the large pelta⟨te⟩ stigma as a support for their hind legs whilst poking their noses into the disk. Surely here is an ⟨i⟩nstance of the effects of direct mechanical action ⟨o⟩n the part of insects, the style having become declinate through this perpetual "sagging" (as we say in Yorkshire, ⟨but a⟩s Shakespeare says it too I suppose the word is ⟨go⟩od English).[7]

When I have seen flies tugging & poking at the ⟨s⟩iphon-like corollas of some Gardoquias, Salvias, Siphocampyli, &c.[8] I have said to myself "Surely these flowers have become arcuate through the long-continued action of the flies."

I have not by me your charming book on the Fertilisation of Orchids,[9] but I thought you had admitted in part the direct agency of insects in modifying the structure of the flowers. When my paper is printed I will take care not to attribute to you any opinions you do not really hold.

Does not Natural Selection balance all internal & external influences, to the good of the individual, as ⟨f⟩ar as may be; and when opposing agencies become too strong for it, still keep up the struggle (as it were) to the last gasp?

There are some structures which (so far as we can see) are neither beneficial nor otherwise to plants and animals in their struggle for life. Such may b⟨e⟩ the sacs on the leaves of Tococas, which perhaps ⟨ ⟩ not have attained to permanence except in a ⟨reg⟩ion where the climate scarcely varies throughout the year.[10]

If a gall were always to appear in the same part of a plant—if it could be permanently tenanted—if it c^{d}. be proved not prejudicial but possibly eve⟨n⟩ beneficial to the plant—I do not see why it shou⟨ld⟩ not become hereditary.

I have many times seen sacs on Tococa-leaves even full-grown ones—that were imperforate an⟨d⟩ therefore untenanted by ants: that alone proves t⟨he⟩sacs to be inherited. I sh^{d}. think instances of th⟨is⟩ might be found on my specimens at Kew.[11] I wish ⟨I⟩ had them here & c^{d}. send you a leaf.

My only doubt is, if the ants really cultivate these sacs, as we cultivate cabbages & turnips, w^{d} not the saccate leaves become plane even in a few generations?

I do not see what influence the secretions of ants c^{d} have on the Tococas, except in the way of manure. Dead vegetable tissues are by their action changed to the colour & consistence of tinder, as my carefully put away linen has many times borne witness.

In fine, you must not let my speculations turn you aside from more important matters; but if at any time you think I may be able to furnish information that you require I will most gladly do it to the best of my ability.

I am Dear Sir | Yours very respectfully | Richd Spruce

Ch. Darwin Esq.

* In Blakea, Bellucia, & other large-flowered genera, but not ⟨ ⟩

DAR 177: 242

CD ANNOTATIONS
1.1 I was ... copy it here. 3.3] *crossed pencil*
6.8 stigma as a ... ability. 15.3] *crossed pencil*
Top of letter: *one word illeg, blue crayon*; 'Manner of fert of Melastom.' *pencil*

1 See letter to Richard Spruce, 1 April 1869.
2 See letter from Richard Spruce, [before 1 April 1869] and n. 2. Spruce refers to the order (now family) Lythraceae. CD had published on *Lythrum* in his paper 'Three forms of *Lythrum salicaria*'.
3 Baccate: having berries.
4 Spruce 1861; see also Spruce 1908, 2: 261–310. Spruce spent 1860 in Ecuador, obtaining seeds and young plants of *Cinchona succirubra* (now *C. pubescens*; the red bark tree) for the Indian Government. *Cinchona* was used to produce quinine, an anti-malarial drug. Specimens of the wood of *Cinchona* are in the Richard Spruce Collection, Royal Botanic Gardens, Kew.
5 El Limon was a group of small cane farms on the western side of the mountain Chimborazo (Spruce 1908, 2: 259).
6 *Blakea* is a genus in the family Melastomataceae.
7 'The mind I sway by and the heart I bear | shall never sag with doubt nor shake with fear.' William Shakespeare, *Macbeth*, 5.3: 9–10.
8 *Gardoquia* is a synonym of *Satureja* (savory), a member, like *Salvia* (sage) of the family Lamiaceae. *Siphocampylus* is a member of the family Campanulaceae.
9 *Orchids*.
10 Spruce refers to the genus *Tococa*, which he discussed in the paper that he sent to CD (see letter from Richard Spruce, [before 1 April 1869] and n. 3).
11 Several herbarium specimens of *Tococa* species collected by Spruce are in the Richard Spruce Collection, Royal Botanic Gardens, Kew. For a database of Spruce's specimens, see *Richard Spruce collection*, Natural History Museum, London, 2007, http://www.nhm.ac.uk/jdsml/research-curation/projects/spruce/.

From Edward Blyth 17 April 1869

21 Chalcot Crescent, | Regent's Pk, *N.W.*,
April 17/69

Dear Mr Darwin,

Observe the above address, for I have moved into more comfortable & roomy quarters. I regret exceedingly to learn ⟨of⟩ your accident, but must nevertheless congratulate you on its being no worse.[1] As for the mandrills, I saw the other day at

Antwerp adult male, female, and their offspring about 3 weeks old (& remarkably large), but the grown ones were so deep in clean straw that I could not have observed their extent of bareness, even had I tried to do so. This, by the way, was an excellent device to conceal the too demonstrative and gorgeous coloured nudities of the mandrills. There is a young one ⟨ ⟩ in the Regent's Park collection, & also a juvenile drill. I will endeavour to ascertain what you desire as soon as I can manage it, but the next private day at the *B.M.* is Tuesday, before which I cannot get to examine any of the contents of the cases in the public rooms.[2]

Have you seen the just-published portion of the *Procs. Z. S.*? It contains a paper by Gray on the *Canidæ*, in which are some remarks on natural selection which will amuse you somewhat.[3]

I have just received this weeks *Land & Water*, & in the Paris correspondent's letter (he is the Hon. H. Bingham) you will see some notice of an article on hybrids in the *Revue des Deux Mondes*, which letter must be worth your looking over.[4] I doubt if you will recognise me under the occasional signature of "Caviller", writing about the *Sequoia*;[5] but I did send a review for publication this week on a manufactured book of travels, entitled "the Ruined Cities of Zulu Land", which I have been demolishing unmercifully so much so that I fancy the general editor (Glass) hardly likes to offend the publisher, Chapman & Hall, by putting it in![6] However, I will shew it to Chapman myself, for I suspect that he has been taken in by those fictitious travels, as the reviewer in last week's *Athenæum* was decidedly, & there is a letter from the author in this week's *Athenæum* which bears out my view of the matter.[7] I believe that it will appear yet. You will have seen the first part of my review of Wallace's new book, which I was obliged to write somewhat hastily, & I have yet to follow it up as I of course mean to do.[8]

Yours very truly, | E. Blyth

DAR 160: 224

1 No letter has been found in which CD told Blyth of his accident, which occurred on 8 or 9 April (see letter to A. R. Wallace, 14 April 1869 and n. 12).

2 No letter to Blyth has been found in which CD enquired about mandrills. CD discussed the colouring of adult male mandrills in *Descent* 2: 292, 296, 310, but did not cite Blyth. Blyth had visited Antwerp in 1868 (see *Correspondence* vol. 16, letter from Edward Blyth, 20 November 1868). Blyth also refers to the gardens of the Zoological Society in Regent's Park, London, and to the British Museum. CD suspected that the hairlessness of the faces and buttocks of some monkeys was due to sexual selection (see *Descent* 2: 376–7). The mandrill is *Mandrillus sphinx*, the drill *Mandrillus leucophaeus*.

3 In his article in the *Proceedings of the Zoological Society of London* (J. E. Gray 1868), John Edward Gray commented (*ibid.*, p. 510):

> How any one can think that the differences between varieties of domestic animals are such as zoologists would use to distinguish genera and species, is a mystery that I cannot understand; and the theory that the variation produced by breeding and weeding, or selection as it is called, is to be regarded as the origin of the difference between natural species, is more astonishing, and can only have arisen for want of careful study of the subject. There are some minds so constituted, even among the well educated, who believe in animal magnetism, metallic tactors, table-turning, phrenology, spiritualism, mesmerism, the great pyramid, natural selection, and mimicry of animals—and some even two or more of these theories in succession, or at the same time.

[4] The article in *Land and Water*, 17 April 1869, p. 255, referred to Armand de Quatrefage's articles on CD's theory (Quatrefages de Bréau 1868–9). H. Bingham has not been further identified.

[5] In *Land and Water*, 17 April 1869, p. 255, Blyth argued that the genus *Sequoia* was identical with the genus *Wellingtonia*.

[6] Blyth refers to Walmsley 1869, published by Chapman & Hall. Glass has not been further identified.

[7] Blyth refers to John Chapman. Walmsley 1869 was reviewed in the *Athenæum*, 10 April 1869, pp. 499–500; the reviewer, John Doran, remarked, 'The . . . book partakes of romance so much that it is difficult to pluck reality out of it.' In a response in the *Athenæum*, 17 April 1869, pp. 537–8, the author wrote that his only object in writing the book had been to promote research.

[8] Blyth refers to Alfred Russel Wallace and to Wallace's *Malay Archipelago* (Wallace 1869a). Blyth's review has not been found.

To W. B. Tegetmeier 17 April 1869

Down. | Beckenham | *Kent. S.E.*
Ap 17. 1869.

My dear Sir

If not inconvenient I sh[d] be grateful for any scraps of information about the plumage of the chickens of Cuckoo breeds & Seabright bantams, if you have been able to obtain such.[1] Also the table of the greyhound births, if you have been able to set the clerk to work.[2] I must, in about a week's time, begin on subjects which lead me much to require the above information.

Have you had any success about eggs?[3]

Forgive me for being so troublesome & believe me | yours very faithfully | Ch. Darwin

P.S. If you sh[d] ever meet any one who breeds Sebastopol geese, c[d] you find out at what age the curled feathers first appear?[4]

LS
Archives of the New York Botanical Garden (Charles Finney Cox collection) (ser. 14)

[1] See letter to W. B. Tegetmeier, 25 February [1869] and n. 1.

[2] CD had asked Tegetmeier to tabulate male and female births per year among greyhounds using the *Coursing Calendar* (see *Correspondence* vol. 16, letter from W. B. Tegetmeier, [before 15 February 1868] and n. 4, letter to W. B. Tegetmeier, 15 February [1868], and letter from W. B. Tegetmeier, [16–20 February 1868]).

[3] See letter to W. B. Tegetmeier, 25 February [1869].

[4] Sebastopol geese have curled feathers on parts of their bodies (see Tegetmeier 1867, pp. 312, 316).

From W. B. Tegetmeier [before 18 April 1869][1]

[1 page missing]

given. The matter was so complicated and irregular that I could not get any one to do it satisfactorily, so I had to set about it myself, and have only been able to give it an hour or two at a time—[2]

The result is enclosed—Table I

In adding up the tables, I had compiled from each volume of the Field. My son[3] who is a good arithmetician observed that the proportion of males and females varied in the 1st and 2nd six months of the year.— (About $\frac{3}{4}$ trs of the animals are bred in the first half of the year and very few in Oct Novr & Decr) so I had the approximate ratios worked out, for the whole year it is nearly 74 males to 68 females—* but as unfortunately some of the provisional slips (from the separate half yearly volumes) were destroyed before my attention was diverted to the difference I can only send you the half yearly list in its imperfect condition. Table II But the missing matter can be compiled again if you wish it.

*	Approximative ratios.	
	Males	Females
Whole years.	74	68
1st half of year	74	65
2d half	74	70

You will bear in mind that these tables are compiled from those accounts of "Greyhound Produce" sent to the Field and are only a portion of the dogs kept for public coursing.

Few dogs are bred in the winter so they would run to disadvantage in the puppy stakes of the following season.—

There are some extraordinary instances of hereditary influence over colours in greyhounds—if you like to have them collected—certain celebrated dogs are known never to have produced any puppies but those of one colour, which was not influenced by the colour of the female—etc

I *Greyhounds produce*[4]

			Males	Females
1857	From Coursing Calendar		504	516
1858	Field Vol.11.	12	245	219
1859	13	14	157	138
1860	15.	16	275	235
1861	17	18	202	207
1862	19	20	312	254
1863	vols. 21.	22	366	316
1864	23	24	265	278
1865	25	26	298	255
1866	27	28	308	279
1867	29	30	378	325
1868	31	32	295	251
			3605	3273

II	Greyhound produce			
	Jan to June		July to Dec[r]	
	Males	Females	Males	Females
1857				
8	150	114	95	105
9				
1860	223	181	52	54
1	153	164	49	43
2	202	175	110	79
3	279	236	87	80
4	211	221	54	57
5	231	197	67	58
6	269	238	39	41
7	330	274	48	51
8				
	2048	1800	601	568

A Memorandum incomplete
DAR 85: B29–33

CD ANNOTATIONS
1.1 given . . . time— 1.3] *crossed pencil*
5.1 Few . . . season.— 5.2] *crossed pencil*; 'stet' *in margin, pencil*
6.1 There are . . . female—etc 6.4] *crossed pencil*
Table I: *cross after* '516', *pencil; tick after* '18', *pencil; cross after* '207', *pencil, two crosses after* '278',[5] *pencil; foot of first column,* '12 years.' *ink; foot of last two colums,* '3273 / 6878' *pencil*
Foot of table II: 'Tegetmeier', *pencil*

[1] The year is established by the relationship between this letter and the letter to W. B. Tegetmeier, 18 April [1869].
[2] See letter to W. B. Tegetmeier, 17 April 1869 and n. 2.
[3] Egbert Tegetmeier.
[4] CD used the data from this table in *Descent* 1: 263–4, 304; he credited Tegetmeier for his assistance in *ibid.*, p. 304.
[5] In a separate note at DAR 85: B34, CD calculated the ratios of male to female births for 1864 and 1867 (two of the years marked with pencil crosses); he used these ratios in *Descent* 1: 265.

From George Maw 18 April 1869

Club House Hotel | Gibraltar
April 18th. 69

My dear Sir

I am glad to say that I found on Friday last nr Andalusia about 6 miles N of this place the Drosophyllum in great abundance—quite half an acre covered with it.[1] I have sent off to day by "Bengalore"[2] P & O Steam Ship two cases of plants to Kew (Due at Southampton Friday the 23rd. inst) & enclosed to your address a box of the

Drosophyllum which I hope will reach you safely I have packed them carefully & as I was fortunate in getting the plant just before the Southampton steamer sails I really hope they will arrive in good order.[3] I fear from what Dr. Hooker tells me you will not be able to cultivate it as they have several times lost it at Kew.[4] It would be worth while trying it under various methods of treatment—stove—greenhouse, cold frame. & some might be put under a bell-glass.

It is a most beautiful & curious plant & the flowers much more conspicuous than I had imagined

I have just returned from a very pleasant visit to Barbary[5] & procured a good many plants which I have sent home by the "Bengalore".

Believe me my dr Sir | Very truly yrs | Geo Maw

I have not been able yet to get any information as to the age at which Merino ram lambs get their horns.[6]

C Darwin Esqre

DAR 171: 104

1 CD had asked Maw to obtain a specimen of *Drosophyllum lusitanicum* (see letter to George Maw, 13 January 1869 and n. 3).
2 The ship was the *Bangalore* (Rabson and O'Donoghue 1988).
3 CD mentioned receiving *Drosophyllum* specimens from Maw in *Insectivorous plants*, p. 332.
4 Joseph Dalton Hooker was director of the Royal Botanic Gardens, Kew.
5 The Barbary States were the North African states of Tripolitana, Tunisia, Algeria, and Morocco.
6 See letter to George Maw, 13 January 1869 and n. 2.

To W. C. Tait 18 April [1869][1]

Down. | Beckenham | *Kent.*
Ap. 18th.

Dear Sir

I am not very well owing to a fall from my horse & so write a line merely to thank you for your very obliging letter of April 13th.[2] I hear the plants are going on *very well*, but I have not been able to go for a week to the greenhouse.

I have this day ordered my Orchis Book to be sent to Messrs. Rossari & will send there my paper on Climbing Plants.[3] I feel sure that there will be no occasion to trouble you for any more of the Drossophyllum.

Whenever I go to London I will inquire of the Secretary of the Zoolog. Soc. whether they wish for any quadrupeds or birds from Portugal.[4]

With many thanks for all your kindness, believe me, | yours sincerely | Ch. Darwin

LS(A)
Alan R. Tait (private collection)

1 The year is established by the relationship between this letter and the letter from W. C. Tait, 13 April 1869.
2 See letter to A. R. Wallace, 14 April 1869 and n. 12.

[3] See letter from W. C. Tait, 13 April 1869 and nn. 1, 2, and 3.
[4] See letter from W. C. Tait, 13 April 1869. The secretary of the Zoological Society of London was Philip Lutley Sclater.

To W. B. Tegetmeier 18 April [1869][1]

Down. | Beckenham. | *Kent.*
Ap 18th.

My dear Sir

Our notes have crossed on the road. Your table about the greyhounds is of the greatest interest to me & I thank you sincerely. The effect of the seasons is very curious on the sexes, but I do not think it would be worth while to go over the missing years.[2]

I should be grateful for any information about the plumage of the chickens of any breed & about the other points of which you have memoranda.

If it would not cause you too much trouble, I should like to hear one or two cases of the strong transmission of colour by certain greyhounds independantly of that of the other sex.

Many thanks for the very curious paper in the Field.[3]

I am ready for the eggs of the Fowls whenever you can get the three lots, & the sooner the better; & if you will let me have a line telling me that you have them I will send a messenger for them. Please let them be named; & you must let me know what you have paid for them.[4]

My dear Sir, | yours very sincerely | Charles Darwin

P.S. | I enclose a cheque for ten guineas, not as a remuneration, for that is quite out of the question, but that you may not have lost heavily by all the time which you have so kindly given me.—

LS(A)
Archives of the New York Botanical Garden (Charles Finney Cox collection) (ser. 14)

[1] The year is established by the relationship between this letter and the letter to W. B. Tegetmeier, 17 April 1869.
[2] See letter to W. B. Tegetmeier, 17 April 1869, and letter from W. B. Tegetmeier, [before 18 April 1869].
[3] The article in the *Field* newspaper has not been identified.
[4] See letter to W. B. Tegetmeier, 25 February [1869].

From A. R. Wallace 18 April [1869][1]

9, St. Mark's Crescent | N.W.
April 18th.

Dear Darwin

I am very glad you think I have done justice to Lyell, and have also well "*exposed*" (as a frenchman would say) Natural Selection.[2] There is nothing I like better than writing a little account of it, & trying to make it clear to the meanest capacity.

The "Croll" question is awfully difficult. I had gone into it more fully, but the Editor made me cut out 8 pages![3]

I am very sorry indeed to hear of your accident, but trust you will soon recover & that it will leave no bad effects.[4]

I can quite comprehend your feelings with regard to my "unscientific" opinions as to man, because a few years back I should myself have looked at them as equally wild & uncalled for. I shall look with extreme interest for what you are writing on Man, & shall give full weight to any explanation you can give of his probable origin.[5]

My opinions on the subject have been modified solely by the consideration of a series of remarkable phenomena, physical & mental, which I have now had every opportunity of fully testing, & which demonstrate the existence of forces & influences not yet recognised by science. This will I know seem to you like some mental hallucination, but as I can assure you from personal communication with them, that *Robert Chambers, Dr. Norris of Birmingham* the well known Physiologist, and *C. F. Varley* the well known Electrician, who have all investigated the subject for years, agree with me both as to the facts and as to the main inferences to be drawn from them, I am in hopes that you will suspend your judgment for a time till we exhibit some corroborative symptoms of insanity.[6]

In the mean time I can console you by the assurance that I *don't* agree with Q. J. of Science about *Bamboo*, & that I see no cause to modify any of my opinions expressed in my Article on the "Reign of Law".[7]

Believe me | Yours very faithfully | Alfred R. Wallace

C. Darwin Esq.

DAR 106: B79–80

[1] The year is established by the relationship between this letter and the letter to A. R. Wallace, 14 April 1869.

[2] Wallace reviewed Charles Lyell's geological works in the *Quarterly Review* ([Wallace] 1869b). See letter to A. R. Wallace, 14 April 1869.

[3] Wallace refers to James Croll; see [Wallace] 1869b, pp. 376–9. The editor of the *Quarterly Review* was William Smith.

[4] See letter to A. R. Wallace, 14 April 1869 and n. 12.

[5] See letter from A. R. Wallace, 24 March 1869 and n. 9.

[6] Wallace refers to Richard Norris and Cromwell Fleetwood Varley. For more on Wallace's interest in spiritualism, see Kottler 1974 and Fichman 2004, pp. 139–210.

[7] See letter to A. R. Wallace, 14 April 1869 and n. 10. Wallace also refers to his review of George Douglas Campbell's *The reign of law* (G. D. Campbell 1867, Wallace 1867).

To George Charles Wallich 18 April [1869][1]

Down. | Beckenham. | *Kent. S.E.*
April 18

My dear Sir

I should for many reasons be very glad to oblige you in any way. But to come to you to be photographed would certainly lose me two whole days & probably three.[2]

Nor, owing to my health, would it be at all certain if I came to London that I sh^d. be able to go to you next morning.— I really cannot spare so much time, with the chance of subsequent unwellness.— I am very sorry but so it must be. I have already refused some half dozen requests, but none with a quarter of the regret that this causes me.—

If you think it worth while (which I much doubt) to come here, I shall be very glad to do all I can to aid you.— You can, if you like sleep here. Orpington Station (on the S.E.R. Charing Cross) is 4 miles from my house, & I could send a vehicle for you & probably also send you back. Or if your time is precious you could start early in morning & be here in very good time.— We could fit up a small dark closet, in which my son formerly photographed.[3] Even if your photograph was not first rate, no one would probably detect it.— But you could easily get to 16 good men, without taking so much trouble for me.—[4] I can only repeat my regret, & say that here I shall be most happy to do anything I can which you may require.—

My dear Sir | Yours very faithfully | Ch. Darwin

P.S If you decide to come, please let me hear.—

American Philosophical Society (374)

[1] The year is established by the publication date of Wallich's *Eminent men of the day* (Wallich 1870), and from the form of the address, which CD began to use in April 1869. The preface of Wallich 1870 is dated May 1870.

[2] CD had corresponded in 1860 with Wallich about the results of deep-sea soundings made by Wallich (*Correspondence* vol. 8). Wallich was preparing a book of photographs of eminent scientists (Wallich 1870). CD's photograph does not appear in the book. No letter from Wallich requesting a photographic sitting has been found.

[3] William Erasmus Darwin had been interested in photography since at least 1857 (see *Correspondence* vol. 6, letter to W. E. Darwin, [before 11 September 1857] and n. 3). William left home for Southampton in 1861 (see *Correspondence* vol. 9).

[4] Wallich 1870 had sixteen plates.

To J. Noordhoek Hegt 19 April 1869

Down. | Beckenham | *Kent. S.E.*
Ap. 19. 1869

Dear Sir

I hope that you will excuse my troubling you once again, & it shall be the last time. I write partly to ask whether there is now any difference in the development of the spurs in the last year's birds of *Pavo Spicifer* & *cristatus*.[1] But there is another point, on which I sh^d be *greatly* indebted for information.

I hear that you possess adult Mandrills of both sexes, & the young; & I want much to know whether the space at the posterior end of the body, which is almost or quite destitute of hair, is larger in the male than in the female. I believe that the naked space is much more brightly coloured in the male. I want also to know whether the surface destitute of hair on the face is larger in the male than in the female. I have consulted various works of Nat. Hist & find a surprizing difference in the descriptions of the colour of the face in the 2 sexes & in the young.[2]

Can you refer me to any accurate description of the sexual differences in colour? I fear it w^d^ be quite unreasonable in me to ask you to give me a very brief description of the difference.

As I sh^d^ wish to quote you on Pavo & about the Mandrill if you are so kind as to assist me, will you be so good as to tell me whether I read your signature correctly as Hyt or Hijt

It w^d^ give me much pleasure to send you, as a mark of my respect, any of my works, such as my Journal of Travel or Origin of Species &c[3]

I hope that you will forgive me troubling & believe me | dear Sir | yours very faithfully | Charles Darwin

PS. Do you possess the *Macacus inornatus* from Borneo, which is said to shed tears when in distress?[4]

LS(A)
Stadsarchief Amsterdam (395: 614)

1 See *Correspondence* vol. 16, letter to Natura Magistra Artis, 28 November 1868, and letter from J. N. Hegt, 11 December 1868. *Pavo spicifer* is now considered to be a subspecies of *P. muticus* (the green peafowl); *P. cristatus* is the common Indian peafowl.

2 See also letter from Edward Blyth, 17 April 1869 and n. 2.

3 CD refers to his *Journal of researches* and *Origin.*

4 *Macacus inornatus* is now *Macaca maura.* For more on the *Macacus inornatus* from Borneo, which was bought by the Zoological Society of London, see *Correspondence* vol. 16, letter from James Murie, 1 May 1868 and n. 1. There is a note by Francis Darwin, dated 20 June 1867, that includes the statements: 'The *Macacus inornatus* from Borneo certainly cries, so that the tears roll down its face when in grief or even when pitied. The woman who sold it said so. Bartlett & Sutton have both repeatedly seen it.' (DAR 189: 138.) Francis referred to Abraham Dee Bartlett, the superintendent of the Zoological Gardens in Regent's Park, London, and Seth Sutton, a keeper there.

From Orange Judd & Co 21 April 1869

Office of the American Agriculturist | 245 Broadway | New York
Ap^r^ 21 1869

Dear Sir.

We have too long delayed answering your esteemed favor of Jany 25th and beg pardon for the seeming neglect.[1] "The Variation of Animals and Plants" has not sold quite as largely as we had hoped, the second edition of our thousand being now about half sold.[2] But the sale appears steady and will we think continue for sometime to come. Our universal, almost custom is to stereotype or electrotype books which makes it much more difficult to change the matter than if kept in type.[3]

Then you are so exceedingly progressive, which of course we like, that it is difficult for publishers to keep up with you. There is one difficulty with regard to publishing the "Origin of the Species".[4] it needs and must have frequent changes and additions which of course are not congenial with the profits of bookmaking generally. Then again our relations with Messrs Appleton & Co are such that while we might consider fairly that they had given up their claim to publish according to the ordinary rules of American Booksellers & publishers ettiquette yet we would not like in any way to annoy them by appearing to others who do not fully understand the whole matter to trepass upon their grounds.[5] We regard on these accounts feeling obliged to decline your kind offer. We are inclined to keep some of the English Edition of the Origin in stock and will sell what we can of the New Edition. Can the additions and alterations in the Variation be put in the form of an appendix or supplement to our next edition? We shall be pleased to receive a copy in sheets when ready. Enclosed we hand you Bill of Exchange on Baring Bros & Co for fifty pounds and only wish it were more,[6] and remain

Yours truly | Orange Judd & Co

Chas Darwin

DAR 173: 36

[1] CD's letter of 25 January 1869 has not been found.

[2] For the publication of the American edition of *Variation* in 1868, see *Correspondence* vol. 16, letter from George Thurber, 18–20 April 1868.

[3] For works expected to require reprinting publishers usually printed by creating fixed metallic plates, or stereotypes; the text for each page was cast from a mould and was therefore virtually unchangeable.

[4] CD evidently asked the publisher whether they would be interested in publishing an American edition of the fifth edition of *Origin*; he finished making corrections for the fifth edition on 10 February 1869 (see 'Journal' (Appendix II)). The publishing firm D. Appleton & Company had stereotyped the first American edition of *Origin* in 1860 (see *Correspondence* vol. 8, Appendix IV), and CD had been frustrated at not being able to incorporate changes from his later English editions into further American editions (see *Correspondence* vol. 14, letters to Asa Gray, 16 April [1866] and 4 August [1866]).

[5] Later editions of *Origin* in the United States continued to be published by D. Appleton & Company (Freeman 1977).

[6] CD recorded the receipt of £50 in his Account books–banking account (Down House MS) for 11 May 1869: 'From M[ss] Orange Judd & Co for American Edit. of Variation under Domestication'. Baring Brothers and Co. was a merchant bank in London.

From Georg Recht[1] 21 April 1869

Munchen
den 21sten. April 1869

Euer Wohlgeboren!

Aus einer heute stattgehabten Conferenz mit Baron von Liebig über den beifolgenden Artikel in der allgemeinen Zeitung habe ich die traurige Gewissheit erlangt, dass mich mein armes & armseliges Vaterlaendchen ald Fantasten betrachtend im Stiche resp. verhungern laesst.[2] Armselig nenne ich Bayern, weil nicht ein einziger Gelehrte darin zu finden ist, der denkt, unsern Staatsmännern aber und unsern Verwaltungsbeamten, deren wir ein ganzes Heer haben, das mehr an dem Marke des Volkes zehrt als unser unsinniges Militaerheer ist das Denken, insbesondere das practische Denken absolut fremd, und sie sind dabei von einem ungeheurn Eigendünkel beseelt. Arm nenne ich aber das Vaterland, weil die allgemeine Verarmung in Folge von Mangel an Industrie & Handel und in Folge von unsern unnatürlichen Bodenverhaeltnissen, die es gestatteten, dass sich alles Kapital aus denselben zurückzog und in auslaendische Papiere unwandelte, Schauder erregend um sich greifft. Soll die Arbeits Kraft nicht verzweifeln und muss der Boden nicht veröden, wenn Kapital und Intelligenz sie verlassen! wenn die Steuerkraft statt ihrer Rente in die Staatskassa fliest! wenn das Volk vom Boden selbst und nicht mehr von der Rente des Bodens lebt! Ein solches Volk fliehen die Wissenschaft und das Denken, weil sie keine Pflege fühlen.

In dieser Verzweiflung wage ich noch ein mal an Euer Wohlgeboren die Bitte, ob Sie nicht eine englische Stüze für mich finden. Vergebens arbeite ich in Bayern in mehreren Schriften für Einführung der englischen Verhaeltnisse, für welche ich lebe und sterbe.[3]

Vorerst bitte ich um Entschuldigung, dass ich deutsch schreibe und meinen Aufsatz Ihnen in deutscher Sprache übersende. Obwohl ich fleisig englisch treibe, so bin ich doch noch nicht so weit, dass ich mich in Ihrer Sprache bewegen könnte, und zum Übersetzen fehlt mir das Geld. Mein letztes Dankschreiben für Ihre Unterstützung und das für Sie bestimmte Exemplar meiner Schrift werden Sie wohl erhalten haben.[4]

In dem diesem Brief beiliegendem Aufsatze suchte ich zu beweisen, dass alle Zweige der Naturwissenschaft in der Mechanik ihre gemeinsame Vereinigung finden, dass aber die analytische Mechanique irrig sei in so ferne, als ihre Grundprincipien in der Natur nicht gegeben sind, desswegen gestattet sie auch keine Folgerungen für das Wesen der Natur, ihre Körper & ihre Erscheinungen.

Ihre Resultate sind nur richtig in Folge von zwei Fehlern, die sich compensiren, naemlich sie betrachtet Druck oder constante Kraft als continuirlichen Stoss, obwohl das Wesen des Letztern Bewegung, das Wesen des Erstern aber Ruhe ist, und dann betrachtet sie die Hindernisse als rückwirkende Kraft, wahrend sie nur eine stets sich erneuernde Hemmung oder das, was man Masse nennt, und als Streben nach Ruhe bezeichnet werden kann in Folge der continuirlichen Erneuerung, die die Wirkungen des Grundgesetzes der Mechanique, naemlich des Gesetzes der Traegheit ausschliesst, während die Kraft das Streben nach Bewegung bildet. Stoff in bestandiger Erneuerung ist Streben nach Ruhe, Elastizitaet in bestaendiger Erneuerung Streben nach Bewegung, und diese beiden Elemente ruhen in jedem Körper. Rotation & Translation zusammen bilden die Entwicklung, und eine Wirkung zweier Körper auf einander ist nur moeglich, wenn sie zuerst zu einen einzigen Koerper verbunden sind, und alsdann der eine das bewegliche, der andere das hemmende Element bildet. Im *stabilen* Gleichgewichte zwischen Kraft und der durch Bewegung zur Kraft gewordenen Hemmung besteht jede gleichmaessige geradlinige Bewegung.

Diese Principien muss die Mechanique adoptiren und die Naturwissenschaften müssen die Gesammtnatur als einen einzigen organischen Koerper betrachten, dann widersteht auch nicht eine einzige Frage mehr in Bezug auf ihre richtige Beantwortung.

Edler Wohlthaeter & Gönner, nehmen Sie sich dieser Principien & meiner an, die Menschheit wird durch sie auf eine bisher nicht geahndete Höhe gehoben werden, und die Mit- und Nach-Welt wird Ihnen Dank wissen.

Ich bitte Sie herzlich um baldige Antwort und erlaube Ihnen *unumschraenkten* Gebrauch von diesem Briefe.

Euer Wohlgeboren | ergebenster | D Recht

DAR 176: 76

1 For a translation of this letter, see Appendix I.

2 Recht refers to Justus von Liebig, president of the Bavarian academy of sciences. Recht had published an article, 'Die Mechanik in ihrer Anwendung auf die Natur' (Mechanics and its application to nature), in the *Allgemeine Zeitung* of 3 April 1869. It has not been found in the Darwin Archive–CUL.

3 The treatises have not been identified.

4 See letter from Georg Recht, 1 March 1869. Recht refers to Recht 1869.

From E. A. Darwin [after 21 April 1869][1]

than Paget & that you may get well soon— I looked upon muscles as of such little consequence compared to all the rest of the machinery that I believe I looked upon Pagets verdict as better than I ought. You'll require B Jones to put you on horseback again.[2]

yours affec | E D

Incomplete
DAR 105: B65

1 The date is established by the reference to James Paget and a horse-riding accident. Paget visited CD on 21 April 1869 (Emma Darwin's diary (DAR 242)), after CD had fallen from his horse (see letter to A. R. Wallace, 14 April 1869 and n. 12).

2 Henry Bence Jones had treated CD successfully for his chronic illness since 1865 (see *Correspondence* vol. 13), and had advised CD to go riding for exercise (see *Correspondence* vol. 14, letter from H. B. Jones, 10 February [1866] and n. 3).

From Alfred William Bennett 22 April 1869

Lancaster Villa | York Road, Upper Holloway N.
April 22nd 1869

Sir

By this post I venture to send you a copy of a paper written for the "Linnean Society", which illustrates an interesting fact connected with appliances for cross-fertilization in the structure of Parnassia palustris.[1]

I have been observing during the past winter the mode of fertilization in our common wild-flowers which habitually flower & bear seed in the winter season when we cannot suppose that any insect agency exists to produce cross-fertilization, such as Lanium album, Capsella Bursa-pastoris[2] &c; & find that the pollen is almost invariably discharged in the bud before the opening of the flower, thus insuring *self-fertilization*; can you direct me to any previous observations on this subject?

Apologizing for the trouble, permit me to describe myself | yours very respectfully | Alfred W Bennett

C. Darwin Esq.

DAR 160: 137

1 Bennett refers to Bennett 1868. The offprint has not been found in the Darwin Pamphlet Collection–CUL, but the article in CD's own copy of the *Proceedings of the Linnean Society of London (Botany)* is lightly scored.

2 *Lamium album* is the white dead-nettle; *Capsella bursa-pastoris* is shepherd's-purse.

From J. N. Hegt [23 April 1869][1]

⟨*letterhead excised*⟩

Charles Darwin Esqr
Beckenham Down | Kent
Dear Sir!

In answer to your kind letter of the 19th instant I have the honour to inform you, that the young Pavo spicifers which we bred last year (and are at present healthy and strong) show as yet no difference whatever in the development of the spurs with Pavo Cristatus, and the white variety of the same age. At present there is only a small knob or indication at the place where the spurs must grow.[2]

As to the second point mentioned in your letter, I am very sorry I can give no information. The monkey tribe does not thrive well in our marshy Country. As far as I know there are living specimens of Mandrils in the Cologne Frankfurt and Antwerp gardens. D^r. Bodinus, D^r. Max Schmidt ⟨*1 line excised*⟩ of Monkeys and never had *Macacus inornatus* alive[3]

If ever you should want any information I always shall be glad to assist you with the little I know. At present we have a splendid collection of wild Cattle of Cranes Pheasants and Aquatic birds. In some branches as Carnivores Pachydermata, Ursus, Cratt. Parrots and Fringillidae we can make a fine show.

I hope you will excuse me if I have sinned to often against the English grammar and remain dear Sir | Yours obediently | J. Noordhoek Hegt. | *adj. Dir.*

DAR 86: A40

CD ANNOTATIONS
1.4 At present ... grow. 1.5] *double scored ink and pencil*; 'Keep' *added ink*
2.1 As to ... remain 4.2] *crossed ink*
Top of letter: 'April 23^d. 1869' *ink*

1 The date is established by the annotation.

2 See letter to J. N. Hegt, 19 April 1869 and n. 1. *Pavo cristatus* has a white form. CD cited Hegt for his remarks on the spurs of peafowl in *Descent* 1: 290 n. 28.

[3] See letter to J. N. Hegt, 19 April 1869; see also letter from Edward Blyth, 17 April 1869. Hegt refers to Heinrich Bodinus. There is a draft of this letter in the Municipal Archive at Amsterdam (Gemeentearchief). In the draft, the last sentence of this letter reads: 'As far as I know there are living specimens of Mandrills in the Cologne gardens D^{r}. Bodinus director in the Frankfurt gardens D^{r}. Max Schmidt, direct. and Antwerp M^{r}. Vekemans direct. We have only a very poor Collection of the Common Species of Monkeys.' The reference is to Jacobus Vekemans.

To P. G. King 24 April 1869

Down. | Beckenham | *Kent. S.E.*
Ap. 24. 1869

My dear King

Although you tell me not to write I must send a line for auld lang syne & to thank you for yr note.[1]

Living the quiet life which I do I often think of old days, & I remember the friendship of no one on board the Beagle with more pleasure than yours.[2]

The cases of the parrot seem very curious, but at present I do not know what species the rosella is; I will however when next in London, shew the paragraph to Mr Gould & find it out.[3] Many thanks for your kind wish to aid me about expression.

I can give you no news about any other of our old shipmates as I have heard of none of them for some months. As for myself my health continues & will ever be weak, but I am better than I was a few years ago. Should you come to England it would be a real pleasure to me to see you.

Believe me my dear King | Yours very sincerely | Charles Darwin

LS
Mitchell Library, Sydney (MLMSS 3447/2, Item 2, pp. 5–8)

[1] See letter from P. G. King, 25 February 1869.
[2] See letter from P. G. King, 25 February 1869 and n. 4.
[3] CD refers to information in an enclosure to King's letter of 25 February 1869 that has not been identified. Rosellas are Australian parrots in the genus *Platycercus*. CD also refers to John Gould.

From John Price 24 April 1869

38 Watergate St Chester
Ap 24. 1869

My dear Darwin/

It is not *seeds* that I have ever tried to grow, but twigs of willow & perhaps *poplar* with 1 or *2* buds on.[1] I am now trying birch (Betula *nana* probably)[2] which *look* very lively. But I never *suceeded* with any thing but a gramineous plant (Cyperus?)[3] of which the Ptarmigan had extracted the very *core*, almost like a *bulb*. They shot up 1 $\frac{1}{2}$ inch & I (sicut mens est meus)[4] **forgot** to water them. This incapacitates me from **any** experiments, or I should have done something in that line, by this time(?) I have had to give up little *half-hour* jobs, after trying to remember them for **30 years** or more, in vain. You *result-mongers* must take the will for the deed.

The disapp[t] to me is a *sore* trial, you may suppose. If I were rich, I s[hd]. keep *a Curate*, in that sense

I *may* send you a few sketches of my Drake with this; & some Card. pratensis for the *Babbies*[5] (alas! **thorough bred Saxons**).

Fra. Parkers are very near neighbours, & good friends[6]

Your's very truly | J. Price

DAR 205.2 (Letters): 248

CD ANNOTATIONS
1.5 This . . . friends 4.1] *crossed ink*
Top of letter: '(Distribution)' *blue crayon*

1 No letter to Price from CD on this subject has been found.
2 *Betula nana* is the dwarf, shrub, or bog birch.
3 *Cyperus* is a genus of sedge commonly called nut-grass.
4 *Sicut mens est meus*: such is my mind (Latin).
5 The sketches have not been found. Price refers to *Cardamine pratensis*, the cuckoo flower.
6 Price refers to CD's nephew Frank Parker and his family.

From Edward Blyth 26 April 1869

21 Chalcot Crescent, | Regent's Pk,
April 26/69.

Dear M[r]. Darwin,

I was unable last week to visit the B.M., but I examined the rhesus monkeys at the Gardens, & there is no difference whatever in the amount of extension of the nude face in the two sexes.[1] In the females of this group, the sexual organs exhibit an immense development at times, as you cannot but know. In the mandrill & drill, which animals exemplify the culmination of the *Macacus* & *Cercopithecus* series, the full grown females are very much smaller than the full-grown males; & in the male mandrill more especially the peculiar characteristics of the series are most highly enhanced; the brilliant colouring of the nude & especially of the sexual organs, most unusual in the class mammalia, but which is also seen in some of the small *Cercopitheci*.[2]

That is a capital figure of the musk-deer in last weeks no. of the Illustrated London News, & I furnished the description of it. I have now just written to the editor of that paper, to request him to figure the Abyssinian wart-hog (*Phacochœrus Æliani*) that has recently arrived, and which is much more different from the Southern species (*P. æthiopicus*) than I had expected.[3] Although a half-grown sow, I can perceive that the curious facial appendages will be much less developed than in the female of the Southern species, and it has immense white whiskers, & very long flowing black hair upon the middle of its back—

Yours very truly, | E Blyth

DAR 85: A107–8

CD ANNOTATIONS
1.1 I was . . . examined] *crossed blue crayon*

1.1 the rhesus . . . sexes. 1.3] *enclosed in square bracket blue crayon*
1.3 In . . . back— 2.8] *crossed blue crayon*
Top of letter: '(Naked skin & Colour)' *blue crayon*; *square brackets in MS*; 'Add that no difference in Hairiness in the 2 sexes.—' *ink*

[1] See letter from Edward Blyth, 17 April 1869 and n. 2.
[2] The mandrill (probably *Mandrillus sphinx* is referred to) and drill (*Mandrillus leucophaeus*) are members of the family Cercopithecidae. The genera *Macaca* (formerly *Macacus*) and *Cercopithecus* are also part of the same family.
[3] A picture of musk-deer recently arrived at the gardens of the Zoological Society of London was published in the *Illustrated London News* for 24 April 1869, p. 408; Blyth's description was on p. 422. The editor of the *Illustrated London News* was John Lash Latey. No picture of a warthog appeared in May 1869. An Abyssinian warthog was purchased by the Zoological Society and arrived on 15 April 1869 (*Proceedings of the Zoological Society of London* (1869): 665). *Phacochoerus aethiopicus* is the desert warthog; *P. aeliani* is now *P. africanus aeliani*, the Eritrean warthog.

To J. N. Hegt 27 April [1869][1]

Down. | Beckenham | *Kent. S.E.*
April 27

Dear Sir

I write one line merely to thank you very much for your note & to say that I hope you will accept from me a copy of my Journal of Travels, which as you are so good a scholar might possibly in parts interest you.— I have told my publisher to send you a copy by Post.—[2] It does not seem possible that the spurs of the two species of Pavo will ever differ in rate of development.[3]

With my best thanks | Pray believe me | Dear Sir | Yours truly obliged | Charles Darwin

P.S. | If hereafter you sh[d] perceive any difference in the development of the spurs, perhaps you will have the great kindness to inform me.—[4]

Stadsarchief, Amsterdam (395: 614)

[1] The year is established by the relationship between this letter and the letter from J. N. Hegt, [23 April 1869].
[2] CD refers to his *Journal of researches*, and to his publisher, John Murray.
[3] See letter from J. N. Hegt, [23 April 1869].
[4] In *Descent* 1: 290 n. 28, where CD cited Hegt's information on the peafowl, he added, 'I presume that I should have been informed if any difference in the rate of development had subsequently been observed.'

To George Maw 27 April [1869][1]

Down. | Beckenham | *Kent. S.E.*
Apr 27th

My dear Sir

I thank you sincerely for your great kindness about the rare Drosophyllum.— The plants have arrived, but in a rather sickly condition: they may, however, recover.[2] A strange coincidence has occurred: after writing to you I received a letter from a stranger in Oporto, offering to send me any plants, so I mentioned the Drosophyllum, which he found immediately & sent me a supply; they arrived in a very bad state, but have recovered enough for me to observe.[3] The species, however, possesses far less interest than I had confidently expected.

Dr Hooker has been here & tells me that you have sent home a good many plants.[4] What an indomitable worker you are in many subjects!

With very sincere thanks for your great kindness | pray believe me | My dear Sir | Yours very faithfully | Ch. Darwin

Royal Horticultural Society, Lindley Library

1 The year is established by the relationship between this letter and the letter from George Maw, 18 April 1869.

2 See letter from George Maw, 18 April 1869.

3 CD refers to William Chester Tait; see letter from W. C. Tait, 26 January 1869, letter to W. C. Tait, 2 February [1869], and letter from W. C. Tait, 2 March 1869. See also letter to George Maw, 13 January 1868.

4 Joseph Dalton Hooker was the director of the Royal Botanic Gardens, Kew.

From Edward Jones to Mary Anne Ruck 28 April 1869

Hendrewallog
April 28th. 1869

Mrs. M L. Ruck
Dear Madam,

your letter of the 19th inst., came to hand duly but as to make an examination to the matter you referred to, I have been for few days looking through the sheep & I have to inform you, that there are good many of the male lambs having a horns to be seen when they first born, those of the femels which shall have a horns at all you can feel them in the skin[1] there are several of the males having a horns half an inch some times above the wool when they first born— The male once will grow sooner than the femels a good deal— Though there are good many of males & femels without have any horns at all, when they grow unto ful size they got no more horns than a head of a finger— My Mother is better but not yet good enough as to leave home for change of air But we expects that she will be able to go in course of a few days, She wishes to send her best respects to you & all the Family as well and so we do the same, my Father as well as myself[2]

I remain Dear Madam | your most obedient servant | Edward Jones

To Mrs M. L. Ruck
Pantlludw | nr. Machynlleth

DAR 83: 182–3

1 CD had evidently asked Mary Anne Ruck, one of whose sons was friendly with Leonard Darwin (see *Correspondence* vol. 16, letter to Horace Darwin, 26 [July 1868] and n. 6), to ask farmers in her neighbourhood in Wales about the growth of sheep's horns.

2 Jones's parents have not been identified.

From John Murray 28 April [1869][1]

50, Albemarle St. | *W.*
Ap. 28

My Dear Sir

It appears that I have sold 400 copies of Muller out of 1000 printed.[2] It is *a very* bad time for Bookselling, this—indeed for every kind of business. I was sorry to hear you had met with an accident— Did you get the last Quarterly? It seems to me the best exposition of the Darwinian Theory I have met with[3]

I propose to print 2000 copies of the new Edition of Origin of Species & hope to reduce the price to the original s14/—[4] Mr Hegt's book is sent off.[5]

My Dear Sir | Yours very faithfully | John Murray

Ch Darwin Esqre

DAR 171: 368

[1] The year is established by the reference to Dallas trans. 1869 (see n. 2, below).

[2] William Sweetland Dallas's translation of Fritz Müller's *Für Darwin* (Dallas trans. 1869, F. Müller 1864a) was published in March 1869 (see letter from Robert Cooke, 12 March 1869).

[3] Murray refers to a review by Alfred Russel Wallace in the *Quarterly Review* ([Wallace] 1869b). On CD's accident, see the letter to A. R. Wallace, 14 April 1869 and n. 12.

[4] Murray refers to the fifth edition of *Origin*.

[5] CD had asked Murray to send J. Noordhoek Hegt a copy of his *Journal of researches* (see letter to J. N. Hegt, 27 April [1869]).

From Spiridion Brusina[1] 29 April 1869

Agram[2]
29 Aprile 1869.

Esimio Signore!

Ajutato da alcuni colleghi sto fondando una socièta di storia naturale. La stessa avrà per iscopo lo studio della Fauna, Flora e Gea del triregno di Dalmazia, Croazia e Slavonia, e degli altri paesi slavo—meridionali Carniola, Serbia, Bosnia, Erzegnovina, Montenero ecc.; indi tendera a digulgare fra noi le cognizioni di quelle scienze naturali la cui importanza di tanto fu elevata dalle classiche opere da Lei pubblicate.

I diplomi da distribuir si fra i membri desideriamo fregiarli di quattro ritratti dei più distinti naturalisti, ed essendo chè fu stabilito che il Suo debba essere fra i quattro, nè ci viesci d'averlo, mi faccio a pregarla d'aver la bontà di volermi mandare una Sua fotografia per poterlo ritrarre.

Domandando Le mille volte perdono d'un tanto disturbo ed anticipatamente ringraziando La ho l'onore di dirmi

l'obbsme dev. serv. | Spiridion Brusina

Custode al Museo Nazionale di stor. naturale

DAR 160: 354

CD ANNOTATIONS

Top of letter: 'Austria | **via Belgium** | 6d for $\frac{1}{2}$ oz | 1s for 1 oz' *pencil*

[1] For a translation of this letter, see Appendix I.

2 Agram is the German name for Zagreb (Croatia).

From J. V. Carus 29 April 1869

Leipzig
April 29th., 1869.

My dear Sir,

During the present bookseller-fair Mr Koch (Schweizerbart) called here and told me, that a new edition of the translation of the "Origin" would be necessary, to be published in the first part of next year. He intends to begin the printing of it in October next. As far as I know there has not been a new edition of that work in England since 1866.[1] Should it have been the case I should be sorry to have overlooked it. At all events I beg to take the liberty of asking your kind help as in the last edition. If you wish to make any alteration or to give some additions you would oblige me exceedingly if you would kindly prepare them so that I may be able to begin the revision and reworking in September. I am afraid I shall not have time to look to it before that time as I am bound to finish my 'History' within the next months.[2]

The Variation under Domestication has also been sold very well.[3] As it is rather expensive for German purses, it is no wonder that the whole of the edition has not yet been sold. But I think it always a very good result, that two thirds of the edition are gone, as Mr Koch tells me. I hope we shall have a new edition in 1871.[4] Mr Koch also asked me, if any of the works were out which you mention in the Introductory Chapter of the Variation[5] I know that is not the case, as you wrote me you were busy on man.[6] When this new work of yours will be ready I should be happy if you would allow me to bring it before the German public. Mr Koch, too, makes a point of it to publish your works. Of course I could not speak with him positively about it, as I don't know what you intend. But I must say it would be very kind of you if you would think of us, if you intend at all to have it translated.[7]

I hope your health is better than it was when I heard last of you.

Believe me | My dear Sir | Yours very sincerely | J. Victor Carus

DAR 161: 70

1 Eduard Koch was the director of E. Schweizerbart'sche Verlagshandlung, publishers of German translations of CD's work. The last German translation of *Origin* had been the third German edition, Bronn and Carus trans. 1867, including changes made in the fourth English edition, which was published in 1866. The fourth German edition was published in 1870 (Bronn and Carus trans. 1870), incorporating changes from the fifth English edition.

2 Carus was at work on his *Geschichte der Zoologie bis auf Joh. Müller und Charl. Darwin* (Carus 1872).

3 Carus had made the German translation of *Variation* (Carus trans. 1868); see *Correspondence* vols. 15 and 16.

4 The second German edition of *Variation* was published in 1873 (Carus trans. 1873).

5 See *Variation* 1: 4–9, where CD wrote of planned 'second' and 'third' works to follow *Variation*. These were never published.

6 Carus refers to *Descent*. No letter in which CD told Carus that he was working on humans has been found.

7 Carus did translate *Descent* (Carus trans. 1871).

From R. F. Cooke 29 April 1869

50A, Albemarle Street, London, W.
April 29 *1869*

My dear Sir

I suppose the enclosed letter is correct & that you wish us to supply a set of the Electrotypes of the work on Orchids, at as reasonable a price as we can?[1]

Yours faithfully | Robt. Cooke

C. Darwin Esq

DAR 171: 369

1 The enclosure has not been found. John Murray published CD's *Orchids* in 1862. No other English-language edition was published until 1877. The letter probably concerns the projected French translation of *Orchids* (see letter from R. F. Cooke, 22 February 1869).

To John Murray 29 April [1869][1]

Down. | Beckenham | *Kent. S.E.*
Ap. 29th

My dear Sir

Some of the Ladies took off the cover of Quarterly & burnt it, & I had no idea who sent it. I thank you sincerely for the present. The article by Wallace is inimitably good & it is a great triumph that such an article shd. appear in the Quarterly, & will make the B. of O. & Owen gnash their teeth.[2] My neighbour, Sir J. Lubbock,[3] who had no idea who wrote it, was very much struck with the article, as extraordinarily well done.—

I am delighted to hear of sale of Fritz Müller.[4]

It is good news that you will print 2000 copies of Origin at 14s, for I think I have brought it up to present standard of Science, as the cost of 6 weeks work.

I am making slow, but sure progress with my present book.—[5]

Thanks for your kind enquiries about my health: my horse rolled over me & hurt me rather seriously, but I have made a surprisingly quick recovery.[6]

My dear Sir | Yours very sincerely | C. Darwin

National Library of Scotland (John Murray archive, Ms. 42152 ff. 201–2)

1 The year is established by the relationship between this letter and the letter from John Murray, 28 April [1869].

2 The *Quarterly Review* was published by John Murray; he had evidently sent the issue containing an anonymous review by Alfred Russel Wallace that discussed CD's theory of natural selection ([Wallace] 1869b; see letter from John Murray, 28 April [1869]). CD also refers to the bishop of Oxford, Samuel Wilberforce, and Richard Owen, both notable opponents of CD's theory. The 'Ladies' would have been Emma Darwin, Henrietta Darwin, and Elizabeth Darwin; it is not known why they burned the cover.

[3] John Lubbock.
[4] See letter from John Murray, 28 April [1869] and n. 2.
[5] CD refers to *Origin* 5th ed. and to *Descent.*
[6] See letter to A. R. Wallace, 14 April 1869 and n. 12.

To James Paget 29 April [1869][1]

Down. | Beckenham | *Kent. S.E.*
Ap. 29th

My dear Paget

I have made a wonderful recovery. Mr Engleheart says all the fluid is gone, only some thickening left.—[2] I have no pain, & very slight tenderness, & can walk & do walk several times a day.— I shall soon take to gentle rides again.

As far as sensations serve, I am inclined to think the fomentations acted admirably.

My general health is also improving.— Do not waste your precious time in answering this, but accept my cordial thanks for your kind advice & believe me

My dear Paget | Yours very sincerely | Ch. Darwin

P.S. If ever you come across an extra blushing damsel do not forget the downward extent of the blush. I am assured that some half naked natives of Eastern races blush over arms & legs!

Also any case in which when screaming or breathing laboriously if platysma myoides contracts skin of neck into transverse folds?—[3]

The Bodleian Library (Auto 4:119a)

[1] The year is established by the reference to CD's horse-riding accident (see letter to A. R. Wallace, 14 April 1869 and n. 12). Paget had visited CD on 21 April 1869 (Emma Darwin's diary (DAR 242)).
[2] Stephen Paul Engleheart was the surgeon in Down.
[3] Paget had offered to make observations of blushing, and of the platysma muscle in persons screaming, in 1867 (see *Correspondence* vol. 15, letter from James Paget, 9 July 1867). On blushing on the arms and legs, see *Correspondence* vol. 16, letter from F. F. Geach, April 1868 and n. 5.

From M. A. Ruck to Francis Darwin [29 April – 27 May 1869][1]

13. Kensington Park Road
Thursday.

Dear Mr Darwin,

Caerdeon is most lovely. It is on the Barmouth Estuary with Cadir in front and delightful heathery hills behind.[2] You could not do better. I spoke to Clement on Sunday last and he was so vague about his plans that I dare not feel sure you could have it and the other place is a thousand times more beautiful—[3]

I am sorry it is not nearer to us but we can easily meet for expeditions in all directions—[4]

Tell your Father that at present the Sheep question stands thus— Horns are always like felt at birth, but do not develope themselves much for a fortnight or more and that earlier in the Males.[5] I have several people on the look out and will send more certain information when the season is more advanced. I hope Horace is well give him my love please and say that we have his things here and will give them to Dickie[6] Tell Mrs. Darwin that I shall be very glad to welcome her in Wales

Believe me | Yours very sincerely | M L Ruck

DAR 83: 190, DAR 84.2: 211

CD ANNOTATIONS
2.1 I am … thus— 3.1] *crossed pencil*
3.1 question stands] *underlined pencil*
3.2 fortnight … out and 3.3] *double scored blue crayon*
3.3 earlier in the Males.] *underlined red crayon*
3.4 I hope … Wales 3.6] *crossed pencil*
Top of letter: 'Mrs Ruck' *red crayon*
End of 2d page: 'give new [quote] | *Mammals*' *blue crayon*; 'New' *pencil*

1 The date range is established by the relationship between this letter and the letter from Edward Jones to M. A. Ruck, 28 April 1869, and by a note dated 'May 29/69/' written on the verso of the second page of the letter. In 1869, the first Thursday after 28 April was 29 April, and the last Thursday before 29 May was 27 May.
2 Ruck refers to Plas Caerdeon, the house where the Darwins stayed from 12 June to 30 July 1869 (Emma Darwin's diary (DAR 242); see also *Emma Darwin* (1904), 2: 228). She also refers to the mountain Cader Idris.
3 Clement has not been identified.
4 The Ruck home in Wales was Plas Pantlludw near Machynlleth.
5 See letter from Edward Jones to M. A. Ruck, 28 April 1869 and n. 1.
6 Ruck refers to Horace Darwin and Richard Matthews Ruck.

To Max Schmidt [29 April 1869][1]

Dear Sir

I have thought that your kindness wd excuse me taking the liberty to trouble you with a question. I have been told, that you have adult Mandrills of both sexes.—[2] Now I want very much to know whether this species at the posterior end of the body, *which is almost or quite destitute of hair*, is larger in ♂ than in ♀? Is not this species [much] brighter colour in the male?— Also is a larger surface on the face of the ♂ destitute of hair than in the ♀. Some allowance wd have to be made for the greater relative size of body in the ♂. The description of the colour of the face of ♂ & ♀ Mandrills, I find to be very different; but I fear you wd think me too [unreasonable] if I were to ask, for a very short description of the colour of the face in both adult sexes with the [permission] to quote you as my authority.— I hope that you forgive me for asking this favour, but I have for some time vainly endeavoured to find an accurate description of the *sexual* differences of colour in the Mandrills— If you cant aid me, could you tell me the name of the [Curator] of ZG.[3] with *much respect*, I beg leave to remain Dear Sir | Yours truly | C. D

[Cologne]

Name of man at *Antwerp*,[4] & introduction to *[*where I know there are mandrills of both sexes*]*, & if you are *[6 words illeg]* as addresses to try.

ADraftS
DAR 96: 65

[1] The date is established by the letter from Max Schmidt, 14 June 1869.
[2] See letter from J. N. Hegt, [23 April 1869] and n. 3.
[3] Schmidt was director of the Frankfurt am Main Zoological Gardens.
[4] See letter from J. N. Hegt, [23 April 1869] and n. 3.

From M. T. Masters 30 April 1869

Gardeners' Chronicle | & Agricultural Gazette Office, | 41, Wellington Street, Strand, W.C.
April 30 *1869*

My dear Sir,

On the chance that you may not have seen it I forward to you by book post a paper on the "Origin of Genera"—[1] If it is of the slightest interest to you, please retain it, but if you have it already return it to me at any time— No doubt you saw Decaisne's letter on the apple last week. He surely cannot mean that there are no intermediates between Malus & Pirus[2]

The other day at Chiswick Dr Hogg, Mr Moore Mr Barron the Superintendent[3] & myself were looking at a tree in full blossom & none of us could tell for certain whether it was an apple or a pear— To be sure we had not the fruit before us—

faithfully yrs. | Maxwell. T. Masters

C. Darwin Esq

DAR 171: 80

[1] Masters refers to Cope 1868. There is an annotated offprint of this paper in the Darwin Pamphlet Collection–CUL.
[2] Joseph Decaisne wrote a reply to an enquiry that the editors of the *Gardeners' Chronicle* had sent to him about the 'Paradise apple'; it was published in *Gardeners' Chronicle*, 24 April 1869, p. 443. *Malus* is the genus of apples; *Pyrus* that of pears. Decaisne had written: 'I have accumulated materials which do not permit me to adopt the theory of the transformation of one species into another, though I admit that species are very variable. The "Malus" remain Apples, as the "Pirus" remain Pears, in spite of the diversity of their forms and the infinite number of their varieties.' The paradise apple is now *Malus pumila* var. *paradisiaca* but a synonym is *Pyrus malus* var. *paradisiaca.*
[3] Robert Hogg, Thomas Moore, and Archibald F. Barron, superintendent of the Royal Horticultural Society's gardens at South Kensington and Chiswick.

From J. J. Weir [1–13] May 1869[1]

6 Haddo Villas | Blackheath SE
May 1869

My Dear Sir

I send by this days post a copy of a short paper on Insects and Insectivorous birds.— The results yet arrived at are meagre, but they point in one direction & support Wallaces theory on the subject.—[2]

I have recently twice visited the South Downs during the lambing season and was very much surprised to see how very far from being white the lambs were even when some weeks old.—[3]

The flocks were of the choicest character, one my Brother in Laws was in the farm just vacated by the late John Ellmans son & the rams used as sires were of great value, yet the lambs reminded one of Jacobs spotted & speckled, some had large patches of color all over the back, & the faces of many were mottled, & the ewes even were often black & white, yet so powerfully has selection acted entirely to produce a perfect adult, that all these would ultimately have white wool and light brown faces (the true South Down has never a black face).—[4]

I was much interested to find from a recent visitor there that in the remote Island of St. Kilda the sheep are still even in the adult state *brown.*—[5]

I may further remark when on the subject of sheep that this season, much below the average number of twin lambs have been produced, the Flock masters think that insufficient food last autumn was the cause.—

I endeavoured to ascertain whether any lambs were ever produced without tails but could hear of none.—

This I think remarkable because the sheep dogs which have had their tails amputated in a similar manner, are now commonly pupped tailless.—[6]

Believe me | My Dear Sir | Yours sincerely | J Jenner Weir

C Darwin Esqr.

DAR 181: 79

CD ANNOTATIONS

1.1 I send ... subject.— 1.3] *crossed pencil*
2.1 I have ... *brown.*— 4.2] 'South down lambs mottled with Black' *added red crayon*
5.1 I may ... tailless— 7.2] *crossed pencil*
Top of letter: 'Embryology *Sheep* of South Downs | Reversion?' *pencil*

1 The date range is established by the relationship between this letter and the letter to J. J. Weir, 13 May [1869].

2 Weir refers to 'On insects and insectivorous birds; and especially on the relation between the colour and the edibility of Lepidoptera and their larvae' (Weir 1869), which was published in the April 1869 issue of *Transactions of the Entomological Society of London*. Alfred Russel Wallace had theorised that bright colour in caterpillars was protective (see letter from A. R. Wallace, 10 March 1869 and nn. 2 and 3).

3 Weir refers to hills around Lewes in East Sussex, where the Southdown breed of sheep originated. CD had discussed the colour of Southdown lambs in *Variation* 2: 30.

4 John Ellman (1753–1832) was well known for his work in developing the Southdown sheep at Place Farm, Glynde, near Lewes. His son, John Ellman (1787–1867), took over the farm in 1829 (Farrant 1978). The son of the second John Ellman whom Weir refers to may have been Robert Harvey Ellman. The early breed standard described the face as long and speckled (*ibid.*, p. 78), but the modern breed standard refers to the face as mouse-coloured (http://www.southdownsheepsociety.co.uk/breed.htm, accessed 2 November 2007). Weir also refers to the Jacob sheep, an ancient, multi-horned breed with a piebald fleece. Weir's brother-in-law has not been identified.

5 Weir refers to Soay sheep, native to the island of Soay in the St Kilda archipelago of western Scotland. It is considered one of the oldest breeds (Ponting 1980, p. 69).

6 CD had recently corresponded with William Chester Tait concerning Tait's observations of the inheritance of taillessness in dogs (see letters from W. C. Tait, 26 January 1869 and 19 February 1869).

From A. W. Bennett 3 May 1869

Lancaster Villa | York Road, Upper Holloway | London N.
May 3rd 1869

Dear Sir

Tempted by the kindness of your reply to my previous note, I venture to send you by this post my notes made during the past winter & spring on the fertilization of certain plants flowering at those seasons of the year, with the accompanying drawings.[1] I should not think of publishing the observations as a series until they have extended over a much wider range, or I have had the opportunity of confirming them by a second series;[2] but it would be a great assistance to me if the accuracy of any of them were tested by an independent observer; & I venture to hope you may be able to give me some valuable advice as to the best direction in which to prosecute them, or even as to whether any scientific end will be gained by following them.

My observations so far have led me to the following conclusions (subject of course to correction from a wider series of observations) as to the different classes into which winter-flowering plants may be divided.

1. Those which flower & fructify abundantly through the whole year equally in Dec. & Jan to the summer. These I find to be *self-fertilized* & generally in the lead *at whatever time of the year they flower*; & this I take to be the circumstance which adapts them to so long a period of flowering not requiring insect-agency for their fertilization, e.g. *Lamium album*, *Veronica Buxbaumii*, *Euphorbia Poplus*.[3]

2. Those which normally flower in the spring; but in mild seasons are found not infrequently in the winter; these are *cross-fertilized*; e.g. *Sinapis arvensis*, *Fumaria officinalis*[4]

3. Very early spring-flowering plants; these are generally either showy or scented, attracting the earliest spring insects, & are *cross fertilized*. e.g. *Viola odorata*, *Galanthus nivolis*.[5]

4. Plants flowering in the depth of winter, but natives of warmer climates, where there are probably plenty of insects in the winter; these are *cross-fertilized*; e.g. *Chimonanthus fragrans* & *Eranthis hyemalis* where the provisions for crossing are even more perfect that in Parnassia, & which appears never to fructify in our gardens.[6]

The most interesting observations to myself have been those on the three species of *Euphorbia*, *Peplus*, *Heliscopia*, & *amygdaloides*; the two former, genuine winter-flowering plants being apparently self-fertilized, i.e. the pistil from the stamens in the same *head of flowers*; while in the latter, which never flowers before April, the pistil appears to be fertilized from the *male* heads of flowers.[7]

You will find *Eranthis hyemalis*, the Christmas Rose[8] among the plants in which some observations have been made

I am specially curious about the fertilization of *Ruscus*.[9] In most unisexual winter flowering plants, e.g. *Corylus avellana*,[10] there appears to be special provision for fertilization in the enormous preponderance of male flowers over female, & the manner in which the former are hung, so that the pollen is freely dispersed in clouds by every breath of wind. In the *Ruscus* on the contrary, which is generally considered to be *diœcious*, there appears to be no such preponderance, & the excessively rigid stems, growing under the shelter of shrubs & under-wood, would seem to be peculiarly inaccessible to the action of the wind. I found abundance of flowers out last winter in the Surrey heaths during the last week of December, & there were a fair number of the previous year's berries still remaining on the female plants.

I should be much obliged if you would kindly return the notes & drawings when you have had leisure to glance over them, & must really apologize for troubling you with so long a letter as this

& remain | yrs respectfully | Alfred W. Bennett

C. Darwin Esq

DAR 76: B176–7

CD ANNOTATIONS

3.1 1. Those … *Poplus*. 3.5] *double scored pencil*

Top of letter: 'See 2[d] paper Lamium album' *pencil, ringed pencil*; 'Gorse' *pencil, ringed pencil*; 'Does this not indicate sterile at one period & fertile at next' *pencil*

1 CD's letter has not been found, but it was evidently a reply to the letter from A. W. Bennett, 22 April 1869. The notes and drawings that Bennett refers to have not been found.

2 Bennett published 'On the fertilisation of winter-flowering plants' on 4 November 1869 in the inaugural issue of *Nature* (Bennett 1869). He discussed observations made in the winter of 1868 to 1869. CD's annotated copy of the article is in DAR 139.13: 1.

3 *Lamium album* is white dead-nettle, *Veronica buxbaumii* (now *V. persica*) is bird's eye speedwell, and *Euphorbia peplus* is petty spurge (see Bennett 1869, p. 11).

4 *Sinapis arvensis* is charlock; *Fumaria officinalis* is common fumitory (see Bennett 1869, p. 11).

5 *Viola odorata* is the sweet violet; *Galanthus nivalis* is the common snowdrop.

6 *Chimonanthus fragrans* (now *C. praecox*) is wintersweet or Japanese allspice; *Eranthis hyemalis* is winter aconite (see Bennett 1869, p. 12). For Bennett's observations on *Parnassia palustris* (grass of Parnassus), see the letter from A. W. Bennett, 22 April 1869 and n. 1.

7 *Euphorbia helioscopia* is sun spurge; *E. amygdaloides* is wood spurge (see Bennett 1869, pp. 12–13).

8 See above, n. 6; 'Christmas rose' is now the common name for *Helleborus niger*.

9 *Ruscus* is a genus of dioecious evergreen shrubs formerly in the family Liliaceae but now in the family Ruscaceae.

10 *Corylus avellana* is hazel.

To J. V. Carus 4 May 1869

Down. | Beckenham | *Kent. S.E.*
May 4. 1869

My dear Sir

I am pleased to hear about the Origin, for I am now completing a new edit., which will be published in a month's time or less.[1] I have gone very carefully thro' the whole, trying to make some parts clearer, & adding a few discussions & facts of some importance. The new ed. is only 2 pages, at the end, longer than the old; though in one part 9 pages in advance, for I have condensed several parts, & omitted some passages. The translation, I fear will cause you a good deal of trouble; the alterations took me 6 weeks, besides correcting the press; you ought to make a special agreement with M. Koch.[2] Many of the corrections are only a few words, but they have been made from the evidence on various points appearing to have become a little stronger or weaker.[3]

Thus I have been led to place somewhat more value on the direct & definite action of external conditions,—to think the lapse of time as measured by years not quite so great as most geologists have thought,—& to infer that single variations are of even less importance, in comparison with individual differences, than I formerly thought.[4] I mention these points because I have been thus led to alter in many places *a few words*; & unless you go thro' the whole new edition, one part will not agree with another, which w[d] be a great blemish. I c[d] lend you the corrected sheets, which w[d] shew where changes had been made, but then some corrections have been added in the proofs, which w[d] not appear in the above sheets. Please sometime to inform me whether you wish for the loan of the corrected sheets, & whether you w[d] like a *bound* copy, or *clean loose* sheets for y[r] translation.

I am pleased to hear about the sale of "Domestic Animals,"[5] Altho' I am never idle my new book makes very slow progress; & I will assuredly inform you as soon as it goes to the printers.[6]

Believe me | my dear Sir | yours very sincerely | Charles Darwin

LS
Staatsbibliothek zu Berlin – Preußischer Kulturbesitz, Slg. Darmstaedter Lc 1859 Darwin, Charles, Bl. 41/42

1 See letter from J. V. Carus, 29 April 1869. CD refers to the fifth edition of *Origin*. For more on the publication of the edition, see the letter to J. V. Carus, 5 February 1869 and n. 5.
2 Eduard Koch.
3 For CD's list of additions and corrections, see *Origin* 5th ed., pp. xi–xiii.
4 On the direct and definite action of external conditions, see *Origin* 5th ed., pp. 165–8. CD also refers to evidence for the age of the earth from astronomical calculations (see, for example, letter from James Croll, 4 February 1869, and *Origin* 5th ed., pp. 348–54). For CD's discussion of 'individual differences', see *Origin* 5th ed., pp. 104–5; see also letters to A. R. Wallace, 22 January [1869] and 2 February [1869].
5 Carus reported on sales of *Variation* in his letter of 29 April 1869.
6 CD refers to *Descent*.

To Charles Lyell 4 May [1869][1]

Down. | Beckenham. | *Kent. S.E.*
May 4th

My dear Lyell

I have been applied to for some photographs (carte de visite) to be copied to ornament the Diplomas of Hon. members of a new Nat. Soc in *Servia*!!![2] Will you give me one for this purpose? I possess only a full-length one of you in my own album, & the face is too small, I think, to be copied.—

I hope that you get on well with your work & have satisfied yourself on the difficult point of glacier-lakes.—[3] Thank Heaven I have finished correcting the new edit. of Origin, & am at my old work of Sexual selection.—[4]

Wallace's article struck me as *admirable.* How well he brought out the revolution which you effected some thirty years ago.[5] I thought I had fully appreciated the revolution, but I was astounded at the extracts from Cuvier.—[6] What a good sketch of nat. selection, but I was dreadfully disappointed about man: it seems to me incredibly strange & weak; & had I not known to the contrary, would have sworn it had been inserted by some other hand.—[7] But I believe that you will not agree quite in all this.

My dear Lyell | Ever yours sincerely | C. Darwin

I have got over my crushing by my horse wonderfully—3 weeks, instead of 3 months as Paget expected.[8]

American Philosophical Society (369)

1 The year is established by the relationship between this letter and the letter from Spiridion Brusina, 29 April 1869.
2 See letter from Spiridion Brusina, 29 April 1869. 'Servia' was an alternative spelling of 'Serbia'.
3 CD refers to Lyell's reluctance to accept the theory proposed by Andrew Crombie Ramsay of the glacial origin of rock basins (Ramsay 1862). For more on the controversy, and Lyell's alternative explanation, see *Correspondence* vol. 13, letter to Charles Lyell, 21 February [1865] and nn. 7 and 8, and letter from J. D. Hooker, 2 May 1865 and n. 10, and Davies [1969], pp. 304–7.
4 CD refers to *Origin* 5th ed. and to *Descent*.
5 Wallace reviewed Lyell's *Principles of geology* and *Elements of geology* (C. Lyell 1865 and 1867–8) in the *Quarterly Review* ([Wallace] 1869b). The 'revolution' was Lyell's refutation of catastrophism in geology and proposal of what came to be known as uniformitarianism (see [Wallace] 1869b, pp. 363–4).
6 In [Wallace] 1869b, pp. 360–3, Georges Cuvier's approach to geology was reviewed. See letter to A. R. Wallace, 14 April 1869 and n. 2.
7 See [Wallace] 1869b, pp. 391–3. See also letter from A. R. Wallace, 24 March 1869 and n. 10.
8 James Paget had visited CD on 21 April 1869 (see letter from E. A. Darwin, [after 21 April 1869] and n. 1).

From J. D. Caton 5 May 1869

Ottawa Ill.
May 5. 1869.

Charles Darwin Esq.
Dear Sir.

Your esteemed favor met with delay on the way & I have taken some time since to make inquiries.[1] I have carefully examined my herd of Elk[2] (26 in number) and am satisfied that there is no difference in color, depending on Sex or even age, at least at this season of the year, when they are just ready to shed their winter coat— The two darkest colored in the herd are females, but the difference in shade among the whole lot is quite immaterial. I shall continue my observations when they are in summer coat & will advise you further.[3]

Among my deer & Elk I have never observed any female to manifest any preference for any male at any season of the year. Indeed I have never been able to detect any two deer, who manifested a preference for each others society over that of others— During the rutting season the female deer seems always endeavoring to escape from the male. Neither myself nor any of my men have ever seen a female stand quietly for the male If that ever occurs it must be in the night— When observed, the males success has been purchased by a hard chase, but this has been seen very rarely.

Soon after I received your letter I was called upon by a Mr Reed[4] who desired to sell me some live specimens from the Rockey Mountains, where he said he had spent the last twenty five years hunting & trading— He is, I judge, as good an observer as most hunters— He had the impression that in the wild state the male Elk is darker than the female, but after studying my herd for some time, he expressed a doubt of the accuracy of his recollection, as in truth he had never carefully examined them with express reference to that point

He said he had often witnessed battles between Bull bisons, but he could recollect no movement or circumstance which would suggest that the mane or long hairs of the neck & shoulders served as a shield or protection in their encounters.[5] He had never observed any scars nude the mane, though as he never examined for them such might have escaped his observation. He had never observed one to strike the other where the long hairs grow. He says the battle is conducted in precisely the same way as between bulls of domestic Cattle. Neither party gets much opportunity to goad the other, till the vanquished turns to flee— I hope to visit the plains this season, when I may be able to obtain more satisfactory information on the subject.[6]

I confess myself quite unable to discover any advantage to any of the deer family resulting from the many points to the horns.[7] On the contrary they seem to be a manifest disadvantage, or a protection to those whome they may attack— Once my son 13 years old was attacked by an old buck & thrown on his back, and I am satisfied his life was only saved by reason of the many points of the horns of the deer. In the first charge but two penetrated the clothes so as to break the skin, whereas a single spike would have pierced him through at once—[8] While the boy was on his back he fought him with his feet, & received several severe punctures in the back part of his thighs, but he soon seized him by the prongs & was enabled to deaden the force of his thrusts till a favorite dog came to his rescue. As it was, he received many deep wounds in the back part of his thighs, from the many parts of the bucks horns. Had the whole force which inflicted these, been expended upon a single point, fatal consequences would probably have resulted— Clearly the many points are an advantage to the adversary. Again, it has often, occured with both the deer and the Elk, that in the rush of battle their horns have become, inextricably interlocked, & both died together. The mode of joining battle, with both species is after approaching each other slowly till quite near by a terrific plunge towards each other, with their heads lowered so as to bring their faces nearly parallel with the ground, so that they can hardly see each other at the moment of collision. It is in this first shock, undoubtedly, that the horns become interlocked— We can appreciate, that in such an encounter the many antlers serve as a sort of shield. Without them, no doubt fatal consequences would much more frequently occur—[9] So far as I know no animal which joins battle with a violent plunge presents to the adversary a single point, although my very limited information on the subject, disqualifies me to speak with confidence on the subject. The common ram who always joins battle in this way, never presents the points of his horns to the adversary, but depends on the battering powers afforded by the base of the horns and I judge the same thing is true with the Rockey mountain sheep (Ovis Montana Rich.) judging from indications on a single specimen which I have.[10] I make these suggestions, with the few facts stated as tending to throw some light on the subject of your inquiry, not as fully answering it, appreciating that you will be much better able to draw correct conclusions from the facts than myself. Still I cannot help remarking, that it is not difficult to appreciate the wisdom, of a provision, the tendency of which is to ameliorate the severity of a weapon, which might otherwise be too fatal when used in warfare as I have described

Can you inform me whether the secondary or succentorial hoofs are common to all of the varieties of the Antelope and the deer of the old world, and should not its presence or absence, be considered an important circumstance in discribing, their characteristics or distinguishing marks? It is a point on which authorities accessable to me, are quite generally silent, embarressing me much. I am studying with as much care as I am capable of, and comparing the Gazelle of Palestine and the American or Prong horned Antelope as it has been called living specimens of which I have in my parks.[11] While there are many points of close resemblance, the former has these hoofs very prominent, while they are entirely wanting in the latter, as much as in the horse. Indeed, so far as my limited observation goes this is the only one of the larger or cloven footed ruminants of this country, in which this hoof is wanting—

I have been much interested in making observations, which were suggested by the perusal of Chap 18. "Animals & plants under domestication" in "Sterility from changed conditions of life" some of which have suggested the possibility, that the *fault*, if I may use the expression, for the sterility is more with the female than with the male, at least in some cases. For instance, I have a pair of wild turkeys now nearly two years old. Last year neither showed any inclination to breed, while tame ones of same age were prolific.[12] This year the wild male couples with the tame females, though not as freely as the tame male usually does, while the wild female manifests no inclination for the male as yet— Still this may be exceptional, for she is manifestly in some respects an exceptional bird. She has the rare appendage for a female turkey, of a *beard*, which is now three inches long. This however is not a certain evidence of sterility for I once had a tame one with a like appendage and she was more than usually fertile. These are the only two instance I have ever observed of this appendage to the female turkey— The fact also that both are of the same parentage, may have its influence, and my desire to get a pure wild progeny has prevented me from allowing her to associate with the tame male— They are not closely confined but are allowed the range of one of the parks of about forty acres— I have lately procured four other wild female turkeys which will extend my means of observation.[13] They are however but eight months old, so I can hardly expect much from them this Season.— They however do not avoid the society of the male as the older female does— Indeed she declines to associate with any of the turkeys, & seems to prefer the society of the common hen, or barnyard fowl

I am also making pretty careful observations of the deer & Elk in the same direction— Undoubtedly in my parks they are not as fertile as in a wild state. But I will not trouble you with extended observations on the subject fearing that my minute detail may have already wearied you.

You cannot do me a greater favor than by giving me an opportunity of laying my observations before one who is so much more capable than myself to draw correct conclusions from established facts— I shall be ever most happy to hear from you, and if I may be of the least service to you in the great work you have in hand it will be a source of the highest pleasure

With great respect I am very truly yours | J. D. Caton

P.S. Dr. John Paul, Secy. Ottawa Academy of Natural Sciences,[14] wishes me to present you his high consideration of your eminent service to Science, & to enquire if an assortment of fossils collected in this vicinity would be acceptable to you & in what way they should be forwarded.

DAR 83: 170–1

CD ANNOTATIONS

1.2 I have … winter coat— 1.4] 'no difference in colour of sexes in Elk' *added blue crayon*
2.4 seems always … the male. 2.5] *underl blue crayon*
3.3 He is, … point 3.7] *crossed blue crayon*
4.2 would suggest … shoulders 4.3] *underl blue crayon*
5.2 On the … of the deer. 5.6] *double scored blue crayon*
5.12 Had … resulted— 5.13] *scored blue crayon*
5.20 We can … occur— 5.22] *double scored blue crayon*
6.1 Can you … marks? 6.4] *scored blue crayon*
7.7 This year … fertile. 7.13] *scored blue crayon*; 'Audubon'[15] *added blue crayon*
7.13 These are … pleasure 9.5] *crossed blue crayon*
9.1 You … pleasure 9.5] *scored blue crayon*
10.1 With … forwarded. 11.4] *crossed pencil*
11.1 P.S. … forwarded. 11.4] *scored blue crayon*

1 See letter to J. D. Caton, 20 March 1869.
2 Caton had kept elk (*Cervus canadensis*) in an enclosure since the early 1860s; see Caton 1877a, p. 78.
3 Caton later observed a difference in male and female coat colour; see *Descent* 2: 289, and Caton 1877a, pp. 145–6.
4 Mr Reed has not been identified.
5 CD mentioned this observation as coming from an 'experienced hunter' known to Caton in *Descent* 2: 267.
6 Caton refers to the great plains of central North America.
7 In Caton 1877a, pp. 230–1, Caton noted CD's question about the utility of branched antlers in Cervidae, and discussed the 'mode of warfare' among deer species.

8 In *Descent* 2: 254, CD mentioned that Caton agreed that a single point would be more dangerous than branched antlers. See also Caton 1877a, p. 230. Caton's son was Charles Edward Caton.

9 For CD's discussion of battles between male deer with branching horns, see *Descent* 2: 252–6.

10 *Ovis montana* is now *Ovis canadensis* (the bighorn sheep). In *Descent* 2: 249–50, CD noted that the common ram used the base of the horns as the point of contact in a charge.

11 Caton refers to what is now called the Palestine mountain gazelle, *Gazella gazella* subsp. *gazella*, and to *Antilocapra americana*, now called the pronghorn; for his discussion of their hooves, see Caton 1877a, pp. 36–7.

12 See *Variation* 2: 145–72 (for the section on birds, see pp. 153–8). Caton also mentioned the pair of wild turkeys in his letter of 2 [October] 1868 (*Correspondence* vol. 16). CD added Caton's observation in *Variation* 2d ed., 2: 91.

13 In *Variation* 1: 292, CD mentioned that male wild turkeys were known to breed with domestic hens. Caton described his breeding of wild turkeys on forty acres through eleven generations in Caton 1877b; he also discussed his experiments on crossing wild turkeys with the domestic breed. Caton 1877b was reprinted in Caton 1880, pp. 176–88.

14 John Paul was the secretary, or president, of the Ottawa (Illinois) Academy of Natural Sciences (*Ottawa Free Trader*, 28 September 1872).

15 For John James Audubon's description of the turkey, see Audubon 1831–[9], 1: 33–4, and Audubon 1839, pp. 195–7.

From T. H. Farrer 5 May 1869

3 Gloucester Terrace | Regents Park
5 May /69

My dear Mr Darwin

You asked me last year what I meant by the "degraded" structure of Viola Odorata.—[1] I think now, looking at the flowers again, the word was justified. In "Viola Tricolor" admirably described in Hildebrand, there is an apparatus for cross fertilization by insects almost as wonderful and complete as that of the Orchises.[2] In Viola Cornuta, it is even still more elaborate.

In Viola Canina there is the same structure: but in almost every particular, less complete.— The flower less conspicuous: less open: the lower petal less coloured & less displayed: the anthers less syngenesious: good pollen less universal & smaller in quantity: and above all the style & stigma less adapted either in position or form to catch the pollen from the proboscis of the bee. Consequently Viola Canina does sets some seeds from open flowers—but not constantly or frequently as in Viola Tricolor.

In Viola Odorata the deficiencies of V. Can. are much increased. The nature & position of the footstalks: of the sepals: the colour & position & shape of the petals make the flower one of the most inconspicuous there is—and the way to the nectary is no "royal road"— The anthers seldom have good pollen. And the form &c of style & stigma though similar in generally to that of V. tricolor has lost the peculiarities & use which give it a meaning. Consequently it is very seldom that open flowers of Viola odorata set their seed.

Is there not here a gradation or **de**gradation, shewing how an elaborate structure originally full of function & purpose becomes comparatively meaningless & useless.

Very truly yours | T H Farrer

DAR 111: A79–80

CD ANNOTATIONS
1.2 In . . . wonderful 1.4] *double scored blue crayon*; '& Fourier??'[3] *added blue crayon*
Top of letter: 'Viola' *blue crayon*

1 See *Correspondence* vol. 16, letter from T. H. Farrer, 26 October 1868, and letter to T. H. Farrer, 29 October [1868]. Farrer had referred to the efficient contrivances of *Viola cornuta* (horned violet) compared with the degraded ones of *V. odorata* (sweet or common violet). CD had written that, never having seen *V. cornuta*, he was not sure what part was degraded in *V. odorata*.
2 For Friedrich Hildebrand's description of how the floral morphology of *Viola tricolor* (pansy) prevents self-fertilisation, see Hildebrand 1867, pp. 53–6.
3 The reference has not been identified.

From Charles Lyell 5 May 1869

May 5, 1869.

I am pleased at the impression which the historical part of Wallace's review made on you.[1] It reminds me of Cuvier's daughter, a charming and intelligent girl, telling me she had been reading my book (vol. i. of 'Principles') to her father, and that they had been struck with the complete antagonism of my views to those which he had propounded in his 'Theory of the Earth.'[2]

I was always made to feel myself a welcome guest at Cuvier's *soirées*, but he never alluded to my book, and but for Mademoiselle Cuvier's saying she had been reading it to him in their carriage as they drove out, I should never have known he had seen it.

P. S I quite agree with you that Wallace's sketch of Natural Selection is admirable.[3] I wrote to tell him so after I had read the article & in regard to the concluding theory I reminded him that as to the origin of man's intellectual & moral nature I had allowed in my first edition that its introduction was a real innovation interrupting the uniform course of the causation previously at work on the earth. I was therefore not opposed to his idea that the Supreme Intelligence might possibly direct Variation in a way analogous to that in which even the limited powers of man might guide it in Selection as in the case of the breeder and horticulturist. In other words as I feel that Progressive Developement or Evolution cannot be entirely explained by Natural Selection I rather hail Wallace's suggestion that there may be a Supreme Will & Power which may not abdicate its functions of interference but may guide the forces & laws of Nature.[4] This seems to me the more probable when I consider not without wonder that we should be permitted to give rise to a monstrosity like the pouter pigeon & to cause it to breed true for an indefinite number of generations, certainly not to the advantage of the variety or species so created.

At the same time I told Wallace that I thought his arguments—as to the hand, the voice, the beauty, the symmetry, the naked skin & other attributes of man implying a preparation for his subsequent developement—might easily be controverted. That a parrot endowed with the powers of Shakespeare might dictate the "Midsummer

Nights Dream', & that Michael Angelo if he had no better hand than belongs to some of the higher apes might have executed the statue of Lorenzo di Medici.[5]

In reply to this & other analogous comments Wallace said "It seems to me that if we once admit the necessity of *any action* beyond "natural selection" in developing man we have no reason whatever for confining that action to his brain. On the mere doctrine of chances it seems to me in the highest degree improbable, that so many points of structure all tending to favour his mental developement should concur in man & in man alone of all animals. If the *erect posture*, the *freedom of the anterior* limbs from *purposes of locomotion* the *powerful* & *opposable* thumb, the *naked skin*, the *great symmetry of force* the *perfect organs of speech* & his mental faculties—*calculation of numbers, ideas of symmetry*, of *justice*, of *abstract reasoning* of *the infinite* of a *future state* & many others can not be shown to be each & all *useful* to man in the very lowest state of civilization, how are we to explain their coexistence in him alone of the whole series of organized beings? Years ago I saw a Bushman boy and girl in London & the girl played very nicely on the piano. Blind Tom the idiot *negro* had a *musical ear or brain* superior perhaps to that of any living man.[6] Unless Darwin can shew me how this rudimentary or latent musical faculty in the lowest races can have been developed by survival of the fittest—can have been of *use* to the individual or the race, so as to cause those who possessed it to win in the *struggle for life*, I must believe that some other power caused that developement— & so on with every other especially human characteristic. It seems to me that the "*onus probandi*"[7] will lie with those who maintain that man, body *&* *mind*, could have been developed from a quadrumanous animal by Natural Selection."

As to the scooping out of lake-basins by glaciers, I have had a long, amicable, but controversial correspondence with Wallace on that subject, and I cannot get over (as, indeed, I have admitted in print) an intimate connection between the number of lakes of modern date and the glaciation of the regions containing them.[8] But as we do not know how ice can scoop out Lago Maggiore to a depth of 2,600 feet, of which all but 600 is below the level of the sea, getting rid of the rock supposed to be worn away as if it was salt that had melted, I feel that it is a dangerous causation to admit in explanation of every cavity which we have to account for, including Lake Superior.[9] They who use it seem to me to have it always at hand, like the 'diluvial wave, or the wave of translation,' or the 'convulsion of nature or catastrophe' of the old paroxysmists.[10]

I have just got a letter from Professor Leslie, and an important paper by him in the American 'Philosophical Society' for 1862, and another on a projected map, 'intended to illustrate five types of earth-surface in the United States,' published in 1866.[11] He was formerly a catastrophist, but of late years he seems to have anticipated Geikie and Croll[12] in regard to sub-aerial denudation, giving, like them, too little to the sea. But he is a man intimately acquainted with the Appalachians, and he gives his reasons for not believing that the ice-sheet has had any hand in eroding the Appalachians.[13] It has polished the surface, and carried erratics so far as mid-Pennsylvania, and no farther; but the surface erosion is just as great in Southern

Pennsylvania and Virginia, &c., which was not reached by the ice, and where there is not a single glacial scratch or groove. He says that the large map which he has planned will make the ice-scooping of lakes in the United States appear as absurd as if applied to tropical Africa or the Albert Nyanza Lake.[14]

Believe me ever affectionately yours, | Charles Lyell.

LS incomplete
K. M. Lyell ed. 1881, 2: 441; DAR 85: A100–1[15]

CD ANNOTATIONS
Above para. 3: 'After *[when]* I have said women voice??' *pencil*
End of para. 5: 'Sexual selection | H Spencer[16] corrected ask | questions—' *blue crayon, crossing text*

1 See letter to Charles Lyell, 4 May [1869] and n. 5. Lyell refers to Alfred Russel Wallace.
2 Lyell evidently misremembered the event. Clémentine Cuvier died in 1827, but the first volume of *Principles of geology* (C. Lyell 1830–3) did not appear until 1830. Lyell met Clémentine and her half-sister, Sophie Duvaucel, in 1823 (K. M. Lyell ed. 1881, 1: 125, misspells her name). It may have been Duvaucel, the daughter of Georges Cuvier's wife by a former marriage, who later read Lyell's book to Cuvier. Lyell also refers to *Essay on the theory of the earth* (Cuvier 1815).
3 See [Wallace] 1869b, pp. 382–91.
4 Lyell's view is similar to that expressed by Wallace in [Wallace] 1869b, p. 393.
5 See [Wallace] 1869b, pp. 391–3. Lyell refers to William Shakespeare and Michelangelo.
6 The reference is to Thomas Greene Wiggins (see Southall 1999).
7 *Onus probandi*: burden of proof (Latin).
8 See letter to Charles Lyell, 4 May [1869] and n. 3, and C. Lyell 1865, pp. 168–74.
9 Maggiore is a lake, partly in Novara and Varese provinces, Italy, and partly in Ticino canton, Switzerland. Lake Superior is the world's largest freshwater lake, bordering on Minnesota, Michigan, and Wisconsin in the United States and Ontario in Canada (*Columbia gazetteer of the world*). See also C. Lyell 1865, p. 173.
10 Paroxysmist: 'a person who holds or advocates the theory of catastrophism' (*OED*).
11 Lyell refers to J. Peter Lesley and to Lesley 1862 and Lesley 1866.
12 Archibald Geikie and James Croll.
13 See Lesley 1866, pp. 310–12. Lesley concluded that erosion of the Appalachian mountain chain had resulted from subaerial denudation alone.
14 Lake Albert Nyanza borders on the Democratic Republic of the Congo and Uganda in east central Africa (*Columbia gazetteer of the world*).
15 The text of the letter as transcribed here is reproduced in K. M. Lyell ed. 1881; however, part of the original letter, paragraphs 3 to 5, is in DAR 85: A100–1. Paragraphs 3 to 5 have been transcribed from the original text.
16 Herbert Spencer.

To T. H. Farrer 6 May [1869][1]

Down. | Beckenham | *Kent. S.E.*
May 6th—

My dear Mr Farrer

Thanks for your note. I doubted about **de**gradation, & had not slightest objection to gradation, because I did not like the idea, & knew nothing about the facts.

I disliked & still dislike the idea because I can see no reason for it. V. cornuta (I believe) & certainly V. canina & odorata produce perfect flowers adapted for a cross,

& small closed flowers to insure a stock of seed through self-fertilisation. V. tricolor which it seems stands near one end of the series does not produce the small closed flowers, & is only very moderately fertile when the visits of insects are prevented.[2] Hence I can see no meaning in degradation, but every step in progress w^d be advantageous.

Yours very sincerely | Ch. Darwin

—But to invert a saying of Agassiz "Nature often lies" & does things quite contrary to theory.—[3]

Linnean Society of London (LS Ms 299/1)

1 The year is established by the relationship between this letter and the letter from T. H. Farrer, 5 May 1869.
2 See letter from T. H. Farrer, 5 May 1869 and n. 1. CD refers to different species of *Viola*.
3 Louis Agassiz, responding to reports of observations opposed to his theory about when saurians were first created, claimed he did not believe them, 'for Nature never lied' (See *Correspondence* vol. 6, letter to Asa Gray, 1 January [1857]).

To J. D. Hooker [before 7 May 1869][1]

With Sclavonic races, (or indeed in any other race as Dutch) is the beard often of a different tint from the hair on the head; & when it differs, is it lighter in tint? or generally reddish?

C. Darwin

DAR 94: 133

1 The date is established by the relationship between this letter and the letter from J. D. Hooker, 24 June 1869, and by the date of Hooker's departure for Russia, 7 May 1869 (see L. Huxley ed. 1918, 2: 85; see also letter from J. D. Hooker, 11 March 1869 and n. 4).

From George Bentham 7 May 1869

25. Wilton Place. | S.W.
May 7 1869

My dear Mr Darwin

You will receive from Mr Kippist[1] two papers read at the Linnean Society which the Council have referred to you for your report with reference to their value for publication. I am very sorry that you should be bothered when you have so much valuable work on hand but one of these is Mr Spruce's on the vesicles inhabited by ants in Tococa etc which has already passed through your hands about which we have some doubts and are anxious for your opinion.[2] We have many important papers on hand which will tax all our powers to bring out in tolerable time and are just now printing a very long one of Spruce's on the Amazonian Palms[3] at the same time I believe that Mr Spruce thinks highly of this one referred to you and we should be very sorry to reject it if *really good*—but we cannot afford to print what is not so.

the other one on the fecundation of Duvernoya in S. Africa is very short the chief doubt is about the illustration— a coloured plate would be an expensive matter[4]

Yours very sincerely | George Bentham

DAR 160: 162

[1] Richard Kippist was the librarian of the Linnean Society.

[2] Richard Spruce had sent his paper to CD with his letter of [before 1 April 1869]. The paper, 'Ant-agency in plant-structure' was read 15 April 1869 (see letter from Richard Spruce, [before 1 April 1869] and n. 3). Spruce refers to the genus *Tococa* (family Melastomataceae).

[3] Bentham refers to Spruce 1869.

[4] The published version of Mary Elizabeth Barber's paper, 'On the fertilization and dissemination of *Duvernoia adhatodoides*' (Barber 1869) had diagrams of the flower, but no coloured plate.

From T. H. Huxley 7 May 1869

My dear Darwin

Do you recollect recommending that the 'Nassau' which sailed under Capt. Mayne's Command to Magellan's straits some years ago should explore a fossiliferous deposit up the Gallegos river?[1]

They visited the place the other day as you will see by Cunningham's letter which I inclose—and got some fossils which are now in my hands[2]

The skull to which Cunningham refers consists of little more than the jaw—but luckily nearly all the teeth are in place—and prove it to be an entirely new ungulate Mammal with teeth in uninterrupted series like *Anoplothermi*—about as big as a small horse[3]

What a wonderful assemblage of beasts there seems to have been in South America! I suspect if we could find them all they would make the classification of the Mammalia into a horrid mess.

Ever | Yours faithfully | T. H. Huxley

Please return Cunningham letter.

Jermyn St.
May 7th 1869

DAR 166: 319

[1] Huxley refers to the memorial that CD enclosed with a letter to him of 4 July [1866] (*Correspondence* vol. 14); although the memorial has not been found, there is a draft of it (see *Correspondence* vol. 14, letter to the Lords of the Admiralty, [2–4 July 1866]). Richard Charles Mayne was captain of HMS *Nassau*, which surveyed the Straits of Magellan from 1866 to 1869 (*ODNB*). Huxley also refers to the discovery of fossil bones of mammals at the base of a cliff along the Río Gallegos in Patagonia (see *Correspondence* vol. 14, letter from B. J. Sulivan, 27 June 1866 and n. 7).

[2] Robert Oliver Cunningham's letter has not been found; he was the naturalist on HMS *Nassau* (see *Correspondence* vol. 14, letter from J. D. Hooker, 4 February 1866 and n. 4). Some of Cunningham's description of the find is reproduced in Flower 1873 (see n. 3, below).

3 Huxley refers to the genus *Anoplotherium*. In his presidential address to the Geological Society of London for 1870, Huxley proposed the genus name *Homalodotherium* for the fossil discovered by Cunningham (see *Quarterly Journal of the Geological Society of London* 26 (1870): lvii). The species was later described by William Henry Flower as *Homalodontotherium cunninghami* (see Flower 1873). The accepted name is now *Homalodotherium cunninghami*.

From Harrison William Weir 7 May 1869

Weirleigh | Brenchley. Kent
May 7th. 1869

Dear Sir

I have raised some hundreds of daisies from *seed* this last year. I send you some *curious* results— the *green* ones *always* remain *green*. They are *all* from different roots, and shew about the same stage of growth

I heard of a *tame* Rabbit the other day that had **18** young at a birth— I have one with **11** all doing well— the 18 were also *all* brought.— the greatest number *wild* Rabbits have as far as I can learn is *6*.[1]

In greatest haste | Yours Very Truly | Harrison Weir

To C. Darwin Esq L.L.D. etc

DAR 181: 71

CD ANNOTATIONS
1.1 I have … growth 1.3] *crossed pencil*
2.1 I heard … *6.* 2.3] *scored pencil*

1 In *Variation* 2:111, CD compared the relative fertility of tame and wild rabbits.

To Spiridion Brusina 8 May [1869][1]

Down Beckenham | Kent
May 8th

Esteemed Sir

I have much pleasure in sending photographs of Sir C. Lyell & Professor Owen, who I think are generally considered our most distinguished natural philosophers.[2] I also enclose, according to your desire a photograph of myself.

With my best wishes for the success of your institution & with much respect I have the honour to remain

yours faithfully | Charles Darwin

LS
Archives of the Croatian Academy of Sciences and Arts (XV-48/AI 340)

1 The year is established by the relationship between this letter and the letter from Spiridion Brusina, 29 April 1869.

2 See letter from Spiridion Brusina, 29 April 1869; see also letter to Charles Lyell, 4 May [1869]. CD also refers to Richard Owen.

From David Taylor Fish to *Gardeners' Chronicle* 8 May 1869

8 May, 1869

Since writing my chapter on worms, I am indebted to the courtesy of the author, for the paper read by Charles Darwin, Esq., before the Geological Society, and published in their Transactions for 1837, p. 505.[1] Its perusal has scarcely modified the views I have already expressed, but it reveals a marvellous coincidence, inasmuch as this careful observer notices the two other modes of elevation of surface adverted to in my chapter, and dismisses them as inefficient, while he clearly and emphatically credits the worms with the whole of the work. And yet it would appear, from pages 506 and 507, that the soil found above the marl, &c., was not of the same quality as that found below it. I presume Mr. Darwin would attribute the difference to the fact of the earth having passed through the stomachs of the worms. In fact, this is obviously his theory, for he adds, that the surface soil would be more properly called animal than vegetable mould.[2] I, however, would accept the fact of the difference in quality as a strong proof that the bottom soil had not been brought up by the worms, but brought down rather by the plants. The greater part of it has an atmospheric origin. Plants incarnate the sun himself in a visible body; this body is nurtured and sustained mainly by invisible food, produced by the atmosphere, and when it perishes part of it goes back to the air; but a large portion of it is spread out over the surface of the earth in the form of vegetable mould. And doubtless in this emptying of the invisible air down, as it were, in a solid mass of decomposing carbon annually, and hourly, in fact, over the surface of the world, through the instrumentality of every living plant, we have agents of sufficient energy, force, and persistency to cover the earth with vegetable mould without the assistance of manure. The amount of carbon utilised and converted from gases into solids in a single year is well nigh incalculable, and would account for all the elevations of surface so graphically described in the admirable paper of Mr. Darwin. It would be gratifying to many of your readers to know if this distinguished geologist still adheres to his original estimate of the usefulness and power of worms as earth-lifters and surface-elevators. Of this he may be assured, that I had no idea of running counter to any theory of so eminent and careful a naturalist when I began my chapter on worms. Even now I would gladly retreat, if I could consistently with the two rules that I have laid down for myself as a writer:—Is it true? then it ought to be spoken. Is it right? then it ought to be done. But what will Mrs. Grundy say?[3] Well, what matters? No great man will ever respect a little one the less for speaking his mind freely, even though it may be in opposition to him. And truth has been said by some to have fallen from heaven down into the gutter, and the humblest—those closest to the gutter—sometimes find it first. *D. T. Fish.*

Gardeners' Chronicle and Agricultural Gazette (1869): 501

1 Fish's work on worms was 'A chapter on worms', *Gardeners' Chronicle* (1869): 417–18, 444, printed in the 17 and 24 April issues. Fish refers to 'Formation of mould'. CD cited Fish's article in *Earthworms*, p. 6. CD replied to Fish's letter in his letter to the *Gardeners' Chronicle*, 9 May [1869].

2 See 'Formation of mould', p. 508 (*Collected papers* 1: 52).
3 Mrs Grundy is a character in Thomas Morton's *Speed the plough* (1798); she is 'proverbially referred to as a personification of the tyranny of social opinion in matters of conventional propriety' (*OED*).

From Asa Gray and Jane Loring Gray 8 and 9 May 1869

Florence
May 8, 1869.

My Dear Darwin

We have just learned with great sorrow, that you have met with a painful accident and have been suffering much. I hope the suffering is over before this, and that the injury is not serious, not one to leave any enduring effects.[1] Of the particulars I know nothing. We heard of it thru. one of our young ladies who called on the Misses Horner, yesterday.[2] I hope to see them soon, to learn more Pray take good care of yourself. We have much need of you yet.

We have been great travellers since you have heard from us, went up Nile to 2nd Cataract—had a pleasant winter, an agreeable time generally, got fond of our life and surroundings, so completely extra-European, specially fond of Arabs, &c, and the remains of old Egypt were profoundly interesting.[3] I cannot say that Mrs. Gray is much the stronger for it,—at least since a nasty voyage from Alexandria to Messina, which prostrated her much;—but she has been able to see and enjoy Naples and part of its surroundings, Rome, Assizi, & Perugia fairly well, and we are now three days in Florence. We are going to Venice, the Lakes, and to Vienna, where we hope to pass the early part of June, & to meet some travelling friends there. I wonder if Hooker may be of the number. I write to St. Peterburgh to enquire.[4]

I doubt if one could make sure of having direct news of you before we reach Vienna,—where, if we could find a line from Mrs. Darwin, telling us that you are all right again we should be very happy.

Mrs. Gray is inclined to write a line, so I will leave half the sheet, on the chance of her finding time to fill it. Hoping to see you in England near the close of summer.

I remain, Ever Yours | A. Gray

May 9— 1869–

My dear Mr. Darwin,

We are so grieved to hear of your accident! To one who has always to suffer so much, it seems hard to have any new cause added— We can only hope it will not aggravate old symptoms, & that we may soon hear that you are better—

I enclose the few notes I made for you on the Nile— I am afraid you will think us very stupid people not to have done more. But it is surprising when one's attention is drawn to it, how little we see *what makes* the different expressions in faces—[5] We are satisfied with what we think the expression is— Then not understanding the language was greatly in our way, for the Arabs are great & excited talkers, & roll out consonants & their rs in a most emphatic way, & what seemed a quarrel or fierce discussion, turned out only an animated account of adventures— But they are a charming people! So docile & gentle, so courteous & dignified, such born gentlemen— Though our crew lo⟨ok⟩ed very odd to us at first as sailors, in their long, blue gowns & white turbans, one soon came to know each face, to admire such limbs as rarely greet European eyes— It seemed as if I had never seen walking & running before!— And the men keep up such perpetual ablutions as part of their religion, they never seemed like common sailors in coming in contact with them, & it was pleasant to have their willing help on all occasions. They are very much like children, pouting when vexed, & smirking like any conscious girl when they knew their beauty was admired; & some are so handsome!— Two of our crew were Ababdeh, a tribe of the desert towards the Red Sea— They were much darker than the Cairines & river natives, & quite different in feature, but the most gentle of them all, & such sweet smiles showing such White teeth!—[6] As for blacks, the negroes from the interior were blacker than anything I ever saw before— A deep, lustreless, solid black.— But our crew burned & grew dark as whites do on our voyage— We thought when we first came back to white skins, how ugly they were! This lovely, golden bronze is so beautiful—

I thought you would have been interested in seeing an old picture here of Fra Angelico's of the deposition from the cross— The Madonna has the distress muscles very carefully painted. The other women only the up & down wrinkles—[7] Pray give my love to Mrs. Darwin & your daughters.[8] And heartily hoping you are relieved, | believe me very sincerely & cordially | Yours, | Jane L Gray

[Enclosure]

In Alexandria I saw the Arab commissionaire or dragoman of an old gentleman, who seemed a German, use exactly the gestures of *No. 13*, when the gentleman would go in a direction different from that which he pointed to—[9]

Passing slowly a common country cargo boat, this old man on board stood looking at us, with brow wrinkled & mouth compressed & upper lip raised— An expression, as I read it, of dislike & contempt—almost hatred—

A black towing a country trading-boat, when there came some obstruction in passing us, had the *distress* muscles, (the forehead transversely wrinkled in the middle,) very strongly marked, as he stood on the bank watching us—[10]

I should say yes, certainly, to *no 6*—after watching a good deal—[11]

(17) We all decided that the head was never "shaken laterally in negation" & that they did not understand what it meant— Nodding in affirmation was rare, it was more often a sign of approval or greeting.—[12]

(2) One day the rais (Reis, i.e. captain) of the small boat, in bringing up his boat nearly ran it into the large one, the men chaffed him a good deal, & my brother said he *blushed* quite to the back of his neck.[13] He was a man full of gestures & very emphatic— When he wished to say he had nothing to do with a thing, that he washed his hands of it, he would hold both hands each side of his head, opened out flat—

One of the little village girls who carried our water-jar on excursions, a girl about 11 or 12, had one day some remark made to her by a young man of another party who passed us, evidently something which displeased her— She drew herself up, raised her head, compressed her lips dropping the corners, & fell back by one of the gentlemen of our party— Her expression was of great contempt— And the whole action very dignified.—

(16) I heard the hiss to keep quiet, but as often from our dragoman & whites as from the natives—

Sunday, May 9,

Dear Darwin,

I saw Miss Horner[14] to-day, and she told me of a letter just received from Mrs. Darwin, which announced that, under Dr. Paget's care, you were all right again, and already resuming your rides,[15] Very glad we are.

This sheet is one upon which Mrs. G. had noted a few mem. on expression[16]—of no great account,—but showing that we often thought of you in Egypt.

A.G

DAR 165: 167–8

CD ANNOTATIONS
9.1 I thought ... painted. 9.3] *scored ink*
Enclosure:
1.1 Arab] *underlined ink*
1.2 *No. 13*] 'shrugging shoulders' *interl ink*
3.1 A black ... us— 3.3] *scored ink*
4.1 I should ... deal—] 'good spirits' *end of para ink*
9.1 Sunday, ... Egypt. 11.2] *crossed ink*
Top of enclosure: 'Prof. & Mrs Gray. *Egypt* May 8th 1869' *ink*; 'Africa' *pencil*
End of enclosure: 'Mrs G. says in old Picture of Fra Angelicos of the deposition from the cross, the Madonna has the distress wrinkles by compress painted' *ink*

1 CD had been thrown by his horse; see letter to A. R. Wallace, 14 April 1869 and n. 12.

2 Susan and Joanna Baillie Horner had lived in Florence since soon after 1864 (Susan Horner collection, Harold Acton Library of the British Library of Florence). CD had once been well acquainted with the Horner family (see *Correspondence* vol. 2). The Grays' 'young ladies' included one or two nieces of Jane Loring Gray's, and the daughter or daughters of family friends (Dupree 1959, p. 337).

3 The Grays travelled from England to southern Europe in mid-November, and sailed to Egypt, arriving in Alexandria 12 December 1868 (J. L. Gray ed. 1893, 2: 566–72). For accounts of the Grays' travels in Egypt, including the three-month trip on the Nile, see *ibid.*, 2: 575–88.

4 The Grays sailed for Messina, Italy, on 19 March 1869, and travelled within Italy, and then to points further north in Europe (J. L. Gray ed. 1893, 2: 586–91). On Joseph Dalton Hooker's journey to St Petersburg from 7 May to 23 June, see the letter from J. D. Hooker, 11 March 1869 and n. 4, and L. Huxley ed. 1918, 2: 85–9. Hooker returned via Stockholm, and evidently did not meet the Grays.

5 Asa Gray was familiar with CD's queries on expression (see *Correspondence* vol. 15, letter from Asa Gray, 26 March 1867). CD probably asked the Grays to observe expression during their travels when they visited Down in October 1868 (*Correspondence* vol. 16, letter to John Tyndall, 20 October 1868). CD acknowledged 'Professor and Mrs. Gray' for their observations made while on the Nile in *Expression*, p. 22. For a copy of CD's printed queries on expression, see Appendix VI.

[6] The 'Abābda were a Bedouin tribe in Egypt (Aharoni 2007, pp. 77, 198). Residents of Cairo are called Cairenes. For accounts of nineteenth-century women travelling in Egypt, see Starkey and Starkey eds. 2001.

[7] In *Expression*, p. 185, CD wrote that a lady had informed him that in Fra Angelico's 'Descent from the Cross' in Florence, the expression of the grief muscles on the forehead was clearly exhibited in one of the figures 'on the right-hand'. The 'Deposition from the Cross', or 'Descent from the Cross', by Fra Angelico is in the Museo di San Marco, Florence (Lloyd 1992). See Hertz 1981, plate 35, for a detail of this figure in the painting (she is in fact on the left).

[8] Emma Darwin, Henrietta Emma Darwin, and Elizabeth Darwin.

[9] For the numbered queries on expression referred to, see Appendix VI. Question 13 concerned shrugging. CD gave Jane Loring Gray's observation in *Expression*, p. 268.

[10] CD recorded Jane Loring Gray's observation in *Expression*, p. 187.

[11] Question 6 concerned the expression of good spirits.

[12] CD recorded this as an observation of 'Professor and Mrs. Gray' in *Expression*, p. 275.

[13] CD cited this observation as Jane Loring Gray's in *Expression*, pp. 316–17. The brother Gray referred to was probably Charles Greely Loring.

[14] Susan Horner.

[15] James Paget treated Darwin after his accident; see letter to James Paget, 29 April [1869].

[16] Asa Gray wrote his note on the back of his wife's enclosure.

To T. H. Huxley 8 May [1869][1]

Down. | Beckenham | *Kent. S.E.*
May 8

My dear Huxley

Thanks for the enclosed letter which I return.[2]

I remember the circumstances & hoped for a better haul, but am delighted to hear of the one very curious fossil—[3]

ever yours very sincerely | Ch. Darwin

LS
Imperial College of Science, Technology, and Medicine Archives (Huxley 5: 268)

[1] The year is established by the relationship between this letter and the letter from T. H. Huxley, 7 May 1869.

[2] See letter from T. H. Huxley, 7 May 1869 and n. 2.

[3] See letter from T. H. Huxley, 7 May 1869 and n. 3.

From T. H. Farrer 9 May 1869

3 Gloucester Terrace | R.P.
9 May /69

My dear Mr Darwin.

Many thanks for your lesson—[1] I think I see what you mean—that each plant lives as it best can, and survives because it has organs which suit it best; and consequently it is wrong to call any plant which lives & thrives "degrade⟨d"—⟩ Still when one finds an organ in one species fully developed & in full function: and the same organ with a similar structure in another species half developed & functionless, may one not (passing over the question whether the less perfect form is derived from the more perfect—or vice versâ, or both from some other form) say that the less perfect "*organ*" is "*degraded*?

I think that the poor Cephalanthera comes in for the term more than once in the book which is my great delight[2]

Very truly yours | T H Farrer

DAR 164: 52

[1] See letter to T. H. Farrer, 6 May [1869].

[2] Farrer alludes to CD's description of *Cephalanthera* as a degraded form of *Epipactis*, in which he described the abortion of the rostellum as a 'degradation of structure' (see *Orchids*, pp. 93–4, 106).

To *Gardeners' Chronicle* 9 May [1869][1]

May 9.

As Mr. Fish asks me in so obliging a manner whether I continue of the same opinion as formerly in regard to the efficiency of worms in bringing up within their intestines fine soil from below, I must answer in the affirmative.[2] I have made no more actual measurements, but I have watched during the last 25 years the gradual, and at last complete, disappearance of innumerable large flints on the surface of a field with very poor soil after it had been laid down as pasture. I have also purposely covered a few yards square of a grass-field with fine chalk, so as to observe the worms borrowing up through it, and leaving their castings on the surface, which were soon spread out by the rain. The Regent's Park in early autumn is a capital place to observe the wonderful amount of work effected under favourable circumstances by worms, even in the course of a week or two. My observations in Staffordshire were chiefly made on poor, sandy grass-land, and I think that Mr. Fish will find that the proportion by weight or measure of carbon in poor soil is but small, and that the decay of the Grass will account for but a small proportion of the matter successfully deposited on the surface.[3] Except when peat or peaty soil is forming, the carbon-compounds seem soon to be decomposed and disappear. Judging from the quick rate at which I have proved that the surface becomes covered with fine soil, if the mere decay of the Grass were as effective as Mr. Fish thinks, many feet in thickness would be formed in the course of a few centuries—a result which would be as surprising as delightful to the dwellers on poor land, or indeed on any land, which is never overflowed by mud-bearing water. In ordinary soils the worms do not burrow down to great depths, consequently fine vegetable soil is not accumulated to any inordinate thickness. *Charles Darwin.*

Gardeners' Chronicle and Agricultural Gazette (1869): 530

[1] The year is established by the publication date of the letter in *Gardeners' Chronicle*, 15 May 1869.

[2] The reference is to David Taylor Fish. See letter from D. T. Fish to *Gardeners' Chronicle*, 8 May 1869.

[3] See 'Formation of mould', p. 509 (*Collected papers* 1: 52).

To George Bentham 10 May [1869][1]

Down. | Beckenham | *Kent. S.E.*
May 10th

My dear Mr. Bentham

I have received the 2 papers, reported on them to the best of my ability & returned them to the Soc.y—[2]

And now I want to trouble you on a little point, which I would not do if Hooker[3] was at home. You will remember some years ago considerable interest about a strange monstrous form of Begonia frigida.—[4] Fritz Müller has found wild (& this seems to me rather remarkable) an analogous monstrous form, as you will see in enclosed letter.[5] Will you read it, & be so kind as to determine whether it is worth communicating to Linn. Soc.—[6] I am unable to form *any* opinion on subject— As Hooker wd. probably like to have plants, will you be so good, when next you go to Kew, to hand the attached seeds to Oliver or some one.— They ought to be labelled "Begonia from S. Brazil producing monstrous flower like that of B. frigida: Fritz Muller."—[7]

Pray forgive me for asking this favour & believe me | Yours very sincerely | Ch. Darwin

If Letter of no use be so kind as to return it.—

Royal Botanical Gardens, Kew (GEB/1/3)

1 The year is established by the relationship between this letter and the letter from George Bentham, 10 May 1869.

2 See letter from George Bentham, 7 May 1869 and nn. 2 and 4.

3 Joseph Dalton Hooker.

4 See *Variation* 1: 365–6.

5 CD enclosed the letter from Fritz Müller, 14 March 1869.

6 Müller's letter was read at a meeting of the Linnean Society on 3 June 1869, and appeared in the *Journal of the Linnean Society* (*Botany*) 11 (1871): 472–4.

7 CD refers to Daniel Oliver. The Royal Botanic Gardens, Kew, Inwards book, 1869, p. 147, No. 142, records the receipt of 'Seeds of Begonia spe. semperflorens / with monstrous flowers' from Müller via CD.

From George Bentham 10 May 1869

25. Wilton Place. | *S.W.*
May 10 /69

My dear Mr Darwin

Many thanks for Fritz Muller's letter which I will have read at the next meeting of the Society— it will probably go into the first part of the Journal after Spruce's Palm paper which will take up at least two parts[1]

I have given the seeds over to Oliver labelled as you desire[2]

Ever yours sincerely | George Bentham

DAR 160: 163

[1] See letter to George Bentham, 10 May [1869] and n. 6. Bentham refers to Richard Spruce's paper '*Palmae Amazonicae*' (Spruce 1869).
[2] See letter to George Bentham, 10 May [1869] and n. 7.

From John Hellins 10 May 1869

Co Prison. Exeter
May 10. 69

Dear Sir

I have noted the sexes of the Lepidoptera already bred by me this year, & will continue to do so, & forward you the results as before.[1]

Yours faithfully | J. Hellins

DAR 166: 138

[1] For Hellins's earlier list of the proportions of the sexes in Lepidoptera, see *Correspondence* vol. 16, letters from John Hellins, 20 April 1868 and 16 December 1868.

To Linnean Society, President and Council 10 May 1869

Report on Mrs Barber's paper.

Mrs Barber's paper seems to me worth publishing, because the dependence of the fertilization of a plant on one kind of insects alone, though not an unknown case, is very rare.[1] Nor has any instance been recorded, as far as I can remember, of the access of other insects, being prevented by a mechanical obstacle requiring strength to be overcome. All the figures would of course be desirable and ornamental, but It would quite suffice to give on wood two of the adjoining flowers, one with a bee entering and one without. The outline fig 4 should also be given, as this shews better than fig 1 the contracted tube. It would be well to give the outline fig 3. if anyone can explain to the artist the structure of the fold embracing the base of the pistil, which is not clear to me.[2]

Charles Darwin

To the Pres. & Council of Linnean Soc.
Down. May 10th. 1869.—

LS(A)
Linnean Society of London

[1] CD refers to Mary Elizabeth Barber. See letter from George Bentham, 7 May 1869 and n. 4.
[2] See Barber 1869, p. 470.

To Linnean Society, President and Council [10 May 1869][1]

Report on Mr Spruce's paper.[2]

I feel considerable difficulty in expressing an opinion on this paper, from not knowing how few Botanists have generally supposed that the sacs &c are independent growths in the plants. If this be so, as I suppose from the terms used, it seems to me highly desirable that the facts, here recorded and observed by a naturalist on the spot, should be published; more especially as M[r] Spruce shews under what conditions and contingencies the sacs are formed. The presence of sacs &c in plants belonging to various widely different orders, is also a very interesting circumstance. But the point which M[r] Spruce seems to think the most important, namely that the sacs in the course of ages have become inherited, is manifestly very improbable, and appears to rest on hardly any evidence. The evidence advanced is, firstly, the generality of the presence of sacs, accompanied however by an almost equal generality in the presence of ants, in certain species; and secondly, that he saw in one case (p. 6) sacs beginning to be formed long before any ants had visited the plant. But it seems to me not a little doubtful that it could have been positively ascertained that ants had never visited such plants; and D[r] Hooker informs me that none of the Melastomas have ever produced sacs in the hothouses at Kew.[3] I would suggest that the paper should be printed, with two or three illustrations, on condition that the author is willing to alter the title and to strike out several short passages (marked with pencil) in which the sacs &c. are spoken of as being inherited. The author might of course state that he had been led to suspect that these structures are now inherited, and this might lead to further investigations; but to speak positively on the subject of inheritance without much fuller evidence would be in my opinion injurious to the reputation of the author and of the Society.

To the President | & Council of the | Linnean Soc[y].

L(A)
The Linnean Society of London

1 The date is established by the relationship between this letter and the letter to George Bentham, 10 May [1869].

2 Richard Spruce had sent CD a paper on the modification of plant structure by ants for submission to the Linnean Society (see letter from Richard Spruce, [before 1 April 1869]. For a transcription of the paper, which was not published by the society, and an account of events surrounding its submission, see Spruce 1908, 2: 384–412. There is a draft of this report in CD's hand in DAR 96: 75.

3 Joseph Dalton Hooker was director of the Royal Botanic Gardens, Kew. No correspondence between CD and Hooker on this subject has been found, but Hooker visited Down on 25 April 1869 (Emma Darwin's diary (DAR 242)).

From Louis Rérolle 10 May 1869

Lyon,
the 10 May 69

Sir

I have the honour to inform you that the French translation of your Orchis book will be published by Mr Reinwald, at Paris.[1] The treaty is decisively concluded with him since eight days, but I could not let you know it earlier. The size of the book will be that of the French translation of your work on the variations of animals and plants, published at the same library by Mr Moulinié.[2] It will appear five or six months after I shall have sent to Paris my manuscript, and Mr Reinwald is to write immediately to Mr Murray,[3] concerning the acquistion of the woodcuts.

I shall now, Sir, persue the work of revision which I have undertaken, examining with much care the Orchid-flowers, under the direction of Mr Faivre;[4] Save the Mormodes Ignea, Angræcum Sesquipedale, Cephalanthera grandiflora, and perhaps one or two others, I think I shall be able to see and study by myself every flower you have described.[5] I will do my best to respond well to the honour you did me, in giving me your authorization.

As you intend to join to the French edition a series of notes, be so kind as to send me them as soon as possible, indicating at the same time the place you wish to give them, either in the course, or at the end of the book.[6] The simplest manner to send them to Lyons is to give them to the post, under bonds, as a manuscript or a commercial paper.

According to the advice of Mr Faivre, I will perhaps write three or four pages of introduction, these I will send you, desiring to write nothing which you would object to.

Allow me, Sir, to thank you for your kindness in sending me your very interesting work on climbing plants,[7] and to assure you of the deep respect with which I have the honour to be | Your most obedient servant | Louis Rérolle

DAR 176: 130

1 Rérolle refers to *Orchids* and the publisher Charles-Ferdinand Reinwald (C. Reinwald). See also letter from Ernest Faivre, 15 March 1869.

2 The French translation of *Variation* by Jean Jacques Moulinié (Moulinié trans. 1868) was also published by C. Reinwald.

3 John Murray.

4 Ernest Faivre.

5 *Mormodes ignea* is a South American orchid; *Angraecum sesquipedale* is from Madagascar; *Cephalanthera grandiflora* (now *C. damasonium*) is native to southern, central, and western Europe.

6 For more on CD's additions, see the letter to J. D. Hooker, 8 March [1869], n. 5.

7 *Climbing plants.*

From W. C. Tait 10 May 1869

Oporto
May 10th 1869.

Charles Darwin Esq.r
Dear Sir,

I was sorry to see from your letter of the 18th. ultmo. that you had met with an accident and hope that ere this you have quite recovered from its effects.[1]

I have received the "Fertilization of Orchids" and pamphlet on climbing plants and am very much obliged indeed to you for them.[2] I have looked into them here and there and find them very interesting I hope soon to read them attentively such subjects as these make me feel inclined to study botany— Curiously I yesterday afternoon when out for a walk found an orchis (of what species I do not know) which practically illustrated the mode of adhesion of the pollinia— How curious that a vegetable should be dependant on an animal— I suspect that many other instances will be discovered—

With respect to sheep I have written to a Portuguese agricultural periodical without as yet having received any answer regarding the rate of development of their horns.[3]

I have examined a few sheep on the Candal side of the river Douro[4] and found the rams horned and some of the ewes had very small and shrivelled horns of about an inch or two in length—

Another shewed only two white spots where the horns usually spring from and some did not seem to shew signs of horns—

I shall not relinquish the subject until I get the opinion of experienced people—

I suppose you are trying to prove the elimination under domestication of the horns in the ewes of the merino breed— I think the effect of the pasturage should always be kept in view— We see the effect of soil on the vine for Port Wine cannot be produced at Oporto even with the proper vine but only (as yet) in a certain district in the neighbourhood of Regoa and Lamego—[5]

Tail-less Dogs— My pointer bitch only brought forth 6 pup 3 of them have stumps of tails like the father and mother two have $\frac{3}{4}$ of a tail and another an apparently entire tail— 5 were dogs and one with $\frac{3}{4}$ of tail was the only bitch. Here we have a new breed of dog! I do not know how I am to prove that it proceeds from mutilation with inheritance but the length of the stump seems to make it somewhat probable—[6]

I have examined minutely with glasses of different powers the Drosophyllum Lusitanicum but cannot see any motion—[7]

The purple disc on the top of the pedicle seems to become darker after an insect has clung to it for some time and they also appear drier the liquid forms into white crystals. Three of the flowers have not come to maturity and one of the two remaining plants seems to be drying but we have really had such shockingly bad weather that all plants seem to have suffered—

The glands on the flowers differ somewhat from those on the leaves looking more like vegetable hairs—

It would be interesting to know whether the plants cross through *insect* agency I should think it scarcely possible and a curious paradox. Also whether the plant can exist without insect food or be proved to be another plant dependant on animal life.

I beg to enclose Brotero's description of the plant which may prove useful to you.[8] There is a kind of beetle which swarms on flowers and may carry pollen about whilst feeding on some other part of the flower—

I am dear Sir | Yours very truly | William C. Tait.

[Enclosure][9]

Description of Drosophyllum Lusitanicum from Brotero's Flora Lusitanica page 215.

"4 Spergula droseroides.

S Radice caulescenti: caulibus superne ramosis; foliis subulatis, carinatis, apice spiraliter tortis, pilsis, pilis rorido-glandulosis: seminibus pedicellatis

Lusit Herva Pinheira orvalhada

Drosera Lusitanica Linn.

Chamaeleontioides. Grisl Vir Lusit[10] N 1351.

(Grisleyus in Virid Lusit nonnumquam eandem unicam plantam sub duplice nomine indicavit sic Apium hortense, Seleri dictum N 122 Seleri, sive Apium palustre sativum N 1309.

Quibus ambobus denotare voluit, Apium graveolens Linn var β)

Hab. in Sabulosis aridis trans Tagum circa Seixal et Arrentella, circa Torres Vedras, Montejunto, Chaõ de Maçãas, Redinha, et alibi in collibus siccis ex Olisipone usque Aveiro ad quinque leucas ab Oceani littoribus Fl æstate Perenn et quasi suffrutex.

Radix perennis, ramosa, superne crassitudine pennæ anserinæ, sensim quotannis extus terram sese erigens, seu duas ad quatuor uncias caulescens, foliis emortuis superne cincta: ibi nova folia, quasi ut in Palmis, progerminant, cum ad terram nulla, nisi á seminibus germinatione anno primo. Caulis in ætate plantæ tenera unicus, postea duo tres quatuorve ex apice radicis caulescentis, semipedales aut paulo altiores, teretes, paucifolia, erecti, superne ramosi, ramis alternis, erectis, unifloris, inferioribus sæpe altioribus, omnibus, uti caulio, pilosis, pilis apice capitato-glandulosis succumque glutinosum exsudantibus. Folia radicalia, et quæ in apice radicis caulescentis, in orbem congesta, erectiuscula, caule paulo breviora, subulata, carinata, super planiuscula apice spiraliter retorta, ad oras præsertim piloso glandulosa: caulina similia, alternum sursum versus caulis apicem sensim breviora. Flores quatuor ad septem, caulem et ramos terminantes. Calyx pentaphyllus, foliolis ovato-lanceolatis, acuminatis, intus glabris, extus ad oras piloso-glandulosis, corolla brevioribus.

Corolla petula lutea obovata.

Stamina constanter decem. Pistilla quinque Capsula calyce fere duplo longior, glabra, glabra, unilocularis, quinquevalvis, valvis ovato-lanceolatis. Semina conica unam lineam longa, nigricantia, basi lata subconcava, ibique in media umbilicata, decem ad duodecem, nonnulla abortientia, omnia in fundo capsulæ, singulaque pedicello parvo imposita. Integumenta duo, membranacea.

Albumen semini conforme, sordide albidum, cartilagineum, apice ubi embryo, excavatum, Embryo inversus, minutus, albus, conicus, in apice angusto et excavato albuminis sedens— Cotyledones obovatæ, crassæ, hinc ad plumulum subconcavæ, inde convexæ. Plumula conica, alba, apice bifida, cotyledonibus dimidio brevior: radicula brevissima, acutiuscula.

Planta, carpologice considerata, sui generis; fructus enim nec Droseræ nec Spergulæ: loco tamen habitationibus sabuloso et arido, caule ramoso et foliato, numeroque staminum perpenso, potiori jure Spergulis, quam Droseris eam associandam esse, maxime in tyronum commoditatem, crediti. Succus glandularum papyrum colore saturate purpureo tingit.'

Portuguese names— Pinheiro [moncoso]
Caça [moscas]
Herva dos [Impigemo]
Herva Pinheira Orvalhada

DAR 58.1: 19–20, DAR 178: 47

1 See letter to W. C. Tait, 18 April [1869] and n. 2.
2 See letter to W. C. Tait, 18 April [1869] and n. 3.
3 See letter from W. C. Tait, 5 March 1869 and n. 5.
4 The Douro (Spanish 'Duero') river flows from north central Spain into Portugal to the Atlantic Ocean at Oporto. Candal is a suburb of Oporto on the left bank of the Douro.
5 Regoa (now Peso da Régua) and Lamego are towns on the Douro river in the centre of the port wine grape growing region of Portugal.
6 See letters from W. C. Tait, 26 January 1869 and 19 February 1869, and letter to W. C. Tait, 24 February 1869.
7 Tait had sent CD specimens of *Drosophyllum lusitanicum* (see letter from W. C. Tait, 5 March 1869).
8 Tait refers to Félix de Avellar Brotero and Brotero 1804, 2: 215–17. Aside from minor errors, Tait's transcription of Brotero's diagnosis is correct.
9 For a translation of the Latin part of the enclosure, see Appendix I.
10 Grisley [1661].

To J. J. Weir 13 May [1869][1]

Down, Beckenham, Kent., S.E.
May 13

My dear Sir

Many thanks for your paper, which I read with *extreme* interest. Your verification of Wallace's suggestion seems to me to amount to quite a discovery.[2] Here is an odd chance, about a fortnight ago I twice noticed the blackness of the lambs in some flocks which I believed to be Southdown, and looked out for the shepherd to ask him whether the lambs would lose their dark tints, but could not see anyone to ask. And now you have answered the question with respect to a flock of a very high character.[3] It is a capital case like the kittens of black cats showing traces of stripes. It will work in usefully for me some day. In fact I have been perpetually quoting you in my MS. of late. I am hardly a believer in inheritance of mutilations, and not of dogs' tails, so I feel no surprise at all about the sheep.[4] With many thanks.

Yours very sincerely | C. Darwin.

Copy
DAR 148: 321

1 The year is established by the relationship between this letter and the letter from J. J. Weir, [1–13] May 1869.
2 See letter from J. J. Weir, [1–13] May 1869 and n. 2.
3 See letter from J. J. Weir, [1–13] May 1869 and n. 3.
4 See letter from J. J. Weir, [1–13] May 1869.

From John Beck 15 May 1869

Copthorne Cottage near Shrewsbury.
May 15th 1869.

My dear Sir—

I am anxious to offer for your perusal, at your leisure—the accompanying Manuscript, sent through the Book Post, on "the Real Presence" &c. &c.—a most mercifully to myself and Country Spiritually revealed to me from Heaven— I believe—full entire exposition—most wonderfully minute & clear literal revealment of the truth reality—as to "the Real Presence" in the Bread and Wine—in the Holy Sacrament of "The Lord's Supper" and shall be obliged by your returning it to me by the Book Post in the course of a month or later than that, if more convenient to you, for which I inclose a Postage Stamp![1]

I would especially direct your attention to that especial infinite mercy of the Allmighty—through the Saviour to all of us, who are most mercifully granted true repentance, and are earnestly striving for newnesss of life (whatever our past sins may have been) in the operation upon us of "The Hedge set about the Vinequard" (fo 12 & 14)[2] and in those most mercifully blessed words of Holy Scripture in Isaiah 1. 18 (fo 16)

"Come now—and let us reason together, saith the Lord: though your sins be as scarlet—they shall be as white as snow,—though they be red like crimson,—they shall be as wool—"[3]

With my best wishes & humble prayer for your welfare & happiness here & eternally in Heaven—& for your restoration to health—& for each blessing—& for Mrs Darwin & family—& upon my MSS—& for increase of grace & of the Holy Spirit most mercifully—for us all—with peace & joy in believing, I remain my dear Sir— Yours faithfully | John Beck

C. Darwin Esq^{e}.

DAR 160: 104

1 The manuscript referred to has not been found; it was evidently returned by CD. CD had received an earlier letter from Beck on religious subjects (*Correspondence* vol. 12, letter from John Beck, 6 October 1864).

2 Beck probably refers to Mark 12, which begins with a reference to a man who 'planted a vineyard and set an hedge about it' (Mark 12:1) and Mark 14.

3 The passage printed here in bold is in red ink in the original letter.

From George Maw 17 May 1869

Benthall Hall, | n^{r} Broseley.
May 17— 69

My dear Sir

I dried a few specimens of the drosophyllum & as the plants did not arrive in very good order, I think the dried specimen may perhaps better exhibit the characters of the plant; I will forward one to you by rail tomorrow.[1]

When I found it it was just coming into flower & had evidently not attained its full stature The foliage dries well but the bright yellow color of the flowers is very fugitive I fear it will be a tiresome plant to cultivate my live specimens arrived in bad condition but I am glad to find they are breaking nicely in the stove & I hope to save six or seven plants. Should yours not live I will find you another specimen when they are a little more recovered.

I got a number of very pretty Sedums most of which were new to me & are I think not in general cultivation[2]

I spent a day or two in the granite district north of Madrid & obtained a number of interesting plants, all of which appear to be doing well & I think I shall save the greater part of those I sent home from Barbary[3] & Gibraltar.

Believe me to be | very truly y^{rs}. | Geo Maw

C Darwin Esqr

DAR 171: 105

[1] See letter from George Maw, 18 April 1869, and letter to George Maw, 27 April [1869]. Maw refers to *Drosophyllum lusitanicum.*

[2] *Sedum* (stonecrop) is a genus of succulent-leaved plants of the family Crassulaceae.

[3] The Barbary States were the North African states of Tripolitania, Tunisia, Algeria, and Morocco.

To Charles Lyell 20 May 1869

Down. | *Kent. S.E.*
May 20 1869

My dear Lyell

I have been much pleased to hear that you have been looking at my S. American book, which I thought was as completely dead & gone as any pre-Cambrian fossil.[1] You are right in supposing that my memory about American geology has grown very hazy. I remember however a paper on the Cordillera by D. Forbes, with splendid sections, which I saw in M.S., but whether referred to me or lent to me, I cannot remember. This w^{d} be well worth your looking to, as I think he both supports & criticizes my views.[2] In Ormerod's index to the Journal, which I do not possess, you w^{d} no doubt find a reference;[3] but I think the sections w^{d} be worth borrowing from Forbes.

Domeyko has published in the Comptes Rendus papers on Chili, but not, as far as I can remember, on the structure of the mountains; Forbes however w^{d} know.[4] What you say about the plications being steepest in the central & generally highest part of the range is conclusive to my mind that there has been the chief axis of disturbance.

The lateral thrusting has always appeared to me fearfully perplexing. I remember formerly thinking that all lateral flexures probably occurred deep beneath the surface, & have been brought into view by an enormous superincumbent mass having been denuded. If a large & *deep* box were filled with layers of damp paper or clay, & a blunt wedge was slowly driven up from beneath, w^{d} not the layers above it &

on both sides become greatly convoluted, whilst those towards the top w^{d} be only slightly arched?

When I spoke of the Andes being comparatively recent, I suppose that I referred to the absence of the older formations. In looking to my volume, which I have not done for many years, I came upon a passage (p. 232) which w^{d} be worth your looking at if you have ever felt perplexed, as I often was, about the sources of volcanic rocks in mountain chains.[5]

You have stirred up old memories, & at the risk of being a bore, I sh^{d} like to call y^{r} attention to another point which formerly perplexed me much, viz. the presence of basaltic dykes in most great granitic areas; I cannot but think the explanation given at p. 123 of my *Volcanic I's* is the true one.[6]

My dear Lyell | Yours affectionately | C. Darwin

LS(A)
American Philosophical Society (370)

1 Lyell's discussion of *South America* may have been in the now missing section of his letter of 5 May 1869.

2 CD refers to 'On the geology of Bolivia and southern Peru' by David Forbes (Forbes 1860). For Forbes's references to CD, see Forbes 1860, pp. 10, 11, and 29.

3 George Wareing Ormerod's revised *Classified index to the Transactions, Proceedings, and Quarterly Journal of the Geological Society of London* was published in February 1870. The first edition appeared in 1858 (Woodward 1907, pp. 205–6).

4 Ignacy Domeyko published 'Mémoire sur la constitution géologique de Chile' in *Annales des Mines* (Domeyko 1846).

5 CD had written that it was difficult to trace the streams of porphyries to their eruptive sources and conjectured that it was because some original points of eruption tended to become points of injection (*South America*, p. 232).

6 CD hypothesised that the dikes were formed by 'fissures penetrating into partially cooled rocks of the granitic and metamorphic series, and by their more fluid parts, consisting chiefly of hornblende, oozing out, and being sucked into such fissures' (*Volcanic islands*, p. 123).

From Henry Maudsley 20 May 1869

32 Queen Anne St | Cavendish Square W
20 May 1869.

Dear Sir,

As my present opportunities did not admit of my giving you satisfactory information in respect of the queries in your letter of 7th. Feb^{y}., I sent the letter on to my friend D^{r}. C. Browne, Medical Superintendent of the West Riding Asylum, who has upwards of 1000 insane patients under his constant observation, and whose acquirements are of a high order.[1]

He has been kind enough to send me some of the results of his experience this morning, and I now forward them, together with his note—

I trust that you may find the results useful, and I am sure D^{r}. Browne will be very glad to give you any further assistance which it may be in his power to give.

I am, my dear Sir | Yours faithfully | H Maudsley

James Crichton-Browne.
Wellcome Library, London. Image ref. V0028544.

[Enclosure 1]

West Riding Asylum, | *Wakefield.*
19th. May. 1869.

My dear Dr. Maudsley,

Along with this I send you some notes on 'Expression' for Mr. Darwin. They are exceedingly crude & deal only with the two questions specially referred to in his letter to you.[2] I have not had time to say anything upon the numerous topics suggested by the printed slip.[3] To tell you the truth I have been greatly puzzled to know how to reply even to the two queries. I do not exactly understand what sort of information Mr. Darwin wishes—whether statistics cases, or generalizations founded upon experience. I have thought it safer to stick to facts.

Should the sort of information which I have sent prove of any service to Mr. Darwin I can supply him with much more of the same description

In great haste | Yours very faithfully | J. Crichton Browne

Dr. Maudsley | &c &c

[Enclosure 2]

Notes on the Expression of the Emotions amongst the insane and idiotic.

I have repeatedly seen the hair of the insane bristle—or as it is called stand on end—under the emotion of sudden & extreme terror.[4] In a woman named Jane Leech (Ward 23) who is now under my care, the hair of the front of the head used to become partially erect whenever morphia was administered by hypodermic injection, the skin of the face becoming at the same time deadly pale & the limbs being forced in a sort of tetanic spasm. This patient's dread of the operation is not due to the very trifling degree of pain which it occasions but is founded on the delusion that it is a method of introducing into her system a deadly poison by which her bones are to be softened & her flesh turned into dust.[5] The bristling of the hair however—so common amongst the insane & observed in this case, is not always associated with the emotion of horror, or any allied Sentiment. It is perhaps most frequently seen in chronic maniacs who rave incoherently, & have destructive impulses.[6] It is present in a remarkable degree in a patient of mine named Ruth Lockwood (Ward 30) whose, photograph I forward & who is affected in this way.[7] She is always excited & violent, & noisy, but passes through various stages of exacerbation & mitigation, the state of her hair being a sure & convenient criterion, of her mental condition. When she is at her worst "each particular hair" does literally "stand on end", "Like quills upon the fretful porpentine",[8] & when she is at her best her hair is more amenable to the brush & comb although it never reaches anything like an ordinary amount of subordination. The photograph was taken when it was in its most tractable humour.

In another patient of mine John Grace (Ward No 1) the hair becomes erect immediately prior to the recurence of manical paroxysms to which he is liable. It rises up from his forehead like the mane of a shetland pony.[9] The same condition of the hair has been recorded in acute mania, hysterical mania, acute melancholia, & hypochondriasis.

Woman with bristling hair,
probably Ruth Lockwood.
DAR 161: 323/6v

Man who died of horror. CD annotated the picture in pencil: 'Man who died of Horror and often attempted suicide'.
DAR 161: 323/7

Photographs of patients at the West Riding Asylum, pasted to the second enclosure to the letter from Henry Maudsley, 20 May 1869.
Reproduced by permission of the Syndics of Cambridge University Library.

The following case from my note book illustrates the physiognomy of Terror in an insane woman. —Grace Mellor— a patient in Ward 21. at 35. is subject to attacks of intense fear & phrenalgia—in which she fancies herself in hell.[10] While in such a paroxysm she screams out "This is hell! This is hell! Oh dear! Oh dear! There are chariots of fire! There is a black woman! I can't get out! Oh-dear! Oh dear! There are rats & horses! dont tickle me to death with feathers This is hell." &c. When so screaming her movements are those of alternate tension & tremor. For one instant she clenches her hands, holds her arms out before her in a stiff semiflexed position, then suddenly bends her body forwards sways rapidly to & fro, draws her fingers through her hair, clutches at her neck & tries to tear off her clothes, these movements being tolerably symmetrical. Her hair at the back of her head which has been cut short, but which is brushed down when she is calm, & stands on end—that in front being dishevelled by the movements of her hands. Her countenance is very expressive of mental agony. The skin is flushed over the face & neck, down to the clavicles, & the veins of the forehead & neck stand out promently like thick cord. The sterno-cleid-mastoid-muscles[11] also stand out prominently as if swollen & the skin in front of them is much wrinkled. The lower lip droops & is somewhat everted. The mouth is kept half open the lower jaw projecting. Two deep parallel lines pass, in a double curve from the outer edge of each ala nasi[12] to a point—about a quarter of an inch to the outer side of each angle of the mouth. The nostrils are raised & distended. The cheeks are hollow. The eyes are widely opened & beneath each there is a raised semilunar patch (which I very frequently observe in all varieties of Melancholia) as if the loose skin of the cheek beneath the lower eyelid were puffed up or slightly œdematous. The pupils are large. There are several strongly marked furrows (divergent on to the brow) at the inner termination of each eye-brow. These are evidently produced by the powerful & persistent action of the corrugatores supercili.[13] The forehead is wrinkled transversly in wavy folds.

Opposite is the photograph[14] of a man who died of "*horror*' & in whom the bristling of the hair was only present to a very trifling extent. He believed that he had not only brought ruin upon himself—but upon all with whom he was brought into contact—& that "an *awful doom* (awful because of its very vagueness') awaited him. He was perpetually endeavouring to "fly into the arms of death to avoid the terrors of his countenance" & attempted suicide by hanging—by scratching the veins of his arm, by drowning, by beating his head upon the wall & by throwing himself before a loaded waggon. When he died no organic disease was found in his body, only emaciation, & redness of the grey matter of the brain.

II.

I have often had occasion to observe that under the influence of rage & resentment—insane & idiotic persons, uncover & obtrusively display the teeth or *"girn"* as the Scotch have it. I could quote many examples of this.

1. I was recently consulted about a Lady A. Y. subject to incontrolable outbursts of anger—brought on by—jealousy & delusive suspicions, & actually saw her in one of

these orgasms. At first she vituperated & foamed at the mouth. Next she approached close to her husband with compressed lips & a virulent set frown.[15] Then she drew back her lips & showed her teeth, until the corners of the upper lips became almost right angles at the same time aiming at her husband a vicious blow.

2. John Smith No. 1 Ward is impulsive & readily irritated. Although an old soldier he cannot be made to conform to discipline but whenever requested to conform to the regulations of the establishment, gives way to discontent terminating in fury.[16] I have repeatedly witnessed these attacks. When they take place in my presence, they are directed against me. He commences by asking if I am not ashamed of the institution, if I feel no remorse for my treatment of him? He then proceeds to swear & blaspheme, to pace up & down, to toss his arms about & to menace me & the other officers in his neighbourhood. At last as his exasperation culminates, he rushes up towards me, by a peculiar sidelong movement, shaking his doubled fist in my face, & threatening to destroy me. At that time his upper lip is drawn up on each side by the levator anguli oris,[17] so that his large canine teeth are exhibited, his expression being that of great ferocity, & his curses being hissed out through his set teeth.

3. Almost the same description as that given above in the case of John Smith might apply to another patient of mine named William Wheatley (Ward 14.) who however, spits & foams in a singular manner when he is put out of temper (as he often is) & who moreover dances when he is in this condition jumping rapidly & perpendicularly upwards upon both feet, & hopping from one foot to the other as he shrieks out his maledictions in a shrill falsetto.[18] This man has two of his fingers webbed together to their very points.

4. Jonathan /Kilburn/ (Ward 4) an epileptic idiot (photograph enclosed) has just mind enough to feed himself which he does voraciously.[19] He is incapable of independent movement & spends his whole day seated on a low stool or chair, holding an old cigar box containing a piece of tin. To these toys he is much attached becoming exceedingly irate, if any one attempts to take them from him. His temper is generally sullen & morose—but is easily lighted into fierceness, by any interference. If touched or disturbed, he slowly raises his head, from its habitual downcast position & fixes his eyes on the offender with a tardy, but angry scowl. If the annoyance is repeated he draws back his thick lips & reveals a prominent row of hideous fangs. (Large canine being especially noticeable) & then makes a quick cruel clutch with the open hand at the offending body. The rapidity of this clutch is marvellous in a being ordinarly so torpid who takes about fifteen seconds to turn his head from left to right when attracted by any noise. If when he is thus incensed a handkerchief, book or any other article be put into his hand he will drag it to his mouth & bite it.

Phthisis terminated the life of a young woman, who had fostered the growth of an inherited germ of disease by intense religious excitement, and hemiplegia and coma were the concluding stages of the case of a woman, who, though represented as having been only nine months insane at the date of her death, had probably been deprived of reason for above ten years.[20] *Before her insanity was officially recognised, she had been in the habit of escaping from home, and living in solitude in the woods, feeding upon wild fruits, or what she could occasionally beg at a cottage, and sleeping amidst the brushwood. She had frequently lived in this manner for a fortnight at a time, seeking no converse with human beings. While upon one of these expeditions she was delivered of twins. At this time she was without attendance or help. She sought out a sheltered hollow for her delivery, and there, reverting to a primitive instinct, she actually gnawed through the umbilical cords with her teeth. The twins were alive when found two days after birth, but the mother was in a very exhausted state, having had no food or covering since her delivery. When brought to this asylum, she was the most perfect specimen of brutalized humanity that I have ever encountered. The most casual observers on seeing her, had powerfully suggested to them her striking resemblance to a wild animal. Her hair had all been cut or broken off short, while, over the summit of each parietal bone, the skin was quite denuded, owing to frequent rubbing, not with fingers but with the inner side of her hand and wrist. The forehead was small, receding, and deeply wrinkled. The cheek bones were naturally prominent, and appeared even more so, in consequence of the emaciation of the face, which also strongly delineated massive temporal muscles. The eyelids were without lashes, and so opened as to expose a large rounded portion of the eyeballs, thus giving the countenance a staring appearance. The width between the eyes was considerable. The bridge of the nose was depressed, the alœ wide-spread, the nostrils forward opening. The lips were thin, and always retracted so as to display a grinning set of teeth. The lower jaw was protuberant, and the chin was always held forward. The complexion was sallow, and the features exhibited no play of expression. The limbs were wasted and the fingers were fixedly contracted or flexed, and had the aspect of claws. The habits of this woman, as well as her bodily configuration, had many points in common with those of wild creatures. She was averse to a sitting posture, and preferred to lie upon the ground, or to run about in a half-stooping attitude. As she lay she drew up her legs, and twisted herself into the most singular contortions. When stared at she winked forcibly and repeatedly. She used her teeth for tearing and gnawing, as well as mastication, and she bit and licked the hands of those who fed her, as a mark of her fondness for them. She never expressed coherent thoughts, but used a few words strung together without apparently attaching any meaning to them. She constantly emitted undescribable gibbering noises. Her only attachment was for the nurses who fed her, from whom she would snatch the food brought with sudden avidity, turning away to devour it in secret. Her greatest pleasure seemed to consist in chewing a hard sea-biscuit. When awake she was invariably restless, picking, rubbing, rocking her body, and moving about. Sometimes for a whole night she would pace round her room sideways, placing her feet heavily down with the most rythmical regularity, and clasping a pillow in her arms. She was destructive to furniture, especially to crockery, which she delighted in breaking; but she did not injure her own clothing. When out in the airing court, she would pull up the flowers and grass and toss them wildly above her head. And yet this woman, presenting so many animal traits, had been of average intelligence previous to her insanity, and had adequately discharged the duties of her station in life.*

J. C. B.

DAR 161: 308, 323/6–9; DAR 171: 93

CD ANNOTATIONS

Enclosure 2:

1.1 Notes … forward 2.14] *crossed pencil*

2.1 I have … terror. 2.2] 'Hair end in terror.' *left margin pencil*

2.2 sudden … terror. 2.2] *underl blue crayon*

2.4 partially erect] *underl pencil*

2.4 morphia] 'morphia' *added above, pencil*
2.4 hypodermic] 'hypoderm' *added above, pencil*
2.5 the skin . . . spasm. 2.6] 'at same time face deadly pale, & limbs tetanic' *left margin pencil*
2.6 This patient's . . . dust. 2.9] 'not from pain but from hallucination at its result.' *left margin pencil*
2.9 The bristling . . . impulses. 2.12] 'Bristly hair common in chronic maniacs independently of emotions—' *left margin pencil*
2.11 the emotion . . . Sentiment.] *underl blue crayon*
2.14 She is always . . . condition. 2.16] *underl pencil*
2.15 the state . . . criterion 2.16] *scored blue crayon*
2.16 When she . . . porpentine", 2.18] *scored pencil*; 'very good' *left margin pencil*
2.19 although . . . subordination. 2.20] 'permanently affected' *left margin pencil*
2.19 brush & comb] 'brush & comb' *added above, pencil*
3.3 The same . . . hypochondriasis. 3.5] 'Is it always accompanied by passion or fury' *left margin pencil; scored pencil*
3.4 acute melancholia] *underl pencil*
3.5 hypochondriasis.] 'hypo-' *below, pencil*
4.1 The following . . . in such 4.3] *double scored pencil; cross in margin red crayon*
4.3 & phrenalgia] *underl pencil*; 'phrenalgia' *added above, pencil*
4.3 such] 'such' *added below, pencil*
4.4 a paroxysm . . . folds. 4.27] *scored pencil*
4.7 tension & tremor.] 'tension & tremor' *added above, pencil*
4.7 For one . . . position, 4.9] *underl pencil*
4.10 these movements . . . symmetrical. 4.11] *underl pencil*
4.11 Her hair . . . head] *del pencil*
4.12 has been] *del pencil*
4.12 cut short] *underl, circled pencil, line to caret after* 'Her hair' 4.11
4.12 but] *del pencil*
4.12 brushed] 'brushed down' *added above, pencil*
4.12 stands on end] *underl pencil*; ';' *added after* 'end', *pencil*
4.13 that in front] 'that in front' *added above, pencil*
4.13 dishevelled . . . hands.] *underl pencil*
4.14 agony] *double underl pencil*
4.14 flushed . . . neck,] *underl pencil*
4.15 clavicles,] *underl pencil*
4.16 The sterno-cleid-mastoid-muscles . . . wrinkled. 4.17] *circled pencil, line to caret after* 'symmetrical' 4.11
4.16 The sterno-cleid-mastoid-muscles . . . everted. 4.18] 'Cyclop. of Anat. vol. 3. p. 566 Bring Head down to chests'[21] *left margin ink*; '(which seem to bend to head ['hair' *del*] on the chest)' *left margin pencil*
4.16 sterno-cleid-mastoid-muscles] *illeg added above, pencil*
4.17 droops] 'droops' *added above, pencil*
4.18 projecting] 'projecting' *added below, pencil*
4.18 Two . . . mouth. 4.20] 'the cheeks are *hollow & [*interl after del illeg*] deeply furrowed | curved line runs from the wings of the *[*nostrils*]* to the corner of the mouth' *added below, pencil*
4.19 pass, . . . mouth. 4.20] *crossed pencil*
4.19 ala] 'ala' *added above, pencil*
4.20 outer . . . each] *underl pencil*
4.20 The nostrils . . . hollow. 4.21] *scored pencil*; 'there' *interl pencil after* 'nostrils'
4.20 The nostrils . . . distended. 4.21] *underl pencil*
4.21 The cheeks are hollow] *del pencil*
4.21 widely opened] *underl pencil*
4.22 beneath . . . patch.] '*[*swelling*]*' *added above, pencil*
4.22 raised . . . œdematous. 4.24] *crossed pencil*
4.22 semilunar] *underl pencil*
4.24 slightly œdematous.] *underl pencil*
4.25 furrows] 'furrows' *circled and interl pencil*

4.26 & persistent] '& persistent' *added above, pencil*
4.27 The forehead ... folds.] '& strongly furrowed or drawn down at the near end of the *[2 words illeg]* by the same powerful & persistent contraction of the corrugator supercilii' *added after, pencil*
5.6 by scratching ... brain. 5.9] *crossed blue crayon*
6.1 I have ... bite it. 10.14] *crossed pencil*
6.1 I have ... of this. 6.3] 'grinning in rage' *left margin pencil*
6.1 rage & resentment 6.2] 'rage and resentment' *added above, pencil*
7.1 I was ... blow. 7.6] 'paroxysm of rage & also jealousy | *[illeg]*' *left margin pencil*
7.3 vituperated . . . until 7.5] *underl pencil*
7.5 until] 'until' *added above, pencil*
7.6 right angles at] 'at' *left margin pencil*
8.1 No. 1 Ward] *opening parenthesis before* 'No.'; *closing parentheses after* '1' and 'Ward'
8.1 impulsive] 'impulsive' *added above, pencil*
8.3 fury.] 'fury' *added above, pencil*
8.7 pace] 'pace' *added above, pencil*
8.9 & threatening 8.10] '& threatening' *added above, pencil*
8.10 At that ... teeth. 8.12] *scored pencil*; 'very good' *left margin pencil*
9.3 spits & foams] *underl pencil*
9.4 & who ... feet, 9.5] *scored pencil*; 'jumping up' *left margin pencil*
9.5 & hopping ... points. 9.7] *scored blue crayon*
10.1 4. . . . bite it. 10.14] 'Photo not sent | An epileptic idiot' *left margin pencil*
10.6 fierceness] 'fierceness' *added above, pencil*
10.7 raises his head] *underl pencil*
10.7 fixes] 'fixes' *added above, pencil*
10.8 tardy] 'tardy' *added above, pencil*
10.8 If the ... fangs. 10.9] *scored pencil*
10.8 annoyance] 'annoyance' *interl pencil*
10.8 he draws ... reveals 10.9] *underl pencil*
10.9 & reveals] '& reveals' *added above, pencil*
10.10 cruel clutch] 'cruel clutch' *added above, pencil*
10.10 hand] 'hand' *added above, pencil*
10.11 rapidity] *underl pencil*
10.11 clutch] 'clutch' *added above, pencil*
10.13 incensed] 'incensed' *added above, pencil*
10.14 mouth ... it.] 'mouth & bite it' *added below, pencil*
11.9 She sought ... teeth. 11.10] *scored pencil*
11.22 The lips ... forward. 11.23] *scored pencil*
Top of enclosure: '(C.)' *pencil, square brackets in ms*
End of enclosure: 'D^{r}. J. C. Browne | West Riding Asylum | Wakefield' *ink*

1 CD's letter to Maudsley has not been found. CD believed that infants and the insane were most likely to give natural expression to strong emotion, and applied to Maudsley, a leading practitioner in the care of the mentally ill, for assistance (*Expression*, pp. 13–14). Maudsley refers to James Crichton-Browne.
2 See n. 1, above. From the account given in *Expression* of CD's correspondence with Maudsley and Crichton-Browne, it is clear that the questions he asked were whether extreme fear induced erection of the hair, and whether 'retraction of the lips and uncovering of the teeth ... as if to bite the offender' was common in mental patients during paroxysms of rage (*Expression*, pp. 244, 294).
3 Maudsley had evidently forwarded a copy of CD's printed questionnaire on the expression of the emotions (see Appendix VI).
4 CD quoted this observation in *Expression*, p. 295.
5 CD quoted Crichton-Browne's account of Jane Leech almost verbatim in *Expression*, p. 295.
6 CD cited this observation in *Expression*, pp. 295–6.
7 CD cited the case of Ruth Lockwood in *Expression*, p. 296, and referred to photographs of two women sent to him by Crichton-Browne to illustrate erect hair. One photograph, which is probably that of

Lockwood, is attached to Crichton-Browne's letter (DAR 161: 323/6v), and reproduced on p. 235. The other is in DAR 53.1: C68, and is reproduced as a woodcut in *Expression*, p. 296; it was probably sent as an enclosure with the letter from Crichton-Browne, 6 June 1870 (see *Correspondence*, vol. 18).

8 Crichton-Browne quotes the ghost of Hamlet's father:

> But that I am forbid
> To tell the secrets of my prison-house,
> I could a tale unfold whose lightest word
> Would harrow up thy soul, freeze thy young blood,
> Make thy two eyes, like stars, start from their spheres,
> Thy knotted and combined locks to part
> And each particular hair to stand on end,
> Like quills upon the fretful porpentine.

Shakespeare, *Hamlet* 1.5 (London: Isaac Iaggard and Ed. Blount, 1623).

9 CD quoted this observation in *Expression*, p. 296. John Grace has not been further identified.

10 CD cited Crichton-Browne's account of Grace Mellor in *Expression*, pp. 292–3. Phrenalgia was a term used to describe severe depression. Grace Mellor has not been further identified.

11 The sternocleidomastoid muscles on either side of the neck connect the sternum, the clavicle, and the mastoid process of the temporal bone and are used to turn and nod the head.

12 The *ala nasi* is the wing-like cartilage at either side of the nose.

13 The *corrugator supercilii* muscles run from either side of the nose to the middle of the eyebrow.

14 DAR 161: 323/7. The photograph is reproduced on p. 235.

15 CD cited this case in *Expression*, p. 244. A.Y. has not been further identified.

16 CD cited this case in *Expression*, pp. 244–5. John Smith has not been further identified.

17 The *levator anguli oris* is a muscle connecting the jawbone above the canine tooth to the angle of the mouth.

18 CD cited this case in *Expression*, p. 245. William Wheatley has not been further identified.

19 CD cited this case in *Expression*, p. 245. The photograph was not enclosed; see CD annotations. Jonathan [Kilburn] has not been further identified.

20 The remainder of the enclosure, from the beginning of this paragraph, with the exception of Crichton-Browne's signature, is a cutting from a printed notice of 'deaths' and 'recoveries'.

21 R. B. Todd ed. 1835–59; this states (vol. 3, p. 566): 'The two sterno-mastoid muscles acting together directly bend the head on the chest'.

To J. J. Weir 20 May [1869][1]

Down, Beckenham, Kent. S.E.
May 20.

My dear Sir

I feel that I have a sort of prescriptive right to bother you for information. Mr. Gould told me that *male* nightingales immigrate before the *females*; and that he had ascertained this was the case with the snipe and he believed that it was general with migratory Birds.[2] Do you know anything on this head? There is a man in or near Brighton, who sometimes writes in the Ibis, and who I imagine may be a bird-stuffer, and who seems to have paid special attention to migratory Birds.[3] Do you know to whom I refer, for if I could discover his name and address, I would write to him, supposing that you are not able to give me information on this head. I fear that the migratory birds are not largely caught by the Bird-catchers; and the South coast would be the best place for observation

Believe me, my dear Sir | Yours very sincerely | Ch. Darwin

I could no doubt find out the Brighton man's address through Sclater.[4]

Copy
DAR 148: 322

[1] The year is established by the relationship between this letter and the letter from J. J. Weir, 26 May 1869.

[2] CD recorded John Gould's remarks in a note dated 24 March 1868 (DAR 86: C18).

[3] George Swaysland was a naturalist with premises at 4 Queen's Road in Brighton (*Post Office directory of the six home counties* 1870).

[4] Philip Lutley Sclater.

To James Crichton-Browne 22 May 1869

Down,
May 22, 1869.

Dear Sir

I do not know how to thank you enough for your MS. observations on expression.[1] They contain exactly and fully the information which I wanted; and besides being of the greatest use to me, are most interesting and so graphic as to be almost painful.

You kindly say that you are willing to give me further information. I may premise that my object is to make out to a limited extent the causes of the movement of certain muscles under various emotions in man and the lower animals. Your answers are distinct and amply sufficient about the hair "standing on end"; and before receiving them, I felt doubtful whether this expression was not a mere poetic licence. With mammals of all orders, the hair (and in birds the feathers) are erected both under fear and anger; and I think especially when these emotions are combined. Now from your account it seems that with man the hair is erected under both terror and rage;[2] but I understand from one of your remarks, and from what you say about the hair under acute melancholia and hypochondria that the erection often occurs without any emotion either of terror or rage being felt; Is this the case? As any habitual expression seems ultimately to affect the features, is it conceivable that frequent paroxysms of terror and rage might give a tendency to the hair to stand on end without at the time any such emotion being excited? I suppose however that this notion is too improbable.

Your remark on the sterno-cl.-mast.-muscle[3] leads me to ask for information on one point, about which I have long been extremely curious. Duchenne (whose great work on expression perhaps you know) gives photographs of persons with the platysma myoides contracted by galvanism, which wrinkles the skin of the neck transversely, and draws a very little downwards the cheeks near the corners of the mouth.[4] These photographs when tested by showing them to many persons without any explanation and asking what they meant, seem well to exhibit extreme fear; but whether the contraction of the platysma plays so important a part as Duchenne thinks, I wish to ascertain.[5] Would you kindly observe patients suffering under extreme fear, with widely distended eyes, open mouth and erect hair, whether you can observe the contraction of this muscle? Gratiolet says it contracts under severe dyspnœa: a good observer denies this: another surgeon says he has seen it contracted during violent screaming by a patient who was quite insensible from an injury on the head.[6]

Altogether this muscle has quite perplexed me. If it does contract under terror, one might suspect that it partook of the contraction of the homologous scalp muscles.[7] Your description of the grinning and exposure of the canine teeth under furious rage is excellent.[8] I presume that you would not object to my quoting, if I require it, some of your words, as well as giving your general results. I fear that you will think me quite unreasonable, but I should very much like to hear whether the "grief muscles" (see query 5) act frequently and in a prolonged manner with patients who are suffering from extreme depression and anxiety.[9] I should expect that it would be oftenest observed in women. I enclose a poor photograph of a girl who could voluntarily (and this is not common) contract these muscles; but as I have no other copy, having sent the remainder to distant countries I must beg you to return it some time hence.[10] I have noticed the "grief muscles" in action with girls who were pretending to be in distress and by girls who could not voluntarily bring them into action.

I must not write any more, or I shall weary you; but I will add that I should be grateful for any hints from so excellent an observer on any points, such as Query 10 or 9 or 2.[11] Mr. Paget has been observing for me with respect to blushing, but has never seen a blush extend lower down than the clavicles, except a few irregular blotches.[12] I am however assured that some savages who go nearly naked blush over a much more extended surface.[13] I am led to suppose that you must long have felt some interest in Expression; and this leads me to ask whether you have seen Duchenne's great work in folio with numerous photographs.[14] If you would like to see it I would send it by rail, and you could keep it for a couple of months; but the plates are hardly intelligible without reading the text, and this perhaps would be hardly worth your while, even if you had spare time. With the most cordial thanks for your great kindness, and apologies for the length of this letter

I remain dear Sir | Yours very faithfully | Charles Darwin

P.S. In about a month's time a new Edit. of my Origin of Species will appear and for the chance of your liking to have a copy I will direct the publisher to send one[15]

Copy
DAR 143: 327

1 See letter from Henry Maudsley, 20 May 1869, enclosure 2.

2 CD cited Crichton-Browne as an authority on this point in *Expression*, p. 295.

3 CD refers to the sternocleidomastoid muscle (see letter from Henry Maudsley, 20 May 1869, enclosure 2, and n. 11).

4 CD refers to Guillaume Benjamin Amand Duchenne, and Duchenne 1862. There is an annotated copy of Duchenne 1862 in the Darwin Library–CUL (see *Marginalia* 1: 209–10). See *Expression*, pp. 298–303, for CD's discussion of the platysma myoides, a sheet of muscle in the neck extending from the collar bone to the lower part of the cheek.

5 See *Expression*, pp. 299–300, for CD's account of this experiment. *Expression*, p. 299, fig. 20, is a woodcut based on Duchenne 1862, 'Atlas', fig. 61; *Expression*, plate VII, fig. 2, is a reproduction of fig. 64 from Duchenne 1862. CD's unbound copy of the 'Atlas' is in the Darwin Library–CUL. The tabulations of

results of this experiment are in DAR 186: 27–9. See Duchenne 1862, 'Atlas', pp. 44–5, for Duchenne's discussion of the relative role of the muscles of face and neck in expressions of terror. CD cited Charles Bell's belief that corrugation of the neck muscles was strongly indicative of fear only when accompanied by widely staring eyes and an open mouth in *Expression*, p. 298.

6 CD refers to Louis Pierre Gratiolet; see Gratiolet [1865], p. 167. There is an annotated copy of Gratiolet [1865] in the Darwin Library–CUL (see *Marginalia* 1: 345–7). The 'good observer' was probably Charles Langstaff; see *Correspondence* vol. 16, letter to W. E. Darwin, 16 April [1868] and n. 6. The surgeon may have been James Paget, who had offered to make observations on the condition of the platysma in patients under the influence of chloroform (see *Correspondence* vol. 15, letter from James Paget, 9 July 1867). In *Expression* CD cited both Langstaff and William Ogle on this point, but CD only approached Ogle later, in 1870 (see *Expression*, p. 301, and letter to William Ogle, 9 November 1870, *Calendar* no. 7364).

7 CD discussed this point in *Expression*, p. 302.

8 See letter from Henry Maudsley, 20 May 1869, enclosure 2.

9 For CD's printed questionnaire on the expression of the emotions, see Appendix VI. The first part of question five reads, 'When in low spirits, are the corners of the mouth depressed, and the inner corner of the eyebrows raised by that muscle which the French call the "Grief muscle"?' CD defined the corrugator muscle and the occipito-frontalis when 'in conjoint yet opposed action' as the 'grief muscles' (see *Expression*, pp. 179–80; see also *Expression*, p. 25, for a diagram including these muscles). The phrase 'muscle de la douleur', referring to the corrugator supercilii, was used by Duchenne in his *Mécanisme de la physionomie humaine* (Duchenne 1862, pp. 35– 53). For the difference between CD's and Duchenne's understanding of the muscles used in creating this expression, see *Expression*, p. 181 n. 3; see also *Correspondence* vol. 15, letter to J. P. M. Weale, 27 August [1867].

10 The enclosure has not been found, but see *Expression*, pl. 2, fig. 3, facing p. 180, for a detail of the photograph; see also *Expression*, p. 182. CD had apparently sent copies of it with his questionnaire on expression of emotion (see *Correspondence* vol. 15, letter to J. P. M. Weale, 27 August [1867]).

11 See n. 10, above. Question 2 refers to blushing, question 9 to expressions of contempt, and question 10 to expressions of disgust.

12 CD cited Paget on the extent of blushing in *Expression*, p. 313. No letter from Paget containing this information has been found, but he had been observing people blush on CD's behalf for several months; see *Correspondence* vol. 15, letter from James Paget, 9 July 1867, and this volume, letter to James Paget, 29 April [1869].

13 See *Correspondence* vol. 16, letter from F. F. Geach, April 1868. CD cited Frederick F. Geach in *Expression*, p. 317. See also letter to James Paget, 29 April [1869].

14 CD refers to the 'Atlas' of Duchenne 1862; see n. 5, above.

15 The fifth edition of *Origin* was published in the week beginning 22 June 1869 (letter from R. F. Cooke, 22 June 1869).

To George Maw 22 May [1869][1]

Down. | Beckenham | *Kent. S.E.*
May 22[d]

My dear Sir

I am much obliged for your kindness in sending me the splendid specimen on the Drosophyllum, which I have received safely.[2]

I will not trouble you to send me any living plants, as I have done all I could, but to my great regret have not succeeded in making out much.

With my best thanks | Yours sincerely | Ch. Darwin

Royal Horticultural Society, Lindley Library

1 The year is established by the relationship between this letter and the letter from George Maw, 17 May 1869.
2 See letter from George Maw, 17 May 1869 and n. 1.

To J. D. Caton 24 May 1869

Down— Beckenham Kent— S. E.
May 24 1869.

Dear Sir—

I thank you very sincerely for your long letter, the whole of which has interested me much, & for the extremely kind manner in which you offer to give me any further assistance[1]

I have been extremely glad to read your various details, almost all of which will be of more or less use to me—

With respect to the prong-horned Antelope, the *absence* of secondary hoofs is given as a generic character by Dr J E Gray;[2] & from this circumstance I suppose these hoofs are present in most other genera—

They are present throughout the whole deer family. I am surprised to read what you say about the rarity of the tuft of hair in the turkey hen; for Audubon expressly states that this is always acquired by the hen, tho in a less degree at a greater age than in the male—[3] Please to give my best thanks to Dr Paul. The sympathy of a brother naturalist is the highest reward which any one can wish for Will you tell him that I will not trouble him to send me the fossils as I c^d^ not at present make any use of them[4]

With my sincere thanks for all your kindness, I remain, dear sir | With much respect yours very faithfully | Charles Darwin—

Copy
DAR 143: 255

1 See letter from J. D. Caton, 5 May 1869.
2 See letter from J. D. Caton, 5 May 1869 and n. 11. For John Edward Gray's description of the pronghorn, see J. E. Gray 1852, p. 116.
3 John James Audubon wrote that the tuft of hair on the breast was present on all turkey hens, but only when full-grown (Audubon 1831–[9], 1: 15). In *Descent* 2: 180, n. 29, CD cited Audubon but also included Caton's observation that females rarely acquired the tuft of hair. Approximately ten per cent of wild female turkeys are now thought to have the pendulous 'beard' (*Birds of the world* 2: 374).
4 CD refers to John Paul. See letter from J. D. Caton, 5 May 1869 and n. 14.

From J. J. Weir 26 May 1869

6 Haddo Villas | Blackheath SE
26 May 1869

My Dear Sir

I have now made my visit to the Birdcatchers, and find that there is no doubt as to the male of the Nightingale arriving in this country first.—[1]

One man told me he had received several males this Spring, and it was not till a fortnight after the first, that he had any females, and then he had two, and he said that the same rule held good for the Blackcap.—[2]

This latter bird is one that is easily recognized in the male sex & I have invariably noticed that the early birds are males, the head of the female has a brown cap, the other warblers are not easily distinguished, but it has always appeared to me that they arrive in full song, and are afterwards joined by the females.—

One of the Chaffinch "Peggers" has this Spring "pegged" in one day the hitherto unparalleled number of 70 birds.—

A match came off a few days since between two experts one caught 62 & the other 40 birds.—[3]

There seems no doubt that the males are considerably in excess of the females, and the jealousy of the males makes them fall such an easy prey to these men.—

Chaffinches have but one nest during the year to each female, unless the first has been destroyed.—

I feel very unwilling to differ from you, but as to the fertility of canary mules I think you must have been misinformed, the universal testimony of Fanciers including our mutual friend Wicking is that they are barren.—[4]

I have heard of but one case of a linnet[5] mule breeding ♂ but the testimony was not that of an experienced man.—

I placed two mule hens in my aviary one a Goldfinch & the other a Siskin mule, both Goldfinches ♂ & a Siskin ♂ are with them but I notice no advances on either side.

On the other hand I placed one canary ♀ in the aviary to see out of all the species, Linnets, Goldfinches, Siskin, Greenfinch, Chaffinch's & others which she would choose, there was never any doubt the Greenfinch carried the day & this morning I observe two young ones hatched.[6]

You would have been pleased to see how so dull a colored bird as the Greenfinch (chloris), made the most of his generally concealed brighter feathers when courting.— The tail was spread, so as to exhibit the bright yellow outer webs of the feathers, and the wings were drooped to make the most of their golden edgings, the bird singing all the time, however he proved irresistible—

It appears to me that birds which breed but once are more quarrelsome than those which have more

Incomplete
DAR 86: A52a-c, DAR 84.1: 143

CD ANNOTATIONS
1.1 I have ... nest 7.1] *crossed ink*
2.1 One man ... for 2.3] *scored blue crayon*
2.3 the Blackcap.—] *double scored blue crayon*
4.1 One ... birds.— 4.2] *scored blue crayon; top of page* 'Keep for Birds | S Selection' *blue crayon, del pencil*; '& for Origin crosses' *pencil del pencil*; 'Jenner Weir' *pencil*
6.1 There ... females,] *underl blue crayon; double scored blue crayon*
7.1 Chaffinches ... destroyed.— 7.2] *crossed blue crayon*

8.2 Fanciers ... barren.— 8.3] *crossed ink; cross in right margin, blue crayon;* 'Keep' *top of page, blue crayon, circled blue crayon, del pencil*
11.1 On ... hatched. 11.4] *crossed pencil;* 'p 99' *or* 'p 97' *added and circled pencil;* '(Keep for chr' *added blue crayon, square bracket in original ms*
12.1 You would ... irresistible— 12.5] *crossed blue crayon*
13.1 It appears ... more 13.2] *crossed blue crayon*
Top of letter: 'Scraps Sexual selection' *ink*

1 See letter to J. J. Weir, 20 May [1869] and n. 2. The area around Club Row, Sclater Street, and Brick Lane in Spitalfields, London, was known for its bird dealers and animal traders.
2 The blackcap is *Sylvia atricapilla.* In *Descent* 1: 259, CD cited Weir for this information.
3 The chaffinch is *Fringilla coelebs.* In *Descent* 2: 53, CD explained the practice of 'pegging', a method of using caged and stuffed birds to catch male songbirds, and also cited Weir for the information about the number of chaffinches caught in a day (see also *Descent* 1: 307, n. 47).
4 In *Variation* 1: 295, CD had written that the canary had been crossed with nine or ten other species of Fringillidae, of which some hybrids were almost completely fertile. Matthew Wicking had supplied CD with pigeons (see *Correspondence* vol. 6, letter to W. B. Tegetmeier, 11 February [1857] and n. 4).
5 The linnet is now *Carduelis cannabina.*
6 The goldfinch is now *C. carduelis*; the siskin is now *Carduelis spinus*; the greenfinch is *Carduelis chloris.* In *Descent* 2: 115, CD cited Weir for the information on the canary choosing the greenfinch male.

To J. J. Weir 27 May [1869][1]

Down, Beckenham, Kent.
May 27

My dear Sir

Many thanks for your answer about the migratory birds and I have also written to Mr. Swaysland at Brighton.[2] All the pieces of information in your last letter are just the fish which I like to catch in my net. I don't remember what I have said about Canary hybrids, but thought I had only stated that the Canary crosses easily with many other species; I will however keep your note and look to my authorities if I live to come to the subject.[3] I am glad to hear about the Canary at once selecting as its mate the greenfinch, but I am surprised it did not choose the siskin.[4]

With many thanks believe me | Yours very sincerely | Ch. Darwin

Copy
DAR 148: 323

1 The year is established by the relationship between this letter and the letter from J. J. Weir, 26 May 1869.
2 CD's letter to George Swaysland has not been found, but see the letter from George Swaysland, 12 June 1869.
3 See letter from J. J. Weir, 26 May 1869 and n. 4.
4 See letter from J. J. Weir, 26 May 1869 and n. 6.

From Adolf Reuter 30 May 1869

Sans Souci near Potsdam
the 30. 5. | 1869.

Sir,

You will be very astonished to receive a letter from a person, which has not the honour to be acquainted with you in any way, but I hope also you will excuse me, when I tell you the reason for what I took me liberty, to address this letter to you.—

Reading your very interesting and celebrated book on the species formation and animals and plants variation in the domestication,[1] I thought that it would be perhaps agreeable to you, when I did send you also some similar examples, confirming your great Ideas:—

I am Gardener longer then 25 years, I was nearly 5 years in other' countrys, at Liege in Mr. Jacob-Makoy's nursery, at Gand in Mr. van Houtte's;[2] some time in France and longer then a year I was employed at Kew in the Royal Bot. Garden; I was always so happy to be working in the propagating houses.

Returned of my voyage, I have been employed 13 years in our Royal nursery for trees and shrubs &c. and in the same time I was obliged to give lessons our young gardeners, so that I did find very much occasion to study the nature of plants and collect experiences; now I am two years employed as Gardener of His Majesty the King.[3]

Sir, I know it very well that I am immodest to molest such a Gentleman as you with my simply knowledge, but because you are relating often in your celebrated books of English and french Gardening, so I thought it would be perhaps interesting for you to receive also some communication of German Gardening.

I should be glad indeed, when you would be so kind by giving me the permission to send you some examples of my small experience and so I hope to have the honour to receive some words like an answer.—

With my very best respects and assuring you of my attraction at all times | I am dear Sir | Your obedient servant | Ad. Reuter

Your honourable answer may be addressed so

Ad. Reuter | Gardener of His Majesty | SansSouci | near Potsdam (Prussia.)

DAR 176: 124

1 Reuter refers to *Variation*.

2 Reuter refers to Lambert Jacob-Makoy and Louis Benoit Van Houtte. Gand is the French spelling of Ghent.

3 Wilhelm I, king of Prussia.

From J. J. Weir June 1869

6 Haddo Villas | Blackheath SE
June 1869

My Dear Sir

I have again been amongst the Farmers and at the risk of being tedious have ventured to give a few notes more.[1]

The lambs have now obtained their normal color & I could see no traces of the mottled faces before adverted to.—

I found upon enquiry that the dark color of the lambs is considered a good sign, such being considered stronger; a very inferior breed of sheep I saw, had several quite black, the wool as well as the face being of that color, but no South Down Farmer would have had such on his land.—

It would therefore appear probable that the "South Down" was originally nearly black & that the color is now confined to the face & legs which are in the best flocks light brown.

The Short Horned cattle were at one time the most remarkable in this country for the large

Incomplete
DAR 181: 80

CD ANNOTATIONS
1.1 I have ... more. 1.2] *crossed pencil*
2.1 The lambs] 'South Down' *added pencil*
2.1 The lambs ... brown. 4.3] 'June' *added pencil*
5.1 The ... large 5.2] *crossed pencil*
Top of letter: 'J. J. Weir | Keep Embryology (1st sheet)' *pencil*

1 See letter from J. J. Weir, [1–13] May 1869.

From James Crichton-Browne 1 June 1869

West Riding Asylum | Wakefield
June. 1st. | 1869.

Dear Sir

Permit me to thank you very sincerely for your gratifying letter of the 22nd ultimo, & to assure you that it gives me sincere pleasure to afford any little assistance in my power to so distinguished & revered a naturalist as yourself.[1] I only wish that the numerous & harassing duties of my office here allowed me a little more leisure for such congenial pursuits so that I might collect & utilize the mass of interesting material which is as it were going to waste around me in this huge hospital for want of accurate observation & which might be of immense value if submitted to a judgement like yours. My limited opportunities however such as they are are diligently employed & the fruits of my researches are very much at your service, if you care to avail yourself of them. You will confer a real favour upon me by making me of use so that you need not hesitate to command my assistance.

Tomorrow I start for Scotland where I shall remain a few days, but immediately on my return home I shall send you some information upon the points adverted to in your letter & printed slip.[2] Just touching upon one or two of these in the meantime I may say—

1st. With reference to the erection of the hair in acute melancholia & hypochondriasis. There is in both these conditions, a persistent state of painful emotion (either terror or anxiety) varied by paroxysms of more intense suffering. It is in these *paroxysms* that the bristling of the hair is most frequently noticed, although in the intervals it is also occasionally present. It is attributable, I think when seen in the intervals, partly to the sub-acute emotional perturbation then present & partly to that influence of *habit* to which you refer, (derived from the acute paroxysms), which creates a tendency to certain actions & modes of action, & which is operative I am satisfied even upon the hairs. The wife of a medical man, with whom, a patient of mine a *lady* suffering from acute melancholia, (characterized by convictions of unpardonable sin, fear of death for herself her husband & her children), said to me the other day, without the faintest suggestion on my part,—"I think M^rs. H. is going to begin to improve because her hair is beginning to get smoother. I always *notice* that our patients" (a considerable number of lunatics have resided with the speaker) "begin to get better whenever their hair ceases to be rough & untidy & unmanageable" This is a curious empirical observation.[3] There can be no doubt that when convalescence sets in muscular spasm is relaxed, the balance of the capillary circulation is restored, while the cutaneous secretions are re-established.

2nd. With reference to the action of the platysma myoides I have frequently remarked transverse wrinkles on the neck, such as you mention, under excitement but under what sort of excitement, I am not just now prepared to say. The matter shall be carefully considered. The wrinkles to which I refer are very peculiar & quite different from the ordinary folds of the skin of the neck seen in stout people. They are small, parallel, close together, have a direction downwards & forwards & are restless or unstable having sometimes a sort of vibratory or vermicular movement. I am almost sure that I have seen them in terror. I remarked them last, in a man now under my care, labouring under General Paralysis, (in which there is imperfect co-ordination of muscular action, some feebleness, loss of all precision & a loss of all power of directing the diffusions of nervous energy). In this man, to whom I allude the action of the platysma myoides is singularly well marked. Whenever he is asked to show his tongue, & opens his mouth & protrudes that organ the platysma contracts & seems to rise as it were under the skin. It draws up the skin from the lower part of the neck & the upper part of the chest that from the latter situation being visibly drawn up over the clavicle, while the posterior edge or border of the muscle forms a prominent ridge— extending from the clavicle to the angle of the lower jaw. The skin over the neck anteriorly is puckered presenting numerous small transverse wrinkles & a few small longitudinal ones at the anterior & posterior margins of the muscle. The movements of the muscle are all distinctly upwards, it must be remembered however that the lower jaw is already depressed.

3rd. With reference to the grief muscles.[4] They do act frequently & in a prolonged manner amongst the insane & I shall be able to send you many cases illustrative of this. I now return the photograph of the girl who could voluntarily induce the expression of grief which you were good enough to enclose in your letter.[5] It is exceedngly good. Along with it I send 5 photographs of an idiot girl under my care who has many singular ways of expressing her emotions. When pleased she flaps her hands in front of her like little wings, & when displeased, throws the head back in the most extraordinary manner (shown in photograph 5) so that the occiput rests upon the dorsal vertebræ between the scapulæ. It is an anatomical puzzle to me how this is accomplished. May I beg your acceptance of these photographs of this little Yorkshire Cretin.[6]

Will you also look over another packet of photographs which I send you, & let me know if there are any that you would like to possess. I must ask you to return these as they are out of my album, but of most of them I believe I can procure copies. They are all patients of my own.

I have not seen Duchenne's photographs & would much wish to have an opportunity of studying them. If you will be good enough to forward them by mail, as you suggest, I shall take great care of them & return them punctually.[7]

I shall thank you most cordially for the promised copy of the new edition of your book.[8] It will be an old friend endeared by many new & excellent qualities.

Excuse this most hurried note & Believe me to be | With profound respect, | yours faithfully | J. Crichton Browne

Charles Darwin Esq | &c &c

DAR 161: 309

CD ANNOTATIONS
1.1 Dear ... say— 2.4] *crossed blue crayon*
3.1 1st.] 'Dr J. Crichton Browne,— W. Riding Asylum Wakefield.—June 1st 1869.—' *top of page, ink*
3.2 persistent] *underl blue crayon*
3.3 terror or anxiety] 'terror or anxiety' *added above ink*
3.3 intense] 'intense' *added above ink*
3.5 It is attributable ... upon the hairs. 3.19] 'The Hair erect chiefly in the paroxysm, but reduced almost permanent by habit & by a certain state of terror or anxiety being persistent.' *top of page, ink*
3.6 partly ... refer, 3.7] *scored blue crayon*
3.12 Mrs. H] 'Mrs. H' *added above ink*
3.13 her hair ... observation. 3.16] 'A curious empirical confirmation of less erection of Hair coincident with improvement of Health.' *ink*
3.13 her hair ... patients" 3.14] *scored blue crayon*
3.13 always] '*always*' *added above ink*
3.15 (a considerable ... unmanageable" 3.15] *double scored blue crayon*
4.1 2nd. . . . depressed. 4.21] '*Platysma myoides.*—' *ink*
4.1 2nd. . . . depressed. 4.21] *crossed pencil*
4.3 just] 'just' *added above ink*
4.7 I am almost ... terror. 4.8] *underl ink*
5.1 3rd. . . . Cretin. 5.11] '"Grief Muscles"' *blue crayon*
5.1 3rd. . . . of this. 5.3] *scored blue crayon*
5.3 I now ... good. 5.5] *crossed blue crayon*
5.5 Along] 'Along' *added above ink*
5.5 Along ... Cretin. 5.11] *crossed pencil*
5.7 like little wings] 'like little wings' *added above ink*
5.8 (shown ... Cretin. 5.11] *crossed blue crayon*

1 See letter to James Crichton-Browne, 22 May 1869.
2 Crichton-Browne had received a copy of CD's printed Queries about expression (see first enclosure to letter from Henry Maudsley, 20 May 1869, and Appendix VI). CD's initial letter of enquiry to Maudsley has not been found.
3 CD recounted Crichton-Browne's observations about the erection of the hair in *Expression*, pp. 295–7; he quoted the remark about 'Mrs H.' Mrs H. and the 'medical man' and his wife have not been identified.
4 On 'grief muscles', see letter to James Crichton-Browne, 22 May 1869 and n. 9.
5 See letter to James Crichton-Browne, 22 May 1869 and n. 10.
6 These photographs have not been found in the Darwin Archive–CUL; apparently CD returned them (see letter to James Crichton-Browne, 8 June 1869).

[7] Crichton-Browne refers to Guillaume Benjamin Amand Duchenne and the 'Atlas' of Duchenne 1862. See letter to James Crichton-Browne, 22 May 1869.
[8] *Origin* 5th ed.

From George Cupples [1 June 1869][1]

Sheep–

from Robert Thomson, Esqr. | Seggie House | near St Andrews Fifeshire.[2]

Further returns for previous years, will probably be got from this gentleman and his neighbours.

In the year 1867

	out of	318	Lambs
	He had		{ 147 ewe lambs
			171 wether (or wedder)
in 1868			
	out of	207	Lambs
			{ 102 ewes
			105 wedders
for 1869			

The return will be got.

DAR 85: B56

CD ANNOTATIONS
End of letter: 'sent by G. Cupples— | Jun 1— 1869.—' *ink*

[1] The date is established by the annotation.
[2] CD discussed the proportions of the sexes in sheep in *Descent* 1: 304–5.

To Asa Gray 1 June [1869][1]

Down. | Beckenham | *Kent. S.E.*
June 1st

My dear Gray

I was uncommonly glad to get your letter of May 8th. & it was extremely good in M^rs. Gray to write to me so nice a note.[2] Very many thanks to her & to you for the answers about Expression. I well know how extremely difficult it is to observe. One of her answers about the "grief muscles" in a *negro* is of especial value, as I have failed on this head with all the more distinct races of man, & began even to doubt whether it could be general.[3] The case of the shrugging the shoulders will also be useful, & that about the head not being shaken laterally for negation, is very disagreeable & surprising.[4]

Very many thanks for all your kind expressions about my accident: my horse fell & partly rolled over me. Paget thought I sh^d require three months to recover, but I was nearly well in 3 weeks.[5] I have, however, had more pain lately & not been very brisk. My good & dear wife, in consequence, is going to take me, nolens volens, on the 10th to a house, which we have hired in N. Wales for 6 or 7 weeks.[6]

We all heartily wish you could have given a better account of Mrs. Gray's strength. It is very disheartening that the voyage shd. have driven away so much of the good effects of your grand trip up the Nile.—[7]

I have little to say about my own doings. My regular work has been much interrupted by preparing a new Edit. of the everlasting old Origin, which consumed 8 weeks, & by preparing notes for a French Edit. of my Orchis book. In these notes I give references to all papers on subject, with very brief abstracts; so that I have had to quote you incessantly. I think I shall also publish these notes in English.[8] By the way I tried to get Orange Judd & Co to publish the new Edit. of the Origin; but the Appletons were too strong for me. They say courtesy prevents them: fear of their pockets, I presume, is nearer the truth. What an evil this stereotyping scientific works is.—[9] I ought to tell you that they sent me £50 for the "Variation under Domestication", which I think very handsome & which I owe to you.—[10]

I have not heard from Hooker, but I see in Gard. Chronicle, that when elected as President for next Congress, he was received with *great* applause.—[11] We have seen hardly a soul for a long time except the Huxleys, & two detachments of Nortons.[12] I then verified a grand generalisation, which I once propounded to you, that all persons from the U. States are perfectly charming.

Farewell, my dear Gray, I often think with pleasure of your visit here.[13] Give from all of us our kindest remembrances to Mrs. Gray & believe me | Ever yours very sincerely | Ch. Darwin

P.S. | You know how many odd points I want observed: look whether the beards of Germans, when differing in tint from Hair of head, are of a lighter or redder tint? Does Beard & Hair often differ in tint?[14]

Gray Herbarium of Harvard University (86a)

1 The year is established by the relationship between this letter and the letter from Asa and J. L. Gray, 8 and 9 May 1869.

2 Letter from Asa and J. L. Gray, 8 and 9 May 1869.

3 See letter from Asa and J. L. Gray, 8 and 9 May 1869 and n. 10. On 'grief muscles', see the letter to James Crichton-Browne, 22 May 1869 and n. 9.

4 See letter from Asa and J. L. Gray, 8 and 9 May 1869 and nn. 9 and 12.

5 James Paget treated CD after his accident; see the letter from Asa and J. L. Gray, 8 and 9 May 1869.

6 The Darwins left on 10 June 1869 for Caerdeon, Barmouth, and returned on 31 July (see 'Journal' (Appendix II)).

7 See letter from Asa and J. L. Gray, 8 and 9 May 1869.

8 CD finished the fifth edition of *Origin* on 10 February 1869 (see 'Journal' (Appendix II)). For the French edition of *Orchids*, see the letter to J. D. Hooker, 8 March [1869] and n. 5, and the letter from Louis Rérolle, 10 May 1869. For CD's quotations from Gray in his English publication of his orchid notes, see 'Fertilization of orchids' and *Orchids* 2d ed.

9 See letter from Orange Judd & Co., 21 April 1869 and n. 3.

10 On the American edition of *Variation* and the publisher's payment to CD, see the letter from Orange Judd & Co., 21 April 1869.

11 On Joseph Dalton Hooker's election as a president of the second congress of the Russian International Horticultural Exhibition, see the *Gardeners' Chronicle*, 29 May 1869, p. 582.

12 According to Emma Darwin's diary (DAR 242), the Huxleys visited on 25 May 1869. The family of Charles Eliot Norton had rented a house in nearby Keston in 1868 (*Correspondence* vol. 16); they were now staying in London (J. Turner 1999, pp. 222–6). Charles's wife, Susan Ridley Norton, visited on 27 March, with a 'Miss Norton' visiting the preceding day (Emma Darwin's diary (DAR 242)); the Nortons were travelling with two of Charles's sisters, as well as Susan's sister (J. Turner 1999, p. 222). The Nortons also visited on 13 May 1869 (Emma Darwin's diary (DAR 242)).

13 Asa and Jane Loring Gray visited the Darwins from 24 to 30 October 1868 (Emma Darwin's diary (DAR 242)).

14 For CD's interest in different tints of hair and beards, see *Descent* 2: 379–80. CD's notes on the topic are in DAR 85: 42. See also letter to J. D. Hooker, [before 7 May 1869].

To W. C. Tait 1 June [1869][1]

Down. | Beckenham | *Kent. S.E.*
June 1st.

Dear Sir

I am very much obliged for your kindness in bringing me the specimens; but I am so much overworked & have so little strength that I shall not be able to examine them for a considerable time.[2] You can send them if you please addressed to me at

Orpington Station,
Sth E. Railway[3]
(To be called for)

If you stay some little time in England & would feel inclined to come to dinner & sleep here I should be much pleased to thank you in person for all your kindness; but at present I feel so far from well, & having lately had so many visitors, that I am sure the excitement of more conversation would quite knock me up.[4]

I hope that you will enjoy your visit in England & I remain, Dear Sir | yours faithfully | Ch. Darwin

LS
Alan R. Tait (private collection)

1 The year is established by the relationship between this letter and the other letters from Tait in 1869, for example the letter from W. C. Tait, 13 April 1869.

2 No letter from Tait has been found announcing his visit to England. The specimens would have been of *Drosophyllum lusitanicum* (see letter from W. C. Tait, 13 April 1869 and n. 1).

3 South East Railway Company.

4 There is no record of Tait's visiting CD in 1869.

To Stephen Paul Engleheart 2 June [1869][1]

Down
June 2d

My dear Sir

Will you do me a favour by observing during next few weeks any women just towards the end of parturition especially in a bad labour; you will, I believe, observe that at the last & worst efforts & pains, they close the orbicular muscles round the eyelids, like a Baby when crying; now what I want to know is whether this muscular contraction cause a few tears to flow. Of course the woman must not be *crying* from suffering in the ordinary sense. I want to hear about the same point during violent *retching*, when nothing is actually thrown from the stomach, which c^d. get into the nose, & thus irritate the membrane, & cause tears.— I sh^d be very grateful for any information on above point.—[2]

As I am writing I will ask one other question: have you heard any credible account of good being derived in dyspepsia & nervous weakness from "Pulvermachers Volta-Electric Chain bands"; I see he quotes, Sir C. Locock, H. Holland, Ferguson &c, &c.—[3]

Are all the statements quackery & lies, or w^d it be worth my trying as an experiment?

Forgive me troubling you & believe me | Yours sincerely | Ch. Darwin

American Philosophical Society (398)

[1] The year is established by the reference to Pulvermacher's volta-electric chain bands (see n. 3, below).

[2] CD was gathering information for his work on *Expression*.

[3] Isaac Louis Pulvermacher's electric belts were advertised in *The Times* from 1852 onwards, at first as the 'patent portable hydro-electric chain'; an advertisement for the 'volta-electric chain band' with testimonials from the people CD mentions (Charles Locock, Henry Holland, and William Fergusson) appeared on 8 March 1869, p. 12. An article on Pulvermacher's hydro-electric chain bands appeared in the *Lancet* 58: 388–9 (25 October 1851).

To Adolf Reuter 2 June 1869

Down. | Beckenham | *Kent. S.E.*
June 2. 1869

Dear Sir

I am much obliged for your very kind letter.[1] I am sure that an observer who shews such zeal & love for science as you do, & who has had such wide experience, must have collected very valuable facts & conclusions.

The observations which you are so good as to offer me would no doubt be of much value; but I am bound to tell you that from having so many other subjects to attend to, & from my health being weak, I have unwillingly resolved no longer to pursue the subject of variation under domestication, except to correct any errors which I may have published.[2]

Secondly, I am a very poor German scholar, & if your notes were written in German I sh^d have great difficulty in understanding them; & although you write so well in English, it w^d probably cause you some trouble to write in this language.

Under these circumstances I w^d suggest that you sh^d publish your observations in some German periodical & inform me or kindly send me a copy. Your observations w^d thus be of use to every one, & I might not be able to use them for years.

With my sincere thanks for your sympathy & interest in my works I remain | Dear Sir | yours faithfully | Ch. Darwin

LS
Staatsbibliothek zu Berlin – Preußischer Kulturbesitz, Slg. Darmstaedter 34.7

1 See letter from Adolf Reuter, 30 May 1869.
2 CD refers to his work on *Variation*. Reuter had offered to send information about German horticulture (see letter from Adolf Reuter, 30 May 1869).

From J. V. Carus 3 June 1869

Leipzig,
June 3. 1869.

My dear Sir,

I must beg your pardon for answering your kind letter so late.[1] But as there there is no hurry for the new german edition, there was no necessity for troubling you again. Now, as the time of the publication of the new English edition is coming on, I answer first your question about this. If you would be kind enough to send me *a bound copy* I should be very obliged. There is no occasion to send me the corrected sheets, as I shall of course go through the whole new edition line by line.[2]

I don't know if the late Mr Bronn has ever written to you about the german title of the "Origin". His translation of this word by "Entstehung" is not so correct as "Ursprung" would be.[3] I used the last word in the few introductory lines I added to the first edition I published after Bronn.[4] The publisher has no objection to alter the word in the title.[5] I am rather doubtful if it would be well done; for the book is now known in Germany as "Entstehung". Of course there is plenty of time to think of this, yet I thought it better to mention it to you. Perhaps you will be so good as to tell me some time what you think best.

I see from your letter that you changed your residence. As I don't know if it is only a temporary change I direct these lines to Beckenham, hoping that they will reach you.[6]

Believe me | My dear Sir | Yours very sincerely | J. Victor Carus

DAR 161: 71

1 See letter to J. V. Carus, 4 May 1869.
2 Carus refers to the fifth edition of *Origin*, which was published in the week beginning 22 June 1869 (letter from R. F. Cooke, 22 June 1869). The German edition corrected from this edition was published in 1870 (Bronn and Carus trans. 1870).
3 Heinrich Georg Bronn translated the first and second German editions of *Origin* (Bronn trans. 1860 and 1863); his work formed the basis of Carus's later German editions of *Origin*.
4 Carus refers to Bronn and Carus trans. 1867.
5 The publisher was Eduard Koch of E. Schweizerbart'sche Verlagshandlung.
6 CD's postal address had changed from Bromley to Beckenham in April 1869.

To Joseph Henry Gilbert 3 June [1869][1]

Down. | Beckenham | *Kent. S.E.*
June 3

Dear Sir

I am much obliged for your kind invitation, but I am sorry to say that the state of my health prevents me ever attending so pleasant an excursion as you propose.[2]

Pray believe me | dear Sir | yours faithfully | Ch. Darwin

LS
Rothamsted Experimental Station (GIL9.6)

1 The year is established by the position of this letter in a chronological set in a bound volume of Gilbert's correspondence at the Rothamsted Experimental Station.

2 Gilbert had probably invited CD to visit the grass experiments at Rothamsted; in a letter dated 5 June 1881 (*Calendar* no. 13196), Gilbert invited CD again:

> It is just at this season that our "Grass-Experiments" are at their best for inspection; and although the agricultural result is not very promising, I think the differences among the 20 plots, now in the 26th season of experiment, are perhaps as striking as ever, as illustrating the vast influence of external conditions on the character and results of the struggle between the numerous components of an established mixed herbage.

CD discussed the results of growing a single species of grass or many different genera of grasses in *Origin*, pp. 113–14.

From Robert Elliot to George Cupples 4 June 1869

Laighwood, Dunkeld
4th June 1869

Dear Sir

On my return home from the North, I am last night in receipt of yours of the 31st. Ulto. and have pleasure in replying to your inquiries.—[1]

I have now been a Breeder of Sheep for the last 43 years; for the last 13 years, I have had fully 2.000 Lambs annually; and for many years in my life, when a younger man assisted my friends in reducing the Male Lambs into Wethers, some times to fully the extent of the number named annually, and this in the Counties of Roxburgh, Selkirk and Dumfries, and now in the County of Perth, and I can without hesitation say that the numbers of Male and Female Lambs on an average have been Equal, some years the Male will be in greater number, some years the females, but on the average Equal.— This is not only my own opinion but the opinion of all the shepherds I have had, or met with, when performing the cutting operation, and the opinion of all Farmers I have met with, who had experience; the Males have a worse life than the females, more males die than females, up till 3 years old, when the Wethers are sold,[2] and after that I cannot compare the classes, but at birth the numbers are Equal; in cutting the Lambs, we have been in the habit of laying the points of the tails of the Males to one side & the points of the tails of the females to the other, and on the average I have found them equal.—

The same with Cattle.— I think on this point you now have no doubt, but may take the matter for granted, after I have had the opportunity of testing the matter in so many localities, so many numbers and all kinds of sheep— Leicester, Cheviot Blackfaced &c—

As you wish the names of some extensive flock masters and Cattle Breeders however, to ascertain their opinion also, I may mention

William Aitchison Esq of Brieryhill by Hawick[3]

James Oliver Esq., Howpasley, by Hawick.—[4]

John Graham Esq of Shaw by Lockerby.[5]

James Brydon Esq Kinnellhead Moffat. or Langehar[6]

—Mitchelson Esq, Old Fascally by Pitlochry,—[7]

Fletcher N. Menzies Esq. the Secretary to the Highland Society, George IV Bridge Edin^r^—[8]

Allan Pagan Esq Innergeldie Crieff[9]

M^r^. M'Combie M. P. Tillyfour Aberdeen.[10]

I could give you a list of a thousand, but for what good?, You may take what I have stated as a matter of fact.

I am Dear Sir | Yours truly | Robert Elliot

George Cupples Esq | The Cottage | Guard Bridge | Fifeshire

DAR 86: A30–1, A62

CD ANNOTATIONS
1.1 On … experience; 2.10] *crossed pencil*
2.1 the Males] *square bracket added before, pencil*
3.1 The same … Fifeshire 15.1] *crossed pencil*
Top of letter: 'Old letter O | about males dying most' *pencil*

CD note:

Mr Rob Elliot of Laighwood Dunkeld has bred for 43 years, about 2000 Lambs annually & otherwise much concerned with sheep—is convinced *(& they are counted) [*interl*] that on average *(& with cattle) [*interl*] sexes ['about' *del*] equal.— but fluctuate during different years. "The males have a worse Life than the ♀^s^, up till 3 years old"— when the whethers are sold,— he does not know— If not for *castration* very good for me

1 Cupples had offered to find statistics of the proportions of the sexes in various domestic animals for CD (see letter from George Cupples, 22 March 1869).

2 In *Descent* 1: 305 n. 41, CD mentioned that it was Elliot who first drew his attention to the premature deaths of male sheep.

3 Cupples added, 'also Thomas C. Borthwick Es^q^ Hop*[illeg]* by Langholm (written to)'. Borthwick has not been identified.

4 Cupples added, 'written to, June 29/69'.

5 Cupples added, 'written to, June 29/'. John Graham has not been identified, but see also the letter from John Graham to George Cupples, 6 July 1869.

6 Cupples added, 'written to June 29th'. James Brydon has not been identified.

7 Cupples added, 'written to, June 29th.' Mitchelson has not been identified.

8 Fletcher Norton Menzies. Cupples added, 'written to June 28/69'.

9 Cupples added, 'written to June 28/69'. Allan C. Pagan has not been further identified but a letter from him to Cupples survives (see letter from A. C. Pagan to George Cupples, 16 July 1869).

[10] Elliot refers to William McCombie. Cupples added, '(written to June 23/69) also Thos. Elliot, Esq. Hyndhope, by Jedburgh (written to June 29).' Thomas Elliot has not been further identified. He was recommended as a source of information in the letter from William Aitchison to George Cupples, 25 June 1869.

From Ponderer to the *Athenæum* [before 5 June 1869][1]

Darwin's Elephants.— In the last edition of Mr. Darwin's 'Origin of Species,' he makes the following statement (chap. iii., page 74, line 18):—"The elephant is reckoned the slowest breeder of all known animals, and I have taken some pains to estimate its probable minimum state of natural increase. It will be under the mark to assume that it begins breeding when thirty years old, and goes on breeding till ninety years old, bringing forth three pair of young in this interval. If this be so, at the end of the fifth century there would be alive fifteen million elephants descended from the first pair."[2] Perhaps some of your readers will be able to enlighten my dull intellect as to the process of reasoning by which this result is obtained. According to Mr. Darwin's theory, each pair brings forth a pair when it is thirty, when it is sixty, and when it is ninety. Hence if there be one pair in the first year, there will be one pair born in the thirtieth year; these two pairs will produce two pairs in the sixtieth year, and these four will produce four pairs in the ninetieth. After that we have only to add the numbers born in the three preceding periods to find out how many are born in each period; because after they have attained the age of ninety years they cease to breed. This method of reasoning gives the number of pairs born in each period of thirty years as 1, 1, 2, 4, 7, 13, 24, 44, 81, 149, 274, 504, 927, 1,705, 3,136, 5,768, 10,609, 19,513; the last number being born in the period commencing with the five hundred and tenth year. Therefore the number of elephants alive at that time would be 42,762 pairs, that is, 85,524 elephants, less the number that would have died by reason of their age. But Mr. Darwin says that there would be fifteen millions. On what does he base his calculation? | PONDERER.

Athenæum, 5 June 1869, p. 772

[1] The date is established by the date of publication of this letter in the *Athenæum*, 5 June 1869.
[2] 'Ponderer' refers to *Origin* 4th ed.; see also the first edition of *Origin*, p. 64, where the passage exists in almost the same form. It was altered in the sixth edition (*Origin* 6th ed., p. 51) so that the conclusion read, 'if this be so, after a period of from 740 to 750 years there would be nearly nineteen million elephants alive, descended from the first pair.'

From J. D. Hooker 6 June 1869

Stockholm
June 6 /69

My dear Darwin

Here we are, homeward bound, but moored here by the amazing kindness of the people for a longer time than we expected; enjoying ourselves very much indeed.[1]

We went to St Petersburgh in two stages, stopping only at Berlin & there for 3 nights; our Hotel was opposite the Pertz's,[2] who were most attentive & hospitable.— I saw also a good deal of Peters,[3] who is a capital fellow, & has, as you know, been much in England. At the Museum there was a very badly stuffed specimen of a new & small species of Gorilla from **East** Africa, & that it was the only novelty I saw. At St. Petersburgh I found many friends, including some cousins of my Mothers, who had married into Russian families, & one of whom, being an ADC to the Emperor, was immensely useful to us; nothing indeed can exceed the kindness of these Scandinavians & Scythians—[4]

At the Academy I was much interested with old Brandt the Zoological Director who declares that all the Bos'ses *longifrons* & Co are one species—that there is but one fossil Elephant, & that the Dinotherium is simpliciter a mammoth![5] he believes in you with a vengeance, & I hope I do not misinterpret his ideas— I saw the *Rhinoceros tichorhinus* with *skin* & *hair* on head & feet found in East Russia—[6] I was not aware, or had forgotten, that this animal had been found with the soft parts preserved.. The Mammoth is certainly a magnificent thing: the skin is preserved in huge masses. But the *Syremi Stelleri*[7] (I cannot spell the name) of which they have a complete skeleton is even more interesting; the texture of all the bones is like the hardest ivory & the proportion of bone to size of animal I should think exceeds that of any other animal: this & its curious organization rendered it to my eyes the most curious thing I ever saw. The Birds & beasts at the Academy are most admirably stuffed & set up, & the series of varieties of Rodents &c is most instructive in the variability point of view. With the mineralogical Gallery at the School of Mines I was disappointed— Von Baer was not at St. P., he has gone to live at Dorpat, old Admiral Lutke & Bunge the Botanist were by far the most interesting men I saw—[8]

With St. P. I was a good deal disappointed, it is huge, tasteless, & void of all national architecture, except the churches which are Sublime, & the choral services celestial: beside these our emasculated Anglican services, with its halting imagery & puling intonation is contemptible;—if we are to have music & gesticulation & incense & gold & jewels, give them me hot & strong, & the Russo-Greek X is the place for my money. The altar Screens & chapels are literally *ablaze* with jewels & every jewel given, is a full & perfect sacrifice for some real stunning crime sin or misdemeanor committed by this most immoral people.

The Palaces are gorgeous, but one gets quite sick of French decoration, & endless cabinets of diamonds & rubies.— The streets are Enormous & horribly paved, distances are tremendous, living very expensive & the place terribly unhealthy.— I never saw such sickly children & I am assured that the mortality exceeds the births by 6000 per annum immigration of French & Germans keeps up the popn.

The Exposition was very fair, but the arrangements extremely bad.— The Emperor[9] was most polite, received a lot of us at his palace of Tsarkoe Selo[10] showed us himself over the private apartments that were of Historic interest, gave us two dejeuners, & at the end decorated a dozen or so of the savants & expositors— as I declined a Decoration, he has sent me a pair of beautiful Jasper Vases from a private mine he has in Tomsk, which he reserves for such purposes. I was sorely puzzled what to do about the Decoration, not wishing to be rude on one hand, & on the other, anxious to avoid it, lest my motives in coming, after the refusal of Low to send me, should be misunderstood—so I said that as we could not wear foreign decorations at our Court, I would decline, adding that to English men of science they were not of the same Value as to Foreigners.—[11] His functionaries were most civil about it— & he consequently sent me the Vases &c to my two compatriots, Murray & Hogg,[12] each a malachite table.

With Moscow we were enchanted—& could have spent weeks there with pleasure —it is as eminently national as S^t. P. is the contrary.

We left last Tuesday by steamer sailing through the wooded glaciated isles of the Gulf of Finland & Baltic, all the way to this most lovely city—the most enchanting I ever was in. Andersson[13] was on the Quay to meet us & has persuaded me to stay till the meeting of the Academy on Wednesday, when we leave for Copenhagen, & shall thence go to Hamburgh & home by steamer next week.

Here I have seen much of the officers of the Spitzbergen Expeditions—most interesting men— they hope next year to go to Nova-Zembla.[14]

Andersson would go again to the Galapagos for 6 months with a schooner from Guayaquil[15] if he could raise £500— I should think we could do this it would be a splendid thing to do & he is a capital man.

The weather has been cold & comfortless but suitable on the whole for travelling. Mrs Hooker[16] has stood it perfectly & enjoyed herself amazingly, she sends love to you all—please add mine.

Ever my dear Darwin | Yours affec | J D Hooker

Clement Wedgwood & his wife[17] arrived before we left & we spent a very happy day altogether

DAR 103: 14–17

1 For Hooker's trip to St Petersburg, see the letter from J. D. Hooker, 11 March 1869 and n. 4.

2 Georg Heinrich and Leonora Pertz.

3 Probably Wilhelm Peters.

4 Hooker's mother was Maria Hooker, née Turner. Her cousins Anne, Maria Justine, and Gertrude Rigby had married in Estonia George de Wahl, Robert de Rosen, and Theophile de Rosen respectively; Gertrude's daughter married General Manderstjerna, aide de camp to Tsar Alexander II, and the rest of Gertrude's children also married and stayed in Russia (L. Huxley 1918, 1: 17, 2: 86). Estonia had been under Swedish control from the sixteenth to the eighteenth century but was currently ruled by Russia. De Wahl, the de Rosens, and General Manderstjerna have not been further identified.

5 *Bos longifrons* was one of three fossil forms of *Bos* identified as distinct species by most palaeontologists at the time (see *Variation* 1: 81–3). Johann Friedrich von Brandt had published on the mammoth *Elephas primigenius* (now *Mammuthus primigenius*, the woolly mammoth) and on the relationship between *Deinotherium* and the elephant family (Brandt 1866 and 1869).

6 Hooker may refer to a *Rhinoceros tichorhinus* (now regarded as a synonym of *Coelodonta antiquitatis*) discovered by Pyotr Simon Pallas in Siberia, a part of which, with skin and hair on the sides of its head, was kept at the Museum of Natural History at St Petersburg (Roderick Impey Murchison, 'Habitation and destruction of the mammoths', *Edinburgh New Philosophical Journal* 40 (1845–6): 349).

7 Hooker refers to *Sirene stelleri*, now *Hydrodamalis gigas*, Steller's sea cow.

[8] Hooker refers to Karl Ernst von Baer, Feodor Petrovich Lütke, and Alexander von Bunge. Dorpat is now Tartu, in Estonia.
[9] Tsar Alexander II.
[10] Tsarskoe Selo or Tsarskoye Selo.
[11] Hooker had originally planned to attend the horticultural congress as an official delegate, but the Treasury refused to send him and he attended independently (L. Huxley 1918, 2: 85). Hooker refers to Robert Lowe, the chancellor of the Exchequer. The *Gardeners' Chronicle* reported that none of the English delegates were given decorations, on Hooker's advice (*Gardeners' Chronicle* (5 June 1869): 609).
[12] Andrew Murray and Robert Hogg.
[13] Nils Johan Andersson.
[14] Nova Zembla (Novaya Zemlya): two islands in the Arctic Ocean between the Barents and Kara Seas. Spitsbergen was a starting point for Arctic exploring expeditions.
[15] Guayaquil is in Ecuador.
[16] Frances Harriet Hooker.
[17] Clement Francis and Emily Wedgwood.

To John Traherne Moggridge 7 June 1869

Down Beckenham
June 7. 1869

My dear M^r. Moggridge

I heard of your illness but fear that you were much worse than I had supposed.[1] It is extremely kind of you to have taken so much trouble in sending me a copy of your observations with such beautiful illustrations.[2] I have read them with much interest, and as I went on I wished that you would draw up a paper on the subject, so that I was glad when I saw that this is your intention.[3] I do not know when I shall come to this subject, nor how to treat it when I do. You have made a good point in drawing attention to the difference between the variability of distinct individuals and between parts of the same plant.

These latter cases feebly come in fact under what I have called bud-variation.[4] Your observations on the Arbutus when much pruned will I think prove very interesting. Pray attend to how far there is any correlation in the variations of the leaves, flowers and fruit. Could you get specimens growing at the utmost verge of height, or in unusually damp places &c. &c., so as to ascertian how far altered conditions add to the variability of the plant. H. Müller of Lippstadt has examined epipactus and platanthera, and Meehan of the U. States has examined Mitchella as well as epigæa, and some Germans have examined mosses, under the same point of view as you are examining arbutus &c.[5]

Some time ago my horse fell and partly rolled over me, and I have in consequence not been quite so well as usual.[6] On the 10th we all start for N. Wales where we shall stay for nearly 2 months, and I hope to return refreshed for my work.[7]

I sincerely hope that your health may greatly improve during the summer, and with M^rs. Darwin's very kind regards, I remain, with many thanks

Yours very sincerely | Charles Darwin

Copy
DAR 146: 377

[1] See letter from F. A. von Hartsen, 5 February 1869 and n. 8.
[2] Moggridge's observations have not been found in the Darwin Archive–CUL. They concerned *Arbutus* and *Ophrys* (see letter from J. T. Moggridge, 16 September [1869]).
[3] Moggridge may have intended to submit a paper to the Linnean Society (see letter from J. T. Moggridge, 12 December 1869). No paper by Moggridge on *Ophrys* or *Arbutus* was subsequently published in the *Journal* or *Proceedings* of the Linnean Society.
[4] CD discussed bud-variation in *Variation* 1: 373–411.
[5] CD refers to Hermann Müller and H. Müller 1868; Thomas Meehan and Meehan 1868a and 1868b; and Alexander Walther, Ludwig Molendo, and Walther and Molendo 1868. There are annotated copies of H. Müller 1868 and Meehan 1868a and 1868b in the Darwin Pamphlet Collection–CUL, and of Walther and Molendo 1868 in the Darwin Library–CUL (see *Marginalia* 1: 843).
[6] See letter to A. R. Wallace, 14 April 1869 and n. 12.
[7] The Darwin family set out for Plas Caerdeon near Barmouth in Wales on 10 June, and returned to Down on 31 July (Emma Darwin's diary (DAR 242)).

From Henry Walter Bates 8 June 1869

Royal Geographical Society | 15, Whitehall Place, S.W.
June 8 1869

My dear Mr Darwin

I am ashamed to say that my brother Entomologists have almost nothing to offer you in statistics of sexes of bred insects.

Last evening there was a Meeting of the Society & I took the opportunity of asking all the practical Entomologists individually if they had done anything towards answering the appeal made to them some months ago & I found they had done nothing.[1] There was only one exception & I am promised the results of enumeration by one gentleman a Mr Buckler, which as soon as I receive I will forward to you.[2]

I trust this glorious weather finds you well & in happy working condition

Yours sincerely | H W Bates

I have had a fearfully busy season—in Geography[3]

DAR 160: 87

[1] Bates had first asked the members of the Entomological Society of London on CD's behalf for information on the proportion of the sexes in insects in February 1868 (see *Correspondence* vol. 16, letter from H. W. Bates, 18 February 1868).
[2] See memorandum from William Buckler, [after 8 June 1869]. CD gave Buckler's results from rearing seventy-four species of Lepidoptera in 1869 in *Descent* 1: 312.
[3] Bates was assistant secretary of the Royal Geographical Society (*ODNB*).

To James Crichton-Browne 8 June 1869

Down,
June 8, 1869

Dear Sir

I am grateful for the extremely kind manner in which you offer me further assistance, and for the very valuable aid already given.[1] And this is all the more kind, as

I am well aware how much your time must be occupied with your arduous duties. The information about the hair standing on end is now quite ample.[2] I am also glad to hear that you have seen the "grief muscles" in action, for these cannot often be observed.[3] But I may add that I have just had a letter from Egypt from a trustworthy observer that he has seen this expression in a negro.[4] I shall be particularly anxious to hear at some future time about the platysma which has been my *bête noire* for a year or two.[5] It is indeed wonderful what aid you have already given me, and I think will give me. I may remark with respect to what you say about the neck of one of your patients (though to you the remark will probably be superfluous) that there is a muscle, the name of which I have just now forgotten, running from near the clavicle to the corners of the mouth; I see this muscle in strong action, producing prominent diverging lines on the neck of one of my sons, when he draws strongly backwards the corners of his mouth in producing certain notes on the flute.[6] I am very much obliged for your present of the photographs of the cretin; they are dreadful to look at.[7] Can the throwing back of the head be an extreme exaggeration of the attitude assumed during laughter? I have noticed the head thrown a little backwards when a person out of complaisance pretends to laugh and be amused. Sometime ago I went into several shops in London to try to buy photographs of the insane, but failed;[8] so you may believe with how much interest I have carefully looked at your excellent ones, and made some notes. I do not, however, think that your kind offer of copies would be of any special use to me. They shall be returned together with Duchenne by rail; and are sent off to-day.[9]

I have arranged the plates of D. with the text, as I think this is the most convenient plan; but you can take them out and arrange them as you like. If you have the inclination and time, I should be delighted if you would write any remarks on the plates or separately,—such as whether you thought any decidedly good or bad etc. Some French authors sneer at Duchenne, but I am inclined to think highly of this work, and Huxley tells me that he has made out much about the muscles of the hand by the aid of galvanism.[10] I shall not want the book returned for a couple of months and I enclose the best address for its return. On the 10th I and my family start for North Wales to stay there for about 7 weeks, as my health which is always very poor has lately been worse. My address there is C. Darwin, Caerdeon, Barmouth, N. Wales.

Pray believe me, my dear Sir with cordial thanks | Yours sincerely | Charles Darwin

June 8th, 1869

Copy
DAR 143: 328

[1] See letter from Henry Maudsley, 20 May 1869, and letter from James Crichton-Browne, 1 June 1869.
[2] See n. 1, above.
[3] See letter from James Crichton-Browne, 1 June 1869.
[4] The observer was Jane Loring Gray (see letter from Asa and J. L. Gray, 8 and 9 May 1869).

5 See letter from James Crichton-Browne, 1 June 1869. CD had been asking his correspondents to observe the action of the platysma myoides muscle since at least 1868 (see *Correspondence* vol. 16). He discussed the action and significance of this muscle in *Expression*, pp. 298–303.

6 It is not known which of CD's sons is referred to.

7 See letter from James Crichton-Browne, 1 June 1869.

8 CD was in London from 16 to 24 February 1868 (Emma Darwin's diary (DAR 242)).

9 CD had offered to send Crichton-Browne his copy of the 'Atlas' of Duchenne 1862 (see letter to James Crichton-Browne, 22 May 1869, and letter from James Crichton-Browne, 1 June 1869).

10 No letter from Thomas Henry Huxley on this subject has been found. Huxley lectured on the teleology and morphology of the hand at Glasgow in 1876, and on the hand at the Working Men's College in 1878 (L. Huxley ed. 1900, 1: 456, 489–90).

From William Buckler [after 8 June 1869][1]

Larvæ	W. Buckler bred in 1869	♂	♀
3	Argynnis Euphrosyne[2]	2	1
3	Arge Galathea[3]	3	
6	Thecla rubi[4]	3	2
1	Thecla Quercus[5]		1
1	Hesperia sylvanus[6]	1	
28	Acherontia atropos[7]	5	12
2	Sphinx ligustri[8]		1
2	Chærocampæ elpenor[9]		1
5	Sesia myopæformis[10]	1	1
6	Sesia cynipiformis[11]	1	2
2	Hepialus hectus[12]	1	1
1	Hepialus humuli[13]		1
3	Zygæna trifolii[14]	1	
5	Zygæna filipendulæ[15]	1	3
4	Lithosia rubricollis[16]		2
3	Arctia mendica[17]	3	
1	Arctia lubricipedæ[18]	1	
4	Arctia urticæ[19]	1	3
7	Demas coryli[20]	5	
1	Eriogaster lanestris[21]		1
4	Bombyx castrensis[22]	1	
25	Bombyx trifolii[23]	7	6
1	Odontopera bidentata[24]		1
1	Amphydasis prodromaria[25]		1
1	Boarmia Roboraria[26]		1
1	Dasydia obfuscata[27]	1	
1	Hemythia thymiaria[28]		1
1	Acidalia incanaria[29]	1	

6	Hybernia aurantiaria[30]	2	2
4	Eupithecia consignata[31]	1	
1	Eupithecia castigata[32]	1	
1	Lobophora hexapterata[33]		1
5	Pelurga comitata[34]	1	
1	Dicranura bifida[35]	1	
3	Dicranura Vinula[36]	1	2
5	Ptilophora plumigera[37]		1
1	Notodonta dictæoides[38]		1
1	Cymatophora or[39]		1
1	Acronycta tridens[40]	1	
1	Acronycta alni[41]	1	
1	Acronycta Rumicis[42]		1
1	Hydræcia micacea[43]	1	
1	Caradrina blanda[44]	1	
4	Agrotis segetum[45]		1
2	Agrotis exclamationis[46]	1	
40	Agrotis ripæ[47]	9	19
3	Noctua plecta[48]	2	
1	Noctua triangulum[49]		1
1	Noctua brunnea[50]	1	
112	Noctua Dahlii[51]	51	48
1	Taeniocampa gothica[52]		1
1	instabilis[53]	1	
2	Cerastis vaccinii[54]	1	1
8	Xanthia Citrago[55]	3	2
14	Xanthia cerago[56]	4	4
12	Xanthia silago[57]	4	3
1	Cosmia trapezina[58]	1	
4	Epunda nigra[59]	1	1
4	Cucullia Scrophulariæ[60]	1	
27	Cucullia Lychnitis[61]	3	6
11	Heliothis peltigera[62]	4	4
1	Plusia chrysitis[63]	1	
1	Plusia interrogationis[64]	1	
1	Euclidia glyphica[65]		1
4	Pyralis glaucinalis[66]	1	1
1	Aglossa pinguinalis[67]	1	
18	Scoparia muralis[68]	7	6
4	Chilo Phragmitellus[69]	1	1
2	Nephopteryx angustella[70]	2	
6	Rhodophea formosella[71]	2	
1	Carpocapsa pomonana[72]	1	

1	Tortrix on Yellow rattle unnamed[73]		1
1	Crambus? unnamed[74]		
50	Yponomeuta vigintipunctella[75]	26	17
		179	169

AMemorandum
DAR 85: B127

CD ANNOTATIONS
1.1 W. ... 1869.] 'Of Emsworth Hants' *ink*
Bottom of first column: '495' *pencil*; 'larvæ' *ink*
End of table: '74 species counted' *ink, circled ink*
End of table:
'179
168
347 180' *pencil*

1 The date is established by the relationship between this letter and the letter from H. W. Bates, 8 June 1869.
2 *Argynnis euphrosyne* is now *Boloria euphrosyne*, the pearl-bordered fritillary.
3 *Arge galathea* is now *Melanargia galathea*, the marbled white.
4 *Thecla rubi* is now *Callophrys rubi*, the green hairstreak.
5 *Thecla quercus* is now *Neozephyrus quercus*, the purple hairstreak.
6 *Hesperia sylvanus* is now *Ochlodes sylvanus*, the large skipper.
7 *Acherontia atropos* is the death's-head hawk-moth.
8 *Sphinx ligustri* is the privet hawk-moth.
9 *Chaerocampa elpenor* is now *Deilephila elpenor*, the large elephant hawk-moth.
10 *Sesia myopaeformis* is now *Synanthedon myopaeformis*, the red-belted clearwing.
11 *Sesia cynipiformis* is now *Synanthedon vespiformis*, the yellow-legged clearwing.
12 *Hepialus hecta* is the gold swift moth.
13 *Hepialus humuli* is the ghost moth.
14 *Zygaena trifolii* is the five-spot burnet.
15 *Zygaena filipendulae* is the six-spot burnet.
16 *Lithosia rubricollis* is now *Atolmis rubricollis*, the red-necked footman.
17 *Arctia mendica*, the muslin moth, is now *Diaphora mendica*.
18 *Arctia lubricipeda*, the white ermine, is now *Spilosoma lubricipeda*.
19 *Arctia urticae* is now *Spilosoma urticae*, the water ermine.
20 *Demas coryli* is now *Colocasia coryli*, the nut-tree tussock.
21 *Eriogaster lanestris* is the small eggar.
22 *Bombyx castrensis* is now *Malacosoma castrensis*, the ground lackey.
23 *Bombyx trifolii* is now *Lasiocampa trifolii*, the grass eggar.
24 *Odontopera bidentata* is the scalloped hazel.
25 *Amphidasis prodromaria* is now *Biston strataria*, the oak beauty.
26 *Boarmia roboraria* is now *Hypomecis roboraria*, the great oak beauty.
27 *Dasydia obfuscata* is now *Gnophos obfuscata*, the Scotch annulet.
28 *Hemythia thymiaria* is now *Hemithea aestivaria*, the common emerald.
29 *Acidalia incanaria* is now *Idaea dilutaria*, the silky wave.
30 *Hybernia aurantiaria* is now *Agriopis aurantiaria*, the scarce umber.
31 *Eupithecia consignata* is now *E. insigniata*, the pinion-spotted pug.
32 *Eupithecia castigata* is now *E. subfuscata*, the gray pug.
33 *Lobophora hexapterata* is now *L. halterata*, the seraphim.
34 *Pelurga comitata* is the dark spinach.

35 *Dicranura bifida* is now *Furcula bifida*, the poplar kitten.
36 *Dicranura vinula* is now *Cerura vinula*, the puss moth.
37 *Ptilophora plumigera* is the plumed prominent.
38 *Notodonta dictaeoides* is now *Pheosia gnoma*, the lesser swallow prominent.
39 *Cymatophora or* is now *Achlya flavicornis*, the yellow horned moth.
40 *Acronicta tridens* is the dark dagger.
41 *Acronicta alni* is the alder moth.
42 *Acronicta rumicis* is the knot grass.
43 *Hydraecia micacea* is the rosy rustic.
44 *Caradrina blanda* is now *Hoplodrina blanda*, the rustic.
45 *Agrotis segetum* is the turnip moth.
46 *Agrotis exclamationis* is the heart and dart.
47 *Agrotis ripae* is the sand dart.
48 *Noctua plecta* is now *Ochropleura plecta*, the flame shoulder.
49 *Noctua triangulum* is now *Xestia triangulum*, the double square spot.
50 *Noctua brunnea* is now *Diarsia brunnea*, the purple clay.
51 *Noctua dahlii* is now *Diarsia dahlii*, the barred chestnut.
52 *Taeniocampa gothica* is now *Orthosia gothica*, the Hebrew character.
53 *Taeniocampa instabilis* is now *Orthosia incerta*, the clouded drab.
54 *Cerastis vaccinii* is now *Conistra vaccinii*, the chestnut.
55 *Xanthia citrago* is the orange sallow.
56 *Xanthia cerago* is now *X. aurago*, the barred sallow.
57 *Xanthia silago* is now *X. togata*, the pink-barred sallow.
58 *Cosmia trapezina* is the dun-bar.
59 *Epunda nigra* is now *Aporophyla nigra*, the black rustic.
60 *Cucullia scrophulariae* is now *Shargacucullia scrophulariae*, the water betony.
61 *Cucullia lychnitis* is now *Shargacucullia lychnitis*, the striped lychnis.
62 *Heliothis peltigera* is the bordered straw.
63 *Plusia chrysitis* is now *Diachrysia chrysitis*, the burnished brass.
64 *Plusia interrogationis* is now *Syngrapha interrogationis* the scarce silver Y.
65 *Euclidia glyphica* is the burnet companion.
66 *Pyralis glaucinalis* is now *Orthopygia glaucinalis.*
67 *Aglossa pinguinalis* is the large tabby.
68 *Scoparia muralis* is now *Cryphia muralis*, the marbled green.
69 *Chilo phragmitellus* is now *C. phragmitella.*
70 *Nephopteryx angustella* is now *Nephopterix angustella.*
71 *Rhodophaea formosella* is now *Pempelia formosa.*
72 *Carpocapsa pomonana* is now *Cydia pomonella*, the codling moth.
73 Yellow rattle is *Rhinanthus minor*; larvae of *Endothenia marginana*, a moth of the family Tortricidae (tortrix moths) are known to feed on the seedheads of yellow rattle.
74 Buckler later described twenty-two species of *Crambus* (family Pyralidae) in *The larvae of the British butterflies and moths* (Buckler 1886–1901, 9: 284–318).
75 Buckler evidently refers to a species of small ermine moth, *Yponomeuta vigintipunctatus.*

From George Swaysland 12 June 1869

4 Queens Road | Brighton
June 12th 1869

Sir

In reply to your letter I beg to say that the birds that arrive for the first few days are all males, I have never seen the female before the males of any birds males &

Females come together after the first few days, but it is not the case when they go away, I one Spring shot 39 Rays Wagtails before I saw a female, I have several times known the male Whitethroat build the nest before the females arrive[1]

I have seen clods of earth on the birds feet such as Wheatears Winchats Wagtails particulary on fogey mornings and certain they had only just crossed the water[2]

I have never noticed any substance in crops or stomachs of Birds when they arrive as they cross the water in the night or before they feed in the morning the greater part of the Wagtails arrive between 7 or 8 Oclock in the morning I have shot and caught birds for over 40 years and do so now if you should come to Brighton I could tell you a great deal more

I am | Sir | Your Most Ob[ent] | Servant | Geo Swaysland

DAR 205.2 (Letters): 252

CD ANNOTATIONS
Top of letter: 'Keep for Distribution Means of; p. 13' *pencil*; '& Origin' *ink*

1 CD's letter to Swaysland has not been found. CD had asked John Jenner Weir for Swaysland's name and address (see letters to J. J. Weir, 20 May [1869] and 27 May [1869]). CD cited Swaysland's information on Ray's wagtail (now the yellow wagtail, *Motacilla flava*) in *Descent* 1: 260.

2 CD cited Swaysland's information on earth on birds' feet in *Origin* 6th ed., p. 328. Wheatears belong to the genus *Oenanthe*; the whinchat is *Saxicola rubetra*; wagtails belong to the genus *Motacilla*.

From Max Schmidt[1] 14 June 1869

Frankfurt [a]/M.
14 Juni 1869.

Sehr geehrter Herr!

Mehrfache geschäftliche Reisen und Überhäufung mit Arbeiten haben mich bis jetzt verhindert Ihre geehrte Zuschrift vom 29 April zu beantworten und ich bitte dies freundlichst entschuldigen zu wollen.[2] Leider bin ich auch jetzt nicht in der Lage, Ihnen die gewünschte Auskunft zu geben, da unser Garten nicht ein Paar ausgewachsene Mandrillen sondern nur einen männlichen Drill (Mormon leucophaeus)[3] besitzt, welcher sich bereits seit 11 Jahren hier befindet und da er ganz jung zu uns kam, wohl jetzt 12 Jahre alt sein dürfte. Sollte Ihnen mit einer genauern Beschreibüng dieses Thieres gedient sein, so stelle ich mich gern zu Ihrer Verfügung und werde dieselbe sofort abfassen, da ich gegenwärtig wieder etwas mehr Muse habe.

Ob der zoologische Garten in Antwerpen *ein Paar* Mandrille hat, weiss ich nicht, doch wäre es vielleicht die einzige aller ähnlichen Anstalten, bei welcher dies wahrscheinlich ist. Jedenfalls würde ich rathen, sich an Hrn. Direktor Vekemans dortselbst zu wenden und bin überzeugt dass auch bei diesem Herren Ihr Name die beste Empfehlung sein wird.[4] Sollten Sie ausserdem sich auf mich beziehen wollen, so steht dies in Ihrem Belieben.

Sollte ich Ihnen fernerhin irgendwie nützlich sein können, so bin ich mit Vergnügen dazu bereit und bitte Sie, ganz über mich zu verfügen

Mit ausgezeichneter Hochachtung | ergebenst | D^r. Max Schmidt | Director des zoologischen Gartens.

DAR 177: 57

CD ANNOTATION
2.3 Vekemans] *underlined pencil*

1 For a translation of this letter, see Appendix I.
2 See letter to Max Schmidt, [29 April 1869].
3 The drill is now *Mandrillus leucophaeus.*
4 No correspondence between CD and Jacobus Vekemans has been found.

From Gerolamo Boccardo[1] 15 June 1869

Genova
15 Guigno | 1869

Miò Illustre Signore

Accolga, La prego, con la sua usata benevolenza il nuovo mio libro, intitolato "Sismopirologia–Terremoti, Vulcani e lente Oscillazioni del suolo", il cui solo merito è forse quello di tributare ad ogni tratto a V. S. l'ossequio che profondi io Le professo.[2]

Non mando in Ighilterra altra copia che questa, perché temerei che alla "Sismopirologia" incorresse come già alla mia "Fisica del Globo", di cui ho fatto omaggio a Sir Carlo Lyell ed a parecchio Società Scientifiche, Senza riceverne tampoco un cenno che mi assicurassse che il mio dono fosse pervenuto.[3]

Voglia conservarmi, illustre Signore, la sua preziosa indulgenza e mi creda pieno di rispetto | Sua Devmo | G. Boccardo

DAR 160: 234

1 For a translation of this letter see Appendix I.
2 Boccardo refers to Boccardo 1869 (in this version, 'Sismopirologia: earthquakes, volcanoes, and slow oscillations of the ground': the word 'slow' (lente) is apparently not in the printed title of the book). No copy has been found in the Darwin Libraries at CUL or Down.
3 See *Correspondence* vol. 16, letter from Gerolamo Boccardo, 23 April 1868. Boccardo refers to Charles Lyell and Boccardo 1868 (Physics of the world).

From Fritz Müller 15 June 1869

Itajahy, S^a Catharina, Brazil
June 15. 1869.

My dear Sir.

I have been prevented from answering sooner your kind letters of March 14th and 18th by a botanical excursion, I made last month.[1]

I do not know how to express my warm gratitude for all the pains, you have taken in the publication of the Translation of my little book, to which, after all, you attach far too much importance.[2] The three copies have arrived here safely; many thanks for them. I have been much gratified by the appearance of the book in its English dress, as well as by the Translation, which appears to me to be very good. I have to thank you also for the reviews in the "Athenæum" and the "Scientific Opinion", which you have kindly sent me. I am sorry to learn from these reviews, that some expressions of mine have shocked the religious feelings of one of the reviewers.[3] I should indeed have suppressed or modified some passages, before offering my book to English readers.—

As to *Faramea*, I am quite sure, that the two forms are really the long-styled and short-styled form of a single species.[4] The plant is very common in my own forest, and after examining numerous specimens, I cannot find any difference between the two forms excepting those in the sexual organs. The size of the flowers varies much, but is not characteristic of either form. Some long-styled flowers, which I had fertilized artificially with pollen from the opposite form, are yielding fruits.— A second very beautiful species of the same genus which grows near the mouth of the Itajahy, is probably also dimorphic; I have hitherto examined the flowers of only one plant; these were longstyled and the styles as long as in the longstyled form of the first species.—

Many thanks for the *Eschscholtzia*—seeds, which will be sown in a few weeks.— It is curious to see, on what trifling circumstances fertility sometimes depends.[5] Thus I have in my garden several specimens of an endemic *Irideous* plant (*Cypella*?), which I suspected to be self-sterile.[6] I fertilized repeatedly numerous flowers with pollen of the same and of distinct plants, without obtaining a single pod. The unprotected flowers were regularly visited by insects; by these also, for months, did not yield a single fruit, when unexpectedly all the flowers, about 20 in number, which had opened on a certain day, produced fine pods. The plants have continued to flower after this time for many weeks, but without producing any more pods.—

That curious grass with spirally contracting barbs, the *Streptochæta*, appears to be extremely rare here.[7] I have looked in vain for more specimens at different localities; that part of my forest, where I had found it first, had been cut down in the meantime; a single specimen, which I had planted in my garden, seems to thrive well and will, I hope, furnish me with an opportunity of settling the points to which you allude in your letter.

From seeds of a *long-styled* white *Oxalis*, legitimately fertilised with pollen of the longer stamens of the *mid-styled* form I raised some plants; 8 of these have hitherto

Incomplete
DAR 110: B115

CD ANNOTATIONS
1.1 I have … readers.— 2.10] *crossed pencil*
3.1 As … species.— 3.10] 'faramea (Rubiaceæ)' *added ink*
4.1 Many … pods.— 4.9] *crossed blue crayon*
4.6 by these … pods.— 4.9] *double scored pencil*
5.1 That … hitherto 6.2] *crossed pencil*
Top of letter: 'Eschotzia | Dimorphic Plant in *[*growth*]* | Begonia Hooker' *pencil del ink*; 'Faramea' *blue crayon*; '(Claparède)' *ink del ink*; '(Peridromia)'[8] *blue crayon*

1 See letters to Fritz Müller, 14 March 1869 and 18 March 1869. In a letter of 12 June 1869, Müller told his brother, Hermann Müller, that he had travelled to the furthest part of his land to get supplies of garlic vine (*Pseudocalymma alliaceum*) to use as rope for his garden fences (Möller ed. 1915–21, 2: 158).

2 Müller refers to the translation of his book *Für Darwin* (F. Müller 1864a) into English as *Facts and arguments for Darwin* (Dallas trans. 1869; see letter to Fritz Müller, 14 March 1869).

3 The review in the 21 April 1869 issue of *Scientific Opinion* (Anon. 1869b) was positive, but the one in the 27 March 1869 issue of the *Athenæum* was critical, and the author remarked, 'It is to be feared that Dr. Müller's moral code is rather Crustacean than Christian' ([Leifchild] 1869). The reviewer, John R. Leifchild, had written a review of *Origin* for the *Athenæum* that CD found unfair 'under a theological point of view' (see *Correspondence* vol. 7, letter to C. S. Wedgwood, [after 21 November 1859]). In his letter to his brother Hermann, Müller confided that he thought the remark about his moral code a rather good joke (Möller ed. 1915–21, 2: 158).

4 See letter to Fritz Müller, 14 March 1869 and n. 9.

5 CD refers to *Eschscholzia californica* (California poppy); see letter to Fritz Müller, 14 March 1869.

6 *Cypella* is a South American genus of the family Iridaceae. Müller later published observations on two species of *Cypella* in 'Biologische Beobachtungen an Blumen Südbrasiliens' (F. Müller 1883).

7 See letter from Fritz Müller, 12 January 1869 and nn. 3–5, and letter to Fritz Müller, 14 March 1869 and n. 13.

8 CD's notes are for his reply to Müller of 8 September [1869]; he refers to Joseph Dalton Hooker and Eduard Claparède. Müller discussed a monstrous begonia in his letter to CD of 14 March 1869. The butterfly genus *Peridromia* is now *Hamadryas* (*LepIndex*).

From William Aitchison to George Cupples 16 June 1869

Linhope
16th June | 1869

Sir

I regret my absence from here should have left yours of the 12th so long without an answer.—[1]

I am no stranger to Mr Darwin's scientific fame & shall be glad to extend to him, through you, any information in my power about sheep— We sheep Farmers always count the number of our Lambs someway about the 20th or 25th May when we castrate the Tup Lambs, but I suspect none of us know the exact proportional deaths of the sexes previous to that period, but there is no doubt the males are most liable to die after birth—[2] notwithstanding I have always been of opinion that the balance upon the whole is in numbers with the female sex—

In evidence of this I shall give you a statement of the number of Lambs I had on one of my largest Farms during the last two years at the time of castration about the 20th May—[3]

1867	Ewe Lambs	——	57·14
	Tup Lambs	——	55·17
1868	Ewe Lambs	——	62.17
	Tup Lambs	——	52.1
1869	Ewe Lambs	——	57·9
	Tup Lambs		49·13

You are at perfect liberty to further communicate with me about these matters—
I am Sir | Your obdt. Servt. | Wm Aitchison

DAR 86: A54–5

CD ANNOTATIONS
2.3 the 20th ... know the 2.4] *scored blue crayon*
2.5 but there ... always 2.6] *scored blue crayon*
Top of table: '(Scores)' *pencil*
End of table: 'Cheviot Breed at period of castration' *pencil*

1 Aitchison had been suggested to Cupples by Robert Elliot as a breeder worth contacting for information about the proportions of the sexes (see letter from Robert Elliot to George Cupples, 4 June 1869). This letter was enclosed in the letter from George Cupples, 20 June 1869.

2 CD cited Aitchison for this information and for returns on sheep in *Descent* 1: 305.

3 In the table, the first column (before the point) represents groups of twenty, the second (after the point) units up to nineteen. See letter from William Aitchison to George Cupples, 30 June 1869. Sheep were traditionally counted in a vigesimal counting system. Cupples added conventionally written figures at the side of the table.

From Eduard Koch[1] 17 June 1869

Stuttgart
den 17. Juni 1869.

Hochgeehrter Herr!

Durch Herrn Professor Carus—Leipzig werden Sie bereits erfahren haben, dass von der "Entstehung der Arten" die vierte Aufl. in nächster Zeit begonnen wird. Wie mir Herr Carus schreibt waren Sie so freundlich ihm die Aushaengebogen der fünften engl. Auflage zuzusenden.[2]

Mit dem Absatz des "Variiren" bin ich sehr zufrieden und kann Ihnen mitteilen dass in etwa 2 Jahren eine zweite Auflage nöthig sein wird.[3]

Professor Dub—Berlin hat sich mit mir in Verbindung gesetzt wegen Herausgabe einer auch für die grössere Menge leicht verstaendlichen Darstellung Ihrer Lehre und werden wir zu diesem Zweck das Werk mit einigen Illustration verzieren wodurch manches für den Laien klarer werden wird.[4]

Für das bei dieser Gelegenheit an den Tag gelegte Eintreten für meine Rechte an der deutschen Uebersetzung sage ich Ihnen meinen verbindlichsten Dank und werde ich das möglichste Vorbereiten Ihrer Bücher als Pflicht der Dankbarkeit gegen Sie betrachten.

In neuerer Zeit ist eine Broschüre von Prof. Jaeger hier erschienen über "die Stellung der Darwin'sche Theorie zur *Moral* u *Religion*"[5] Sollten Sie dasselbe noch nicht erhalten haben so kann ich es zuschicken und verbleibe

Hochachtungsvoll u Ergebenst | E Schweizerbart'sche Verlagshdlg Eduard Koch

DAR 169: 44

1 For a translation of this letter, see Appendix I.

2 Julius Victor Carus had written to CD about making a fourth German edition of *Origin* using the fifth English edition (see letter from J. V. Carus, 3 June 1869 and n. 2).

3 Koch refers to the German translation of *Variation* (Carus trans. 1868). A second edition was published in 1873 (Carus trans. 1873).

[4] Julius Dub; his popular exposition of Darwinism was Dub 1870. There is a copy in the Darwin Library–Down.

[5] Koch refers to Gustav Jäger and to his *Die Darwin'sche Theorie und ihre Stellung zur Moral und Religion* (Jäger [1869]). There is an annotated copy in the Darwin Library–CUL (see *Marginalia* 1: 429).

From John Russell[1] to George Cupples 18 June 1869

Pitbladdo | Cupar Fife
18th June 1869

Dear Sir,

I am favoured with the note you referred to when I saw you yesterday.

I fear I can give you little, or I should say no information that can be of service to you in the enquiry you desire to make. As a matter of course in most of flocks, many Lambs die either at birth or shortly afterwards, and shepherds as a rule are not over communicative to their employer on these mishaps— I never ask my shepherd to give me exact numbers until the males have been castrated which is done in the month of May; this year on the 12th of May from 167 Ewes there survived Males 124. Females 112—[2]

Although this is no data I have always found that at weaning time there were more Males than Females.

It quite escaped me to talk about the Terrier Yesterday I do not wish "Dan" back as he leads the big dog astray. If you can get the "Cocker" we spoke of *cheap*, and let "Dan" assist the purchase it might be a good arrangement. I wont pay a large price for any dog at present, as I have been spending too much of my spare cash on horses &c of late!!!

Excuse this hurried note as I could not have written you for a day or two had I let this post escape

Yours truly | John Russell

DAR 86: 56–7

CD ANNOTATIONS
2.3 and shepherds . . . mishaps— 2.4] *scored blue crayon*
2.4 I never . . . castrated 2.5] *double scored blue crayon*
2.6 this year . . . 112— 2.7] *scored left margin, double scored right margin, blue crayon*
Top of letter: 'Sheep' *blue crayon*

[1] John Russell has not been further identified. Cupples had evidently contacted him about CD's request for information about the proportion of the sexes in sheep.

[2] CD aggregated the information he received on sexes of sheep in *Descent* 1: 304–5.

To the *Athenæum* 19 June 1869

Caerleon, North Wales,[1]
June 19, 1869.

I am much obliged to your Correspondent of June 5 for having pointed out a great error in my 'Origin of Species,' on the possible rate of increase of the elephant.[2] I inquired from the late Dr. Falconer with respect to the age of breeding, &c., and understated the data obtained from him, with the intention, vain as it has proved, of not exaggerating the result.[3] Finding that the calculation was difficult, I applied to a good arithmetician; but he did not know any formula by which a result could easily be obtained; and he now informs me that I then applied to some Cambridge mathematician.[4] Who this was I cannot remember, and therefore cannot find out how the error arose. From the many familiar instances of rapid geometrical increase, I confess that, if the answer had been thirty or sixty million elephants, I should not have felt much surprise; but I ought not to have relied so implicitly on my mathematical friend. I have misled your Correspondent by using language which implies that the elephant produces a pair of young at each birth; but the calculation by this assumption is rendered easier and the result but little different. A friend has extended your Correspondent's calculation to a further period of years. Commencing with a pair of elephants, at the age of thirty, and assuming that they would in each generation survive ten years after the last period of breeding—namely, when ninety years old—there would be, after a period of 750 to 760 years (instead of after 500 years, as I stated in 'The Origin of Species'), considerably more than fifteen million elephants alive, namely, 18,803,080.[5] At the next succeeding period of 780 to 790 years there would be alive no less than 34,584,256 elephants.

CHARLES DARWIN.

Athenæum, 26 June 1869, p. 861

[1] CD and his family were staying at Caerdeon in North Wales from 12 June to 30 July 1869 (Emma Darwin's diary (DAR 242)).
[2] See letter from Ponderer to the *Athenæum*, [before 5 June 1869].
[3] CD refers to Hugh Falconer. No correspondence with Falconer on this subject has been found.
[4] The arithmetician and Cambridge mathematician have not been identified.
[5] See letter from Ponderer to the *Athenæum*, [before 5 June 1869] and n. 2.

From George Cupples 20 June 1869

The Cottage | Guard Bridge | Fifeshire. N.B.
June 20 /69.

My Dear Mr Darwin,

On hearing from you I at once wrote to Mr Elliot again, inclosing exact copy of your letter (the one intended for him), and have been waiting to hear from him.[1] I have not yet had any reply, but meanwhile have written to some other sheepowners not mentioned in his list— Two of these are among the very foremost breeders in the Border districts—and I send you an answer received from one of them yesterday—along with a minor one from a gentleman near here. Mr Aitchison of Linhope writes very obligingly, at all events—[2] I do not need to have any more letters returned to me, as I shall keep notes so far as necessary for re-writing to these correspondents. Perhaps you might give me a brief hint or two for further guidance, so as to extract more definite information. I intend to write to *all* the persons named by Mr Elliot—and to others—both for sheep and cattle. As to horses, I daresay you can manage best in England. *Swine* are bred by far most largely, I suppose, in England and Ireland. With regard to dogs, greyhounds and foxhounds are the only chance; I should say.

If the statistics are really reliable as to the number of births among sheep, I think there is no doubt I could get returns of many thousands for the present year—probably in some cases for previous years also. Writing the letters gives me *no* trouble.[3]

All that I have got about the births among dogs may be thus summed up.

Mr Barr stated that out of a certain considerable number of litters, there was an excess of females (precise amount of excess not given, I think.) From what he said at the time, I infer that the *breeds* of dogs in question were various (not deerhounds alone.)[4]

Captain Graham, of Redrock, Dursley, Gloucestershire[5]—states his decided impression that females exceeded in his experience.

My own experience was similar.

Mr Wright, of Yeldersley House, near Derby,[6] had similar experience—but entered his *caveat* against any conclusion to that effect—as a large induction through a number of years and circumstances would be required.

Every dog-breeder seems to think females in excess (because bitches are less valued?)[7] As in *table-turning* the involuntary sympathetic nervous action works, so perhaps here! I have lately a letter from the Headforester of the Breadalbane deer-forest[8]—in which he says "I never kept a journal of the litters of deer-hounds, as to number of males and females, but I may mention in trusting to my memory, that in general there are more females than males in a litter of deerhounds, and I believe it is the case in all other animals of the Dog tribe—and it is the same with Sheep, but as to cattle I am not sure about them"—(!!!)

Valeat tantum.[9]

I shall be sending letters on sheep & cattle to you from week to week—and only hope there may be some actual information in the result. I have lost a litter of puppies, but expect more during the season.

Mrs Cupples has not yet returned. She has only told us that she was delighted with her visit to you, and with your kindness and Mrs Darwin's. We expect to hear all in a few days.[10]

Pray excuse haste | ever most truly yours | George Cupples

DAR 86: 58–9

CD ANNOTATIONS

1.1 On ... years also. 2.3] *crossed pencil*
1.9 Perhaps ... information. 1.10] *scored blue crayon*
8.1 *Every* ... Headforester 8.3] *scored pencil*
10.1 I shall ... result. 10.2] *double scored blue crayon*
10.1 I shall ... Cupples 12.1] *crossed pencil*

11.1 Mrs Cupples … Mrs Darwin's. 11.2] *scored blue crayon*
Top of letter: '**Sheep** | Sexual selection' *pencil*

1 CD's letter to Cupples has not been found. Cupples refers to Robert Elliot (see letter from Robert Elliot to George Cupples, 4 June 1869).
2 See the letter from William Aitchison to George Cupples, 16 June 1869, and the letter from John Russell to George Cupples, 18 June 1869.
3 CD thanked Cupples for having procured returns for sheep from Scotland in *Descent* 1: 305 n. 41.
4 See *Correspondence* vol. 16, letter from J. G. R. Barr to George Cupples, [after 11 May 1868?], and this volume, letter from George Cupples, 11 March 1869.
5 George Augustus Graham.
6 See *Correspondence* vol. 16, letter from John Wright, 11 June 1868.
7 CD reported Cupples's opinion in *Descent* 1: 304.
8 Peter Robertson.
9 A variation of *Valeat quantum valere potest*: 'take it for what it's worth'. 'Appropriate when passing information of doubtful authenticity' (Ehrlich 1986).
10 No further information on Anne Jane Cupples's visit to the Darwins has been found.

To J. V. Carus 21 June [1869][1]

Down, Beckenham | Kent [Caerdeon]
June 21st

My dear Sir

I have despatched by Book-Post, this day "viâ France" (as postage is less) the new Edit. of the Origin, & very truly obliged I am to you for your offer to look through the whole of the new Edit.—[2] I have had to make some corrections even since the sheets were printed; & I have been led into one dreadful blunder, (at p. 75), by the false calculation of a mathematical friend.—[3]

I am very decidedly of opinion that it w[d]. be mistake to change "enstehung" into "ursprung", although the latter may be much more correct. If this change were made, purchasers w[d] think it a *new* Book & w[d]. complain; I have known trouble thus caused in England.

I have not changed, & never shall, my House; but the Post-Office have changed the head Post-Town. which for the future will be Beckenham instead of Bromley.—

I am writing this in N. Wales, where I have taken a house for a couple of months for a change & rest, as I have lately not been so well as usual.—[4]

Believe me | My dear Sir | Yours very sincerely | Ch. Darwin

Staatsbibliothek zu Berlin – Preußischer Kulturbesitz, Slg. Darmstaedter Lc 1859 Darwin, Charles, Bl. 19/20

1 The year is established by the relationship between this letter and the letter from J. V. Carus, 3 June 1869.
2 See letter from J. V. Carus, 3 June 1869 and n. 2. CD refers to the fifth edition of *Origin*.
3 See letter from Ponderer to the *Athenæum*, [before 5 June 1869], and letter to the *Athenæum*, 19 June 1869. The error was on page 74 of the fifth edition of *Origin*.
4 CD and his family stayed at Caerdeon in North Wales from 12 June to 30 July 1869 (Emma Darwin's diary (DAR 242)).

From Robert Elliot to George Cupples 21 June 1869

Laighwood, Dunkeld
21st. June 1869

Dear Sir

I am in receipt of yours of the 11th. Inst. but absence from home in Dumfriesshire, & Kirkcudbrightshire prevented me from replying to it until now.—

We all know very well who Mr. Darwin is, although some of us do not think ourselves bound to agree with him in all his theories, if I could assist him however in his researches I would willingly do so; I cannot however give you any more correct statistics regarding the number of males and females among sheep and Cattle than I have done, as I keep no record, and when Mr. Darwin speaks of 104.5 male births to 100 female births in the human species,[1] there may be some slight difference among sheep, but I do not think so, but without closer observation I cannot speak to the exact fact; when in Dumfriesshire I put the question to an extensive farmer, both on lowland and hills, and he said on the average he believed they were equal.—

As regards the point on which Mr. Darwin wishes more information as to whether more male lambs die than females, independently of castration, I have not found more male *Lambs* die, before weaning, than females excepting from Castration, but I have always found males, when *Mogs*, that is from the time they are weaned until they are one year old, more liable to die from *Braxy* and other diseases than females.— Afterwards when females begin to breed they are more liable to casualties than males, that is that more males die before they are one year old or one and a half than females and that independently of castration, and more females than males from one cause and another after they are two year old.—

The number of births of Horses 25,000 tabulated by Mr. Darwin[2] seems a large number, but it is a small number in comparison to the Lambs I have had through my hands first and last.—

You ask the Breed of the sheep I speak of, my opinion is you have that information in my former Letter;— they are now Blackfaced, Mountain sheep, formerly Leicesters and Cheviot.

I am Dear Sir | Yours truly | Robert Elliot

George Cupples Esq | The Cottage

DAR 86: A60–1

CD ANNOTATIONS

1.1 I am … equal.— 2.9] *crossed pencil*
4.1 The number … last.— 4.3] *crossed pencil*
5.2 they are … Cheviot. 5.3] *scored pencil*
Top of letter: 'On Males | Lamb dying, when 1 year old' *pencil*
End of letter: 'Mr Aitchison says on death ♂ *most likely* to die **after birth** & before castration[3] | Oliver says ♂ most apt to die[4]' *ink*

[1] CD gave this statistic in *Descent* 1: 300.

[2] The tabulation was made by William Bernhard Tegetmeier (see *Correspondence* vol. 16, letter to W. B. Tegetmeier, 21 February [1868], and *Descent* 2: 303).

[3] See letter from William Aitchison to George Cupples, 16 June 1869.
[4] See letter from James Oliver to George Cupples, 3 July 1869.

From R. F. Cooke 22 June 1869

50A, Albemarle Street, London, W.
June 22 *1869*

My dear Sir

You will have received a copy by Saturday post of the new edition of the Origin.[1]

I merely write a line to say all your presentation copies have been sent out & that we shall publish the work this week.

We sent by yr directions a set of Electros in Jan[y] last, to Bologna. Are we to make any charge for it & if so *to whom*?[2] At present we have only a memorandum in our books of their being sent.

Yours faithfully | Rob[t] Cooke

Charles Darwin Esq

DAR 171: 370

[1] Cooke refers to the fifth edition of *Origin*. The Saturday before 22 June 1869 would have been 20 June.
[2] The electrotypes were presumably sent in connection with a prospective translation into Italian of *Variation* (see *Correspondence* vol. 16, letter from Giovanni Canestrini, 14 May 1868).

To J. D. Hooker 22 June [1869][1]

Caerdeon, Barmouth | N. Wales
June 22[d]

My dear Hooker

It was very good of you to write to me from Stockholm, telling me all the things about which I liked to hear.—[2] I had heard nothing of your doings, except, as stated in Gard. Chronicle, that the announcement of your being President of the next Congress, was received with loud & general applause; & this applause made me applaud the meeting.—[3]

I suppose the Emperor could have given you nothing which you would have liked better than the Vases, which even I sh[d] like to see: he must have heard of your Crockery madness!—[4]

We have been here for 10 days: how I wish it was possible for you to pay us a visit here: we have a beautiful House, with a terraced garden, & a really magnificent view of Cader, right opposite.[5] Old Cader is a grand fellow & shows himself off superbly with every changing light. We remain here till end of July, when the H. Wedgwoods have the house.—[6] I have been as yet in a very poor way: it seems as soon as the stimulus of mental work stops, my whole strength gives way: as yet I have hardly crawled half a mile from the House, & then been fearfully fatigued.— It is enough to make one wish oneself quiet in a comfortable tomb.

I suppose that you have read Bentham's address: it has interested me greatly, as I particularly wished to hear how Botanists agreed with Zoologists about Distribution.[7] Everything Bentham says, always seems to me remarkably *wise*; & I have an instinctive feeling that every hint from him is well worth pondering over.— Nevertheless I must still think more of the importance of isolation in preserving old Forms, than he seems inclined to do.—[8] Does not this address make you wish to write another essay like your former splendid ones?[9]

I am glad to hear about Andersson & Galapagos: I w^d. gladly subscribe, & if necessary w^d go as far as 50£.— He surely ought to visit Cocos Isl^d.—[10]

Has anyone ever examined the Revillagagos or some such name Arch. off Mexico? I remember looking in old days with longing eyes at this group on the Chart.—[11]

I fear that you forgot all about the colour of the Beards of the Sclavonic races relating to the colour of the Hair.—[12]

Give our very kind remembrances to M^rs. Hooker,[13] & I am very glad to hear that she enjoyed your trip, which seems to have been in every way a most successful affair—

Yours most affect. | C. Darwin

DAR 94: 134–6

[1] The year is established by the relationship between this letter and the letter from J. D. Hooker, 6 June 1869.

[2] See letter from J. D. Hooker, 6 June 1869.

[3] On Joseph Dalton Hooker's election as a president of the second congress of the Russian International Horticultural Exhibition, see the *Gardeners' Chronicle*, 29 May 1869, pp. 582 and 586.

[4] See letter from J. D. Hooker, 6 June 1869. Hooker had been an avid collector of Wedgwood ware (see *Correspondence* vol. 10, letter from J. D. Hooker, [27 or 28 December 1862]).

[5] CD refers to Cader Idris, a mountain. The house was Plas Caerdeon.

[6] CD refers to Hensleigh Wedgwood, his cousin and brother-in-law, and his family.

[7] George Bentham's anniversary address to the Linnean Society was devoted to the subject of geographical distribution (Bentham 1869b). There is a lightly annotated copy of the address in the Darwin Pamphlet Collection–CUL.

[8] Bentham had written, 'Isolation may be an important element in the fixation of a race, but not an indispensable one, and may take only a subordinate part in its maintenance when fixed' (Bentham 1869b, lxxix).

[9] CD refers to Hooker's introductory essay to the flora of Tasmania (J. D. Hooker 1859), his outline of the distribution of Arctic plants (J. D. Hooker 1860), and his paper on insular floras (J. D. Hooker 1866).

[10] See letter from J. D. Hooker, 6 June 1869; CD refers to Nils Johan Andersson. CD had written in 1845: 'I have little doubt that Cocos isl^d., north of the Galapagos Archipelago, from its insulated position, & judging from the Galapagos Arch. would have a most peculiar flora & fauna' (*Correspondence* vol. 3, letter to Edward Forbes, 13 May [1845]; see also *Correspondence* vol. 6, letter to Charles Lyell, 11 February [1857]). Cocos island is off the coast of Costa Rica.

[11] The Revillagigedo islands are off the west coast of Mexico. CD had suggested that a naturalist should visit these islands in his letter to Edward Forbes, 13 May [1845] (*Correspondence* vol. 3), and his letter to Charles Lyell, 11 February [1857] (*Correspondence* vol. 6).

[12] See letter to J. D. Hooker, [before 7 May 1869].

[13] Frances Harriet Hooker.

To Eduard Koch 22 June [1869][1]

Caerdeon Barmouth | North Wales.
June 22

Dear Sir

I am much obliged for your note & am particularly pleased to hear that D[r] Dub is going to publish with you, as I felt doubtful whether his work w[d] interfere with Prof. V. Carus' translation[2]

I sh[d] be very much obliged for Prof Jaeger's brochure; if sent by post please to address it as above as I shall not return to my home at Down until the end of July.[3]

Dear Sir | yours faithfully | Ch. Darwin

P.S. I send yesterday the new English ed. of the Origin to Prof. Carus—[4]

LS
Württembergische Landesbibliothek, Stuttgart (Cod. hist. 4° 333a, 77, 3)

1 The year is established by the relationship between this letter and the letter from Eduard Koch, 17 June 1869.
2 See letter from Eduard Koch, 17 June 1869 and nn. 2 and 4. CD refers to Julius Victor Carus and Cristoph Julius Dub.
3 See letter from Eduard Koch, 17 June 1869 and n. 5. CD refers to Gustav Jäger and Jäger [1869]. The Darwins returned to Down on 31 July 1869 (Emma Darwin's diary (DAR 242)).
4 CD refers to the fifth edition of *Origin*. See letter to J. V. Carus, 21 June [1869].

From Alexander F. Boardman 23 June 1869

Brunswick, Maine, US.
June 23 1869

Mr Charles Darwin
Dear Sir

Having seen, yesterday, what appears to me to be a very extraordinary animal and of a higher grade than any of which I have seen an account, I have thought that even such a rough description of it as I could give might be interesting and possibly useful to you.

You may perhaps have had something like it in England, but judging from what I have read, I should think not.

He was advertised & exhibited as a young Gorilla from four to five years old, bought of the natives in Africa about nine months ago.[1] The natives had had him some considerable time.

The most remarkable thing about him to me was the intellectual shape of his head.

His forehead is high & full, benevolence, causality & comparison being particularly developed.[2]

Through from ear to ear it is quite thick but behind the ears is moderately developed, concientiousness & caution appearing small. As a whole the brain was *very similar* in shape to the human, and in benevolence causality & comparison decidedly superior in shape to some heads of common occurence in human beings of good capacity & intelligence. The animal is about the same height perhaps a ⟨little⟩ less than the usual height of of child of four years— He has a regular knee pan— His heel is decidedly human looking, while the rest of the foot is monkeyish— His finger nails however are decidedly human looking, being in fact better shaped & handsomer than usual in humans.

His arms & legs were very human looking, tho having much less flesh on them—

His ears were decidedly human in appearance. He wore shoes & stockings & was otherwise dressed very much like a child and, it being a damp day, wore a shawl also which when removed it would immediately replace covering his head also all but the face. It was very tractable & quiet, not being confined in any way, would allow itself to be examined & handled very freely by perfect strangers when none but strangers were near. When requested so to do by his owner, would lean forward, put out its lips and kiss him—

His health was apparently not good and his owner did not expect to be able to keep him alive long— His eye teeth were larger but not very much larger than the human.

His resemblance to man struck every one as being very much greater than they had ever seen in any other animal—

Many thought he might be a mixture of human & ape, but he seemed to me only a much higher order of ape than I had ever seen or heard of. I am not skilled in such matters however— ⟨His⟩ nostrils occupied the same place that they do ⟨in⟩ man, but they had no projection over them.

Many remarked that if he only had a nose he would have a very human countenance.

This different position of the nostrils, with the full high forehead, took away much of the monkey look. His face was mostly free from hair and the hair on his legs & arms was no thicker or longer than on on many men and was brownish in color.

I have just shown what I have written to you to one of my neighbors who saw the above described animal & he says that he considers it a fair unexaggerated description.

I tried to get an instrument for measuring heads but before I could find one he had left town, remaining here only one afternoon & evening. Somebody however I presume has measured or will measure his head properly, and his skull of course will be preserved & taken care of.

Hoping that your health is better than when I last heard from you,[3] and that many, many years of happiness & usefulness are yet in store for you | I remain | Yours respectfully | Alex F. Boardman

DAR 160: 229

1 Boardman's description is of a chimpanzee. As late as 1893, some circuses were still referring to chimpanzees as gorillas (Chindahl 1959, pp. 142–3). CD had been gathering information on expression in chimpanzees (see, for example, *Correspondence* vol. 15, letter from Seth Sutton, 8 August 1867).

2 Boardman was analysing the chimpanzee's head using terms from phrenology. According to phrenology, as elaborated by Franz Joseph Gall, the shape of the skull reflects the 'organs' or faculties of the brain. See *EB* 21: 534–40.

[3] CD's letter to Boardman has not been found.

To R. F. Cooke 23 June [1869][1]

Caerdeon, Barmouth | N Wales
June 23[d]

My dear Sir

I distinctly told Prof. Canestrini that you w[d]. charge 10£ for the stereotypes, as you had done for other foreign editions.—[2] You had better write direct to him—or to his publisher.— I thought that you had been paid long ago.—

Thanks for the copy received of Origin & for the copies despatched abroad & elsewhere.— Pray do not forget my suggestion of a copy for Review to "Scientific Opinion"

My dear Sir | Yours sincerely | Ch. Darwin

National Library of Scotland (John Murray archive, Ms. 42152 ff. 203–4)

[1] The year is established by the relationship between this letter and the letter from R. F. Cooke, 22 June 1869.

[2] See letter from Robert Cooke, 22 June 1869 and n. 2. CD refers to Giovanni Canestrini.

From James Croll 23 June 1869

Edinburgh
June 23[d]. 1869.

Dear Sir,

Please to accept of my warmest thanks for the copy of the new edition of your Origin of Species which you were so kind as to present me with.[1] I trust that you will not consider me vain for saying that I am very much gratified with the complimentary way in which you have referred to my papers.[2] I am particularly interested in the section on alternate glacial periods in north and south.[3] I had no idea of this application of the theory when engaged on the subject or I might have brought out the point more clearly than I have done, that when the northern hemisphere, for example, is under a glacial period the line of highest temperature will not be at the equator but will lie a very considerable distance to the south of the equator. And again when the glacial period is transferred over to the southern hemisphere the line of greatest heat will move over to as great a distance to the north of the equator. Your idea that the temperate climate plants will move up the mountain side while this line of greatest heat is being transferred from the one hemisphere to the other is most ingenious as it is natural.

I am, | Yours very truly | James Croll

Charles Darwin Esq. M.A; F.R.S.

DAR 161: 265

[1] The fifth edition of *Origin* was published in the week beginning 22 June 1869; presentation copies were sent out earlier (see letter from Robert Cooke, 22 June 1869).

[2] CD added a number of references to Croll's work on subaerial denudation and glacial periods to *Origin* 5th ed. (see *ibid.*, pp. 349, 352, 379, 451).

[3] See *Origin* 5th ed., pp. 450–61. CD used Croll's hypothesis of alternating glacial periods in the northern and southern hemispheres to help to explain the survival of tropical species, and the distribution of temperate ones.

From A. R. Wallace 23 June [1869][1]

9, St. Mark's Crescent N.W.
June 23rd.

Dear Darwin

Thank you very much for the copy of your fifth Edition of the "Origin". I have not yet read all the additions, but those I have looked at seem very interesting, though somewhat brief,—but I suppose you are afraid of too great & rapid growth.

A difficult sexual character seems to me the *plumules* or battledore scales on the wings of certain families & genera of butterflies—almost invariably changing in form with the species and genera in proportion to other changes,—and always constant in each species yet confined to the males, & so small & mixed up with the other scales, as to produce no effect on the colour or marking of the wings. How could sexual selection produce them?[2]

Your correspondent Mr. Geach[3] is now in England, & if you would like to see him I am sure would be glad to meet you. He is staying with his brother,[4] address, *Guildford*, but often comes to town.

Hoping that you have quite recovered from your accident[5] & that the *great work* is progressing, | Believe me | Dear Darwin | Yours very faithfully

Alfred R. Wallace—

P.S. You will perhaps be pleased to hear that German French & Danish translations of my Malay Archipelago are in progress.[6]

A R W.

DAR 106: B81–2

[1] The year is established by the reference to the fifth edition of *Origin*, which was published in June 1869.

[2] The 'battledore' scales were later found to be scent-producing scales (see letter from Fritz Müller, 19 October 1877 (*Calendar* no. 11191), which was published in *Nature* 17 (1877): 78).

[3] Frederick F. Geach had sent CD responses to his Queries about expression in 1867 and 1868 (see *Correspondence* vols. 15 and 16).

[4] Robert Edgar Geach.

[5] See letter to A. R. Wallace, 14 April 1869 and n. 12.

[6] Wallace refers to Wallace 1869a; the German, French, and Danish translations are Wallace 1869c, Wallace [1870?], and Wallace 1870–1, respectively.

To ? 23 June 1869

"I have never once expressed a wish for aid or for information, which has not been granted, as far as possible, in the most liberal spirit."
(Fertilisation of Orchids. Introduction.—)[1]
Charles Darwin

Down, June 23$^{d.}$ 1869.—

The Pierpont Morgan Library, New York (MA 6512)

[1] The quotation is from *Orchids*, p. 4.

From J. D. Hooker 24 June 1869

Royal Gardens Kew
June 24 /69.

Dear Darwin

A few lines just to say that we are back—as yesterday—having come round by Stockholm, Upsala, Copenhagen, Hamburgh, Hanover Utrecht, Amsterdam, Hague, Leyden & Rotterdam, inspecting the Bot. Gardens & their museums throughout—[1] I got *very* tired of it—thought it was excessively interesting—but the constant packing & moving got odious. Such lots of people asked for you— Even at the Hague I found a young Frenchman busy making notes on the Pictures, so I pointed out the *Dodo* to him & he immediately asked me whether it was alluded to in Darwins last book on Animals & Plants, *which he had read.*[2] Miquel we staid 2 days with, in his House—so nice & pleasant: he has 4 very fine daughters such fine pleasant English like girls—[3] two of them will visit us in September— M. is quite a convert;—he is a very intelligent man, but in poor health. Œrsted[4] at Copenhagen also talked heaps about you— he too is a very able man, & good naturalist.

We have good news of Willy, who has been up to near the scene of the late murders in a Govt. steamer with Capt. Haultaine, the Defence Minister—[5] Hector[6] is looking out for a settlement for him— he is perfectly well & of course happy— I do hope he will take to a settler's life— I am so sure it is the only thing that can suit him for the next 5 years— Hector truly says that he is many years younger than his age— Everyone is charmed with his manner & general conduct—

Miquel has been telling me how the Flora of Sumatra & Borneo are identical, & of Java quite different—just as Wallace shows for the Animals.[7]

How I wish I could join you in Wales, but it is impossible I have a pile of letters that appalls *myself*, & I am not easily frightened—plus a large unopened box of documents & pamphlets accumulated during my absence I too sometimes wish myself in a tomb—though I hold that the balance of life is always on the side of enjoyment, & that the bitterness of the bitterest loss is an insufficient measure of the enjoyment we had in the object lost.—

I am always rejoiced when you like Benthams addresses—[8]

I read it all in mss, & modified some very heterodox passages about Insularity & its effects— you have hit—the flaw in the address— Indeed I do wish I could write another Essay on Islands,[9] & do not give up the hope— I think if Bentham had read Wallace's volumes he would have been more cautious—but he had no time— he however modified extensively what he had written on the strength of what I told him—that is to say he struck out several passages & put others more guardedly.—

Your offer to aid Andersson is a noble one:—I too have often gazed at the Cocos & Revillagigedos & others more isolated still—there away—[10] if Andersson gets a schooner he could do all.

I am woefulley disappointed to hear of your health— is it the sequelle of your fall that you now suffer from?—[11]

Now as to Beards, we never forget them & began to count—in Russia, but soon gave it up, as there was no exception to the rule of the Beard Moustachies & Whiskers being paler than the hair, usually ruddier also.[12]

The Copenhagen & Stockholm Prehistoric Museums are perfectly wonderful, we have nothing at all like them for richness of material & admirable instructive arrangement.— The Danish one is en suite with an Ethnographical Museum of the very highest value, extent, & beauty, which is now under rearrangement, & that again graduates into an illustration of Danish Arts & customs, carried down to the present century. The Dutch Ethnographical & Antiquities collections are very far behind hand indeed—in every respect.

I was charmed to find my former visit to Leyden in 1845.! well remembered by Schlegel & the 2 or 3 survivors of the 10–12 naturalists I then knew there—but what an old man it made of me![13]

Ever yrs affec | J D Hooker

DAR 103: 18–21

CD ANNOTATIONS
9.1 Now as to Beards,] *after opening square bracket pencil*
10.4 & that … 1845.! 11.1] *crossed pencil*

[1] Hooker and his wife, Frances Harriet, were returning from a horticultural congress in St Petersburg (see letter from J. D. Hooker, 6 June 1869, and L. Huxley ed. 1918, 2: 85–9). Upsala and Leyden are now spelled Uppsala and Leiden, respectively.

[2] Hooker refers to *Variation*. The dodo is not mentioned in it. The picture that Hooker pointed out to the Frenchman, who has not been identified, was probably a painting by Roelandt Savery of the animals, including a dodo, listening to Orpheus; it is now in the Mauritshuis, The Hague (s Gravenhage).

[3] Hooker refers to Friedrich Anton Wilhelm Miquel; his daughters have not been identified.

[4] Hans Christian Oersted.

[5] William Henslow Hooker had been sent to New Zealand; in February 1869, he went in the government steamer *Sturt* to Wanganui, and then Taranaki, where they had heard that some settlers had been killed by Maoris (Yaldwyn and Hobbs eds. 1998, p. 212). The event had been reported in *The Times*, 17 May 1869, p. 10. Hooker also refers to Theodore Minet Haultain.

[6] James Hector.

[7] Hooker refers to Alfred Russel Wallace and Wallace 1869a, 1: 227–30.

[8] Hooker refers to George Bentham and Bentham 1869b. See letter to J. D. Hooker, 22 June [1869].

[9] Hooker 1866a.

10 Hooker refers to Nils Johann Andersson. See letter to J. D. Hooker, 22 June [1869] and nn. 10 and 11.
11 See letter to A. R. Wallace, 14 April 1869 and n. 12.
12 CD cited Hooker for this information in *Descent* 2: 319.
13 Hooker refers to Hermann Schlegel; he had visited botanists in Europe in early 1845 (L. Huxley ed. 1918, 1: 177–89).

To Alexander Goodman More 24 June 1869

Down, Beckenham, Kent.
June 24, 1869

Dear Sir

I have not heard for some time how your health is; if still weak pray forgive me for troubling you, and burn this note.[1] But if you are well and inclined to take a little expedition, I should be very grateful if you would endeavour once again on a fine day to observe during half an hour what insects visit *Epipactis palustris*, and especially how they make their entrance into and exit from the flowers. Perhaps you would be so good as to read my description of the supposed manner.[2] This is the point about which I am most curious, and secondly the kind of insect (which I should be glad to see and get named) for I have now clearly ascertained that *Ep. latifolia* is regularly and exclusively fertilised by wasps.[3] The motive for my applying this year is that I have drawn up some notes and corrections for a French Translation of my Orchis book, and these notes I shall also publish in English.[4] Pray forgive me for troubling you, and if in your power I believe that you will kindly feel inclined to aid me.

Dear Sir | yours very faithfully | Ch. Darwin.

Copy
DAR 146: 408

1 CD's last extant letter to More is that of 1 October 1862 (see *Correspondence* vol. 10). More's letters to CD have not been found.
2 See *Orchids*, pp. 99–102. More had sent specimens of *Epipactis palustris* to CD and assisted him in his research (see *Orchids*, pp. 99 n., 101 n.).
3 CD recorded his observation of *Epipactis latifolia* being pollinated by wasps in a note in DAR 70: 56, dated 13 August 1863. For his published references to wasps pollinating *E. latifolia*, see 'Fertilization of orchids', p. 150 (*Collected papers* 2: 146–7), and *Orchids* 2d ed., pp. 101–2.
4 CD refers to Rérolle trans. 1870 (French translation of *Orchids*), and 'Fertilization of orchids'. CD's account of More's latest experiments on *Epipactis palustris* is in 'Fertilization of orchids', p. 149–50 (*Collected papers* 2: 146). See also *Orchids* 2d ed., pp. 98–9. More removed part of the labellum of the flower, to ascertain how important it was to pollination.

From William Aitchison to George Cupples 25 June 1869

Linhope
25th June | 1869

Dear Sir

In reply to yours of the 23[d] I have to inform you I have nothing on my Hill Farms but Sheep of the Cheviot Breed—

It was at the time of Castration I gave you the number of Lambs I had on my Penchrise Farm & now add the numbers I had at that time on this Farm of Linhope—[1]

		Scores Ends
1868	of Ewe Lambs	33. 0
	of Tup Do.	30. 8
Total		63. 8
1869	of Ewe Lambs	30. 3
	of Tup Do.	29. 7
Total—		59. 10

I observe I have sold at the time of Castration 305 Lamb skins from Penchrise & from this Farm at that time 156 Skins—

The above added gives you the total number of Births but not of each Sex, which I fear you will not can procure

I have not yet received a statement of the Lamb Skins sold from my Farms this year—

I should think Mr Elliot Esqre, Hyndhope by Jedburgh a likely person to apply to for information—[2] He is an extensive & distinguished Farmer.

I am | Yr's faithfully | Wm Aitchison

DAR 86: A63–4

CD ANNOTATIONS
Top of letter: 'Sheep' *blue crayon*

1 For Aitchison's earlier returns, see the letter from William Aitchison to George Cupples, 16 June 1869. As in that letter, Aitchison gave the figures here in a base twenty notation, and Cupples translated them in pencil (see letter from William Aitchison to George Cupples, 16 June 1869 and n. 3).

2 See letter from Robert Elliot to George Cupples, 4 June 1869, n. 10. Thomas Elliot has not been further identified.

From Robert Elliot to George Cupples 25 June 1869

Laighwood, Dunkeld
25th June 1869

Dear Sir

I am in receipt of yours of the 23rd. Inst. to get the exact number of Lambs born on large Hill farms is impossible, the numbers can only be taken at the time of Castration.— I have kept no record of the numbers, but as I mentioned to you previously, I have obtained the exact numbers annually, and have found as near as possible that they are equal on the average, that is equal for all practical purposes, I cannot speak to one or two per cent.[1] I know no one likely to have kept a record of the numbers at Castration time, the numbers are taken accurately at Weaning time

& some may have retained a record I have not.— I am only told the numbers for what number of females I keep and keep all the males.—

This year I have 11 Female Calfs and 4 males, last year I had a few more males than females. I may mention that our ratcatcher says that there are 5 male rats for each female, and that he has found that to be the case in all rat nests he has found with young ones in them, thinking this extraordinary I have examined all old ones killed, some hundreds, since he said so, and have found that they were chiefly males.—[2]

I am | Dear Sir | Yours truly | Robert Elliot

George Cupples Esq
The Cottage | Guard Bridge | Fifeshire

DAR 86: A65

CD ANNOTATIONS
1.1 I am … males.— 1.9] *crossed blue crayon*
2.1 This … males,] *scored blue crayon*
Top of page: 'Male Rats' *pencil*; '& Cattle' *blue crayon*

[1] See letters from Robert Elliot to George Cupples, 4 June 1869 and 21 June 1869.
[2] CD cited Elliot for the information on proportions of the sexes in rats in *Descent* 1: 305.

To Edwin Ray Lankester 25 June 1869

Down, Beckenham, | Kent.
June 25th, 1869.

I have read several of Mr. E. Ray Lankester's papers in various Journals, and have been struck by the ability and knowledge shown in them, and by his zeal for Science. Therefore in my opinion he would be well-qualified for giving Lectures on Biology, and for leading students along the right path.[1]

Charles Darwin.

Copy
DAR 221.4: 219

[1] Lankester was a candidate for the Lee's Readership in anatomy at Christ Church, Oxford; he was unsuccessful (see Lester 1995, pp. 22–3). No letter from Lankester requesting a recommendation has been found.

To A. R. Wallace 25 June [1869][1]

Caerdeon, Barmouth | N. Wales
June 25

My dear Wallace

We have been here a fortnight & shall remain here till beginning of August.—[2] I can say nothing good about my health, & I am so weak that I can hardly crawl half a mile from the House; but I hope I may improve & anyhow the magnificent view of Cader is enjoyable.—[3]

I do not know that I have anything to ask M^{r}. Geach: nor do I suppose I shall be in London till late in the Autumn, but I sh^{d}. be particularly obliged, if you have any communication with M^{r}. Geach, if you w^{d}. express for me my *sincere* thanks for his kindness in sending me the very Valuable answers on Expression.— I wrote some months ago to him in answer to his last letter.[4] I w^{d}. ask him to Down, but the fatigue to me of receiving a stranger is something which to you w^{d} be utterly unintleligible.

I think I have heard of the scales on Butterflies; but there are lots of sexual characters, which quite baffle all powers of even conjecture.[5]

You are quite correct, that I felt forced to make all additions to Origin as short as possible.—[6]

I am indeed pleased to hear & fully expected that your Malay work w^{d}. be known throughout Europe.—[7]

Oh dear what w^{d}. I not give for a little more strength to get on with my work.—

Ever yours | C. Darwin

I Wish that you c^{d}. have told me that your place in the new Museum was all settled.[8]

British Library Board (Add 46434 f. 186–7)

1 The year is established by the relationship between this letter and the letter from A. R. Wallace, 23 June [1869].

2 CD and his family stayed at Caerdeon in North Wales from 12 June to 30 July 1869 (Emma Darwin's diary (DAR 242)).

3 CD refers to Cader Idris, a mountain.

4 CD refers to Frederick F. Geach. See letter from A. R. Wallace, 23 June [1869] and n. 3. CD's letter to Geach has not been found.

5 See letter from A. R. Wallace, 23 June [1869] and n. 2.

6 CD refers to the fifth edition of *Origin*. See letter from A. R. Wallace, 23 June [1869].

7 CD refers to Wallace 1869a; see letter from A. R. Wallace, 23 June [1869] and n. 6.

8 In 1869, the government was planning to open a branch of the South Kensington Museum at Bethnal Green to combine art and natural history. Wallace was hoping to be appointed as a director. In the event, when the museum opened in 1872, it was managed from South Kensington. (Fichman 2004, p. 60.)

From W. W. Reade 28 June [1869][1]

C/o Charles Heddle *Esq.* | Sierra Leone—
June 28. '68

Dear Sir

Mr Tickel, the civil commandant of Badagry writes to inform me that two rams of the Guinea breed born on Feby. 10th. showed horns on 6th. of March.[2] This single observation from a man who is not an observer cannot be depended on of course, though useful to compare with other observations of a like kind. Alas! science has not penetrated to the Coast— If you care sufficiently about the matter to do so you might easily obtain a young ram & ewe from this colony for about £2 or less & my host M^{r}

Heddle the leading merchant here, & a man of intellect (though not scientific) has promised to ship them for you, should you require them— His address is above—

I have just returned from the interior after an absence of five months & return tomorrow with the hope of discovering Source of Niger which is not far from here but in a very awkward locality— I shall be absent probably 3 months & if I can be of any use to you pray write & instruct me— I must confess that your queries about expression are too difficult for me to answer. I do not possess the faculty of seizing a fugitive change of feature especially when unprepared for it.

I saw a blue-eyed negress in the interior pure African, & not abnormal in respect of complexion &c. She was brownish but not more so than average of her tribe— Hilly country.

I remain | Yours truly | W. Winwood Reade

Charles Darwin Esq

DAR 86: A32–3

CD ANNOTATIONS

1.1 Mr. Tickel . . . like kind. 1.4] *scored blue crayon*

1.5 If . . . Hilly country. 3.3] *crossed blue crayon*

Top of letter: '(Preliminary discussion Sexual Selection)' *blue crayon*

Verso of last page: 'On the Guinea [*above del* 'coast'] coast of Africa there is a breed of sheep, in which the ewes are hornless, like those of the new breed; & Mr. Winwood Reade informs me that in one case observed, a young ram was born on Feb 10 & showed horns on March 6th.; so that in this instance, the development of the horns,, compared with one /such/, ['seems later' *del*] occurs at a later period of life than in the Welsh sheep, in which both sexes are horned.'[3] *ink*

1 Reade dated this letter 1868 in error. Reade travelled to Africa in mid-1868 (*Correspondence* vol. 16, letter from W. W. Reade, 19 May 1868), and wrote to CD on 17 January 1869 that he was about to travel in search of the source of the river Niger, from which expedition he had evidently just returned. See also Reade 1873, vol. 2.

2 Reade refers to Thomas Tickel. CD had asked Reade to investigate the age at which rams' horns appeared in his letter of 21 May [1868] (*Correspondence* vol. 16). CD reported this observation, which he credited to Reade, in *Descent* 1: 289–90 n. 27. Badagry is now spelled Badagri.

3 This text is a draft of *Descent* 1: 289–90 n. 27.

From Edward Lacy Garbett to the *Athenæum* 29 June 1869

Mornington Road,
June 29, 1869.

Without pretending to an exact solution of Mr. Darwin's elephant problem, I would submit that his first calculation must have been the truest, if, indeed, it made the rate of increase sufficiently high; for the period of doubling seems to me some where about 23 years, whereas his present estimate in the *Athenæum*, letting 18,803,080 become no more than 34,584,256 in 30 years, requires as much as 34 years for doubling.[1]

Remembering that if the births of the first generation extend through 60 years, those of the second will be spread over twice that period, those of the third over thrice, and so on, we shall find the births in the first three centuries as follows:—

Total of each generation.	Dates between which they are born.	No. born in 1st century.	No. born in 2nd century.	No. born in 3rd century.
6	30 – 90	6	–	–
18	60 – 180	6	12	–
54	90 – 270	3	30	21
162	120 – 360	–	54	67
486	150 – 450	–	81	162
1,458	180 – 540	–	81	405
4,374	210 – 630	–	–	937
13,122	240 – 720	–	–	1,640
39,366	270 – 810	–	–	2,187
No. surviving each century		15	258	5,419

Now, as the first parents are not here counted, not surviving the first century, nor even their youngest calf living through the second, in the third at least the regular birth and death rates must be established. The number alive after two centuries, we see, is but 17 times that at the end of the first; while the number alive after three centuries is 21 times those surviving the second. But a 21-fold increase per century is doubling in every 22.7 years; for as Log. 31 is to 100, so is Log. 2 to 22.7. This rate will give in 5 centuries about 2,400,000 elephants, and in one more century 50,000,000. Even the second century's 17-fold increase would approach Mr. Darwin's first results. His present correction, therefore, allowing not even an 8-fold centurial increase, must, I think, involve some great error.

Edward L. Garbett

Athenæum, 3 July 1869, pp. 18–19

[1] See letter from Ponderer to the *Athenæum*, [before 5 June 1869] and n. 2, and letter to the *Athenæum*, 19 June 1869.

To Edward A. Smith [before 30 June 1869][1]

Please to keep carefully the accompanying engraved plates as they are taken from valuable engraved works.[2]

Please to return these memoranda.

The wood-cuts must be fac-similes, done neatly & lightly

(Pl. 25 from Milne-Edwards) fig. 1 the anterior part of a crab, as much as is included in the pencil outline.[3]

(Annals & Mag. of Nat. Hist. Pl. XI) fig *3a*; I think a line drawn across the limb, above & below, where cut off, looks best. Also Fig 9 & 10 from same plate. These 3 figures to be arranged close together on the same block, & in a fitting position[4]

Four wings of Butterfly, exact copies of AA' BB'—to be placed much closer together on same block[5]

(Tab. X from Landois) fig 1 & 2 on same block to be put closer together; in fig 2. all the skin beneath the teeth (as shewn by the pencil line) need not be engraved; alter the letter of reference in fig 1 from s into r.

Fig 3 & 4 to be placed on same block nearer to each other: attend to the dotted line which runs from r to the **second** nervure, & is not shewn plainly in the engraving; attend to letters of reference.—

Fig 5 on a block by itself. Cut off top of.

Fig 6 & 7 on same block; in fig 7 omit some skin on each side, as shewn by pencil lines.

Fig 8 on block by itself. Give all the letters of reference changing rl into r.[6]

Ch. Darwin

Down, Beckenham, Kent.—

The *Plates* can be returned to
C. Darwin
6 Queen Anne St
Cavendish Square[7]
To be left till called for

LS(A)
DAR 81: 35–6

[1] The date is established by the relationship between this letter and the letter to E. A. Smith, 30 June [1869]. The addressee was previously thought to be Edgar Albert Smith, a son of Frederick Smith's, but is now thought more likely to be Edward A. Smith, an entomological illustrator, also a son of Frederick Smith's (Census returns, and Peter Dance, personal communication).

[2] Smith was making woodcuts for *Descent*.

[3] See *Descent* 1: 330 fig. 4 and Milne-Edwards 1834–40, 4: fig. 25. Versions of the woodcut are in DAR 81: 61 and 71. CD's annotated copy of Milne-Edwards 1834–40 is in the Darwin Library–CUL (see *Marginalia* 1: 579–81). CD has circled the part of the drawing of a crab that he wanted Smith to reproduce.

[4] See *Descent* 1: 329 fig. 3 and *Annals and Magazine of Natural History* 11 (1853): fig. 1 (facing p. 25). The plate is from Lubbock 1853a. Versions of the woodcut are in DAR 81: 57 and 60. The figure is of parts of a specimen of *Labidocera*, a copepod.

[5] See *Descent* 2: 133 fig. 52. The woodcut was taken from a drawing by Roland Trimen (see *Correspondence* vol. 16, letter from Roland Trimen, 13 January 1868). A version of the woodcut is in DAR 84.1: 169. The woodcut is complete, whereas Trimen's drawing has had one wing cut out of it.

[6] The woodcuts referred to were taken from Landois 1867, table 10. CD's annotated copy, which includes the pencil marks referred to in this letter, is in the Darwin Pamphlet Collection–CUL. For figs. 1 and 2, see *Descent* 1: 357 fig. 13; there are corrected proofs in DAR 81: 39, 40. For figs. 3 and 4, see *Descent* 1: 353 fig. 10; there are corrected proofs of it in DAR 81: 44 and 45. For fig. 5, see *Descent* 1: 354 fig. 11; there is a corrected proof in DAR 81: 62. For figs. 6 and 7, see *Descent* 1: 379 fig. 24; there are corrected proofs in DAR 81: 41, 43. For fig. 8, see *Descent* 1: 380 fig. 25; there is a corrected proof in DAR 81: 63.

[7] Six Queen Anne Street, London, was the address of CD's brother, Erasmus.

From William Aitchison to George Cupples 30 June 1869

Linhope
30th June | 1869

Dear Sir

In reply to yours of the 28th I have to express my regret that I have not been more explicit in my enumerations—[1]

The 33 convey scores which amount to 660 Lambs & the 30.8 to 608 Lambs & which will give an explanation of all the figures I have sent you—

I note what you say about Mr Elliot' opinion but my long experience has brought me to an opposite conclusion, for every year at the time of castration (about the 22d May) I have had more female than male Lambs, which is no doubt partly balanced by Tup Lambs being more liable to die after birth—[2]

I am Dear Sir | Y'rs faithfully | Wm Aitchison

DAR 86: 66

CD ANNOTATION
3.3 which ... birth— 3.4] *scored pencil*

[1] See letter from William Aitchison to George Cupples, 16 June 1869.
[2] See letter from Robert Elliot to George Cupples, 4 June 1869. CD cited Aitchison's confirmation of Elliot's observation about the greater number of premature deaths among male lambs in *Descent* 1: 305 n. 41.

To Edward A. Smith 30 June [1869][1]

Caerdeon | Barmouth | N. Wales
June 30th—

Dear Sir

Absence from home has prevented me from sooner thanking you for the proofs of woodcuts, which will do very well & some seem excellently executed.[2]

In Fig 2a. I think the fig of the female would be a good deal improved, if made rather less dark; she now looks much more massive than the male.— Please observe that the ♂ must be removed.—

I must have these proofs or others returned to me: Be so kind as to attend to this.—

Please, also, be so good as to return the M.S. list of specimens or instructions for the wood cuts; for I have forgotten the names of some of the species figured.[3]

Please address the returned M.S. to address above given which will be my address for all this month.

I am much obliged for your assistance: I shall probably, but not certainly, require a few more cuts; but my work has lately been much delayed by ill-health.—

Dear Sir | Yours faithfully | Ch. Darwin

P.S. | *I think* I formerly returned a proof for correction, & a revise was never sent me.— When I get my M.S. list, I can look over the proofs, & ascertain whether I have copies of all.—

American Philosophical Society (372)

[1] The year is established by the address. CD and his family stayed at Caerdeon in North Wales in June and July 1869 (Emma Darwin's diary (DAR 242)).

[2] The woodcuts were for *Descent.* See also letter to E. A. Smith, [before 30 June 1869].

[3] CD's list of woodcuts for *Descent* is in DAR 82: B29

From W. B. Tegetmeier 30 June 1869

The Field. | *346, Strand,* | *London, W.C.*
June 30 *1869*,.

My dear Sir

I have just forwarded by rail a basket of pile game eggs.— they are from a pile cock & pile hens.—[1] they were specially put together for me by the Rev[d] F Watson of Kelvedon,[2] a very well known breeder of game. so they may be depended on. I hope they are not too late.— I have a quantity of notes for you which I hope to be able to

AL incomplete
DAR 142: 49c

CD ANNOTATIONS
Top of letter: 'Pile game | Rev. F. Watson | Breed' *pencil*

[1] See letter to W. B. Tegetmeier, 25 February [1869].

[2] Frederick Watson.

To J. J. Weir 1 July [1869][1]

Caerdeon, Barmouth, N. Wales
July 1

My dear Sir

You will see by this address where we are and where we remain all this month. My health got so bad that I could do nothing at Down and we came here for a change of air; but though the views are delightful and air admirable, as yet it has done me no good, and I am fit for nothing.

Very many thanks for your interesting letter. I am glad to have the confirmation of the lamb case and to hear about the insects and Swifts.[2] How wonderfully complex the action and reaction of the elements and all organic beings are, and what a fine field for observation.

Darwin family at Caerdeon, 1869. Left to right: Henrietta Emma Darwin, Francis Darwin, Leonard Darwin, Horace Darwin, Elizabeth Darwin.
CUL MS 8904.4: 1153. By permission of the Syndics of Cambridge University Library.

I have had an answer from M^r. Swaysland of Brighton and he most fully confirms about the males of our migratory Birds; he has never known an instance of the females arriving first.[3]

Once again let me thank you for your interesting letter, and believe me | My dear Sir | Yours very sincerely | Ch. Darwin.

Copy
DAR 148: 324

[1] The year is established by the address. The Darwins left on 10 June 1869 for Caerdeon, and returned on 31 July (see 'Journal' (Appendix II)).
[2] See letter from J. J. Weir, June 1869. The section of the letter in which Weir discussed insects and swifts is now missing.
[3] See letter from George Swaysland, 12 June 1869.

To R. F. Cooke 2 July [1869][1]

Caerdeon | Barmouth | N. Wales
July 2nd

My dear Sir

Very many thanks for your kind present of the Q. Review. There are two articles which concern me, besides others I daresay of general interest. If the article on

Design is written, as I daresay is the case, by an impartial and able man it will interest me greatly.[2] Again many thanks.

Yours sincerely | C. Darwin

We are far from shops and are run short of note-paper, so excuse this.

Copy
DAR 143: 272

[1] The year is established by the address. The Darwins left on 10 June 1869 for Caerdeon, and returned on 31 July (see 'Journal' (Appendix II)).

[2] The July 1869 issue of the *Quarterly Review* contained articles reviewing Alfred Russel Wallace's *Malay archipelago* (Wallace 1869a) and George Henry Lewes's 'Mr. Darwin's hypotheses' (Lewes 1868; see *Quarterly Review* 127 (1869): 68–98, 134–76). The author of the article that discussed design in nature (*Quarterly Review* 127 (1869): 134–76) was James Bowling Mozley (see *Wellesley index*).

From Ernst Haeckel[1] 2 July 1869

Jena
2. Juli 69.

Hochverehrter theurer Freund!

Vor einigen Tagen erhielt ich die fünfte Auflage Ihres "Origin of Species" und danke Ihnen herzlich für diesen neuen Beweis Ihrer Güte.[2] Ich freue mich sehr dass Ihre epochemachendes Werk immer noch so zahlreiche neue Abnehmer mit jedem Jahre findet, trotzdem es schon in so vielen tausend Exemplaren und in so zahlreichen Übersetzungen durch die ganze Welt verbreitet ist. Besonders freue ich mich, dass Sie noch die grosse Genugthuung erleben, die ungeheure Wirkung des colossalen Fortschritts, den Ihr Werk in der Biologie hervorgerufen hat, mit eigenen Augen zu sehen, und daraus zu ermessen, wie gross erst sein Ruhm in späterer Zukunft sein wird.

Von meiner "Natürlichen Schöpfungsgeschichte" bereite ich jetzt die II. Auflage vor, welche im nächsten Jahre erscheinen wird.[3] Die Zahl der Schriften, welche die einzelnen Seiten Ihrer Theorie ausbauen, wächst mit jeder Woche, und ebenso auch die Zahl Ihrer Anhänger. Unter den deutschen Naturforschern der *jüngeren* Generation giebt es jetzt keine irgend bedeutenden Gegner mehr. Besonders freut es mich aber, dass auch höchst verdiente ältere Naturforscher sich zur Descendenz-Theorie bekehren, wie z. B. *P. Harting* in Utrecht und *M. Sars* in Christiania, welche mir Beide in letzter Zeit geschrieben haben, und ihre volle Übereinstimmung ausgedrückt haben.[4]

In einigen Wochen werden Sie meine "Entwickelungsgeschichte der Siphonophoren" erhalten, welche jetzt endlich fertig gedruckt und von 14 Tafeln begleitet ist.[5] Gegenwärtig arbeite ich seit bald einem Jahre sehr eifrig an den *Kalkschwämmen* (Calcispongiae)[6] einer höchst interessanten Thiergruppe, welche auch für Ihre Theorie von sehr grosser Bedeutung ist. Die organische "Species" ist hier bei manchen Grantien quasi "*in statu nascenti*" zu beobachten. Um die Naturgeschichte dieser Thiere möglichst vollständig zu bearbeiten, werde ich gegen Ende Juli auf einige

Wochen nach *Norwegen* (Christiania, Bergen, Drontjhem)[7] gehen, wo es sehr viel Kalkschwämme zu geben scheint. Ich werde Ihnen eine vorläufige Mittheilung über meine Resultate in einigen Wochen zuschicken. Eine umfassende "*Monographie der Kalkschwämme*" denke ich dann im nächsten Jahre zu veröffentlichen.[8] Um den systematischen Theil derselben recht vollständig zu machen, erbitte ich nun Material von allen Seiten. Sollten Sie oder einer Ihrer Freunde mir Kalkschwämme schicken können (wenn auch nur die *ganz gemeinen* Specien der britischen Küsten, Grantia ciliata,[9] Leucosolenia botryoides etc) so würde ich dafür sehr dankbar sein. Bei *der ausserordentlichen Variabilität der Kalkschwämme* kann ich *nicht genug Individuen* von verschiedenen Standorten vergleichen. Grade von den britischen Küsten habe ich noch sehr wenig erhalten.—

Hoffentlich befinden Sie und Ihre liebe Familie sich wohl. Mir geht es recht gut. Mein Junge ist jetzt 9 Monate alt und fängt schon an, viel Pithecoides zu verlieren, ja selbst bereits zu philosophieren![10]

Mit den herzlichsten Grüssen | Ihr treu ergebener | Haeckel

DAR 166: 52

CD ANNOTATIONS
Top of letter: 'Calc-sponges | Ask Huxley'[11] *pencil*

1 For a translation of this letter, see Appendix I.
2 *Origin* 5th ed. was published in June 1869 (letter from R. F. Cooke, 22 June 1869).
3 The second edition of Haeckel's *Natural history of creation* appeared in 1870 (Haeckel 1870a). CD's annotated copy of Haeckel 1868 (the first edition) is in the Darwin Library–CUL (see *Marginalia* 1: 358–60). His copy of Haeckel 1870a is in the Darwin Library–Down.
4 Haeckel refers to Pieter Harting and Michael Sars.
5 Haeckel refers to his study of the development of siphonophores (Haeckel 1869a). CD's annotated copy is in the Darwin Library–CUL (see *Marginalia* 1: 355). Siphonophores are hydroids of the order Siphonophorae.
6 The current name for the class of calcareous sponges is Calcarea.
7 Christiania is now Oslo; Haeckel also refers to Trondheim.
8 *Die Kalkschwämme* (Calcareous sponges) was published in 1872 (Haeckel 1872). CD's annotated copy is in the Darwin Library–CUL (see *Marginalia* 1: 357–8).
9 *Grantia ciliata* is now *Sycon ciliatum*.
10 Haeckel refers to his son, Walter Haeckel. In an earlier letter, Haeckel had described his son's 'atavistic movements' (see *Correspondence* vol. 16, letter from Ernst Haeckel, 9 November 1868).
11 Thomas Henry Huxley. See letter to T. H. Huxley, 9 July [1869].

From John Lubbock 2 July [1869][1]

High Elms
2 July

Dear Mr. Darwin

In the "Origin" 4 Ed. P 532 you say that Agassiz regards it as a law of nature that the embryos indicate the structure of their Ancestor, so that the embryos of existing species resemble the mature forms of extinct ones.[2]

Can you give me the reference for this?

Hoping you are all well, & enjoying yourselves in Wales.[3]
Yours very sincerely | John Lubbock

C Darwin Esq

DAR 170: 73

CD ANNOTATIONS
1.1 In … nature] *double scored ink*
1.1 that the embryos 1.2] 'of existing species' *added ink*
1.2 their Ancestor … ones. 1.3] 'early *[precursors]* in the same class, their progenitors,— I sh[d] cover this.—' *interl ink*
Top of letter: 'See to this & write to Victor Carus'[4] *ink*

1 The year is established by the relationship between this letter and the letter to John Lubbock, 3 August 1869.
2 Lubbock refers to Louis Agassiz.
3 CD was on holiday at Caerdeon, Barmouth, Wales (see 'Journal' (Appendix II)).
4 CD refers to Julius Victor Carus.

From John Scott 2 July 1869

Roy Bot Gardens | Calcutta
2[d] July 1869

Dear Sir,

It is long since I wrote you[1]—so utterly had I failed in collating the information you required—and now when I do write I have only a few imperfect scraps. I have written many parties in different parts of India to make and communicate to me observations on expression but all have been fruitless or at most afforded me only a few general remarks.[2] I can therefore only afford you a few additional ones of my own.

1— The following on your query *3.* were made by me in Sikkim last September.[3] I purposely raised a quarrel between two Mechis[4] by making a present to one of the whole of a few pence which I had promised to divide between them, and telling the other that he might get his share as he best could. I having previously cautioned the recipient against giving it up. This had the desired effect; both ultimately got into a violent passion— As they warmed on the argument, I observed that their bodies became less erect, (withdrawn below) the head pushed forward strong grimaces, shoulders raised, arms rigid, bent at elbows, hands turned into body, their backs facing opponents, and somewhat spasmodically closed (not clenched—the third joint unbent the middle and dictal phalanges being simply pressed on palm while the thumb is bent upon the proximal phalange of the first finger and not placed slanting across and in opposition to the three fingers. They continued thus for several minutes approaching and withdrawing from each other; and frequently lifting their arms as if they would strike. This they never did. This they never did, and I always observed that in the action of lifting their hands were always *opened*, thumbs and fingers slightly bent and spread in a sort of grasping position. I did not once see them strike at each other with closed fists.[5]

The Lepcha (Sikkim)[6] manifests indignance and violent passion very similarly; though perhaps less demonstratively, they being naturally I believe a milder race. In their cases I observed that the arms were scarcely bent at elbow, but placed nearly parallel to body, rigid the hands pushed somewhat backward and but loosely closed much in the Mechi fashion, with the thumbs placed parallel to the metacarpas and bent upon the proximal phalange of the first finger.[7] I have also repeated observations with such slight difference of results from those I have previously recorded as might be expected from differences of temperament. In these observations, however, I must tell you, that I have always made it my endeavour to confine to those who have had little to do with Europeans.

Were I to give you results, from the Bengali Manjees (boatmen) who have had a training from our sailors you would have but a repetition of what you could see at home— Natives are apt pupils in all sorts of vices.

(2). With regard to your query (17) I have abundant evidence that the lateral shake of the head is somewhat indifferently given in negation and affirmation. The vertical nod is much less commonly used by natives—though exclusively in affirmation. From repeated observations on this however I am convinced that in affirmation a single shake only is given, the head being first indifferently bent to the right or left and then given an oblique jerk: while in negation the head is shaken several times and I think usually in a **nearly** *vertical* position. I have paid particular attention to this since you wrote me and I now think that negation and affirmation is thus distinguished by them.[8]

(3) I have only lately been afforded illustrations of your query (5). The first case was of a poor Hindustani who after a long illness and want of means had to sell a favorite goat. He was deeply affected in receiving the money, and looked alternately at it and the goat. On the goat being taken from his hand and bound, it tried hardly to get back to him, he then went up to it, when it at once leaped upon him and licked his hands. Thus he stood silently looking from the money to the goat—then as if with an effort he put the money in his cloth. Thus far his eye was unfixed and wavering, though his mouth was partially closed with the corners very decidedly depressed Now however that he seemed to have determined to part from the goat—I observed as he looked down upon it an obliqueness of the eyes, the lower eyelids wrinkled below at the inner angle, the eyebrows somewhat obliquely set with a marked puckering or swelling at their inner angles and over the nose, but though I looked very carefully I could not observe the transverse wrinkling on the middle of forehead. The poor man stood thus for a minute or so, then heaving a deep sigh, burst into tears, held up his hands blessed the goat, then turned, and went away without once looking round again.[9]

The other observations under this head was on a young Dhangar woman whose first and only child was fast dying, while she sat nursing it below a trellis covered with Pumpkins. I having observed her sitting thus pensively with the child in her lap went quietly up to her and got quite unobserved behind the Pumpkin screen. She seemed really affected, and I very distinctly observed the *transverse wrinklings* on the middle of the forehead, the eyebrows slightly raised at the inner end, the upper eyelid somewhat drooping and the lower wrinkled at the inner angle. Mouth slightly open, lips protruding a little and arched so that the corners were much depressed. After having watched her for sometime I went from behind the screen of Pumpkins, and spoke to her she started up burst into tears and cried most bitterly for me to save her child. This is by far the best illustration I have been able to make under this head—she was about 16 years of age and the wife of one of our garden coolies a Dhang⟨ar⟩[10]

⟨*2 pages excised*⟩

observations. I have tried in vain to get the breed with black comb and wattles. I see in your reference to the Sooty fowls of India, that the hens alone are said to exhibit the peculiarities.[14] I have lately got a dull greyish white Cock, (about 2 months old) comb and wattles, brownish red, ear-lappets of a bluish tinge, legs stoutish and as is also the beak of an olivy-brown colour, skin and bones black. The natives assure me that this bird will retain its present colour of plumage. It agrees however in many respects with the silk-fowl as described by you and may possibly be a cross between that and the Sooty race. I am indeed sorry that I cannot get a thoroughly pure breed. I do not know a single fancier of fowls here, & it is a mere chance to get such from the native dealers. They are most unscrupulous and deceitful race. If I succeed in getting the above breed however I shall have very great pleasure in making the experiment and reporting result.

(6) I have particularly attended to your query about differences in the tint of beard and hair of the head: but can detect no such characteristics in any of the races which have come under my observation in India.[11] We of course see occasional differences in individual cases (just as at home) in which the beard is lighter, (I really do not remember having seen it darker among natives) than the hair of the head, but I cannot hear of its being common to any race. Some of the Northern races here have little or no beard while some of those of Eastern Bengal—as do also many of the Malays—practise depilation.

The only instance in which I really thought I had found real differences was with some of the Arabs which frequent our bazaars &c. Those with light brown or greyish hair have a carroty-red beard I was soon undeceived however. I find that it is a custom to *dye* the *whiskers* with the bruised leaves of the Henna—Lawsovia alba.[12]

I shall be glad if any of the above scraps are of service to you. It has been a pain to me to have failed so long & so signally in collating the information you required. I still hold my old position in the garden here and for which I have you to thank.[13] With best wishes for your continued health.

I am Sir very respectfully yours John Scott

Incomplete
DAR 85: A31, DAR 177: 119

CD ANNOTATIONS
1.1 It . . . Europeans. 3.10] *crossed pencil*
2.2 Mechis] *underl blue crayon*
2.7 less erect] *underl blue crayon*
2.7 the head pushed forward] *underl blue crayon*
2.8 their backs] 'of Hands' *interl*
3.1 Lepcha] *underl blue crayon*
3.8 I must . . . Europeans. 3.10] *double scored blue crayon*
4.1 Were . . . home— 4.3] *crossed blue crayon*
5.1 (2).] *red crayon opening square bracket*
5.1 (2). . . . a Dhang⟨ar⟩ 7.13] *crossed pencil*
6.2 Hindustani] *underl red crayon*
6.4 it tried . . . a Dhang⟨ar⟩ 7.13] 'John Scott' *pencil*
6.8 corners very decidedly depressed 6.9] *underl red crayon*
6.11 wrinkled below] *underl red crayon*
6.11 the eyebrows . . . nose, 6.12] *scored red crayon*
6.15 burst into tears] *underl red crayon*
7.5 the *transverse* . . . end, 7.6] *scored red crayon*
7.8 lips . . . depressed.] *double scored red crayon*
7.13 Dhang⟨ar⟩] 'from Nagpore—.' *ink*
8.1 observations. . . . result. 8.12] *crossed pencil*
9.1 (6) . . . alba. 10.4] 'Man sexual selection —July 2^{d} 1869' *pencil*
9.3 occasional . . . lighter, 9.4] *scored red crayon*
9.4 (I . . . natives) 9.5] *underl red crayon*
9.6 Some . . . health. 11.4] *crossed pencil*
Top of letter: 'John Scott | This is first letter' *ink*

CD note:

'Abstract | *p. 1–3 Passion & indignation two Races from Sikkim; do not clench fists, do not hold themselves erect, but do not strike each other— grimaces.— good & /recent/ account [*crossed pencil*] | p (4) on the shaking of Head in affirmation & negation— not very /firmly/ fixed, /likely/ | I suspect the lateral is primordial— & the vertical a principle of opposite /movement/ *p 5 & 6 on cor⟨ner⟩ of mouth depressed & oblique eyelids— ⟨ ⟩ wrinkled forehead [*crossed pencil*]'

1 Scott's last extant letter to CD is that of 4 May 1868 (*Correspondence* vol. 16).
2 CD evidently sent Scott a version of his queries on expression (see Appendix VI) in 1867 with a letter that has not been found (see *Correspondence* vol. 15, letter from John Scott, 24 September 1867 and n. 16). Scott sent some replies and CD asked further questions in 1868 (see *Correspondence* vol. 16, letter from John Scott, 4 May 1868, and letter to John Scott, 3 June 1868).
3 Question 3 on CD's list of Queries about expression concerned the facial expression and bodily postures of indignation or defiance.
4 The Mechi, or Mech people, lived in Sikkim, Bengal, Assam, and Nepal during the nineteenth century (Risley 1891, 2: 86–91; Bhatt and Bhargava eds. 2005, p. 29).
5 CD included some of Scott's observations on the quarrel in *Expression*, p. 249.
6 The Lepcha people are thought to be the original inhabitants of Sikkim (Risley 1891, 2: 6–14, *EB*).
7 CD cited Scott for this observation in *Expression*, p. 249.
8 CD had asked for further observations of lateral shakes of the head in his letter to Scott of 3 June 1868 (*Correspondence* vol. 16). CD recorded Scott's conclusion in *Expression*, p. 276.
9 Question 5 on CD's list of Queries about expression concerned the facial expression of grief. Scott's account was recorded in *Expression*, p. 188.
10 CD gave Scott's account of the Dhangar woman in *Expression*, p. 187.
11 CD's query to Scott regarding beard and hair colour has not been found, but see the letter to Asa Gray, 1 June [1869] and n. 14.
12 *Lawsonia alba* is a synonym of *L. inermis* (henna).

13 For more on CD's role in helping Scott to emigrate to India and become curator at the Royal Botanic Gardens, Calcutta, see *Correspondence* vols. 12 and 13.

14 In *Variation* 1: 230, 256, CD mentioned the sooty fowl of India, in which only the female had black skin and bones. In his letter to Scott of 3 June 1868 (*Correspondence* vol. 16), CD had asked about the plumage of the male sooty fowl.

To W. B. Tegetmeier 2 July [1869][1]

Caerdeon Barmouth | N. Wales
July 2[d]

My dear Sir

I received this morning your note in which you tell me that you have kindly procured for me the eggs of Pile Games, & I have written to Down to my servants to send to the Station at once & do the best with the eggs.—[2]

My health became so bad that I c[d]. do no work at home, so we have all come here where we shall remain all this month.[3]

You say that you have got a good many notes for me, & I thank you *most sincerely* for your kindness: will you be so good as to send them *here*?

I congratulate you about the Crested Turkey; it will be a wonderful feat if you recover the breed.[4] I suppose both sexes are crested.

I am far from well today & will write no more, except once again to thank you for your kindness.—

Yours very sincerely | Ch. Darwin

Archives of the New York Botanical Garden (Charles Finney Cox collection) (ser. 14)

1 The year is established by the relationship between this letter and the letter from W. B. Tegetmeier, 30 June 1869.

2 See letter from W. B. Tegetmeier, 30 June 1869 and n. 1.

3 CD was holidaying at Caerdeon, Barmouth, Wales (see 'Journal' (Appendix II)).

4 Tegetmeier's information on the crested turkey was evidently in a now missing part of his letter of 30 June 1869. See letter from W. B. Tegetmeier, 20 July 1869. A crested turkey had been exhibited at the Zoological Gardens, London (see *Correspondence* vol. 16, letter from Edward Blyth, 3 August 1868).

From James Oliver to George Cupples 3 July 1869

Howpasley
3[d] July 1869

Sir

I have been from home for the last three weeks on account of my health, and only came home this afternoon, when I found yours of the 29 ult. awaiting me.

I have now had considerable experience in the breeding and rearing of Cheviot Sheep but in a hill country only, where it is impossible to tell the number of either of the sexes when lambed, or born as few of them require any assistance and we never have them in hand until the time of castration, however I do not think there can be much difference in the number of either of the sexes when lambed, but if the lambing season is cold and wet the male sex are more liable to die from having a greater tendency to a stoppage of the urine—[1] My numbers this season at the time of castration were male 826 female 864

I remain Sir | Your obed^{t}. servant | Ja^{s}. Oliver

George Cupples Esqre.

DAR 86: 67

[1] Oliver refers to what is now known as obstructive urolithiasis, a urethral blockage that is more common in male lambs, especially those that have been castrated.

From William Erasmus Darwin 5 July [1869][1]

Southampton
July 5

My dear Father,

I went to Alum Bay Hotel yesterday, and walked to Colwell Bay where after some little search I found a considerable number of Epipactis P. tho' with not v. many flowers out.[2] I send you in little box the result of my watching; the flies seemed to visit the flowers very seldom, it was a very bright hot afternoon tho' with perhaps a little too much breeze, but after watching about $2\frac{1}{2}$ hours I only saw 4 or 3 flies of the kind I enclose visit.

after watching a group of 3 plants for some time I saw two or possibly three flies (one of which I send without the pollen masses) crawl into some flowers that were considerably open, that is so much open that the labellum was about in the position of the labellum in Fig B Page 94 of orchis Book.[3] I observed that the flies in *every** instance after crawling down to collect honey apparently from the stigma and beneath it instead of backing out, turned round and came out, and I saw that they could turn round just underneath the pollen masses or anther without touching them. These seemed to be essentially crawling flies, and I caught one which I send just to show which flies I then fancied could *not* be the agents.

These flies have not weight enough to depress the lip at all as I could plainly see; tho' from trial I found the lips were depressed very easily, but the spring back seemed to me extremely slight, in one or two instances of flowers that certainly had never been visited, I found on pushing down gently the lip with a bit of grass that I had to give it a lift to get it into position again.

In looking about for flowers with the labellum in the position of Fig A p. 94 I came upon one just in that position, namely the edges of the labellum near the hinge were just inside the body of the flower. the two frilled edges of the distal portion of the labellum were in contact; and on holding the flower opposite the eye and looking over the centre of the groove formed by the Junction of the two edges of the labellum one could just fairly see the yellow anther case but no lower The moment I came

on this flower I found a fly apparently of the same sort crawling out of the flower head formost without depressing the lip (i.e. not backing) with the pollen masses attached to his back (this is the fly I send you). If I had only had the luck to be a minute sooner I should have seen him enter.

I then went to some other flowers, and saw a small beetle or two crawl in with no effect also an ant, and a long bodied thin fly alighted but went off again— I saw no bee come near them within a yard or so. I saw another of same sort of fly on a flower (of the Epipactis) with what I am nearly sure sure was the remains of the pollen masses on his back in just the same position, but I was just too late with my bottle to catch him. on one flower I saw the pollen masses sticking to an upper petal, in another case to a lower sepal, in another to the top of an unopened bud.

I also found a longish hair or a leg sticking across a flower which drew out the pollen masses with it; I brought it away as carefully as possible, but it has vanished.

on going back to some flowers I had previously looked at, *and in which there were no traces of pollen masses sticking to them or protruding*; I found an ant in each of two flowers, one of which ants I send. in one flower a single pollen mass was hanging down out of position, as if it has just been drawn out, and in the other the ant was sucking hard away and the pollen masses were behind him in the channel of the lip, and looked quite fresh.

I am almost certain that the ants did not back out but turned round. The flowers in which I found them were about in a state intermediate between A and B. I could easily go next Sunday and have another look, or get you some flowers, I might also have another look at the spring in hinge of labellum which then seemed so slight, but in some flowers I have brought away with me it seems a little greater.

You will find in box another much slighter fly which I caught in a flower not that he is worth sending.

I shall be down about the 13th or 14th.[4]

Your affect son

W. E Darwin

I took notes on the spot

I send Sat. Rev.[5] please keep it as there is an Article I want to read when down with you as I have not read it

* I don't remember seeing a fly back out

DAR 162: 100

CD ANNOTATIONS
1.1 I went … 94 4.1] *crossed pencil*
4.8 could just … read it 14.2] *crossed pencil*
Top of letter: 'WED | Epipactis palustris' *pencil*

1 The year is established by the relationship between this letter and the letter to W. E. Darwin, 7 [July 1869].

2 Alum Bay and Colwell Bay are on the western side of the Isle of Wight. William refers to *Epipactis palustris* (marsh helleborine), a native British orchid commonly pollinated by bees, wasps, and syrphid flies.

3 William refers to *Orchids*.
4 CD was holidaying at Caerdeon, Barmouth, Wales (see 'Journal' (Appendix II)).
5 *Saturday Review*.

From Louis Agassiz 6 July 1869

Museum of Comparative Zoology, | at Harvard College, | Cambridge, Mass.,
7th. 6th. 1869

My dear Sir,

It gives me great pleasure to introduce to you my only son Alexander Ag⟨as⟩siz,[1] the more so as I beleive you will find him more tractable than myself upon questions which you are especially interested about.[2]

Ever truly yours | Ls Agassiz

Ch. Darwin Esqr.

DAR 159: 10

CD ANNOTATIONS
End of letter: '115 Jermyn St | S.W.'[3] *pencil*

1 Alexander Agassiz travelled to Britain in September 1869 (G. R. Agassiz ed. 1913, p. 97).
2 For Alexander Agassiz's views on Darwinism, see Winsor 1991, pp. 148–63.
3 Asa and Jane Loring Gray stayed at 115 Jermyn Street (a private hotel) when they arrived in London (see letter from Asa Gray, 23 August 1869; *Post Office London directory* 1869); Alexander Agassiz may have stayed there too.

From John Graham to George Cupples 6 July 1869

Shaw. by Lockerbie
6th July 1869

Sir

With your letter of the 29 ultimo, I am duly favoured, and in answer am indeed sorry I can give you little information upon which you can rely regarding the subject of enquiry. viz.

1st The comparative numbers of the sexes born in any year or years from any flock or herd?—

2nd Whether there is any excess of mortality in either sex over the other at birth or afterwards?—

1st From observation could not say, on an average of years, that equality of numbers of either sex vary much in sheep or Cattle.—

2nd Mortality at birth, and more afterwards, will I think be in excess in males, to a certain degree,—lambs when young, during cold weather are subject to affections in the urinary organs often proving fatal, to which the female are not liable—[1] altogether males both sheep and Cattle are more tender and require better keep and attention until they reach eighteen months.—

Born here		
1869	Bull calves 9	Heifer do— 12
1868	— — 9	— — 7

No record of lambs kept.—
Pray excuse my delay in answering your letter.—
I am yours truly | John Graham

George Cupples Esq | Guard Bridge

DAR 86: 68–9

CD ANNOTATIONS
0.1 Shaw … fatal, 5.3] *crossed pencil*
5.3 altogether … more 5.4] *double scored blue crayon, pencil*
Top of letter: 'Cattle' *blue crayon*

[1] See letter from James Oliver to George Cupples, 3 July 1869, n. 1.

To the *Athenæum* 7 [July] 1869[1]

Caerdeon, Barmouth,
June 7, 1869.

I have received a letter from Germany on the increase of the elephant, in which a learned Professor arrives at a totally different result from that of Mr. Garbett, both of which differ from that of your Correspondent "Ponderer."[2] Hence you may perhaps think it worth while to publish a rule by which my son, Mr. George Darwin,[3] finds that the product for any number of generations may easily be calculated:—

"The supposition is that each pair of elephants begins to breed when aged 30, breeds at 60, and again, for the last time, at 90, and dies when aged 100, bringing forth a pair at each birth. We start, then, in the year 0 with a pair of elephants, aged 30. They produce a pair in the year 0, a pair in the year 30, a pair in the year 60, and die in the year 70. In the year 60, then, there will be the following pairs alive, viz.: one aged 90, one aged 60, two aged 30, four aged 0. The last three sets are the only ones which will breed in the year 90. At each breeding a pair produces a pair, so that the number of pairs produced in the year 90 will be the sum of the three numbers 1, 2, 4, *i.e.* 7. Henceforward, at each period, there will be sets of pairs, aged 30, 60, 90 respectively, which breed. These sets will consist of the pairs born at the three preceding periods respectively. Thus the number of pairs born at any period will be the sum of the three preceding numbers in the series, which gives the number of births at each period; and because the first three terms of this series are 1, 2, 4, therefore the series is 1, 2, 4, 7, 13, 24, 44, &c. These are the numbers given by 'Ponderer.' At any period, the whole number of pairs of elephants consists of the young elephants together with the three sets of parents; but since the sum of the three sets of parents is equal in number to the number of young ones, therefore the whole number of pairs is twice the number of young ones, and therefore the whole number of elephants at this period (and for ten years onwards) is four times the corresponding number in the series. In order to obtain the general term of the series, it is necessary to solve an easy equation by the Calculus of Finite Differences."[4]

Charles Darwin.

Athenæum, 17 July 1869, p. 82

[1] CD evidently wrote 'June' by mistake. He arrived in Wales on 12 June 1869 and stayed until 30 July (Emma Darwin's diary (DAR 242)). See also letter to the *Athenæum*, 19 June 1869.

[2] The letter from Germany has not been found; CD refers to Edward Lacy Garbett. See letter from Ponderer to the *Athenæum*, [before 5 June 1869] and n. 2, and letter from E. L. Garbett to the *Athenæum*, 29 June 1869.

[3] George Howard Darwin.

[4] The calculus of finite differences is concerned with changes in a dependent variable due to discrete changes in the independent variable. Its equations are finite analogues of differential equations.

To W. E. Darwin 7 [July 1869][1]

Caerdeon Barmouth | N. Wales
June 7th.

My dear W.

Very many thanks for your excellent observations. I will send the Fly to B. Museum & get it named—[2]

If you are not wearied out, I certainly shd be very glad if you cd. pay another visit. I still think some larger insect must visit the plant; probably wasps.—

I feel sure about the elasticity of Hinge in the specimens which I examined.—[3] May not the dry Weather have acted injuriously on the plant?—

When you come here, I must ask you about one point— viz whether the fly with pollen-masses attached to it, crawled out *over* the distal & flexible part of the labellum but without depressing this part, or crawled out (as I supposed) nearer the centre of the flower, ie between the 2 upper petals.

I am ashamed that you shd. have had to watch so long.—

Very many thanks, my dear old fellow, | In Haste for Post | yours affect. | C. Darwin.

DAR 210.6: 130

[1] The month and year are established by the address. CD spent from 12 June to 30 July 1869 at Caerdeon; he evidently wrote 'June' by mistake.

[2] See letter from W. E. Darwin, 5 July [1869]. In *Orchids* 2d ed., pp. 99–100, CD described William's observations and mentioned two species of fly that he had observed, *Sarcophaga carnosa* (now *S. carnaria*, a flesh fly) and *Coelopa frigida* (the seaweed fly).

[3] In his letter of 5 July [1869], William had noted that the elasticity in the hinge of the labellum of *Epipactis palustris* seemed very slight.

To William Ogle 7 July [1869][1]

Caerdeon, Barmouth | N. Wales
July 7th

My dear Sir

You must let me write half-a-dozen words to say how much I admire your paper on Salvia.[2] It seems to me excellently worked out, & contains much that is new. I shd. have committed a heavy sin, if I had been the means of stopping its publication.—[3]

I suspect that that the Salvia with shorter pistil will prove merely an arrested form, for, as far as my experience goes, it is not likely that two such wonderful means, as complex mechanical contrivances & Dimorphism, shd have been co-used for the same end.—[4]

Judging from what I know occurs with some plants, for instance Verbascum, I apprehend that the Malvaceæ, which do not secrete nectar, are visited by pollen-collecting Bees.—[5]

I congratulate you on publishing so striking a paper & remain, My dear Sir | Yours very faithfully | Ch. Darwin

I am not certain of your private address, & have not *here* any address-book.—[6]

DAR 261.5: 3 (EH 88205901)

[1] The year is established by the reference to Ogle's paper on *Salvia* (see n. 2, below).

[2] Ogle's paper 'The fertilisation of *Salvia* and some other flowers' appeared in the 1 July 1869 issue of *Popular Science Review* (Ogle 1869). CD's annotated copy of the paper is in the Darwin Pamphlet Collection–CUL.

[3] CD had lent Ogle his copy of a paper on *Salvia* by Friedrich Hildebrand (Hildebrand 1866). When Ogle returned the paper he mentioned that he had been unaware that the main facts on the mechanism of fertilisation in the species had already been described, but concluded that he might still add some of his own observations (see *Correspondence* vol. 16, letter from William Ogle, 2 September 1868).

[4] Ogle had observed two forms of *Salvia patens*, one with a shorter style (see Ogle 1869, p. 269). He noted the the longer styled blossoms were far more numerous and speculated that the second form was an accidental deformity.

[5] Ogle had suggested that self-fertilisation was easily accomplished in mallows (Malvaceae) that did not have nectaries, and suggested a causal connection between the two facts (Ogle 1869, pp. 270–1). In 'Specific difference in *Primula*', p. 453, CD had observed that although the flowers of *Verbascum* did not have nectaries, they were frequently visited by pollen-eating bees

[6] CD addressed the letter to Ogle at 'St. George's Hospital, Hyde Park Corner, London'.

From William Ogle [after 7 July 1869][1]

34 Clarges St. | Piccadilly

My dear Sir,

I was very much gratified by receiving your kind letter, and reading the complimentary terms in which you speak of my article on Salvia.[2] I am so utterly ignorant of most botanical literature, and have such a limited smattering of botany that I could not but fear that any observations I might make would turn out either erroneous, or, if correct, mere feeble repetitions of what others had described already & better.

One however of the benefits which your works have conferred is that those who are little skilled in systematic botany can yet examine flowers with the certainty of rational enjoyment, and a fair chance of hitting upon some unnoticed adaptation of structure—

I am on the point of starting myself to spend my summer holiday at Barmouth & Dolgelley,[3] and trust you will pardon me if I give myself the pleasure of calling upon you.

Believe me | Y^rs. sincerely | W. Ogle.

DAR 173: 2

1 The date is established by the relationship between this letter and the letter to William Ogle, 7 July [1869].
2 See letter to William Ogle, 7 July [1869] and n. 2.
3 CD was holidaying at Barmouth, Wales (see 'Journal' (Appendix II)). The small market and resort town of Dolgellau is within ten miles of Barmouth.

To J. D. Hooker 8 July [1869][1]

Caerdeon Barmouth N. Wales
July 8th

My dear Hooker.

I have been very idle not to have thanked you long ago for your very pleasant letter of June 24th. which I could see you wrote to cheer me up, though at the time overburdened with correspondence. I liked your letter very much indeed, but do not write again, as it made me feel guilty.— Very many thanks about the Beards of the Russians.—[2] I cannot gain any strength & have given up the attempt to walk & have now got a pony lent me by Miss Lloyd (who is staying close by with Miss Cobbe) & this will perhaps do me some good & anyhow pass the time till I get home.—[3]

My object in writing to you now is to tell you that I have had a letter from Dr. Habel of N. York, who has staid 5 months at the Galapagos: I have told him,, if he collected plants, to communicate with you, & I hope I did rightly.[4]

From reading several essays by Nägeli & others I have come to think that it is all important with respect to the principles of Variability to learn all that we can about the polymorphic or protean genera, such as Rubus, Hieracium &c &c.[5] Keep this, if you can, a little in your mind.

The New Zealand genera are interesting under this point of view. I want especially to hear of as many cases, as occur, of single species, or sections of genera, being tolerably fixed in form, whilst the other species are eminently variable.— From the time before writing the Origin, these genera have always seemed to me very perplexing.

We shall be at Down at very end of month.[6]

I am truly glad that you receive such comfortable accounts from N. Zealand about Willie.[7]

My dear old Friend | Yours affect | C. Darwin

PS. My note was written before receiving yesterday the paper on Snakes, which I have been very glad to read.[8] I am strongly inclined to be a believer, not in fascination but in extreme fear which leads the wretched creature to fall into the snakes power, in same manner as many people feel inclined to throw themselves down a precipice. Similar accounts have been published about the Rattle-snake in U. States, which I have believed, though ridiculed. I go further & am much inclined to suspect, that the noises of the Rattle-snake & of the Puff adder, which are by no means dissimilar, serve to paralyse their prey with fear.—

I think the paper ought to be published,—if not in Linnean Journal, in Popular Science Jrnl or in the Annals or somewhere.[9]

Yours affect | C.D

DAR 94: 137–9

[1] The year is established by the relationship between this letter and the letter from J. D. Hooker, 24 June 1869.
[2] See letter from J. D. Hooker, 24 June 1869.
[3] CD refers to Mary Charlotte Lloyd and Frances Power Cobbe. Lloyd lent her pony, 'Whitey', to CD (Lucas 2007, p. 322).
[4] The letter to CD has not been found, but Simon Habel did write to Hooker and the herbarium at Kew holds five specimens of plants collected by Habel in the Galapagos (letter from Simon Habel to J. D. Hooker, 30 June [1869], Royal Botanic Gardens, Kew, DC 204, f. 609; for the list of specimens, see the Kew Herbarium Catalogue, http://apps.kew.org/herbcat/).
[5] In *Origin*, p. 46, CD had discussed 'protean' or 'polymorphic' genera, in which the species presented 'an inordinate amount of variation', and said, 'hardly two naturalists can agree which forms to rank as species and which as varieties.' Carl Wilhelm Nägeli had argued that natural selection acted only on physiological, not on morphological characteristics of plants, which, Nägeli argued, were controlled by a principle of 'perfectibility' ('*Vervollkommnung*'; see Nägeli 1865, p. 29, and letter to J. D. Hooker, 13 January 1869 and n. 1). CD evidently hoped that the study of polymorphic genera would provide clear examples of the action of natural selection on morphological features.
[6] CD returned to Down on 31 July 1869 ('Journal' (Appendix II)).
[7] William Henslow Hooker. See letter from J. D. Hooker, 24 June 1869 and n. 5.
[8] The paper on snakes was written by Mary Elizabeth Barber (see letter from J. D. Hooker, 17 July 1869 and n. 9).
[9] CD refers to *Journal of the Linnean Society of London*, *Popular Science Review*, and *Annals and Magazine of Natural History*.

From George Robert Gray 9 July 1869

British Museum
9th. July 1869

Dear Mr Darwin

I should feel greatly obliged if you would kindly favour me with a testimonial of my abilities &c. as a Naturalist it might prove highly useful to me at the present moment[1]

I remain in haste | faithfully Yours | George Robert Gray

C. Darwin; Esqre

DAR 96: 69v

[1] Gray became senior assistant keeper at the British Museum at the end of July 1869 (*ODNB*).

To T. H. Huxley 9 July [1869][1]

Caerdeon | Barmouth. N. Wales
July 9th.

My dear Huxley

I wd. not trouble you if I could think of anyone else to apply to; but ask Mrs. Huxley[2] to answer me.—

Häckel is hard at work at *Calcareous* Sponges & is actually starting soon to Norway to study them.[3] He says he badly wants British specimens. Now can you suggest anyone or two men to whom I could apply with fair chance of success?— I know so few naturalists now-a days.— Häkel forgets to say whether he requires the specimens in spirits or dry. Can you answer this? I feel bound to do anything Häckel asks me.—

We shall remain here all this month; but I have been much disappointed at gaining no strength & failing to climb even a hill, & I had longed once again to set foot on summit of a mountain.—

What a splendid success, as I hear on all sides, your Comte article has been.—[4] Give my kindest remembrances to Mrs. Huxley & all your dear little ones.— Ever Yours | C. Darwin

Imperial College of Science, Technology, and Medicine Archives (Huxley 5: 271)

1 The year is established by the relationship between this letter and the letter from Ernst Haeckel, 2 July 1869.

2 Henrietta Anne Huxley.

3 See letter from Ernst Haeckel, 2 July 1869.

4 CD refers to 'The scientific aspects of positivism' (T. H. Huxley 1869b), which appeared in the 1 June 1869 issue of the *Fortnightly Review.* Huxley had also criticised the positivist philosophy of Auguste Comte in an earlier article (T. H. Huxley 1869a). See letters to T. H. Huxley, 10 March 1869 and 12 March [1869], and letter from T. H. Huxley, 11 March 1869.

From W. C. Tait 11 July 1869

Oporto
July 11th. 1869.

Charles Darwin Esqr.
Dear Sir,

I went to Coimbra a few days ago and met at the botanical gardens there a german gardener at the head of the establishment a Mr. Edmund Goetze.[1]

He most obligingly showed us over the gardens and conservatories and in course of conversation I asked him whether he knew the Drosophyllum Lusitanicum as I had seen mentioned in a book that it grew near Coimbra— He said that he had sent some plants to Dr. Hooker at Kew Botanical Gardens where Mr. Goetze had been for some time and that Dr. Hooker wanted them for you to experimentalize upon but that all the plants he sent died, further that Dr. Hooker had written that you had already received some plants from Portugal which plants were looking very sickly—[2]

Mr. Goetze had sent some seed also to Dr. Hooker and these had germinated. Mr. Goetze told me that he had always experienced great difficulty in growing the Drosophyllum Lusitanicum which in a short time after transplantation invariably died he thinks the best mode of cultivating it is by seed, What however may interest you more is that he suspects this species has spontaneous movements of the leaves moving with the sun—[3]

I did not notice this but now the last of my plants is dead after flowering and producing 3 seeds— A colony of ants had established themselves at the roots and seemed to attend to some aphides which the also appeared to have placed on the stalk and some of the leaves. By the aid of a hand lens I have noticed a great resemblance between the glands of the Drosophyllum and those of many other plants such as the rose and on the calix(?) of the moss rose I have actually seen a gnat which had been caught by the secreted fluid— No answer have I received from the portuguese paper about the rate of development of sheep's horns and I have not yet been able to meet the only person who I think could give me reliable information on this subject.[4] I keep it before me and sooner or later shall send you an answer—

In the Geranium I have noticed a contrivance for crossing analogous to that of the orchis No doubt you have observed it already but if not the enclosed rough sketch will explain what I mean.[5]

In the earlier stage of the flower the stigmas are over the pistil and this latter is still up, but shortly after the stigmas bend down and the ends containing the pollen fall, the pistil now being over the stigmas opens out like an umbrella and is ready to receive the pollen brought by bees from other flowers— I have taken great interest in this case—

I am trying to make as complete as possible a list of all the endemic species in Portugal and also of those which are found only in Spain and Portugal and I expect to find that the endemic ones have little adaptability for long journeys and for crossing with idividuals at a distance.

There is here a variety of the fox, as I see by a Portuguese Pamphlet.[6] I have never compared the skin with the English species and I should like to exchange a skin. A friend of mine here would give me a young fox cub—

It must be a great satisfaction to you to see your theory now becoming more and more generally accepted I think that what makes people most prejudiced against it is that it must apply to man in common with other animals— Their pride and preconcieved opinions receive a smart shock and it is not always that truth is sought for its own sake and in an unprejudiced or unbiassed manner, Max Müller seems to have given natural selection a kind of application to language[7] but I do not see why it should not be applied also to customs, instincts, opinions, politics, social institutions, competition in commerce and professions, and why in fact it should not be an universal law.

I see that you are about to publish a work on instinct this is to me a most interesting subject of observation and on which I might perhaps gather together a few curious instances.

I have an idea that reason is a kind of consciousness–super-consciousness and receives its value from a knowledge or consciousness of the laws of nature.

An uncle of mine in Lisbon many years ago found on the bank of the habour the carcase of an animal in a species of grey mud it was quite soft but in a beautiful state of preservation This was a kind of petrifaction and he feels almost sure that it was an extinct species; Unfortunately he kept only a small portion which is now a hard petrifaction— My uncle remembers that it was a mammal.

Believe me to remain | Dear Sir, | Yours very truly, | William C. Tait.

DAR 178: 48

1 Edmund Goeze was the director of the Coimbra botanic garden (*GEPB*).
2 See letter from J. D. Hooker to Emma Darwin, 29 March 1869 and n. 7.
3 In *Movement in plants*, p. 450, CD discussed the unusual lack of heliotropism in *Drosera rotundifolia*, which he believed belonged to the same family as *Drosophyllum*, but did not mention the phenomenon in *Drosophyllum lusitanicum*.
4 See letters from W. C. Tait, 13 April 1869 and 10 May 1869.
5 The enclosure has not been found.
6 The fox variety referred to has not been identified. The red fox (*Vulpes vulpes*) is common throughout Europe and the British Isles.
7 Tait refers to Friedrich Max Müller. For Max Müller's theory of the origin of language, see *Correspondence* vol. 10, letter from Asa Gray, 4 and 13 October 1862 and n. 2. See also Max Müller 1861 and 1864.

From W. E. Darwin 12 July 1869

Bank, Southampton,
July 12 *1869*

My dear Father,

I send some Epipactis by post.[1]

I have not time to write particulars.

I have seen bees visit about 14 to 20 flowers, and have caught one with pollen about root of trunk. I **did not** see one actually withdraw pollen.

When a bee alights on a fully open flower it rests in a more sloping position from the lip sinking more than in case of a flower just at perfection

the bee remains quite outside the flower & pushes his head in till the top of the Anther just rests between the antennæ, and keeps boring its proboscis down with a jogging motion. A shows the angle of trunk & body of bee sucking an open flower B a perfect flower.

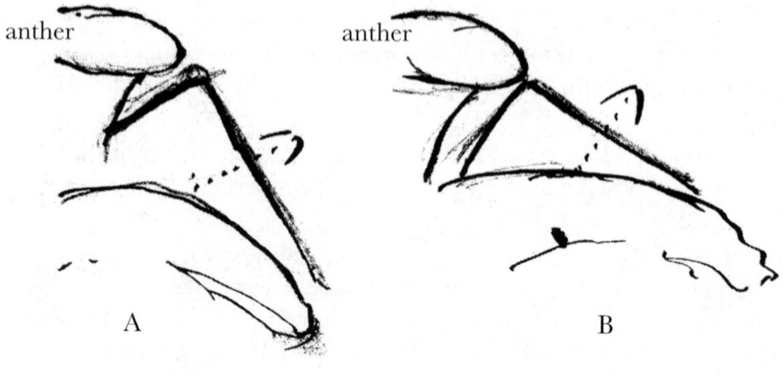

I am tolerably sure (but it is most difficult to be quite so) that in case of bee sucking as at B that it flew off in rather more vertical a direction than in A. see dotted line & ↑ in each case whether open or not the lip flew up and appeared to hit the bee on the belly, and in one or two cases of perfect flowers the bees appeared to fling themselves off with a quick jerk kicking out with their legs.[2]

I found that bending a rush point  and passing down under Anthers, and bringing it out with a jerk I very often drew out the pollin

I send you the Epipactis in case you want to try any more as to withdrawing pollen. I will tell the rest when I come on Thursday.[3]

Your affect son | W.E.D

DAR 162: 101

CD ANNOTATION
0.2 July . . . Thursday 6.2] *crossed pencil*

[1] William had been observing the insects that visited the native British orchid *Epipactis palustris*, while he was on the Isle of Wight. See letter from W. E. Darwin, 5 July [1869], and letter to W. E. Darwin, 7 [July 1869].

[2] CD discussed William's observations of hive-bees visiting flowers of *Epipactis palustris* in *Orchids* 2d ed., p. 99.

[3] According to Emma Darwin's diary (DAR 242), William arrived at Caerdeon, Barmouth, Wales, on Friday 16 July 1869.

To G. R. Gray 12 July 1869

Caerdeon | Barmouth.
July 12. 1869.

Dear Mr. Gray

As you have always given me information & aided me in every way in the kindest manner at the British Museum, it would give me much pleasure to send you a Testimonial of your abilities as a Naturalist.[1] But you are aware that I am not an ornithologist or an Entomologist; & it is impossible for me to express any opinion regarding the merits of your works on these subjects; but I may add that I have heard those who are capable of judging speak in high terms of your knowledge of the species of Birds, & especially of your knowledge of the literature of ornithology.—[2]

Pray believe me | Dear Mr Gray | Yours very faithfully | Ch. Darwin

To G. R. Gray Esq

British Library Board (Egerton 2348: 232–3)

[1] See letter from G. R. Gray, 9 July 1869. Gray had assisted CD in completing the volume on birds (*Birds*) for *Zoology of the voyage of HMS Beagle* (*Zoology*) after John Gould left for Australia in 1838 (Lightman ed. 2004).

[2] Although Gray was in charge of the bird collection at the British Museum, his personal interest was in entomology and he wrote on both ornithological and entomological topics (Lightman ed. 2004).

From Alexander Wallace 15 July [1869][1]

Trinity House, | Colchester.
July 15

Dear Sir

In reply I send the following[2]

Yrs truly. | A Wallace.

1869.

Bombyx Pernyi[3] sent to me from N. China in the cocoon 347 moths emerged

Of which 123 were ♀ & 224 ♂: out of about 4000 cocoons.

In 1868, of 46 Yama Mai[4] cocoons which emerged, 25 were ♂ & 21 ♀

in 1869, of 52 Yama Mai cocoons

27 ♂ & 25 were ♀

these are all I have at your disposal[5]

AW.

DAR 86: A98

CD ANNOTATIONS

4.1 Pernyi] *underl pencil*; '?' *added pencil*

6.1 Yama Mai] *underl pencil*

[1] The year is established by the last year given for the emergence of the moths.

[2] CD's query to Wallace has not been found.

[3] *Bombyx pernyi* is now *Antheraea pernyi*, the Chinese oak silk-moth.

[4] *Bombyx yamamai* is now *Antheraea yamamai*, the Japanese oak silk-moth.

[5] CD published Wallace's results in *Descent* 1: 313.

From T. H. Huxley 16 July 1869

Jermyn S
July 16th. 1869

My dear Darwin

Haeckel wrote to me some time ago about his calcareous sponges.[1] I was unable to help him & I do not know anyone who is likely to be able to aid him except Bowerbank—[2] I forwarded a letter from him to Bowerbank but I do not know with what effect

To tell you the truth, what with Fossils & Ethnology & the great question of 'Darwinismus' which is such a worry to us all, I have lost sight of the collectors & naturalists "by grace of the dredge", almost as completely as you have[3]

I am very sorry to hear you have not been picking up strength— But it very often happens that the effect of sea air is cumulative and one picks up more in the last week than in all the others—

We have not been particularly flourishing— My poor wife[4] has had to undergo an operation which has put her in great pain, but she is picking up again now & will I hope take a fresh start. I myself, (who boasted so much) have been pulled down by an attack of acute dyspepsia—what they call Gastralgia—& kept my bed a day or two—but I am on the turn too—

The chicks have all colds or coughs—so I think the sooner we are out of this the better

It was miserably cold up to within a few days— now it is tropically hot—

With kindest remembrances to Mrs Darwin & all your circle

Ever Yours | T. H. Huxley

Hooker has come back, radiant as I hear—but I have only seen Madame—who is decidedly radiant also[5]

DAR 166: 320

[1] Huxley refers to Ernst Haeckel. See letter to T. H. Huxley, 9 July [1869], and letter from Ernst Haeckel, 2 July 1869.

[2] James Scott Bowerbank.

[3] See letter to T. H. Huxley, 9 July [1869]. Huxley refers to the practice of dredging for marine specimens.

[4] Henrietta Anne Huxley.

[5] Huxley refers to Joseph Dalton Hooker and Frances Harriet Hooker, who had spent six weeks travelling in Russia and northern Europe. See letter from J. D. Hooker, 24 June 1869.

From Allan C. Pagan to George Cupples 16 July 1869

Laurel House | Crief
16 July 1869

Dear Sir

I now append you a list of Lambs on my Farms for 8 years—[1] If necessary my Books can be produced, I give you this for Mr Darwins private information as of course no one has a right to know what my Farm produces but my self.

If you require any more information I will be happy to give it

Yours Faithfully | Allan C Pagan

[Enclosure]

Account of Tup & Ewe Lambs on Innergeldie Farm for **8 years** — on Cheviot and Blackfaced Hirsels

	Male	Female	Male	Female	
	Cheviot Stock		Blackfaced Stock		Total
1862	481	494	449	477	1901
1863	505	494	541	547	2087
1864	523	511	536	553	2123
1865	516	499	538	581	2134
1866	506	502	562	617	2187
1867	526	513	588	560	2187
1868	545	536	581	578	2240
1869	502	551	600	553	2206
	Total 4104	4100	4395	4466	Total 17065
	Male	Female	Male	Female	

Allan C Pagan | Innergeldie | 15 July 1869

DAR 85: B54–5

[1] In *Descent* 1: 305, n. 41, CD thanked 'Mr. Payan' for his information on sheep. See also letter from Robert Elliot to George Cupples, 4 June 1869.

From W. B. Dawkins 17 July 1869

Gungrog. | Welsh Pool.
17 July 1869

My dear Sir,

Your very kind note was forwarded from London here, and reached me this morning.[1]

I should very much like to see the bones etc that you have obtained from Ruthin, and would run over to see the fissure, if you could put me in the way of doing so.[2] I am staying here for 14 days, after which I return home to

Norman Road, Rusholme,
Manchester, where,

if you will be kind enough to send the remains, I will try to make out the species.

That Ruthin district is very interesting, and ought to furnish large numbers of cave-beasts. The Cefn cave contained the rare *R. leptorhinus*.[3]

I am delighted to see how fast Geolog. evidence is being discovered in favor of the 'Development Theory'. This last week, I read a paper of Prof. Owen's in which three species of S. American "Hippidian" (Ow) fill up in great measure the blank between Horse and Hipparians This discovery makes Mr. Gaudry's genealog of Horse very complete.[4]

The small sketch of the British mammals, (Pleistocene) is very incomplete: but at some future time I hope to fill it in, and to finish the details.[5]

I am | My dear Sir | Yours truly | W. Boyd Dawkins

Charles Darwin Esq. F.R.S

DAR 162: 123

[1] CD's letter to Dawkins has not been found.

[2] Dawkins refers to prehistoric bones that had been collected from a cave near Corwen in the hills south of the town of Ruthin in Wales and given to CD (for more on CD's acquisition of the bones, see Lucas 2007, pp. 325–6).

[3] Cefn cave is one of a group of Palaeolithic and later caves in the lower Elwy valley near Denbigh. Dawkins had written a paper on the dentition of *Rhinoceros leptorhinus* (Dawkins 1867).

[4] In his paper 'On fossil remains of equines from Central and South America' (Owen 1869), Richard Owen proposed the group name 'Hippidion' for three species of extinct American horses (Owen 1869, p. 572). 'Hipparion' was a group of extinct old-world horses. Albert Gaudry had discussed the group in 'Sur les Hipparions' (Gaudry 1865).

[5] Dawkins refers to Dawkins and Sanford 1866–72.

To Albany Hancock 17 July [1869][1]

Caerdeon, Barmouth | N. Wales
July 17th

My dear Sir

I am going to beg a favour of you, which I know it is a mere chance whether you can grant.— Ernst Häckel is working zealously on *Calcareous* Sponges, & is now starting for Norway solely to carry out his investigations.[2] He says he is in great want of British specimens; & I know not to whom to apply, nor does Huxley, from whom I heard this morning.[3] Can you spare him any out of your collection? Do not trouble yourself to write *unless you can aid us.*— I shall be at home "Down, Beckenham Kent" last day of this month

Forgive me for asking this favour & believe me with the most sincere respect.

My dear Sir | Yours very faithfully | Ch. Darwin

E. C. Hodgkin (private collection)

[1] The year is established by the relationship between this letter and the letter from Albany Hancock, 25 July 1869.

[2] See letter from Ernst Haeckel, 2 July 1869.

[3] CD refers to Thomas Henry Huxley. See letter from T. H. Huxley, 16 July 1869.

From J. D. Hooker 17 July 1869

Royal Gardens Kew
July 17th /69.

Dear old Darwin

It is mild weather—therm 81. in my room, & I am indisposed to describe *Cucurbitaceæ* for the W. Africa flora, the interest of which vanished after I had done these Genera for Gen. Plants. & I have just been idling over Fritz Muellers work, & that makes me wonder how you are & that makes me write this letter!—[1] I think here is causation enough to please Herbert Spencer. I find the first 96 pp. of F. Mueller very dull, not remembering enough of Crustacea—but Chapt X on Principles of Classification & XI on Evolution strike me as being remarkably good in matter, though the manner of them is not unexceptionable He certainly hits Agassiz hard, & his analysis of the great God J. Muller is searching, & very reverential though damaging.[2] The same thoughts anent embryogenic resemblance &c had often crossed my mind.

The obs. 5 in note of p. 119 is a very good one, & exceedingly suggestive.[3]

I have had a queer Strasburgh Mathematician here with me this morning, about Phyllotaxy &c & we have had a long chat, during which he has expounded certain queer aspects of scientific theories—eg. that the original primordial cell—from which all organized creatures were developed, was that of *Man*—in as much as it has attained it's highest development in Man.[4] I told him that Pangenesis would demand this, for the original cell must have contained the original *Gemmules* which enter into the composition of every cell of Man.—[5]

Hector writes me that he thinks of putting Willy on a farm with a friend of his own, a goodly man with a good wife, who has leased Gov[r]. Gray's farms on Kawa Island—near Auckland.[6]

Hector tells me that they have found *lots of Saurian bones* in the brown coal of N.Z.[7]

I have just got the 1st part of the Transact[s]. of the N.Z. Institute—a most excellent volume. Colensos account of the natives seems to me to be most admirable & very interesting—& there are various other (unread) articles with interesting titles.[8]

I agree with you about fascination, & oddly enough had already written to Mrs Barber, the same view as to not doubting the facts, but the explanation. I have sent it to Pop. Sc. Rev.[9]

I saw Huxley yesterday at a dinner party, looking well I thought, but complaining of getting *bloated.* I see he has been at Comte again but I have not read his Article.[10]

Pray let some one write & tell me how you are. I do wish you would get better. I suppose the book on Man is at a stand still.—[11]

Ever yrs affec | J D Hooker

DAR 103: 22–4

1 Hooker was working on the section on Cucurbitaceae for Daniel Oliver's *Flora of tropical Africa* (Oliver 1868–77); the section appeared in the second volume, published in 1871. The section on Cucurbitaceae for *Genera plantarum* (Bentham and Hooker 1862–83) was published in vol. 1, part 3, in 1867. Hooker also refers to Fritz Müller's *Facts and arguments for Darwin* (Dallas trans. 1869).

2 In chapter ten, Müller discussed the views of Johannes Peter Müller on the relationship between type, as discovered through comparative anatomy, and developmental history (Dallas trans. 1869, pp. 98–9). He then gave a detailed critique of Louis Agassiz's view of embryogenesis (*ibid.*, pp. 99–109). In the next chapter, Müller discussed embryonic and larval development from an evolutionary standpoint (*ibid.*, pp. 110–21).

3 In Dallas trans. 1869, pp. 119–20, Müller noted that in insect orders with few genera the differences among the members were much greater than in large orders like the Coleoptera whose members were easily identified as belonging to the order. He compared this case to that of ferns of certain genera poor in species whose members differed more from each other than any two forms in a genus with numerous species.

4 The Strasbourg mathematician has not been identified.

5 Hooker refers to CD's hypothesis of heredity, published in *Variation* 2: 357–404. CD had not speculated about the origin of life or primordial cells.

6 Hooker's son William Henslow Hooker had travelled to New Zealand for his health. James Hector was trying to find a position for William and had discussed having him work on a farm on Kawau Island (see Yaldwyn and Hobbs eds. 1998, pp. 117–19). Hooker also refers to George Grey, the former governor of New Zealand, who had retired to Kawau Island in 1868 (*ibid.*, p. 248).

7 Bones of plesiosaurs and mosasaurs were discovered in the Waipara River valley, north Canterbury, following a flood (Yaldwyn and Hobbs eds. 1998, p. 119).

8 Hooker refers to the first volume of *Transactions and Proceedings of the New Zealand Institute* (1868), which was issued in May 1869, and to a paper by William Colenso, 'On the Maori races of New Zealand' (Colenso 1868), that appeared in the journal.

9 Mary Elizabeth Barber had evidently sent Hooker a paper on snakes that he passed on to CD (see letter to J. D. Hooker, 8 July [1869]). The paper was not published in *Popular Science Review.*

10 Hooker refers to Thomas Henry Huxley and T. H. Huxley 1869b. See letter to T. H. Huxley, 9 July [1869] and n. 4.

11 CD had been unwell for most of the family holiday in Wales (see 'Journal' (Appendix II)). Hooker refers to *Descent.*

To W. C. Tait 17 July [1869][1]

Caerdeon, Barmouth | N. Wales
July 17

My dear Sir

Your obliging letter of the 11th has been forwarded to me here.[2] The case of the Geranium is not new; there are many plants, often called Dichogamous, in which the anthers shed their pollen before the stigma is ready, and a few converse cases in which the stigma is mature before the pollen is ready. All these dichogams absolutely *require* cross-fertilisation. When I left home a month ago *one* alone of the Drosophyllums was still healthy.[3] D[r] Hooker[4] begins to have some doubts whether Drosophyllum is so closely allied to Drosera. Pray do not give yourself any more trouble about the development of the Horns in Sheep. I have acquired some facts from Saxony which must suffice.[5]

I am not intending at present to publish on Instinct, though I have much MS. almost ready for press;[6] and I doubt whether my strength will last out to publish much more on any subject. We have taken this house for 8 weeks for change of air, but it has done me very little good, and I hope to be at home again by the end of this month.

Pray believe me, | My dear Sir, | Yours very faithfully | Ch. Darwin

[...]

P.S. 2[d] [Asks whether Tait saw CD's] little Book on the Fertilisation of Orchids by Insect Agency [and offers to send him a copy.][7]

Copy and catalogue entry (incomplete)[8]
DAR 147: 545; Sotheby's, London (19 July 1990)

1 The year is established by the relationship between this letter and the letter from W. C. Tait, 11 July 1869.
2 See letter from W. C. Tait, 11 July 1869.
3 Tait had sent CD several specimens of *Drosophyllum lusitanicum*, but they arrived in poor condition (see letter to W. C. Tait, 12 and 16 March 1869).
4 Joseph Dalton Hooker.
5 See letters from W. C. Tait, 10 May 1869 and 11 July 1869. CD had received information on Saxon merino sheep from his German translator, Julius Victor Carus (see letters from J. V. Carus, 22 January 1869 and 2 February 1869).
6 CD refers to *Descent*.
7 The reference is to *Orchids*. The second postscript is from the sale catalogue. The rest of the letter is transcribed from a copy in DAR 147 which does not include the postscripts.
8 The original letter is complete and is five pages long.

To Fritz Müller 18 July [1869]

Caerdeon, Barmouth N. Wales
July 18

My dear Sir—

I received your last letter shortly before leaving home for this place.[1] Owing to this cause, & to having been more unwell than usual, I have been very dilatory in writing to you.— When I last heard about 6 or 8 weeks ago from Mr. Murray, 400 copies of your Book had been sold, & I daresay 500 may now be sold.[2] This will quite repay me, if not all the money; for I am sure that your book will have got into the hands of a good many men capable of understanding it: indeed I know that it has. But it is too deep for the general public. I sent you 2 or 3 Reviews: one of which in the Athenæum was unfavourable; but this Journal has abused me, & all who think with me, for many years.—[3] I enclose two more notices not that they are worth sending: Some other brief notices have appeared.[4]

The case of the Abutilon sterile with some individuals is remarkable: I believe that I hd. *one* plant of Reseda odorata which was fertile with own pollen; but all that I have tried since are sterile except with pollen from some other individual. I planted the seeds of the abutilon; but I fear that they were crushed in the letter.—[5] Your Escholtzia plants were growing well, when I left home, to which place we shall return by the end of this month, & I will observe whether they are self-sterile.—[6]

I sent your curious account of the monstrous Begonia to Linnean Soc. & I suppose it will be published in the Journal.—[7] I sent the extract about grafted Orange trees to Gardeners' Chronicle, where it appeared.—[8] I have lately drawn up some notes for a French Translation of my Orchis-book: I took out your letters to make an abstract of your numerous discussions, but I found I had not strength or time to do so, & this caused me great regret. I have alluded to your work, as you will see in my Paper, which will also be published in English, & which I will send it you.[9]

With the most cordial feelings believe me, my dear Sir, | Yours very sincerely | Charles Darwin

P.S. By an odd chance, since I wrote the beginning of this letter, I have received one from Dr. Hooker, who has been reading Für Darwin: he finds that he has not knowledge enough for the first Part; but says that Chpt X & XI "strike me as *remarkably* good"— He is also particularly struck with some of your highly suggestive remarks in the note to p. 119—[10] Assuredly all who read your book will greatly profit by it, & I rejoice that it has appeared in English.

Endorsement: 'Received Septbr.9.69.'
British Library (Loan 10:29) (by kind permission of English Heritage)

1 See letter from Fritz Müller, 14 March 1869.

2 The reference is to Dallas trans. 1869. See letter from John Murray, 28 April [1869].

3 See letter from Fritz Müller, 15 June 1869 and n. 3.

4 The reviews have not been identified.

5 Müller had evidently enclosed seeds of an *Abutilon* species (later named *A. darwinii*) found by him near the Capivary river with his letter of 14 March 1869. Müller published the results of his hybridisation studies with *Abutilon* in 'Bestaubungsversuche an Abutilon-Arten' (Pollination experiments on species of Abutilon; F. Müller 1872–3). CD reported the results of his own experiments with *A. darwinii* in *Cross and self fertilisation*, pp. 333–5. *Reseda odorata* is mignonette. CD reported the results of his experiments with this species in *Cross and self fertilisation*, pp. 119–23.

6 CD refers to plants raised from seeds of *Eschscholzia californica* (California poppy; see letter to Fritz Müller, 14 March 1869 and n. 11.)

7 See letter to George Bentham, 10 May [1869] and n. 6.

8 Müller evidently enclosed a brief article from *Colonie Zeitung* with his letter of 14 March 1869. A translation of a portion of the article appeared in *Gardeners' Chronicle*, 26 June 1869, p. 686. The excerpt does not mention CD or Müller by name, but mentions that the newspaper was published at 'Santa Catharina, Brazil'. The article discussed the influence of the stock of *Citrus decumana* (now *C. maxima*, the pomelo) on the scion of an orange variety known as Laranga selecta.

9 CD refers to Rérolle trans. 1870 (French translation of *Orchids*), and 'Fertilization of orchids'.

10 CD refers to Joseph Dalton Hooker. See letter from J. D. Hooker, 17 July 1869 and nn. 2 and 3.

From Adolph Reuter 18 July 1869

Sanssouci near Potsdam
the 18. 7. 1869

Sir,

I am uncapable to express you my feeling, when I did have the honour to receive your very amiable letter and I beg your pardon, that I have retarded so long-times my answer, because one of my colleagues is dead and I am now obliged to conduct two gardens till to the next October.[1]

I assure you my dear Sir, I am not only very glad—no I am proud indeed, that you have been so kind to give me the permission, by entering with you in correspondence and I should be only sorry when it would be seeming as when I would publish my small German knowledge— no Sir not at all, my intention is *only* to send it to *you*, perhaps that then one or the other object would be interesting for you and worth to find a small place in your publications.[2]

I should be glad, when I could give you more then only some proofs of my practical experience, but you know very well, the Gardener has much business and so you may excuse me when I talk about an object not quite scientifically and sometimes also more as it would be necessary.—

For the first, I send you some arguments about the more or less constancy of trees and shrubs variety's seedling's and also some remarkable facts of some woody climbers, changing their character in an older state; with the next letter I suppose to deliver to you interesting examples; where grafted trees have changed their characteristic etc.—

But forgive me my very dear Sir, when I am unpolite, molesting you so long-times.—

I wish, that your weakness may not derange you longer and my letter may find you healthy and happy.—

With my greatest respects and the assurance of my attraction at all times | I am dear Sir | Your very obedient servant | Ad. Reuter

[Enclosure]

The *propagation* by *seeds* of different trees and shrubs' *varietys* and *their resultance*:

1. *Acer Pseudoplatanus foliis variegatis*: Travelling with Mr. Linné, the late Director of our Royal Gardens, I did find near Aachen (Aix-la-Chapelle) in the Garden of a medicine a very fine looking exemplar of Acer Pseudoplat. foliis variegatis.[3] The

Leaves enclosed with the letter from Adolph Reuter, 18 July 1869. DAR 176: 125.
By permission of the Syndics of Cambridge University Library.

tree was nearly 1 foot big and the ground round the Acer covered with white spotted seedlings, what is most remarkables indeed, because the possessor did affirm me that it were the same appearance every year.

2. *Acer Pseudoplatanus foliis purpureis*,[4] which has the *under* site of the leaves red coloured, exist also in our Botan. Garden at Berlin as a beautiful tree. Mr. Bouché, Inspector of the Garden, was so kind to show me last autumn some pots full of seedlings, where all the young plants did have the *same* peculiarity like the *mother*-tree.

3. *Berberis vulgaris foliis purpureis*,[5] from this fine coloured shrub exist at Sanssouci a strong exemplar and I was sawing very often the fruits, but (between 100 plants) I

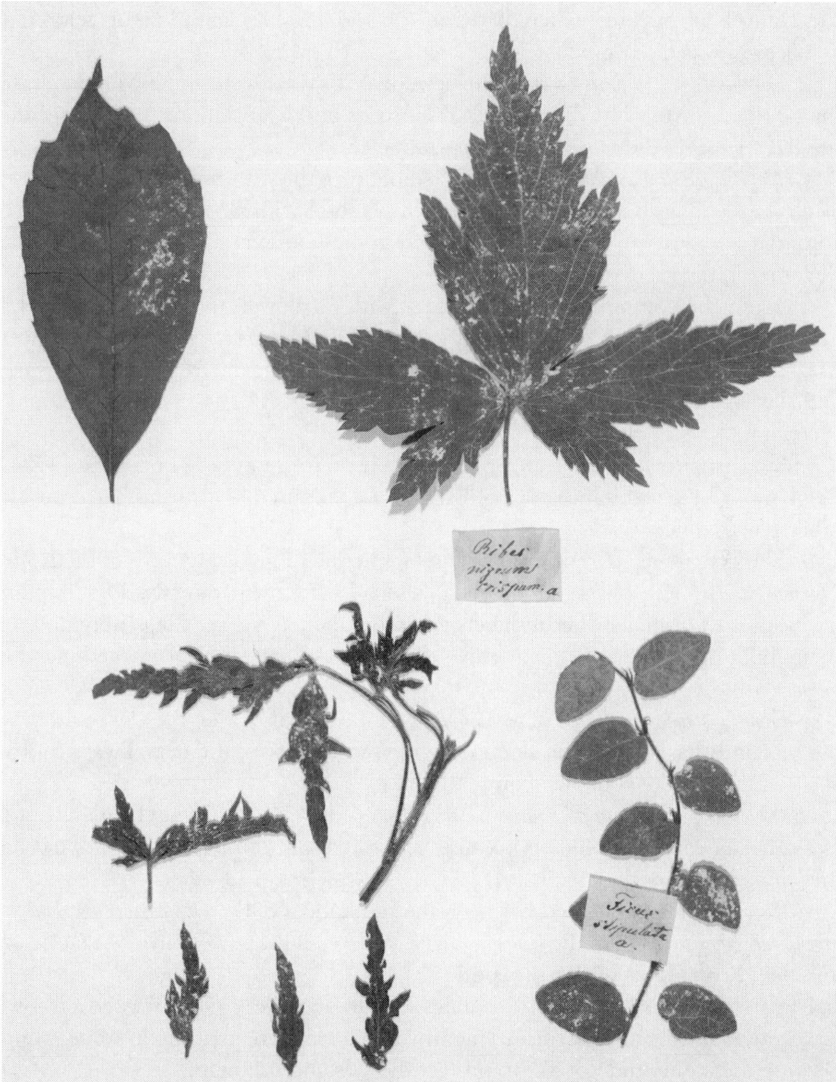

Leaves enclosed with the letter from Adolph Reuter, 18 July 1869. DAR 176: 125. By permission of the Syndics of Cambridge University Library.

was only so happy to find 1–2 *red* exemplars; all the others were *green* leaved like the mother plant.—

An argument, that no rule is without exception may be, that we did receive very often Seedlings for our nursery of trees etc. from Mr. Grundel at Offenbach and

Mr. Dauveli at Orléans,[6] where between 100 and more *red* leaved plants seldom one *green* leaved was discovered.—

4. *Corylus Avellana foliis laciniatis* (urticaefolia), I saw this shrub (at Magdeburgh in the Garden of Mr. Häneb) propagated by seeds and it was about $\frac{1}{3}$ of the seedlings *constant*, the other did have the *common* character.[7]

5. *Corylus Avellana foliis purpureis* has sometimes the affection so *generate* by seeds *mostly red* seedlings but also sometimes $\frac{1}{3}$ *red* plants, $\frac{1}{3}$ with *red brown* leaves and $\frac{1}{3}$ with common *green* leaves. The first case passed in the Garden of Mr. Vogel at Potsdam, the other in my garden at Sanssouci.[8]

6. From the common *Daphne Mezereum* (with *red* flower and berry), we did have in our nursery of trees etc also a variety with **yellow** *flower* and *fruit* — Daphne Mezereum fructa lutea.[9] More then six times I was sawing the fruits of the last variety and allways I was so happy to bring up again plants *like* the *mother* shrub (without any exception.).—

7. *Fagus sylvatica foliis purpureis*[10] Seedlings I have seen not only in Germany also in Belgium and the resultance was: $\frac{1}{3}$ with *fine red* coloured, $\frac{1}{3}$ with *rust-brown* and the other $\frac{1}{3}$ with quite *green* leaves.

8. Some years ago I did receive by a Gentleman a great quantity of seeds, from *Fraxinus excelsior monophylla* (simplicifolia), which I was sawing directly. The resultance has been extraordinarily because between more then 3–400 seedlings were not more then about 30 exemplars, looking like the mother-plant, *all the other* seedlings were *common* ashs.—

9. *From Quercus pedunculata var. pyramidalis*[11] I did saw also often the glans and mostly the seedlings has been again fine looking pyramidal trees, the rest allways more or less slightly or quite complete common Oaks.

10. *Quercus pedunculata foliis argent. variegatis*,[12] I did also propagate by glans but the essays has been *not* succeeding because between about 100 plants only *two* did have the leaves a *little* spotted.

11. Very interesting indeed has been the resultances of the *Quercus pedunculatæ foliis argent. marginalis*,[13] one of the finest varietys, where all the leaves have *small white edges* and the **glans** also is **white striped**.

I was so glad to receive 30 Seedlings but all and every plant was covered with leaves so *white* as *snow* and curious fact the little oaks were expiring in some months, because the nature did not like to protect the existence of them.

12. *Ribes nigrum crispum* (aconitifolium) (spectabile) I propagated 8 years ago also by seeds and obtained $\frac{1}{3}$ common black currants, $\frac{1}{3}$ like the mother plant and $\frac{1}{3}$ with *frizzled* leaves, which was called by Prof. Koch *Ribes nigrum apiifolium*.[14]

The last again to propagate by seeds, was quite impossible because my seedling did never show a flower, he was a regular monstrosity and I was allways obliged to propagate by cuttings.

13. For to be sure, that berry varietys could be also propagated sometimes by seeds, I slected *Ribes rubrum var. cerasiferum* (Grosseille à cerise).[15] In the fourth year all the

seedlings did show their fruits and alltogether so also the leaves were looking like the mother plant.

14. *Rubus laciniatus* I was very often sawing nd between all the seedlings and the mother was never a difference, so that I believe it will be perhaps a *pure* species!—[16]

15. *Thuja orientalis var. aurea*[17] is also sometimes propagated by seeds and a great quantity of the young plants has then the same originellity as the old plant.

Thuja occidentalis var. Wareana[18] was also propagated in our Botan. Garden at Berlin by seeds and all the seedlings did have a similar appearance as the mother tree.

17. *Taxus hibernica* (fastigiata.)[19]

A friend at England was so kind to give me one day a small portion of Seed, which rised in the second year, but also not one plant did have the character like T. hibernica. The were alltogether looking like the *common* Yew with the exception of *3* exemplars, growing more pyramidal like *Taxus baccata var. erecta.*

At least I mention yet two examples of *2 very known species of Spiraea*, which I was propagating by seeds: Spiraea callosa and Douglasii.[20]

18. *Spiraea Douglasii* one of our oldest red flowering Spiraees, I was allways obliged to propagate by cuttings, because between more then 400 seedlings scarsely 10 exemplars has been, which did have the same figure like Sp. Douglasii, the seedlings were representing the forms: Billardi, eximia, carpinifolia etc.[21] so that I believe, we are not more in the possession of the *right* old Spiraea Douglasii or what also possible — it is only a variety.—

The same resultance was stated at Muskau in Prussia (Possession of His Roy. H. the Prince of Low-land.)[22]

19. *Spiraea callosa* this nice and very much estimated Spiraea was propagated all and every year in our nursery in great quantity and allways the curiosity has been, that some exemplars (between 5-600) did have *red grape*-like flowers and not like all the other *umbelliferous*, also the leaves were differently looking.

Because the variety was trained at Sanssouci so Prof. Koch did give her the signification Spiraea *Sanssouciana*. Some times later we did receive from England Sp. Nobleana, trained at Bagshot (?) and Sp. Regeliana from Frankfurth trained in Mr. Rinz's nursery, but all the three varietys seemed me to have the same mother— that is to say—Sp. Callosa!—[23]

Climbers changing their character in an elder state:—

1. *Hedera Helix* the common Ivy with all his varietys is allways changing the physicomie, when he is beginning to show his flowers and it is a very curious fact that propagated in this period by cuttings the Ivy will not accept again his original character. We find this form announced in the Catalogues of the nursery men as *Hedera Helix arbuscula.*[24]

2. *Lonicera brachypoda* has as a creeper allways more tender leaves but because the plant in our country will be very often attacked by the frost, so we find her sometimes cultivated in pots on sticks, 2-3' long, where the leaves are changing their character in a manner that they are looking like evergreen as the leaves of a Viburnum Tinus.—[25]

3. In a similar way as the first it is also the case with *Decumaria barbara.*—

4. *Cissus acuminata*, a fine looking climber from the Himalaya is allways in my country a very sensible plant and is to be covered in the winter. Some plants standing on a wall are at once producing on the top very notable leaves (Cissus acuminata. β.) what is indeed very curious, because I never saw such a fact at the similar Cissus quinquefolia (Ampelopsis hederacea).[26]

5. *Ficus stipulata (a)* is no dought the most interesting example of all because planted on the wall one our houses (for forcing beans) would not change her character till to the time where she was showing the fruits, then the now leaves at once were beginning to be larger and longer, so that no person would know again the Ficus stipulata[27]

In our Botan. Garden at Berlin is such a shrub propagated by a fruit-branch in a pot, which is every year full of fruits and which will never return again to his true character.

The fact is almost incredible, but Mr. Bouché,[28] Inspector of the Roy. Botan. Garden at Berlin is capable to affirm the truth.—

[Enclosure]

[Reuter enclosed a number of leaves: see plates on pp. 324 and 325.]

DAR 176: 125

CD ANNOTATIONS
Enclosure:
2.1 *Acer*] 'acer' *added above pencil*
34.1 In ... character. 34.3] *scored pencil*
Top of enclosure: '*[illeg]* Plants & *[illeg]*' *ink*
End of enclosure: 'Herr Reuter | of Sanscouci, Potsdam | **Inheritance**. Difference (like Ivy) between leaves of Branches & flowering Branches can be propagated & do not *[*revert*]*' *ink*

1 See letter to Adolph Reuter, 2 June 1869.
2 In his letter of 2 June 1869, CD advised Reuter to publish his observations on horticulture.
3 'Foliis variegatis': with variegated leaves (Latin) *Acer pseudoplatanus* is the sycamore maple; there is a cultivar 'variegatum'. Reuter also refers to Peter Joseph Lenné.
4 'Foliis purpureis': with purple leaves. The cultivar *Acer pseudoplatanus* 'Atropurpureum' has leaves that are dark green above with purple undersides.
5 *Berberis vulgaris* is the common barberry; there is a cultivar 'variegata'.
6 Offenbach is a city in Hesse, Germany, on the river Main near Frankfurt; Mr Grundel has not been identified. Orléans is a city in north-central France in the Loire valley. Mr Dauveli has not been identified.
7 'Foliis laciniatis (urticaefolia)': with slashed leaves (nettle-like leaves). *Corylus avellana* is the hazel tree; the modern cultivar is called *C. avellana* 'heterophylla' (the cutleaf hazel).
8 There are now several red or purple cultivars of *Corylus avellana*; in some varieties, like 'Rote Zeller' the anthocyanins (red pigments) are affected by temperature and humidity.
9 The more common form of *Daphne mezereum* ('rubrum') has pink flowers and red berries. The form Reuter describes as 'fructa lutea' (yellow fruit) is the modern 'album', which has cream flowers and yellow fruit.
10 *Fagus sylvatica* is the beech tree; 'purpurea' is the most common purple-leaf variety.

[11] *Quercus pedunculata* is now *Q. robur* (the English oak or pedunculate oak); the variety 'pyramidalis' is now 'fastigiata' (cypress oak).
[12] 'Foliis argent. variegalis': with silver variegated leaves.
[13] 'Foliis argent. marginalis': with silver-edged leaves.
[14] *Ribes nigrum* is the blackcurrant. 'Crispum (aconitifolium) (spectabile)': curled (with aconite-like leaves) (showy). *Ribes apiifolium* is now considered to be a synonym of *R. nigrum*. Reuter also refers to Karl Heinrich Emil Koch and Koch 1869–73, 1: 661.
[15] *Ribes rubrum* is the redcurrant. 'Cerasiferum': bearing wax (Latin). 'Grosseille à cerise' (cherry currant) is not a currently recognised variety, however, there is a western North American species, *Ribes cereum*, the wax currant.
[16] *Rubus laciniatus* is the cut-leaved blackberry and is a distinct species.
[17] *Thuja orientalis* is now *Platycladus orientalis*, the oriental arborvitae; 'aurea nana' is a modern golden-leaved variety.
[18] *Thuja occidentalis* is the eastern white cedar; 'Wareana' is a cultivar with blue-green leaves.
[19] *Taxus hibernica* is now *T. baccata* 'hibernica' (the Irish yew); *T. baccata* 'fastigiata' is a synonym. *T. baccata* 'erecta' is a cultivar of the English yew.
[20] *Spireaea* is meadowsweet. *S. douglasii* is sometimes called western spiraea.
[21] *Spiraea* x *billardii* is a hybrid of *S. douglasii* x *S. salicifolia*; *S. eximia* and *S. carpinifolia* are synonyms of *S. salicifolia* (willow spiraea).
[22] Muskau Park is a large English-style park founded in 1815 in Prussia (the park is now in both Germany and Poland). It was taken over by Prince Frederick of the Netherlands and extended from 1852 to 1881 (*Oxford companion to gardens*).
[23] *Spiraea sanssouciana* is a synonym of *S. douglasii* (see *Berliner allgemeine Gartenzeitung* 25 (1857): 214 for Koch's description). *Spiraea nobleana* is a synonym of *S. douglasii*; Charles Noble was the owner of a nursery in Bagshot (*Post Office directory of the six home counties*). Reuter also refers to the nursery in Frankfurt am Main founded by Sebastian Rinz.
[24] The cultivar *Hedera helix* 'arbuscula' ('small tree') has not been identified.
[25] *Lonicera brachypoda* is now *L. japonica* (Japanese honeysuckle). *Viburnum tinus* is laurustinus.
[26] Reuter refers to *Cissus acuminata* A. Gray (now *Vitis acuminata*) and to *C. acuminata* Thwaites. *Cissus quinquefolia* or *Ampelopsis hederacea* is now *Parthenocissus quinquefolia* (American ivy).
[27] *Ficus stipulata* is now *F. pumila* (climbing fig).
[28] Carl David Bouché.

To W. B. Dawkins 19 July [1869][1]

Caerdeon, Barmouth, North Wales
19 July

[... admitting that he had disobeyed his instructions and dispatched a box of bones to him by rail, giving an account of the discovery of the bones at Perth y Chwaril on the Rhagatt estate[2] and his promise to Miss Lloyd[3] to obtain from Dawkins the English names of the principal bones, with postscript and emendation in text in another hand outlining details to journey to Rhagatt.]

Incomplete[4]
Phillips, Son & Neale (24 October 1985)

[1] The year is established by the relationship between this letter and the letter from W. B. Dawkins, 17 July 1869.
[2] See letter from W. B. Dawkins, 17 July 1869 and n. 2. The bones were found on land belonging to the farm of Perthichwareu in Iâl, part of the Rhagatt estate (Lucas 2007, pp. 320, 325).

[3] Mary Charlotte Lloyd was the daughter of the owner of Rhagatt (Lucas 2007, p. 320).
[4] The original letter is complete and is described in the sale catalogue as being three-and-a-half pages long.

From E. L. Garbett to the *Athenæum* 19 July 1869

7, Mornington Road,
July 19, 1869.

Either of my last methods, applied to Mr. Darwin's supposition of twin births alone at intervals of 30 years, would give the same results as the series 1, 1, 2, 4, 7, 13, &c. (continued by the scale 2+0+0-1), whose terms quadrupled give, as Mr. George[1] observes, the elephants alive, on that supposition, in the decade after each bearing year.[2] In the periods of 20 years, however, between these exceptional decades, their numbers are only the doubles of these terms. The general term of this progression being 1:839, shows the period of doubling to be about 34 years, while, on my truer supposition of single births at intervals of 12 years, the increase is about a sixteenth faster. The mistake that led me at first to a doubling in 23 years may be the same as the original calculator's. At least it would fully account for his result.

E.L.G.

Athenæum, 24 July 1869, p. 115

[1] George Howard Darwin.
[2] See letter to *Athenæum*, 7 [July] 1869 and n. 2.

From John Lubbock 20 July [1869][1]

15, Lombard Street, E.C.
20 July

Dear Mr. Darwin

Mr Powell has asked me to subscribe to build him a house at Down & I feel rather in a difficulty.[2]

It seems that about £1700 is required. The Ecclesiastical Commissioner would give £700, he thinks he could get £290 from the Queen Anne's Bounty Fund, £150 from a building society, & his own friends would do something.[3]

What do you think about it?

Believe me | yours most sincerely | John Lubbock

C Darwin Esq | &c &c &c

DAR 170: 74

[1] The year is established by the relationship between this letter and the letter to John Lubbock, 3 August 1869.
[2] Henry Powell became vicar of Down sometime earlier in the year. John Brodie Innes, the former vicar of Down, had encouraged the curate, John Warburton Robinson, to resign in December 1868 (see *Correspondence* vol. 16, letter from J. B. Innes, 4 December 1868).

3 The Ecclesiastical Commissioners administered church revenues; a common fund instituted in 1850 gave grants for new parsonages. Queen Anne's Bounty was a fund that loaned money for building and mortgages (*Church dictionary*). The lack of a suitable parsonage was cited by John Brodie Innes as the principal cause for failing to find or retain suitable candidates for the curacy of Down (see *Correspondence* vol. 16, letter from J. B. Innes, 4 December 1868).

From James Rait to George Cupples 20 July 1869

Castle Forbes, Whitehouse, Aberdeenshire,
20 July 1869

George Cupples Esq | The Cottage, Guard Bridge | Fifeshire—
Sir

I have now to state more fully in reply to your letter of 5th. inst. in which in behalf of Mr Darwin you propose certain questions relative to the comparative numbers of the sexes born from any flock of sheep or herd of Cattle, in any one or more years, also whether there is noticeable any excess of mortality in either sex over the other, at birth or afterwards (independently of Castration) that Cattle or sheep are not bred at Castle Forbes to any extent.[1] I have, however, laid these subjects before certain gentlemen on whose veracity and accuracy thorough reliance may be placed, and from them have derived the following information.

I. Anthony Cruickshank Esq. Sittyton, Aberdeenshire in a letter to me dated "7th. Mo. 20th. 1869 For a period over 20 years we have slightly had an excess of Bull over Heifer Calves, with the exception of last year, when we had 60 Bulls + 68 Heifers, & as far as we know, are alive at present: in 1867 60 Bulls & 54 Heifers and about this proportion of sexes in former years as near as we can arrive at. We don't know the proportion of deaths at calving or dead at birth, no memo being kept of these. Our deaths on the whole range from 3 to 7 p. cent at and after birth"

II. Robert Bruce Esq. Newton, Kinloss, Morayshire, gives as follows

"Year 1864—	5	Bulls—	4	Heifers	in all	9	
1865—	6	" —	6	"	—	12	
1866—	5	" —	5	"	—	10	
1867—	6	" —	4	"	—	10	
1868—	5	" —	7	"	—	12	
1869—	6	" —	6	"	—	12	
Total in 6 years	33	bulls—	32	Heifers	—	65	Calves

Never lost a calf."

III. David Gibb Esq.[2] Bridge of Dye, Kincardineshire, states "I have 5 bull & 11 heifer calves this year, and last year I had equal numbers, and the year before that 9 bull and 8 heifer calves. As to sheep I have generally more females than males. Exact numbers I cannot state but say 30 annually. If you wait a month I shall give you the exact numbers for this year"

I may add that Mr Cruickshank is the largest holder of Cattle with whom I am acquainted, and perhaps the most extensive breeder of these in the north of Scotland. Mr. Gibb has for many years had upwards of 4000 sheep.

Trusting that these statistics may be useful as far as they go I submit them. I may mention that there is an impression among cattle breeders in Aberdeenshire that as a cow is served early or late, at the time she comes into season, there is greater probability of the one sex than the other, and that in the case of poultry the age of the male serves in some measure to determine the preponderance of the respective numbers of the sexes.

I am, | Sir, | Yours most obdt. servant | James Rait

DAR 86: 70–1

CD ANNOTATIONS
2.1 I. . . . birth" 2.7] *scored blue crayon, scored pencil*
4.1 "I . . . calves. 4.3] *scored blue crayon, scored pencil*
4.4 Exact . . . sexes 6.6] *crossed blue crayon*
Verso of first page: 'Cattle' *blue crayon*

[1] Cupples's letter to Rait has not been found. Rait was the forester at Castle Forbes (Brown 1861, p. 630).
[2] Robert Bruce and David Gibb have not been further identified.

From W. B. Tegetmeier 20 July 1869

The Field. | *346, Strand,* | *London, W.C.*
July 20 *1869*

My dear Sir

I send you by this post a Field with an engraving of the crested Turkey and enclosed are one or two of the crest feathers.[1]

I am afraid I shall not do any good with the breed this season though I have nine eggs now incubating.

I enclose you some notes as to hereditary colours in Greyhounds they are by a very good authority. (Mr C M Browne) on whom I can rely[2]

Also about the colour of the young of Sebright bantams.[3]

Also a letter from Major Holland an amateur writer who figures—'serene gossip' —The letter was not written for you to see but as it contains a fact? new to me I send it to you Please return it—as I should not like the letter which is rather equivocal to get into strange hands.[4]

I am going I hope to America in October, to stay about two months.— I shall ask you to give me a letter of credit if you are so disposed.

I know the value of an introduction from you in the United States. for Prof Shepherd[5] of Amherst College Mass.. once sent me a most elegant flower vase, in return for two or three of your letters which he requested me to give him—. If I take over

Feathers of crested turkey enclosed with the letter from W. B. Tegetmeier, 20 July 1869. DAR 193: 39.
By permission of the Syndics of Cambridge University Library.

all the letters I have received from you and exchange them for green backs I shall return a millionaire.

Sincerely trusting you are improving in health | Believe me truly yours | W B Tegetmeier

Please *dont* trouble yourself to reply unless you have any thing special to write to me about

[Enclosure]

⟨Effor⟩t a fawn dog by Larriston ⟨ ⟩ ⟨fa⟩wn got most of his puppies ⟨of th⟩at color. He was ⟨ ⟩ J⟨ ⟩ last ⟨ ⟩ the ⟨ ⟩ in, the ⟨ ⟩ Bedlam⟨i⟩te bring ⟨ ⟩ on owner over sixteen hundred pounds ⟨ ⟩ his services.[6]

'Bedlamite'.
John Henry Walsh ('Stonehenge'), *Greyhound* (London, 1875), p. 220.
By permission of the Syndics of Cambridge University Library.

⟨Mr⟩ Randells Black dog Ranter by Bedlamite ⟨out of Black⟩ fly was another Stud Greyhound ⟨who⟩ got most of his puppies, if not all, ⟨b⟩lack.[7]

Retribution by David (Lord Uffington's) ⟨was⟩ a handsome brindle and got his puppies mostly of that color.[8]

Mr Browne's Bedlam⟨ite⟩, no ma⟨tter⟩ ⟨wh⟩at color the bitches he served, he ⟨almost⟩ ⟨inv⟩ariably ⟨got⟩ all his puppies black. ⟨ ⟩ preponderance of ⟨ ⟩ on the sire a⟨nd⟩ Brewer[9] own brother ⟨ ⟩ them mostly fawns.

At the Scottish National Club t⟨ ⟩ March 25/65 Mr J L Ewing h⟨ad⟩ a dog Leonardo running marked blac⟨k⟩ and tan for all the world like ⟨a⟩ black and tan English terrier. There was no reason either on sire or dam side for the color.[10]

DAR 178: 81

Mr. Long's 'DAVID.'[1]

'David'.
John Henry Walsh ('Stonehenge'), *Greyhound* (London, 1875), p. 238.
By permission of the Syndics of Cambridge University Library.

CD ANNOTATIONS
Enclosure
1.1 ⟨Effor⟩t … services. 1.3] 'Pedig' *pencil del pencil*; 'Important | Pedigree & *[*Profile*]* | Preponderant *[*transfer*]*' *pencil*
1.3 services.] *closing square bracket red crayon follows*
2.1 Randells … ⟨b⟩lack. 2.2] 'Ranter' *pencil*
3.1 Retribution … color. 3.2] 'Transmission of colour' *pencil*
5.1 At … color. 5.3] *crossed, double scored pencil*; 'good | Black & Tan *Greyhound*' *pencil*
Top of enclosure: *2 words illeg, pencil*
End of enclosure: 'M[r] C. M. Browne (case sent on by M[r] Tegetmeier | man to be trusted' *pencil*

1 CD's copy of the article on the crested turkey (*Field*, 17 July 1869, p. 46) is in DAR 193: 39 together with three feathers (see plate on p. 333). See also letter to W. B. Tegetmeier, 2 July [1869] and n. 4.
2 C. M. Browne has not been identified.
3 Most of the enclosure has not been found, but it was evidently written by Edward Hewitt. CD returned part of the enclosure with his letter to W. B. Tegetmeier, 21 July [1869] in order to clarify the meaning

of a word. In Tegetmeier's *The poultry book* (Tegetmeier 1867, pp. 241–7), most of the information on Sebright bantams had come from Hewitt.

4 CD evidently returned the letter (see letter to W. B. Tegetmeier, 21 July [1869]). Major Holland has not been identified.

5 Charles Upham Shepard.

6 The references are to the greyhound dog 'Effort', his sire 'Larriston' and his maternal grandsire 'Bedlamite' (see J. H. Walsh 1875, p. 457).

7 The references are to the greyhound dog 'Ranter', his sire 'Bedlamite' and dam 'Black Fly' (see J. H. Walsh 1875, p. 475). Mr Randell has not been further identified.

8 The references are to the greyhound dog 'Retribution' and his sire 'David' and William Augustus Frederick Craven, Viscount Uffington (see J. H. Walsh 1875, p. 456; see also http://www.greyhound-data.com, accessed 23 June 2008).

9 'The Brewer' was a brother of 'Effort' (see above, n. 6; see also J. H. Walsh 1875, p. 451).

10 'Leonardo' is described in pedigree records as black. His sire, 'Loadstar', was fawn and his dam, 'Joan of Arc', was white and red (http://www.greyhound-data.com, accessed 23 June 2008). His owner and breeder, James L. Ewing, has not been further identified.

To W. B. Tegetmeier 21 July [1869][1]

Caerdeon, Barmouth | N. Wales
July 21st

My dear Sir

I have to thank you for *many* things including the grand illustration & account of the Crested Turkey.[2]

I am glad to hear about the transmission of colour in the Greyhound, & especially about the black & tan colours appearing without cause.—[3]

What a curious letter is Major Hollands! As you no doubt know Strelecki maintained the very same law about the Australian women, but it has not been confirmed by further observations[4] Yet it seems hardly credible that the same belief shd. arise in Borneo & Australia without foundation.

I can generally read every word which you write, but I cannot read the word before *spangled* in the enclosed: to save you trouble I enclose an envelope addressed; & please rewrite the word, & add answer to the P.S.

Can you do me the real kindness to make an effort & to find out soon the character of plumage of any Cuckoo sub-breed & of any Spangled Polish breed, in which the adult sexes are alike.—[5] If I could find out these, they wd. suffice for me: I am almost certain that I have noticed the cuckoo marks in young chickens, but I fear to trust my memory.—

When near the time of starting for America, let me hear & you may feel sure that I will do what I can. But I know personally no one there, except Prof Asa Gray, & he will not be returned from Europe so soon— I have formerly corresponded with Dana of New Haven & Wyman of Boston; but I hardly know how far I shd be justified in giving a letter of introduction to them.—[6] We have corresponded only on Science; but they feel friendly towards me.— Anyhow I will do what I can.

Your's very faithfully & | truly obliged | Ch. Darwin

I cannot say much about my confounded health.—

[Enclosure][7]

Sebright bantams

Sexes alike at 7 and 8 weeks both being *mossily*? spangled not definitely laced[8]
The darkest birds when in the early plumage being the best laced eventually
From Mr E Hewitt[9] a very successful breeder of this variety.

P.S. Do you suppose that the spangles in the young Sebrights would be circular or crescent shaped, for the latter ⟨*illeg*⟩ change into lacing

Archives of the New York Botanical Garden (Charles Finney Cox collection) (ser. 14)

1 The year is established by the relationship between this letter and the letter from W. B. Tegetmeier, 20 July 1869.
2 See letter from W. B. Tegetmeier, 20 July 1869 and n. 1.
3 See letter from W. B. Tegetmeier, 20 July 1869 and n. 2.
4 Major Holland has not been identified. See letter from W. B. Tegetmeier, 20 July 1869 and n. 4. In *Physical description of New South Wales and Van Diemen's Land* (Strzelecki 1845), p. 347, Paul Edmund de Strzelecki maintained that Australian women who had had intercourse with European men became infertile with their own race. CD scored the passage in his copy of the work, which is now in the Darwin Library–CUL (see *Marginalia* 1: 790–1). See also *Descent* 1: 220–1, n. 10, where CD gave evidence that the claim had been disproved.
5 CD discussed cuckoo sub-breeds of fowl in *Variation* 1: 244. The name comes from the cuckoo-like appearance of the plumage of these varieties.
6 See letter from W. B. Tegetmeier, 20 July 1869. CD refers to James Dwight Dana and Jeffries Wyman.
7 The enclosure is in Tegetmeier's hand, except for the postscript, which is in CD's hand, and is part of an otherwise missing enclosure to Tegetmeier's letter of 20 July 1869. CD has underlined 'mossily' and added the question mark following it.
8 'Spangled' refers to a crescent-shaped marking at the feather-tip, while 'laced' refers to an edging of colour around the feather. In Tegetmeier 1867, p. 148, Tegetmeier referred to feathers with a 'mossy appearance' as being caused by 'two colours running into one another'.
9 Edward Hewitt.

To the editors of the *Annals and Magazine of Natural History* 23 July 1869

[Caerdeon, Barmouth]

Gentlemen,

Having drawn up some notes for a French translation of my work 'On the various contrivances by which British and Foreign Orchids are Fertilized by Insects' (1862), it has appeared to me that these notes would be worth publishing in English.[1] I have thus been able to bring up the literature of the subject to the present day, by giving references to, together with very brief abstracts of, all the papers published since my work appeared. These papers contain, on the one hand, corrections of some serious errors into which I had fallen, and, on the other hand, confirmations of many of my statements. I have also been able to add, from my own observations and those of others, a few new facts of interest. A heading is given to each note, which will show the nature of the correction or addition, without any reference to my book; but I have added in a parenthesis the page to which the note ought to be appended.[2]

Gentlemen, | Your obedient Servant, | CHARLES DARWIN.

Down, Beckenham, Kent.
July 23, 1869.

Annals and Magazine of Natural History 4th ser. 4 (1869): 142

1 CD refers to *Orchids* and Rérolle trans. 1870. For more on CD's additions, see the letter to J. D. Hooker, 8 March [1869], n. 5.

2 This letter accompanied 'Fertilization of orchids'.

From Jules Joseph Fleury 23 July 1869

London,
July 23rd 1869.

Sir,

A man of the name of Edward Darwin, married, in 1823, a french woman called Orricette Parfaite Adèle Boucher, or Bouché, de Forges.[1]

Should you be a relation of M. Edward Darwin, you may perhaps inform me ⟨ ⟩ when the marriage ceremony took place.

I will feel obliged if you can give me an answer at your earliest convenience.

I remain, | Sir, | Your's truly | The Consul General for France | Fleury

Mr. Darwin | Down Bromley

DAR 164: 136

1 CD's half uncle, Edward Darwin, was never married (Freeman 1978). Orricette Parfaite Adèle Boucher has not been further identified.

To J. D. Hooker 24 July [1869][1]

Caerdeon, Barmouth N. Wales
July 24th

My dear Hooker.

We shall be at home this day week, taking two days on the Journey & right glad I shall be.[2] The whole has been a failure to me but much enjoyment to the young, except to poor George[3] who has been a good deal unwell & has just got over a severe abscess in the throat, & is much pulled down. My wife has ailed a good deal nearly all the time; so that I loathe the place with all its beauty.—

I was glad to hear what you thought of F. Müller, & I agree wholly with you: your letter came at nick of time for I was writing on the very day to Müller, & I passed on your approbation of Chapt X & XI.—[4] Some time I shd like to borrow the Transact. of N. Zealand Institute so as to read Colenso's article.—[5] You *must* read Huxley v. Comte: he never wrote anything so clever before, & has smashed everybody right & left in grand style.[6] I had a vague wish to read Comte & so had George, but he has entirely cured us of any such vain wish.

There is another article, just come out in last North British, by some great mathematician, which is admirably done: he has a severe fling at you, but the article is directed against Huxley & for Thomson.[7] This review shows me, not that I required being shown, how devilish a clever fellow Huxley is, for the Reviewer cannot help admiring his abilities— There are some good specimens of mathematical arrogance in the Review, & incidentally he shows how often astronomers have arrived at conclusions, which are now seen to be mistaken; so that geologists might truly answer that we must be slow in admitting your conclusions. Nevertheless all Uniformitarians had better at once cry "peccavi";[8] not but what I feel a conviction that the world will be proved rather older than Thomson makes it, & far older than the Reviewer makes it.—[9] I am glad I have faced & admitted the difficulty in last Edit. of Origin, of which I suppose you received, according to orders, a copy.[10]

You ask about my man-book; everything has been of late at a stand still with me, for I have not had strength to do hardly anything: the new Edit of Origin took 8 weeks, & notes for French Edit. of Orchids nearly 2 weeks; I shall publish these notes in English likewise.—[11]

In the last Quarterly there is a long article on "Design", & I sh[d] like to hear who the author is, if you sh[d]. ever hear.—[12] I have read only a small part & do not think much of what I have as yet read.

Did you ever study Drosophyllum, as you proposed to do when at Down, & does it prove closely allied to Drosera?[13]

It will be very comfortable to you, when you know that your Willie is settled with a good man with a quiet & healthy occupation.—[14]

My dear old friend— Yours affect. | C. Darwin

DAR 94: 140–2

1 The year is established by the relationship between this letter and the letter from J. D. Hooker, 17 July 1869.

2 CD arrived home from Wales on 31 July 1869 (see 'Journal' (Appendix II)).

3 George Howard Darwin.

4 CD refers to Fritz Müller's *Facts and arguments for Darwin* (Dallas trans. 1869). See letter from J. D. Hooker, 17 July 1869, and letter to Fritz Müller, 18 July [1869].

5 The reference is to William Colenso. See letter from J. D. Hooker, 17 July 1869 and n. 8.

6 Thomas Henry Huxley had written an article criticising the philosophy of Auguste Comte (see letter to T. H. Huxley, 9 July [1869] and n. 4).

7 CD refers to a review article, 'Geologic time' , written by Peter Guthrie Tait ([Tait] 1869; see *Wellesley index*). Hooker 1868 was briefly considered in Tait's essay (see [Tait] 1869, pp. 413–14). Tait discussed several of William Thomson's works that challenged the school of uniformitarian geology on its estimate of the age of the earth, and responded to Huxley's critique of Thomson's own calculations of the earth's age.

8 *Peccavi*: I have sinned (Latin).

9 On Thomson's and Tait's varying estimates of the age of the earth, see [Tait] 1869, p. 432.

10 In [Tait] 1869, p. 429, Tait had quoted CD's conjecture, based on his estimate of time needed for the denudation of the Weald, that far more than 300 million years must have elapsed since the latter part of

the secondary period (*Origin*, p. 287). In response to criticisms in the *Saturday Review*, 24 December 1859, however, CD halved his estimate in the second edition of *Origin*, and removed the discussion altogether from the third edition. In *Origin* 5th ed., pp. 349–54, CD discussed recent estimates by James Croll of the time elapsed since the the first Cambrian formation. *Origin* 5th ed. was published in June 1869 (letter from R. F. Cooke, 22 June 1869).

11 CD refers to *Descent*, *Origin* 5th ed., Rérolle trans. 1870, and 'Fertilization of orchids'. See also 'Journal' (Appendix II) for more on CD's work schedule and health.

12 The author of the unsigned article 'The argument of design', which appeared in the July 1869 issue of *Quarterly Review* ([Mozley] 1869; see *Wellesley index*), was James Bowling Mozley.

13 CD had recently obtained specimens of *Drosophyllum lusitanicum* and had begun studying it in the light of his earlier research on species of *Drosera* (see letter to W. C. Tait, 12 and 16 March 1869).

14 CD refers to William Henslow Hooker. See letter from J. D. Hooker, 17 July 1869 and n. 6.

To Mary Charlotte Lloyd [24 July 1869][1]

Saturday

My dear Miss Lloyd

I have thought that you w^d. perhaps like to see the enclosed from Mr Boyd Dawkins.[2] Althought the bones themselves are not interesting, yet if he is correct that they indicate the work of Prehistoric man, they promise well, as every scrap of information about our old ancestors is valuable.

Pray believe me | Yours very truly obliged | Ch. Darwin

P.S. | If I do not hear to the contrary, I will send for the dear old Horse tomorrow at 12.—[3]

Huntington Library, Art Gallery and Botanical Gardens (CB 383)

1 The date is established by the relationship between this letter and the letter to W. B. Dawkins, 19 July [1869] and by the reference to borrowing Lloyd's pony the next day (see below, n. 3). The Saturday following 19 July 1869 was 24 July. CD had already left Wales by the following Saturday.

2 CD refers to William Boyd Dawkins. The enclosure has not been found but was evidently a list of extinct species whose bones had been found on property belonging to Lloyd's family (see letter to W. B. Dawkins, 19 July [1869] and nn. 2 and 3).

3 CD refers to Lloyd's pony, 'Whitey' (see letter to J. D. Hooker, 8 July [1869] and n. 3).

From Albany Hancock 25 July 1869

S^t. Mary's Terrace | Newcastle on Tyne
25th July 1869.

My dear Sir,

My having been from home must account for the delay that has occurred in replying to your note of the 17th Inst. I have just returned after an absence of upwards of a week.[1]

It would give me great pleasure to assist you to the required Calcareous Sponges, but I unfortunately have no collection of them—[2] I possess, however a few of the British Clionæ and could supply two or three species of these curious excavating forms if you think they would be of any service.[3]

The gentleman best able to aid you is the Rev. A. M. Norman,[4] Newbattle—Fence-Houses, Durham— He has an extensive collection of British Sponges, and would, I, am sure, have much pleasure in giving every assistance in his power.— I have given his address above, that you can if you think proper apply to him direct;—or should you prefer it I will make the application in your name—

With every wish to aid you in this matter—I am | My dear Sir, | Your's ever truly | Albany Hancock.

Ch. Darwin Esqre

DAR 166: 97

[1] See letter to Albany Hancock, 17 July [1869].
[2] See letter to Albany Hancock, 17 July [1869].
[3] Hancock refers to sponges of the genus *Cliona*, whose members burrow into coral, calcareous rock, or shells of both living and dead molluscs (see Hancock 1849).
[4] Alfred Merle Norman.

From Charles Voysey 27 July 1869

Healaugh Vicarage | Tadcaster
July 27. 1869

Sir

I have been reading with intense interest & pleasure your new work on Variation &c I little expected to have any fact drawn from my personal observation worthy of your notice, until I came to page 314 (Chap. XXIV), where you remark of the potato that it is probably as "tender in England as when first introduced".—[1]

I have a partially sheltered garden, but not sufficiently so to have shielded my early potatoes in the Spring of 1865 & 1866—in which great damage was done to a crop of Lunn Kidneys out of Leicestershire.

The crop of 1865 was remarkably weak & lanky owing to premature budding in the previous autumn. But I had an idea that the plants might be strengthened by invariable selection of the largest & soundest tubers & by richer tillage of the soil. *Less & less damage has been done every succeeding year by frost*; for the last two years I have not had one plant destroyed, and *this* year, in sheltered & unsheltered parts of my garden alike, I have not had even a single blistered leaf.

To give further value to this fact, I must add that my potatoes were out of ground a full month before those of my neighbours, and it is the wonder of the district that while repeated severe frosts injured later crops in all directions, mine wholly escaped. The stems were splendid both in thickness & height (2 ft. 3 to 2 ft. 8) & the produce has been magnificent.—

I am perhaps not hasty in inferring that the *largest* tubers will furnish the *least tender* plants, provided the land be in rich culture.—

I am Sir | very faithfully Y^{rs}. | Charles Voysey

Charles Darwin Esqre.

DAR 180: 16

CD ANNOTATION
Top of letter: 'Acclimatisation by selection' *pencil*

[1] In *Variation* 2: 313–14, CD had argued that attempts at acclimatisation would only succeed if plants were propagated by seed and if hardier individuals were preserved.

To Dudley Campbell 29 July [1869]

Caerdeon Barmouth | North Wales.
July 29.

My dear Sir

I have had pleasure in appending my name as you desire it, for I have no doubt that the Westminster Review has aided materially in spreading liberal opinions.[1] In order to save writing a separate letter I have taken the liberty to enclose a cheque for £5 as my subscription.[2]

The date of this letter will explain the cause of my delay in answering it.[3]

I am much obliged for your kind enquiries about my health, which I am sorry to say remains much the same

My dear Sir | yours very faithfully | Ch. Darwin

P.S | I return home tomorrow morning

Endorsement: '29 July 1869'
LS(A)
National Library of Scotland (MS 5509: 92)

[1] The request from Campbell and enclosure have not been found.
[2] A payment of £5 to Campbell on 29 July 1869 is recorded in CD's Account books–banking account (Down House MS), and the same amount is recorded as paid to *Westminster Review* under the heading 'Gifts–Annual Subscriptions' in CD's Classed account books (Down House MS).
[3] CD probably refers to the place rather than the date. He had been away from home since 10 June 1869 and arrived back on 31 July 1869 (see 'Journal' (Appendix II)).

To Alfred Merle Norman 29 July [1869][1]

Down Beckenham | Kent
July 29

Dear Sir

I hope that you will excuse the liberty which I take in begging a favour of you. Prof. Haeckel is carefully studying *calcareous* sponges & is very anxious for British specimens.[2]

Mr. Hancock informs me that you are the most likely person to possess, & to feel inclined to give under these circumstances, some specimens.[3]

If this is in your power I sh^{d} be grateful if you w^{d} send me any, however few, specimens addressed to

"C. Darwin Esqre
care of Down Postman
Bromley Kent."

With apologies for troubling you I remain dear Sir | yours faithfully | Ch. Darwin

LS
University of Chicago Library

[1] The year is established by the relationship between this letter and the letter from Albany Hancock, 25 July 1869.
[2] See letter from Ernst Haeckel, 2 July 1869.
[3] See letter from Albany Hancock, 25 July 1869.

From Louis Rérolle 30 July 1869

Lyons,
the 30 July 69

Sir

Since the last letter I had the honour to write you, after having received the foot-notes you sent me, I went on translating your Orchis book.[1] Now, the two first chapters are entirely ready and copied out, together with the appended notes, I shall immediately despatch them to the publisher, who has agreed with me to print them before he has received the others.[2] I have, besides, looked at some Orchid-flowers that you describe in other chapters, of which I have finished some passages; so that I still hope that the whole will be ended in November next, and owing to the printing being already begun, the publication will take place in february 1870.[3]

I generally take care to have my French version as near your English text as possible, making only the slight modifications which are required by the peculiar idiom of our language, and in all cases, endeavouring to express as strictly as possible your idea. Allow me to ask you some explanations for a few points of detail.

In the case of the Epipactis palustris, you give to the terminal portion of the labellum the epithet of *distal*, which I cannot understand.[4]

In your notes, you speak of little insects which you call *thrips*; though I suppose it is what we name in French *pucerons*, as I am not sure of it, I shall ask you to send me the Latin or French name.[5]

I suppose it is needless to repeat some epithets like these "*pouch-formed* rostellum, *saddle-formed* disc, *boat-formed* disc, etc." which you frequently repeat; in French, we have not so many compound adjectives as in English, and we are obliged to say "le rostellum en forme de poche, le disque en forme de selle, etc", which being repeated makes the phrases languid.

I hesitated in translating the word *Bank*. For instance, in the list given in the p. 40 of your work, when you say "six plants from a much exposed bank" do you mean a green bank, a grassy mound, a river bank, or simply a locality?

Would you be desirous, Sir, to look over one or several chapters of my translation before printing, I would soon send you the M.S.

Allow me, Sir, to assure you of the deep respect with which I have the honour to remain | Your very obedient servant | L. Rérolle

P.S. Would you be so kind as to answer me, I pray you to direct your letter to the address: quai Joinville, 9, Lyon.

DAR 176: 131

1 Rérolle refers to his letter of 10 May 1869 and to *Orchids*.
2 The publisher of Rérolle's translation (Rérolle trans. 1870) was Charles-Ferdinand Reinwald.
3 The translation appeared in June 1870 (see *Correspondence* vol. 18, letter from Louis Rérolle, 15 June 1870).
4 See *Orchids*, pp. 94–5.
5 'Thrips' is both the common name given to any member of the insect family Thripidae and one of the genus names within that family. CD probably used the term in the former sense. 'Pucerons': aphids.

From C. S. Bate 31 July 1869

Plymouth
July 31. 69

My dear Sir

I find your letter upon my return home after a short absence.—[1] In *carcinas mænas* the male carries the female about hugging her with the dorsal surface against his ventral surface—seeking apparently some quiet place of concealment, this continues longer or a shorter time until the female moults, when immediately copulation takes place & continues some day or two probably until the integument of the female becomes hard & calcareous:— Copulation takes place face to face—[2]

Crustacea—not parasitic or entomostracous are I think generally larger than the female[3] Certainly so in the Brachyura to a marked degree—generally—But not conspicuously so in the Anomoura and Macroura—[4]

In the Amphipoda & Isopoda the males are the larger except in Parasitic genera[5]—in which they are generally smaller— In the Ento-mostracous Crustaceæ I am less certain But I think variable

I hope that my information is of use as it always pleases me to find it so

Believe me yours sincerely | C. Spence Bate

DAR 82: A72–3

CD ANNOTATIONS
1.1 I find … calcareous:— 1.6] *crossed pencil*
Top of letter: 'Size' *pencil*

1 CD's letter to Bate has not been found.
2 *Carcinus maenas* is the common shore crab. For Bate's earlier letter on a male of this species seen fighting a crab of a different genus, see the letter from C. S. Bate, 1 February 1869.
3 In *Descent* 1: 332, CD cited Bate on the relative sizes of the sexes in Crustacea. Entomostraca was the name used to refer to all crustaceans other than Malacostraca (Leftwich 1973).
4 Brachyura (true crabs) and Anomura (a group that includes hermit crabs) are now infraorders of the order Decapoda, while Macrura was the order that included shrimp and long-tailed lobsters. All the above belong to the superorder Eucarida.
5 Amphipoda and Isopoda are now orders within the superorder Peracarida.

From A. M. Norman 2 August 1869

⟨*address excised*⟩

Dear S⟨ir⟩

It always gives me pleasure to aid as far as lies in my power any naturalist engaged in the investigation of any group of the British Marine Invertebrata.

Allman wrote to me a month ago making the same request as that ⟨*2 or 3 words excised*⟩ letter ⟨*2 words excised*⟩ ⟨mak⟩ing and I promised to send what calcareous sponges I could spare to Prof Haeckel.[1] That promise as yet remains unfulfilled though not forgotton. I have during the past month had every moment occupied in moving into a new house which I have just built, and which is hardly yet completed and when I tell you that I have been obliged to be constantly looking after every species of workman in the house—have been laying out the garden & grounds—being a bachelor have had to buy every article of furniture myself from blacking brushes to Piano—and besides this have had all my collection (some 90 cabinet drawers, about 4000 bottles of spirit specimens & innumerable boxes) to pack and transport—as well as usual Parish work, you will understand that it has not been possible for me as yet to fulfil the promise.[2] I enclose a letter which I have written to Prof Haeckel, which you can please read and post.[3] It will show you what help I can give and what not Bowerbank[4] is the only other person who *could* help him where I am unable to do so.

It may interest you to hear that I have received the Crustacea for examination which have been procured during the first part of the cruise of the 'Porcupine' off the West of Ireland.[5] They are an extremely interesting lot of things. For reasons already given I have not yet uncorked a single bottle, but a hand lens through the glass enables me to observe a very large number of northern things in the bottle—The following are from the greatest depths—

Lat 51.22 N Long 12.26 W. 808 faths Temp 41.4

A young specimen of the fine crab Geryon tridens Kröyer[6] (only previously procured from fishermen's deep water lines on a limited part of the Norwegian coast) **eyes well developed**.

Lat 56.10 N Long 13.36. W. *1263* fath Temp 37.3

Rhoea spinosa Sars (? or an allied species)[7]
Nebalia—apparently a new species.

Lat 54.54 N. Long 10.59. W. *1366.* fath

Phoxus— sp.[8]
Tanaidæ— sp
Isopod— a new sp
Diastylis— spe
Cytheria— sp[9]

Lat 56.24 N. Long 11.37. *1380* fath Temp. 37.2

Cumacea— sp
Tanaidæ— sp
New isopod— same as in last
Cythere—
Praniza-like form— too large to be the larval state of Anceus, perhaps the larval form of Tanais[10]

Lat 55.11 N. Long 11.31. W. 1443 fath

Tanais— sp.
Cypridina (Norvegica?)[11]

Besides the things named these bottles contain some very fine Copepoda—species wholly unknown to me—being free swimming species they may be presumed to have been taken into the dredge as it was being drawn up, but the species are so different from those previously known to us as British that I am inclined to think that they are forms which do not come to the surface, but live in the greater depths, and if this be so—here we have a *source of food for the abyss-inhabiting fishes.*[12]

Believe me, dear Sir: | Yours faithfully | A. M. Norman

August 2. 1869.

I have posted to you a copy of the Shetland Dredging Report just received from the Printer.[13]

[Enclosure]

Burnmoor Rectory | Fence Houses | Co. Durham

Dear Sir.

I have received two letters—one from Prof. Allman the other from Mr Darwin—asking me if I could assist you in your studies of the calcareous Sponges, by supplying you with specimens of the British Species.[14]

It will afford me, I assure you, great pleasure to do this as far as lies in my power. Of the 12 British species, I can give you the following

Grantia compressa (several varieties)
Grantia ciliata[15] (two forms)
Leucosolenia botryoides
Leucosolenia coriacea[16]
Leuconia nivea
— fistulosa[17]
Leucogypsia Gossei[18]

Of Leucosolenia contorta[19] I have no specimen; and of the remaining four species I have only single examples. If however *external examination* will be of service to you I will gladly sent you these for that purpose, and they can be returned when you have finished with them. They are all so small that it would destroy them to attempt to make sections; and one is the type specimen of *Leuconia pumila*.[20] On hearing from you whether you would wish to see these specimens and how you would wish the sponges forwarded the parcel shall be sent.

Believe me | Yours faithfully | (The Rev) A. M. Norman

DAR 172: 73; Ernst-Haeckel-Haus (Bestand A-Abt. 1: 1–52/22)

1 See letter to A. M. Norman, 29 July [1869]. Norman refers to George James Allman and Ernst Haeckel.
2 Norman had been appointed rector of the newly created parish of Burnmoor in County Durham in 1866; the building of the new rectory began in 1868. He was a leading authority on, and collector of, North Atlantic marine invertebrates; his collection was ultimately acquired by the British Museum. (Mills 1980.)
3 See enclosure and also letter to Ernst Haeckel, 3 August [1869].
4 James Scott Bowerbank.
5 HMS *Porcupine* made three deep-sea dredging expeditions off the coast of Ireland in 1869; the first of these took place from 18 May to 13 July (Mills 1980).
6 *Geryon tridens* is now *G. trispinosus* (family Geryonidae, deep-sea crabs).
7 *Rhoea spinosa* is now *Apseudes spinosus* (family Apseudidae).
8 The amphipod genus *Phoxus* is no longer recognised because the name *Phoxus* had already been assigned to a genus of Coleoptera, but Norman evidently refers to a species of the modern family Phoxocephalidae.
9 The bivalve mollusc genus *Cytheria* (an incorrect subsequent spelling of *Cytherea* Lamarck) is no longer recognised because the name *Cytherea* had already been assigned to a genus of Diptera, but Norman evidently refers to a species of the modern family Veneridae.
10 Norman refers to the genera *Praniza* and *Anceus* (now *Gnathia*) of the isopod family Gnathiidae and to the genus *Tanais*, formerly considered to be an isopod genus but now placed in the order Tanaidacea.
11 The ostracod *Cypridina norvegica* is now *Vargula norvegica* (family Cypridinidae).
12 The food source for deep-water fish was a matter of contemporary scientific debate, but Norman never published this suggestion; for more on this, see Mills 1980.
13 Norman refers to Norman 1868. Between 1861 and 1867 Norman collaborated with John Gwyn Jeffreys on a series of deep-sea dredging expeditions around the Shetland Islands, supported by the British Association for the Advancement of Science; the marine organisms recovered were described by Norman, Bowerbank, and George James Allman.
14 See above, n. 1.
15 *Grantia ciliata* is now *Sycon ciliata*.
16 *Leucosolenia coriacea* is now *Clathrina coriacea*.
17 *Leuconia fistulosa* is now *Leucandra fistulosa*.
18 *Leucogypsia gossei* is now *Leucandra gossei*.
19 *Leucosolenia contorta* is now *Clathrina contorta*.
20 *Leuconia pumila* is now *Leucandra pumila*.

To Walter Elliot 3 August [1869][1]

Down. | Beckenham | *Kent. S.E.*
Aug. 3[d]

My dear Sir

Your former great kindness in giving me valuable information, makes me think that you will, if in your power, answer me a question.[2]

I want to know what mammals are polygamous, & I cannot find any statement about the wild Boar in Europe. In India do you happen to know whether during the breeding season, the Boar is accompanied by two or more sows?[3]

I hope that you will excuse me for troubling you & believe me | My dear Sir | Yours faithfully | Ch. Darwin

Brown University Library

[1] The year is established by the relationship between this letter and the letter from Walter Elliot, 7 August 1869.

[2] CD cited Elliot a number of times in *Variation* for providing him with information on horses and poultry in India. Their last known correspondence was in 1858; see *Correspondence* vol. 7.

[3] Elliot resided in India from 1820 until his return to England in 1860 (*ODNB*). CD was researching polygamy in relation to sexual selection, and cited Elliot for information on polygamy in boars in India in *Descent* 1: 267.

To Ernst Haeckel 3 August [1869][1]

Down, Beckenham, Kent
Aug 3[d]

My dear Häckel

Many thanks for your very kind letter of July 2[d].[2] I first wrote to Huxley, to whom I found that you had applied;[3] & then to Albany Hancock, who says he sh[d]. have been most happy, but has never collected calc. sponges. He says he could give you 2 or 3 species of the burrowing Cliona; if these w[d]. be of service.—[4] I then applied to the Rev[d]. A. M. Norman & enclose his answer.—[5] You had better write direct to him, as this w[d]. please him; but you could (if you have to write to me) enclose one for him & I could stamp & forward it.—

Do not forget to say how specimens could be forwarded. The man who has most specimens is "J. S. Bowerbank Esq[r] F.R.S, East Ascent, St. Leonard's-on-sea";[6] & I daresay he w[d]. send you specimens, if you were to write to him; but I do not much like to do so.—

What an indomitable worker you are!— I hope you will reap the richest reward possible by making more fine discoveries. You will soon be ranked (not that you will care much about that) amongst the few first naturalists in Europe: already I see your works more & more quoted, in England, Italy & U. States.

Once again many thanks for your very pleasant letter & believe me | Yours most sincerely | Ch. Darwin

Ernst-Haeckel-Haus (Bestand A-Abt. 1: 1–52)

[1] The date is established by the relationship between this letter and the letter from Ernst Haeckel, 2 July 1869.

[2] Letter from Ernst Haeckel, 2 July 1869.

[3] CD refers to Thomas Henry Huxley, the letter to T. H. Huxley, 9 July [1869], and the letter from T. H. Huxley, 16 July 1869.

[4] See letter to Albany Hancock, 17 July [1869], and letter from Albany Hancock, 25 July 1869 and n. 3.

5 See letter to A. M. Norman, 29 July [1869], and letter from A. M. Norman, 2 August 1869 and enclosure.

6 CD refers to James Scott Bowerbank. See also letter from A. M. Norman, 2 August 1869.

To John Lubbock 3 August 1869

Down
Aug 3^{d}. 1869

My dear Lubbock

I am at home again & thank Heavens at work again;[1] & I want very much to plague you with 2 or 3 questions.

I write to mention, (& perhaps give woodcut) as curious sexual difference, the right-hand antenna in ♂ *Labidocera, Darwinii* as described by you in Annal & Mag Vol. XI. (1853) Pl. I. Now will you tell me in a few words the character of the left-hand antenna, which resembles both antennæ in ♀.— May I say that these antennæ consist of "simple tapering joints less in number than in the right-hand antenna?"[2] Or are the joints equally numerous, though different in form?

In the Crustacean which has the right-antenna modified, is one of the hinder thoracic legs *generally* or only *sometimes* converted into a forceps? & is this forceps on the *same* right side of the body with the modified right antenna?

Lastly, I can find no account of any secondary sexual differences in the Myriapoda, except that it is said (I know not do you know? whether on good authority) that in Lithobius the *female* has a pair of pincers or forceps at entrance of vulva.—[3] Do you know whether the males ever differ from ♀^{s} in size, colour, jaws for fighting, or prehensile organs?

I have looked at the sentence in the Origin about which you wrote, but I find that I do not so expressly imply that Agassiz believes in descent or derivation of species: I have, however, by mere good luck altered the sentence in the new Edit. of Origin.—[4]

Have you settled anything about Mr Powell.[5]

Pray forgive me for troubling you & believe me | My dear Lubbock | Yours most truly | Ch. Darwin

P.S. One other question: in such Entomostraca as you know are the males larger than the females, as is the case with the stalk-eyed & sessile-eyed crustaceans?[6]

DAR 263: 66 (EH 88206510)

1 The Darwins arrived home from Wales on 31 July 1869 (see 'Journal' (Appendix II)).

2 CD discussed sexual differences in the antennae of the copepod species *Labidocera darwinii* in *Descent* 1: 329, and reproduced part of the plate that accompanied Lubbock's description of the organism in *Annals and Magazine of Natural History* (Lubbock 1853).

3 *Lithobius* is a genus of centipedes in the arthropod subphylum Myriapoda.

4 See letter from John Lubbock, 2 July [1869]. In *Origin* 4th ed., pp. 532–3, CD wrote:

> As the embryonic state of each species and group of species shows us more or less completely the structure of their less modified ancient progenitors, we can see why ancient

> and extinct forms of life should resemble the embryos of our existing species, their descendants. Agassiz believes this to be a law of nature; but I am bound to confess that I only hope to see the law hereafter proved true.

He altered this in *Origin* 5th ed., p. 534, to read:

> As the structure of the embryo generally shows us more or less plainly the structure of its less modified and ancient progenitor, we can see why ancient and extinct forms so often resemble the embryos of existing species in the same class. Agassiz believes this to be a universal law of nature; and I hope to see it hereafter shown in most cases true.

5 CD refers to Henry Powell. See letter from John Lubbock, 20 July [1869] and n. 2.
6 Entomostraca was formerly used to refer to all crustaceans other than Malacostraca (Leftwich 1973).

To M. C. Lloyd 4 August [1869][1]

Down. | Beckenham | *Kent. S.E.*
Aug. 4th

My dear Miss Lloyd

I have thought that you wd. like to see the enclosed letter from Mr B. Dawkins, who has evidently been much pleased by his visit, though in some geological respects not so interesting as he expected.[2] There is something wonderfully curious in coming across remains of an old savage progenitor, with their ground-down teeth.—[3]

Pray believe me | yours truly obliged | Ch. Darwin

American Philosophical Society (373)

1 The year is established by reference to William Boyd Dawkins's visit to north Wales; see n.2, below.
2 Dawkins visited Lloyd's family home at Rhagatt on the Gower peninsula in north Wales from 29 to 30 July 1869. He was investigating the site where a group of prehistoric bones, sent on to him by CD, had been discovered. (Dawkins 1870, Lucas 2007, p. 326; see also letter to W. B. Dawkins, 19 July [1869], and letter to M. C. Lloyd, [24 July 1869] and n. 2.) Dawkins's letter to CD has not been found, but presumably reported the initial results of his investigation; see n. 3, below.
3 Dawkins commented in his later published account of the excavations that all the adult human teeth discovered at Rhagatt were ground perfectly flat (Dawkins 1870, p. 444).

From J. D. Hooker 5 August 1869

Royal Gardens Kew
Aug 5 /69.

Dear Darwin

I thought you would be interested to know that Dr Cunningham has brought the upper & lower jaws of an Anoplotherium from the Gallegas river— Huxley shewed me them today.[1]

I went to Brighton yesterday & saw Hallett's wheat crops.— his results by Selection are very striking—[2] He finds by experiment that having advanced to a certain point in improving the quality in a desired direction he can get no further, & that then the power of varying in other directions is much reduced— thus, the Le Couteur wheat which,—itself a very highly improved strain, he has tried in vain to break—

he can however from common wheats raise a strain far superior to LeCouteurs,—[3] I was much pleased with the man & his methods;—though not altogether a Scientific observer & experimenter,—I think he is a sound & good one in many respects.— He is drawing up a paper for the British Association.[4]

Has this been proved *on wheats* before?

Huxley is arranging all his Darwinian papers for republication—[5] he is a *most* wonderful man—& I am beginning to feel quite infantine beside him!— I have read his answer to Congreve,—it is crushing, but too discursive as it struck me—[6]

I have not seen the N. B. which has a "fling at me" as you call it—[7] tell me "on your soul" should I see it?— I have not the smallest curiosity to do so except you think it merits it in any way, great or small. I have no time for mere literature & no mood for worry: & I still dream at times that I have the address to deliver!.[8]

What did you think of Govt. offering, & Sabine's accepting, K.C.B.—[9] I do think that the chucking the bauble into the poor man's Coffin—so to speak—is in the most atrocious bad taste—& worse than that is the exhibition of eternal childhood in the aged philosopher's clutching at it— I should have refused it & written to the Times! I am sure, if he deserved it at all, he did so 20 years ago, when it would have been a gracious acknowledgment of public services.

I am anxious to hear how you are now that you are at home? & Mrs Darwin & George.[10]

Ever yrs affec | J D Hooker

DAR 103: 25–6

CD ANNOTATIONS
2.2 He finds ... to break— 2.5] *crossed pencil*
Top of letter: '(Palaeontology)' *pencil*
End of letter: '(Lyell)' *ink*

1 Hooker refers to Robert Oliver Cunningham and Thomas Henry Huxley. See letter from T. H. Huxley, 7 May 1869 and nn. 2 and 3.

2 Frederic Francis Hallett had been experimenting on wheat since the 1840s but CD apparently did not read his principal paper, 'On "pedigree" in wheat as a means of increasing the crop' (Hallett 1861) until Hallett sent it to him in 1875 (letter from F. F. Hallett, 21 May 1875 (*Calendar* no. 9988)).

3 CD cited John Le Couteur's experiments in the first edition of *Variation* (*Variation* 2: 200); he expanded this in the second edition to refer also to Hallett's experiments (*Variation* 2d ed., 2: 184). There is a copy of Le Couteur 1836 (*On the varieties, properties, and classification of wheat*) in the Darwin Library–CUL.

4 Hallett read a paper titled 'On the laws of the development of cereals' at the British Association for the Advancement of Science meeting at Exeter on 18 August 1869; an abstract was published in the *Report of the thirty-ninth meeting of the British Association for the Advancement of Science*, Transactions of the sections, p. 113. There is copy of the paper in the Darwin Library–CUL sent to CD by Hallett in 1875 (letter from F. F. Hallett, 18 May 1875 (*Calendar* no. 9982)); it appears to have been privately printed.

5 Hooker refers to Thomas Henry Huxley and his volume of collected papers, *Lay sermons, addresses and reviews* (T. H. Huxley 1870). In it he reprinted two reviews of *Origin* from the *Westminster Review*, April 1860 (T. H. Huxley 1860), and the *Natural History Review*, 1864 ([T. H. Huxley] 1864).

6 Hooker refers to T. H. Huxley 1869b, in which Huxley responded to an article by Richard Congreve attacking his views on Auguste Comte. See also letter to J. D. Hooker, 24 July [1869] and n. 6.

7 On the criticism of J. D. Hooker 1868 in an article in the *North British Review*, see the letter to J. D. Hooker, 24 July [1869] and n. 7.

[8] Hooker refers to his presidential address to the previous year's meeting of the British Association delivered in Norwich on 19 August 1868 (J. D. Hooker 1868).

[9] Edward Sabine's appointment as a Knight Commander of the Order of the Bath was announced in the *London Gazette* on 27 July 1869.

[10] The Darwins had returned to Down from a holiday in Wales on 31 July 1869 (see 'Journal' (Appendix II)). CD, George Howard Darwin, and Emma Darwin had all been unwell (see letter to J. D. Hooker, 24 July [1869]).

From John Lubbock 5 August [1869][1]

15, Lombard Street, E.C.
5 Aug.

My dear Mr. Darwin

The left hand antenna of Labidocera consists of simple tapering joints as you say, but the number is the same on both sides.[2] Indeed the right hand one in some species seems to have fewer segments in consequence of coalescence.

One of the hinder, generally the hindermost leg is often converted into a forceps, & this always on the right side.[3] I have examined hundreds, I might say thousands, & never met with an exception—that is I never saw the left leg modified instead of the right.

I cannot answer your question about the Myriapods offhand, but will gladly lend you the Suites a Buffon if you care to have it.[4]

One pair of the legs is I know in many species sexually modified.

In the Entomostraca the males are very frequently smaller than the females, this is especially the case in the parasitic Copepods.[5]

I am afraid I shall not get to see you before Exeter as I am off to open a tumulus on Thursday next[6]

Believe me | Yours most sincerely | John Lubbock

DAR 82: 84–5

[1] The year is established by the relationship between this letter and the letter to John Lubbock, 3 August 1869.

[2] See letter to John Lubbock, 3 August 1869.

[3] CD cited this information in *Descent* 1: 329.

[4] See letter to John Lubbock, 3 August 1869. Lubbock refers to Walckenaer and Gervais 1837–47. In *Descent* 1: 339–40, CD mentioned *Lithobius*, the female of which has prehensile appendages for holding the male during intercourse, as one of the few well-marked instances of sexual differences in the *Myriapoda*; he gave Walckenaer and Gervais 1837–47, 4: 19, as the authority.

[5] See letter to John Lubbock, 3 August 1869 and n. 6. In *Descent* 1: 332, CD said that Entomostraca were an exception to the general rule in crustaceans that males were larger than females, but did not cite Lubbock.

[6] Lubbock attended the meeting of the British Association for the Advancement of Science held in Exeter from 18 to 25 August 1869 as a vice-president of the biology section (*Report of the thirty-ninth meeting of the British Association for the Advancement of Science, held at Exeter*). On 14 August, he took part in the excavation of a prehistoric structure recently discovered at Cwm on the Gower peninsula, Wales (Lubbock 1870).

From John Lubbock [after 5 August 1869][1]

High Elms

Dear Mr Darwin

Thanks for your kind note.[2]

I am not yet quite sure whether we shall have people with us tomorrow or not. If not I will come up & see you.

I send the Suites a Buffon on Myriapods.[3]

Any way I hope you will come to luncheon the Sunday following.

Believe me | Yours most sincerely | John Lubbock

DAR 170: 70

1 The date is established by the relationship between this letter and the letter from John Lubbock, 5 August [1869].

2 No letter from CD in reply to Lubbock's letter of 5 August [1869] has been found.

3 Walckenaer and Gervais 1837–47; see letter from John Lubbock, 5 August [1869] and n. 4.

From Walter Elliot 7 August 1869

Wolfelee | Hawick | N.B.
7 August 1869

My dear Sir

In reply to your inquiry about the Indian Wild Boar, I believe the male in the breeding-season consorts with several sows.[1] Except in the breeding season, the old Boars are solitary, but in beating them out of cover, I have frequently found an old Boar with the Sounder.[2] When not so associated the old Boar goes out to feed alone for I have often, where the cover was too extensive to beat, gone out early in the morning before dawn, & watched for these solitary males returning from feeding, as they always show the best sport. They are the most savage & powerful & when they join the sounder, they drive the young boars away, wh. are then found two or three together, harboring apart but returning to the Sounder after the breeding period when the old one reverts to his solitary habits. I cannot say positively that this is the universal & constant habit of the animal, but it is a very general one, as far as my observation goes.

I dont think I ever thanked you for the copy of the "Variation of Animals & Plants under Domestication", wh you were kind enough to send me.[3] It came during a prolonged absence from home & I found it on my table when I returned. I had hoped to do so in person, when I was at High Elms about two months ago, but Mr John Lubbock, who had promised to take me over to Down, did not come home till too late to do so & I was obliged to go back to Town next morning[4]

He promised however to tell you of my regret at the disappointment.

Believe me | Yours very truly | Walter Elliot

Charles Darwin Esqr

DAR 86: A72–3

CD ANNOTATION
Top of letter: 'Polygamy' *blue crayon*

[1] See letter to Walter Elliot, 3 August [1869]. CD cited Elliot for this information in *Descent* 1: 267, in a discussion of polygamy in relation to the principles of sexual selection.
[2] 'Sounder': a herd of wild swine (*OED*).
[3] Elliot's name appears on CD's presentation list for *Variation*; see *Correspondence* vol. 16, Appendix IV.
[4] John Lubbock and his family, of High Elms, Down, were near neighbours and friends of the Darwins. Elliot and Lubbock had both attended the meeting of the British Association at Dundee in 1867 (see *Anthropological Review* 6 (1868): 17).

To J. D. Hooker 7 August [1869][1]

Down. | Beckenham | *Kent. S.E.*
Aug 7th

My dear Hooker

There never was such a good man as you for telling me things which I like to hear. I am not at all surprised that Hallett has found some vars. of wheat could not be improved in certain desirable qualities, as quickly as at first.[2] All experience shows this with animals; but it wd. I think, be rash to assume, judging from actual experience, that a little more improvement cd. not be got in the course of a century; & theoretically very improbable, that after a few 1000s rest, there wd not be a start in the same line of variation. What astonishes me, as against experience, & what I cannot believe, is that varieties already improved or modified do not vary in other respects: I think he must have generalised from 2 or 3 spontaneously fixed varieties. Even in seedlings from same capsule some vary much more than others: so it is with sub-varieties & varieties.—

It is a grand fact about Anoplotherium;[3] & shows how even terrestrial quadrupeds had time formerly to spread to very distant regions. At each epoch the world tends to get peopled pretty uniformly; which is a blessing for Geology.—

The article in N. British is well worth reading scientifically: George D. & Erasmus were delighted with it.[4] How the author does hit! It was a euphemism to speak of a fling at you: it was a kick.— He is very unfair to Huxley & accuses him of "quibbling" &c—[5] Yet the author cannot help admiring him extremely.— I know I felt very small, when I finished the article. You will be amused to observe that geologists have all been misled by Playfair, who was misled by 2 of the greatest mathematicians![6] and there are other such cases; so we could turn round & show from Reviewer how cautious Geologists ought to be in trusting mathematicians.—

There is another *excellent*, original article, I feel sure by McClennan, on Primeval man;—well worth reading.—[7]

I do not quite agree about Sabine:[8] he is unlike every other soldier or sailor I ever heard of, if he would not put his second leg into the tomb with more satisfaction as K.C.B. than as a simple man.— I quite agree that Government ought to have made him long ago: but what does Government know or care for Science.— So much for your splenditious letter.—[9]

We came home this day week; & good Lord, how pleased I was: walking has done me the greatest good & I am above my usual standard of strength. George is nearly well, & Emma better, but not quite right.—[10]

I have 3 plants of Drosophyllum in splendid health & flowering. I sh[d] much like to hear what you decide on its affinities with Drosera.—[11] W[d]. you like one or 2 of my plants— I suppose we c[d] somehow pack them up?—

Whether you will care to read the enclosed $\frac{1}{2}$ letter I am sure I do not know: you can keep the seed: the monstrosity is like that which you described formerly in Gardeners Chronicle.[12]

Thanks for sheets from Proc. Z. Soc: I knew Elephants did not produce twins; but I stupidly said 3 pairs meaning 3 males & 3 females.[13] The paper has otherwise been *very* useful to me.— If you write to D[r] Campbell (*but not otherwise*) pray ask him whether he knows whether the wild Boar in India is polygamous: I have just written to Sir W. Elliot to ask.—[14]

Lastly have you the recent census of India: perhaps Thomson has: would Royal Society have it? I want to see proportion of men to women; but then I must hear whether the census is esteemed trustworthy.—[15]

Do you know anything about Lyell,[16] & what he is doing.—

This is a shamefully long letter.

Ever yours | C. Darwin

P.S. Do tell me what I ought to subscribe to the Faraday Memorial: I have never seen an advertisement or list of Subscriptions.—[17]

DAR 94: 144–8

1 The year is established by the relationship of this letter and the letter from J. D. Hooker, 5 August 1869.

2 CD refers to Frederic F. Hallett. See letter from J. D. Hooker, 5 August 1869 and n. 2.

3 See letter from J. D. Hooker, 5 August 1869 and n. 1.

4 CD refers to George Howard Darwin, Erasmus Alvey Darwin, and Peter Guthrie Tait's anonymous article in the *North British Review* ([Tait] 1869). See letter from J. D. Hooker, 5 August 1869; see also letter to J. D. Hooker, 24 July [1869] and n. 7.

5 Tait suggested that a 'spirit of quibbling' was evident in Thomas Henry Huxley presidential address to the Geological Society (T. H. Huxley 1869c) when Huxley discussed William Thomson's arguments on the age of the earth ([Tait] 1869, p. 426).

6 According to Tait, John Playfair's arguments in favour of a uniformitarian view of an essentially unchanging solar system were based on erroneous calculations by the mathematicians Joseph Louis Lagrange and Pierre Simon, marquis de Laplace ([Tait] 1869, pp. 414–15). Huxley had argued that geologists should be cautious in trusting the conclusions of physicists (T. H. Huxley 1869c, pp. xlvii–lii).

7 CD refers to John Ferguson McLennan and [McLennan] 1869 (*Wellesley index*).

8 CD refers to Edward Sabine. See letter from J. D. Hooker, 5 August 1869 and n. 9.

9 Letter from J. D. Hooker, 5 August 1869. 'Splenditious' is an invented Darwin family word (see *Correspondence* vol. 13, letter to J. D. Hooker, 1 June [1865] and n. 3).

10 See letter from J. D. Hooker, 5 August 1869 and n. 10. CD refers to George Howard Darwin and Emma Darwin.

11 See letter to J. D. Hooker, 24 July [1869] and n. 13.

12 CD evidently forwarded a portion of a letter from Fritz Müller, probably the missing part of the letter from Fritz Müller, 15 June 1869 (see letter from J. D. Hooker, 13 August 1869, and letter to Fritz Müller, 8 September [1869]). CD refers to Hooker's article 'The monstrous Begonia frigida at Kew, in relation

to Mr. Darwin's theory of natural selection', *Gardeners' Chronicle*, 25 February 1860, pp. 170–1. Müller had sent seed of a monstrous begonia with his letter of 14 March 1869 or later; see also letter to George Bentham, 10 May [1869] and nn. 5 and 7.

13 In a paper published in the *Proceedings of the Zoological Society of London* (A. Campbell 1869), Archibald Campbell stated that twin births were unknown in elephants (*ibid.*, p. 138). A reference in *Origin* to the rate of breeding of elephants had recently been misunderstood as suggesting that elephants could give birth to twins; see letter from Ponderer to the *Athenæum*, [before 5 June 1869] and n. 2. See also letter to the *Athenæum*, 7 [July] 1869. CD quoted a sentence on polygamy in elephants from A. Campbell 1869 in *Descent* 1: 267.

14 See letter to Walter Elliot, 3 August [1869] and n. 3.

15 A report on the census of the Central Provinces of India had been published in 1867, and one on the census of Oudh in 1869 (*A classified list, in alphabetical order, of reports and other publications in the record branch of the India Office, December 1892* (London: E. Spottiswoode, Queen's Printer, 1894), pp. 24–5). Thomas Thomson, a friend of Hooker's, had been superintendent of the Calcutta botanic garden and professor of botany at Calcutta Medical College.

16 Charles Lyell.

17 A subscription fund for a memorial to Michael Faraday, who had died in August 1867, was established at a public meeting at the Royal Institution on 21 June 1869 (*The Times*, 22 June 1869, p. 5). See also *Correspondence* vol. 16, letter from J. D. Hooker, 13 February 1868 and n. 6.

From T. H. Farrer 8 August 1869

Abinger Hall, | Dorking.
8 August /69

My dear Mr Darwin

You remember mentioning the separate stamen of papilionaceous flowers as a point worth looking into.[1] I have had too much to do to pay proper attention to it, but cannot help thinking that its function is to give the insects proboscis access to the nectar. In many cases where it does not exist. eg. Sarothamnus—Ulex—Genista Ononis—Lupin—the staminal tube is a tight fitting, thin tough, jacket to the pistil, with no space for nectar at foot between it and the pistil, & no space for a proboscis. In some at any rate of these flowers—the bees, which visit them constantly, certainly dont get what they want from the inside, but from the outside of the staminal tube. On the other hand in Lathyrus Pisum, Phaseolus, Robinia, Faba, where the 10th stamen is quite or partly separate, it is stiff & expands at bottom so as to leave between it & the pistil a cavity, in which there is nectar. And where the stamen is not quite separate all the way, it is separated at the bottom so as to leave a passage into the nectary.[2]

There are many other interesting points in these flowers illustrating your views about fertilization—eg the position of the flower before at and after blossoming—Invariably, as far as I have seen, the flower when in bloom is so placed as to make the wings & keel (or in one or two cases; the vexillum), a tempting lighting place for insects.— Almost, though not quite, as invariably, the bud and the pod are in quite different positions.— What can be more curious than to see the Common Pea[3]—the bud pendent: with the peduncle bent—straightening & lifting itself with the blossom—and again drooping in the pod? or again, the Lupin with a precisely opposite motion?

Except perhaps the still more curious case of the pendent racemes of Laburnum or Robinia, where the peduncle of each flower takes a half twist as it approaches blossoming, so as to bring the wings and keel into the same position as they usually occupy on an upright spike.

Again the brush on the style of Lathyrus: Vicia & Faba in its different positions, obviously serves to sweep out the pollen from the different shaped keels.

But I shall get tedious—and hope to be able when once well settled here to follow this up.

Hooker has promised to come and see us in our new abode in the course of the autumn, when I return from an official cruise amongst Lighthouses[4] and Lockyer has promised to come & set me up a Telescope—[5] I fear there is no chance of getting you to come and meet them: it would not be till the latter end of September or beginning of October.

We had a visit from the Leith Hill place party yesterday—but I have not yet been here long enough to go & see & envy their lovely view.[6]

Believe me | Very truly yours | T H Farrer

Charles Darwin Esq[r] FRS

DAR 164: 53

1 See *Correspondence* vol. 16, letter from T. H. Farrer, 21 November 1868 and enclosure and n. 4.

2 *Sarothamnus* (now *Cytisus*), *Ulex*, *Genista*, *Ononis*, *Lupinus*, *Lathyrus*, *Pisum*, *Phaseolus*, *Robinia*, and *Faba* were all genera in the family Leguminosae (now Fabaceae).

3 *Pisum sativum.*

4 Farrer refers to Joseph Dalton Hooker. Building work on Farrer's new house, Abinger Hall, Dorking, Surrey, was completed in 1872 (Nairn and Pevsner 1971). Farrer was permanent secretary of the Board of Trade, and had a long-standing professional interest in marine trade in particular (*ODNB*).

5 Farrer refers to the astronomer Joseph Norman Lockyer.

6 CD's brother-in-law, Josiah Wedgwood III, and his family lived at Leith Hill Place, Dorking, Surrey (Freeman 1978).

To Thomas Spencer Cobbold 9 August [1869][1]

Down, Beckenham, Kent,
August 9.

Dear Sir,—

In looking over some bottles with specimens in spirit from S. America and adjoining seas, collected by me nearly forty years ago, I find a few parasitic worms, which it has occurred to me you might possibly like to have. Should this prove the case, be so kind as to inform me and they shall be sent to you. I have looked at only one lot, viz. from the *Rhea*, or American Ostrich, and these seemed not in a bad state;[2] 2nd, worms from stomach of a Porcupine;[3] 3rd, from the mouth of a Snake;[4] 4th, from the wild *Cavia Cobaya*—these might be compared with any worms from the domestic Guinea-pig, which some authors think (I believe falsely) to be descended from the

C. Cobaya.[5] Also three sets from fish; but as I was very ignorant when I collected them, these perhaps are Lerneæ or their allies.[6] Should you care to have these specimens, I will give exact locality and date at which they were preserved.

Pray believe me, | Dear Sir, | Yours faithfully, | CHARLES DARWIN.

Cobbold 1885, p. 174

1 The year is established by Cobbold's statement that he received this letter in the autumn of 1869 (Cobbold 1885, p. 174)

2 When he later sent this specimen to Cobbold, CD labelled it 'From the stomach of a Rhea, Bahia Blanca, North Patagonia, 1832'; Cobbold identified it as *Filaria horrida* (Cobbold 1885, pp. 175–6). CD's 'Catalogue for specimens in spirits of wine' (Down House MS, Notebook 63.2, entry 400) lists an 'Intestinal worm taken out of the stomach of an Ostrich' at Bahia Blanca in September 1832 (R. D. Keynes ed. 2000, p. 333).

3 This is likely to be a mistaken transcription for 'Porpoise'. Included with the specimens that CD sent to Cobbold were some labelled 'Worms from stomach of a Porpoise off the I. of Chiloe, Jan. 1835' and identified by Cobbold as *Ascaris simplex* (now *Anisakis simplex*; see Cobbold 1885, pp. 176–7). See also CD's 'Catalogue for specimens in spirits of wine' (Down House MS, Notebook 63.2, entry 1159; Keynes ed. 2000, p. 357).

4 CD labelled this specimen 'Worm from mouth of a Snake allied to *Coluber*; Maldonado, Rio Plata, May 1833'; Cobbold identified it as *Distoma incerta* (Cobbold 1885, pp. 177–8). CD's 'Catalogue for specimens in spirits of wine' (Down House MS, Notebook 63.2, entry 645) lists a 'Small worm from mouth of Coluber' collected in Maldonado in May or June 1832 (R. D. Keynes ed. 2000, p. 340).

5 CD labelled the specimen 'Worm from the duodenum of *Cavia Capybara*, floating amidst the green digesting mass: Rio Plata, May 1833' ('Catalogue for specimens in spirits of wine', Down House MS, Notebook 63.2, entry 633 (R. D. Keynes ed. 2000, p. 340)). When CD sent Cobbold the specimens, he included a note that read, 'N.B. I see I made a mistake and spoke in my letter of *Cavia Cobaya*'. Cobbold identified the worm as *Oxyuris obesa* (now *Protozoophaga obesa*; see Cobbold 1885, p. 176). *Cavia capybara* is now *Hydrochaeris hydrochaeris*, the capybara.

6 Cobbold identified one specimen labelled by CD, 'Worm buried in the tail of a *Gadus*; T. del Fuego, Mar. 1833', as *Lernea branchialis* (now *Lernaeocera branchialis*). The remaining specimens remained unidentified and were labelled: 'Worm from under branchial covering of fish; T. del Fuego, 1834', 'Worm on scale of fish; T. del Fuego, Jan. 1833', and 'Worm from branchiae of *Lota* or *Gadus*; when caught protruded a transparent case full of eggs; T. del Fuego, Jan. 1834' (Cobbold 1885, p. 178; see also CD's 'Catalogue for specimens in spirits of wine', Down House MS, Notebook 63.2, entries 521, 552, 805 and 836 (R. D. Keynes ed. 2000, pp. 336–7, 345–6).) *Lota* is the genus of burbots; *Gadus* is the genus of codfishes.

From John Blackwall 10 August 1869

Hendre House,
August 10th, 1869.

Dear Sir,

In replying to your letter of the 5th instant,[1] I have to remark that immature male spiders usually appear to resemble in colour the females of their respective species of the same age, and that they do not seem to differ materially from them in size till the digital joint of their palpi, by its increasing tumidity, indicates a near approach to maturity, and affords a decided characteristic of sex. Whether these remarks are applicable to the entire order of the Araneidea or not, of course I am incompetent to state; the very remarkable differences in size and colour of the sexes of some species belonging to the genus Nephila may, perhaps, be thought to render it doubtful.

Lycosa lugubris,[2] Sparassus smaragdulus,[3] Philodromus dispar,[4] Linyphia pratensis[5] and Linyphia fuliginea[6] are species the sexes of which resemble each other in colour when young, but differ very decidedly in that respect when adult. I may mention here that young spiders of many species undergo modifications of colour in their progress towards maturity (especially on the occasion of changing their integument), and that such modifications seem to be common to both sexes.

Before your letter came to hand I had collected cocoons of the varieties of Theridion lineatum[7] for the purpose of observing whether the young on quitting them present any characteristics by which they may be distinguished from each other or not; my present opinion is that they do not, but should the result prove this impression to be inaccurate, I will immediately communicate with you on the subject.

In all my investigations relative to spiders I never heard them produce any sound by means of an organ or organs adapted to that especial purpose; the fact that they do so, if well ascertained, is as you remark highly interesting, and would be still more so should the function be limited to the male sex.

One of my correspondents, the Rev. O. P. Cambridge, has received from Ceylon two species of spiders each of which has only four eyes, descriptions of them will be given in the next number of the 'Journal of the Linnean Society.'[8]

I was sorry to learn from Miss Holland[9] that you had met with an accident;[10] hoping that you no longer suffer from the effects of it,

I am, dear Sir, | very truly yours | John Blackwall.

DAR 82: 82–3

CD ANNOTATIONS
1.2 usually] *underl blue crayon*
1.3 same age,] *underl blue crayon*
2.6 be ... sexes.] *underl blue crayon*

1 CD's letter has not been found.
2 *Lycosa lugubris* is now *Pardosa lugubris*, a thin-legged wolf spider.
3 *Sparassus smaragdulus* is now *Micrommata virescens*, the giant crab spider.
4 *Philodromus dispar* is a species of running crab spider.
5 *Linyphia pratensis* is now *Linyphia hortensis*, a species of dwarf and sheetweb weaver spider.
6 *Linyphia fuliginea* is now *Microlinyphia pusilla*, a species of dwarf and sheetweb weaver spider.
7 *Theridion lineatum* is now *Enoplognatha ovata*, a cobweb weaver spider.
8 Blackwall refers to Octavius Pickard Cambridge and Cambridge 1869. The specimens had been sent from Ceylon (Sri Lanka) by George Henry Kendrick Thwaites. Cambridge named them *Miagrammopes thwaitesii* and *Miagrammopes ferdinandi*.
9 Miss Holland has not been identified but may be one of the Holland sisters, Mary, Lucy, Elizabeth and Hannah Holland, of Knutsford, Cheshire, who were friends of Amy Wedgwood (see *Correspondence* vol. 8, letter to Miss Holland, [April 1860]).
10 CD had been injured in an accident while riding his horse in April 1869; see letter to A. R. Wallace, 14 April 1869 and n. 12.

To T. H. Farrer 10 August [1869][1]

Down. | Beckenham | *Kent. S.E.*
Aug. 10th.

My dear Mr Farrer

Your view seems most ingenious & probable; but ascertain in a good many cases that the nectar is actually within the staminal tubes. One can see that if there is to be a split in the tube, the law of symmetry, would lead it to be double & so free one stamen.[2]

Your view, if confirmed, would be extremely well worth publication before Linnean Soc.[3] It is to me delightful to see what appears a mere morphological character proved to be of use; it pleases me the more as Carl Nägeli has lately been pitching into me on this head:[4] Hooker, with whom I discussed subject, maintained that uses wd. be proved for lots more structures, & cheered me by throwing my own orchids into my teeth.[5]

All that you say about changed position of peduncle in bud, in flower & in seed is quite new to me, & reminds me of analogous cases with tendrils. This is well worth working out, & I daresay the brush of the stigma.—[6]

With respect to the hairs or filaments (about which I once spoke) within different parts of flowers.— I have a splendid Tacsonia with perfectly pendent flowers, & there is only a microscopical vestige of the corona of coloured filaments; whilst in most common passion-flowers, the flowers stand upright & there is the splendid corona which apparently wd catch pollen.—

On lower side of corolla of Foxglove there are some fine hairs, but these seem of not least use,—a mere purposeless exaggeration of down on outside,—as I conclude after watching the Bees at work, & afterwd covering up some plants; for the protected flowers rarely set any seed, so that hairy lower part of corolla does not come into contact with stigma, as some Frenchman[7] says occurs with some other plants, as Viola odorata & I think Iris.

I heartily wish I could accept your kind invitation, for I am not by nature a savage, but it is impossible—[8]

Forgive my dreadful handwriting, none of my women-kind are about to act as amanuenses—

Pray believe me | Yours very sincerely | C. Darwin

I am going to publish some notes on Orchids & will send you a copy: I give, but with utmost brevity, your correction of my two errors.—[9]

Linnean Society of London (LS Ms 299/8)

1 The year is established by the relationship between this letter and the letter from T. H. Farrer, 8 August 1869.

2 See letter from T. H. Farrer, 8 August 1869. Farrer cited CD on symmetry in the double aperture in Farrer 1872, p. 501.

3 Although Farrer wrote up his observations on fertilisation in papilionaceous flowers in the autumn of 1869 and sent them to CD, he did not send them to the Linnean Society; they were published in *Nature* in 1872 (Farrer 1872; for the history of the paper, see *ibid.*, p. 478).

4 Carl Wilhelm von Nägeli had criticised CD's theories in Nägeli 1865; see letter to J. D. Hooker, 13 January 1869 and n. 1.

5 See *Correspondence* vol. 16, letters to J. D. Hooker, 25 December [1868] and 29 December 1868; see also this volume, letter from J. D. Hooker, 15 January 1869.
6 See letter from T. H. Farrer, 8 August 1869.
7 The French botanist has not been identified.
8 See letter from T. H. Farrer, 8 August 1869.
9 CD cited Farrer in 'Fertilization of orchids', pp. 144, 146, for the information that the pollinia of both *Ophrys muscifera* and *Peristylus viridis* underwent a movement of depression.

From T. H. Farrer 12 August 1969

Abinger Hall, | Dorking.
12 Aug/69

My dear Mr Darwin

I will certainly under such encouragement try to work out the papilionaceous flowers when I return from a three weeks semi offical cruise in the Trinity House Steamer.[1] I have a few notes.

Are you sure about the hairs in the Foxglove.[2] I watched some large humble bees—who are very fond of them—and it seemed to me that the hairs on the under lip of the corolla served as an excellent foothold. The flower bends directly the bee alights, and he has quite a struggle to get perpendicularly up in which as it seemed to me he was much helped by the hairs under his feet. This may be fancy but I thought I saw it—

I feel pretty sure about the hairs on the style of Vicia & Lathyrus.[3]

But of all this more by & by.

I think I must when I come back run over for a day and ask you to shew me your crossing & covering up processes.

Fanny begs me to say that you are the last person she credits with savagery, & only wishes you were strong enough to come and enjoy a song[4]

Very truly yours | T H Farrer

Charles Darwin Esq[r] FRS

DAR 164: 54

1 See letter from T. H. Farrer, 8 August 1869 and n. 4, and letter to T. H. Farrer, 10 August [1869]. Trinity House corporation in London was the official regulator of British shipping, its responsibilities including the erection and maintenance of lighthouses on the coasts of England and Wales (*OED*).
2 See letter to T. H. Farrer, 10 August [1869].
3 See letter from T. H. Farrer, 8 August 1869, and letter to T. H. Farrer, 10 August [1869].
4 Farrer refers to Frances Farrer; see letter to T. H. Farrer, 10 August [1869].

From J. D. Hooker 13 August 1869

Kew
Aug 13 /69.

Dear old Darwin

I was delighted to hear of your being back & better.[1]

I did not mean to imply that Hallett affirmed that all variation stopped, far from it—[2] he maintains the contrary, but, if I understand him aright—he soon arrives at a point, beyond which any further accumulation in the direction sought is so small & so slow, that practically, a fixity of type (not absolute fixity however) is the result. Also that coincident with this point is that the plant is also very slow to vary in other directions than that it was bred to accumulate. This, I supposed, correlation would account for—viz. that while you are *knowingly* accumulating in one direction correlation obliges you *unknowingly* to be accumulating in others.

I suppose I must read the N. B.[3] but I never read now, & am getting very tired of the struggle, not for life, thank God, but of life. & am getting overweighted with duties of the Colonial & Foreign Office which want endless supplies of seeds & forest trees, &c, that I alone can procure, & I only through personal correspondence, with people, who would snap their fingers at Official requests.

The D. of A. has further **requested** me to superintend the publication of a Flora Sylvatica of India, that will give me a lot of trouble. I think he is paying me off for my kick at Nat: Theology![4] Next Tuesday I go to the Brit. Ass. which grieves me.[5] then for 2 days to Sir Thos Acland, & for as many to Symond of Pendock, where I hope the fair Cinthea will amuse me, for I am very dull & down in the mouth—[6] Especially because poor Willy is coming home from N.Z.—like a bad shilling to its passer!— he is so young of his age, childish & purposeless, that Hector can get no one to take it him—[7] He has also been threatened with more hemorhage, & has had a little more indeed, but of this I have said nothing to any one but my wife.[8] He will arrive about beginning of winter & I shall have to send him off again at once—

I have written to Mr Merivale at the F. O. & asked him to send you a copy of the India Census if he can.[9]

I have figured Drosophyllum & will make a few observations on it's affinities it is clearly Drosereceous, I went over the matter with Oliver—[10] I observe one remarkable character & I think *unique* in plant, that the leaf is **revolute** in Vernation please look to this— our specimens have gone to the dogs, I cannot conceive why— they had my special attention— Do not pack & send yours' as I am going away. I should like to run down for a Sunday after Exeter & have a chat with you about it—[11]

I saw Huxley the other day as bright as a lark—about to publish his "Darwiniana"[12]—he is a *prodigy*

What a capital letter of Fritz Mueller you sent me— how admirably he plans his observations— the *Begonia* is growing.[13]

I saw Lyell 3 weeks ago[14] he is I believe in Scotland—what doing I know not.

I see no necessity for your subscribing to the Faraday Monument— I *had* to as P. B. S.—& to do the maximum of £5·5·0[15] Also I had to do the Ross memorial, which was £10.—[16] These are the luxuries of Presidentship.

There will be a collection for the Faraday monument at Exeter, & I will pay for you if you really wish to, & say how much. but I see no call *whatsoever*. & disapprove this eternal touting for dead bones.— £2·2·0. would be ample any-how—

I have letters from Sydney with account of Ld Howes Island, the Bot^{y}. of which approaches Norfolk Island. There are 4 Palms on the Island (as many as in all N.S.W.!)

which is only $4\frac{1}{2}$ miles long! *Not a Proteaceæ*! & very few *Leguminosæ*, *Myrtaceæ* or *Rutaceæ*! I suspect that the Palms are Australian types.[17] What is to be done with such Islands, they fit into no theory.

Ever yrs affec | J D Hooker

Poor Moggridge has just called— he looks *dreadfully* ill—& has been very bad indeed.—[18]

DAR 103: 27–9, DAR 100: 156

CD ANNOTATIONS
3.1 I suppose … of life. 3.2] 'X (Tait)'[19] *added pencil*
4.6 Especially … passer!— 4.7] 'X' *added pencil*
6.2 I observe … to this— 6.4] 'X' *added pencil*
6.5 I should … it— 6.6] 'X' *added pencil*
10.1 I see … any-how— 11.3] *crossed pencil*
12.1 I have … theory. 12.5] 'Geograph Distrib' *added pencil*
14.1 Poor … indeed— 14.2] *crossed pencil*

1 See letter to J. D. Hooker, 7 August [1869]. The Darwins arrived home from Wales on 31 July 1869 (see 'Journal' (Appendix II)).
2 Hooker refers to Frederic Francis Hallett and his paper 'On the laws of the development of cereals', read at the British Association for the Advancement of Science meeting at Exeter on 18 August 1869. See letter from J. D. Hooker, 5 August 1869 and n. 4.
3 *North British Review*; see letter to J. D. Hooker, 7 August [1869] and n. 4.
4 George Douglas Campbell, eighth duke of Argyll, was secretary of state for India. Hooker published the first of seven volumes of *The flora of British India* (J. D. Hooker 1875–97) in 1875 under the authority of the secretary of state for India in council, and dedicated it to Campbell. In his presidential address to the British Association in 1868, Hooker had described natural theology as 'a science falsely so-called', which is 'to the scientific man a delusion, and to the religious man a snare' (*Athenæum*, 22 August 1868, p. 248); Campbell was well-known as a supporter of natural theology.
5 The meeting of the British Association was held in Exeter from 18 to 25 August 1869.
6 Hooker refers to Thomas Dyke Acland and to William Samuel Symonds, rector of Pendock, Worcestershire, and his daughter, Hyacinth Symonds; see also letter from J. D. Hooker, [25 January 1869].
7 Hooker refers to his son, William Henslow Hooker, who had been sent to New Zealand for his health, and to James Hector; see letter from J. D. Hooker, 17 July 1869 and n. 6.
8 Frances Harriet Hooker. William Hooker had suffered a burst blood vessel in the lungs in 1868 (see *Correspondence* vol. 16, letter from J. D. Hooker, 30 August 1868).
9 See letter to J. D. Hooker, 7 August [1869] and n. 15. Herman Merivale was permanent under-secretary at the India Office. Hooker presumably wrote 'F. O.' (Foreign Office) in error.
10 See letter to J. D. Hooker, 7 August [1869] and n. 11. Hooker refers to Daniel Oliver.
11 See n. 5, above. Hooker was unable to make the suggested visit to Down; see letter from J. D. Hooker, 7 September 1869.
12 See letter from J. D. Hooker, 5 August 1869 and n. 5.
13 See letter to J. D. Hooker, 7 August [1869] and n. 12.
14 See letter to J. D. Hooker, 7 August [1869]. Hooker refers to Charles Lyell.
15 See letter to J. D. Hooker, 7 August [1869] and n. 17. Hooker was president of the British Association. An upper limit of five guineas had been set for individual subscriptions to the Faraday Memorial Fund, set up in memory of Michael Faraday (*The Times*, 22 June 1869, p. 5).
16 Hooker probably refers to a memorial collection for James Clark Ross.
17 Lord Howe Island is in the Pacific Ocean, about halfway between the mainland of Australia and Norfolk Island, which is itself about 1500 km east of Australia. Charles Moore, director of the Royal Botanic Garden, Sydney, undertook a survey of the vegetation of Lord Howe Island in June 1869 and

listed thatch palm (*Howea forsteriana*), curly-leaved palm (*Howea belmoreana*), umbrella palm (*Hedyscepe canterburyana*), all in the family Arecaceae, and a fourth unnamed dwarf species that he himself did not see (C. Moore 1869 and 1870). Moore also listed four species of Leguminosae, two of Myrtaceae, and two of Rutaceae; in his opinion the flora was closer to that of Norfolk Island than to mainland Australia (C. Moore 1870). Proteaceae is the family of proteas; Australia has more species than any other country. Leguminosae (now Fabaceae) is the bean family; Myrtaceae is the myrtle family; Rutaceae is the rue or citrus family.

18 John Traherne Moggridge suffered chronic ill health (R. Desmond 1994). See also letter to J. T. Moggridge, 7 June 1869.

19 See letter to J. D. Hooker, 14 August [1869] and n. 6. CD refers to Peter Guthrie Tait.

To J. D. Hooker 14 August [1869][1]

Down. | Beckenham | *Kent. S.E.*
Aug. 14—

My dear Hooker

I write one line to tell you that you are a real good man to propose coming here for a Sunday after Exeter—[2] Do keep to this good intention.

It is very bad news about y^r poor boy & we are both grieved to hear it.[3] I am not at all surprized that it makes you out of spirits. You must feel much perplexed what to do, & I shall be glad to talk about your plans with you when you are here. I am sure Exeter & your other visit will do you good.[4] I often wonder how you stand all y^r multifarious work.

I quite agree about the folly of the endless subscriptions for dead men; but Faraday is an exception & if you will pay 3 guineas for me, it will save me some trouble;[5] but it will be best to enclose a cheque, which as you will see, must be endorsed. If you read the North. Brit. Rev. you will like to know that George has convinced me, from correspondence in style & spirit, that the article is by Tait, the co-worker with Thompson.[6]

I was much surprized at the leaves of Drosophyllum being *always* rolled backwards at their tips, but did not know that it was a unique character.[7]

Thanks for all the news in y^r letter yours affectionately | Ch. Darwin

LS
DAR 94: 149-50

1 The year is established by the relationship of this letter and the letter from J. D. Hooker, 13 August 1869.

2 See letter from J. D. Hooker, 13 August 1869 and n. 11.

3 CD refers to William Henslow Hooker. See letter from J. D. Hooker, 13 August 1869 and n. 7.

4 See letter from J. D. Hooker, 13 August 1869 and nn. 5 and 6.

5 CD refers to the memorial for Michael Faraday. See letter from J. D. Hooker, 13 August 1869 and nn. 15 and 16.

6 See letter to J. D. Hooker, 7 August [1869] and n. 4. George Howard Darwin correctly attributed the article in the *North British Review* ([Tait] 1869) to Peter Guthrie Tait; Tait and William Thomson had published a joint work, *Treatise on natural philosophy* (Thomson and Tait 1867).

7 See letter from J. D. Hooker, 13 August 1869 and n. 10.

From George St Clair 14 August 1869

Banbury | Oxon.
14th Aug. 1869.

Dear Sir,

In your last book you say you have heard of a family of six-toed cats" (Vol. I. p. 48.)[1] I have *seen* two members of such a family, in this town.

I learn that the first was brought here from Brighton by M^rs. Grimbly, of the High Street, & afterwards came into the possession of M^r. H. Walford, High Street.[2] It is said to have had six toes on each of its two front paws. It is now dead, but has left a score of descendants, of which one or two in every litter has had extra toes. One now in the possession of M^r. H. Walford I have seen. It has six claws on each front paw; & differs from others of the family (from *all* others of the family as far as Mr W. knows) in having a sixth claw on the left hind foot also & in possessing a bushy tail. The feet attract notice by their largeness, as soon as you ⟨see⟩ the cat. The family are said to be goo⟨d⟩ ratters. I saw another six-toed cat ⟨of⟩ this family in the possession of the Rev. H. A. Nash, formerly of this town & now of London.[3] I have also the names of thr⟨ee⟩ other persons in this town said to possess specimens; & I think I could procure one for you, if you like. If you do *not* wish for one, nor place any value on these particulars, pray pardon me for troubling you.

I am, Dear Sir, | Yours faithfully | George St. Clair.

Chas. Darwin, Esq. F.R.S.

DAR 177: 6 (fragile)

CD ANNOTATION
Top of letter: '⟨*one word missing*⟩ w 6 Toes' *ink*

1 *Variation* 1: 48.
2 Mrs Grimbly has not been identified; in 1869, James Grimbly, grocer and provision and wine merchant, lived at 17 High Street, Banbury. There was also a Mrs Grimbly at 15 Horsefair, Banbury. Henry Walford, watchmaker and silversmith, lived at 68 High Street (*Post Office directory of Northamptonshire etc.* 1869).
3 St Clair refers to Henry Allden Nash.

From John Beddoe 21 August 1869

Clifton,
Aug 21. 1869

Dear Sir

I enclose the results of several more years of observation on the point of conjugal selection, respecting which you wrote to me—[1] I am sorry that I have been so long in sending these particulars, but must beg you to excuse my tardiness, as I have been intensely occupied in various ways, & have had no one by me to whom I could entrust the business— You will see that the tendency of these later figures is to confirm the former ones—[2]

In this later investigation I have excluded the women between 20 & 25, & between 45 & 50, who were included in the former one—[3] I do not think the hair grows notably darker after 25 in most people—[4] But red hair does become less vividly red—

To test the matter further, I will divide the women into two classes, viz, those above & those below 35, & let you know the result—[5]

Your apology for troubling me in this matter really made me feel ashamed— I think there can be but few of us humbler cultivators of natural science, who would not feel it an honour to be permitted to contribute his stone towards the building of your great edifice— At any rate, it will always be a pleasure to me to do for you anything that you may esteem a service— I hope before long to send you a copy of a statistical memoir of mine "on the stature & bulk of Man in Britain", which is being printed in the Anthropological Memoirs—[6]

I am, dear Sir | Yours faithfully | John Beddoe M.D.

[Enclosure][7]

Women of the lower class, from 25 to 45 years of age—natives of Bristol and the neighbourhood for the most part—

	Red	Fair	Brown	Dark	Black	Total
Single Women	11.5	17.5	51.5	44	5.5	130
per cent	8.8	13.4	39.6	33.8	4.2	
Married Women	23	83½	218½	310½	42½	678
	3.4	12.3	32.2	45.8	6.2	

DAR 85: A21–3

CD ANNOTATIONS

1.1 I enclose … business— 1.5] *crossed pencil*
1.3 but must … tardiness,] *scored red crayon*
2.2 I do … red— 2.3] *scored red crayon*
3.1 I will … result— 3.2] *double scored pencil*
4.1 Your … Memoirs— 4.7] *crossed pencil*
End of table: '23
83½
218½
310½
42½
6780' *ink*

1 CD's letter has not been found. He had evidently written with questions concerning Beddoe's paper 'On the supposed increasing prevalence of dark hair in England', in which Beddoe suggested sexual selection as a mechanism for a perceived change in the predominant hair colour in the human population (Beddoe 1863).

2 From his analysis of data on the hair colour and marital status of 737 female patients at Bristol Royal Infirmary aged between 20 and 50, Beddoe had concluded that fewer light-haired women than dark-haired women married, and suggested that this would eventually result in greater prevalence of dark hair in the general population (Beddoe 1863, p. 312).

3 See n. 2, above.
4 Beddoe had qualified his published conclusions with the caveat that hair darkens with age (Beddoe 1863, p. 312 n.).
5 See enclosure to letter from John Beddoe, [22 August – 12 September 1869].
6 Beddoe 1869.
7 There is an analysis of the combined data from Beddoe 1863 and this table in CD's handwriting in the Darwin Archive–CUL (DAR 85: A24); there are also calculations of percentages based on Beddoe's tables made by George Howard Darwin, some with CD annotations (DAR 85: 15, 16, 19, 20, 25–28). Although CD cited Beddoe's work on human stature in *Descent*, he did not use any of the data on hair colour; see also CD annotations to letter from John Beddoe, 13 September 1869.

From John Beddoe [22 August – 12 September 1869][1]

Professor Christison,[2] | Arroquhar, by Glasgow

My dear Sir

Your letter has followed me hither, where I am enjoying a holyday—[3] And by the same post I have received the result of the last examination of data—[4] It is vexatiously subversive of my former inferences, & really looks very much as if the gradual darkening of colour were the real cause, or main cause at least, of the whole set of phenomena—[5] Now that I have got a trustworthy amanuensis[6] I will have the 35–45 women abstracted from the books which furnished the original data for my paper of 1863;[7] and when I get the result you shall have it at once—[8]

Yours very faithfully | John Beddoe

[Enclosure]

July 13th 1863 to Aug 6. 1869

Women between 35 and 45									
Single					Married				
Red	Fair or Lt. B.	Brown	Dk. B.	Black	Red	Fair or Lt. B.	Brown	Dk B.	Black.
3½	2	8½	13	5	14½	28½	95½	60	26½
Total 32					Total 225				
Women between 25 and 45									
11½	17½	51½	44	5½	23	83½	218½	310½	42½
Total 130					Total 678				

DAR 85: A13–14

1 The date is established by the relationship between this letter and the letters from John Beddoe, 21 August 1869 and 13 September 1869.
2 Robert Christison.
3 CD's letter has not been found.
4 See letter from John Beddoe, 21 August 1869.
5 See letter from John Beddoe, 21 August 1869 and nn. 2 and 4.
6 See letter from John Beddoe, 21 August 1869.

7 Beddoe 1863; see also letter from John Beddoe, 21 August 1869 and n. 2.
8 See enclosure to letter from John Beddoe, 13 September 1869.

From Federico Delpino[1] 22 August 1869

Firenze
22. Agosto 1869

Celeberrimo uomo!

Animato dalla cortesia con cui la S.V. Si compiacque aggradire due miei scritti editi nel 1867, mi prendo la libertà di comunicar le tre altri miei lavori successivi.[2] Accetti in omaggio il tenuissimo dono.

Le dottrine che harmo la fortuna di essere propugnate dalla prima Intelligenza del secolo nostro, trovano anche in me un difensore, scarso di mezzi sì ma pieno di buona volontà. Ho cercato di contribuire per quanto stava in me alla illustrazione delle dottrine della Dicogamia e della variabilità delle specie.

Duolmi per altro di non potere consentire in alcune idee generali. Forse per errore di mente ma certo con profonda convinzione io sono teleologo. Io credo che quel principio razionale di cui l'uomo si Sente animato, governi non meno tutti gli altri esseri, comprese le piante. E credo che la causa prima di tutte le variazioni, sia in bene che in male, con progresso o con regresso, consista appunto in questo principio intrinseco, innato in tutti gli esseri.[3]

Comunque sia, dalle mie ricerche botaniche mi risulta che la variabilità delle specie non merita più di essere chiamata teoria: tutte le parti della scienza concorrono, almeno secondo me, a metterne in chiara luce l'assoluta verità di fatto.

Onore dunque alla Intelligenza che seppe dare tanto impulso al pensiero umano. Onore al primo naturalista del Secolo, del quale pregio protestarmi

Ossequentissimo discepolo | Federico Delpino

Assistente di botanica al museo di Storia naturale in Firenze.

DAR 162: 143

1 For a translation of this letter, see Appendix I.
2 Delpino refers to 'Pensieri sulla biologia vegetale, sulla tassonomia, sul valore tassonomico dei caratteri biologici, e proposta di un genere nuovo della famiglia delle labiate' (Thoughts on plant biology, taxonomy, the taxonomic value of biological characters, and the proposal of a new genus of the labiate family; Delpino 1867c), the first part of *Ulteriori osservazioni sulla dicogamia nel regno vegetale* (Further observations on dichogamy in the vegetable kingdom; Delpino 1868–75), and 'Sulla darwiniana teoria della pangenesi' (On the Darwinian theory of pangenesis; Delpino 1869a). Delpino had sent CD two papers on dichogamy in 1867 (Delpino 1867a and 1867b; see *Correspondence* vol. 15, letter from Federico Delpino, 5 September 1867 and n. 2).
3 For more on Delpino's teleological approach to CD's theory, see Pancaldi 1991, pp. 107–36.

From Asa Gray 23 August 1869

115, Jermyn St, London
23[d] Aug, 1869

My Dear Darwin

We are back in England at length, I am glad to say,[1]—the Hookers' being away from Kew I can learn nothing of you.[2] Pray tell me how you are.—& where. The last I knew you were down in Wales.[3]

We expect to be here till near the end of the week, and then probably run down into Somersetshire on a short visit.[4] Mrs. Gray (pretty well for her) sends best regards.

Ever, | A. Gray

DAR 165: 169

[1] Jane Loring and Asa Gray arrived in London on 20 August 1869 (J. L. Gray ed. 1893, 2: 591). On their travels in Europe and Egypt, see the letter from Asa Gray and Jane Loring Gray, 8 and 9 May 1869.

[2] Joseph Dalton Hooker, and presumably his wife Frances Harriet Hooker, left the Royal Botanic Gardens, Kew, on 17 August 1869 for several destinations (see letter from J. D. Hooker, 13 August 1869).

[3] The Darwins returned to Down from Wales on 31 July 1869 (see 'Journal' (Appendix II)).

[4] The Grays visited Richard William Church, rector of Whatley, Somerset (J. L. Gray ed. 1893, 2: 591–4).

From W. C. Tait [23 August 1869]

the mother" I have them all preserved in spirits.

I shall try to obtain for you some seed of the drosophyllum and shall also try to obtain some plants to accustom them to growing in large flower pots before sending them to you.[1] Perhaps the plants prefer some peculiar species of fly. The greater part of the flies which I saw sticking to the plants at the locality were of the shape more or less of the house fly but much smaller.  I hear from the gardener at the botanical gardens Coimbra[2] that a french botanist is now employed in studying the drosera plants. Hoping that your health may improve and that you may enjoy the making of fresh discoveries in natural history

I remain, Dear Sir, | Yours very sincerely, | William C. Tait.

Incomplete
Postmark: 23 | Agosto | 1869
DAR 178: 49

[1] Tait had previously sent CD specimens of *Drosophyllum lusitanicum* but they had arrived in poor condition; he had been advised that sending seed might be more successful (see letter to W. C. Tait, 12 and 16 March 1869, and letter from W. C. Tait, 11 July 1869).

[2] Edmund Goeze; see letter from W. C. Tait, 11 July 1869 and n. 1.

To Federico Delpino 24 and 25 August [1869][1]

Down. | Beckenham | *Kent. S.E.*
Aug 24th

Dear Sir

I am very much obliged to you for your kind present of your three publications.— I will send the duplicates "from the Author" to the Linnean Society, where they are most likely to be appreciated.[2] It is a bitter grief to me that I cannot read Italian; for I am sure that I shd. feel the *greatest* interest in your works; but my wife will translate parts to me.— I am particularly pleased that you have discussed pangenesis, though I find that you are by no means favourable to this hypothesis. I will send it to a person, whom I employ, to be carefully translated; & after reading the translation will endeavour to find some Scientific Review in which it may be published.—[3]

With my best thanks & much respect, I remain | Dear Sir | Yours faithfully | Ch. Darwin

P.S. Since the above was written I have just received your very kind letter of Aug 22[d] with all its too honouring & complementary remarks on my works.— I again return you my very sincere thanks.—

Aug 25th—

American Philosophical Society (376)

[1] The year is established by the relationship between this letter and the letter from Federico Delpino, 22 August 1869.

[2] See letter from Federico Delpino, 22 August 1869 and n. 2.

[3] CD refers to Delpino 1869a. A translation of the article was published in parts in the 29 September, 6 October, and 13 October 1869 issues of *Scientific Opinion* (Delpino 1869b).

To W. B. Tegetmeier 25 August 1869

Down. | Beckenham | *Kent.*
Aug. 25. 69

My dear Sir,

I am much obliged for your note.[1] But when I wrote last I returned to you a memorandum with information from M[r]. Hewitt about the chickens of Sebright bantams; for I could not read one word about the spangling of the feathers; I further asked whether the Spangles were crescent-shaped or rounded.—[2]

Many thanks for the feathers of the Silver Spangled Polish, but I do not understand whether the general plumage of the chickens resembles that of the adult birds as nearly as do the crest feathers.[3]

I hope that you will succeed in finding out for me the character of the plumage of "cuckoo chickens" when about 7 or 8 weeks old.[4]

It is very important to me to learn whether in breeds in which the adult sexes resemble each other in plumage, the chickens nearly approach in plumage to the adults;[5] but I am well aware that I have no claim to give you so much trouble.—

Many thanks for the Review; the fact would have been of much interest to me, if M[r] Williams had stated to what species his experiments referred.[6]

Believe me, my dear Sir, | Yours very faithfully | Ch. Darwin

P.S. I will not accept your very kind offer of lending me the Polish chickens.

When you say that "the notes give the information you require" I suppose you refer to the notes on the paper with the feathers; but I am more concerned with the general plumage of the body.

LS
Archives of the New York Botanical Garden (Charles Finney Cox collection) (ser. 14)

[1] Tegetmeier's note has not been found.
[2] See letter to W. B. Tegetmeier, 21 July [1869] and enclosure. CD refers to Edward Hewitt.
[3] There are two sets of Polish chicken feathers sent by Tegetmeier in the Darwin Archive–CUL; one is of crest feathers of twelve-week old chicks (DAR 142: 51) which CD annotated as 'silver spangled', and the other is headed 'Genealogy of cross-bred chickens' in Tegetmeier's hand (DAR 142: 52).
[4] See letter to W. B. Tegetmeier, 21 July [1869] and n. 5.
[5] CD discussed the plumage of immature and mature birds in a number of breeds of fowl, including Sebright bantams and cuckoo sub-breeds, in the context of the inheritance of sexual characters in *Descent* 1: 294. He expanded the discussion in *Descent* 2, chapter XVI; see especially pp. 208–16.
[6] Neither Mr Williams nor the review has been further identified.

From Frederick Smith 26 August 1869

Ilfracombe
August. 26th. | 1869.

My dear Sir

The bee to which you allude from Australia—the male of which of a bright brassy colour and clothed with rich fulvous pubescence—the female being of a steel-blue colour—is the Lestis bombylans.[1] The female is as brilliant as Musca vomitoria[2]—the common blue-bottle—having—frequently bright tints of vivid green—

Your conjecture as to the Male of Anthophora being of a rich fulvous brown—the female is quite black.—[3]

The Males in the genus Bombus vary as you say—and the same may be said of their parasites the Apathi— the Males of A. campestris vary extremely—so much so—that Mr. Kirby has described three or four as distinct species;[4] some males are gaily coloured black and yellow being banded and tufted with the latter colour—others are clothed principally with yellow—with narrow bands of black intermixed on the abdomen—and from these standards—every shade of colouration occurs between them and a totally black variety— The female varies from a black and yellow banded insect to a totally black one—and this species does not confine its parasitism to one species— it has been obtained from nests of two or three species of Bombus.

The greatest sexual differences in coloration certainly obtains among the solitary bees—as in Anthophora and Xylocopa— in the latter genus many species consist of black females and bright yellow males

believe me | Yours very truly | Fredk. Smith

Chas. Darwin Esqre.

DAR 82: 1–2

[1] CD's letter to Smith has not been found. Smith refers to *Lestis bombylans* (now *Xylocopa bombylans*, the Australian green carpenter bee). CD cited Smith's description of this bee in *Descent* 1: 336.
[2] *Musca vomitoria* is now *Calliphora vomitoria*.
[3] Smith refers to *Anthrophora retusa*, the potter flower bee. CD cited Smith's description of this bee in *Descent* 1: 336.
[4] Smith refers to the *Bombus* subgenus *Apathus* (now *Psithyrus*) or cuckoo bumble-bees, which are kleptoparasites on other bumble-bee species. (Kleptoparasitism is a form of parasitism in which an animal habitually steals food from members of another species (*OED*).) *Apathus campestris* is now *Bombus*

(*Psithyrus*) *campestris*. In *Monographia apum Angliæ* (Kirby 1802, 2: 331, 333–5, 371), William Kirby had described five species that Smith later synonymised with *Apathus campestris* (see Smith 1853–9, 1: 384).

To W. C. Tait 27 August [1869][1]

Down, Beckenham, Kent. S.E.
Aug. 27th,

My dear Sir

I write merely to thank you for your very kind offer to send me more specimens and seed of Drosophyllum;[2] but I will not trouble you, for they would be superfluous. Strangely 3 of your plants, after being sickly, have *perfectly* recovered and are growing *vigorously* and have produced plenty of seed! They catch abundance of insects; but I cannot make out anything peculiar about their structure or functions. I suppose it is my soil; for they cannot succeed at Kew.[3] I am going to give one or two of my plants to Kew and seed. Dr Hooker has made out that the genus is certainly closely allied to Drosera, though having such different habits.[4] The case of the Lizard is very interesting and shall certainly be quoted, if I ever live to come to embryology.[5]

With my best thanks. Pray believe me | yours sincerely and obliged | Ch. Darwin.

Copy
DAR 147: 546

1 The year is established by the relationship between this letter and the letter from W. C. Tait, [23 August 1869].
2 See letter from W. C. Tait, [23 August 1869].
3 See letter from J. D. Hooker to Emma Darwin, 29 March 1869 and n. 7.
4 CD refers to Joseph Dalton Hooker. See letter from J. D. Hooker, 13 August 1869 and n. 10.
5 CD presumably refers to information contained in the missing portion of the letter from W. C. Tait, [23 August 1869]. There are no references in CD's published work to information from Tait on lizards.

From T. S. Cobbold 1 September [1869][1]

Westward Ho! Hotel | North Devon
Septr. 1st

Dear Sir,

I have this morning received your note (which has tardily followed me viâ Exeter).[2]

I have learnt that the box of specimens was duly delivered at 84 Wimpole Street; & from the "list" I quite think I may congratulate myself upon the possession.

I hope to utilise a mass of materials (by me) this winter, in which case your kind addition will be a useful adjunct.

At present, I am here studing rock-succession & collecting fossils (in view of my Spring series of Swiney Lectures).[3]

I had hoped to have read a paper on the "facts of Succession" at our Exeter meeting; which as you will have heard was on the whole successful—despite certain stupid intrusions.[4]

I am, Dear Sir | Yours very truly | T. Spencer Cobbold

Charles Darwin Esq. F.R.S—

DAR 161: 190

1 The year is established by the relationship between this letter and the letter to T. S. Cobbold, 9 August [1869].
2 See letter to T. S. Cobbold, 9 August [1869]. CD had offered to send Cobbold some parasitic worms collected on the *Beagle* voyage.
3 Cobbold was appointed Swiney Lecturer on geology at the British Museum in 1868 (*ODNB*).
4 The annual meeting of the British Association for the Advancement of Science was held at Exeter from 18 to 25 August 1869. CD received an account of the meeting in the letter from J. D. Hooker, 7 September 1869.

From William Bowman 3 September [1869][1]

North End | Hampstead
Sept. 3rd

My dear Mr Darwin,

Prof. Donders of Utrecht is with me, and if you should be at home and able to receive a call from us on Tuesday next, we would do ourselves the pleasure of coming to Bromley but I write beforehand to ascertain—thinking it very likely you may be from home at this season.—[2]

I know of no one abroad more instructed in all recent progress or more able to appreciate your works—

I hope next year he will pass a month with me but on this occasion his stay is very short & he leaves on Wednesday— I have asked him to think over the points you referred to me from Sir C. Bell's works—[3]

Yours very truly | W Bowman

C. Darwin Esqr

American Philosophical Society (B/D25.313)

1 The year is established by the reference to Frans Cornelis Donders's visit (see n. 2, below).
2 Donders's visit to Down is mentioned in CD's letter to J. D. Hooker, 10 September [1869]. The Tuesday after 3 September 1869 was 7 September.
3 CD had asked Bowman about a passage in Charles Bell's *Anatomy and philosophy of expression as connected with the fine arts* (Bell 1844) in his letter of 2 April [1868] (*Correspondence* vol. 16). CD had previously asked Bowman for observations pertaining to Bell's account of weeping (see *Correspondence* vol. 15, letter to William Bowman, 30 July [1867]).

To Frans Cornelis Donders[1] 6 September 1869

Down, Beckenham | Kent
6 Sepr 1869.

"1st Are any facts known which support Sir C Bell's statement that the contraction of the muscles round the eyes prevents the eye-balls and surrounding parts from becoming gorged with blood, when the circulation is impeded by screaming or or other similar exertions?[2]

2^ndly^ Is it conceivable that the strong pressure of the muscles round the eyes can in any way stimulate the lachrymal glands, so as to lead to the secretion of tears? I am aware that strong voluntary contraction of these muscles does not produce this effect, or only in a very slight degree: yet I cannot avoid the suspicion that there is some relation between their contraction and the secretion of tears.[3]

Ch. Darwin

Copy[4]
American Philosophical Society (B/D25.314)

1 The correspondent is identified as Donders on the basis that the letter is part of the collection of CD's letters to Donders at the American Philosophical Society; and on the assumption that CD sent the note to Donders prior to their meeting on 7 September (see letter from William Bowman, 3 September [1869] and n. 2).
2 CD refers to Charles Bell's account of the physiology of weeping (see Bell 1844, pp. 105–6).
3 For some of CD's investigations into the effects of contraction of the eye muscles, see *Correspondence* vol. 16, letters to W. E. Darwin, [15 March 1868] and 8 April [1868].
4 The memorandum is a copy in an unknown hand. A note at the top of the letter states that the original was in Emma Darwin's hand, and that the signature, address, and date were in CD's hand.

From Walter Elliot 7 September 1869

Wolfelee | Hawick | N.B.
7th. Sepr. 1869

Dear Mr Darwin

In case you have not seen the Proceedgs. Asiatic Socy Bengal, I extract the following notice of the *polygamous* habits of the Asiatic Elephant, by Lieut' Johnstone, Superintendt Kheddas, Cuttack Tributary, Mahals, which are particularly trustworthy:[1]

"The average number of a herd of elephants is about 20 or 25, they have a female at their head, who leads the way;—in the cold weather, three or four herds often join together; & when a female is in heat a male joins them, otherwise the males remain apart."

"often a large male is seen attended by 3 or 4 smaller ones."

"Males when with the herd, never help to defend the others from any outward attack, but are generally the first to run." Procgs. for 1868. p. 128 (May)

He also mentions that, in a herd of 30 to 40, it is not unusual to find only one really old female, the average duration of life being 80 to 100 years. He has also observed that, a large proportion of the males are puny & stunted, tho' why, he does not understand. "Of those that attain a large size (say 1 in 10) the smallest are always killed or turned out of the herd by the larger ones, wh. tends to keep up the size of the breed".[2]

By some accident this No was not sent to me last year, in regular course; & I have only just recd it.

Believe me Yours very truly | Walter Elliot

Charles Darwin Esq

DAR 86: A74

CD ANNOTATIONS
1.3 which … trustworthy: 1.4] *underl blue crayon*
5.4 "Of … breed". 5.6] *double scored blue crayon*

[1] Elliot refers to James Johnstone and Johnstone 1868, pp. 128–9. CD cited the paper in *Descent* 1: 268, n. 7. Cuttack was a tributory state (or 'Mahal') in east India (see Mohapatra 2005). It is now a district in the state of Orissa (*Columbia gazetteer of the world*).
[2] Johnstone 1868, p. 128. CD cited the paper in *Descent* 1: 268, n. 5.

From J. D. Hooker 7 September 1869

Royal Gardens Kew
Sept. 7. 69

Dear Darwin

At last I sit down to write to you—after a fortnight of the West country, one week of restless excitement at Exeter, & another of quiet in country houses, 1st. Sir Thos Acland, 2 Symonds of Pendock, 3d. Douglas Galtons Mothers near Droitwich—[1]

Exeter was a very good meeting & Stokes made a first rate President,—whereby he agreeably disappointed his friends.[2] The "punctum saliens"[3] of the whole meeting was decidedly Huxley's answer to Dr McCann. he literally poured boiling oil over the bumptious man— you have read it of course.— McCann's was the most conceited dogmatic sermon you ever heard preached, & was delivered with a sort of swagger that rendered it offensive to Huxley in the last degree.—[4] Next in point of interest was Lubbock's answer to the D. of Argyll, which I thought uncommonly good, but the bird was hardly worth the shot, & the discussion was singularly poor.[5] & unsuggestive considering who the speaker was. Stokes address was very sound science for the most part; but no approach to novelty or originality in matter & very little in dealing with it— the last part was singularly weak—& mindless.[6] All had expected an address on his own special pursuits in science.

Phillips lecture on Vesuvius I thought very poor, both in manner matter & style. Lockyers was far better, but he tried too much, & became obscure.[7] There were few foreigners & fewer nobs. & the whole meeting was a very quiet one, except at the end, when the struggle between Edinburgh & Liverpool took place—a struggle which bodes no good to the Association, as showing how little the General Committee regards the matured opinion of the Council in grave matters that require calm & unprejudiced consideration[8]

I had hoped to have asked whether I could have spent a Sunday with you soon, but Prof Miquel of Utrecht is coming tomorrow to stay a fortnight with us, bringing 2 of his daughters—very nice (& pretty!) girls whose acquaintance we made in Holland— he is an old friend & has been twice in England before[9]

At Mr Galton's I met your connections (or relations) Mrs & Miss Wheler, the latter of whom I met at your house last winter or Autumn I think.[10]

Do not bother to answer this, but by a line saying how you are.

Ever affec Yrs | J D Hooker

DAR 103: 30–1

1 Hooker had attended the meeting of the British Association for the Advancement of Science, which was held at Exeter from 18 to 25 August 1869. He refers to Thomas Dyke Acland, William Samuel Symonds, and Douglas Strutt Galton and his mother, Isabella Galton, of Hadzor House, Droitwich.

2 George Gabriel Stokes was president of the British Association in 1869. Hooker had remarked on Stokes's upcoming presidency in his letter to CD of 30 August 1868 (*Correspondence* vol. 16).

3 *Punctum saliens*: the central or most important point (*OED*).

4 James McCann delivered a paper titled 'Philosophical objections to Darwinism or evolutionism' in the biology section of the British Association meeting. It was followed by remarks from Thomas Henry Huxley. McCann's paper was particularly critical of claims made in Huxley's 1868 address, 'The physical basis of life' (later printed as T. H. Huxley 1869a). A report of McCann's paper and Huxley's reply was printed in the *Athenæum*, 4 September 1869, p. 309. McCann's paper was published under the title *Anti-Darwinism*, together with Huxley's reply, and a rejoinder from McCann (McCann 1869). A copy is in the Darwin Pamphlet Collection–CUL.

5 John Lubbock delivered a paper that responded to criticisms by George Douglas Campbell, the duke of Argyll (Lubbock 1869b). Campbell had published four articles in *Good Words* that took issue with Lubbock's address, 'On the origin of civilization and the primitive condition of mankind' (G. D. Campbell 1868, Lubbock 1867). A report of the discussion appeared in the *Athenæum*, 4 September 1869, pp. 309–10.

6 In the conclusion of his presidential address, Stokes remarked on the 'profound mystery' of the phenomena of life and mind (Stokes 1869, pp. civ–cv).

7 Hooker refers to John Phillips and Joseph Norman Lockyer. Lockyer spoke on the 'Physical constitution of the stars and nebulæ' (*Report of the meeting of the British Association for the Advancement of Science, held at Exeter in 1869*, lxxxvii–lxxxviii).

8 Hooker alludes to a dispute over the proposed location of the 1870 British Association meeting. In the event, the meeting was held in Liverpool (*Report of the meeting of the British Association for the Advancement of Science, held at Exeter in 1869*, p. lxxxviii).

9 Hooker refers to Friedrich Anton Wilhelm Miquel; his daughters have not been identified.

10 Hooker refers to Elizabeth Ann Wheler and Lucy Elizabeth Wheler. According to Emma Darwin's diary (DAR 242), Lucy Wheler visited Down from 27 to 31 October 1868.

From John Blackwall 8 September 1869

Hendre House,
September 8th, 1869.

Dear Sir,

In fulfilment of my promise, I write a few lines to inform you that the result of my observations on the young of the varieties of Theridion lineatum on quitting the cocoons that I had collected for the purpose of ascertaining whether on that occasion and for some time afterwards they would present any characteristics by which they might be distinguished from each other or not, completely confirms my opinion previously expressed, that no such characteristics can be perceived; and this remark applies also, as might be expected, to the young spiders in their customary haunts.[1] It would be well to extend the observations through the several stages of growth in the young of these varieties, but the undertaking is surrounded by so many and great difficulties as to render it scarcely practicable.

The colours on the abdomen of Epëira diadema[2] present considerable variety in adult specimens, but the design formed by their distribution is pretty constant; the young, however, on quitting the cocoon differ decidedly from the parents and cannot be distinguished from each other.

An additional motive for bestowing my tediousness upon you has been induced by meeting with a passage in Westring's "Araneæ Svecicæ" having relation to the stridulous sound produced by several species of Theridia.[3] At the conclusion of the description of Theridion serratipes,[4] p. 175, he adds the following "Obs. Maris abdomen circa nervum, quo thoraci affixum est, valvula denticulata munitum, cujus usum sæpe animo perpendi, et tandem mihi fortuito contigit comperire hanc valvulam instrumentum stridoris esse, et quod thoracis basi, sub oculo armato transverse subtilissime rugulosa, ab animalculo applicatum, stridorem fere ut in *Cerambycinis*, *Reduvio personato* et multis aliis insectis reddebat. Hunc sonum audiebam, cum aranea, acu affixa, pulvillo acuum vel segmento papyri usque ad pectus apprimebatur, et tunc aranea, aut sua sponte, aut digitis erga anum attacta, abdomen sursum deorsum movebat, hujusque basin vel valvulam serratam erga basin thoracis affricabat. *Femina* organis iis caret. Idem post tempus plures mares *Theridiorum* stridentes mihi obvenerunt."[5]

I may remark also that M. Westring terminates his description of the male of Theridion castaneum,[6] p. 184, with this statement. "Instrumenta stridentia in basi abdominis et thoracis ut in *Th. serratipede*, *4-punctato*, *hamato*, *guttato* et *albomaculato* adsunt."[7]

Hoping that the above particulars may interest you, | I am, dear Sir, | very truly yours | John Blackwall.

DAR 82: A80–1

CD ANNOTATIONS
3.2 "Araneæ Svecicæ"] *underl blue crayon*
3.4 Theridion serratipes, p. 175,] *underl blue crayon*
4.2 castaneum] *underl blue crayon*
Top of letter: 'I must add that it is a singular circumstance, that the young of Theridion, do not acquire their *[*darker*]* [*interl*] colour until late in life; a circumstance which I shd have expected only in secondary sexual characteristics'[8] *ink*; '(& Stridulation)' *blue crayon*

1 See letter from John Blackwall, 10 August 1869. The spider *Theridion lineatum* is now *Enoplognatha ovata* (family Theridiidae, cobweb weavers).
2 *Epeira diadema* is now *Araneus diadematus*, the cross orbweaver.
3 The reference is to Niklas Westring and Westring 1861.
4 *Theridion serratipes* is now *Steatoda phalerata.*
5 The Latin may be translated as follows: 'Observation. The abdomen of the male is next to a sinew, by which it is attached to the thorax, and which is protected by a tooth-edged flap, the use of which I have often debated; at length it happened by chance to become clear to me that this flap is the instrument of stridulation, because when applied by the insect to the very finely horizontally wrinkled base of the thorax under the armoured eye, it yielded a noise almost as in the *Cerambycinis*, the *Reduvio personato*, and many other insects. I have heard this sound when a spider, affixed with a pin either to a pin-cushion or to a sheet of paper, was pressed on the breast. Then the spider, either spontaneously or from the touch of my finger against its rear, moved its abdomen up and down, and the base of this or the serrated flap was rubbing against the base of the thorax. The *female* lacks those organs. Likewise, in time many stridulating males of the *Theridiorum* have come before me.' Westring refers to the longhorn beetle

tribe Cerambycini. 'Reduvio personato' may refer to the family Reduviidae (assassin bugs) in the order Hemiptera.

6 The species described in Westring 1861 is *Theridium castaneum* (now *Steatoda castanea*). *Theridium* was an invalid synonym of *Theridion*.

7 The Latin may be translated as follows: 'Instruments of stridulation are present in the base of the abdomen and thorax as in *Th. serratipede*, *4.-punctatum*, *hamatum*, *guttatum*, and *albomaculatum*.' CD referred to this discussion in *Descent* 1: 339 n. 16. *Theridion guttatum* is now *Crustulina guttata*; *T. albomaiculatum* is now *Chrysso albomaculata*; *T. 4.-punctatum* and *T. hamatum* have not been identified.

8 In *Descent* 1: 338, CD reported on Blackwell's authority that in spiders both sexes usually resembled each other when young, but underwent great changes in colour as they matured.

To Fritz Müller 8 September [1869][1]

Down. | Beckenham | *Kent. S.E.*
Sept. 8

My dear Sir

I write now to ask you to observe the Papilio or Peridromia feronia, which I believe inhabits St. Catharina— see my Journal of Travels p. 33— I want to know whether *both* sexes equally make the ticking noise.— I sh[d]. think the structure at base of wings w[d]. be worth your investigation.—[2]

I received your letter of June 15th & am glad to hear the confirmation of the dimorphic case of Faramea.—[3]

The plants of Eschotzia from your seed (N.B. Asa Gray has been here, & says your plant & mine are both strictly E. Californica)[4] have, when artificially & spontaneously self-fertilised, produced pods! but they are very small (& few in number) compared with the crossed pods.[5] They are not yet ripe, so I know nothing about seed.— I shall be very curious to hear how my plants behave with you.—[6]

I sent your account of Begonia to Hooker, who was much struck by it.[7]

I have cases with Reseda odorata as capricious in respect to fertilisation as yours of Cypella(?).—[8]

Claparède has published what appears a splendid memoir on the Acarids; & he applies an argument like yours about the breathing organ of Crustaceans in my favour.—[9]

I will send in a few days a copy of a small paper on Orchids.[10]

In Haste pray believe me | Yours ever sincerely | Ch Darwin

British Library (Loan 10:30) (by kind permission of English Heritage)

1 The year is established by the relationship between this letter and the letter from Fritz Müller, 15 June 1869.

2 CD had remarked on the clicking noise of *Papilio feronia* (now *Hamadryas feronia*, the blue cracker; *Peridromia feronia* is a synonym) in *Journal of researches*, p. 38. He had inserted a note in the second edition on a drum-like structure at the base of the forewings of the insect (*Journal of researches* 2d ed., p. 133). For more on sound production in this species, see Yack *et al.* 2000. Müller had received a copy of the second edition of *Journal of researches* from CD in 1866 (see *Correspondence* vol. 14, letter from Fritz Müller, 13 February 1866).

3 See letter to Fritz Müller, 14 March 1869 and n. 9, and letter from Fritz Müller, 15 June 1869.

4 According to Emma Darwin's diary (DAR 242), Asa Gray had visited CD on 28 August 1869. *Eschscholzia californica* is the California poppy. CD had worried that the seeds he obtained from a commercial grower were of a different species of *Eschscholzia* (see letter to Fritz Müller, 14 March 1869 and n. 11).

5 See letter to Fritz Müller, 18 July [1869] and n. 6.

6 CD had sent seeds of *Eschscholzia californica* after his letter to Fritz Müller of 14 March 1869.

7 See letter from J. D. Hooker, 13 August 1869 and n. 13.

8 On *Reseda odorata*, see the letter to Fritz Müller, 18 July [1869] and n. 5. On *Cypella*, see the letter from Fritz Müller, 15 June 1869 and n. 8.

9 CD's annotated copy of Edouard Claparède's 'Studien an Acariden' (Claparède 1868) is in the Darwin Pamphlet Collection–CUL. In discussing the parasitic lifestyle of different groups of acarids, Claparède referred to Müller's account of the breathing apparatus in different families of crabs (see Claparède 1868, pp. 535–7; see also F. Müller 1864a, pp. 24–6, and Dallas trans. 1869, pp. 35–7). Claparède argued that differences in the grasping organs of the parasitic acarids he studied demonstrated that the adaptation for a parasitic lifestyle had developed independently and thus there was no single ancestor for all parasitic forms. In modern taxonomy, Acari is now a subclass of the class Arachnida.

10 'Fertilization of orchids'.

From James Orton 8 September 1869

Vassar College | Poughkeepsie, New York.
Sept. 8. 1869.

Charles Darwin Esq &c | Down, Bromley, Kent.
My dear Sir—

I take the liberty of sending you the first lower molar of a Horse, the skull of which (almost entire) I found fossil in a deep deposit in the Valley of Quito, 10000 ft. above the sea. The remains of the Horse, as I mentioned in a former letter, were accompanied by relics of a Mastodon, Deer, Llama, Auchenia &c.[1] I am curious to know the *species* of the Horse; how it differs from other sp. of Horse found elsewhere in South America and lately by Prof Owen in the tertiary of Mexico.[2] The entire specimen (and I believe this is the first equine skull found in South America) would be a better guide; but perhaps this tooth will show its affinities. If you have not the time to examine it, will you please send it to Prof. Owen.

I am with high respect | Your obedient Servant | James Orton.

PS. My book on "The Andes and the Amazon" will be out in November, when I shall have the honor of sending you a copy.[3]

DAR 173: 38

CD ANNOTATION
Top of letter: 'Name will *[illeg]* suffice' *pencil*

1 Orton's letter of 4 January 1869 described his scientific expedition to South America, where he had found mammalian fossils. *Auchenia* is a genus of the family Camelidae; it formerly included extant species like llamas and alpacas but is now restricted to one fossil species.

2 For Richard Owen's paper on fossil horses, see Owen 1869. See also the letter from W. B. Dawkins, 17 July 1869.

3 Orton 1870 was dedicated to CD (see letter from James Orton, 4 January 1869, and letter to James Orton, 23 January [1869]). CD's annotated copy is in the Darwin Archive–CUL (see *Marginalia* 1: 644).

From Edward Burnett Tylor 8 September 1869

Linden | Wellington Som.
Sep. 8. 69

My dear Sir

I understand from Sir John Lubbock that you have a copy of M[r] Haliburton's curious paper, published in some Nova Scotia transactions, on Superstitions connected with Sneezing.[1] If you can find this paper without difficulty, I should be much obliged if you would lend it to me. I would not trouble you with this request, but I do not know how otherwise to get at this paper, which would complete a collection of accounts I have on the subject.[2] With many apologies

Believe me Dear Sir | Yours very faithfully | Edward B. Tylor

Chas. Darwin Esq

DAR 178: 201

[1] Tylor refers to Robert Grant Haliburton and Haliburton 1863. A copy inscribed by the author is in the Darwin Pamphlet Collection–CUL. CD had lent his copy to John Lubbock, in 1867 (see *Correspondence* vol. 15, letter from John Lubbock, 28 September [1867] and n. 2).

[2] Beliefs associated with sneezing are discussed in Tylor 1871; Haliburton 1863 is cited on 1: 90 and 93.

To Gustav Jäger 9 September 1869

Down. | Beckenham | *Kent. S.E.*
Sep. 9 1869

Dear Sir

I lately received, but I do not know by whom sent, a copy of yr "Darwinsche Theorie," & I hope that you will allow me to express the admiration & interest which it has excited in me.[1] I must also thank you very sincerely for the manner in which you speak of my works. I am well aware that your generosity leads you to form *much* too high an estimate of what I have done; but it is deeply gratifying to receive the sympathy of so experienced an observer as yourself. As I am a very poor German scholar, I have as yet read only a portion; but I have already found many observations & incidental remarks of especial interest to me.

The case of the male silver pheasant who was rejected by the females when despoiled of his ornaments will be very useful to me to quote in what I am now writing. I wish you had specified a little more in detail in what manner the plumage was injured.[2] If you have yourself observed any analogous facts with mammals or birds, & would be so obliging as to communicate them I sh[d] feel particularly grateful.

Pray believe me dear Sir, with my best thanks & sincere respect | yours very faithfully | Charles Darwin.

P.S. I will venture to trouble you with one other question, on the chance of your having bred last year the *Pavo spiciferus*: I am very anxious to know whether the spurs in this Species are developed earlier or later in life, or at about the same period, as in the male of Pavo muticus.[3]

It w^{d} be necessary to compare birds of the 2 species hatched last summer; for the spurs are quite small during the first year.

I fear that there is no chance of y^{r} possessing both sexes of the mandrill baboon,—

LS
Dr. Hildegard Jaeger (private collection)

1 There is an annotated copy of Jäger [1869] in the Darwin Library–CUL (see *Marginalia* 1: 429). This copy was evidently sent by Eduard Koch, who had informed CD about Jäger's work (see letter from Eduard Koch, 17 June 1869, and letter to Eduard Koch, 22 June [1869]). Darwin later received another copy from Jäger himself (*Correspondence* vol. 18, letter to Gustav Jäger, 17 February 1870); this copy is now in the library of the Linnean Society. CD's letter did not reach Jäger and was eventually returned after several months; CD then resent it, enclosed in his letter to Jäger of 17 February 1870 (*Correspondence* vol. 18).

2 Jäger had described a male pheasant who, after his ornamental plumage was spoiled, was superseded by a rival in competition for females (Jäger [1869], p. 59). CD cited this example in *Descent* 2: 120.

3 The Burmese peafowl (*Pavo spiciferus*) is now *P. muticus spicifer*. CD discussed the development of spurs in *P. cristatus* (the Indian peafowl) and *P. muticus* (the green peafowl) in *Descent* 1: 290, n. 28.

To J. D. Hooker 10 September [1869][1]

Down. | Beckenham | *Kent. S.E.*
Sept. 10th

My dear Hooker

We are all very sorry you cannot come here at once, but shall be *very*, **very**, ***very***, glad whenever you can.[2] We have had our Dutchman, Prof. Donders, who came with Bowman to luncheon: he is a very pleasant, jolly man & good "Darwinian"(!).[3] He was speaking about Miquel. By the way, if *you* N. B would like it, & Miquel not dislike it, you could propose to bring him here for a Sunday, but it would rather spoil your visit selfishly for us, & in this case you must tell the truth & tell M. how little I can see of anyone.—[4]

Thanks for news about Exeter; much of which was new to me, but as far as I have heard & read I agree to every word you say.[5]

But you do not mention the most charming (& that is the correct word) speech of all & that was yours on resigning the Presidentship.—[6] By Jove you ought to have been an embassador, a foreign councillor or some such swell to have made speeches when monarchs are received!—

I received the Indian Census, for which many thanks; but now I know not how far to trust it—[7]

Yours affectionately | C. Darwin

I am unusually well & been working unusually hard.—

P.S. Henrietta[8] will be away on Sat. 18 & at home on Sat. 25. & she will be much pleased if the latter will suit you equally well, but most likely you will not be able to take your choice of times being so busy—

DAR 94: 151–2

1 The year is established by the relationship between this letter and the letter from J. D. Hooker, 7 September 1869.
2 See letter from J. D. Hooker, 7 September 1869. The third 'very' is underlined three times.
3 CD refers to Frans Cornelis Donders and William Bowman. See letter from William Bowman, 3 September [1869].
4 CD refers to Friedrich Anton Wilhelm Miquel. See letter from J. D. Hooker, 7 September 1869.
5 See letter from J. D. Hooker, 7 September 1869.
6 Hooker's resignation speech preceded the address of the new president for the coming year, George Gabriel Stokes, on 18 August (*Report of the meeting of the British Association for the Advancement of Science, held at Exeter in 1869*, p. lxxxvii). A short excerpt from Hooker's speech was published in *The Times*, 20 August 1869, p. 7.
7 See letter to J. D. Hooker, 7 August [1869] and n. 15, and letter from J. D. Hooker, 13 August 1869.
8 The reference is to Henrietta Emma Darwin. The postscript is in Emma Darwin's hand.

From John Beddoe 13 September 1869

Clifton,
Septr 13 '69

Dear Sir

Enclosed is the *last* instalment of my data respecting conjugal selection—[1] The dark shades do in this case predominate in the married women; but the numbers of the single are so small that I fear nothing can be made of them—

Believe me | Yours faithfully | John Beddoe

[Enclosure]

In- & Out-patients British Infirmary[2]

	Single					Married				
	Red	Fair	Brown	Dark	Black	Red	Fair	Brown	Dark	Black
Women between 35 & 45	1	3	3	7·5	2·5	9	12.5	39	$68\frac{1}{2}$	6
Women between 45 & 50	1.5	·5	6	3	1	5·5	12.5	49	$85\frac{1}{2}$	$11\frac{1}{2}$
Total	2.5	3·5	9	10.5	3·5	14·5	25	88	154	17·5
	29					299				

DAR 85: A17–18

CD ANNOTATION
Top of letter: 'I must give up whole case.—' *blue crayon*

1 Beddoe had last supplied CD with data in his letter of 21 August 1869.
2 Beddoe was physician to the Bristol Royal Infirmary.

From Edward Blyth 13 September 1869

21 Chalcot Crescent, | Regent's Pk,
Sept 13 /69

Dear Mr Darwin,

I find that I have an unanswered letter from you dated so long back as May 28th—[1] Meanwhile, I have been much away, & have had much trouble to undergo, entirely through the misconduct of others. You ask me about the mandrils at Antwerp, & I have since passed more than a fortnight at Antwerp,[2] as you will probably have seen by the last three Nos. of the *Field*, for which paper I write now instead of *Land and Water*, to my own great pecuniary advantage.[3] One reason that I did not reply to you before was that I rather expected to revisit Antwerp, but I found that the mandrils, parents & infant, were all dead, & I could get no satisfactory answer to your queries. When I saw the animals they were half-concealed in deep straw, which answered the purpose of a garment. My *impression* is that both wild boar and elephant are polygamous, but I could not undertake to speak positively.[4] However, I think that there is hardly a doubt that such is the case. I was exceedingly ill treated by a vulgar upstart Glass, who calls himself "manager" of the *Land & Water* concerning ⟨th⟩e *[getter] up* of the paper. His object was to put J K. Lord in my place, & he succeeded, for I could brook his impertinence no longer, & besides I had no end of trouble in getting paid by him for my contributions.[5] Now I receive considerably *more than double* from the *Field*.

Trusting that this will find you better in health, I remain | Yours Sincerely | E. Blyth

DAR 160: 225

1 CD's letter to Blyth has not been found.

2 The mandrill is *Mandrillus sphinx*. See letters from Edward Blyth, 17 April 1869 and n. 2, and 26 April 1869.

3 Blyth's article, 'The Zoological Garden at Antwerp', appeared in three parts in the *Field* under the pseudonym 'Zoophilist' (28 August, 4 and 11 September 1869, pp. 175, 191–2, 215–16).

4 CD described the polygamy of the Indian elephant and wild boar during the breeding season in *Descent* 1: 267.

5 John Keast Lord had been a member of the staff of *Land and Water* since the journal was launched in 1866 (*ODNB*). Glass has not been further identified, but see also the letter from Edward Blyth, 17 April 1869.

From George Cupples 13 September 1869

The Cottage, | Guard Bridge | Fifeshire. N.B.
Septr 13 /69—

My Dear Mr Darwin,

I lately saw an old man in this neighbourhood (above 90, but in full possession of his memory and characteristic Scotch shrewdness) who was for more than 40 years a *mole*-and-rat-catcher. He states that he has often heard it alleged among the fraternity of the trapping line of business that there is a decided majority of *males* in the nests of young *rats*, and his own experience disposes him to corroborate this. As to *moles*, he had not happened to think of the question (if I understood him rightly)—but I only saw him for a short time, and can easily see him again. He spoke of a professional friend who is to be in this part of the country ere long, and who could doubtless throw light on these points.[1]

I did not get in all my returns on sheep—and nothing on cattle. If you have much time before publication or completion, I could obtain further statistics on any head that might appear worth while to you—[2]

Incomplete
DAR 86: 75

CD ANNOTATIONS
1.1 I lately … rat-catcher. 1.3] *crossed blue crayon*
1.3 He states … corroborate this. 1.5] *scored blue crayon; closing square bracket blue crayon*
1.5 As to … these points. 1.9] *crossed blue crayon*
2.1 I did … to you— 2.3] *crossed blue crayon*
Top of letter: 'G. Cupples | Rats' *blue crayon*

1 CD was collecting information on the proportion of the sexes for his chapters on sexual selection in *Descent*. On rats and moles, see *Descent* 1: 305.

2 Cupples had been compiling information on sheep and cattle bred in Scotland (see letter from James Rait to George Cupples, 20 July 1869).

From V. O. Kovalevsky 13 September 1869

Paris
13 September 1869

My dear Sir!

It is such a long time that I had no news from You, nor gave You any of me that I fear I am quite forgotten in Your amiable house at Down.[1]

I left Russia in March and passed the whole summer in Germany at Heidelberg devoting myself entirely to Geology and have now a reasonable hope to regain the time lost by me these five years, while affairs dragged me away from regular studies. My wife was happy enough to be accepted in the University as a regular student and worked so well that in eighteen months or two years she will be able to pass her examination as "Doctor der Mathematik und Physik".[2]

In the vacation I travelled about five weeks in Switzerland, am now in Paris and hope to be in London, though my chief attraction in England will be Down. If you have not quite forgotten Your visitor of 1866 & editor of Your book[3] be so kind as to write me two lines to Paris (W. Kowalevsky, Paris. Boulevard Montparnasse, Rue Vavin, Hotel Heranger N 4.) to inform me wether You will be able to receive our visit, I am very Impatient to see You once more and my wife is the more so to make Your aquaintance. The best wish of my brother was also to go this year to England to visit You, but he was send by the University of Kieff to the Caspian Sea and the project was to be remitted to the next year.[4] I present also my humble respects to M[rs] Darwin and the Ladies[5] and am Sir

Yours most truly | W. Kowalevsky

P.S. The Russian Edition of Your work was not quite finished in March, when I got away, but now it is all ready and perhaps I could bring You a copy of my edition if I receive it by time from home.—[6]

DAR 169: 78

[1] Kovalevsky visited CD at Down in August 1867, not 1866 as Kovalevsky writes in this letter (see *Correspondence* vol. 15, Appendix II).

[2] Kovalevsky's wife was Sofia Vasilyevna Kovalevskaya. For more on her career in mathematics, see Koblitz 1983.

[3] Kovalevsky had translated *Variation* into Russian (V. O. Kovalevsky trans. 1868–9).

[4] Kovalevsky refers to Alexander Onufrievich Kovalevsky. Kovalevsky had sent CD some of his brother's embryological papers in 1867 (see *Correspondence* vol. 15, letter from V. O. Kovalevsky, 15 March 1867 and n. 6).

[5] Emma Darwin, Henrietta Emma Darwin, and Elizabeth Darwin.

[6] The Russian translation of *Variation* (V. O. Kovalevsky trans. 1868–9) was published in seven parts between 1867 and 1869 and also issued in book form (see Freeman 1977). CD's copy of the book is in the Darwin Library–Down. Kovalevsky and his wife visited Down from 30 September to 1 October 1869 (Emma Darwin's diary (DAR 242)).

To Edward Blyth 14 September [1869][1]

Down. | Beckenham | *Kent. S.E.*
Sept[r] 14th

My dear M[r]. Blyth.

Very many thanks for your attempts to find out about the Mandrills:[2] I have written to several Zoolog. Gardens on the continent, but with no success.—[3] I have got from Sir W. Elliot & others some information about Indian Boar & Elephant confirming your impression.—[4]

I am very sorry to hear that you have been so badly treated by the manager of L. & W.[5] He has cut his own throat; for to my mind your articles were incomparably the best which appeared. I left off taking in L. & W. a little time ago, for I found I had not time to read so many periodicals; but I have doubted since whether I was wise, & this was exclusively on account of your articles.

Anyhow the case is now decided. I must think about the Field.[6]

With many thanks | Yours sincerely | C. Darwin

McGill University Libraries, Rare Books and Special Collections Division

[1] The year is established by the relationship between this letter and the letter from Edward Blyth, 13 September 1869.

[2] See letter from Edward Blyth, 13 September 1869.

[3] See letter from J. N. Hegt, [23 April 1869], and letter to Max Schmidt, [29 April 1869].

[4] CD cited William Elliot on the wild boar in *Descent* 1: 267. See letters from Walter Elliot, 7 August 1869 and 7 September 1869.

[5] CD refers to *Land and Water*; see letter from Edward Blyth, 13 September 1869.

[6] Blyth had started writing for the *Field.*

From Frederick Smith 14 September 1869

Brit. Museum.
14 Sept[r]. 1869.

My dear Sir

I regret that I am not prepared to decide as to whether the sexes stridulate in Mutilla I can only affirm that the female does so—and that so sharply as to attract your notice immediately—[1] I have taken more specimens of the Male than the female but have not certainly noticed in the former sex.— One of the musicians that I have heard is Trox sabulosus— this insect I once observed on the wing— it settled in the grass within a yard of me—and on doing so I was attracted by the shrill tone it produced— I caught it in my hand and so loud was the squeak-like noise that a gamekeeper who was standing by asked if I had caught a mouse—[2] All the species of Necrophorus stridulate loudly[3]—but the most amusing performance is that the little weevil Acalles roboris—[4] it is about this size.  You will recollect Mr Wollastons paper in the Zoologist on the Musical Curculio—that he found I think in Madeira—this was a Sp. of Acalles[5] Wollaston wrote to me when I was at Deal where the Acalles roboris is plentiful asking me to observe whether this little sp.[c] was musical—I obtained about 20— I had to place them in a small pill box—then shake them up—and on applying the box to my ear—their music was distinctly audible— I tried—various experiments placing 6. 4. 3. 2. 1 in the box— the single beetle was very difficult to hear— the tone was shrill but very low but I did hear it— the tone was produced by rubbing the abdomen against the elytra—there being a very smooth space on which it operated— Wollaston has described it in his paper

I shall probably see you soon when I might be able to say more— next week I shall I expect be in the Isle of Wight

Yours very truly | Fred[k]. Smith

DAR 82: A5–6

CD ANNOTATIONS
1.1 I regret . . . sex.— 1.4] *crossed pencil*
1.5 Trox sabulosus] '**Trox**' *added pencil*
1.7 I caught . . . gamekeeper 1.8] *scored blue crayon*
1.8 All . . . roboris— 1.10] *crossed pencil*
1.15 I tried . . . Wight 2.2] *crossed pencil*
Top of letter: 'Get reference to Wollastons Paper.— | Sexes of Mutilla to examine—[6] | How are colours in Mutilla?' *blue crayon*

1 *Mutilla* is a genus of parasitic wasp of the family Mutillidae (velvet ants). In *Descent* 1: 366, CD reported that both sexes of *M. europaea* stridulated. Smith's observation of stridulation in females of the genus had been given in the letter from E. W. Janson, 25 May 1868 (*Correspondence* vol. 16).

2 CD included Smith's account of the beetle *Trox sabulosus* (the sand-loving trox) in *Descent* 1: 380.

3 *Nicrophorus* (*Necrophorus* is an incorrect subsequent spelling) is a genus in the family Silphidae (carrion beetles). CD discussed the stridulating organs of *Necrophorus* in *Descent* 1: 378–9.

4 *Acalles* is a genus in the family Curculionidae (snout beetles, weevils). CD mentioned the stridulation of *Acalles* in *Descent* 1: 384.

[5] The reference is to Thomas Vernon Wollaston's paper 'On certain musical Curculionidae' (Wollaston 1860); it was published in *Annals and Magazine of Natural History*, not the *Zoologist*. The paper is cited in *Descent* 1: 378.

[6] CD reported his own observations of the stridulating mechanism in *Mutilla europaea* in *Descent* 1: 366.

To Samuel Hubbard Scudder 15 September [1869][1]

Down. | Beckenham | *Kent. S.E.*
Sept. 15

Dear Sir

I am much obliged to you for sending me your "Revision &c" & paper on Lepidoptera, & for various other papers on former occasions, many of which I have read with very great interest—[2]

I thank you also for your present of Harris' Ent.— Correspondence, in which I daresay I shall find points of interest;[3] for I possess two Editions of his "Insects of New England", which I have read with profit—[4]

With my best thanks & sincere respect.— | I beg leave to remain | Dear Sir | Yours very faithfully | Ch. Darwin

Boston Society of Natural History

[1] The year is established by the reference to Harris 1869; see n. 3, below.

[2] CD refers to Scudder's *Revision of the large, stylated, fossorial crickets* (Scudder 1869a); CD's copy is in the Darwin Pamphlet Collection–CUL. Scudder had recently published several papers on Lepidoptera (Scudder 1868, Scudder 1869b, and Scudder 1869c). A copy of Scudder 1869c is in Darwin Pamphlet Collection–CUL. Beginning in 1861, Scudder had sent numerous publications and reports to CD (see *Correspondence* vol. 10, letter to Andrew Murray, 10 April [1862] and n. 2, and Scudder 1861). They are in the Darwin Pamphlet Collection–CUL.

[3] Scudder edited the *Entomological correspondence of Thaddeus William Harris* (Harris 1869). CD's lightly annotated copy is in the Darwin Library–Down (see *Marginalia* 1: 361).

[4] CD's annotated copy of *A treatise on some of the insects of New England, which are injurious to vegetation* (Harris 1842) is in the Darwin Library–CUL (see *Marginalia* 1: 361–2). CD's copy of a later edition of the book (Harris 1862) is in the Darwin Library–Down.

From J. T. Moggridge 16 September [1869][1]

2 Montague Villas | Richmond | (*Surrey.*)
Sept. 16

Dear Mr. Darwin

Thank you very sincerely for the pamphlet containing supplementary *Notes on the Fert: of Orchids*, which arrived here yesterday—[2]

I am much pleased that my observations are found worth a place, & that they recieve such kind mention—[3]

The history of the fertilization of the tiny *Herminium monorchis* is very wonderful, & admirably observed.[4]

Dr. H. Müller's papers on *Platanthera* & *Epipactis* appear to be of first rate importance, & I look forward with great interest to reading them in full.[5]

I have lately read with delight the paper on *Epigæa* by Dr. Meehan— It would appear that, in respect of variability, this Ericaceous plant is almost comparable to *Arbutus*—.[6]

Some of his generalisations appear to me to want qualification; as, for instance, when, speaking of plants generally, he says that—"there is as much variation in the wild plant, as in those under the best gardener's skill"—[7]

I do not remember to have seen evidence to shew that any constant proportion exists between the amount of variability in the wild & the cultivated states—

I must now thank you for your kind & much-valued letter of June 7, recieved in reply to my own containing an imperfect abstract of some observations on *Arbutus*, *Ophrys*, etc.—[8]

I hope on my return to Mentone to follow out your hints, especially that with reference to the lopped trees.—

I have already noted in each case, the altitude & exposure to which each tree is subject, & find, as one might expect, a sure relation between small leaves & a burning exposure & large leaves & a shady one. Whether the leaves & fruits or flowers are correlated, remains to be seen.

I have gained some evidence this spring, to shew that there are in England two varieties of *Ophrys apifera*, which tend to form colonies in which one is found to the exclusion of the other.—[9]

Thus I have recieved 17 specimens from Walton-on-the Naze, & 7 from Colston Basset (Notts.), all of which present the long-petaled form (A); while out of 20 from Dover, 16 were short-petaled (as. B.), & the remainder were either diseased or monstrous, & produced in the same flower long & short petals—

A.

B.

I have heard from time to time from Dr. Hooker, that you are recovering from the effects of that most unfortunate accident with your horse.[10]

What a pity it is that now-a-days there are no Fates to propitiate with sacrifices, or such accidents would, I am sure, be made impossible—

I know that you will be glad to hear that I am now much stronger, & look forward to starting for Mentone on October 15— —

It would give me sincere pleasure if I could be of use to you in any way there—

We leave Richmond for London on Oct. 8th., where we shall stay a week—

Believe me | yrs. most sincerely | J. Traherne Moggridge.

DAR 171: 212

1 The year is established by the reference to 'Fertilization of orchids' (see n. 2, below).

2 'Fertilization of orchids' was published in the September 1869 issue of *Annals and Magazine of Natural History*.

[3] CD cited Moggridge 1864 in 'Fertilization of orchids', pp. 143, 145 (*Collected papers* 2: 140–1, 142).
[4] See 'Fertilization of orchids', pp. 145–6 (*Collected papers* 2: 142–3). *Herminium monorchis* is the musk orchid.
[5] CD had mentioned Hermann Müller's paper on Westphalian orchids (H. Müller 1868) with its discussions of *Platanthera* and *Epipactis* in his letter to J. T. Moggridge, 7 June 1869. He cited the paper in 'Fertilization of orchids', pp. 147, 150 (*Collected papers* 2: 144, 147).
[6] The reference is to Thomas Meehan and Meehan 1868a. See letter to J. T. Moggridge, 7 June 1869 and n. 5.
[7] Meehan 1868a, p. 153.
[8] See letter to J. T. Moggridge, 7 June 1869 and n. 2. Moggridge's letter has not been found.
[9] CD and Moggridge had corresponded at length on the bee orchis (*Ophrys apifera*); see *Correspondence* vol. 14. Walton-on-the-Naze is in Essex. Colston Basset is in Nottinghamshire.
[10] Moggridge refers to Joseph Dalton Hooker. For more on CD's accident, see the letter to A. R. Wallace, 14 April 1869 and n. 12.

From J. D. Hooker 17 September 1869

Royal Gardens Kew
Sept 17 /69

Dear Darwin

I see nothing to prevent my being with you on afternoon of tomorrow week—to wh. I look forward—

Thanks for the supplementary which I shall quote in the Brit. Fl.[1]

Ever yrs | J D Hooker

The Dutch could not come through illness in family.[2]

DAR 103: 32–3

[1] Hooker was preparing *The student's flora of the British Islands* (J. D. Hooker 1870). See letter from J. D. Hooker, 11 March 1869 and n. 6.The 'supplementary' Hooker refers to is probably new material added to the French edition of *Orchids* (Rérolle trans. 1870) and separately published in English ('Fertilization of orchids').
[2] CD had suggested that Hooker invite Friedrich Anton Wilhelm Miquel to Down. See letter to J. D. Hooker, 10 September [1869].

From T. H. Farrer 18 September 1869

Abinger Hall, | Dorking.
18 Sept/69

My dear Mr Darwin

Many thanks for your notes and for the kind notice you have taken of the tiny facts I sent you.[1]

I have found out myself this year how comparatively difficult it is to observe physiological functions when owing to bad weather or otherwise the plants are not in full vigour. The pollinia of Orchids did not function half so well in the last cold June as in the hot one of 1868.

I am much interested in what you say about bees getting nectar from below the surface.[2] In the papilionaceous flowers—Ulex, Lupin, &c I see them at work where

I can find no nectar— But I find no open staminal tube without nectar. And now I am going to trespass terribly on your good nature & your time.

I have put together a few notes upon a few Papilionaceous flowers and am venturing to send them to you. If you have not time to look at them please send them back: and if you find my scribbling hard to read, I will gladly have it copied. Should you be able to look at the paper I shall be very much obliged if you will tell me whether you think all or any part of it worth printing: and whether it would be wise to keep it for another summer.[3] I am pretty sure of the facts I have stated, but alas! I was in London entirely till July and afterwards too busy, when away from London, with settling here & with Office matters, to do what I should have liked.[4]

The fact is, one needs to go on to the hill side & look & look, & come home and think it over and go & look again before one can be certain of any thing.

It is too late now this year for real flowers and drawings in books are of little use.

Will you kindly return me the paper addressed to me at the Board of Trade and marked "Private", where it will need no postage stamps.

And pray do not trouble yourself with it to the detriment of more important work or to your own fatigue.

Hooker is coming to us on the 5th October—and I shall victimize him with Papilionaceous flowers— What a work of labour is his Genera Plantarum.[5] It is Cyclopean like Johnsons Dictionary[6]

If railways will suit, I might come over for an hour or two some day, and thus save you the trouble of writing. Is there any station on the S.E.R.—eg. Godstone Road, Caterham, &c that is within fly-distance of Down?[7]

Believe me | Sincerely yours | T H Farrer

Charles Darwin Esqr FRS

DAR 164: 55

CD ANNOTATIONS

5.1 The fact ... look again 5.2] *scored blue crayon*

1 Farrer's observations of the movement of the pollinia in *Ophrys muscifera* (now *Ophrys insectifera*, the fly orchid) and *Peristylus viridis* (now *Coeloglossum viride*, the frog orchid) are cited in 'Fertilization of orchids', pp. 144, 146 (*Collected papers* 2: 141, 143). See *Correspondence* vol. 16, letters from T. H. Farrer, 17 May 1868 and 4 June 1868.

2 In 'Fertilization of orchids', pp. 142–3 (*Collected papers* 2: 140), CD added further observations confirming his speculation in *Orchids*, pp. 50–1, that in many orchid species, insects punctured the inner lining of the nectary, withdrawing nectar that had been secreted between the inner and outer membranes of the nectary.

3 Some notes by Farrer on the floral morphology of Papilionaceae are at the Linnean Society (MSS Case B, Ms 508). He later published on fertilisation in the Papilionaceae in *Nature* (Farrer 1872).

4 Farrer was permanent secretary of the Board of Trade (*ODNB*).

5 Joseph Dalton Hooker was co-author, with George Bentham, of the multi-volume work *Genera plantarum* (Bentham and Hooker 1862–83).

6 Farrer refers to Samuel Johnson's *A dictionary of the English language* (Johnson 1755).

7 The station nearest Down village on the South East Railway was Orpington. CD's reply to Farrer has not been found, but Farrer may have visited Down on 8 October 1869 (see letter from T. H. Farrer, 9 October 1869).

To J. D. Hooker 18 [September 1869][1]

Down. | Beckenham | *Kent. S.E.*
18th

My dear Hooker

We shall all be right glad to see you this day week; let us know Trains to Orpington & in all probability we can send to meet you.—

I much want a query answered.— Consult any of your colleagues learned in physiology.— When a large piece of bark is removed from a tree, does the bark ever regrow in isolated points, *separate* from the growing margins of the surrounding bark. I fancy I have heard that this is sometimes the case.— It bears on Pangenesis & on little powers of repair by plants.[2]

If you can answer in affirmative pray send me a line,—if I do *not* hear, I shall understand your answer is a negative—.

Yours affect, | C. Darwin

DAR 94: 153–4

[1] The month and year are established by the relationship between this letter and the letter from J. D. Hooker, 24 September 1869.

[2] CD discussed the regrowth of bark in his letter to *Scientific Opinion*, [before 20 October 1869], drawing an analogy with the regrowth of amputated limbs in some animals. CD added a section on the regrowth of amputated parts to his discussion of pangenesis in the second edition of *Variation* (*Variation* 2d ed., 2: 357–9), but did not discuss plants.

From J. T. Moggridge 18 September 1869

2 Montague Villas | Richmond | (Surrey)
Sept. 18 | 1869.

Dear Mr. Darwin

Thank you for your kind note recieved this morning.[1]

The plant you enclose is *Lathyrus Ochrus* DC., & I fully hope that I shall be able to get some seeds for you before long, as it is not a rare species—[2]

I enclose a fragment from a more advanced specimen, also gathered at Cannes; in order to shew you that leaflets (to the number of 2–4) are developed towards the summit of the stem—

I notice that it is common to find (as in the fragment sent), three leaflets, of which the two upper ones form a pair, while, the lower one is solitary—[3]

This species has much in common with *L. Clymenum* L., a plant which abounds in the South, & of which I could easily obtain seeds, if you wished for it—[4]

Believe me | yrs. very sincerely | J. Traherne Moggridge.

DAR 171: 213

[1] CD's letter has not been found.

2 *Lathyrus ochrus* is cyprus vetch. It is found near the Mediterranean, where Moggridge was expecting to return shortly (see letter from J. T. Moggridge, 16 September [1869]).
3 On CD's interest in phyllotaxy, see *Correspondence* vol. 11. CD's notes and an unpublished draft on the subject are in DAR 51a: 1–32.
4 *Lathyrus clymenum* is Spanish vetchling.

From John Murray 18 September [1869][1]

50A, Albemarle St. | W.
Sept. 18

Dear Mr Darwin

The enclosed paper will show that I am about to start a new Monthly Literary Review to convey early & trustworthy intelligence to English Readers on Literature Science & Art & to establish, if possible, a higher tone of Criticism than that now prevailing—[2]

I sh[d] like very much to announce among my Literary News in the 1st number, a short notice of the New Work upon w[ch] you are now engaged.[3] Will you kindly draw up a short account of it for me? The first number will be circulated far & wide & I shall send out many thousand copies—so that this at least will be a favorable mode of advertising— It need not commit you at all as to the time of publication, or of going to press even. I hope to be able to announce at the same time Mr Grotes Aristotle—[4]

I hope you have recovered from the accident w[ch] I was sorry to hear had befallen you[5]

I am My Dear Sir | Yours very faithfully | John Murray

Charles Darwin Esq

DAR 171: 371

CD ANNOTATIONS

Top of letter: 'Good wishes | puzzled— | alter | It ['is clear' *del*] is too clear that ['it' *del*] as it stands it was *[written]* by me' *pencil, del pencil*

1 The year is established by the reference to the *Academy* (see n. 2, below).
2 The enclosure has not been found. Murray refers to the *Academy: a Monthly Record of Literature, Learning, Science, and Art*, published from 9 October 1869. The new journal covered the same subject areas as the *Athenæum*. On the history of the journal, see North 1997, 1: 82–3.
3 The notice appeared in the *Academy* 1 (1869): 15–16. It announced a three-part work: 'I. *The Descent of Man*; II. *On Sexual Selection*; and III. *On Expression of the Emotions*.'
4 George Grote's *Aristotle* was not published until after his death (Grote 1872).
5 CD had been thrown from his horse. See letter from A. R. Wallace, 14 April 1869 and n. 12.

To John Murray [after 18 September 1869][1]

Down. | Beckenham | *Kent. S.E.*

My dear Mr Murray

I heartily wish your Periodical all success: I wish it had been weekly, as then perhaps it would have killed the Athenæum by a lingering death, & that to me would have been a pleasing sight.—[2]

I have been sorely puzzled what to write about my book, & know not in the least whether the enclosed will do, or is too long.[3]

Pray ask your Editor to strike out or alter to any extent.[4] It looks now too clearly as coming from me. Perhaps if the Editor w^{d}. put in here & there. "We hear" or "we are informed"—it w^{d} improve matters, but I leave it to his or your discretion.— Pray do not publish anything which seems absurd, superfluous or unnecessary.—

I shall be in London later in the Autumn, & want to consult you about woodcuts &c[5]

Your's very sincerely | Ch. Darwin

The M.S. is legible, but badly written, as I dictated it to one of my boys.—[6]

National Library of Scotland (John Murray archive, Ms. 42153 ff. 39–40)

1 The date is established by the relationship between this letter and the letter from John Murray, 18 September [1869].
2 CD refers to the *Academy* and the *Athenæum*. See letter from John Murray, 18 September [1869] and n. 2.
3 See letter from John Murray, 18 September [1869] and n. 3. The enclosure has not been found.
4 The editor of the *Academy* was Charles Edward Cutts Birchall Appleton.
5 CD was in London from 1 to 9 November (Emma Darwin's diary (DAR 242)).
6 Of CD's sons, Horace Darwin was the most frequently at home; however Francis and George were also at Down House during the second part of September (Emma Darwin's diary (DAR 242)).

To Albert Günther 21 September 1869

Down. | Beckenham | *Kent. S.E.*
Sep. 21 1869

My dear Dr Günther

I am going to beg, I hope for the last time, a great deal of assistance from you. I enclose a whole string of queries, but most of these may be answered by a few words written after them. My only excuse for troubling you so much is that four-fifths of the facts which I give, are quoted from you & your works.[1] To save you as much trouble as possible, I enclose an envelope directed.

In the second place, you most kindly offered to help me about illustrations.[2] I now enclose a list somewhat different from my original one.[3] Will you look it over, as some of the drawings cannot be made, unless you can supply the specimens. Mr Ford[4] w^{d} do the drawings *incomparably* better than any one else; but if his time is taken up with more important & original work, could you find any other artist who w^{d} make accurate & neatly finished drawings. If Mr Ford c^{d} pledge himself to complete the drawings within 2 months, this w^{d} do; otherwise I must search for some one else, but this w^{d} be a grievous loss to me. I am not quite certain when I shall go to press, for it depends upon whether I shall add another subject, & this I cannot at present decide;[5] but I sh^{d} so much regret being delayed by the want of the drawings, that I must get them done within the time specified. It will be a real kindness if you will assist me on this head.

Pray forgive me for asking so many favours & | believe me | yours sincerely obliged | Ch. Darwin

[Enclosure]

Queries for Dr Günther

(1) In my notes from y^r^ conversation I speak of the dorsal fin of Mollienesia Petenensis as occellated with **bright** spots;; but you in the Zoolog. Trans. merely speak of a "row of large rounded spots" Which of these is correct?[6]

(2) Agassiz writes to me that certain chromids, as Pomotis, Hygrogonus &c sit on their nest; do you happen to know whether it is the male, or the female, or indifferently both sexes, which thus sit?[7]

(3) I understood you to say that the female of Solenostoma is brighter coloured than the male; is this the case, for you do not say so in the Fishes of Zanzibar?[8] Do you believe Kaup's statement that the males have marsupial sacks & carry the eggs?[9] What must I say on this head?

(4) Is it known whether any Lizards utter any sound during their courtship?

(5) You gave me cases, with Lizards, of the dorsal crest being confined to the males; & other cases in which the crest was larger in the males than in the female;[10] Now are there other cases in which the dorsal crest is equally developed in both sexes?

(6) Chamælio bifurcus.[11] What is its native Country? Is it brightly coloured, especially the projections on the head of the male?

(7) Chamælio owenii.[12] What is its native Country? Is it brightly coloured?

Has the female any rudiment of the triple horn? Is the horn in the male flexible or rigid?

Is the triple horn formed by the modification of **one** or **more** scales?

(8) Can you remember whether any amphibian or reptile is marked with true *ocelli* of bright colours; that is a spot of colour surrounded by a ring of another colour?

(9) With Lizards are the males generally or often or rarely larger than the females?

LS(A)
Shrewsbury School, DAR 82: B14–15

[1] CD cited Günther extensively on fishes, amphibians, and reptiles in *Descent* (see *Descent* 2: 9–10, 12, 18–19, 22, 30, 33–4, 37). He made particular use of Günther's descriptive catalogue of central American fish (Günther 1864–6), and his monograph on the reptiles of British India (Günther 1864). There is an annotated copy of Günther 1864 in the Darwin Library–Down (see *Marginalia* 1: 352). An annotated copy of Günther 1864–6 is in the unbound pamphlet collection in the Darwin Archive–CUL.

[2] CD had discussed the illustrations for *Descent* with Günther during a visit to London in March 1868. See *Correspondence* vol. 16, letter to Albert Günther, 23 March [1868] and n. 2.

[3] The enclosed list has not been found. CD had enclosed a slip (also not found) in the letter to Albert Günther of 23 March [1868] (*Correspondence* vol. 16).

[4] George Henry Ford.

[5] CD originally planned to include material on the expression of emotions in *Descent*.

[6] CD refers to Günther 1864–6, p. 485. *Mollienesia petenensis* (now *Poecilia petenensis*) is the peten molly.

[7] See *Correspondence* vol. 16, letter from Louis Agassiz, 22 July 1868. The former family Chromidae includes the present families Cichlidae and Pomacentridae (Pauly 2004). *Pomotis* is a now invalid genus that included species now recognised as sunfishes as well as some cichlids. *Hygrogonus* is now *Astronotus*.

[8] CD refers to *The fishes of Zanzibar* (Playfair and Günther 1866), p. 137; *Solenostoma* is now *Solenostomus*).

[9] Günther had reported Johann Jakob Kaup's observation in Playfair and Günther 1866, p. 138.

[10] In his letter of 19 December 1867 (*Correspondence* vol. 15), Günther wrote that in lizards with a dorsal crest, the feature was more developed in the male.
[11] CD refers to *Chamaeleo bifurcus* (now *Furcifer bifidus*, the two-horned chameleon).
[12] CD refers to *Chamaeleo owenii* (now *Chamaeleo oweni*, Owen's three-horned chameleon).

To T. H. Huxley 21 September [1869][1]

Down. | Beckenham | *Kent. S.E.*
Sept 21

My dear Huxley

A naturalist, James Orton of U. States, has found in a thick deposit, at height of 10,000 feet, in Quito, associated with remains of Mastodon, Llama &c the skull of a Horse.—[2] He has sent me from U. States "the first lower molar", & asked me to show it to Owen to learn its relations to the fossil Horses of Mexico & La Plata. This I cannot do, for I can hold no communication with Owen.—[3] I wish for several reasons to oblige Mr. Orton, who is going to publish an account of his Researches in Nat. Hist in S. America.[4]

Can you aid me & him? Would it give you much trouble to compare the tooth with any American fossil in College of Surgeons or elsewhere?[5] I know it is an unreasonable request.— I send the specimen by this Post: Mr Orton does not ask for its return, & unless he claims it, I shd think it wd be well worth depositing in your Museum[6] or British Museum, as an authentic specimen.

Mr Orton ought to have sent the specimen to Leidy.[7]

Forgive me for being so very bothersome & | believe me | Ever yours | Ch. Darwin

Imperial College of Science, Technology, and Medicine Archives (Huxley 5: 273)

[1] The year is established by the relationship between this letter and the letter from James Orton, 8 September 1869.
[2] See letter from James Orton, 8 September 1869 and n. 1.
[3] See letter from James Orton, 8 September 1869 and n. 2. CD had fallen out with Richard Owen following Owen's anonymous review of *Origin* ([Owen] 1860, see *Correspondence* vols. 8 and 9).
[4] See letter from James Orton, 8 September 1869 and n. 3.
[5] CD refers to the Hunterian Museum in the Royal College of Surgeons.
[6] CD refers to the Museum of Practical Geology, which was attached to the Royal School of Mines in Jermyn Street, London, where Huxley was employed.
[7] Joseph Leidy.

From M. T. Masters 21 September 1869

Gardeners' Chronicle | *& Agricultural Gazette Office,* | *41, Wellington Street, Strand, W.C.*
Sept. 21 *1869*

My dear Sir/

Many thanks for your note and the accompanying pamphlets.[1]

At the Hort. Soc. meeting to day Mr Fenn (Rectory Woodstock) showed some Potatos which would I think have interested you had you seen them viz. 1. results of grafting potatos—all failures except one which *may* be all right 'Milky white' grafted into a fluke— result an inferior potato having the characters of both parents[2]

2^{d}. A large series of tubers the result of a *single cross* between a round red potato & a white round. The offspring are of all sorts sizes shapes colors degrees of hardihood etc—but what is most curious is that he has got among them a *small round* mottled tuber which is almost precisely identical with some so called wild tubers grown in Peru—a nice case of reversion!—[3] The half bloods exhibit traces of the blood of both parents in very varying degrees— I asked him to forward you samples— He assures me he took great pains to prevent crop fertilisation with other sorts but as I understand the experiments were conducted en pleine terre foreign pollen may have been introduced without his knowledge In any case the reversion fact remains

Believe me | faithfully yrs | Maxwell. T. Masters

DAR 171: 81

[1] CD's note to Masters has not been found. The pamphlets may have been offprints of 'Notes on the fertilization of orchids'.

[2] Robert Fenn was a steward to Woodstock Rectory, Oxfordshire (R. Desmond 1994). His experiments with potatoes, presented at the Royal Horticultural Society, were reported in *Gardeners' Chronicle* 39 (1869): 1011–12. CD had discussed hybrid potatoes produced by grafting in *Variation* 1: 395–6. See also *Correspondence* vol. 16, letter from Friedrich Hildebrand, 2 January 1868, and letter to Friedrich Hildebrand, 5 January [1868].

[3] CD discussed reversion arising from crossing in *Variation* 2: 39–47.

To [?] 21 September [1869]

Down. | Beckenham | *Kent. S.E.*
Sept. 21

Dear Sir

I am much obliged for your kindness in sending me the curious facts about your Cats.[1] I have so many letters to write that I hope you will excuse my entering on the subject in detail.—

With my best thanks | I beg leave to remain | Dear Sir | Yours faithfully | Ch. Darwin

Endorsement: 'Rec^{d} 22/9/69 J. J. L.'
National Library of Australia (MS 760/2/571)

[1] CD had discussed cats in *Origin* and *Variation.*

From Hermann Müller 22 September 1869

Lippstadt
22. Sept. 1869.

My dear Sir

I am very much obliged to you for your kindness in having sent me the translation of my brothers book and recently your memoir on Orchids, which I have read all two with the greatest interest.[1] I beg leave to thank you for the kind and very honorable mention which you make of my observations on Orchids.[2]

Last spring I have succeeded to observe repeatedly the fertilisation of Orchis mascula by different species of Bombus. I have described in detail my observations in a report which will be printed in the "Verhandlungen des naturhist Vereins etc" and which I shall send you as soon as I have received separata.[3]

All my leisure-time during the last summer has been bestowed upon insects visiting and fertilizing flowers. Until next summer passed I hope to can finish preliminarely and publish these observations

With sincere thanks and respect I remain my dear Sir | yours very faithfully | H Müller.

DAR 171: 294

[1] The references are to Fritz Müller's *Facts and arguments for Darwin* (Dallas trans. 1869) and 'Fertilization of orchids'.

[2] Müller's observations are cited in 'Fertilization of orchids', pp. 142, 147, 150, 152, 156 (*Collected papers* 2: 139, 144, 147, 149, 153).

[3] Müller's paper 'Die Anwendung der Darwin'schen Lehre auf Blumen und blumen-besuchende Insekten' (The application of Darwinian theory to flowers and flower-visiting insects; H. Müller 1869) was published in the *Verhandlungen des naturhistorischen Vereines der preussischen Rheinlande und Westphalens.* CD's annotated copy is in the Darwin Pamphlet Collection–CUL. In 'Fertilization of orchids', p. 142 (*Collected papers* 2: 139), CD had noted that flowers of *Orchis mascula* (the early purple orchid) were visited by *Bombus muscorum* (the large carder bee), adding that insects were seldom seen visiting this species. For Müller's observations on visits to *O. mascula* by three different species of *Bombus* (bumblebees) as well as other insects, see H. Müller 1869, pp. 52–5.

From Albert Günther 23 September 1869

British Museum
23.9.69

My dear Sir

In order to show you how much pleasure I have in answering your queries, I express a hope that this is merely the first batch of questions[1]

I am glad to say that Mr Ford will be able to complete your drawings within 2 months, & he has been at work at them today.[2] With regard to the heads of Salmon, a figure of a *male S. lycaodon* can be copied from Richardson,[3] but then we have no female of this species. Therefore I should advise you to take a British Salmonoid which will answer your purpose perfectly.[4] Again, you require the figures of males & females of exactly the same size, but then one sex is sometimes the larger, as in *Callion. draco*,[5] & therefore your figures would be faulty in this respect. I wait for your instructions.

It is but a short time ago that you told me that a good wife is the greatest blessing for a man.[6] How often did I remember these words, when I hurried away from trouble & work home where peace & happiness awaited me. Alas! it was a short dream in life's long night—my wife who was the greatest blessing to me every day of our married life, has been taken away from me. Last month she gave birth to a strong & healthy child, a boy; all appeared to go well, when the excessive heat came, & with it a fever which hurried her off into an early grave.[7]

A time & a trial like this make a man think of affairs beyond the grave; and that men must be strong who can rest satisfied with the glimpses or the semblance of truth gained by man's speculation. I confess I am weak enough to require better comfort, & this I can only find in faith into truth revealed

Yours ever faithfully | A Günther

DAR 165: 243

1 CD had enclosed a list of queries in his letter to Albert Günther of 21 September 1869.

2 Günther refers to George Henry Ford. See letter to Albert Günther, 21 September 1869.

3 *Salmo lycaodon* is now *Oncorhynchus nerka*, the sockeye salmon. It is described in J. Richardson 1829–37, 3: 222–3 as *S. paucidens* but not figured. John Richardson describes it as being similar to *S. scouleri* (now *Oncorhynchus gorbuscha*, the pink salmon) and refers to the plate for that species (pl. 93), which includes a figure of the head of a male.

4 Illustrations of the heads of male and female common salmon (*Salmo salar*, the Atlantic salmon) appear in *Descent* 2: 4 and 5. They were drawn by Ford from specimens at the British Museum. CD discussed the change in colour and jaw shape in male salmon during the breeding season, remarking that these changes were thought to be permanent in *Salmo lycaodon*, native to western North America. CD's source for this information, John Keast Lord, was evidently unaware that unlike Atlantic salmon, sockeye salmon die shortly after spawning (for more on salmon breeding systems, see Fleming and Reynolds 2004).

5 Günther probably refers to *Callionymus dracunculus*, which had been recognised as the female of *C. lyra*, the dragonet. In *Descent* 2: 7–8, CD discussed sexual dimorphism in *C. lyra* and included illustrations of both sexes.

6 CD may have said this to Günther in person. They met in London in March 1868 (see letter to Albert Günther, 21 September 1869 and n. 2).

7 Günther's wife, Roberta, died on 2 September 1869. Their son, Robert William Theodore Gunther, was born on 23 August 1869. (*ODNB* s.v. Günther, Albert Charles Lewis Gotthilf, and Gunther, Robert William Theodore.)

From Adolf Reuter 23 September 1869

Sanssouci near Potsdam
the 23. 9. 69.

Dear Sir!

Enjoyed very much indeed by your honourable and kind answer, I take me again the liberty to address a letter to you, in the hope not to be immodest to molest you my dearest Sir so very often—[1]

I tell you my sincere thank for your great amability and I feel me very much honoured that you have accepted my simple knowing.

Reading nearly every day your celebrated book,[2] I send you with this letter again some examples about the variation of the plants, which will be perhaps interesting and also useful for you—as a small stone to your great building.—

Excuse me, when you will find also closed two drawings by me, I am sorry, that I am not capable to make it better, but I think it will be sufficient to explicate it in the right manner.

But I will not longer molest you, and so I pray you only to accept this letter in the same indulgent manner as the first.

With the best wishs for your health and the assurance of my greatest adoration to you, I have the honour to remain | Dearest Sir | your | very obedient servant | Adolph Reuter.

[Enclosure 1]

I.

The propagation by the root of a plant with white spotted leaves:

Reading in your book the example of the white spotted Symphytum officinale propagated by the roots,[3] I did believe that it may be perhaps also interesting for you to hear, that I was propagating Rubus fruticosus foliis variegatis in the same manner.[4]

Between about 30 plants has been again not one variegated plant,— the were all-together green leaved, looking like the common Black-berry and so I was always obliged to use the propagation by cuttings or grafting.—

II.

The species and varietys of the Vinca minor and Vinca major:

You told in your book, that the varietys and species of the Vinca are very scarcely fructificating[5]— indeed but it is curious what this plant is so very changeable. We are in the possession of 3 species with different varie

a, Vinca *herbaria* (without any variety)
b, Vinca *minor* foliis argent. variegatis.
" " flore coeruleo
" " " purpureo.
" " " albo[6]
" " inter media/ remarquable variety, because not creeping like the others, she is quite closed and very much to propogate by separation.
c: Vinca *major*
" " foliis reticulatis.
" " " argent. variegatis.
" " " aur. variegatis.[7]

The last has been discovered by me, because I did find a shoot with yellow spotted leaves in a plant of Vinca major fol. argent. variegatis, which I used for a cutting, but believing, that he would be again in short time quite green or white spotted like the mother-plant.—

Sketch of *Robinia rubra*, enclosed with the letter from Adolf Reuter, 23 September 1869. DAR 176: 126/3.
By permission of the Syndics of Cambridge University Library.

Sketch of *Pirus malus*, enclosed with the letter from Adolf Reuter, 23 September 1869. DAR 176: 126/4.

My intension was to keep the yellow spotted variety and I was also so happy and more as hundred plants of her are now ornating our gardens.—

You will find closed some leaves of the Vinca major fol. aur. varieg., some leaves of the mother—Vinca major fol. argent. varieg. and also some fruits of the common Vinca major et minor.

III

A very good *argument*, that the *varietys are often more hardy or tender as the mother species* may be mentioned. Syringa persica, which as a young plant will be often attacked in **our** country by the frost but never the more strongly growing variety Syr. pers. laciniata.

The also closed variety—Syringa persica mimosaefolia is always very sensible and it is nearly quite impossible to cultivate[8] the last in the open ground she is mostly only a green-house plant for **our country**

IV.

It is indeed a *curious fact*, that some *varietys of trees* and *shrubs* will often *show again* past 10–15 years *their original type* from some years:

a, *Carpinus Betulus quercifolia*, from this fine looking tree is standing in the garden of Sanssouci an exemplar ⟨a⟩bout 18 high, which has large branches with small leaves like an oak and other with leaves like the quite common Carpinus Betulus.[9]

b, *Platanus orientalis cuneata*, a tree about 15' high was always covered with foliis cuneatis, but this year alonoe the branches are mixed with leaves like the common Platanus orientalis.[10]

c. *Robinia Pseud. Acacia inermis rubra*, a beautiful variety, where the branches are forming a regular closed sphere from dark green colour is also planted at Sanssouci and the tree may be 12 years old. Last year on the highest top (c'.) of the crown was growing a shoot now longer as S[1] and—representing the quite common Robinia Pseud. acacia.— (x the place where the Rob. Ps. inermis rubra has been grafted.)[11]

d. Another fact not less interesting as the first is *an apple tree* (Golden Pipping) (inoculated on the ground x.) which has a fine looking crown, formed by his branches with the exception of one (α), which has a quite pyramidal form.—[12]

What may be the reason?–

Perhaps it will be his parents original character or what will be also not quite impossible, the character of the savage stock.

For to be sure, that the pyramidal character will be constant so I have grafted from him and it will be certainly quite interesting when we obtain in this manner a pyram⟨idal⟩ growing apple tree.—

V.

I send you also for the curiosity *3 different kinds of leaves from* the common *Symphoria racemosa*/Symphoricarpor race⟨mosa⟩ which I did find in a hedge formed by this shrub.

I think it will show us the origin ⟨ ⟩ or the affinity with the Caprifoliaceae to which family the plant is belonging.[13]

VI.
The influence of the wild stock on the graft!—

a, Some years ago I did find between my *grafted Fraxinus excelsior aurea* (variety with prettily yellow coloured wood) an exemplar, which did have *not* the same fine colour as all the others, it was quite different looking—regular gray yellow.

What was the cause?—

The wild stock from this ash tree was Fraxinus pubescens and not like the others—Fraxinus excelsior.[14]

b, *Acer Negundo californicum*, van Houtte (Negundo fraxinifolia californica) a very good variety of the common Acer Negundo, not only because the tree is growing more strongly, so the wood of him is also quite white gray coloured with a tender powder.

For this reason I did propagate by inoculation about 80 exemplars of the common Acer Negundo (5' high), but the results has been not the best, because all my grafts did accept the character of the wild stock and so I was always obliged to use the cuttings and layers propagation.[15]

VII.
The influence of the graft on the root stock.

a, *Ptelea trifoliata foliis aur. variegatis*, from this shrub I did receive some eyes, which I grafted on the common Ptelea trifoliata, but it was singular, that also not one of the eyes was growing, however was it not wonderfull?[16] that some months afterwards the shrub was covered with yellow spotted leaves like the mother shrub from whom I did receive the eyes:—

I think—it was the same phenomen⟨on⟩ as with the variegated ash tree, which has been described in your book.[17]

b, *Acer colchicum rubrum Hort.* (Acer laetum C. A. Mey) a beautiful and very much estimated tree has been propagated in our Garden every year and mostly by inoculation or copulation.[18]

Because the sleeping eyes[19] often has b killed in the winter-time or also yet in ⟨the⟩ spring when she begin to grown so I did observe twice a very curious fact.

Once at an copulated plant the wild stock (Acer platanoïdes) did accept the **regular** character of the Acer colch. rubrum because the graft was not growing.[20]

The other time perhaps 30 inoculated Acer platanoi trees where the eyes have been destroyed by the frost did change their character in this manner, that all the leaves of them has been coloured and formed *not* regularly but *similarly* like the Acer colchicum rubrum.

c, A valuable example of the affinity between *Prunus spinosa* and *Amygdalus communis* will be the result of an Amygdalus communis inoculated on a Prunus spinosa.[21]

The first (the Amygd.) was not easily growing because it was forgotten to take of the young shoots of the wild stock and so did find, that one of the last, sitting just under 〈 〉 inoculation did have the leaves in the middle of the Prunus spinosa and Amygdalus communis.

This fact is the most remarquable appearance for me in all the time, that I have been observing, working in the nature and living between the plants. Reuter.

[Enclosure 2][22]

[Sketch of *Robinia rubra*]

[Enclosure 3]

[Sketch of *Pirus Malus*]

DAR 176: 126

CD ANNOTATIONS
Enclosure 1
1.1 *The* . . . grafting.— 3.3] 'Var not propagated by Reuter' *in margin, blue crayon*
5.1 You told . . . minor. 7.3] *crossed pencil*
8.1 A very . . . **country** 9.3] 'good Hardy & tender vars of syringa' *in margin, blue crayon*
11.1 It is indeed . . . Betulus. 11.3] *crossed blue crayon*
12.1 b, *platanus* . . . orientalis. 12.3] *illeg in margin, blue crayon*
14.1 d. . . . reason?— 15.1] 'Sport— apple' *blue crayon*
17.1 I send . . . belonging. 18.2] *crossed blue crayon*
19.1 *The* . . . *graft!*—] *triple scored blue crayon*
End of letter: 'Sept. 25 — 1869 | Sanssouci' *blue crayon*

CD note:

Reuter

(1) Rubus var – not propagated by roots 3) Syringa vars different degrees of Hardiness

p 4 Sport in foliage in Platanus & Robinia. – both cases of reversion. apple *on measure of growth* ['say Herr Reuter has given me *[second]* case' *added*]

p 5. var graft-hybrid?

7) – variegated Ptelia *[1 word illeg]* died, but which parent affected graft-hybrid between Prunus & Amygdalus but not *[1 word illeg]* enough to *[1 word illeg]*

1 CD's letter has not been found. His last extant letter to Reuter is that of 2 June 1869. Reuter replied to this letter on 18 July 1869.

2 Reuter refers to *Variation*.

3 *Symphytum officinale* (common comfrey) is a perennial herb. Reuter appears to have meant: 'not propagated by the roots'. In *Variation* 1: 384 n. 67, CD reported that the *Symphitum* with variegated leaves could not be propagated by division of the roots. See also CD's note.

4 The cultivar *Rubus fruticosus* 'variegatus' is a blackberry with cream and pink marbling on the edges of the leaves.

5 CD reported that *Vinca minor* (the lesser periwinkle) was said 'scarcely ever to produce fruit in England' (*Variation* 2: 170). *Vinca major* is greater periwinkle.

6 Reuter refers to cultivars of *Vinca minor* with variegated silver (white) leaves or with blue, purple, or white flowers.

7 Reuter refers to cultivars of *Vinca major* with veined leaves, or silver (white) or golden variegated leaves.

8 *Syringa persica* is Persian lilac. *Syringa persica* 'laciniata' is now thought to be a distinct species, *S. laciniata* (the cutleaf lilac), of which *S. persica* is a hybrid (see Fiala 1988, p. 52). The variety 'mimosaefolia' has not been identified.

9 *Carpinus betulus* is the hornbeam; 'quercifolia' is a variety with oak-leaf-shaped leaves.

10 *Platanus orientalis* is the oriental plane tree; *foliis cuneatis*: with wedge-shaped leaves.

11 See the second enclosure (plate on p. 400). *Robinia pseudoacacia* is the black locust or false acacia; 'inermis' is a globe-shaped variety. The variety 'inermis rubra' has not been identified.

12 See the third enclosure (plate on p. 401). 'Golden pippin' is a variety of *Malus domestica*, the apple tree.

13 *Symphoricarpus racemosa* is now *Symphoricarpos albus*, the common snowberry.

14 *Fraxinus excelsior* is the European ash; 'aurea' is a cultivar name. *Fraxinus pubescens* is now *Fraxinus pennsylvanica* (the green ash).

15 *Acer negundo* (*Negundo fraxinifolia* is an invalid synonym) is the box elder or ashleaf maple; 'californicum' is a variety with pink flowers. 'Propagation by inoculation' is bud-grafting, the union of one bud and a small piece of bark from the scion with a rootstock (*OED* s.v. inoculation).

16 *Ptelea trifoliata* is the common hop-tree; 'aurea' is a cultivar with bright yellow leaves that turn yellow-green with age.

17 In *Variation* 1: 394, CD described how the buds of a golden variegated ash affected the root stock of a common ash, which thereafter produced shoots with variegated leaves both above and below the point of insertion.

18 *Acer laetum* is a synonym of *A. cappadocicum* (the Caucasian maple); 'rubrum' is a cultivar with dark red leaves that turn green, then red and gold in autumn.

19 Sleeping eye: undeveloped bud.

20 *Acer platanoides* is the Norway maple.

21 *Prunus spinosa* is the blackthorn; *Amygdalus communis* is now *Prunus dulcis* (the sweet almond).

22 For the second and third enclosures, see plates on pp. 400 and 401.

From J. D. Hooker 24 September 1869

Royal Gardens Kew
Sept 24 /69

Dear Darwin

I will arrive at Orpington at 5"44 PM. tomorrow—& bring the bark story with me[1]

Ever yours | J D Hooker

DAR 103: 34

1 See letter to J. D. Hooker, 18 [September 1869] and n. 2.

To Albert Günther 25 September 1869

Down. | Beckenham | *Kent. S.E.*.
Sept 25/69

My dear Dr. Günther

I have just received your letter, & am astonished & deeply grieved at its contents.[1] You have my most entire sympathy. Your words so full of tenderness & resignation brought tears into my eyes.

There is at present no comfort for you, & I can only wish you fortitude to bear the greatest loss, which a man can ever have to bear.—

I will write in few days & in the meantime I thank you for all your very kind assistance, so kindly rendered.—

Believe me, Yours very sincerely | Ch. Darwin

Shrewsbury School

[1] See letter from Albert Günther, 23 September 1869 and n. 7.

From George Henslow 25 September 1869

St John's Parsonage | *St John's Wood*
Sep 25 /69

Dear Mr Darwin.

Many thanks for the pamphlet sent, & duly received—on additions to fertility of Orchids;[1] & which has interested me much. I must apologize for not having returned a paper on Phyllotaxis. I cannot understand part of it at all; viz: about the "*rectangle*" & the letters to the fig I found are *wrong* in some places.[2] Though the rest of the paper is interesting.

I beg to send you a copy of my Expression on same subject.[3]

Hoping you are in good health | believe me | yrs very truly | Geo Henslow

Ch Darwin Esq

DAR 166: 169

[1] The reference is to 'Fertilization of orchids'.

[2] Henslow refers to C. de Candolle 1865. See C. de Candolle 1865, p. 208, for the 'rectangle' and p. 205 for the figure. CD's annotated copy, a separately paginated offprint, is in the Darwin Pamphlet Collection–CUL. For CD's own interest in phyllotaxy, see *Correspondence* vol. 11.

[3] Henslow had presented a paper on phyllotaxy at the Linnean Society in April 1868 (Henslow 1868). See also *Correspondence* vol. 16, letter from George Henslow, 13 April 1868.

To Ellen Frances Lubbock [26 September 1869?][1]

Down. | *Kent. S.E.*
Sunday

My dear Lady Lubbock

I shd have liked extremely to have come to luncheon & accepted your kind invitation, but I have been talking so much to Dr. Hooker that I have no strength left.—[2] Dr Hooker desires me to thank you & to say that he will call in the afternoon.

Pray believe me | Yours truly obliged | Ch. Darwin

Houghton Library, Harvard University (Autograph File, D)

[1] The date is conjectured from the style of the address, with '*Bromley*' deleted from the printed letter-head, which is similar to that used by CD between April 1869 and May 1871, and from the fact that 26 September is the only Sunday in this period on which it is likely that Joseph Dalton Hooker was visiting Down on his own.

[2] Hooker arrived at Down on 25 September (see letter from J. D. Hooker, 24 September 1869).

From Albert Günther [before 27 September 1869][1]

1. I have described *Mollienesia petenensis* with a dorsal fin ornamented with 'a row of large rounded spots' because this is sufficient for specific description.[2] Ichthyologists rarely describe bright-colored ornamentations, because they are most subject to variation as regards form, number intensity & color (what is blue in one individual, may be green or even red in another, as in many Labridæ),[3] & because they disappear or change in spirit, & therefore, are of no specific value. I perfectly recollect that, when I received the examples of *Moll.*, spots of dark & bright (bluish) colors were mixed; after the specimens had been in spirit for some time, the bright shades disappeared, & now I could not exactly state their distribution. The large round spots may have been dark, blackish, or violet; this I cannot decide; but they were certainly surrounded by a lighter ring (true ocelli), as it is also indicated in the figure.[4]

2/ I have also heard that certain American Chromids sit on a nest, but I do not know which sex; nor do I recollect that any exact observations have been published about this.[5] Knowing several American Ichthyologists, shall I write to one of them?

3. Since I saw you last, I have subjected Solenostoma to a careful examination. We have about a dozen females of about the size of the specimens figured in the Fish. of Zanzibar, & one male only, of about $\frac{1}{3}$ the size of the females.[6]

a./ It may be concluded from this disproportion of male & female specimens, either that males are scarcer than females, or that males always are of much smaller size than females. I am inclined to believe the latter, because of those Cyprinodonts[7] in which the male is much inferior in size to the female, we always receive many more females than males, simply for that reason, that smaller fish escape more easily through the meshes of the net or to the eye of the collector than larger ones.

b./ The male example we possess, is much more uniformly colored, less spotted, less ornamented, than the females, which were put into spirits at the same time.[8]

c/ Kaup's assertion that the males have marsupial sacs, is not founded on actual observation; he fancied this to be the case, because it is so in the true Lophobranchs.— In Solenostoma it is the female, which carries the ova in a pouch formed by the ventral fins; in the male (Kaup evidently never saw a male) the ventral fins are quite free & do not form a pouch.[9]

4/ I cannot find any positive evidence of a Lizard uttering a sound during courtship.

5/ In male & female *Iguana tuberculata* I find the dorsal crest equally developed; slight differences in the length of the lobes are more properly accounted for by the age of the individuals.[10]

6/ *Chamæleo bifurcus* from Madagascar. Nothing positive known about the colors. The projections certainly not of a bright colour[11]

7/ *Chamæleo owenii* from West Africa. Its colors are known to me from examples in spirit only, & from those they do not appear to be brighter than in other species; they vary much, some specimens being spotted, others uniform. The female has not a trace of a horn. The horns of the male are osseous, in fact excrescences of the facial bones, each horn being covered by a smooth sheath which is part of the general integument. In fact the horns of this species are in structure identical with the horns of those Ruminants which do not shed their horns.[12]

8/ Nothing is more common than true ocelli in Reptiles & especially in Frogs & Toads; but the ocelli are more frequently black with a white, yellow, red or blue ring, than light-colored with black ring.

9/ In Lizards the females are more frequently the larger sex, than the males; the fact is, I do not at this moment recollect one species, in which the male is the larger.[13] You must recollect that lizards propagate before they have ceased growing, & therefore it would be hazardous for a Museum-naturalist to decide from his examples which is the larger sex in each species. In our European lizards the females are the larger; and very large examples of foreign species are generally females.

DAR 82: B16–20; DAR 84.2: 168

CD ANNOTATION
2.3 Knowing ... them?] *triple scored pencil*; 'Yes' *pencil*

1 The date is established by the relationship between this letter and the letter to Albert Günther, 27 September [1869].
2 See letter to Albert Günther, 21 September 1869 and n. 6.
3 Labridae is the family of rainbowfishes and wrasses.
4 Günther refers to plate 86 in Günther 1864–6. CD included Günther's description of *Mollienesia petenensis* (now *Poecilia petenensis*) in *Descent* 2: 9.
5 See letter to Albert Günther, 21 September 1869 and n. 7. For more on nesting behaviour in these fishes see Tweddle *et al.* 1998.
6 The reference is to Playfair and Günther 1866. See letter to Albert Günther, 21 September 1869 and n. 8.
7 Günther refers to the family Cyprinodontidae, pupfishes. CD discussed their sexual differences in *Descent* 2: 7, 9.
8 CD reported Günther's information on the coloration of *Solenostoma* (now *Solenostomus*) in *Decent* 2: 22.
9 The former order Lophobranchii is roughly equivalent to the present suborder Syngnathoidei (pipefishes and seahorses), which includes the genus *Solenostomus* (ghost pipefish). Günther refers to Johann Jakob Kaup. See letter to Albert Günther, 21 September 1869 and n. 9. In *Descent* 2: 22, CD cited Günther for the information on the marsupial sac in female *Solenostoma*.
10 CD cited Günther's information on crest size in *Iguana tuberculata* (now *Iguana iguana*) in *Descent* 2: 32.
11 CD described and illustrated the large bony projections on the upper part of the skull in the male *Chamaeleon bifurcus* (now *Furcifer bifidus*) in *Descent* 2: 34–5.
12 CD described the three horns of the male *Chamaeleon owenii* (now *Chamaeleo oweni*) in *Descent* 2: 35–6.
13 CD appears to have misread Günther on this point. In *Descent* 2: 32, he reported on Günther's authority that male lizards are generally larger than females.

To Albert Günther 27 September [1869][1]

Down. | Beckenham | *Kent. S.E.*
Sept 27th

My dear Dr. Günther

Your answers to all my many queries are so full, clear & explicit that I have not a word more to ask, & they will be of the *greatest* use to me.—[2]

I would not on any account ask you to write to America about which sex of the Chromids sits on the nest; but *if you write on any other subject*, I sh[d] be very much obliged if you would ask.—[3]

I am delighted to hear that M[r] Ford will undertake the drawings. I had forgotten about differences of size in sexes; no doubt the female must be in due proportion smaller.—[4]

Any species of Salmon, which presents during the breeding season the difference in jaws will do very well.—[5]

Will you ask M[r] Ford whether he has done the drawings of the feather of Polyplectron & Peacock.— I hope so, otherwise I shall have a terrible hunt to get out the specimens again.—[6]

With the most cordial thanks & sincere sympathy | Your's very truly | Ch. Darwin

P.S. | *Private* | I wish you had told me some news about your child, & how you succeed.[7] You have indeed had a hard & cruel lot to bear.—

Shrewsbury School

1 The year is established by the relationship between this letter and the letter to Albert Günther, 21 September 1869.

2 See letter to Albert Günther, 21 September 1869, and letter from Albert Günther, [before 27 September 1869].

3 See letter from Albert Günther, [before 27 September 1869] and n. 5.

4 George Henry Ford. See letter from Albert Günther, 23 September 1869.

5 See letter from Albert Günther, 23 September 1869 and nn. 3 and 4.

6 Illustrations of a tail feather of a peacock and of tail-coverts of two species of *Polyplectron* (peacock-pheasants) appear in *Descent* 2: 137, 139. There is a note with detailed instructions for Ford on the illustration of these specimens in DAR 84.1: 170.

7 See letter from Albert Günther, 23 September 1869 and n. 7.

To Adolf Reuter 27 September 1869

Down. | Beckenham | *Kent. S.E.*
Sep. 27. 1869

Dear Sir

I am very much obliged for your extremely courteous letter, & for your valuable notes together with the specimens.[1]

It is very good of you to take so much trouble for my sake.

I shall give several of your notes in any future edition of my work. The facts about grafting & budding are perhaps the most important.[2] But the case of the great difference in hardiness in the varieties of the Syringa Persica has particularly struck me. The varieties themselves are also very curious.[3]

Dr Hooker of Kew happened to be staying here, & I shewed him your specimens, several of which were new to him.[4]

With my best thanks for your very great kindness I beg leave to remain dear Sir | yours very faithfully | Charles Darwin

LS
Staatsbibliothek zu Berlin – Preußischer Kulturbesitz, Slg. Darmstaedter Lc 1859 Darwin, Charles, Bl. 224/225

[1] See letter from Adolf Reuter, 23 September 1869.
[2] Reuter's experiments on grafting potatoes are mentioned in *Variation* 2d ed., 1: 422, but this information is cited from a published source. Reuter did not mention potatoes in his notes with his letter of 23 September 1869.
[3] CD had mentioned *Syringa persica* (Persian lilac) in *Varation* 2: 164 in a discussion of the effects of climate on fertility. See letter from Adolph Reuter, 23 September 1869 and n. 8.
[4] Joseph Dalton Hooker had arrived at Down on 25 September (see letter from J. D. Hooker, 24 September 1869).

From Alphonse de Candolle[1] 28 September 1869

Genève
28 Sept. 1869.

Mon cher Monsieur

Je vous envoie des graines qui ont un véritable interet au point de vue dont nous nous occupons tous les deux et je viens vous proposer de les semer pour suivre à une expérience commencée et pour la vérifier vous même.

J'ai fait venir en 1868 des graines de quelques espèces *spontanées* de quatre localités et même cinq localités, très distantes, en Europe, savoir: *Palerme* et *Montpellier*, *Kharkoff*[2] et *Moscou*, *Edinburgh*; afin de les semer à Genève et de voir si la même espèce etait devenue plus précoce ou plus tardive pour avoir eu des ancêtres exposés pendant des siècles à des climats très-differents.[3] J'ai choisi des espèces faciles à reconnaitre et n'offrant pas des variétés bien distinctes. Malheureusement mes correspondants n'ont pas tous recueilli toutes les espèces demandées et il m'a fallu restreindre l'experience à peu d'espèces. Malheureusement aussi le jardinier au quel j'avais cru pouvoir me fier pour certains détails, n'a pas élevé convenablement toutes les plantes semées. Néanmoins il est arrivé dans mes semis de 1869, sur trois espèces, bien comparables, que les *Senecio vulgaris* et *Sisymbrium officinale* originaires du nord, ont été les plus prompts à se developper et à fleurir. Au contraire dans le *Trifolium repens*, les plantes du nord ont été plus lentes, mais elles etaient aussi plus faibles, plus petites dans toutes les parties, et l'on peut dire qu'elles avaient l'apparence d'une variété *minor*.[4]

J'ai semé en 1869 à Genève des *Senecio vulgaris* des 4 origines, et à cette seconde génération, la précocité des plantes du nord s'est maintenue.

Combien de temps cela durera-t-il? Voila qui n'est pas sans interêt.

La rigueur de notre climât et le peu de serres et jardiniers dont je dispose ne me permettent pas de recommencer un semis dans cette saison. Peut-être seriez vous disposé à le faire chez vous et si vous préferez, vous pouvez vous entendre avec le Dr Hooker, par example, pour faire un semis qu'on observerait soigneusement.[5]

Je vous envoie:

Sisymbrium officinale de 1868, recueilli à	Edinburgh Moscou Kharkoff Palerme
Trifolium repens, de 1868, recueilli à	Moscou Palerme
Senecio vulgaris, de 1868, recueilli à	Moscou Palerme
Id. de 1869, recueilli à Genève, originaires de	Edinburgh Moscou Montpellier Palerme

Ces derniers sont les plus intéressants.

Vous savez que plusieurs indices font croire que la même espèce ressent differemment les effets de la temperature selon le climat. Ainsi nos arbres d'Europe se feuillent à Madère sous des températures analogues à celles de l'ètè de l'Angleterre, et restent sans feuilles sous la température d'hiver semblable à celle de nos printemps. J'avais cru en trouver la cause dans un repos plus complet des sucs dans les régions froides, repos qui prépare une végétation vigoureuse au printemps des pays du nord; repos qui est imparfait ou qui manque à peu près completement à Madère, à Naples, etc. Mr Lindseer jeune astronome de St Petersbourg, a publié depuis dans les Mémoires de l'Académie de St Petersb^{g} un mémoire sur les époques de foliaison, floraison, etc, de diverses espèces en divers points de l'Europe, et il arrive à la conclusion d'une qualité differente de la même espèce au nord et au midi pour ressentir la chaleur, les espèces etant plus hâtives au nord, et il croit que cela tient à l'hérédité.[6]

Probablement les deux causes hérédité et état des plantes en hiver concourent, au moins dans les espèces vivaces et ligneuses, mais les experiences que je viens de faire sur des espèces annuelles montrent bien l'effet de l'hérédité seule. Le *Trifolium repens* a été different— Cela vient-il de l'état des graines ou de la difference de variété? Je ne sais.

J'aurais aimé avoir des espèces plus surement indigènes, les Senecio, Sisymbrium et Trifolium pouvant bien provenir de graines importées avec les marchandises, graines, etc. Cependant j'ai eu soin de recommander qu'on les recueillit hors des jardins, loin des habitations, ce qui a été fait. Ces espèces ont été choisis parce qu'il y a vraiment très peu d'espèces communes à des pays aussi éloignés en Europe, et en même temps faciles à trouver et sans variétés differentes.

J'ai envoyé mes Senecio de la recolte de 1869 de Genève, à *Florence*, où j'espère les trouver semés lorsque j'irai dans cette ville en automne. Mon projet est effectivement d'aller passer 2 mois dans le nord de l'Italie, depuis le milieu d'octobre prochain et je resterai au moins 15 jours à Florence—tout à vos ordres si vous desirez quelque renseignement.

Recevez, mon cher Monsieur, l'assurance de mes sentiment très dévoués | Alph. de Candolle

DAR 261.11: 32.i

CD ANNOTATIONS
2.12 Au … faibles, 2.13] *scored blue crayon*
3.1 J'ai … maintenue. 3.2] *scored blue crayon*
7.10 de diverses … chaleur 7.12] *scored blue crayon*
9.3 Cependant … fait. 9.4] *scored blue crayon*

1 For a translation of this letter, see Appendix I.
2 Kharkov (now Kharkiv) is a city in eastern Ukraine, in the Donets valley (*Columbia gazetteer of the world*).
3 CD had written extensively on the effect of changes in climate and other conditions of life on various domestic plants and animals (see *Variation* 1: 271–92).
4 *Senecio vulgaris* is common groundsel, a winter annual. *Sisymbrium officinale* is hedge-mustard. *Trifolium repens* is white clover.
5 CD forwarded Candolle's letter and most of the seed to Joseph Dalton Hooker (see letter to J. D. Hooker, 2 October [1869]).
6 Candolle refers to Carl Linsser and his monograph 'Die periodischen Ersheinungen des Pflanzenlebens in ihrem Verhältniss zu den Wärmeerscheinungen' (The periodic appearance of plant life in relation to heat-phenomena; Linsser 1867). For a discussion of Linsser's work on the effects of temperature on plant development, see Thornthwaite 1943.

From T. H. Huxley 28 September 1869

My dear Darwin

I got back to the mill & received your letter yesterday.—[1] (*Item* the tooth) I will do my best to make out the tooth—though I should tell you before hand that I am not disposed to place much weight upon conclusions that may be based upon a single tooth of a horse

Bad man that you are, you do not say whether you are the better for your trip or not.[2]

As usual, your abominable heresies were the means of getting me into all sorts of hot water at the Association—

Three parsons set upon you—and if you were the most malicious of men you could not have wished them to have made greater fools of themselves than they did[3] They got considerably chaffed & that was all they were worth—

Ever | Yours faithfully | T. H. Huxley

Jermyn S
Sep. 28. 1869

DAR 166: 321

1 See letter to T. H. Huxley, 21 September [1869].
2 The Darwins stayed at Caerdeon, Barmouth, Wales, from 12 June to 30 July 1869 (Emma Darwin's diary (DAR 242)).

[3] Huxley refers to three papers by Anglican clergymen delivered in the biology section of the British Association for the Advancement of Science. The papers were by Philip Freeman, Francis Orpen Morris, and James McCann. The papers and Huxley's remarks were summarised in the *Athenæum*, 4 September 1869, p. 309. For Huxley's dispute with McCann, see the letter from J. D. Hooker, 7 September 1869 and n. 4.

To W. B. Tegetmeier 28 September [1869][1]

Down. | Beckenham | *Kent. S.E.*
Sept. 28. Tuesday

My dear Sir

I am very much obliged. I write in haste, to say that I will send our Carrier on *Thursday* morning to the Field office, as that will be **very much** the safest plan.—[2] But if you send them by Railway pray inform me to what Station addressed, as we have now several stations & the poor creatures may be left long uncared for.—[3] Please put a handful of grain in Basket

[A postscript expresses the hope that the recipient will be able to answer a question.]

Incomplete[4]
Christie's, 6 June 2001

[1] The year is established by the style of the address, with '*Bromley*' crossed through and 'Beckenham' written in, which CD used between April 1869 and May 1871; during this period, 28 September was on a Tuesday only in 1869.

[2] George Snow operated a carrier service on Thursdays between Down and London (*Post Office directory for the six home counties* 1870). Tegetmeier was an editor of the *Field* with offices on The Strand, London.

[3] The railway station at Orpington, four miles from Down House, had opened in 1868 (see *Correspondence* vol. 16, letter to W. D. Fox, 14 May [1868]). CD had been corresponding with Tegetmeier about plumage in young chickens and had recently declined Tegetmeier's offer to lend him some Polish chickens. See letter from W. B. Tegetmeier, 20 July 1869, and letters to W. B. Tegetmeier, 21 July [1869] and 25 August 1869.

[4] The original letter is complete and is described in the sale catalogue as being two and a half pages long. The first page, up to 'Please put a', has been transcribed from a facsimile; the rest of the sentence is from Christie's transcription.

From Ernst Haeckel[1] 29 September 1869

Jena
29. Sept. 69

Hochverehrter theurer Freund!

Gestern Abend erst bin ich von einer zweimonatlichen Reise nach Norway zurückgekehrt, und fand hier Ihren freundlichen Brief vom 5. August vor, der leider erst nach meiner Abreise hier eintraf.[2] Ich bin Ihnen für Ihre liebenswürdige Gefälligkeit und für die viele Mühe, die Sie sich wegen meiner Calcispongien gemacht haben, äusserst dankbar; aber ich bereue es fast, Ihnen durch meine Bitte so viel Mühe und Zeitverlust gemacht zu haben.[3] Hoffentlich werden die Resultate meiner

Arbeit so befriedigend, und besonders für den "Darwinismus" so brauchbar, dass Ihre Bemühungen wenigstens nicht umsonst gewesen sind.— Das gütige Anerbieten des Mr. Norman, mir die englischen Kalkschwämme zu senden, nehme ich um so dankbar an, als ich erst sehr wenige von den englischen Kalkschwämmen erhalten konnte.[4] Bis jetzt hatte nur Mr. Allman die Güte, mir die Spongien aus der Edinburgh-Collection zu senden.[5] Dagegen hat Mr. *Bowerbank*, an welchen ich schon im Sommer desshalb *zweimal* schrieb, keinen meiner beiden Briefe beantwortet.[6] Er glaubt wahrscheinlich *allein* das Monopol des Verständnisses für Spongien zu besitzen und hält andere Arbeiten für überflüssig.

Beifolgend sende ich Ihnen die vorlaüfige Mittheilung über das bisherige Resultat meiner spongiologischen Untersuchungen; die Stellen darin, welche Sie vielleicht interessieren, habe ich *roth* angestrichen (p. 218, 220, 222, 233) 243).[7]

Einen anderen Aufsatz lege ich noch bei, welcher die Beschreibung einer neuen Medusen-Form (Crambessa) und einigen fossilen Medusen enthält.[8] Es ist eine rein descriptive Arbeit, ohne besondere Ideen. Ferner darf ich Sie wohl bitten, das beigelegte *zweite Exemplar* von beiden Arbeiten (Spongien und Medusen) an Mr. Norman zu senden, dessen nähere Adresse ich nicht sicher weiss. Da Sie mir auch die gütige Offerte machten, einen Brief an Mr. Norman befördern zu wollen, lege ich diesen mit ein.

Meine Entwickelungsgeschichte der Siphonophoren, die ich bereits Ende Juli an Sie durch Buchhändler abschickte, werden Sie inzwischen hoffentlich erhalten haben.[9] Von Interesse dürften darin für Sie vielleicht folgende Theile sein: IV (p. 38), VII (p. 73), VIII (p. 80) und X (p. 92, 97, 100, 103).

Meine Reise nach Norwegen war interessant, obwohl nicht sehr angenehm. In Christiania[10] freute ich mich sehr, in dem ausgezeichneten Zoologen *Sars* einen begeisterten *Darwinisten* kennen zu lernen,[11] der auch speciell meine Hypothese von dem Ursprung der Echinodermen billigte.[12] Er ist schon sehr alt, aber geistig noch sehr frisch. Von Christiania ging ich durch das innere Norway (*Valders*) nach dem *Sogne-Fjord. Ich sah wohl mehr als 12 oder 16 Ponys mit ausgezeichneten Zebra-Streifen* an Beinen und Schultern. Sie sollen dort *sehr häufig* sein.[13] (N. B. Hier in Jena läuft auch ein *Esel* mit solchen Beinstreifen herum (aber kein Professor, wie Sie vielleicht denken könnten!) Die längste Zeit fischte ich in *Bergen*, wo ich auch *Mr* Gwyn-Jeffreys begegnete, und in dem ausgezeichneten Naturforscher *Danielson* einen Verehrer Ihrer Theorie fand.[14] Ebenso traf ich in *Hamburg* fast alles *Darwinistisch.* Ihr Porträt hängt in dem berühmten Aquarium.—[15]

Nochmals vielen, vielen Dank, theurer Freund! Hoffentlich befinden Sie sich recht wohl. In unveränderlicher Treue Ihr Haeckel

N. B. Die in Norway beobachteten Kalk-Spongien sind sehr wichtig für Ihre Theorie

DAR 166: 53

[1] For a translation of this letter, see Appendix I.
[2] See letter to Ernst Haeckel, 3 August [1869].

[3] Haeckel had asked CD to help him obtain specimens of calcareous sponges. See letter from Ernst Haeckel, 2 July 1869, and letter to Ernst Haeckel, 3 August [1869].

[4] Haeckel refers to Alfred Merle Norman. See letter from A. M. Norman, 2 August 1869.

[5] George James Allman was professor of natural history and keeper of the natural history museum at Edinburgh University (*ODNB*). See also letter from A. M. Norman, 2 August 1869.

[6] James Scott Bowerbank had a large private collection and was working on a multi-volume monograph on British sponges (Bowerbank 1864–82; *ODNB*).

[7] Haeckel refers to his paper, 'Ueber den Organismus der Schwämme und ihre Verwandtschaft mit den Corallen' (Haeckel 1869b; later translated as 'On the organisation of sponges, and their relationship to the corals', Haeckel 1870b). CD's lightly annotated copy is in the Darwin Pamphlet Collection–CUL.

[8] Haeckel refers to 'Ueber die Crambessiden, eine neue Medusen-Familie aus der Rhizostomeengruppe' (On the Crambessidae, a new medusa family of the rhizostomean group; Haeckel 1869c). CD's copy is in the Darwin Pamphlet Collection–CUL. The genus *Crambessa* and the family Crambessidae are now considered invalid; the family corresponds roughly to the modern family Catostylidae in the order Rhizostomeae.

[9] Haeckel refers to *Zur Entwickelungsgeschichte der Siphonophoren* (On the developmental history of the siphonophores; Haeckel 1869a). See letter from Ernst Haeckel, 2 July 1869 and n. 5.

[10] Christiania is now Oslo, the capital of Norway.

[11] Michael Sars was professor of zoology at Christiania University.

[12] For Haeckel's view on the descent of echinoderms, see Haeckel 1868c, pp. 420–1 and tables I and III.

[13] CD had discussed stripes in the Norwegian horse in *Variation* 1: 58.

[14] Haeckel refers to John Gwyn Jeffreys and Daniel Cornelius Danielssen.

[15] The Hamburg aquarium opened in 1864 and was regarded as one of the finest in Europe (Taylor 1876, p. 17).

From Victor Masson[1] 29 September 1869

Paris, Place de l'Ecole-de-Médecine,
le 29 sept. 1869.

Monsieur

Nous ignorions absolument qu'une cinquième édition de votre *Origine des Espéces* eût été publiée.[2]

Il va de soi en effet que notre intérêt et celui de Madame Royer sont en cette affaire les mêmes que le vôtre, et que nous n'avions nulle raison pour faire quelque chose qui pût à la fois vous être désagréable, et diminuer la valeur de l'ouvrage que nous publions.[3]

Madame Royer qui n'avait pas depuis longtemps eu le plaisir de recevoir de vos nouvelles, a fait l'achat de votre récent ouvrage pour s'en inspirer dans son travail.[4] Elle n'avait aucun motif pour ne pas acheter de même, si elle en avait su l'existence, la cinquième édition de *l'origine des Espèces*, et en conférer le texte avec celui de sa traduction.

Nous écrivons par ce courrier à notre correspondant de Londres de nous envoyer un exemplaire, que nous remettrons à Madame Royer à son retour d'Italie où elle est en ce moment.

S'il est dans votre nouvelle édition quelque point que vous regrettez surtout de voir omis dans l'édition française, nous n'aurions quant à nous aucune répugnance à faire les frais d'un appendice de quelques pages, et nous vous prions de nous le signaler.

N'ayant jamais eu l'honneur jusqu'à présent d'etre en relations directes avec vous, nous regrettons vivement de vous avoir, bien involontairement, contrarié de telle sorte que votre première lettre contienne des reproches bien peu en rapport avec notre caractère, nos habitudes, et nous osons le dire, notre réputation dans le public savant.

Recevez Monsieur l'assurance de nos sentiments | très distingués | Victor Masson fils

M. Charles Darwin | Down | Beckenham | Kent. (S.E.)

DAR 171: 64

1 For a translation of this letter, see Appendix I.

2 The fifth edition of *Origin* was published in June (see letter from R. F. Cooke, 22 June 1869). CD had written to Masson to complain that a third French edition of *Origin* had been prepared without reference to the fifth English edition (see letter to Victor Masson, [after 29 September 1869]); however, no letter to Masson on this subject has been found.

3 Clémence Auguste Royer was the French translator of *Origin*. Two editions had been published (Royer trans. 1862 and 1866). A third edition dated 1870 evidently appeared in 1869 (see letter to J. J. Moulinié, 23 October [1869]).

4 Masson refers to *Variation*, which Royer discussed in her preface to Royer trans. 1870 (see letter to J. J. Moulinié, 23 October [1869] and n. 5).

From W. C. Tait 29 September 1869

Oporto
Sept 29th 1869.

Charles Darwin Esqr. | Beckenham

Dear Sir,

I duly received your letter of the 27th. ultmo. and am very glad to hear such good news of the plants—[1]

The Portuguese flora is I believe rather interesting as I believe there are many endemic plants of curious habits— Mr. Goetze[2] mentioned to me a very curious parasitical fungus which grows on the roots of a cistus I have seen it growing in abundance near a village called Oliveira d'Azemeis[3] and it is eaten by the country people—

I have not seen or heard of one single species of bird confined to this country or even to the peninsula but was much astonished a short time ago to see barred plumage on a young specimen of a Lestris (Lestris Richardsonii I believe) so much resembling that of the Sparrowhawk.[4] The connexion between the predatory habits of the two birds and the similar plumage is naturally striking the more so as they are classified so widely apart.

With regard to the Lizard there are several links between the Seps chalcides and the blindworm[5] One little known is the Sepsina angolense and in Cuviers Animal Kingdom there is a plate containing the figure of a reptile with 2 flaps on its sides and which from my remembrance of it seemed to represent the disappearing hind legs—[6]

The law of the correlation of growth may perhaps be in a measure connected with the law of the tendency to the equal diffusion of force from its centre regulated by the medium through which it travels—[7]

These subjects interest me very much but of course to a business man the time for study is very short and subordinate and any systematic investigations must necessarily go on very slowly—

Perhaps you may not know that if the spurs of a cock are cut off and inserted into the head of the bird they grow and look like horns— The Portuguese sometimes do this while the birds are young— The first specimen which I saw puzzled me very much—

Believe me to remain | Dear Sir, | Yours very truly, | William C. Tait.

DAR 178: 50

[1] See letter to W. C. Tait, 27 August [1869]. Tait had sent plants of *Drosophyllum lusitanicum* from Portugal.

[2] Edmund Goeze.

[3] Oliveira de Azeméis is in the district of Aveiro, in northern Portugal.

[4] *Lestris richardsonii* is now *Stercorarius parasiticus* (family Stercorariidae), the arctic skua or parasitic jaeger. Juvenile birds show varying degrees of barring in their plumage. *Accipiter nisus* is the Eurasian sparrowhawk; the under-parts are barred at all ages.

[5] *Seps chalcides* is now *Chalcides chalcides*, the Italian three-toed skink. The slowworm or blindworm is *Anguis fragilis.*

[6] Tait may refer to a species now known as *Typhlops angolensis*, a species of a blindsnake that superficially resembles blindworm. Tait also refers to Georges Cuvier and *Le règne animal* (Cuvier 1817), which was translated into English as *The animal kingdom* in several different English editions, some of which had additions to the text and illustrations. Tait probably refers to the illustration of *Bipes lepidopodus* (now *Pygopus lepidopodus*, the common scaly foot), a snake-like lizard from Australia that lacks forelimbs and has flap-like hind limbs (see Cuvier 1834–6, 2: pl. 26 bis).

[7] CD had discussed correlation of growth, including correlated variation, or the tendency, when one part of an organism is modified through continued selection, of other parts to be unavoidably modified, in *Origin* and *Variation.* See especially *Variation* 2: 319–21.

To Victor Masson [after 29 September 1869][1]

[Down]

Sir

I am much obliged for your letter of 29th last.—[2] As I had spent two months in correcting the 5th Edition of my Origin I was very much annoyed that the labour shd not have been taken advantage of in the new French Edition.— And I must take the liberty of saying that I still think that you or Mme Royer ought to have written to me before bringing out a new Edition as has been the habit of all other Publishers:—[3]

But I am sorry that I wrote while I still felt rather angry & offer to you my apologies, which I hope that you will accept.—[4] Your offer to print some appended pages is very obliging, but though I have made more additions, some of extended length I do not think that it wd be worthwhile to you any more on this, for I have taken struck out other discussions of similar length; the correction which I care most about, are mere *[1 word illeg]*, depending on how far the evidence has come *[*or*]* demands on other factors

*[*Yours very sincerely*]*

Draft
DAR 96: 61

[1] The date is established by the relationship between this letter and the letter from Victor Masson, 29 September 1869.
[2] Letter from Victor Masson, 29 September 1869.
[3] CD refers to Clémence Auguste Royer. See letter from Victor Masson, 29 September 1869 and nn. 2 and 3.
[4] CD's letter to Masson has not been found.

To T. H. Huxley 1 October [1869][1]

Down | Beckenham | *Kent. S.E.*
Oct 1.

My dear Huxley

M. Kowalewsky, who translated into Russian one of my books, has been staying here.[2] He wishes **very much** to hear one of your Lectures: can you give him an order for admittance? He remains in London only until Oct.— 10th.— He is Brother of the naturalist of the same name.—[3]

I suppose he can look over your Museum.—[4]

Yours affectly | C. Darwin

Many thanks about the Horses tooth.[5]

He is studying Geology & intends travelling with his young & pretty wife[6] in Central Asia!—

Imperial College of Science, Technology, and Medicine Archives (Huxley 5: 275)

[1] The year is established by the reference to Vladimir Onufrievich Kovalevsky's visit (see n. 2, below).
[2] According to Emma Darwin's diary (DAR 242), Kovalevsky visited from 30 September to 1 October 1869.
[3] Alexander Onufrievich Kovalevsky.
[4] CD probably refers to the Museum of Economic Geology at the Royal School of Mines, where Huxley was professor.
[5] See letter from T. H. Huxley, 28 September 1869.
[6] Sofia Vasilyevna Kovalevskaya.

To W. B. Tegetmeier 1 October [1869][1]

Down. | Beckenham | *Kent. S.E.*
Oct 1

My dear Sir

I write one to thank you & to say that the chickens arrived at Down last night all safe & lively.— I am surprised to see the feathers on back of Cock tinged with reddish or fulvous, & shall watch their adult plumage with interest.—[2]

Pray tell me what they cost you—with carriage, porterage, basket &c.—
In Haste | Yours very faithfully | Ch. Darwin

Archives of the New York Botanical Garden (Charles Finney Cox collection) (ser. 14)

1 The year is established by the relationship between this letter and the letter to W. B. Tegetmeier, 28 September [1869].
2 See letter to W. B. Tegetmeier, 28 September [1869] and n. 3.

To J. D. Hooker 2 October [1869][1]

Down, Beckenham | Kent
Oct 2[d]

My dear Hooker.—

I am sorry to bother you with the enclosed long letter; but you will see that De Candolle asks me to communicate with you.— If you wish I will send you half all the many packets of seeds.—[2]

The experiment seems to me quite an *excellent* one; & I sh[d] enjoy trying it, & will do my best, but I have so many other things in hand, & so little strength & time to watch & record daily the state of some score of lots of plants.— Yet, I will do my best.— If anyone in your establishment, w[d] do the work effectually I think it w[d] be Wisest to hand the whole lots over to you.—

Yours affect | C. Darwin
Please return D.C.[s] Letter.—

DAR 94: 155

1 The year is established by the relationship between this letter and the letter to Alphonse de Candolle, 11 October 1869.
2 CD evidently enclosed the letter from Alphonse de Candolle, 28 September 1869. A copy of Candolle's letter along with notes on the results of experiments made at Kew is in the archive at the Royal Botanic Gardens, Kew (Directors' Correspondence, vol. 135, doc. 367a–e).

To Ernst Haeckel 3 October [1869][1]

Down. | Beckenham | *Kent. S.E.*
Oct. 3[d]

My dear Häckel

I received this morning your letter of the 29th. & am heartily glad that you have had so successful an expedition.[2] You give a very pleasant account of all your doings. It is, also, very satisfactory to hear from you of so many naturalists who believe in our views. I am particularly glad to hear about Sars,[3] for whom I have always felt an unbounded respect.

I received some time ago your Siphonophoren, & have admired the exquisite Plates; but I have not yet had time to read any as yet, though it now stands as the top book of a pile next to be read. I am very much obliged to you for telling me of the pages, in the work & in your memoir on Sponges, for, as you know, I am a very poor German scholar, & alas shall never improve.[4]

Do not work yourself to death or sickness.

Let your wife[5] sing that into your ears night & day.— I have forwarded your letter & papers to Mr. Norman.—[6]

My dear Häckel | Yours most truly | C. Darwin

Ernst-Haeckel-Haus (Bestand A-Abt. 1: 1–52/23)

1 The year is established by the relationship between this letter and the letter from Ernst Haeckel, 29 September 1869.

2 See letter from Ernst Haeckel, 29 September 1869.

3 Michael Sars.

4 CD refers to *Zur Entwicklungsgeschichte der Siphonophoren* (On the developmental history of siphonophores; Haeckel 1869a) and 'Ueber den Organismus der Schwamme und ihre Verwandtschaft mit den Corallen' (Haeckel 1869b), which was later translated as 'On the organisation of sponges, and their relationship to the corals' (Haeckel 1870a). See letter from Ernst Haeckel, 29 September 1869 and nn. 7 and 9.

5 Agnes Haeckel.

6 Alfred Merle Norman. See letter from Ernst Haeckel, 29 September 1869 and n. 4.

To James Orton 7 October [1869][1]

Down. | Beckenham | *Kent. S.E.*
Oct. 7th

My dear Sir

I received your note & the Horse's tooth about 10 days ago.

Although I have never had any quarrell with Prof. Owen, he has used such language about me that I can hold no communication with him.[2] The best which I could do for you under these circumstances was to send the specimen to Prof. Huxley, who is quite as able & a more careful observer than Prof. Owen.—[3] He writes to me that he will do his best, but he has not much confidence in the case of the Horse in identification from a single tooth.[4] Unfortunately he is so incessantly engaged that he has not yet written to me again: but as soon as I hear, I will at once write to you.— You do not ask for the return of the tooth, so I have *provisionally* deposited in the Museum, where it will be a valuable addition.—[5] Your own palæontologist Prof. Leidy[6] wd. be as good a judge about the Horses skull (which is a treasure) & teeth as anyone in Europe.—

I am glad to hear that your work will be published this autumn,[7] & I beg leave to remain | My dear Sir | Yours faithfully | Ch. Darwin

History of Science Collections, University of Oklahoma Libraries

1 The year is established by the relationship between this letter and the letter from James Orton, 8 September 1869.

2 See letter from James Orton, 8 September 1869 and n. 2. For more on the relationship between CD and Richard Owen, see, for example, *Correspondence* vol. 8, letter to Asa Gray, 25 April 1860.

3 CD asked Thomas Henry Huxley to examine the horse tooth in his letter of 21 September [1869].

4 See letter from T. H. Huxley, 28 September 1869.

[5] CD refers to the British Museum.
[6] Joseph Leidy.
[7] The reference is to Orton 1870; see letter from James Orton, 8 September 1869.

From Frederick Smith 8 October 1869

British Museum
8th. Octr. 1869.

My dear Sir

I have carefully watched the Mononychus *pseudacori* and am satisfied that both sexes stridulate— I have placed them singly in pill boxes with glass lids and if they are placed in a strong light—sunlight or lamp light they become musical— I send a few in a quill so that you may experiment if you have leisure to do so—[1]

I also feel certain that both sexes of the Musk-beetle *Aromia moschata*—are musical — the same I believe of all the species of Necrophorus—[2] Among the Orthoptera on the contrary as far as I know only the Males stridulate—

I kept the Acheta in June last in my garden and the male somewhat annoyed my neighbours by his loud piping sound. This was A. campestris from Surrey— the same of course is the rule with all the Crickets as well as the Cicadas[3]

Yours sincerely | Fred^{k} Smith

DAR 82: A3

CD ANNOTATIONS
1.1 Mononychus] *underl pencil*; 'n?' *added pencil*
1.1 and . . . stridulate— 1.2] *double scored pencil*
1.1 I have . . . do so— 1.4] 'Add after or before Frank's statement'[5] *blue crayon*
2.2 the same . . . Cicadas 3.3] *crossed pencil*
Top of letter: 'Mutilla[4] | Mononychus' *pencil*; '(Mononychus)' *ink*; *square brackets in original MS*

[1] *Mononychus pseudacori*, the single-clawed weevil, is now *M. punctumalbum*, the iris weevil (family Curculionidae). In *Descent* 1: 382, CD cited Smith for the information on stridulation in both sexes.
[2] *Aromia moschata* is in the family Cerambycidae (longhorn beetles). The genus *Necrophorus* belongs to the family Silphidae (carrion beetles).
[3] *Acheta campestris* is now *Gryllus campestris*, the field cricket. It was in the family Achetidae but now belongs to the family Grillidae (crickets, grasshoppers, and katydids).
[4] In *Descent* 1: 366, CD discussed stridulation in *Mutilla europaea*, the velvet ant. See also letter from Frederick Smith, 14 September 1869.
[5] In *Descent* 1: 382, CD cited Frederick Smith's observations on *Mononychus pseudacori* after his son Francis Darwin's observations on *Geotrupes stercorarius.*

From Federico Delpino[1] 9 October 1869

Firenze
addè 9. 8bre 1869

Celeberrimo uomo!

Ho ricevuto i lavori "On the specific difference between Primula veris etc," "Offspring from the illegitimate unions of dimorphic and trimorphic plants etc." notes on the fertilization of orchids.[2] Io ne rendo alla S. V. le più sentite grazie. Si sto studiando col massimo impegno, come si deve per ogni cosa che esce dalla penna e dalla mente del più grande naturalista di questo secolo.

Mi permetto alcune rispettose osservazioni. Una infinità di dati mi persuadono che per molti fiori tropicali (e probabilmente per molti delle Orchidee) larghissima e importante parte alla fecondazione debbono prendere gli uccelli mellisugi (Trochilus, Ornismya, Nectariania, Melitreptus e simili).[3] O io verso in grave errore, o non mi so spiegare come la S. V. non abbia prego detti pronubi in Speciale considerazione.

Così, contemplando la struttura dei fiori di Corianthes, vedendo la enorme nettaroconea, e la enorme secrezione di liquido dolciastro, dissi tra me: questo fiore deve essere fecondato da un trochilide specialissimo. Ora mi accadde in seguito che sfogliando la opera di Gould sui trochilidi, vidi per lo appunto una tavola ove si scorge un trochilo singolare con becco curiosamente ricurvo visitare i fiori di corianthes, e sorbire il copioso liquido della nettaroconea.[4] All' incontro la spiegazione del Dott. Crüger mi sembra pochissimo fiducievole. La natura non fa glandule melliflue, nè perfettamente lavorate nettaroconche, semplicemente per inumidire le ali di una Euglossa.[5] Io mi rimetto ad ulteriore guidizio della S. V.

Mi riuscì oltremodo istruttivo lo apprendere che alcuni insetti vanno sui fiori di orchidee non già per suggere miele ma per rodere speciali escrescenze.[6] Questa era una idea che già da un pezzo mi girava per la mente. Piacemi che i fatti corrispondano a questa idea. Tra le orchidee nostrali senza dubbio deve succedere la stessa cosa pel genere Serapias (S. lingua, cordigera, neglecta ecc.)[7] Nel fondo del fiore non vi è miele; ma invece vi si trova una grossa *escrescenza tinta in purpureo* che deve appunto fornir cibo ai pronubi.

Ho ricevuto un numero del giornale ove è comparsa una parte del mio lavoro sulla pangenesi.[8] La pangenesi può essere considerata sotto due aspetti; sotto l'aspetto monistico e sotto l'aspetto dualistico.

Sotto l'aspetto monistico mi pare, salvo errore, che non possa concepirsi colle leggi chimiche.

Sotto l'aspetto dualistico, la pangenesi è una mirabile, una grande teoria, che aggruppa felicemente, a mio parere, tutti quanti i fenomeni della genesi degli organismi. Ammettendo la esistenza di gemulle di sostanza immateriale, si sfuggono le objezioni della chimica, si spiega la metagenesi, la generazione alternante, la dislocazione, la moltiplicazione, la diminuzione degli organi, i fenomeni di atavismo ecc. ecc. Io sono compreso di ammirazione per la S.V. che ha saputo formulare una così grande teoria, e sarò lieto se potrò in breve pubblicare una scrittura non indegna dell 'alto argomento, sotto il titolo —*Apologia della pangenesi sotto il punto di vista dualistico*—[9]

Gradisca il tenuissimo dono di due miei opuscoli con cui ho cercato di utilizzare per la geografia botanica le nozioni sulla dicogamia e di estendere l'albero genealogico delle Marantaceo, nel cui stilo credo di aver ravvisato la parte omologa al rostello (!) delle orchidea.[10]

Ho l'onore di dichiarararmi della S.V. | Ossequentissimo discepolo
Federico Delpino

DAR 162: 144

[1] For a translation of this letter, see Appendix I.
[2] Delpino refers to 'Specific difference in *Primula*', 'Illegitimate offspring of dimorphic and trimorphic plants', and 'Fertilization of orchids'.
[3] Delpino refers to the genus *Trochilus* and the now invalid genus *Ornismya* in the family Trochilidae (hummingbirds), the genus *Nectarinia* in the family Nectariniidae (sunbirds), and the genus *Melithreptus* in the family Meliphagidae (honeyeaters).
[4] Delpino refers to John Gould and Gould 1861, 1: plate 3, which depicts *Eutoxeres aquila* (the white-tipped sicklebill) at a flower of *Coryanthes speciosa*, a species of bucket orchid.
[5] In 'Fertilization of orchids', p. 157, CD had described Hermann Crüger's observations on the visits of bees of the genus *Euglossa* to *Coryanthes*. CD wrote, 'The fluid in the bucket formed by the basal part of the labellum is not nectar and does not attract insects, but serves, by wetting their wings, to prevent them from crawling out except through the small passages close to the anther and stigma.' For more on pollination of orchids by *Euglossa*, see Dressler 1968.
[6] See 'Fertilization of orchids', p. 157.
[7] *Serapias lingua* is the tongue orchid; *S. cordigera* is the heart-flowered serapias; *S. neglecta* is the scarce tongue orchid. For more on pollination in this genus, see Jersáková *et al.* 2006, p. 222.
[8] Delpino refers to the first part of the English translation of his article 'Sulla darwiniana teoria della pangenesi' (On the Darwinian theory of pangenesis; Delpino 1869a), published in the 29 September 1869 issue of *Scientific Opinion* (Delpino 1869b).
[9] The proposed work (Defence of pangenesis from a dualistic point of view) was never published.
[10] The articles referred to are 'Alcuni appunti di geografia botanica a proposito delle tabelle fitografiche del Prof Hoffmann' (Some remarks on geographical botany with reference to Prof Hoffmann's phytographical tables; Delpino 1869c) and 'Breve cenno sulle relazioni biologiche e genealogiche delle Marantacee' (Brief remarks on the biology and genealogy of the Marantaceae; Delpino 1869d). CD's annotated copies of these papers are in the Darwin Pamphlet Collection–CUL.

From T. H. Farrer 9 October 1869

Abinger Hall, | Dorking.
9 Oct/69

My dear Mr Darwin

It may be worth your knowing, for the sake of future visitors, that the trains represented on the SER tables as continuous, are not so—but separate at Beckenham Junction: and do not wait for each other—so that one may have to wait & get on by later trains as I had yesterday. It did not signify to me, as there were later trains to Shere.[1]

I have glanced this morning at Delpinos books— they are charming— You should get them read to you— But his last book will save me the trouble of publishing.[2] It contains very nearly all that I have in my paper and a great deal more— It has not the position of the flower: but the coherence of petals & stamens: function of filaments & brush on style: dusty and moist pollen are all there. He says however that where stamens are monadelphic there is *no nectar* in the flower anywhere. I dont think is so, as it is outside the tube in Anthyllis: on vexillum (I think) in Ulex: and I have seen bees poking their noses into vexillum in Ulex—Ononis—& Lupin.[3]

Can you tell me of any book which will give me (with pictures) the machinery of a bees collecting apparatus?

Can the meaning of the Corona in Passiflora be to bring the bee's back as he struggles irregularly over it *up* and into efficient contact with anthers and style? If it were not there he would scarcely touch them.

I have been looking at Antirrhinum's hairs. Is not the meaning of them that the lower lip with its hairs being pressed against the open anthers, the pollen gets shed upon the hairs. A bee comes in, and carries away this pollen—not on his back but on his belly: he goes to another flower and his belly deposits it on the hairs of that flower: when he leaves the second flower the lower lip springs up and its hairs deposit the pollen (which the bee has left) on the stigma of the second flower. Thus the bee carries the pollen but does not immediately fertilize the stigma. This explains too the continuance of the hairs down the corolla, since the bee in entering will spread the pollen all the way down and carry more away from the long line of hairs than if they were not so continued. It would be a good plant to try under a net, since it must get its own pollen.

I shall at once order one of Delpino's books and send you back your copy.

How endless & fascinating these things are?

I must read Delpino carefully— Possibly a short notice of what he says with any little points that he has not, might be of use—only that I should feel it presumptuous to appear to criticize.[4] Or if there were the opportunity and he would care to have it, I would send him what I have written, begging him to make use of it. He seems an admirable observer and describer.

In one of the books he assails you as a fatalist, and wanders into regions of free will & necessity, into which—human time being, as Huxley says, limited—I do not care to follow him.[5]

Sincerely yours | T H Farrer

Charles Darwin Esqr FRS

The Pocket book was in the parcel[6]

DAR 164: 56

1 Farrer refers to the South East Railway, which had a station in Orpington. The line opened in 1868. The village of Shere is six miles west of Dorking.

2 Farrer evidently borrowed CD's copies of articles by Federico Delpino. Delpino had sent CD two papers on dichogamy in 1867 (Delpino 1867a and 1867b; see *Correspondence* vol. 15, letter from Federico Delpino, 5 September 1867 and n. 2). He also sent three works with his letter of 22 August 1869 (Delpino 1867c, 1868–9, 1869a). Farrer probably refers to the first part of *Ulteriori osservazioni sulla dicogamia nel regno vegetale* (Further observations on dichogamy in the vegetable kingdom; Delpino 1868–75) as Delpino's 'last book'.

3 Many species of *Anthyllis* bear stipular nectaries.

4 Farrer did not publish on Delpino's works.

5 Farrer may refer to a passage in Delpino 1869a, translated in Delpino 1869b, pp. 407–8, where Delpino argued that pangenesis was a monistic theory and that monism denied liberty. Farrer also refers to Thomas Henry Huxley.

6 Farrer had evidently recently paid a visit to Down by train (see also letter from T. H. Farrer, 18 September 1869), and possibly left some belongings behind that had to be sent on by post.

To T. Thompson[1] 9 October 1869

[Down]

To Mr T. Thompson, Secretary of the Westerham F. Club

Sir

Mr W. Reeves in this place has called on me as a County Magistrate to consult me on the best means of obtaining the payment which he states is due to him from the Westerham Club.—[2]

I can testify that the condition of his hand shows that he is unable work, & as he cannot at present write, he has asked me to address you on the subject,—before he takes any further steps.— His case seems a hard one, I am informed that some of your members still receive payment, & if this be correct it w^d^ be difficult to justify, any one member who may have a just claim being refused.

Hoping that you will be able to satisfy W Reeves' claim. I beg leave to remain | Sir | Your obed sert | Ch. R. Darwin

Oct 9—1869

ADraftS
DAR 96: 60

1 The addressee is probably Thomas Thompson, grocer, of Westerham (*Post Office directory of the six home counties* 1870); Westerham is about five miles from Down.

2 CD probably refers to William Reeves (b. 1834/5), but may refer to William Reeves (b. 1814/15). CD was a county magistrate for the Bromley division of Kent (*Post Office directory of the six home counties* 1870, p. 1004). CD had served as treasurer of the Down Friendly Club, a local savings and insurance society, since its establishment in 1850 (see *Correspondence* vol. 4, letter to J. S. Henslow, 17 January [1850] and n. 6).

To T. H. Farrer 10 October [1869][1]

Down. | Beckenham | *Kent. S.E.*
Oct 10

My dear Farrer

I declare I am almost as sorry as if I had been myself forestalled, indeed more so for I am used to it. It is, however, a paramount, though bothersome duty in every naturalist to try & make out all that has been done by others on his subject.

By all means publish next summer your *confirmation* & a summary of Delpino's observations, with any new ones of your own.— Especially attend about the nectary exterior to the staminal tube.— This will in every way be *far* better than writing to Delpino.—[2]

It w[d] not be at all presumptuous in you to criticise Delpino.— I am glad you think him so clever; for so it struck me.—

Look at hind legs yourself of some Humble & Hive-bees; in former take a very big individual (if any can be found) for these are the females, the males being smaller & they have no pollen-collecting apparatus. I do not remember where it is figured, probably in Kirby & Spence,[3]—but actual inspection better.— That is good idea about corona of Passiflora— I remember something analogous, but exactly what I cannot call to mind.— Still better, your idea about Antirrhinum, **if true**—[4] Could you experimentise with any coloured powder?

I rather think Antirrhinum is self-fertile, when insects are excluded,— I know I have the result in my notes; & which I will attend to when I go through them in course of winter.—[5]

Please do *not* return any of my Books, until all are finished, & do not hurry.—[6]

I feel certain that you will make fine discoveries.—

Yours very sincerely | C. Darwin

P.S I sent paper by Post on Martha[7]

Linnean Society of London (LS Ms 299/17)

1 The year is established by the relationship between this letter and the letter from T. H. Farrer, 9 October 1869.

2 See letter from T. H. Farrer, 9 October 1869 and nn. 2 and 3. Farrer had been studying many of the plant species discussed in recent publications of Federico Delpino. Farrer did not publish a summary of Delpino's observations, but did eventually publish a revised version of the paper he evidently sent to CD (see Farrer 1872, p. 478).

3 CD refers to Kirby and Spence 1815–26, but the illustration he refers to has not been found. The description of the absence of the corbicula or pollen-basket in males is in *ibid.*, 2: 118.

4 See letter from T. H. Farrer, 9 October 1869.

5 Most of CD's surviving notes on the self-fertility of *Antirrhinum majus* (the snapdragon) concern experiments with artificially pollinated specimens (DAR 51: B18–21). For CD's published observations on the self-fertility of *Antirrhinum*, see *Cross and self fertilisation*, p. 363.

6 See letter from T. H. Farrer, 9 October 1869 and n. 2.

7 CD refers to 'Ueber die Befruchtung der *Martha (Posoqueria?) fragrans*' by Fritz Müller (F. Müller 1866), in which Müller described the mechanism by which *Martha fragrans* (now *Posoqueria densiflora*) prevented self-fertilisation. CD's copy of the paper is in the Darwin Pamphlet Collection–CUL.

To Alphonse de Candolle 11 October 1869

Down. | Beckenham | *Kent. S.E.*
Oct 11 1869

My dear Sir

I am very much obliged for y[r] extremely interesting letter. Your experiment strikes me as a very valuable & simple one,—so simple that now that it has been once suggested, I am surprized that no one ever thought of trying it before.[1] Nägeli, indeed, in one of his recent essays complains that no one has made this experiment with Alpine & lowland forms of the same species.[2] I forwarded your letter to Dr Hooker, who seems very glad to have the opportunity of making the trial.[3] As I shall have many experiments in progress next spring, which cannot be deferred, & as I have little strength to spare, I believe it would be the wisest plan to send all the seeds to Kew.[4] But I will open some of the packets & see how many seeds there are, so that if numerous I will try a few myself— I will not however sow them till the spring, for I have lately had some experience in raising during the winter hardy annuals in my greenhouses, & I find that they become much drawn up—

With my best thanks for your kindness & the most sincere respect, I remain | My dear Sir | yours very faithfully | Charles Darwin

LS
Archives de la famille de Candolle

1 See letter from Alphonse de Candolle, 28 September 1869.
2 CD refers to Carl Wilhelm von Nägeli's remarks in a paper on varieties of *Hieracium pilosella* (mouse ear hawkweed; see Nägeli 1867, p. 478).
3 See letters to J. D. Hooker, 2 October [1869] and 11 October 1869.
4 CD probably refers to his experimental work on the comparative growth of plants raised from seeds of cross and self fertilised parents (see *Correspondence* vol. 18, letter to T. H. Farrer, 28 May [1870]).

To J. D. Hooker 11 October 1869

October 11, 1869

I am very glad that you are willing to try DeCandolle's experiment which strikes me as a very good one. I enclose all the seeds except a very few taken from the Palermo and Moscow packets of 1868 of [. . .] and Senecio.[1]

Incomplete[2]
Rendell's (Catalogue No. 123)

1 See letter from Alphonse de Candolle, 28 September 1869 and n. 4, and letter to Alphonse de Candolle, 11 October 1869.
2 The original letter is complete and is described in the sale catalogue as being one page long.

From John Fox & Sons 12 October 1869

Farnborough Brewery, | By Chislehurst, | Kent, S.E.
12 Oct 1869

C. Darwin Esq. | Down
Sir.

We beg to acknowledge, with our best thanks, the receipt of your Cheque in payment of the enclosed account[1] *and in all your commands, | We remain, | Your much obliged obed*t. *Servants, | John Fox & Sons. | [3 words illeg]*

P.S. | It is requested that all Money Orders may be made payable at the Chief Office, London; and Cheques crossed "London & Westminster Bank."

DAR 83: 61v

[1] The enclosure has not been found but a payment to Fox of £37 2*s.* 6*d.* on 9 October 1869 is recorded in CD's Account books–banking account (Down House MS).

From T. H. Farrer 13 October 1869

Abinger Hall, | Dorking.
13 Oct /69

My dear Mr Darwin

At Albury—near this,—there is house full of Tacsonias and Passifloras—[1] All the Tacsonias are pendent, have no crown & very long styles and filaments— They never—so the gardener says—fertilize themselves—but are readily & constantly fertilized by him with each other and with the Passifloras— The Passifloras turn their flowers upwards, though themselves pendent: their styles & filaments are short: and they have a corona. And they fertilize themselves—i.e. they are fertilized without the gardeners help— This certainly looks as if in this country the bees could manage the Passiflora—but not the Tacsonia: and as if for the latter humming birds or big moths were needed—which can suck without alighting. The bees are as a fact very fond of Passiflora—and I think when raised on the Corona—easily reach anthers & stigma.

It is a funny notion that a humming birds tail should do the business for Tacsonia.[2]

Many thanks for your kind note.[3] It has long seemed to me, in the business of my own shop, that claims to have originated any thing that is true or successful, are idle ignes fatui: things to be avoided & put aside as soon as possible—like patents.

Your books open up a new field. Old Sprengel is delightful— But where is it to end—and what is to become of Merchant Shipping Bills and Maritime Codes?[4]

I could not help revenging myself on Delpinos philosophy during a sleepless hour one night—and I inclose you the result. What a strange mental phenomenon, that an able man of science should think he he has saved any thing worth saving, by reserving one portion of plant life to the regions of the arbitrary & the unknown![5]

Very sincerely yours | T H Farrer

The paper on "Martha" arrived quite safely.[6]

[Enclosure]

The Biological Teleologist.

Highest of living Creatures, Man,
Exhibits clearly "Will and Plan:
The animal can comprehend
A purpose, and attain an end:
Whilst in its turn the humbler plant
Can feel and satisfy a want—
And thus we find throughout the line
Freedom and Will and High Design—

But here we pause. No living Mind
Informs the mass that lies behind;
And Earth, and Air, and Sun and Sky,
Come—God knows how, and God knows why:
Brute Matter all: whilst we inherit,
With beast and plant, Free will and Spirit;
And carry on with noisy clatter
A ceaseless strife twixt Mind and Matter—

———

Strange Creed! For me, Creations Soul
Part seen, part guessed, informs the Whole.

DAR 164: 57

1 Albury, Surrey, was about two miles west of Abinger Hall. *Tacsonia*, the genus of banana passion fruit (family Passifloraceae), is now subsumed within the genus *Passiflora*, but is still often referred to as a subgenus.
2 For more on pollination in *Passiflora*, see Janzen 1968.
3 See letter to T. H. Farrer, 10 October [1869].
4 Farrer refers to the fact that CD's approach to floral morphology in works such as *Orchids* treated morphological features as adaptations that revealed information about agents and methods of fertilisation. Farrer had borrowed CD's copy of Christian Konrad Sprengel's *Das entdeckte Geheimniss der Natur im Bau und in der Befruchtung der Blumen* (The secret of nature discovered in the structure and fertilisation of flowers; Sprengel 1793). Farrer also alludes to his position as permanent secretary in the Board of Trade and his special interest in marine matters (*ODNB*).
5 Farrer refers to Federico Delpino's teleological interpretation of fertilisation mechanisms (for more on the differences between CD and Delpino on this topic, see Pancaldi 1991, pp. 117–36). CD had sent several works by Delpino to Farrer (see letter from T. H. Farrer, 9 October 1869 and n. 2).
6 See letter to T. H. Farrer, 10 October [1869] and n. 7.

From L. E. Becker 14 October 1869

28 Jackson's Row | Albert Square | Manchester
Oct. 14. 1869

My dear Mr. Darwin

Will you permit me to send you the enclosed abstract of a paper read at the Exeter Meeting of the British Association, respecting the curious variety of campion flowers I took the liberty of sending you a few years ago, and in which you were good enough to express some interest[1]

If you think the subject worthy of your attention I should be greatly obliged if you would kindly inform me whether you think it possible that the fungus could exert what looks like an "active coercive force to bend the structure of the flower to its necessities". From all that I have observed I cannot rest in any other conclusion but in the enclosed abstract, I have merely given the result to which I have been led and not entered into the detailed reasons for it.

If you paid any attention to those plants which I sent you, you may have observed some facts which will bear on this curious question.[2]

On your theory of Pangenesis it seems to me not inconceivable that gemmules of stamens may circulate undeveloped in the pistilliferous plants of campion, and that the presence of the parasite may cause the condition under which these can develop—but without this theory I am totally unable to conjecture in what manner the fungus possibly can cause stamens to grow in a flower that would not naturally have produced them.[3]

Apologising for troubling you with this note I am | yours very truly | Lydia E. Becker

Charles Darwin Esq—

[Enclosure][4]

⟨On alteration⟩ in the structure of Lychnis diurna ⟨observed in co⟩nnection with the development of a ⟨parasitic⟩ fungus

Abstract of Paper read before section D— British Association, Exeter—

Specimens were produced of the common red campion, *Lychnis diurna*, infested with a parasitic fungus allied to the "smut" in wheat which fungus develops its fructification in the anthers of the flower. The campion in its ordinary healthy state has flowers bearing stamens only or pistils only, but about half the plants infested with the parasitic fungus bear flowers containing both stamens and pistils. The writer had never observed bisexual flowers on healthy plants, and attributed the occurrence of that condition in the specimens produced to the action of the parasitic fungus. The diseased plants very rarely produce seed but occasionally late in the season perfect capsules bearing good seed are found on them. A few of the flowers had been submitted to M^r^. Darwin, and he had suggested that the pollen being destroyed by the parasite at an early period, the pistil was developed in compensation. But the writer thought that the influence exerted by the parasite was of a much more subtle and surprising character ⟨than this,⟩ and that instead of causing the develop⟨ment of⟩ a pistil in compensation for destroyed ⟨stamens the⟩ fungus has the power to cause a plant which would naturally have produced pistils only to develope stamens for the accommodation of the parasite— She supposed that the spores of the fungus fell on the stigma of a healthy flower, and infected all the seeds produced by that capsule that of the plants thus raised all which were naturally male plants remain unaffected in structure by the parasite, but they have their pollen destroyed by it— that all which were naturally female plants would develope their pistils to a certain extent, but as the fungus cannot produce spores without anthers to fructify in a pollen to feed on, it compels the campion to develope these for its accommodation, and the effort of so doing exhausts the forces of the plant, and causes the decay of the capsule, if indeed the previous stunting of the styles does not prevent fertilization. The production of healthy capsules late in the season may be owing to decay in the vigour of the fungus when, the pressure being removed, the plant resumes its natural functions. The fact that only about half the diseased plants are bisexual favours the view that the latter are female plants in which the growth of stamens has been induced by the presence of the fungus—

DAR 160: 117, 119

[1] The abstract appeared in *Report of the thirty-ninth meeting of the British Association for the Advancement of Science, held at Exeter* (Becker 1869a). CD and Becker had earlier corresponded on the possible dimorphism in *Lychnis diurna* (now *Silene dioica*, red campion) and CD had evidently alerted Becker to the fact that a fungus was the cause of the purple anthers she had observed (see *Correspondence* vol. 11, letter from L. E. Becker, 23–4 May [1863] and n. 5).

[2] Becker refers to plants she sent with her letter of 21 May [1863] (*Correspondence* vol. 11).

[3] Pangenesis, CD's hypothesis of the transmission and development of hereditary characters, is discussed in *Variation* 2: 357–404. CD suggested that each individual cell threw off minute particles (gemmules) that circulated in the bodily fluids and were capable of generating new cells, or remaining dormant until required. For more on the relation of the anther-smut fungus to male and female flowers of *Silene dioica*, see Giles *et al.* 2006.

[4] The enclosure is damaged. Missing parts have been restored from the published text (Becker 1869a).

To Federico Delpino 14 October 1869

Down
Oct. 14. 1869

Dear Sir

I am much obliged for your courteous and interesting letter.[1] I write now in case you should publish on Coryanthes, to caution you to consult a paper in the Bulletin de la Soc. Bot. de France, Tom. 2. 1855. p. 351, in which you will see that the fluid in the labellum hardly deserves to be called nectar.[2] I believe Crüger to be right, and I send his paper in case you have not seen it.[3]

I have been very much struck by the originality of many of your remarks in one of your papers formerly sent; I do not remember the title, and I have lent all your papers to Mr. Farrer who has been studying the Leguminosæ and writes *enthusiastically* about your papers.[4] I have long believed that humming birds play a great part in the fertilisation of many American flowers such as Salvia, Fuchsia &c. From enquiries which I made at the C. of Good Hope I believe that Strelitzia is fertilised by honey-suckers (Nectarinæ). I have received your two papers and hope to get parts of both translated for my own reading.[5]

If the editor does not send the succeeding numbers of "Scientific Opinion", I will send them to you if you will inform me.— I am not satisfied with the remarks which I have made on the re-growth of limbs.[6]

Permit me to suggest to you to study next spring the fertilisation of the Gramineæ; I feel sure that those botanists are wrong who assert that the fertilisation takes place in the flower-bud; though I suspect this occasionally takes place in Triticum. Zea mays would be well worth studying.[7] A Swedish book very much like Hildebrand's has just appeared, so our subject is receiving much attention; I cannot give you the title as I have lent it to Mr. Farrer.[8]

Believe me, dear Sir | with much respect | Yours very faithfully | Charles Darwin

Copy
DAR 143: 379

[1] See letter from Federico Delpino, 9 October 1869.
[2] The article referred to included a report on an analysis of the fluid in the *Coryanthes* 'bucket' that determined that it was mostly water with a small percentage of mucilaginous compounds (Menière 1855, p. 354).
[3] CD refers to Hermann Crüger and Crüger 1864.
[4] CD sent several of Delpino's works to Thomas Henry Farrer (see letter from T. H. Farrer, 9 October 1869 and n. 2).
[5] See letter from Federico Delpino, 9 October 1869 and nn. 3 and 10.
[6] See letter from Federico Delpino, 9 October 1869 and n. 8. The final two instalments of Delpino 1869b appeared in the 6 October and 13 October 1869 issues of *Scientific Opinion.* CD mentioned Delpino's review and criticisms in *Variation* 2d ed., 2: 350 n. 1, and 380–1. For CD's remarks on the regrowth of limbs in response to Delpino 1869b, see the letter to *Scientific Opinion,* [before 20 October 1869]. For CD's original remarks on regrowth of limbs, see *Variation* 2: 376–7.
[7] *Triticum* (wheat) and *Zea mays* (maize or sweetcorn) are members of the grass family, Gramineae (now Poaceae).
[8] CD refers to Friedrich Hildebrand and Hildebrand 1867a, and to Johan Severin Axell's *Om anordningarna för de fanerogama växternas befruktning* (On the mechanisms for the fertilisation of phanerogamic plants; Axell 1869).

To T. H. Huxley 14 October 1869

Down. | Beckenham | Kent. S.E.
Oct 14. 1869

My dear Huxley

I have been delighted to see y^r Rev. of Häckel, & as usual you pile honours high on my head.[1] But I write now (*requiring no answer*) to groan a little over what you have said about rudimentary organs. Many heretics will take advantage of what you have said. I cannot but think that the explanation given at p. 541 of the last Ed of the Origin of the long retention of rudimentary organs & of their greater relative size during early life, is satisfactory.[2] Their final & complete abortion seems to me a much greater difficulty. Do look in my "Var under Dom." Vol. 2. p 397. at what Pangenesis suggests on this head, though I did not dare to put it in the Origin.[3]

The passage bears also a little on the struggle between the molecules or gemmules. There is likewise a word or 2 indirectly bearing on this subject at p 394–395. It won't take you 5 minutes, so do look at these passages.

I am very glad that you have been bold enough to give y^r idea about nat. selection amongst the molecules, though I cannot quite follow you.[4]

Whenever you can spare time let me hear about the horse's tooth from S. America—[5]

Ever yours most truly | Charles Darwin

Good Heavens what a deal of work there is in you.—

LS(A)
Imperial College of Science, Technology, and Medicine Archives (Huxley 5: 277)

1 Huxley's review of Ernst Haeckel's *Natürliche Schöpfungsgeschichte* (Natural history of creation; Haeckel 1868) appeared in the 9 October and 13 November 1869 issues of the *Academy* (T. H. Huxley 1869d).

2 CD is probably complaining about Huxley's arguments concerning teleology. Huxley discussed rudimentary organs in the context of Haeckel's arguments against teleology (T. H. Huxley 1869d, p. 14). Huxley argued against 'coarser forms' of teleology which posited that organs such as the eye were specially designed for the purpose of seeing, but allowed that there was a 'wider Teleology' based on the fundamental proposition of evolution, which Huxley defined as the lawlike mutual interaction of the forces possessed by molecules (*ibid.*, p. 13). CD refers to *Origin* 5th ed.

3 See *Variation* 2: 397–8.

4 See T. H. Huxley 1869d, p. 14.

5 See letter to T. H. Huxley, 21 [September 1869].

From Margaret Susan Vaughan Williams to H. E. Darwin [after 14 October 1869][1]

Down Ampney | Cricklade

My dear Henrietta

I am afraid you will think I have taken a long time in answering your questions, but we do find it so hard to observe Baby's expression accurately, & besides I seldom *can* attend to his eyebrows, when he is crying and I am wanting to comfort him.[2] I see my own eyebrow corners go up perfectly, and we are almost sure his do the same in a less degree—when he is going to cry—but the transverse wrinkles on the forehead not straight but waving downwards the lines converging towards the nose. I cannot draw a face in the least.

Baby crys with tears now and certainly has done so for a long while, but when he began I cannot tell.

He is just over 6 months old.[3] I wish I could answer Uncle Charles questions better, but if we can observe further more accurately, I will write again— I return the "Young Person".[4]

Thank you for your Photo: I like it very much indeed We have been very busy with our garden since we got home, it seems hard when we work so much, that we produce such *very* small results.

I am so glad I met you and Bessie at Leith Hill, it is such a long while since we met.[5] How did your letter get to me when *you* had only put Down Ampney & some postmaster added Cricklade![6]

With best love to Aunt Emma & Bessie[7] | Your affec. | *M. Vaughan Williams*

DAR 180: 4

CD ANNOTATIONS

3.1 I wish … Bessie 6.1] *crossed pencil*

1 The date is established by the age of Williams's baby (see n. 3, below).

2 CD had kept notes on the development of his own children and noted that very early crying was not accompanied by tears (see *Correspondence* vol. 4, Appendix III). As part of his work on expression, CD had asked people with young babies to observe when they began to shed tears when crying (see, for example, *Correspondence* vol. 15, letter from L. F. Kempson to Emma Darwin, 20 June 1867). CD was also interested in the facial muscles associated with crying. In *Expression*, p. 164, CD wrote that infants did not weep until the age of 'from two to three or four months'.

3 The baby was Hervey Wedgwood Williams, who was born on 14 April 1869 (Birth certificate, General Register Office, Dorking).

4 The reference to the 'Young Person' has not been identified.

5 According to Emma Darwin's diary (DAR 242), Henrietta and Elizabeth Darwin visited Leith Hill Place near Dorking, Surrey, the home of Josiah Wedgwood III, on 7 October 1869.

6 Down Ampney is south of Cirencester in Gloucestershire; Cricklade is a couple of miles south of Down Ampney and is in Wiltshire.

7 Emma and Elizabeth Darwin.

From John Brodie Innes 15 October 1869

Milton Brodie | Forres | N.B.
15th. Octr. 1869.

Dear Darwin,

It seems sad to lose sight of old friends, and this perhaps is one chief reason for my taking up my pen to write a line to you today. We are often reminded of you by our admiration of the most beautiful photograph you were so kind as to give me, which hangs in a post of honour in the drawing room looking at another good friend, now gone to his rest, Archbishop Longley, the photograph being however far behind that of yourself—[1] And I have just been reading a paper about you which has given me much pleasure by Mr. Hutton, at the Liverpool Church Congress.[2] I hope you have seen it.

I wonder how you are getting on in Church matters at Downe I do not hear much from Powell Perhaps he thinks I have not helped him as I should in the matter of the proposed parsonage. When Sir John asked me about it I could do no less or more than simply say how the case actually stood.[3] I conclude you saw my letter. It would be a great advantage to restore and re-seat the Church, and for this I should gladly contribute, and probably had the parsonage question been disposed of as I hoped this would soon have been undertaken—

We jog on in a quiet hum drum sort of existence here. Mrs. Innes' violent headaches being I hope rather less frequent of late, but distressing when they come. Johnny is off again to Cambridge after a short holiday.[4] He spent the early part of the vacation in College and had less out door amusement here than he should, having been laid up with bronchitis caught at the opening of the Inverness Cathedral on a wet day.

I think I told you I had one milk white partidge last year.[5] This year I have seen four among brown ones. If they should escape guns and other enemies it is possible a race may be established. I avoid injuring them myself. I wish I had something of interest to tell you. With our kindest regards to your party

Believe me Dear Darwin | Faithfully yours | J Brodie Innes

DAR 167: 25

[1] Innes refers to Charles Thomas Longley, who was archbishop of Canterbury from 1862 until his death in 1868. The photograph of CD has not been identified.
[2] Richard Holt Hutton delivered a paper on 6 October 1869 to the Church Congress held at Liverpool in which he discussed Darwinism, scepticism, and faith (Hutton 1869). In it he argued that CD's theory not only did not increase any moral difficulty in the apparently cruel competition of nature, but in fact made clearer the intellectual purpose at work in creation (see Hutton 1869, pp. 102–3).
[3] Henry Powell, the curate of Down, had asked John Lubbock to subscribe to build him a house (see letter from John Lubbock, 20 July [1869]).
[4] Innes refers to Eliza Mary Brodie Innes and John William Brodie Innes, who was a student at Cambridge (*Alum. Cantab.*).
[5] No letter has been found in which Innes told CD about a white partridge.

From Robert Swinhoe 16 October 1869

33 Oakley Square. SW.
16 October, 1869.

My dear Mr. Darwin,

I have returned from China to England.[1] I took the route via Japan to San Francisco and across the American continent to New York & thence to Liverpool. I have a book of fine photographs of Japanese which somewhat illustrate expression. This I would much like you to see.[2] If agreeable to you I will send the book to your house.

I trust that you are improved in health.

Your's very truly, | Robert Swinhoe

DAR 177: 332

[1] Swinhoe was acting British consul at Amoy (Xiamen), China.
[2] Swinhoe had earlier published CD's queries on expression in *Notes and Queries on China and Japan* 1 (1867): 105 (see *Correspondence* vol. 15, letter to Robert Swinhoe, [27 February 1867], where the queries are reproduced, and letter from Robert Swinhoe, 5 August 1867).

From T. H. Farrer 17 October 1869

Abinger Hall, | Dorking.
17 Oct | /69

My dear Mr Darwin

I have been looking again at Passiflora Cœrulea—at Tacsonia Mollissima— and another very like the one in your greenhouse, also at Sprengels excellent descriptions of P. Cœrulea—[1]And it strikes me that a comparison of the three suggests more than even Sprengel made out. Perhaps the accompanying notes may interest you especially if you can pull to pieces one of your beautiful pendent flowers.[2]

Many thanks for a paper received today

Sincerely yours | T H Farrer

Hooker is pleased with the enclosed and thinks if I doctor it up that it will be worth reading at the Linnean. I would prefix a short statement of what Sprengel says about it.[3]

DAR 164: 58

[1] See letter from T. H. Farrer, 13 October 1869 and n. 1. Farrer had evidently visited CD earlier in the month and seen a specimen of *Passiflora* in CD's greenhouse (see letter from T. H. Farrer, 9 October 1869). The species was evidently *P. van volxemii* (see letter to T. H. Farrer, 20 October [1869]). *P. caerulea* is the common or blue passionflower and *Tacsonia mollissima* (now *Passiflora tripartita* var. *mollissima*) is the banana passionfruit. Farrer also refers to Christian Konrad Sprengel and Sprengel 1793, pp. 160–6 (see letter from T. H. Farrer, 13 October 1869 and n. 4).

[2] The enclosure, which CD evidently returned, has not been found, but the corrected version, in the hand of a copyist, with corrections and diagrams by Farrer, and dated 13 November 1869, has been transcribed as Appendix IV. The version that CD saw was probably very similar to the notes in Appendix IV, except for the introduction on Sprengel.

[3] The paper was never read at the Linnean Society of London or published in any of their journals.

From G. H. Darwin [18 October 1869][1]

14 Arlington St
Monday Ev^{g}

My dear Father,

I forgot to ask you to forward me the £52"10s which I shall have to give to Tatham on Thursday.[2] I cannot tell until Xmas whether I can afford to pay it myself, as I don't know what the fellowship dividend comes to yet. 52£ will make my balance rather indecently small so cd you have it put to my account before Thursday. I have called on Tatham & have been just introduced to my fellow pupils, of whom I will give my opinion of Saturday.—

Your affectionate Son | George. H. Darwin

DAR 210.2: 10

[1] The date is established by the entry in CD's Classed account books on 19 October 1869 (see n. 2, below). The Monday preceding that date was 18 October.

[2] An entry in CD's Classed account books (Down House MS) on 19 October 1869 records a payment of £52 10*s.* for 'George for Tatham'. The reference is to Charles Meaburn Tatham, with whom George was to study law.

To J. B. Innes 18 October 1869

Down. | Beckenham | *Kent. S.E.*
Octr. 18 | 69

My dear Innes

I was wishing to hear some news of you, & had thought of writing, but I get so many foolish letters from foolish people, that I seldom have the heart to write to my friends.[1] There is hardly any news to tell you of your old Parish.— Mr Powell has taken Mr Engleheart's House & this I am very glad of, as he will now be able to look after the Parish & school, & I daresay he will be active & kind;[2] but I rather doubt whether he is above the average in sense.— We suspect on very slight grounds that he is going to be married; for he has given notice to Amy Duberry that a Lady will soon teach in the Sunday School.— Possibly it may be Mrs Lovegrove.—[3] I hear of no chance of a parsonage being built; Mr Powell wished to get up a subscription, but I doubt whether he will succeed.— I offered 20 £ which he seemed to think very small, but I shall not increase the amount; for I see no reason that the Parsonage sh^d cost 16 or 1700 £, as he proposes.[4]

Mr Engleheart's lungs (& I fear purse also) failed him; & he is to us a fearful loss as a doctor.

I have neither seen nor heard anything of the Lubbocks for an age—but this is not true for I often hear their Harriers in the morning. M^rs. H. Lubbock finds Gorringes so dull that they intend taking a London House, & coming occasionally to the Farm for the hunting in the winter; & I suspect poor Henry Lubbock is ready to hang himself at the thought of his London life.—[5]

I have not seen or heard of Mr Huttons remarks on me at the Liverpool congress; & this I regret, for I suppose it is Mr Hutton, editor of the Spectator, who is a very clever man, who feels a deep interest in religion, though he w^d be considered by all Churchmen as highly Latitudinarian.—[6] The newspapers have lately been abusing, praising & chaffing me at a great rate.—

Incomplete
Cleveland Health Sciences Library (Robert M. Stecher collection)

1 See letter from J. B. Innes, 15 October 1869.

2 CD refers to Henry Powell, the curate, and Stephen Paul Engleheart.

3 Amy Duberry was the Sunday-school teacher in Down (see *Correspondence* vol. 16, letter from J. B. Innes, 18 June [1868]). CD also refers to Henrietta Lovegrove.

4 For Powell's estimate of the cost of building a parsonage, see also the letter from John Lubbock, 20 July [1869].

5 Henry James Lubbock and Frances Mary Lubbock were evidently living at Gorringes, a house in Down (Freeman 1978).

6 CD refers to Richard Holt Hutton and a paper he delivered at the Liverpool Church Congress. See letter from J. B. Innes, 15 October 1869 and n. 2.

From Fritz Müller 18 October 1869

Itajahy, Sa. Catharina, Brazil,
Octr. 18. 1869.

My dear Sir

I received some weeks ago your kind letter of July 18th and am very sorry to learn from it, that you have been more unwell than usual.[1] But I hope, that in the meantime you will have recovered again your health, so precious to your numerous friends and admirers.

I have been trying lately a new series of experiments on our self-impotent *Abutilon.*[2] During the winter (from April to August) all my plants, though continuing to flower, had become quite sterile even with pollen from distinct plants. In the meantime one of the two plants, which I supposed to be mutually sterile, had been destroyed by an inundation of the Itajahy, so that I have been prevented from trying by more experiments the exactness of this supposition.— All my plants are descending from seeds of the same pod. Now among them is one, which is distinguished by pale flowers, by longer styli exserted before the expanding of the corolla and by much smaller stigmas. The pollen of this plant fertilizes the flowers of the others; but it has hitherto not yet yielded a single ripe pod. When fertilized with own pollen, the whole flowers fall off, as do those of the other plants, whilst when fertilized with pollen of a distinct plant, only the corolla falls off and the germen begins to swell a little; but in from 5 to 15 days the young pods are dropped. (they w^{d} require about 30 days to ripen.) The same plant differs also from the rest, when fertilized with pollen of two distinct species of Abutilon (a cultivated one, (*A. b.* striatum?) and an endemic one);[3] in this case the whole flowers fall off a few days after fertilization, while the other plants, when thus fertilized, yield fine pods with numerous seeds, more numerous indeed than in most pods naturally fertilized by insects. For instance 8 pods fertilized by insects contained from 11 to 52 (on an average 35,4) seeds; 7 artificially fertilized pods contained from 43 to 68 (on an average 55,7) seeds; and 3 pods fertilized with pollen of *Abutilon stiatum* (?) contained 46, 49 and 62 seeds (on an average 52,3)— Thus, as in many Orchids, self-impotence in our *Abutilon* appears to be accompanied by a great facility of being fertilized by pollen of distinct species.—[4]

Of the *Eschscholtzia*, raised last year from your seed, a single plant has survived; it is now flowering profusely since the beginning of Septbr., but shows only a few very poor pods, of which probably none will produce good seeds.[5] In february or March I had sown some of the Eschscholtzia seeds, which you have sent me this year, but in consequence of the unfavourable weather, they did not germinate, the rest, sown in July, are growing well, but as yet far from flowering.—[6] I hear from Prof. *Hildebrand* that in *Hypecoum littorale*, as in Eschscholtzia, only crossed flowers are producing well developed seed-capsules.[7]

I think, I told you already, that all the seedlings from a longstyled white Oxalis (*O. Martiana* Zucc., as I hear from Kew) legitimately fertilized with pollen of the long stamens of a midstyled plant are either longstyled or midstyled, without a single short-styled plant. Of 15 plants, which are now flowering, 10 are longstyled and 5 midstyled.—[8]

Of the monstrous *Begonia* mentioned in a former letter, I raised numerous seedlings; some are now beginning to expand their first flowers, but only one of these has inherited the monstrosity of the parent-plant, the rest bearing normal male flowers.—[9]

In one of your letters you called my attention on perennial *Passiflora.*[10] I have in consequence planted several species into my garden; only 2 have as yet flowered, & these are self-sterile. In one species the first flowers, which appeared after transplantion, had contabescent anthers.

As to the fertilization of Passiflora, *Delpino* thinks in his interesting "Ulteriori osservazioni sulla Dicogamia", which I rec^{d} a few days ago) that in the larger species this is effected by humming birds.[11] I had come to the same conclusion by observing how frequently these flowers are visited by humming-birds, while I did not yet see neither humble-bees, nor butterflies, nor any other large insects able to fertilize them.

Incomplete
DAR 76: B178, Linnean Society of London (LS Ms 299/12)

CD ANNOTATIONS
1.1 I . . . admirers. 1.4] '*[*Abution*]*' *blue crayon*
2.1 I . . . plants. 2.3] *crossed blue crayon*
2.3 In . . . insects. 2.18] *scored blue crayon*
3.1 Of . . . year, 3.4] *scored red crayon*
4.1 I think, . . . flowers.— 5.3] *crossed blue crayon*
Top of letter: 'F. Muller | Hybridism in *[illeg]*' *ink*

1 See letter to Fritz Müller, 18 July [1869].
2 See letter to Fritz Müller, 18 July [1869] and n. 5.
3 Müller refers to *Abutilon striatum* (redvein Indian mallow). The endemic species was later named *A. darwinii* (see *Cross and self fertilisation*, pp. 333–4).
4 For Müller's earlier work on self-sterility and crossing in orchids, see *Correspondence* vol. 15, letter from Fritz Müller, 1 April 1868.
5 CD had sent seeds of *Eschscholzia californica* (California poppy) with his letters of 30 January [1868] and 3 April [1868]; Müller had obtained seven plants from these seeds (see *Correspondence* vol. 16, letter from Fritz Müller, 9 September 1868 and n. 11).
6 Müller may refer to seeds of *Eschscholzia californica* sent by CD with his letter of 3 April [1868] (*Correspondence* vol. 16). CD sent more seeds of *E. californica* shortly after his letter of 14 March 1869 (see letter from Fritz Müller, 15 June 1869).
7 Friedrich Hildebrand reported his observations on self-sterility in *Hypecoum* and *Eschscholzia* in his paper 'Ueber die Bestäubungsvorrichtungen bei den Fumariaceen' (On the contrivances for pollination in the Fumariaceae; Hildebrand 1870). In this paper *H. procumbens* and *H. grandiflorum* are the species discussed. No letter from Hildebrand on this subject before October 1869 has been found.
8 *Oxalis martiana* is now *O. latifolia*. This is probably the species that Müller first described to CD in 1867 (see *Correspondence* vol. 15, letter from Fritz Müller, 17 July 1867 and n. 4).
9 See letter from Fritz Müller, 14 March 1869.
10 The last two paragraphs of the letter are reproduced from a copy that was enclosed with the letter to T. H. Farrer, [27 November 1869]. See *Correspondence* vol. 15, letter to Fritz Müller, 2 November 1867 and n. 13.
11 See Delpino 1868–75, 1: 172.

From W. B. Dawkins 19 October 1869

Norman Road | Rusholme, Manchester
19 Oct. 1869.

My dear Sir,

I ought to have written to you concerning the Denbigh-shire caves—to which you gave me an introduction.[1] The first one about which I wrote to you, has been fully explored by the energy of M^rs. Lloyd, and has furnished a large number of human skulls & bones, and a few implements.[2] The skulls belonged to short-headed savages of Huxleys River-bed type.[3] Craniologically, and Ethnologically the discovery is of some value, and I hope shortly to bring it before the Ethnological Society.[4] M^rs.

Wynne of Cefn invited me down to dig & supplied me with a virgin cave containing R. Leptorhinus (Owen) & the Grizzly Bear, and a stone cromlech full of short-heads.[5] I have had a most enjoyable time in Denbighshire, and I owe it to you that I hope to be able to make some addition to our knowledge of Cave-short heads in Britain.

I thought that you would like to hear this upshot of your introduction to the Denbigh-shire folk.

Along with this note the three prehistoric scraps.—one of which British Prehistoric Mammals may perhaps be worth your notice. You will see that in it, I have viewed the Chillingham cattle as the lineal descendants of an old domestic breed, an⟨d⟩ not of the wild British Urus.[6] Since it was written I have obtained more evidence on the point. The *white* cattle with red ears were not only the large valuable breed in Wales in the days of Howel Dha,[7] but also in Ireland, in the days subseq⟨uent⟩ to the great Danish invasion.[8] In Britain there is not the slightest proof of the Urus having existed in the wilds of Britain during the Norman occupation, or afterwards. Indeed one would hardly expect to find so *large* an animal capable of living side by side with civilized & quasi-civilized man in so small an Island as Britain. Moreover so far as I can make out the Urus of Northern Germany was not *white* For these reasons, which are merely given in outline, I am obliged to retract my views of the Urus having lingered in Britain into the Historic Times.—(Quart Journ. Geol. Soc.)[9] Possibly you may care to know this, and to have Irish authority for my bouleversement for the 2[d] edition of 'The Variation'.[10] I shall be delighted to tell you all my evidence.

I am looking forward to your new work, with a feeling akin to thirst in a man, and curiosity in a woman,[11]

I am My ⟨dear⟩ Sir | ⟨Yo⟩urs truly | ⟨W.⟩ B⟨oyd Dawkins⟩

Charles Dar⟨win⟩

DAR 162: 124

1 See letter from W. B. Dawkins, 17 July 1869.

2 Dawkins refers to Gertrude Jane Mary Lloyd. CD had promised her sister-in-law, Mary Charlotte Lloyd, to obtain from Dawkins the English names of the principal bones found at Perthichwareu on her family's estate at Rhagatt (see letter from W. B. Dawkins, 17 July 1869 and nn. 2 and 3).

3 Thomas Henry Huxley referred to skulls of the earliest neolithic humans as 'The River-bed' type because they were found in the lower strata of the ancient river-beds (see T. H. Huxley 1862).

4 Dawkins reported on the discovery of human skulls at Perthichwareu and other Denbighshire caves at the British Association for the Advancement of Science meeting of 1870 (see Dawkins and Busk 1870 and Dawkins 1870).

5 The reference is to Anna Williams-Wynn and to the Plas-yn-Cefn estate (see Lucas 2007, pp. 327–8). Dawkins also refers to *Rhinoceros leptorhinus*, the slender-nosed rhinoceros, and *Ursus arctos*, the grizzly bear.

6 Dawkins refers to his paper 'On the prehistoric Mammalia of Great Britain' (Dawkins 1868, pp. 288–9). Dawkins argued that the Chillingham cattle were related to *Bos longifrons*. CD scored the relevant passages in his copy, which is in the Darwin Pamphlet Collection–CUL. The Chillingham cattle were kept on the Chillingham Park estate of Charles Augustus Bennet, sixth earl of Tankerville (see Ritvo 1992). They closely resembled the fossil form, *Bos primigenius* (the urus), a species domesticated in Switzerland during the Neolithic period. In *Variation* 1: 81, CD cited Ludwig Rütimeyer for the information that the

Chillingham cattle were less altered from the 'true primigenius type' than any other known breed and described the Chillingham cattle as 'semi-wild' descendants of *B. primigenius*, though 'much degenerated in size'; see also *Variation* 2: 119.

7 Hywel Dda was a Welsh king who ruled during the first half of the tenth century (*ODNB*).

8 The Danish Viking invasion of Ireland started in the ninth century; Vikings remained until early in the eleventh century (Somerset Fry and Somerset Fry 1988, pp. 47–57).

9 Dawkins refers to his conclusions in an article, 'On the fossil British oxen. Part I. *Bos urus*, Caesar' published in the *Quarterly Journal of the Geological Society of London* (Dawkins 1866, p. 401).

10 The section on the origin of domestic cattle was altered in *Variation* 2d ed., 1: 85–9, to include Dawkins's hypothesis.

11 Dawkins refers to *Descent*.

To *Scientific Opinion* [before 20 October 1869][1]

I could say something in answer to most of Delpino's ingenious criticisms, but I will confine myself to one point; remarking, however, that I cannot see the force of his objection to the belief that the gemmules have the power of self-division.[2] The analogy of the multiplication within the body of an infected person of the contagious atoms of small-pox or scarlet-fever still seems to me fair.[3] The criticism which at first struck me as much the most forcible, relates to the re-growth of an amputated limb or other organ.[4] I have assumed, from what we know of the ordinary laws of growth, that gemmules become developed only when united with the nascent or very young cells which precede them in the due order of successive development. Now, Delpino remarks that when the limb of a mature salamander is cut off, nascent cells would not exist on the stump, but only mature ones; consequently, as he urges, the gemmules, according to the hypothesis, could not be there developed and form a new limb. But have we any reason for assuming that with animals the same cells or organic units endure throughout life and are not replaced by new cells? All physiologists admit that the whole frame is being constantly renovated. The bones, which it might be thought would be least capable of renewal, undergo before maturity great changes in form and size and in the diameter of the medullary canal; even after maturity their shape does not remain the same. If, then, it be admitted that the tissues throughout the body are constantly being renovated by means of the old cells giving birth through proliferation or division to new cells, the difficulty of the gemmules being developed at any point at which a limb may be cut off, disappears; for some nascent cells would exist there; and we must bear in mind that gemmules may possibly become developed in union with cells which are not strictly the proper ones, as shown in the extreme case of nails occasionally growing on the stumps of the amputated fingers of man. It is probable that the renovation of the tissues by the formation of new cells becomes less and less in old age, and so does the power of regrowth. The curious fact that insects which live but a short time after their final metamorphosis, have singularly little power of regrowth, may be partly explained by the deficiency in their tissues of nascent cells.

In the case of plants, the cells composing the tissues endure for life, and are not renewed. The woody cells in the trunk of a tree originally existed as delicate cells in the young shoot. Now, it is remarkable, considering how low plants stand in the scale of life, what little power of re-growth most of their organs or parts possess, not withstanding their high vitality. Cut off a young shoot, and the next bud below will form a new shoot; cut off an old stem, and hundreds of adventitious buds will be formed. No one, I believe, ever saw a mutilated leaf repaired in the same manner as the limb of a salamander; yet on the cut margins of certain leaves innumerable buds will be quickly developed. In the bark, on the other hand, new cells must be continually forming, for the outer layers are continually being thrown off, and bark can certainly reproduce itself on a decorticated surface (see Professor Oliver in *Transactions of Tyneside Club*, vol. iii. 1855, p. 67),[5] from an isolated point not connected with the surrounding uninjured bark; so that this case is fairly analogous with that of the re-growth of an amputated limb.

I will add only one other remark. Delpino insists that, according to the hypothesis, the same cells must throw off gemmules at many successive periods of their development.[6] But I can see no such necessity. Certain epidermic cells, for instance, form horn; but their conversion into this substance is apparently due to the chemical nature of the contents of the cells; and of the changes consequent on the absorption of certain elements. An atom or gemmule thrown off at an early period from one of these cells would, when properly nourished, first form a cell like the parent cell, and ultimately be converted into horn, without the necessity of throwing off successive gemmules. As, however, a cell is a complex structure, with its investing membrane, nucleus, and nucleolus, a gemmule, as Mr. G. H. Lewes has remarked in his interesting discussion on this subject (*Fortnightly Review*, Nov. 1, 1868, p. 508), must, perhaps, be a compound one, so as to reproduce all the parts.[7] The attack by Delpino on the weak point of the hypothesis, namely, that it requires the support of several subordinate hypotheses, is perfectly just, though perhaps pushed to an extreme. But I can speak from recent experience, that he who has to consider complex cases of inheritance, as limited either separately or conjointly by sex, age, and season, with the inherited characters themselves and the form of inheritance liable to change from crossing and variability, will be able to disentangle the phenomena much more clearly, if he admits for the time our hypothesis with all its imperfections. He will then have fixed in his mind that transmission and development are quite distinct phenomena,—the gemmules being thrown off from all parts of the organization, and transmitted from both parents to both sexes at the earliest age, their development alone depending on the nature of the nascent tissues of the individual, whether permanently or temporarily modified by sex, age, season of the year, or other conditions.

Scientific Opinion 2 (1869): 426.

1 The date is established by the date of publication of the letter in *Scientific Opinion*, 20 October 1869.

2 Federico Delpino's criticisms of pangenesis, CD's hypothesis of heredity, were first published in Italian and subsequently translated into English, appearing in parts in *Scientific Opinion* (Delpino 1869a and

1869b). Delpino had argued that self-division in gemmules would be contrary to their nature as dormant or latent corpuscules analogous to seeds or buds (Delpino 1869b, p. 392). CD scored the passage and marked it with an 'x' in his copy of the article, which is in the Darwin Pamphlet Collection–CUL. CD discussed Delpino's criticisms in *Variation* 2d ed., 2: 380–1.

3 See Delpino 1869b, p. 392.

4 See Delpino 1869b, p. 407. Delpino argued, in the case of limb regeneration, that CD had not explained how the cells or tissue surrounding a wound (which were already developed) could recover the nascent state in which new gemmules could be produced.

5 CD refers to Daniel Oliver and to Oliver 1855.

6 See Delpino 1869b, p. 366.

7 CD refers to George Henry Lewes and Lewes 1868.

From J. V. Carus 20 October 1869

Leipzig,
Octobr 20. | 1869

My dear Sir,

Mr Koch (Schweizerbart) begs me to ask you when your new book on Man will be out. In a newspaper he found a notice, that it is nearly finished.[1] At the same time he told me from one of your letters to him, that you would not pledge yourself to give him the right of translation, as that would depend upon whether the book is worth translating and upon whom he may employ for translating it.[2] With regard to the first of these two conditions I dare say it would be a pity if the German public would not be put in a position to know all you write and have written. I must not say more, as you know how thoroughly I admire your method and way and how profoundly I appreciate the results which you arrived at by so wonderfully patient and many sided researches

Another question is, whom you would like to have as translator. I should not offer myself if I knew that you had any and even the slightest reason to object against my doing it. So pray don't think yourself bound. But if you think that I may do it, I shall be proud to be allowed to do it.

The short notice I read tells that the book will be published next year. I shall come to England in the beginning of March. If your book is not out before that time I may be able to speak with you about it. At all events I shall try to introduce myself to you personally.[3]

Believe me | My dear Sir | Ever yours faithfully | J. Victor Carus

DAR 161: 72

1 Carus refers to Eduard Koch, the publishing firm E. Schweizerbart'sche Verlagsbuchhandlung, and *Descent*. An advance notice of the publication of *Descent* had been published in the 9 October 1869 issue of the *Academy* (pp. 15–16).

2 Carus may refer to the letter to Eduard Koch, 13 December [1867] (*Correspondence* vol. 15).

3 Although Carus was in England in the spring of 1870, he was not able to arrange a meeting with CD (see *Correspondence* vol. 18, letters from J. V. Carus, 16 April [1870] and 18 April 1870).

To T. H. Farrer 20 October [1869][1]

Down. | Beckenham | Kent. S.E.
Oct. 20

My dear Farrer

Your notes strike me as good & I agree with Hooker that they are quite worth publication; but I wish you could first have examined more species.—[2] Especially Passiflora princeps, a splendid species, (as I believe called) in which, as I remember (for I cannot find my notes wh. I suppose I made) there were some (3?) *regular rounded passages* or holes through the corona to the nectary. The nectary, if my memory serves me, was constructed like that of Tacsonia.[3]

My *Tac: van volxenis* has produced, without any aid in fertilisation, plenty of fruit with an abundance of apparently good seed: the fert: is effected by each of the three stigmas slowly curving upwards (flowers being pendent) & in thus moving they generally touch one of the anthers of same flower.—[4]

See Hildebrand in pamphlet last sent, on abortion of stamens in uppermost or final dichogamous flower in Geranium:[5] this seems an *eminently* interesting case of abortion— attend to other such cases— see how is the *last* flower in reference to its capacity for being fertilised.—

Yours very sincerely | C. Darwin—

You ought to set up in cool Hothouse & Greenhouse.— Could you not go to Kew & see if any Passifloras are now in flower?[6]

Linnean Society of London (LS Ms 299/10)

1 The year is established by the relationship between this letter and the letter from T. H. Farrer, 17 October 1869.

2 For Farrer's notes on *Passiflora* and *Tacsonia*, see Appendix IV. Farrer discussed *Passiflora princeps*, but noted that he had not seen many specimens. Farrer received specimens of some passifloras, but not *P. princeps*, from Joseph Dalton Hooker (letter from J. D. Hooker to T. H. Farrer, 23 October 1869, Linnean Society Archives). Farrer evidently added more information on this species after visiting the Royal Botanic Garden, Kew, in November 1869 (the additions are not transcribed in the appendix, but are in the Linnean Society of London, MSS Case 6B, MS. 510).

3 The reference to the structure of the nectary in *Passiflora princeps* and *Tacsonia* has not been found in CD's notes.

4 *Tacsonia van volxemis* is now *Passiflora antioquiensis*.

5 CD had evidently sent Farrer his copy of Friedrich Hildebrand's paper 'Weitere Beobachtungen über die Bestäubungsverhältnisse an Blüthen' (Further observations on the conditions of pollination in flowers; Hildebrand 1869). Hildebrand noted that in the protandrous *Geranium macrorrhizum* the earliest flowers had stamens aborted and concluded that this was because there was no use for fertile stamens when no developed stigmas were available (Hildebrand 1869, p. 480). CD scored this section in his offprint of the article, which is in the Darwin Pamphlet Collection–CUL.

6 See above, n. 2.

From Asa Gray 20 October 1869

Charlton House, Kew
Oct. 20. 1869

My Dear Darwin

If you receive about this time by post some bulbous little things they should be of

Drosera filiformis, which plant in sand & peat, and keep cold till near spring, when you can set them growing in greenhouse. I shall be curious to know if you detect any sensibility in the glandular hairs.[1]

We have filled up the time very busily indeed since our pleasant glimpse of you, and now near the end are busier still, We take steamer of 9th Nov.[2]

If ever you & Mrs. Darwin get up to London in these 2 or 3 weeks please let us know that we may see you before we go.

My wife unites in best regards to you & yours. She—remembering your taste for *American* literature—sends specimens of the coming school-books and nursery rhymes.

Ever Yours | A. Gray

DAR 165: 171

[1] In 1861, CD and Gray had discussed the sensitivity of *Drosera filiformis*, the threadleaf sundew. Gray had been unable to detect the motion of the glandular hairs that CD had previously observed with his specimen (see *Correspondence* vol. 9, letter to Asa Gray, 21 July [1861]).

[2] The Grays had visited Down on 28 August 1869 (see Emma Darwin's diary (DAR 242)). Asa and Jane Loring Gray returned to America in November 1869.

From J. B. Innes 20 October 1869

Milton Brodie | *Forres* | *N.B.*
20th. Octr. 1869

Dear Darwin,

Many thanks for your kind letter. I find that the Guardian which contains Mr. Hutton's paper has not yet gone to the fire, so I send it to you by this post with this line to say don't send it back as I make no use of papers but to burn—[1]

It will be an advantage to have Mr. Powell in Downe rather than at Bromley.[2] I had not heard of any matrimonial intentions. Perhaps he has been visiting you and allowed your example to keep out of sight all the awful warnings; poor Henry Lubbock dragged to London when he wants to hunt; myself tied down when I want to go and see the Sultan with the Empress &c &c[3]—but if he really contemplates the rash act he will surely build a parsonage £200 ready & buy land £700 from Eccles Commission, probably £200 from Church building Society leaving £500 for him to pay beg or borrow; the extreme rate of interest and insurance being no more than he would pay for a bad cottage.[4] I do not think there would be any economy in reducing the size and cost, a very comfortable small house would be well built for £1400, and as the Commission give half up to this, every £50 saved to Powell would be £100 off the building. You are aware that he first agreed with me to provide the £700 at once, and when it appeared his available means were not equal to doing this immediately he engages to do it as early as possible. I have no doubt of its being done, my dread is that somehow the site may be missed, as was so often the case with me. If Miss Wedgwood will take the £200 from me, I will buy it at once, and hold it till wanted.[5] It seems to me so much to Powell's advantage to proceed that if he does not I shall agree with you that he is not above the average in sense. It is rather Irish, but one fancies the large majority are rather below an average— Anyhow please do you and Miss Darwin,[6] who are engaged as Counsel on my side look out that the site does not escape the Church—

We have a sudden outbreak of cold, and the hills are covered with snow. It will not be very pleasant to look on at my tenants' ploughing match today in the Cold North wind, but we hope the Indian Summer is to follow.

With our kindest regards to you all | Believe me | Faithfully Yours | J Brodie Innes

DAR 167: 26

[1] See letter to J. B. Innes, 18 October 1869 and n. 6. Innes refers to Richard Holt Hutton and to the weekly church newspaper, the *Guardian*, which contained the text of Hutton's paper in the 13 October 1869 issue.

[2] The reference is to Henry Powell. See letter to J. B. Innes, 18 October 1869.

[3] See letter to J. B. Innes, 18 October 1869 and n. 5. Innes refers to Henry James Lubbock, Sultan Abdülaziz of the Ottoman Empire and Empress Eugénie of France. The empress paid a visit to the sultan from 16 to 18 October 1869 (*The Times*, 18 October 1869, p. 7).

[4] For more on Powell's attempts to raise money for a parsonage, see the letter from John Lubbock, 20 July [1869] and n. 3. Innes refers to the Ecclesiastical Commission.

[5] Innes refers to Sarah Elizabeth Wedgwood, who had purchased Tromer lodge, a property in the centre of Down village, about one mile from Down House. Innes had already expressed an interest in acquiring a part of the land for a parsonage (see *Correspondence* vol. 16, letter from J. B. Innes, 30 January 1868).

[6] Innes refers to Henrietta Emma Darwin.

From J. J. Moulinié 20 October 1869

Geneva
20th October 1869.

Dear sir,

I learn through a German paper that a new work of your's on Man, I believe, is to be published shortly.[1] This is a good news for the scientific public, and if it be correct, may I solicit the high favour of undertaking a french translation of your new contribution to questions which your preceeding works have discussed in such a remarkable manner, and rendered popular in all countries; I consider it a great honour to continue to be "Darwin's translator", and will therefore be very thankful to you for the permission to act in consequence.[2]

Hoping this will find you in a satisfactory state of health, remember me kindly to Mrs. Darwin and your whole family,[3] and believe me, dear sir | yours truly | J. J. Moulinié

DAR 171: 273

[1] The reference is to a notice in the 9 October 1869 issue of the *Academy* (pp. 15–16) that had evidently been reported on in an unidentified German paper. The 'work' was *Descent* and *Expression.*

[2] Moulinié had translated *Variation* into French (Moulinié trans. 1868).

[3] According to Emma Darwin's diary (DAR 242), Moulinié visited CD on 31 August 1868.

From A. R. Wallace 20 October 1869

9, St. Mark's Crescent, N.W.
Octr. 20th. 1869.

Dear Darwin

I do not know your son's (Mr G. Darwin's) address at Cambridge. Will you be so good as to forward him the enclosed note begging for a little information.[1]

I was delighted to see the notice in the "Academy" that you are really going to bring out your book on Man.[2] I anticipate for it an *enormous* sale, and shall read it with intense interest, although I expect to find in it more to differ from than in any of your other books.

Your reasonable & reasoning opponents are now taking the field. I have been writing a little notice of Murphy's book "Habit & Intelligence",—which with much that is strange & unintelligible contains some very acute criticisms and the statement of a few real difficulties.[3]

Another article just sent me from "the Month"—contains some good criticism How incipient organs can be useful is a real difficulty, so is the independent origin of similar complex organs;—but most of his other points though well put are not very formidable.[4]

I am trying to begin a little book on the "Distribution of Animals", but I fear I shall not make much of it from my idleness in collecting facts.[5] I shall make it a popular sketch first, and if it succeeds gather materials for enlarging it at a future time.

If any suggestion occurs to you as to the kind of *maps* that would be best, or on any other essential point, I sh[d]. be glad of a hint.

I hope your residence in Wales did you good. I had no idea you were so near Dolgelly till I met your son there one evening when I was going to leave the next morning.[6] It is a glorious country but the time I like is May and June—the foliage is so glorious.

Sincerely hoping you are pretty well & with kind regards to Mrs. Darwin & the rest of your family | Believe me | Yours very faithfully | Alfred R. Wallace—

C. Darwin Esq.

DAR 106: B86–7

CD ANNOTATIONS
3.1 I have ... difficulties. 3.4] *scored red crayon*

[1] Wallace refers to George Howard Darwin. The enclosure has not been found, but evidently was a request by Wallace for George's opinion of a book by James Challis, *Notes on the principles of pure and applied calculation: and applications of mathematical principles to theories of the physical forces* (Challis 1869). George's response to Wallace, dated 23 October 1869, is in the British Library (Add.46434, f. 191).

[2] The reference is to *Descent* (see *Academy* 1 (1869–70): 15–16). The notice appeared in the 9 October 1869 issue under the heading 'Scientific notes' and described the book as consisting of three parts, descent of man, sexual selection, and expression of the emotions.

3 Wallace refers to Joseph John Murphy and Murphy 1869. Wallace reviewed Murphy 1869 in the 25 November 1869 issue of *Nature* (Wallace 1869d).

4 The unsigned article, 'Difficulties of the theory of natural selection' (Anon. 1869c) appeared in three parts from July to September in the *Month*, a periodical owned by the Jesuits (*ODNB* s.v. Coleridge, Henry James). For the author's argument concerning incipient organs, see Anon. 1869c, pp. 42–8.

5 Wallace refers to *The geographical distribution of animals* (Wallace 1876).

6 CD had visited Caerdeon, Barmouth, Wales, in June and July 1869 (see 'Journal' (Appendix II)). Although all of CD's sons spent some time in Wales during CD's visit, it is likely that Wallace refers to George, who arrived on 16 June (Emma Darwin's diary (DAR 242)).

To A. R. Wallace 21 October 1869

Down. | Beckenham | *Kent. S.E.*
Oct. 21. 1869

My dear Wallace

I forwarded your letter at once to my son George, but I am nearly sure that he will not be able to tell you any thing; I wish he c^{d} for my own sake; but I suspect there are few men in England who could.[1] Pray send me a copy or tell me where your article on Murphy will be published.[2] I have just rec^{d} "the month" but have only read half as yet. I wish I knew who was the author; you ought to know as he admires you so much; he has a wonderful deal of knowledge, but his difficulties have not troubled me much as yet, except the case of the dipterous larva.[3] My book will not be published for a long time, but Murray wished to insert some notice of it.[4] Sexual selection has been a tremendous job. Fate has ordained that almost every point on which we differ sh^{d} be crowded into this vol.[5] Have you seen the Oct. No. of Rev. des deux mondes? It has an article on you, but I have not yet read it; & another art. not yet read by a very good man on the Transformist school.[6]

I am very glad to hear that you are beginning a book, but do not let it be "little", on distribution &c— I have no hints to give about maps; the subject w^{d} require long & anxious consideration.[7] Before Forbes published his essay on distribution & the glacial period I wrote out & had *copied* an essay on the same subject, which Hooker read.[8] If this M.S. w^{d} be of any use to you, *on account of the references* in it to papers &c I sh^{d} be very glad to lend it to be used in any way; for I forsee that my strength will never last out to come to this subject.

I have been pretty well since my return from Wales, tho' at the time it did me no good.[9]

We shall be in London next month when I shall hope to see you—[10]

My dear Wallace | yours very sincerely | Ch. Darwin

LS(A)
British Library Board (Add 46434 f. 189–90)

1 CD refers to George Howard Darwin. See letter from A. R. Wallace, 20 October 1869 and n. 1.

2 See letter from A. R. Wallace, 20 October 1869 and n. 3. The reference is to Joseph John Murphy and Murphy 1869.

[3] CD refers to an unsigned article in the *Month* (see letter from A. R. Wallace, 20 October 1869 and n. 4). The author argued that the development of certain flies from grub to adult could not be explained by natural selection (see Anon. 1869c, p. 47). The passage is scored in CD's copy of the article, which is in the Darwin Pamphlet Collection–CUL.

[4] See letter from A. R. Wallace, 20 October 1869 and n. 2. CD refers to *Descent.*

[5] For more on CD and Wallace's differences on sexual selection, see *Correspondence* vols. 15 and 16.

[6] CD refers to two articles published in the 1 October 1869 issue of *Revue des Deux Mondes*, one a review of Wallace 1869a (Randau 1869) and the other a review by Gaston de Saporta of three works, including *Variation* (Saporta 1869).

[7] See letter from A. R. Wallace, 20 October 1869 and n. 5.

[8] CD refers to Edward Forbes and E. Forbes 1846. In 1856, CD sent Joseph Dalton Hooker a fair copy of his manuscript on geographical distribution (see *Correspondence* vol. 6, letter to J. D. Hooker, [16 October 1856] and n. 1). The manuscript is in DAR 14; see also *Natural selection*, chapter 11.

[9] CD refers to his stay in Wales during June and July (see 'Journal' (Appendix II)).

[10] CD was in London from 1 to 9 November 1869 (see 'Journal' (Appendix II)).

From Emile Alglave[1] 22 October 1869

Paris
le 22 octobre *1869*

Monsieur

J'ai appris que vous alliez publier dans quelques mois un nouvel ouvrage qui comprendrait trois parties, I, Descent of Man.— II, On Sexual Selection, III Expression of the emotions.[2]

Je desirerais vivement qu'une partie de cet ouvrage peut etre traduite pour la *Revue des cours scientifiques.*[3] Par la même occasion l'editeur de la Revue des cours serait très heureux de publier à sa librairie la traduction de l'ouvrage complet.[4] Cette combinaison faciliterait les deux choses et assurait à votre ouvrage en France une publicite immédiate très considerable, la Revue des cours scientifiques tirant *à elle* seule presque autant que tous les autres periodiques scientifiques francais ensemble. La traduction francaise pourrait etre faite sur les feuilles de l'ouvrage anglais au fur et à mesure du tirage de manière à paraitre presque en même temps, et meme avant par fractions, dans la Revue des cours scientifiques.— Dans tous les cas, si vous n'acceptiez pas cette combinaison pour la publication francaise de l'ouvrage complet, rien n'empêcherait de traduire tout de même pour la Revue des cours scientifiques quelques uns des chapitres de cet ouvrage

Agreez Monsieur l'expression de mes sentiments les plus distingués | Em Alglave | directeur de la Revue des cours scientifiques

DAR 159: A35

[1] For a translation of this letter, see Appendix I.

[2] Alglave had evidently seen the notice that appeared in *Academy* 1 (1869–70): 15–16 of *Descent* and *Expression.*

[3] *Revue des cours scientifiques de la France et de l'étranger* was a weekly journal that reported on the lectures of prominent European scientists. Excerpts from *Descent* eventually appeared in the journal in 1871.

[4] The publisher of *Revue des cours scientifiques* was J. B. Baillière et fils.

To J. V. Carus 22 October [1869][1]

Down. | Beckenham | *Kent. S.E.*
Oct 22[d]

My dear Sir

I thank you for your very kind letter. My book will not be ready for, probably, a year, as my health lets me work so very slowly & I have found sexual selection a most laborious subject when treated in detail. My publisher wished so much to insert a notice of my Book that I could not refuse.[2]

There is no man in Germany I sh[d] so much wish to translate any book by me as yourself, for I sh[d] then feel quite safe.— I demurred to agreeing at once to M. Koch, as I had formerly been more annoyed, than you will readily believe, by thinking that my last Book would prove not worth translating.[3] Moreover it has occurred to me, but on that head I must make enquiries, that I ought to receive some small payment for the right of Translation, as my Books seem to sell well.[4] But as I have said I must make enquiries on that head.— Already one publisher in Germany has made offers to me but I gave him no definite answer.—

I shall be *delighted* to make your personal acquaintance when you come to England.[5]

Believe me | My dear Sir | Yours sincerely | Ch. Darwin

Staatsbibliothek zu Berlin – Preußischer Kulturbesitz, Slg. Darmstaedter Lc 1859 Darwin, Charles, Bl. 43–45

1 The year is established by the relationship between this letter and the letter from J. V. Carus, 20 October 1869.

2 See letter from J. V. Carus, 20 October 1869 and n. 1. CD refers to *Descent* and John Murray.

3 Eduard Koch was the head of E. Schweizerbart'sche Verlagsbuchhandlung. For more on CD's doubts about the translation of his earlier book, *Variation*, see *Correspondence* vol. 15, letter to E. Schweizerbart'sche Verlagsbuchhandlung, [19 March 1867].

4 CD had given his German publisher the right to translate *Variation* and earlier works free of charge but reserved the rights to future works. See *Correspondence* vol. 15, letter to Eduard Koch, 13 December [1867].

5 Carus was unable to visit CD when he was in England in the spring of 1870 (see *Correspondence* vol. 18, letter from J. V. Carus, 18 April 1870).

To J. J. Moulinié 23 October [1869][1]

Down. | Beckenham | *Kent.*
Oct. 23[d]

My dear Sir

My Book will not be published for a very considerable time, but the Publishers wished & insisted on inserting a notice.—[2]

I shall in all probability be delighted if you will translate my new Book, when ready. But I am as yet quite ignorant on the subject, yet it strikes me that I ought to receive some payment from the Publisher in Paris for the right of Translation; & I intend to make enquiries whether this is ever done; but subject to this doubt, I repeat that I shall be very happy if you will undertake the Translation.—[3]

Can you procure for me information on one point: if I give the right of Translation for the first edition, & then make considerable corrections & additions to future editions, am I precluded from giving the right of Translation to some fresh publisher of such new & corrected edition? I ask, because I think I have been badly treated by M. Masson & Mad^elle^ C. Royer (for whom I took great trouble); they have just brought out a 3^d^ French Edit. of my Origin without informing me; consequently the numerous corrections & additions in the 4th & 5th English editions are all lost.[4] Under these circumstances I sh^d^ much wish for a new Edition of the Origin in France, not including Mad^elle^ Royer's Preface (with a second one made as injurious to me as she could) & which first Preface I have been assured has injured the circulation in France.—[5] I believe in half a years time or in a whole year it w^d^ answer to any publisher to bring out a *corrected* French edition, if it is legal in France, or in Belgium.— Pray turn this in your mind; & I w^d^ gladly give you such right of Translation, & inform M. Masson of my motives.—[6]

My dear Sir— | Yours sincerely | Ch Darwin

Bibliothèque de Genève, Ms. suppl. 66, f. 11–12

[1] The year is established by the relationship between this letter and the letter from J. J. Moulinié, 20 October 1869.

[2] See letter from J. J. Moulinié, 20 October 1869 and n. 1.

[3] CD had also inquired about payment from his German publisher (see letter to J. V. Carus, 22 October [1869]).

[4] The references are to Clémence Auguste Royer and Victor Masson and to Royer trans. 1870. CD had evidently already seen a copy of the book and written to Masson informing him that a fifth edition of *Origin* had appeared (see letter from Victor Masson, 29 September 1869). CD's copy of Royer trans. 1870 is in the Darwin Library–Down House.

[5] CD probably refers to comments such as the ones he received from Edouard Claparède about the first edition of Royer's translation of *Origin* (see *Correspondence* vol. 10, letter from Edouard Claparède, 6 September 1862). In the preface to Royer trans. 1870, Royer had criticised CD's hypothesis of heredity, pangenesis, claiming that it had 'done wrong' to his theory of transmutation by natural selection (quoted in Harvey 1997, p. 98).

[6] No letter to Masson has been found, but see the letter to John Murray, 8 November [1869]. Moulinié's translation of *Origin* 6th ed. appeared in 1873 (Moulinié trans. 1873).

From Asa Gray 26 and 29 October [1869][1]

Charlton House, Kew
Oct. 26th

My Dear Darwin

As you are to be up in London next Monday, and we leave it the following Monday, we must contrive to meet in the course of that final week.[2] I will write in time when we can arrange our precious days, unless you can fix upon the day that suits you best and apprize us.

Oct. 29.

Even now I am quite unable to say what day next week we can look in upon you in Queen Anne St.[3]

I will hand the slip with question to Agassiz.[4]

His son Alexander, with wife and children are over here—when last heard of at Dublin, soon to be in London; but we shall very likely miss them.[5] The *Drosera filiformis* will probably yet come.[6]

In haste | Ever Yours | Asa Gray

DAR 165: 172

[1] The year is established by reference to the dates that CD would be in London, and Asa Gray would still be in England (see n. 2, below).

[2] The Darwins went to London for a week on Monday 1 November 1869 (CD's 'Journal' (Appendix II)). The Grays were leaving England on Tuesday 9 November 1869 from Liverpool, and left their accommodation at Kew the day before (J. L. Gray ed. 1893, 2: 597).

[3] While in London, the Darwins stayed with Erasmus Alvey Darwin at 6 Queen Anne Street.

[4] CD's question to Louis Agassiz has not been found; however, see Agassiz's reply in the enclosure to the letter from Asa Gray, 27 February and 1 March 1870 (*Correspondence* vol. 18). CD had evidently asked about a particular fish when attending to its nest. For CD's interest in this topic, see *Descent* 2: 19–22.

[5] Alexander Agassiz and his wife, Anna Russell Agassiz, had two young sons (Dobbs 2005, p. 65). Leaving the children in Malvern, Alexander and Anna travelled to Scotland and Belfast, returning for a stay in London from 11 November to 3 December 1869 (G. R. Agassiz ed. 1913, p. 98).

[6] See letter from Asa Gray, 20 October 1869.

From Carl Vogt[1] 26 October 1869

Genève
ce 26 Oct 1869

Monsieur et cher Maitre!

Monsieur Moulinié vous aura écrit, je pense, au sujet de la traduction française de votre livre sur l'homme, dont nous avons lu l'annonce dans "the Academy."[2] Je n'ai donc pas besoin de revenir là-dessus—je pense que vous avez été content de la traduction qu'il a faite et il mettra autant de soins à ce livre comme à celui sur la domestication.[3]

Aujourd'hui, je reçois une lettre de mon éditeur, Mr. J. Ricker à Giessen;[4] qui me demande si je ne veux pas me charger de la traduction en allemand de votre ouvrage à paraitre sur l'homme—il promet de mettre tous ses soins à une édition digne de vous et je crois que sous ce rapport Mr. Ricker a fait ses preuves—tout comme je ferais en sorte, que la traduction soit accomplie. Malgré mes nombreuses occupations, je donnerais à ce travail de préférence tous mes soins.

Vous me repondîtes, lorsque je vous écrivais au sujet de la traduction de votre livre sur la domestication, que vous étiez engagé avec Mrs. Càrus et Schweizerbart.[5] Je ne sais, si ces engagements continuent pour votre nouveau livre et si je ne crains pas de dire que Mr. Ricker fera mieux et beaucoup mieux que Schweizerbart, je ne voudrais en aucune façon aller sur les brisées de Mr. Càrus, digne et brave garçon, qui a eu le grand malheur d'avoir été écarté de la chaire de zoologie à Leipzig et cela certainement en grande partie parcequ'il a professé hautement vos doctrines au lieu de suivre l'ornière officielle comme son concurrent victorieux Mr. Leukart.[6] Mr. Carus m'avait dit, l'année passée à Norwich,[7] qu'il ne ferait plus de traductions—je pense que c'était dans l'espoir d'être nommé à l'Ordinariat— je ne sais si dans les circonstances actuelles, si fâcheuses pour lui, il n'aura changé de résolution. Mais je crois en tout cas de mon devoir de m'abstenir en cas où il vous aurait manifesté le désir de faire la traduction de votre nouvel ouvrage et si cela était, mettez que je n'ai rien dit.

J'ai été bien jaloux, l'année passée, de Mr. Moulinié, qui a pu faire votre connaissance personelle, tandisque des circonstances indépendantes de ma volonté m'en empêchaient. Mais j'espère que le proverbe allemand "Aufgeschoben ist nicht Aufgehoben" trouvera ici son explication[8] et dans cet espoir je reste | Votre devoué | C Vogt

DAR 180: 13

[1] For a translation of this letter, see Appendix I.

[2] See letter from J. J. Moulinié, 20 October 1869 and n. 1. The reference is to *Descent.*

[3] Vogt refers to *Variation* and to Moulinié trans. 1868.

[4] Vogt refers to the firm J. Ricker'sche Buchhandlung.

[5] See *Correspondence* vol. 15, letter to Carl Vogt, 12 April [1867]. The references are to Julius Victor Carus and the firm E. Schweizerbart'sche Verlagsbuchhandlung.

[6] For more on the appointment committee's comments on the three leading candidates for the chair of zoology at Leipzig, see Nyhart 1995, p. 173; on Rudolf Leuckart's appointment to the post and his approach to zoology, see *ibid.*, pp. 310–11.

[7] Vogt evidently met Carus at the thirty-eighth meeting of the British Association for the Advancement of Science, held at Norwich in August 1868.

[8] According to Emma Darwin's diary (DAR 242), Moulinié visited CD on 31 August 1868. *Aufgeschoben ist nicht Aufgehoben*: put off is not called off.

From T. H. Farrer 27 October 1869

Abinger Hall, | Reigate. (Post Town) | Gomshall (Station) S.E.R.
27 Oct/69

My dear Mr Darwin

I will certainly try to get Passiflora Princeps—

Hooker has sent me a number of flowers from Kew but none of P. pr.—[1] I find no very great differences—but one or two interesting details eg (1) that the length of the upright middle corona is proportioned to the height, position & apertures of the inner one, so as to lead the bee over the knobs of the former to the apertures in the latter:

(2) the apertures in the latter seem also to be correlated to the interior projections, so as to render a straight passage to the nectar impracticable.

(3) as a matter of morphology I think that the process round the column is a first corona: making with the three developed coronas, four in all, whilst the process which forms the nectary in Passiflora, seems to be the base of the next interior corona.

I remarked with interest what Hildebrand says about geraniums— There is a passage about it at p. 28 of his book which you recommended me last year in which he ingeniously suggests that the absence of stamens in the disk flowers of Corymbifers may, be due to this cause—[2] And I see that I made a note on it suggesting that it was doubtful. considering that in such flowers as Senecio—or Calendula there are an immense number of blossoms lasting throughout the year: and that unless we assume fertilization to take place in the same flower-head, (which is contrary to the hypothesis) there can be very few disk flowers which are not contemporaneous at first opening with mature pistils in other flowers.— This does not touch the case of a specially early or late Geranium or Campanula[3]

Sincerely yours | T H Farrer

C Darwin Esq[r] FRS

DAR 164: 59

[1] See letter to T. H. Farrer, 20 October [1869]. For Farrer's notes on *Passiflora princeps*, see Appendix IV.

[2] Corymbiferae was a former suborder of composite plants in which the florets of the disc were tubular and perfect while the ray florets, when present, were ligulate (*OED* s.v. corymbiferous). Farrer refers to Friedrich Hildebrand and Hildebrand 1867, pp. 28–9, where Hildebrand noted that the majority of corymbifers had ray florets that were female and disc florets that were protandrous. He further noted that since the ray florets opened first, there would be no need for these to have viable stamens and that over time the stamens must have first aborted, then disappeared altogether.

[3] *Senecio* (the genus of groundsels and ragworts) and *Calendula* (marigold) were in the family Compositae (now Asteraceae); *Campanula* (bellflowers) is in the family Campanulaceae; *Geranium* (cranesbills) is in the family Geraniaceae.

To J. V. Carus [30 October 1869][1]

P.S. I rec[d] this morning since the above was written a letter from Carl Vogt, expressing his readiness to translate my book at the request of the publisher Monsieur Ricker.[2] I shall write to him tomorrow to say that you have undertaken the work, & I mention the circumstance because if you sh[d] unfortunately for me find yourself, now that you know the size of my book, too much overwhelmed with your own work to complete the translation within any reasonable time, I should not be quite thrown over, as Schweizerbart[3] could perhaps arrange with Prof. Vogt. But I trust there may be no need for any change—

L

Staatsbibliothek zu Berlin – Preußischer Kulturbesitz, Slg. Darmstaedter Lc 1859 Darwin, Charles

[1] The date is established by the relationship between this letter and the letter to J. V. Carus, 30 October 1869.

[2] The letter to which this particular postscript was appended has not been found. CD evidently decided to rewrite his letter to Carus after receiving the letter from Carl Vogt, 26 October 1869, but may have accidentally included this original postscript (which is in Emma Darwin's hand) with the second letter. In the Staatsbibliothek zu Berlin, it has been included with the letter to Carus of 22 October [1869]. CD refers to Franz Anton Ricker.

[3] CD refers to his German publisher, C. Schweizerbart'sche Verlagsbuchhandlung.

To J. V. Carus 30 October 1869

Down. | Beckenham | *Kent. S.E.*
Oct 30 1869

My dear Sir

I rec^{d} this morning a letter from Prof. Vogt proposing to translate my new book for Mr Ricker.[1] At the same time he tells me the very bad news that you are no longer Professor at Leipzig.[2] Under these circumstances Vogt suggests that you w^{d} perhaps have more time than before, & find it more convenient to translate my new book, & in that case he begs me to consider his proposition as never made.

Under these circumstances I shall be delighted if you will accept the right of translation, & you can apply to any publisher whom you like & make the best terms you can. As my books sell pretty well in Germany a publisher ought to give you a considerable remuneration for the right and labour of translation. When I know the size of my book, though I dare say this will not be for 9 or 12 months, I will let you know, to aid you in making any agreement. The book will not be very small & will contain a good many wood cuts.

Besides Mr Ricker mentioned above, Mr Duncker of Berlin, & Mr Brockhaus of Leipzig have applied to me for the right of translation.[3]

Pray believe me | my dear Sir | yours sincerely | Ch. Darwin

LS
Staatsbibliothek zu Berlin – Preußischer Kulturbesitz, Slg. Darmstaedter Lc 1859 Darwin, Charles, Bl. 46/47

[1] See letter from Carl Vogt, 26 October 1869. Franz Anton Ricker was the director of the publishing firm of J. Ricker'sche Buchhandlung. The 'new book' was *Descent*.
[2] See letter from Carl Vogt, 26 October 1869 and n. 6.
[3] CD refers to Alexander Duncker of Alexander Duncker Verlag and Heinrich Brockhaus of F. A. Brockhaus Verlag.

From Eduard Koch[1] 30 October 1869

Stuttgart
den 30· Oktober 1869

Hochgeehrter Herr!

Gleichzeitig mit diesem erlaube ich mir Ihnen ein Exemplar des eben erst fertig gewordenen

Dub, kurze Darstellung[2]

zu übersenden und glaube, dass das Werk sowohl nach Inhalt als auch nach Ausstellung Ihren Beifall haben wird; falls Sie noch mehr Exemplare wünschen so bitte es mir gefaell, mitzutheilen.

Aus einer englischen Zeitschrift "Academy" durchlaeuft jetzt die deutschen Zeitungen die Nachricht, dass im Laufe des naechsten Jahres ein neues Werk von Ihnen die Presse verlassen wird, dessen Inhalt den Menschen betreffen soll.[3]

Herr Professor Carus dem ich desshalb bereits geschrieben habe wird Ihnen wahrscheinlich bereits Mittheilung davon gemacht haben.[4]

In der angenehmen Hoffnung, dass auch dieses Werk wie die früheren in meinem Verlage Deutsch bearbeitet von H Carus erscheinen wird, bitte mich gefäll, zu benachrichtigen, ob ein neues Werk erscheint und welches der Inhalt desselben seyn wird.

Nachdem ich all Ihre Werke aufmerksam durchgelesen habe und bis jetzt nirgends eine Schlussfolgerung betreffs des Menschen darin vorkam, kann ich mir nur denken, dass die neueste Schrift des Herzog von Argyll über den Urmenschen Sie veranlasst hat auch diese Frage zu berühren.[5]

Augenblicklich wird an der vierten deutschen Auflage der "Enstehung der Arten" gedruckt und werde nicht ermangeln Ihnen sofort nach Erscheinen eine von Ihnen noch zu bestimmende Anzahl von Exemplaren zuzusenden.[6]

Indem ich die Gelegenheit ergreife *[Sie]* meine vollkommenste Hochachtung zu versichern | verbleibe ich Ihr | ganz Ergebenster | Eduard Koch
firma: E. Schweizerbart'sche Verl.

DAR 169: 45

[1] For a translation of this letter, see Appendix I.
[2] The reference is to *Kurze Darstellung der Lehre Darwin's über die Entstehung der Arten der Organismen: mit erläuternden Bemerkungen* (Brief description of Darwin's theory on the origin of species, with explanatory remarks; Dub 1870); see also letter to Julius Dub, 20 March 1869.
[3] Koch refers to *Descent* and to the notice in *Academy* 1 (1869–70): 15–16.
[4] The reference is to Julius Victor Carus. See letter from J. V. Carus, 20 October 1869.
[5] Koch refers to George Douglas Campbell and to *Primeval man* (G. D. Campbell 1869).
[6] The reference is to the fourth German edition of *Origin* (Bronn and Carus trans. 1870).

To Carl Vogt 30 October 1869

Down. | Beckenham | *Kent. S.E.*
Oct 30 1869

My dear Sir

I thank you very much for y^r letter & for your proposal about the translation, which I look at as a very great honour;[1] & I feel certain that it w^d have made my book much more widely known— But I have already partly agreed with Carus, & now that I hear for the first time about the loss of the professorship I will write at once & tell him what you have said, & will propose to him to make the best bargain he can with any publisher in Germany whom he may prefer. I am much obliged to you for giving me this information with so much kind sympathy for Carus.

As Leukart is so distinguished a man, I hope that Carus holding my views may be only a very secondary cause of the loss of the professorship.[2]

With my cordial thanks for y^{r} proposal about the translation, so honourable to me, & with the hope that I may some time have the real pleasure of making your personal acquaintance I remain my dear Sir | yours very faithfully | Charles Darwin

P.S. I may add that my book will not be ready for a considerable time, but my publisher insisted on a notice.[3]

LS
Bibliothèque de Genève, Ms. Fr. 2188, f. 298–9

1 See letter from Carl Vogt, 26 October 1869.
2 Julius Victor Carus had been a candidate for a professorship in zoology at Leipzig, but the position went to Rudolf Leuckart (see letter from Carl Vogt, 26 October 1869 and n. 6).
3 CD refers to John Murray and to the notice in the *Academy* 1 (1869–70): 15–16 of *Descent.*

To ? 30 October [1869 or 1870][1]

Down. | Beckenham | *Kent. S. E.*
Oct. 30th.

My dear Sir

The case of the Habenaria is very interesting as shewing how distant plants may be crossed; & I wish I had known of it before drawing up the notes of which I send a copy by this post.[2]

The Lychnis-case well deserves investigation; but I doubt whether the bi-sexual condition can be the result of crossing (tho' this w^{d} please me much;) for Gärtner crossed the two sp. repeatedly, & c^{d} hardly have failed to have observed in the hybrids a tendency to hermaphroditism.[3]

If the smaller grains of pollen are of *equal size & really sound,* I sh^{d} suspect a reversion to a dimorphic condition, like that of Primula. I have long had reason to suspect that some dioicous plants originate from the suppression of the reversed sexes in the 2 forms of dimorphic plants.[4] The enclosed diagram, which is little better than a hieroglyphic, will perhaps explain what I mean.[5]

Your hermaphrodite plants ought all to be transplanted into a garden, carefully observed, the small-grained pollen experimented on, with insects, of course, carefully excluded.

This is what I sh^{d} do if I had the plants, & I think that you w^{d} find it worth while to observe them with care—

Pray believe me | My dear Sir | yours very faithfully | Ch. Darwin

LS(A)
King Edward VI High School, Stafford

1 The date range is established by the form of the address with '*Bromley*' crossed out and 'Beckenham' added, which CD used between April 1869 and May 1871.
2 The notes CD sent have not been identified. CD's notes on the floral morphology of *Habenaria* (bog orchids) are in DAR 205.8: 9–10. The species of *Habenaria* studied by CD are now in the genera *Platanthera* and *Coeloglossum*, both known to hybridise intergenerically in nature (see *Orchids*, pp. 76–8, 83–92, and Dressler 1981, p. 189).

[3] *Lychnis* is the genus of campions. For Karl Friedrich von Gärtner's results in crossing different species of *Lychnis*, see Gärtner 1849, pp. 218–19 (see also *Natural selection*, pp. 393–4).
[4] For more on CD's interest in rudimentary sexual organs found in flowers of *Lychnis dioica* (now *Silene dioica*, red campion), see *Correspondence* vols. 9 and 10. In *Cross and self fertilisation*, pp. 410–11, CD discussed the possible causes for the separation of sexes in *L. dioica*.
[5] The diagram has not been found.

To Francis Henry Salvin?[1] 31 October 1869

Down. | Beckenham | *Kent.*
Oct 31. 69

⟨My dear⟩ Sir,

I am very much obliged to you for your kindness in sending me the extracts about the jackal, which are extremely interesting & curious.[2]

With my best thanks | I beg leave to remain | Yours faithfully | Charles Darwin

LS
McGill University Libraries, Rare Books and Special Collections Division

[1] The conjectured recipient is the author of the extracts on the jackal (see n. 2, below).
[2] CD refers to an article by Salvin, 'A domesticated jackal', that appeared in *Land and Water*, 16 October 1869, p. 249, and a reader's query with the author's response that appeared in *Land and Water*, 23 October 1869, p. 267. CD's annotated copies of the extracts are in DAR 53.2: 118.

From Federico Delpino[1] 1 November 1869

Firenze
1. 9bre 1869

Celeberrimo uomo!

Rendo grazie alla S.V. pella indulgente e cortese lettera ultimamente speditami, nonchè pel grazioso dono dell'interessante memoria del Dott. Crüger. La quale è scritta con tanta chiarezza e tanto discernimento, che ha scosso non poco i miei dubbii intorno alla realtà della fecondazione dei *Coryanthes* per intervento dell' *Euglossae*, e collo stranissimo processo *a bagno freddo* ivi descritto.[2] Non ostante mi resta ad approfondire maggiormente la razionalità delle forme florali dei *Coryanthes* e *Gongora*. Non dissimulo però fin d'ora che mi dà a pensare la circostanza che il liquido, il quale si raccoglie per stillicidio nel labello, *è segregato da due cospicue glandole nettarifere.* La natura è molto economa, e non mi contenta che un liquido, il quale avesse semplicemente la funzione indicata dal D^r. Crüger, debba venir segregato da un apparato secretorio e glandoloso.[3]

Mi è giunto singolarmente gradito il cenno datomi che la *Strelitzia*, giusta le ricerche della S.V., venga al Capo di Buona speranza fecondata da nettarinie.[4] Ciò viene a conferma delle mie congetture. Il Prof. Hildebrand suppose che la fecondazione di questa pianta venga effetuata dagli insetti: ma ciò a me parce meno verisimile, perocchè gl'insetti, anche di grossa taglia, non avrebbero la forza sufficiente per far divaricare i lobi della scatola che racchiude le antere.[5] Farò tesoro di questo suo cenno, e, se la S.V. non dissente, me ne prevarrò ad opportuna occasione.

Il suggerimento, che la S.V. mi diede, di studiare la fecondazione nel grano,[6] mi giunse tanto più gradito, in quanto che nella state passata ho fatto appunto alcuni studi in proposito. Mi pare di aver messo fuori dubbio, 1° che ogni flosculo del grano sta aperto per circa un quarto d'ora, in tempo che la fecondazione stimmatica non è ancora avvenuta; 2°. che in consequenza la dicogamia è possibilissima nel grano; 3°. che per altro ogni fiore del grano è suscettibile di essere fecondato col proprio polline. Presso la segala ho trovato che la fecondazione dicogamica è di gran lunga la più facile. Presso *l'Hordeum vulgare* ho trovato che un terzo dei fiori non si aprono menomamente (e quindi non possono essere fecondati dicogamicamente) mentre gli altri due terzi si aprono e possono essere fecondati dicogamicamente.[7] Presso l'*Hordeum diytichum*[8] (con strana eccezione) la fecondazione omogamica (e non dicogamica) è veramente necessaria, perocchè succede quando ancora le spighe non uscirono fuori dalla guaina fogliare. Ma ecco che, con mirabile alternativa, questa specie d'orzo possiede due specie di fiori: gli uni (pochissimi) che si aprono e sono predestinati per la dicogamia; gli altri (moltissimi) che non si aprono, e sono predestinati per la omogamia. Cosicchè presso questa pianta si ripeta il bene conosciuto fenomeno dei fiori di *Viola*, di *Vicia amphicarpa* ecc. ecc.[9] Così, anche presso *l'Hordeum distichum* rimane intatta la gran legge "no hermaphrodite fertilises itself for a perpetuity of generations."[10] Ora sto scrivendo una piccola memoria in proposito, e appena sarà stampata, mi farò un dovere di rassegnarla alla S.V.[11]

Della "Scientific opinion" non mi è pervenuto che il primo fascicolo, cioè il N°. 48. Ho venuto a sapere per caso che nel N°. 51 la S.V. deve avere inserito una risposta al mio scritto.[12] Io la prego con sommissione a volermene favorire un esemplare: e, se Ella permette, io farò *per deferenza* la traduzione della sua risposta, e la farò inserire sulla *Revista Contemporanea*, che oggi muta nome in *Revista Europea*.[13]

Ho avuto pure il libro svedese di *Axell*.[14] È un opera che mi pare di molto merito e scritta con rara coscienza. Godo assai che Axell faccia un altissima stima degli scritti del più grande naturalista del secolo, di cui ho l'onore di riconfermarmi

Ossequentissimo discepolo | Federico Delpino

DAR 162: 145

[1] For a translation of this letter, see Appendix I.

[2] See letter to Federico Delpino, 14 October 1869. Hermann Crüger's paper described how the pollenmass of *Coryanthes* was secured onto the back of the bee (*Euglossa*) as it tried to force its way out through a passage after falling into a 'bucket' of nectar; when the bee next fell into the bucket of the same or another flower, the pollen was deposited on the stigma as it tried to leave (Crüger 1864, p. 130). See *Origin* 4th ed., pp. 229–30, 'Fertilization of orchids', p. 157 (*Collected papers* 2: 153–4), *Orchids* 2d ed., pp. 173–6, and *Correspondence* vol. 12.

[3] See also the letter from Federico Delpino, 9 October 1869. Visits of bees to flowers of the orchid genus *Gongora* were also described in Crüger 1864.

[4] For CD's remark on the Nectarinae (now the family Nectariniidae, sunbirds) fertilising *Strelitzia*, see the letter to Federico Delpino, 14 October 1869.

[5] Friedrich Hildebrand recorded his opinion on *Strelitzia* pollination in Hildebrand 1869, pp. 508–9.

[6] CD had suggested that Delpino study the fertilisation of 'Gramineæ', the grass family; see the letter to Federico Delpino, 14 October 1869.

[7] On dichogamy, see the letter to W. C. Tait, 17 July [1869]. *Hordeum vulgare* is common barley.

[8] *Hordeum distichum* is an alternative spelling of *H. distichon*, now considered to be a variety of *H. vulgare vulgare*. It was known as two-rowed barley.
[9] Delpino refers to violets and to the vetch *Vicia amphicarpa* (now *Vicia sativa amphicarpa*); see also letters from T. H. Farrer, 8 August 1869 and 12 August 1869.
[10] Delpino refers to CD's conclusion in *Orchids*, p. 359: 'Nature thus tells us, in the most emphatic manner, that she abhors perpetual self-fertilisation.'
[11] Delpino probably refers to his paper on dichogamy in cereals (Delpino 1871); an inscribed copy is in the Darwin Pamphlet Collection–CUL.
[12] Issue 48 (29 September 1869) of *Scientific Opinion* included the first part of Delpino 1869, the translation of Delpino's review of pangenesis. In his letter to Delpino of 14 October 1869, CD had offered to send the remaining parts. CD replied to Delpino's criticism in the letter to *Scientific Opinion*, [before 20 October 1869]; this was published in issue 51 (20 October 1869).
[13] Delpino's translation of CD's letter to *Scientific Opinion*, [before 20 October 1869], appeared in the first volume of *Rivista Europea* (1869): 118–24.
[14] The references are to Johan Severin Axell and Axell 1869.

From R. F. Cooke 2 November 1869

50A, Albemarle Street, London, W.
Nov 2 *1869*

My dear Sir

Mr. Murray thinks it is much better to engrave the "Monkey's" &c for ourselves & not write to Paris, so I return the pages you left with me.[1]

Yours faithfully | Robt. Cooke

C. Darwin Esq

DAR 171: 372

[1] CD had intended to discuss woodcuts for *Descent* with John Murray during his stay in London from 1 to 9 November 1869 (see letter to John Murray, [after 18 September 1869], and CD's 'Journal' (Appendix II)). CD evidently left some pages of drawings with Cooke at Murray's publishing house, probably earlier in the day on 2 November.

From Charles Lyell 2 November 1869

2. Nov— 1869—

I have read the article, & cannot attach the importance—to it which you seem to do.[1] To me the most interesting parts of it are the quotations from Thomson, & what is said about changes in the acceleration of the earth's rotation, or that of the moon by the friction of the tides.[2] I do not yet comprehend the argument but am quite willing to believe that physicists will prove something important on this question; but it was surely a good hit of Huxley's to point out that whatever calculations they have been making about the effects of the tides in the course of geological ages, must be very loose & incomplete, if they took no account of the quantity of water locked up in the shape of ice at different periods, when for example the Miocene forests were flourishing at or near the pole with scarcely any ice, & when at the height of the

Glacial Period, Wales & Yorkshire were covered by a sheet of ice.[3] I am astounded at the coolness with which they talk of the theory of Helmholtz as to solar heat being the only one left to us, viz, that it is a hot body cooling & that 10 or 15 millions is all that can be allowed to the geologist (p 438);[4] & how they can speak of there being no access of energy into the sun than meteors, as if they knew so much of the universe as to warrant such assumptions.[5]

I am said (p 422) to talk about *tens of thousands of millions of years* as the very least period I will accept, but he takes care not to refer to any passage.[6] I suppose you have seen Geikie's calculations that denudation like that of the Mississippi would reduce a mass of rock equalling in volume all the land now above the sea to sand, pebbles & mud in 6.000.000 years—[7] Have you any where in reference to speculations of this kind spoken of the supposed fact that the average depth of the sea is 15 times as great as the average height of the land, the latter being 1000 & the sea 15000.

You have spoken somewhere of the permanence of continents,[8] to which in a general way I have no objection, but I should like you to consider ⟨if⟩ the sea ⟨be less⟩ deep (suppose even the mean depth to be 10.000 ft) would it not have a great bearing on the question of the rate of denudation & permanence of continents. Nothing can be clearer to me than that the continent from which the sedimentary strata of the Alleghanies were derived was placed where the Atlantic is now situated.

The sedimentary strata of the Alleghanies being all Paleozoic are 42.000 f^{t} thick. Do think of this & give me any references to yourself or others which may bear upon it. I am sure it will have much to do with estimates of geological time. The rate of denudation must be balanced by an equal amount of volcanic movement & the changes in species will have reference to these two forces & will require probably as much time as you may at any time have thought.[9]

The permanence of continents depends greatly on the formation of deltas near the coast, or in other words, on rivers clearing themselves of mud, sand & pebbles in the first receptacle they meet with—just as they leave behind all their sediment when they pass through a lake. If therefore the height of the land be to the average depth of the sea as 1 to 15—a moderate quantity of movement in the areas fringing the continents will cause the deltas to emerge & the same matter will do duty several times over in ⟨the cour⟩se of the 6.000.000 years— But the same amount of movement will cause but little of the deep ocean to emerge if the *average* depth be 3 miles— The movements however are sufficient at length to cause the submergence & emergence of the bed of the deep sea, & the changes of level seem to be going on so slowly that I cannot but think the conversion of great oceanic basins into continents & of continents into oceanic basins may have been a work of as much time as your speculations have ever demanded— I do not think that the subaërial denudationists allow sufficiently for the constant repair of the loss of land by the up-lifting of comparatively modern strata thrown down near the coast,[10] the Eocene strata are 12,000 feet high on the Alps & the Miocene from 3000 to 7000 (?)ft— At the end of a great number of periods that chain may be as high as ever.—although the quantity of matter which has passed through the denudation mill may equal in volume all the land now above the sea yet this has been supplied by grinding the same stuff over & over again. I still think that the quantity of movement shown by Moel Tryfaen & by the submergence & re-emergence of a large part of the British Isles while the Mollusca have been scarcely in the least changed implies that an enormous quantity of movement & denudation corresponds to an insignificant amount of organic fluctuation.[11]

Contemporary copy, incomplete
Edinburgh University Library, Special Collections Department, (Lyell 1) [Gen.113.ff.3734–7]

1 Lyell refers to a review article by Peter Guthrie Tait, 'Geologic time' ([Tait] 1869). No letter from CD to Lyell on this subject has been found, but see the letter to J. D. Hooker, 24 July [1869]. The article summarised William Thomson's continuing criticism of uniformitarian thought in geology (see Brush 1996, pp. 36–8, and Burchfield 1990, pp. 80–6).

2 Tait quoted Thomson several times in summarising the argument, supported by mathematical calculations, that ocean tides created enough friction to gradually diminish the rate of the earth's rotation ([Tait] 1869, pp. 416–19, 433–4).

3 Thomas Henry Huxley had responded to Thomson's argument regarding the earth's speed of rotation, suggesting that the rate could actually have increased in the glacial period as ice accumulated at the poles, withdrawing water from the sea (see T. H. Huxley 1869c, p. li).

4 Tait wrote: 'the only theory of solar heat left us is that developed by Helmholtz, which regards the sun as a hot body cooling; the heat having been produced during the falling together of its parts.' Consequently, the earth was also thought to be cooling, and Tait estimated that the earth was only ten or fifteen million years old. See [Tait] 1869, p. 438. For Hermann von Helmholtz's argument on the cooling of the sun, see Helmholtz 1856. See also Thomson 1862b.

5 For a discussion of Thomson's suggestion that the sun's energy was partly replenished by meteors falling into it, see [Tait] 1869, pp. 436–8. See also Burchfield 1990, pp. 28–33. For Lyell's doubt as to whether physicists could determine long-term temperature changes of the sun and earth, see C. Lyell 1872, 2: 213–14.

6 Tait wrote ([Tait] 1869, p. 422): '… *popular* geologists … with no less authorities than Lyell, Ramsay, Darwin, and Jukes at their head, still talk with the wildest looseness about thousands and tens of thousands of millions of years as the very least periods they can accept'. The additional references are to Andrew Crombie Ramsay and John Beete Jukes.

7 Lyell is mistaking Archibald Geikie's support of subaerial denudation (erosion by rivers and glaciers) as the primary erosive force with James Croll's calculations of the erosive power of rivers presented in Croll 1868. CD added Croll's discussion to the fifth edition of *Origin*, pp. 349–50; see also *Correspondence* vol. 16, letter to James Croll, 19 September 1868.

8 In *Origin*, chapter 11, 'Geographical distribution', CD wrote (pp. 357–8): 'It seems to me that we have abundant evidence of great oscillations of level in our continents; but not of such vast changes in their position and extension, as to have united them within the recent period to each other and to the several intervening oceanic islands.'

9 For Lyell on denudation, see C. Lyell 1865, pp. 66–79; see also C. Lyell 1872, 1: 104–5, and 2: 180–97. The Allegheny Mountains extend south-west from New York State to Virginia, and form a portion of the Appalachian mountain range in the eastern United States. For CD's most recent published discussion of the topic, see *Origin* 5th ed., pp. 352–4.

10 On subaerial denudation as the major force of erosion, see *Correspondence* vol. 16, letter to James Croll, 19 September 1868, n. 4. CD had recently added comments on denudation, including a reference to Lyell's views, to the fifth edition of *Origin*, p. 348–9 (Peckham, ed. 1959, p. 478).

11 Marine shells found in 1841 on Moel Tryfan, a mountain in North Wales, had suggested marine submergence to some geologists, including CD (see Herbert 2005, pp. 277, 282, and 'Ancient glaciers of Caernarvonshire'). In 1861, the shells were found to be more modern than had previously been thought, suggesting that uplift had been surprisingly recent (see *Correspondence* vol. 9, letter from Charles Lyell, 30 September 1861, and C. Lyell 1865, pp. 158–9).

To Charles Lyell [3 November 1869][1]

6 Queen Anne St

My dear Lyell

Very many thanks for your long & interesting letter,—the latter part of which perplexed me not a little till I received the missing sheet.—[2] You give me some consolation, but I take the Sun much to heart. I have comforted myself partly in the way which you suggest by our ignorance of the Universe, & by reflecting how little we know of the cause of the incandescent masses of vapour forming comets,—by the sudden blazing up of variable stars—extra planetary meteors &c.—[3]

With respect to long endurance of our existing continents, I formed my opinion chiefly from facts of geographical distribution, to which I allude in Origin—& partly from views given under Coral Reefs.—[4] I quite subscribe to what you say about denudation & from something I somewhere read or heard (perhaps from you) had come to conclusion that there was much grinding up of matter brought down to lower level; but this w[d] not much affect, I suppose, some of the Scotch streams, the carrying work of which I think has been measured.[5]

I am tired with my morning's work in Brit. Museum—so no more.—[6]

Ever yours | C. Darwin

American Philosophical Society (346)

[1] The date is established by the relationship between this letter and the letter from Charles Lyell, 2 November 1869, and by the date of CD's visit to the British Museum (see n. 6, below).

[2] The missing sheet of the letter from Charles Lyell, 2 November 1869, has not been found.

[3] See letter from Charles Lyell, 2 November 1869 and nn. 4 and 5.

[4] For CD's allusion to the endurance of continents in *Origin*, see the letter from Charles Lyell, 2 November 1869 and n. 8. In *Coral reefs*, CD argued that reefs formed as the immense area (or continent) on which they were built subsided over a long period of time; see *Coral reefs*, pp. 142–4, and Herbert 2005, pp. 242–3.

[5] Lyell may have discussed subaerial denudation in Scotland in the missing portion of his letter of 2 November 1869.

[6] CD visited the British Museum on 3 November 1869; there is a dated note made on his visit in DAR 85: A71.

From George Cupples [4–10 November 1869][1]

Deerhound puppies born Sept[r]. 23/69.
2 bitches drowned as soon as born.
1 Dog died at the same time.
Alive and thriving specially well 4 Dogs 1 Bitch.

First weighed Oct[r] 7[h]. when 14 days old[2]

Bitch—— 3 lbs 7 oz
Dog —— 3 – 11 –
Dog —— 3 – 11 –
Dog —— 3 – 5
Dog —— 3 – $3\frac{3}{4}$

Weighed again Oct[r] 14 when 3 weeks old—
Bitch — 3 lbs 13 oz
Dog — 4.–
Dog — 4.7
Dog — 4.7
Dog — 4.–

next weighed Nov[r]. 4 when 6 weeks old[3]
Bitch — 8 lbs. 1 oz
Dog — 9 – 8
Dog — 9 – 9
Dog — 8 – 9
Dog — 8 – $1\frac{1}{2}$
to be weighed again Nov[r]. 11 when 7 weeks old.

From this note it will be seen that the bitch and dogs are nearer each other than in the former case given—[4] There is only *one* bitch—but she was not selected for extra size, only by chance—and the circumstances are a fairer test, as *all* were left to the mother, *none* given out to nurse.

Memorandum
DAR 83: 147

CD ANNOTATIONS
5.3 3 – 11 … $\frac{3}{4}$ 5.6] *scored ink, right margin*; '3.7.7 *[*difference*]* .7 of oz' *left margin ink*
6.3 4.– … 4.– 6.6] *scored ink, right margin*; '4.3.5 *[*Difference*]* 7.5 oz' *left margin ink*
7.3 9 – … 1 $\frac{1}{2}$ 7.6] *scored ink, right margin*; '8.14.9. *[*Diff*]* 13.9 oz' *left margin ink*

1 The date range is established by the date of the last weighing of the deerhound puppies (4 November 1869), and the next projected weighing (11 November).
2 In his letter of 22 March 1869, Cupples had promised to weigh every litter of puppies he could in the coming season.
3 In *Descent* 2: 261, CD reported the weight of the female and the average weight of the males at two, three, and six weeks old.
4 See letter from George Cupples, 11 March 1869.

To Eduard Koch 4 November 1869

Down, Beckenham | Kent [6 Queen Anne Street, London]
Nov. 4th. 1869

Dear Sir

I am much obliged for *D[r]. Dub's* work which I received two or three days ago.[1] When next you communicate with D[r]. Dub will you be so good as to say to him that I congratulate him on the completion of the work, & that I shall hope soon to read the greater part of it.

I have already written to Prof. *Victor Carus* on the *subject of my new book* & told him all that I could.[2]

I am glad to hear of the new edition of the Origin;[3] I sent some time ago *the new English edition* to Prof. Carus; & the corrections will cost him a good deal of labour, for although the book is only a page or two longer I have struck out many passages & added others.[4]

Believe me, dear Sir | yours faithfully | Ch. Darwin

LS
Württembergische Landesbibliothek, Stuttgart (Cod. hist. 4^0 333a, 77, 4)

1 The references are to Julius Dub and Dub 1870; see letter from Eduard Koch, 30 October 1869. There is a copy of Dub 1870 in the Darwin Library–Down.
2 CD refers to Julius Victor Carus. CD discussed *Descent* in the letter to J. V. Carus, 22 October [1869]; see also the letter to J. V. Carus, 30 October 1869.
3 The reference is to the German translation of the fifth English edition of *Origin* (Bronn and Carus trans. 1870); see letter from Eduard Koch, 30 October 1869.
4 CD sent the fifth English edition of *Origin* to Carus on 21 June (see letter to J. V. Carus, 21 June [1869]). See also letter to J. V. Carus, 4 May 1869.

From T. H. Farrer 5 November 1869

The Athenæum
5 Nov /69

My dear Mr Darwin,

If you will put questions "no fellah can answer", you at any rate bear patiently the infliction of guesses.

Here is one at Mimosa.[1] The petioles are long and slender; and a considerable surface of leaf is exposed. A puff from the mouth makes the leaflets fold and the leafstalk bend down much more universally and thoroughly than a touch with the fingers or a sprinkle with water.

Now what if the motion should be a protection against storms of wind. On the first blast the prudent plant not only reefs and furls, but also lowers its sails and makes all snug against the coming storm!!

How and where the plants grow: what is their size and mode of life; in short all that one wants to know wholly fails to appear in the big books at the Linnean.[2]

I have a Passiflora princeps though an imperfect one.[3] You are quite right that there are five corners in the process[4] towards the base of the inner corona which leave apparently large apertures towards the nectary. But the upper part of this corona is so closely applied to the column that a straight object can only go straight down and then it is sure to go—not into the nectary, but into the angle between the column and the innermost process—(qu innermost corona) and to be caught there

Sincerely yours | T H Farrer

C Darwin Esq[r] FRS

DAR 164: 60

1 No query to Farrer regarding *Mimosa* has been found. CD later observed and published on several *Mimosa* species (see *Movement in plants*).
2 Farrer refers to the library of the Linnean Society of London.
3 In his letter of 20 October [1869], CD suggested that Farrer observe the flower structure of *Passiflora princeps* (now *P. racemosa*, the red passionflower).
4 Process: an outgrowth (*OED*).

From J. J. Moulinié 5 November 1869

Geneva
5th Novr. 1869.

Dear sir,

On receipt of your letter,[1] I wrote to Mr. Reinwald, my french publisher, to inform him of the future publication of your new work, and at the same time to have his opinion on the question about the possibility of a new translation correct and complete of the last edition of the "Origin of species."[2] On the first point, he is quite disposed to publish the translation of your work on Man, as soon as it will be ready.[3] On the second, Mr. Reinwald would most gladly undertake also a fresh edition of the translation of the Origin, in the conditions that you desire, and with your consent, and believes that you have all rights to do so, "if you have not already formally transferred to another publisher, the right of translation for subsequent editions."

He adds besides, that desirous to act with moderation and out of consideration for the interests of the first publisher and translator of the Origin, he is ready,—on an *explicit authorisation* of yours, to publish a new translation of the last english edition,[4]—to propose to them some reasonable transaction, and obtain their desistance of the rights they might possibly have. In case of refusal on their part, concludes Mr. Reinwald, he would proceed without hesitation, convinced that a thoroughly new translation, better than the former and containing the last corrections, would certainly be successfull, and supersede the others.

With such good dispositions, on the part of Mr. Reinwald, I hope that all material obstacles if any, may be removed, and will then, as far as it concerns me, most gladly contribute to the realization of your wish, in undertaking the translation of the "Origin", which you so kindly propose to me.[5]

My best regards to Mrs Darwin and family, and believe me, dear Sir, your's grateful and truly devoted | J. J. Moulinié

P.S. Mr. Reinwald's direction is the following:[6]

Incomplete
DAR 171: 274

CD ANNOTATIONS
2.4 to propose … have. 2.5] *scored pencil*
Top of letter: 'A fr address' *pencil*

1 Letter to J. J. Moulinié, 23 October [1869].
2 The references are to Charles-Ferdinand Reinwald, *Descent*, and the fifth edition of *Origin*.

[3] *Descent* was published in 1871, and Moulinié's translation in 1872 (Moulinié trans. 1872).
[4] See letter to J. J. Moulinié, 23 October [1869]. Victor Masson and the publishing house of Guillaumin et cie had published Clémence Auguste Royer's first two French editions of *Origin* (Royer, trans. 1862 and 1866); a third edition was being published by Masson alone without the additions from the fourth and fifth English editions (Royer, trans. 1870).
[5] Moulinié's translation of *Origin* was published in 1873 by Reinwald (Moulinié trans. 1873).
[6] Reinwald's address in Paris was 15 Rue Des Saints-Pères (letter from C.-F. Reinwald, 13 November 1869).

From Frederick Smith 5 November 1869

British Museum
5th. Novr. 1869

My dear Sir

I have just read an article in the Entomologists Monthly Magazine for Novr. 1869. page 130 on Scolytus, the Elm Beetle— the paper is entitled. "Observations on the Œconomy of the British Species of Scolytus—By T. Algernon Chapman, M.D.—Abergavenny—

The Author observes "S. destructor, intricatus and pruni, are able to squeak audibly, by a rapid movement of the abdomen against the elytra, intricatus making the loudest noise—[1]

I just send this in case it may be useful to you—[2]

Yours very truly | Fredk Smith

DAR 82: 4

CD ANNOTATIONS
1.2 Scolytus] *underl pencil*
Top of letter: 'Scolytus' *pencil*

[1] The reference is to Chapman 1869; the author was Thomas Algernon Chapman. *Scolytus pruni* is now *S. mali*, the larger shothole borer.
[2] CD cited the article in *Descent* 1: 379 n. 70, but he mistakenly gave the author's name as 'Algen', and the journal as the 'Edinburgh Monthly Magazine'.

From J. V. Carus 6 November 1869

Leipzig,
Nov. 6. 1869.

My dear Sir,

I hardly know how to thank you for your last letter and the exceedingly kind and decidedly undeserved offer you make me.[1] You are really too good; for all I did, was done for the interests of science. I wished to make my countrymen acquainted wth your wonderful way of looking to facts under the guidance of ideas in the best German form possible.[2] I thought this the more to be my duty, as our Zoologists and Physiologists have almost forgotton to think and to reason on facts and are of opinion that science is nothing but an accumulation of single facts. Every describer of a new form of cells or of a new worm or of a slight alteration of the electric current in nerves is now a great man. Of course all these things must be found out and new facts are to be worked out, but there must be a tie to keep all the details together. But, who ventures to bring the different facts under general headings and to discover (or to try to discover) the common bearing of a larger set of facts, he will entirely be decried as "philosopher". This has been my fault. However, Prof Vogt is wrong. I am still, what I was before, Professor of Comparative Anatomy in Leipzig And yet my position is far from being so agreeable as it was before. I hoped and I had been promised to get the place of the late Professor of Zoology. There I failed.[3] Owing to the great influence of a man, who has, as he says himself, no "Organ" for morphological relations, who looks on Embryology only as on a help for physiological researches, and who overlooks that, even in the purely mathematical sense of the word, form is nothing but a function of the matter, owing to him Leuckart got the place.[4] It is so perfectly natural that one cannot wonder at it, that the mere fact of his being called here from another university gives him the air of being so much the superior.[5] I should not like to compare myself with him, he knows a great deal more than I certainly, but I can say positively he is no comparative Anatomist. For this I am sorry; for as he got the regular Professorship he will get influence on the students, and although I always had a nice class of intelligent young men, yet he is the bigger now.[6]

Excuse me for becoming personal, but I thought that I owed you so much that I am obliged to explain to you why I get sometimes a little bitter. I have now been asked to take up the Natural History, Anatomy and Physiology of Domestic Animals. This I try to do, especially as I hope to get thereby an opportunity of learning more of this subject which, as far as I see already has been neglected by far too much by Zoologists, and also of impressing the minds of my unprejudiced hearers with the force of your arguments and your whole theory. And yet, I want some strength of mind to get through my work now.

It was my duty to tell you all that about my position here. Now I must leave it entirely to you if you will give me the right of translation. As your kind letter was written under the impression of a misunderstanding, I could not accept it purely. At all events I should be happy if you would allow me to translate your book, especially as I intend to give in one of our next terms a course of lectures on Anthropology, which is not taught here at all.

I am happy to think that I shall be able ere long to thank you personally and orally better than I can do it in a letter and to tell you how deeply I feel, how very kind you are.[7]

Believe me | My dear Sir | Yours sincerely | J. Victor Carus

DAR 161: 73

[1] See the letter to J. V. Carus, 30 October 1869; CD had asked Carus to translate *Descent.*

[2] Carus may refer to his outspoken support of CD's views; see letter from Carl Vogt, 26 October 1869.

[3] Carl Vogt had written that Carus's application for the chair of zoology at Leipzig had not been successful (see letter from Carl Vogt, 26 October 1869); CD evidently thought that Carus had lost his existing position (see letter to J. V. Carus, 30 October 1869).

[4] The supporter of Rudolf Leuckart's appointment may have been the anatomist Ernst Heinrich Weber, who was one of the senior professors in the Leipzig medical faculty (Lynn K. Nyhart, personal communication, and Nyhart 1995, pp. 81–3, 173).

[5] Leuckart had been professor of zoology at the University of Giessen.

[6] Leuckart's position in zoology at Leipzig was a full professorship while Carus's professorship of comparative anatomy was not (Nyhart 1995, pp. 81, 96).

[7] Carus planned to travel to England in March 1870 (see letter from J. V. Carus, 20 October 1869).

From George Busk 7 November 1869

32 Harley St
Nov 7 1869

My dear Darwin

The Polyzoan creatures you saw emerging from the cells of the main polypidom appear to me to be parasites— There can be no doubt that the principal growth as seen in Fig 1–5 &c is a species of Sertularia, though I am unable at the moment to give it a name if it have one.[1]

The vacated cells of this growth have been taken possession of by a minute species of ctenostomatous Polyzoa[2]—usually as it would seem or at any rate frequently in pairs as in Fig 3—but also in many cases singly—Fig 2–4— 6, 7. And it is remarkable that the parasite seems to be always affixed at about the same point in the empty sertularian cell (Fig 3 & 4). It is a curious, but as you well know not unexampled instance of "necessary parasitism".[3]

Another instance though of a very different kind of this kind of parasitism among the polyzoa is that which I have described under the name of Bugula Ditrupæ which appears invariably to grow from the upper part of the shell of a species of Ditrupa in Madeira.[4]

Besides this "necessary" guest however—the sertularian is much infested with a Campanularia and a small sponge.[5]

Believe Me | Yours very truly | Geo Busk

DAR 160: 382

[1] The diagram Busk refers to has not been identified, and no letter of enquiry to Busk on this subject has been found. CD had been in London since 1 November (CD's 'Journal', Appendix II), and had been been making observations at the British Museum and the Zoological Gardens, Regent's Park. The former class Polyzoa included the present phyla Ectoprocta (bryozoans) and Endoprocta. *Sertularia* was classified by Busk as a genus of the family Sertularidae in the order Anthozoa Hydroida (Busk 1852, p. 387). It is now a genus in the order Leptothecatae in the class Hydrozoa of the phylum Cnidaria. The polypidom (now polyparium) is the common stem or supporting structure of a colonial invertebrate (*OED*).

[2] Busk classified Ctenostomata as a suborder of the order Infundibulata in the class Polyzoa (Busk 1852, p. 349). It is now an order in the class Gymnolaemata (tubular bryozoans).

[3] Busk describes what was probably a surficial (epizoic) growth of the ctenosome on the empty *Sertularia* cells, known as thecae; the association was unlikely to be parasitic. The term 'necessary parasitism'

was not used by CD or others; Busk may have used the term because the ctenosomes attached to what appeared to be the only viable sites on the hydroid. (Dennis Gordon, personal communication.)

4 Busk refers to his description of *Bugula ditrupae* in *Quarterly Journal of Microscopical Science* 6 (1858): 261. *Ditrupa* is a genus of polychaete worm, now in the family Serpulidae (fanworms).

5 *Campanularia* is another genus in the order Leptothecatae (see n. 1, above).

From A. W. Bennett to *Nature* 8 November 1869

Mr. Darwin has done me the honour of calling my attention to one or two points in my paper, published in your last number, "On the Fertilisation of Winter-flowering Plants."[1] He thinks there must be some error in my including *Vinca major* among the plants of which the pollen is discharged in the bud, as he "knows from experiment that some species of *Vinca* absolutely require insect aid for fertilisation."[2] On referring to my notes, I find them perfectly clear with respect to the time at which the pollen is discharged. My observation, however, so far agrees with Mr. Darwin's, that I find no record of any fruit being produced in January; it was, in fact, the absence of capsules on the *Vinca* which induced me to qualify the sentence on this subject, and to say "in *nearly all* these cases, abundance of fully formed seed-bearing capsules were observed."[3] It is worthy of remark, that the *Vinca* is the only species in my list of apparently bud-fertilised plants not indigenous to this country.[4] The second point relates to the white dead-nettle, with respect to which Mr. Darwin says, "I covered up *Lamium album* early in June, and the plants produced no seed, although surrounding plants produced plenty."[5] This again would agree with my conjecture that it is only the flowers produced in winter that are self-fertilised. I may, however, be permitted to suggest that the test of covering up a plant with a bell-glass is not conclusive on the point of cross-fertilisation, as it is quite probable that with plants that are ordinarily self-fertilised, the mere fact of a complete stoppage of a free circulation of air may prevent the impregnation taking place. Has the experiment ever been tried with grasses, which, according to the French observer, M. Bidard are necessarily self-fertilised?[6]

3, Park Village East, Nov. 8, 1869

Nature 1 (1869): 58

1 CD's letter commenting on Bennett 1869 has not been found.

2 For CD's observations of *Vinca* fertilisation, see *Correspondence* vol. 9, letter to *Gardeners' Chronicle*, [before 15 June 1861] and second letter to *Gardeners' Chronicle*, [before 14 September 1861]. Species in the genus *Vinca* are usually called periwinkles.

3 Bennett 1869, p. 11.

4 *Vinca* is indigenous to continental Europe, northern Africa, and central Asia (Mabberley 1997).

5 CD probably experimented with *Lamium album* (the white dead-nettle) in June 1869 (see letter from A. W. Bennett, 3 May 1869).

6 Bennett refers to Bidard's article 'On the structure of the flower of the *Gramineæ*, the function of the organs of which it is composed, and the phenomena which accompany the act of fecundation' (Bidard 1869).

To Antoinette Louisa Brown Blackwell 8 November [1869][1]

Down Beckenham Kent [6 Queen Anne Street, London]
Nov. 8th

Dear Sir

I am much obliged to you for your kindness in sending me your "Studies in General Science", over which, as I observe in the Preface, you have spent so much time.—[2] In turning over the pages I notice that you quote some statements made by me & very little known to public.[3] I received your work only yesterday

With my best thanks | Dear Sir | Yours faithfully | Ch. Darwin

Blackwell Family Papers (A–77). Schlesinger Library, Radcliffe Institute, Harvard University

1 The year is established by the publication date of Blackwell 1869; see n. 2, below.

2 An inscribed copy of Blackwell 1869 is in the Darwin Library–Down. Blackwell wrote in the preface, pp. v–vii, that the book was 'the product of more than twenty years of an exceedingly busy life' of educational pursuits, work as a church pastor, lecturing, and bringing up children.

3 In Blackwell 1869, Blackwell included quotations from *Orchids*, 'Climbing plants', and *Variation*; she also discussed points in *Origin*.

To John Murray 8 November [1869][1]

Down Beckenham Kent [6 Queen Anne Street, London]
Nov. 8th

My dear Sir

Will you be so kind as to inform me what were the exact terms, by which you granted for me to M. V. Masson of Paris the right of translation of my Origin of Species.—[2] M. Masson has brought out a 3d French Edition, without having had the common courtesy to inform me, so that none of the numerous corrections in the 4th & 5th English Editions have been taken advantage of.—[3]

M. Reinwald of Paris now writes to me that he will at once bring out a new Translation of the Origin "if you have not already formally transferred to another publisher the right of translation for subsequent Editions".[4]

You will now understand my motives. I have other & valid reasons (viz one blasphemous preface, & a second preface abusive of myself) for wishing that a new translation shd appear.—[5]

Lastly, if you have **not** granted for me the right of translation for all subsequent editions, will you have the kindness to send **to me** the proper form for granting permission to M. Reinwald to produce a translation of the 5th English Edition.

I am *very anxious* on this subject, so pray forgive me troubling you & kindly inform me as soon as you can.

My dear Sir | Yours sincerely | Ch. Darwin

National Library of Scotland (John Murray archive, Ms. 42152 ff. 205–6)

1 The year is established by the relationship between this letter and the letter from J. J. Moulinié, 5 November 1869 (see n. 4, below).

2 The references are to Victor Masson and the French translation of *Origin* (Royer trans. 1862).

3 See letter from Victor Masson, 29 September 1869, and letter to Victor Masson, [after 29 September 1869]. The third French edition of *Origin* was published in 1870 (Royer trans. 1870).

4 The reference is to Charles-Ferdinand Reinwald; for the quotation, see the letter from J. J. Moulinié, 5 November 1869.

5 The preface to Clémence Auguste Royer's first translation of *Origin* was strongly anti-clerical (Royer trans. 1862); a lightly annotated copy of the preface is in the Darwin Pamphlet Collection–CUL. The second preface CD refers to is that of Royer's third edition (Royer trans. 1870); this included criticism of CD's hypothesis of pangenesis, which had appeared in *Variation*. For more on Royer's translations, including the prefaces, see Harvey 1997, pp. 62–9, 76–9, 97–9. For more on CD's opinion of the translations, see *Correspondence* vol. 13, letter from C. A. Royer, [April–June 1865], n. 5, and *Correspondence* vol. 15, letter to Charles Lyell, 22 August [1867].

From John Murray 8 November [1869][1]

50, Albemarle St. | W.
Novr 8th

My Dear Sir

I have no recollection of my having taken any part in the arrangement with Masson for the French Translation of the Origin, beyond that of securing for you the copyright in France & depositing a copy. Nor is any trace beyond this, to be found in my books—[2]

Unless you received money from Masson, & signed some Document it seems to me there is nothing to bind you.

As regards any transfer of your rights to Reinwald—in the Fifth Edition, I fear they have fallen through—because the three months prescribed for registry after publication of this Ed^{n} have lapsed, now some time[3]

Still this need not prevent Reinwald publishing, if you have not arranged with Masson as aforesaid: The lapse of time w^{d} now allow Masson to translate even the additions to the 5th. Ed^{n}. I suspect—[4]

Chas Darwin Esqre

AL
DAR 171: 373

1 The year is established by the relationship between this letter and the letter to John Murray, 8 November [1869].

2 The only extant correspondence between Murray and CD regarding the French translation of *Origin* is CD's request to have a copy of the third edition sent to Clémence Auguste Royer in September 1861 (see *Correspondence* vol. 9, letter to John Murray, 10 September [1861]). Victor Masson was one of the publishers of Royer trans. 1862 and 1866, and the sole publisher of Royer trans. 1870.

3 Murray evidently refers to practices resulting from the International Copyright Act of 1852; see Nowell-Smith 1968, pp. 32–3. The fifth edition of *Origin* was published in June 1869 (letter from R. F. Cooke, 22 June 1869). Murray refers to Charles-Ferdinand Reinwald.

4 Reinwald eventually published Jean Jacques Moulinié's translation of the fifth edition (Moulinié trans. 1873).

To J. V. Carus 9 November [1869][1]

Down. | Beckenham | *Kent. S.E.*
Nov. 9th

My dear Sir

I write a line to thank you for your letter & say how very glad I am to hear that I was partly mistaken about the Professorship.—[2] What you say does not make the slightest differences in my wishes.—[3]

Herr. Hoffmann of Stuttgart has applied to me, since I wrote to you, for the right of Translation.— I have simply told Herr Koch that I had communicated with you on the subject of my next book, & it will rest with you to decide whether to prefer him.—[4]

In Haste | My dear Sir | Yours very sincerely | Ch. Darwin

Staatsbibliothek zu Berlin – Preußischer Kulturbesitz, Slg. Darmstaedter Lc 1859 Darwin, Charles, Bl. 48/49

1 The year is established by the relationship between this letter and the letter from J. V. Carus, 6 November 1869.
2 See letter from J. V. Carus, 6 November 1869.
3 CD wanted Carus to translate *Descent.*
4 No letter from Carl Hoffmann, a publisher in Stuttgart, has been found. The most recent extant letter to the publisher Eduard Koch is that of 4 November 1869.

To Frederick Smith 9 November [1869][1]

Down. | Beckenham. | *Kent. S.E.*
Nov 9th

My dear Sir

On my return home I found your note on Scolytus, for which many thanks.[2]

Mr Janson did not send to 6. Queen Anne St the Papilio feronia, (so I suppose he has not specimens) & consequently I did not receive the Mutillas, which you so very kindly offered me.—[3]

My dear Sir | Yours sincerely | Ch. Darwin

P.S | Whenever the last lot of woodcuts are all completed will you ask your Son to be so good as to send me proofs *together with my paper of instructions*, by Post.—[4]

American Philosophical Society (Getz 6979)

1 The year is established by the relationship between this letter and the letter from Frederick Smith, 5 November 1869.
2 See letter from Frederick Smith, 5 November 1869.
3 CD refers to the natural history dealer Edward Westey Janson. In his letter to Fritz Müller of 8 September [1869], CD had asked about stridulation in *Papilio feronia.* Smith earlier wrote to CD about stridulation in *Mutilla,* a genus of wasps (letter from Frederick Smith, 14 September 1869). CD had stayed with his brother, Erasmus Alvey Darwin, at 6 Queen Anne Street, London, from 1 to 9 November (see 'Journal' (Appendix II)).

[4] CD refers to Edward A. Smith. In his letter to E. A. Smith of 30 June [1869], thanking him for completed woodcuts of insects for *Descent*, CD mentioned he would probably need more woodcuts later.

From E. A. Darwin 11 November [1869][1]

Nov 11

A parcel of engravings is come done up but with your paper of instructions outside.[2] Shall I send them on?

E D

Sterling sent today is very interesting but is he quite fair about 'Selection'[3]

DAR 105: B64

[1] The year is established by the reference to Stirling 1869 (see n. 3, below).
[2] The engravings may have been those completed by Edward A. Smith (see letter to Frederick Smith, 9 November [1869]).
[3] The reference is to James Hutchison Stirling's *As regards protoplasm, in relation to Professor Huxley's essay, On the physical basis of life* (Stirling 1869); for Stirling's discussion of CD's theory, including his use of the term 'selection', see pp. 58–64.

From F. C. Donders[1] 12 November 1869

Utrecht
12 Novembre | 1869

Ne croyez pas, Mon chèr & très honoré Monsieur, que j'aie oublié de répondre aux questions que vous m'avez fait l'honneur de m'adresser.[2] Le fait est que j'ai trouvé la reponse plus difficile que je n'avais pensé d'abord. C'est pourquoi J'ai commencé à faire, touchant la première question, une serie d'expériences, qui, comme à l'ordinaire, m'ont entrainé plus loin et m'occupent plus longtemps que je ne m'y étais attendu. Mais, m'apercevant avec quelque frayeur que depuis presque deux mois vous êtes en droit d'attendre ma réponse, j'ai senti le besoin d'écarter des soupçons de nègligence de ma part qui repondrait mal au respect que j'ai pour votre personne et pour vos oeuvres.

Veuillez donc me faire savoir, mon chèr Monsieur, si vous avez hâte d'avoir une solution quelconque de vos questions: dans ce cas je vous donnerai immédiatement un exposé des résultats obtenus jusqu'ici. Si, au contraire, vous n'êtes pas pressé, j'espère que dans six semaines ma reponse pourra être moins imparfaite. En attendant, je vous dois des remerciments pour avoir dirigé mon attention sur une question réellement importante sous différents points de vue.

J'ai été très heureux, Monsieur Darwin, d'avoir eu l'avantage de faire votre Connaissance personelle aussi que celle de Madame Darwin, et très réconnaissant de l'accueil que votre bonté m'avait préparé. Je n'oublierai jamais les quelques heures passées à Down.—[3] | Permettez–moi de vous offrir l'assurance de mes sentiments d'affection et de haute estime | F. C. Donders

DAR 162: 223

1 For a translation of this letter, see Appendix I.
2 Memorandum to F. C. Donders, 6 September 1869.
3 Donders and William Bowman visited Down, probably on 7 September 1869; see the letter from William Bowman, 3 September [1869], and the letter to J. D. Hooker, 10 September [1869].

From Hermann Müller 12 November 1869

Lippstadt. Westfalia
Nov. 12. 1869.

My dear Sir!

I have bestowed the last two summers upon observing the insects frequenting and fertilising our flowers. The bees being of eminent importance for the fertilisation of flowers, I have employed my principal attention in observing, collecting, and determinating them, but although I have extricated positively by far the greatest part of Westfalian species, many forms are still left which I cannot find in conformity with any description in my works on bees. Therefore it is my lively desire to hear the opinion of your highest authority, Mr. Frederick Smith, of the British Museum, o⟨n⟩ my yet ind⟨etermi⟩ned species.

Without the judgement of such an authority I should run the risk to publish some species as new ones, which are already described in any work unknown to me.

The domicil of Mr. Fr. Smith being unknown to me, I beg you, to deliver up to him the enclosed letter and to recommand my purpose.[1]

I am ready to send to Mr. Smith a collection of the Westfalian species of bees as complete as I have collected them, if such a collection would be of any value for the British Museum.[2]

I hope that you will be so kind and to oblige me on this head.

With sincere respect I remain my dear Sir | yours very faithfully | H Müller.

DAR 171: 295

1 The enclosure has not been found. Frederick Smith was an entomologist in the zoology department of the British Museum.
2 Smith agreed to review Müller's bee collection; see *Correspondence* vol. 18, letter from Hermann Müller, 8 March 1870.

To J. D. Hooker 13 November [1869][1]

Down. | Beckenham | *Kent. S.E.*
Nov 13th

My dear Hooker

I heard yesterday from a relation who had seen in a newspaper that you were C.B.[2] I must write one line to say "hurrah", though I wish it had been K.C.B, as it assuredly ought to have been; but I suppose they look at K.C.B before C.B. as a Dukedom before an Earldom.—[3]

We had a very successful week in London,[4] & I was unusually well & saw a good many persons, which when well is a great pleasure to me. I had a jolly talk with Huxley[5] amongst others. And now I am at the same work as before & shall be for another 2 months, namely putting ugly sentences rather straighter; & I am sick of the work & as the subject is all on sexual selection, I am weary of everlasting males & females, cocks & hens.—[6]

It is a shame to bother you, but I sh^{d}. like sometime to hear a little about the C.B. affair.

I have read one or two interesting brochures lately, viz Stirling, the Hegelian, versus Huxley & protoplasm.—[7] Tylor in Journal of Royal Institution on the survival of old thought in modern Civilisation.—[8]

Farewell, I am as dull as a duck, both male & female—

Yours affectionately | C. Darwin

To

D^{r}. Hooker C.B. F.R.S[9]
Dr Hooker K.C.B
(this looks better)

I hear a grand account of Bentham's last address, which I am now going to read.[10]

I find that I have blundered about Bentham's address, Lyell was speaking about one that I read some months ago; but I read half of it again last night & shall finish it.[11] Some passages are either new or were not studied enough by me, before.— It strikes me as admirable as it did on the first reading, though I differ in some few points. Such an address is worth its weight in gold, I sh^{d}. think, in making converts to our views. Lyell tells me that Bunbury has been wonderfully impressed with it, & he never before thought anything of our views on Evolution.—[12]

(I have just read & like very much your review of Schimper)[13]

DAR 94: 156–8

1 The year is established by the relationship between this letter and the letter from J. D. Hooker, 14 November 1869.

2 Hooker had been appointed a Companion of the Order of the Bath on 6 November 1869 (*London Gazette*, 9 November 1869, *The Times*, 10 November 1869, p. 6). CD's relative has not been identified.

3 A Knight Commander, or KCB, was the next highest class of the Order of the Bath after a Companion (Risk 1972). See L. Huxley ed. 1918, 2: 145–7 for a discussion of Hooker's honours.

4 The Darwins were in London from 1 to 9 November ('Journal' (Appendix II)).

5 Thomas Henry Huxley.

6 CD had been going over his chapters on sexual selection for *Descent* since 4 August 1869 (see 'Journal' (Appendix II)).

7 James Hutchison Stirling was a follower of the German philosopher Georg Wilhelm Friedrich Hegel (see Stirling 1865); Erasmus Alvey Darwin recently sent CD a copy of *As regards protoplasm, in relation to Professor Huxley's essay, On the physical basis of life* (Stirling 1869), which included a critique of T. H. Huxley 1869a (see letter from E. A. Darwin, 11 November [1869]).

8 The reference is to Edward Burnett Tylor, and Tylor 1869.

9 Hooker became a fellow of the Royal Society (FRS) in 1847.

10 CD refers to George Bentham's anniversary address given on 24 May 1869 to the Linnean Society of London (Bentham 1869b); there is an annotated copy in the Darwin Pamphlet Collection–CUL.

11 For CD's earlier correspondence with Hooker on Bentham 1869b, see the letter to J. D. Hooker, 22 June [1869], and the letter from J. D. Hooker, 24 June 1869. CD evidently saw Charles Lyell while in London (see n. 4, above).

12 For Charles James Fox Bunbury's initial reaction to *Origin*, see *Correspondence* vol. 8, letter from C. J. F. Bunbury, 30 January 1860.

13 Hooker reviewed the first volume of Wilhelm Philipp Schimper's *Traité de paléontologie végétale* (Schimper 1869–74) in *Nature* (J. D. Hooker 1869).

To *Nature* 13 November [1869][1]

Will you permit me to add a few words to Mr. Bennett's letter, published at p. 58 of your last number?[2] I did not cover up the *Lamium* with a bell-glass, but with what is called by ladies, "net." During the last twenty years I have followed this plan, and have fertilised thousands of flowers thus covered up, but have never perceived that their fertility was in the least injured.[3] I make this statement in case anyone should be induced to use a bell-glass, which I believe to be injurious from the moisture of the contained air.[4] Nevertheless, I have occasionally placed flowers, which grew high up, within small wide-mouthed bottles, and have obtained good seed from them. With respect to the *Vinca*, I suppose that Mr. Bennett intended to express that pollen had actually fallen, without the aid of insects, on the stigmatic surface, and had emitted tubes. As far as the mere opening of the anthers in the bud is concerned, I feel convinced from repeated observations that this is a most fallacious indication of self-fertilisation.[5] As Mr. Bennett asks about the fertilisation of Grasses, I may add that Signor Delpino, of Florence, will soon publish some novel and very curious observations on this subject, of which he has given me an account in a letter, and which I am glad to say are far from being opposed to the very general law that distinct individual plants must be occasionally crossed.[6]

Down, Beckenham, Kent, Nov. 13

Nature 1 (1869): 85

1 The year is established by the publication date of the letter in *Nature*.

2 See letter from A. W. Bennett to *Nature*, 8 November 1869; Alfred William Bennett's letter was published in the 11 November 1869 issue of *Nature*.

3 For an example of one of CD's many notes on the use of netting, see *Correspondence* vol. 10, letter to John Scott, 19 December [1862]. For a description of the netting that CD used, see *Cross and self fertilisation*, pp. 10–11.

4 Bennett mentioned the bell-glass in his letter to *Nature*, 8 November 1869.

5 In both Bennett 1869, and the letter from A. W. Bennett to *Nature*, 8 November 1869, Bennett only mentioned the pollen discharging from the anthers of *Vinca major*. In *Cross and self fertilisation*, p. 362, CD maintained that *V. major* was generally sterile when insects were not present.

6 In his letter of 1 November 1869, Federico Delpino discussed his work on grasses, and mentioned his intended publication (Delpino 1871).

From Charles-Ferdinard Reinwald 13 November 1869

Paris. 15 rue des Sts Pères
Novr. 13th. 1869.

To Charles Darwin E^sq^.
Dear Sir

I duely received your letter of Nov. 10th. and am happy to accept fully your proposal to publish a new french translation of your celebrated work, "*On the Origin of Species,*" after the 5th. english edition with some unpublished corrections, you intend to insert.[1]

I therefore write this day to Col. Moulinié to begin the translation which I should be glad to have revised by M. Carl Vogt.[2] In case M. Vogt should not be ready for the purpose; I would try, under your consentment, to engage another french naturalist for it.

The letter you had the kindness to write to me, will be sufficient to prove the right for a new translation, and I approve entirely your reasons not to compose with Mlle Royer or her editors for it.[3] If it is however in my power to avoid to both of them a loss of money or of unsold copies, you may rely on my best intentions to do it.

I should be obliged to you if you would sent a copy of the 5th. edition to Col. Moulinié, and have the goodness to forward my little note to Mess. Murray, in order to get 2 copies of the same edition, which shall be paid by one of my friends in London. I will try to deposit these copies here at the Ministère de l Intérieur, in order to secure copyright, not according the international treaty with England, but according our french law of copyright.[4]

Any other arrangement you could have to propose, as for a number of french copies, or other charges for the right of translation, shall meet my best exertions to deal honestly and fairly with the celebrated author of the book in question.

I am happy, dear Sir, to be honoured with your confidence in the present occassion and beg to believe me | Dear Sir | your most obedient Servant | C Reinwald

To Charles Darwin, E^sq^ | Down. Beckenham. Kent | Angleterre

DAR 176: 92

1 CD's letter to Reinwald of 10 November 1869 has not been found; however, see the letter to John Murray, 8 November [1869], and the letter from John Murray, 8 November [1869].

2 CD had corresponded with Jean Jacques Moulinié about the translation of the fifth edition of *Origin* (see letter from J. J. Moulinié, 5 November 1869). Carl Vogt also revised the French translation of *Variation* (Moulinié trans. 1868; see *Correspondence* vol. 15, letter from Carl Vogt, 23 April 1867).

3 On Clémence Auguste Royer's translations of *Origin*, see the letter to John Murray, 8 November [1869] and n. 5.

4 The reference is to the French Ministry of the Interior.

From J. D. Hooker 14 November 1869

Royal Gardens Kew
Nov. 14 /69.

Dear Old Darwin

I should have written to you ere this & explained the CB. The only intimation

I have hitherto had is from the *Times*!—[1] a month ago Murchison wrote to me to the effect, that the P.M. had written to him, stating that Lord Russell had recommended me for the Bath, regretting that no K.C.B was vacant, & asking if I would like C.B. The letter went on to ask, if I would accept Knighthood en attendant a K.C.B. after longer service—[2] I wrote by return of post, accepting CB. *for services*, & declining K^t^hood— the latter I did on various grounds, partly—because it signifies nothing,—whilst the CB. recognizes *services*, which is the only recognition I care for—& because if they wanted to Knight me (& I do not *wish* for Knighthood) they might have offered it in an order that indicated special Services—[3]

This said & done, I heard no more of the CB., till I saw it in the papers, & have heard nothing since. Nor had I a word from Lord Russell, though I wrote to thank him, through my Lady.—[4]

Murchison & Lyell[5] have however strongly recommended me to the D. of Argyll for the Knight Commandership of the Star of India—but I do not think there is the least chance of my getting the offer of it.—[6] Lyell has seen the Duke several times about it, who always says that he has "set his subs to seek official information on which to act"! Can you conceive such humbug? he may look long enough & find nothing— The E.I.C. absolutely ignored me & refused me every kind of countenance & assistance. The K.C.S.I. is so rare an honor that I might well be proud to have it—for my India services;[7] but I really do not desire Knighthood, & would infinitely rather be plain Dr Hooker with CB to testify to my having done my duty as well as others who have *that certificate*.— So if it comes I shall be proud of it, if not I shall be as well content. Please say nothing about it— The fact is the D. might do it with a stroke of his pen, but he doesn't like my Darwinism & my Address, & I am right proud of that.![8]

And now to more congenial subjects.

I have just come up from Brodie's pretty place on the top of Box hill—& left my wife & Brian till tomorrow—[9] I like B,—exceedingly; only Odling was there; we talked much of P.R.S. supposing Sabine was to retire or "slip his wind"—& we agreed upon Lyell.[10] I have seen Huxley since who tells me that the subject was mooted at the last X club (I was absent) & Lyell, all agreed, was the man—[11] I know that this is the Botanists opinions too, & the only difficulty is, is he able (physically) to take it for a year even?— I am to make out & shall write to Lady Lyell[12] enquiring—

I am greatly disappointed with "Nature", & so is every one I have talked to about it— I had a long talk with the publisher MacMillan about it today—[13] I would recommend that they drop the articles of the style of the "Academy", which they cannot compete with—& give 1. a Summary of week's Scientific news like the Spectator.,[14] 2 Correspondence— 3 good plain descriptive reviews of books— 4 abstracts of Society's meetings—try in fact to make it more simple & *useful*. There is plenty of room for a *good* journal of the kind— none for an inferior one.— What did you think of Huxleys rhapsody on Goethe's ditto—[15] I must say I could not share in his admiration, & the General public could not make head or tail of it.—

I saw Sylvester today who is raving about you[16]—also Huxley who is full of lecturing & Reptilian bones.

No news from Willy by last mail, but Hector writes that he is well, getting fatt, & that he will look out for a ship for sending him home.[17]

I am so glad that you liked my review of Schimper.[18] I was very glad to write it, I could praise so heartily.—

I am working hard at my British Flora,[19] the number of little points wanting attention is wonderful. I find that to do the descriptive part *well*, is all enough for one effort; & I count on doing the other for a companion volume.

Ever Yr affectionate | Jos D Hooker

DAR 103: 35—8

1 See letter to J. D. Hooker, 13 November [1869] and n. 2.

2 Hooker refers to Roderick Impey Murchison; William Ewart Gladstone, the prime minister; John Russell, first Earl Russell, and two different classes of the Order of the Bath; see letter to J. D. Hooker, 13 November [1869] and nn. 2 and 3.

3 See also L. Huxley ed. 1918, pp. 145–7.

4 The reference is probably to Fanny Russell, wife of John Russell.

5 Charles Lyell.

6 The Star of India was an order founded in 1861; it had three classes: Companion, Knight Commander (KCSI), and Knight Grand Commander (*EB*). George Douglas Campbell, eighth duke of Argyll, was the secretary of state for India.

7 For more on Hooker's frustrations with the East India Company during and after his India expedition, and on Hooker's eventual recognition by the Order of the Star of India, see R. Desmond 1999, pp. 201, 217–19, 250–2. See also *Correspondence* vol. 13, letter from J. D. Hooker, [17 February 1865] and n. 8.

8 Hooker refers to his 19 August 1868 presidential address at the meeting of the British Association for the Advancement of Science at Norwich; see J. D. Hooker 1868, and *Correspondence* vol. 16, letter to J. D. Hooker, 23 August [1868], and letter from J. D. Hooker, 5 September 1868.

9 Benjamin Collins Brodie owned a house on Box Hill in the North Downs of Surrey (*ODNB*). Brian Harvey Hodgson Hooker was a son of Hooker and Frances Harriet Hooker.

10 Hooker refers to William Odling and to Edward Sabine. Sabine been president of the Royal Society of London since 1861 (*ODNB*).

11 The X Club met informally for dinner on the first Thursday of every month, and was established primarily for younger men of science united by friendship, and what one member called a 'devotion to science, pure and free, untrammelled by religious dogmas' (quoted in Barton 1998, p. 411; see also *Correspondence* vol. 13). For the club's efforts to change the Royal Society president in 1869, see Barton 1990, p. 64. Hooker refers to Thomas Henry Huxley.

12 Mary Elizabeth Lyell.

13 The first issue of *Nature*, a 'weekly illustrated journal of science', was dated 4 November 1869. The publisher was Alexander Macmillan. For the criticism by X-Club members, including Hooker, of *Nature*, see Barton 2004.

14 The *Academy* was a monthly review of literature, science, and art, established in October 1869 (see letter from John Murray, 18 September [1869]). On the *Spectator*, see *Correspondence* vol. 12, letter from J. D. Hooker, 29 March 1864, n. 32.

15 The opening article in the first number of *Nature* was entitled 'Nature: Aphorisms by Goethe', and included a commentary by Huxley (T. H. Huxley 1869b). The essay has often been attributed to Johann Wolfgang von Goethe, but was in fact written by the Swiss theologian Johann Georg Christoph Tobler in 1783 (McVaugh 2004, p. 792; see also Miller ed. and trans. 1988, pp. 3–7, and Harrison *et al.* eds. 2000, pp. 846–8).

16 For James Joseph Sylvester's earlier admiration of CD, see *Correspondence* vol. 16, letter from E. A. Darwin, [11 October 1868].

[17] James Hector had taken some responsibility for Hooker's son, William Henslow Hooker, while he was in New Zealand (see letters from J. D. Hooker, 24 June 1869, 17 July 1869, and 13 August 1869).

[18] See letter to J. D. Hooker, 13 November [1869] and n. 13.

[19] Hooker was working on *The student's flora of the British Islands* (J. D. Hooker 1870); see letter from J. D. Hooker, 11 March 1869.

To J. J. Moulinié 15 November [1869][1]

Down. | Beckenham | *Kent. S.E.*
Nov 15

My dear Sir

I am glad to say that I have heard from M. Reinwald that he will publish, & I have told Mr Murray to send you a copy *at once.*

If Prof. Vogt will aid you I sh[d] think it w[d] be a great advantage, & his name is a tower of strength.[2]

I strongly advise you to procure or borrow the 2nd or 3rd edit of M[lle] Royer's translation, in order that the more important terms such as "concurrence" & "selection" (which she did not use in the 1st edit.) may be the same.[3] As your translation will have to compete with M[lle] Royer's, who I am told writes very spirited French, you will I think have to keep this in view in regard to style. I will, in a few days send you a few trifling corrections which do not relate to the early parts, & likewise a prefatory letter.

I sincerely hope that for all our sakes this translation may prove successful & I am much obliged to you for under taking it.

Believe me my dear Sir | yours very faithfully | Ch. Darwin

LS
Bibliothèque de Genève, Ms. suppl. 66, f. 13–14

[1] The year is established by the relationship between this letter and the letter from C.-F. Reinwald, 13 November 1869.

[2] Charles-Ferdinand Reinwald was to ask Carl Vogt for assistance with the French translation of the fifth English edition of *Origin* (see letter from C.-F. Reinwald, 13 November 1869). CD also refers to John Murray.

[3] CD refers to Royer trans. 1862, Royer trans. 1866, and Royer trans. 1870. Clémence Auguste Royer used 'concurrence vitale' for 'struggle for existence'. For Royer's use of 'sélection' in Royer trans. 1866, see *Correspondence* vol. 13, letter from C. A. Royer, [April–June 1865] and n. 3. See also Harvey 1997, pp. 62–79.

From Adolf Bernhard Meyer[1] 16 November 1869

Zimmerstrasse 77. 1 Treppe | Berlin,
Nov. 16. 1869.

Hochverehrter Herr

Ich habe Ihnen noch Dank zu sagen für Ihre so freundliche Zuschrift vom 22.

Sep. vorigen Jahres.[2] Heute war ich so frei Ihnen per Post ein Exemplar der Ubersetzung des Werkes des Herrn Wallace zu übersenden,[3] das ich dem deutschen Publicum vorlege, damit Sie, verehrter Herr, von Neuem sehen wie sehr wir Deutschen die vortrefflichen Werke englischer Naturforscher schätzen und uns beeilen ihnen baldmöglichst Eingang zu verschaffen. Da ich zuerst den malayischen Archipel bereisen will[4] so bot mir dieses geistreiche Werk das bedeutendste Interesse und beim Studium desselben beschloss ich es meinen Landsleuten besser zugänglich zu machen durch eine deutsche Ausgabe.

Allein es ist dieses nicht der Anlass meines heutigen Schreibens, sondern es ist der folgende:

Ich wurde von einem Verleger aufgefordert jene erste Kundgebung über die neue Theorie der "Origin of species", die von Ihnen und Herrn Wallace im Jahre 1858 in dem "Journal of the Proc. of the Linnean Society" gegeben ist, zu übersetzen und dem deutschen Publikum vorzulegen, da diese Briefe und Abhandlungen von so bedeutendem—nicht nur historischen Interesse sind.[5] Sie würden in Form einer kleinen Broschüre erscheinen, der ich gern als Einleitung einen kurzen Abriss Ihrer und Herrn Wallace's Lebensgeschichte gäbe.[6]

Meine Bitte an Sie verehrter Herr geht nun dahin

1) mich zu autorisiren jene Abh. zu veröffentlichen im Deutschen und

2) mir einige Data über Ihre Lebensgeschichte an die Hand zu geben welche das deutsche Publikum im hohen Masse interessiren, da dasselbe bei dem Abstand der englischen Verhältnisse von den deutschen über Vieles ununterrichtet ist. Vor Allem bitte ich um eine *vollständige* Liste Ihrer wissenschaftlichen *Abhandlungen* und Werke.

Meine Vorbereitungen zu den tropischen Reisen sind nun bald beendet, so dass ich nur noch die Abwickelung einiger Familienverhaltnisse abzuwarten habe um fortgehen zu können. Ich hoffe im Frühjahr England zu besuchen. Ich habe mit Freude vernommen dass Ihre Gesundheit sich wieder gekräftigt hat, welche frohe Nachricht hier grosse Genugthuung gewährte.— Eine kleine Abhandlung von mir über die Giftapparate der Schlangen, welche ich Ihnen sandte, kam hoffentlich in Ihren Besitz.[7]

Indem ich Sie, hochverehrter Herr, freundlichst bitte mir bald eine geneigte Antwort zukommen zu lassen und indem ich wegen meines Verlangens Ihre Absolution und Würdigung erbitte

Zeichne ich | hochachtungsvoll | Dr Adolf Bernhard Meyer

DAR 171: 166

CD ANNOTATIONS[8]

Top of letter: Poison— fang— | My Health | Wallace's address | England Spring | Translator of Wallace | Copy of Paper in Linn. Journal

[1] For a translation of this letter, see Appendix I.

[2] CD's letter to Meyer of 22 September 1868 has not been found.

[3] The reference is to Wallace 1869c, a translation of Alfred Russel Wallace's *The Malay Archipelago: the land of the orang-utan, and the bird of paradise* (Wallace 1869a).

[4] Meyer spent at least three years (1870 to 1873) travelling and collecting in the Celebes, visiting New Guinea and the Philippines (*Auk* 28: 519).

[5] Meyer refers to the joint publication of CD and Wallace's papers on natural selection. 'On the tendency of species to form varieties; and on the perpetuation of varieties and species by natural means of selection' was published in the *Journal of the Proceedings of the Linnean Society (Zoology)*, and included the two essays, an abstract of a letter from CD to Asa Gray, and an introductory explanation by Charles Lyell and Joseph Dalton Hooker (see C. Darwin and Wallace 1858).

[6] Meyer published *Charles Darwin und Alfred Russel Wallace. Ihre Ersten Publicationen über die 'Entstehung der Arten' nebst einer Skizze Ihres Lebens und einem Verzeichniss Ihrer Schriften* in 1870; a copy is in the Darwin Pamphlet Collection–CUL.

[7] There is a lightly annotated copy of Meyer 1869a in the Darwin Pamphlet Collection–CUL.

[8] CD's annotations are probably notes for his letter to Meyer of 20 November 1869; the letter has not been found, but see the letter from A. B. Meyer, 24 November 1869.

To F. C. Donders 17 November [1869][1]

Down | Beckenham | Kent S.E.
Nov: 17

My dear Sir

I thank you cordially for your extremely kind letter[2] There is no sort of hurry, for as is always the case, my work has gone on much slower than I expected, & I shall not come to the subject in question for some two months.[3]

I am delighted to hear that you have been making experiments, & shall be very anxious to hear what results you have arrived at.

I feel convinced that the question is all important to any one who is interested in expression.

It gave me sincere pleasure to make your acquaintance and your little visit here was very interesting to us all.[4]

With Mrs. Darwin's kind regards I remain my dear Sir | yours very sincerely | Charles Darwin

Copy
American Philosophical Society (B/D 25.315)

[1] The year is established by the relationship between this letter and the letter from F. C. Donders, 12 November 1869.

[2] See letter from F. C. Donders, 12 November 1869.

[3] CD had asked Donders questions for his research on human expression; see memorandum to F. C. Donders, 6 September 1869. At the time, CD still thought that this research would be included in *Descent*, but it was later published in *Expression*. See *Correspondence* vol. 18, letter to F. C. Donders, 21 June 1870.

[4] See letter from F. C. Donders, 12 November 1869 and n. 3.

From Albert Hugh Jones 17 November 1869

Shrublands, | Eltham, | S.E.
17 Nov 1869

Dear Sir,

On the other side I have much pleasure in sending you particulars as regards sexes of Lepidoptera, reared this season. I am sorry I have so few to record, but hope to do better for you next season.[1]

I am, Dear Sir | yours faithfully | Albert H Jones

Profr Darwin

Lepidoptera reared from Egg.

Eriogaster Lanestris	73	♀	68	♂
Orgyia Pudibunda[2]	17	♀	14	♂
	90	♀	82	♂

Lepidoptera reared from Larvæ collected

Bombyx Castrensis[3]	22	♀	32	♂
	112	♀.	114	♂[4]

DAR 86: 77–8

CD ANNOTATIONS

End of letter:

♂	♀
364	311[5]
114	112
478	423. *ink*

[1] No query to Jones has been found. In 1868, Jones enumerated the proportion of the sexes in nine species (DAR 85: B83); see *Descent* 1: 313. No later correspondence with Jones has been found.

[2] The moth *Eriogaster lanestris* is known as the small eggar. *Orgyia Pudibunda* is now *Calliteara pudibunda*, the pale tussock.

[3] *Bombyx castrensis* is now *Malacosoma castrensis*, the ground lackey.

[4] In *Descent* 1: 313, CD reported that Jones 'reared imagos from 4 species' resulting in 112 females and 114 males.

[5] CD calculated the ratio of 364 males to 311 females as equal to 117 males to 100 females on the back of DAR 85: B83, but it is not clear where these figures came from.

From John Murray 17 November 1869

50, Albemarle St. | W.
Novr 17— 1869.—

My Dear Sir

I have attended to the requests of your note of Monday[1]

At my annual sale last week I disposed of copies of your work

Fritz Muller	8	out of	545	
Variations	18	—	300	
Origin	311	—	1530	—

wch. I hope you will consider not a bad days work—[2]

You have my best wishes for the putting down of your Parisian Blaspheemers[3]

My Dear Sir | Yours very faithfully | John Murray

Chas Darwin Esqre. | &c &c &c

DAR 171: 374

CD ANNOTATIONS
2.2 545] '587' *right margin ink*
2.4 1530] '$\frac{311}{1219}$' *interl ink*

1 The Monday before 17 1869 November was 15 November. No letter to Murray of 15 November 1869 has been found.
2 The references are to William Sweetland Dallas's translation of F. Müller 1864a (Dallas trans. 1869), *Variation*, and the fifth edition of *Origin*.
3 See letter to John Murray, 8 November [1869].

From Camille Dareste[1] 18 November [1869][2]

Paris
18 novembre

Monsieur

Permettez-moi de recourir à votre obligeance, dans une circumstance bien grave pour moi.

Je suis actuellement candidat à une chaire de Physiologie comparée vacante à la faculté des Sciences de Paris. J'ai trouvé beaucoup d'appuis; mais aussi des adversaires qui, comprenant peu la voie Scientifique nouvelle dans laquelle je me suis engagé, ne veulent pas admettre que mes travaux sur la production artificielle des monstruosités puissent rentrer dans le cadre de la physiologie.[3]

Vous êtes, Monsieur, l'un des hommes qui ont le mieux compris l'importance de ces études auxquelles j'ai consacré ma vie; et la manière donc vous m'avez cité dans votre dernier ouvrage me prouve que vous les tenez en haute estime.[4]

Serait-ce un indiscrétion de ma part, de vous prier de m'envoyer une témoignage indiquant ce que vous pensez de mes travaux. Votre nom, et la grande position que vous vous êtes acquise dans la Science européenne donnerait à ce témoignage une grande valeur, et pourrait assurer la réussite de ma candidature.

Pardonnez-moi, Monsieur, de Vous demander ce service; mais la question est très grave pour moi, car notre organisation universitaire est si imparfaite, que nous pouvons avoir en province les ressources de toute sorte que les savants peuvent se procurer à Paris.[5]

Recevez, je vous prie, Monsieur, l'assurance de mon profond respect | Camille Dareste

à Paris: rue de fleurus 37 bis.

DAR 162: 45

1 For a translation of this letter, see Appendix I.
2 The year is established by the relationship between this letter and the letter to Camille Dareste, 20 November 1869.
3 For Dareste's research on animal monstrosities, see Dareste 1862, 1863, and 1867.
4 See *Variation* 1: 262, and 2: 289, 331, 340.

5 Dareste apparently forgot to include the 'ne ... pas' in the sentence that would have indicated the lack of resources in the provinces. Dareste was employed at the University of Lille (*DBF*).

To Christian Gottfried Ehrenberg 18 November [1869][1]

Down. | Beckenham | *Kent. S.E.*
Nov 18.

Dear Sir

I beg leave to thank you for the honour which you have done me in sending me your fine Memoir on the Bacillariæ of Mexico.[2] With my best thanks & the most sincere respect

I beg leave to remain | yours truly obliged | Charles Darwin

LS
Archiv der Berlin-Brandenburgische Akademie der Wissenschaften (ABBAW): (NL Ehrenberg, Nr. 321, Brief von Charles Darwin)

1 The year is established by the publication date of Ehrenberg 1869 (see n. 2, below).
2 The reference is to *Uber mächtige Gebirgs-Schichten vorherrschend aus mikroskopischen Bacillarien unter und bei der Stadt Mexiko* (Ehrenberg 1869).

From Roland Trimen 18 November 1869

Colonial Office, | Cape Town.
18th. November, 1869.

My dear Mr. Darwin,

I am very much obliged for your "Notes on the Fertilization of Orchids", received a fortnight ago, and for your kind notice of my observations in the paper.[1] I have examined several fresh Ophryds since my return hither,[2] but have not observed any contrivance that seems to require special notice.

What an extraordinary and unique device you notice in *Coryanthes*![3] There really seems to be no limit to Nature's versatility: it is impossible to conjecture anything too unusual or unlikely for her to accomplish. Every organism seems to be perpetually on the *qui vive* for the slightest chance of bettering its position, and it is difficult to separate the fact of this constant fitness and aptitude for making the utmost of opportunities from the idea of conscious and deliberate intent and action. One certainly cannot say at what point in the scale intellectual action or purpose absolutely ceases.

I have not forgotten your query about Baboons; but no person that I have spoken to on the subject can say from observation that males fight for possession of females, though at the same time I notice that no one seems to think such a proceeding at all unlikely.[4]

Mention of sexes reminds me that I made a note for you respecting the assembling of many ♂s of *Sphinx Convolvuli* about a single ♀.[5] Unfortunately, I cannot lay my hand on it, so as to refer to the work in which I noticed the case, but I am quite sure of the fact's being on record, and that it was observed in some part of England.

Among Moths whose underside of the wings exceed the upperside in beauty of marking and colouring, I find the South-American *Geometræ* belonging to the genus *Erateina* of E. Doubleday. Of those figured (in Trans. Ent. Soc. Lond., New Series, vol. V, pl. XV & XVI), the more striking in this respect are *E. Regina*, E. *Margarita*, and *E. obscura.*[6] In all these the upper surface is brownish or fuscous with a few white markings, while the under surface is elaborately varied with white, red, & chestnut-red, and in *E. Regina* with yellow also.

About a year ago, I witnessed and made a note of a case of sexual selection in a Moth, *Syntomis cerbera*. The ♀ in this instance had but recently emerged from the puparium, for her wings (she was clinging to the underside of a leaf of a tall grass) were still limp and hanging downward. When I first noticed her, she was closely beset by four ♂s. One of these flew off as I approached, but the remaining three persisted in their endeavours, clustering about in constant motion, and holding on by the ♀'s back and wings. I watched them closely; and though each ♂ repeatedly contrived to clasp the extremity of the ♀'s abdomen with his own, and the ♀ remained perfectly motionless, no junction was effected. After I had watched the group for about 5 or 6 minutes, one ♂ appeared to consider it a hopeless case, for he suddenly flew off. Shortly afterwards, another ♂ also departed rather abruptly. The ♂ that remained persisted for some little time in his efforts, and at length the obdurate ♀ yielded; but the union was incomplete, for, in attempting to assume the ordinary position in line with the ♀ on the narrow grass-leaf, the ♂'s struggles occasioned a separation.

A second union, however, immediately followed, the usual position was safely taken up, and the ♀ for the first time moved her wings into their ordinary place on each side of the body. I compared this successful suitor with many other ♂s taken in the same field on the same day, and am bound to say that there was nothing in his size or colouring to entitle him to preference; but, of course, I don't presume to compare my poor *ocelli* with Mrs. *Cerbera's* argus eyes, and there can be little doubt that she found attractions in him to which his disappointed rivals could lay no claim.

I may mention that there was a great number of these Moths in the field; I counted 24 pairs *in copulâ* on the grass stems, in addition to very many individuals. The Genus *Syntomis* is one of the family *Zygænidæ*, to which our English "Burnet Moths" also belong.[7]

I am sorry to say that I have less time than ever for Nat. History. A cry for what is called "Retrenchment" is now prevalent, and reductions are being made in the Civil Service, the effect of which in our Office is to give me much more work without the consolation of additional pay![8] Alas that the Age of Sinecures is well-nigh passed away! In *my* opinion, Government ought to provide appointments with little work and decent emoluments expressly for naturalists; but it is surprising how few of the people I meet share in this view!!

Hoping that your health has improved, and with kindest regards to Mrs. Darwin and your family, I remain | Very faithfully yours,

Roland Trimen.

DAR 82: 27–9

CD ANNOTATIONS
1.1 I am ... England. 4.4] *crossed blue pencil*
6.1 About ... also belong. 8.4] 'Zygænidæ' *added blue crayon*
8.2 The Genus ... belong. 8.4] *scored blue crayon*
8.3 *Syntomis*] *underl blue crayon*
8.3 *Zygænidare*] *underl blue crayon*
9.1 I am ... family, 10.2] *crossed blue crayon*

1 See 'Fertilization of orchids', pp. 143, 144, and 157.
2 Trimen left England for South Africa in May 1868; see *Correspondence* vol. 16, letter from Roland Trimen, 13 April 1868.
3 See 'Fertilization of orchids', p. 157; see also letter from Federico Delpino, 1 November 1869 and n. 2.
4 No query to Trimen on baboons has been found; however, on 27 December 1867, CD spoke to Trimen about the larger size of canines and manes in baboons of South Africa (DAR 83: 19).
5 In *Descent* 1: 311–12, CD mentioned the grouping of male moths around a female, but did not cite Trimen's example. *Sphinx convolvuli* is now *Agrius convolvuli*, the convolvulus hawkmoth.
6 Trimen refers to plates for the article 'On the genus *Erateina*, Doubl.; with descriptions of some new species' (Saunders 1860). Edward Doubleday was the author of *Erateina.* The lower and upper surface of the three species mentioned are illustrated in plate XV, figures 3 and 3a; plate XVI, figs. 2 and 2a; and plate XVI, figs. 3 and 3a. In *Descent* 1: 397, CD cited Trimen for information on the wings of *Erateina.* The former family Geometrae corresponds roughly to the modern family Geometridae.
7 *Syntomis cerbera* (the heady maiden moth) is now in the family Ctenuchidae. The valid name for the genus *Syntomis* is now *Amata.* Burnet moths (*Zygaena*) are still in the family Zygaenidae.
8 Trimen was a clerk of the second class in the Colonial Office at the Cape of Good Hope; between 1868 and 1870, his pay increased from £190 to £220 (*Colonial office list*, 1868, 1869, and 1870).

To J. D. Hooker 19 November [1869][1]

Down. | Beckenham | *Kent. S.E.*
Novr. 19

My dear Hooker.

Thank you much for telling me all about the C.B, for I wished much to hear. It pleases me extremely that the Government has done this much; & as the K.C.B.^s are limited in number (which I did not know) I excuse it.[2] I will not mention what you have told me to anyone, as it w^d. be Murchisonian. But what a shame it is to use this expression, for I fully believe that Murchison w^d. take any trouble to get any token of honour for any man of science.—[3]

I like all scientific periodicals, including poor "Scientific Opinion", & I think higher than you do of "Nature".[4] Lord what a rhapsody that was of Goethe;, but how well translated— it seemed to me, as I told Huxley, as if written by the maddest English scholar.[5] It is poetry, & can I say anything more severe? The last number of the Academy was splendid, & I hope it will soon come out fortnightly.—[6]

I wish "Nature" w^d search more carefully all foreign Journals & Transactions.— I am now reading a German thick pamphlet by Kerner on Tubocytisus; if you come across it look at the map of distribution of the 18 quasi species, & at the genealogical tree.[7] If the latter, as the author says, was constructed solely from the affinities of the forms, then the distribution is wonderfully interesting; we may see the very steps of the formation of a species. If you study the gen: tree & map you will almost understand the book. The 2 old parent connecting links just keep alive in 2 or 3 little areas; then we have 4 widely extended species, their descendants; & from them little groups of newer descendants inhabiting rather small areas.—

I must enjoy myself & tell you about Mad^elle^. C. Royer who translated the Origin into French, & for whose 2^d^ Edit I took infinite trouble.[8] She has now just brought out a 3^d^ Edit without informing me, so that all the corrections &c in the 4th & 5th English editions are lost. Besides her enormously long & blasphemous preface to 1st Edit, she has added a 2^d^ Preface, abusing me like a pick-pocket for pangenesis, which of course has no relation to the Origin—[9] Her motive being, I believe, because I did not employ her to translate "Domestic Animals".—[10] So I wrote to Paris; & Reinwald agrees to bring out at once a new Translation from the 5th English Edition, in Competition with her 3^d^ Edit.—[11] So shall I not salt her well? By the way this fact shows that "evolution of species" must at last be spreading in France.—

I rejoice that Pres. R. Soc will be offered to Lyell, though I cannot believe that he is strong enough for labour.—[12]

Yours most affect^ly^ | C. Darwin

What a pity it is, as your boy is well, "fat" & happy in N. Zealand, that he cannot find a home for a time at least there.—[13]

DAR 94: 159–61

1 The year is established by the relationship between this letter and the letter from J. D. Hooker, 14 November 1869.

2 See letter from J. D. Hooker, 14 November 1869.

3 Roderick Impey Murchison had recently recommended Hooker for the KCSI (Knight Commander of the Star of India); see letter from J. D. Hooker, 14 November 1869. For more on Hooker's and CD's use of 'Murchisonian', see *Correspondence* vol. 11, letter from J. D. Hooker, 15 September 1863 and n. 16.

4 On the weekly newspaper *Scientific Opinion*, see the letter to J. D. Hooker, [22 January 1869] and n. 8. For Hooker's opinion of the new journal *Nature*, see the letter from J. D. Hooker, 14 November 1869.

5 Thomas Henry Huxley translated an essay mistakenly attributed to Johann Wolfgang von Goethe for the first issue of *Nature* (T. H. Huxley 1869b, see letter from J. D. Hooker, 14 November 1869 and n. 15). The essay was in fact by Johann Georg Christof Tobler. For another translation, see Miller ed. and trans. 1988, pp. 3–5.

6 See letter from J. D. Hooker, 14 November 1869 and n. 14.

7 CD refers to Anton Kerner von Marilaun's pamphlet *Die Abhängigkeit der Pflanzengestalt von Klima und Boden. Ein Beitrag zur Lehre von der Entstehung und Vergreitung der Arten, gestützt auf die Verwandtschaftsverhältnisse, geographische Verbreitung und Geschichte der Cytisusarten aus dem Stamme Tubocytisus D. C.* (Kerner von Marilaun 1869). The map and the genealogical tree are at the end of the article. The tribe *Tubocytisus* has been subsumed within *Cytisus*.

8 On Clémence Auguste Royer's translations of *Origin*, see the letter to John Murray, 8 November [1869] and n. 5. For CD's work on her second edition (Royer trans. 1866), see *Correspondence* vol. 13, letter from C. A. Royer, [April–June 1865], n. 5. See also Harvey 1997, pp. 76–9.

9 CD refers to Royer trans. 1870, Royer's third French translation of *Origin*; the translation did not include the changes in the fourth and fifth English editions of *Origin*. On the prefaces to the editions, see the letter to John Murray, 8 November [1869] and n. 5.

10 *Variation* was translated by Jean Jacques Moulinié, as suggested by Carl Vogt (Moulinié trans. 1868; see *Correspondence* vol. 15, letter from Carl Vogt, 23 April 1867).

[11] CD wrote Charles-Ferdinand Reinwald of Paris on 10 November 1869; the letter has not been found. Reinwald agreed to publish Moulinié's translation of the fifth edition of *Origin* (Moulinié trans. 1873) in his letter of 13 November 1869.

[12] Hooker had mentioned Charles Lyell as a possible president of the Royal Society of London in his letter of 14 November 1869.

[13] CD refers to William Henslow Hooker; see letter from J. D. Hooker, 14 November 1869.

To J. V. Carus 20 November 1869

Down Beckenham | Kent
Nov 20 1869

My dear Sir

If not too late, will you add to the Origin the few enclosed additions which I think are worth inserting.[1] I enclose another application about the translation of my next book—[2]

My dear Sir | yours sincerely | Ch. Darwin

LS
Staatsbibliothek zu Berlin – Preußischer Kulturbesitz, Slg. Darmstaedter Lc 1859 Darwin, Charles, Bl. 50

[1] Carus was translating the fifth edition of *Origin* into German (Bronn and Carus trans. 1870); see letter to Eduard Koch, 4 November 1869. The enclosure has not been found.

[2] The application regarding the translation of *Descent* has not been found. See also letter to J. V. Carus, 9 November [1869].

To George Cupples 20 November [1869][1]

Down. | Beckenham | *Kent. S.E.*
Nov. 20

My dear M[r] Cupples

It is indeed very superfluous on your part to wish that you could have given me more assistance. The facts which you have communicated have been of very high interest to me; & nobody in the whole world could have been kinder in taking so much trouble for me.— The data for all that I have to say about the Scotch deer-hound are, owing to you, really almost sufficient, & much better data than I have got in many other cases.—[2]

I have not D[r] Stirling's paper at hand, for I have lent it, but I think he w[d] be compelled to admit some change in the famous protoplasm in our domestic races, both in regard to the structure of the body & qualities of the mind.—[3]

Yours very sincerely obliged | Ch. Darwin

Det Kongelige Bibliotek, Copenhagen (NKS 4127 II, 4^0)

[1] The year is established by the publication date of Stirling 1869 (see n. 3, below).

[2] For the most recent extant information on Scottish deer-hounds sent by Cupples, see the memorandum from George Cupples, [4–10 November 1869]. There is an empty envelope addressed to Cupples in Emma Darwin's hand, postmarked 12 November 1869 at the American Philosophical Society.

[3] On Stirling 1869, see the letter to J. D. Hooker, 13 November [1869] and n. 7.

To Camille Dareste 20 November 1869

Down | Beckenham | Kent. S.E.
Nov[r]. 20— 1869.

Dear Sir

I am glad that you are a candidate for the Chair of Physiology in Paris—[1] As you are aware from my Published Works I have always considered your investigations on the production of monstrosities as full of interest. No subject is at the present time more important, as far as my judgment goes, than the ascertaining by experiment how far structure can be modified by the direct action of changed conditions; & you have thrown much light on the subject—[2]

I observe that several naturalists in various parts of Europe have lately maintained that it is now of the highest interest for Science to endeavour to lessen, as far as possible, our profound ignorance on the cause of each individual variation; &, as Is: Geoffroi S[t]. Hilaire long ago remarked, monstrosities cannot be separated by any distinct line from slighter variations.[3]

With my best wishes for your success in obtaining the Professorship & with sincere respect—

I have the honour to remain | dear Sir | Yours faithfully | Charles Darwin

To | M. Camille Dareste

Copy
DAR 143: 369

[1] See letter from Camille Dareste, 18 November [1869].

[2] See letter from Camille Dareste, 18 November [1869] and nn. 3 and 4. See also *Variation*, chapter 23, 'Direct and definite action of the external conditions of life' (*Variation* 2: 271–92).

[3] CD refers to the French zoologist Isidore Geoffroy Saint-Hilaire, and to his statement in Geoffroy Saint Hilaire 1832–7, 3: 352; CD cited this comment in *Variation* 2: 254.

To J. J. Moulinié 20 November 1869

Down Beckenham | Kent
Nov 20. 1869

My dear Sir

I enclose the few small additions & corrections for the Origin.[1] I also send my letter which will serve as a sort of preface to y[r] translation, but pray consider it well & suggest any alterations;[2] it should also hereafter be shewn to M. Reinwald for his approval & correction.[3]

My dear Sir | yours sincerely | Ch. Darwin

LS
Bibliothèque de Genève, Ms. suppl. 66, f. 16-17

[1] Moulinié was translating the fifth edition of *Origin* into French (Moulinié trans. 1873; see letter to J. J. Moulinié, 15 November [1869]). The enclosure has not been found.
[2] The letter to Moulinié has not been found. However, since the French translation was not published until 1873 (and included changes from the sixth English edition), CD wrote another letter to Moulinié, dated 23 September 1872, which appeared in Moulinié trans. 1873, pp. ix–x.
[3] Charles-Ferdinand Reinwald was the publisher of Moulinié trans. 1873.

From J. D. Hooker 21 November 1869

Royal Gardens Kew
Nov 21 /69

Dear Darwin

I have been corresponding with MacMillan about Nature, which I find too desultory, & like you, not half full enough in the matter of foreign journals nor indeed of English Societies.[1]

I must get Kerner on Tubocytisus, Oliver does not know it—[2] the worst of such papers is, that it must be a *mere guess* which form to fix on as the type. May not the supposed 2 old parent like tribes, be two modern varieties?— imagine them destroyed, or not discovered, & would not the whole subject have to be treated from quite another assumed type—? I thought of adding a chapter on this subject to the Australia Essay[3] when I wrote it, but could not for the life of me find reasonable evidence of any one species being the earliest—; Something might be done in the case of very old genera like Juglans[4] which we *believe* have spread Westward, whence we may assume that the Eastern forms are the oldest, but that is at once met by the facts, that if the genus has been longest in the East it has had as much time to vary there as to spread to the west!—

What a sell for Mlle Royer serve her right—[5] how *nasty* women are when spiteful, & French women perhaps the worst in the world.

I met young Agassiz at Huxleys two nights ago, but I was recovering from a horrid headache & could not enjoy a *very* pleasant party of Flowers & Maskelynes Young Agassiz seems a nice fellow & his wife too.[6]

Lyell absolutely declines P.R.S. & we look to Grove: I put it to him last night, & he will accept if offered, but will have no fights with Sabine—[7] I assured him that it was only in case of Sabines retirement or being called to his maker (in which case I pity the latter for he wont know what to do with him) that the question of a successor would arise.

It *is* a pity—that Willy cannot get *located* in N.Z. for I shall not know what to do with him when he returns[8]

I think you will like to see Lady L.^{s} letter so I enclose it.[9]

Ever yrs affec | J D Hooker

Pray do not C.B. your letters to me— I can't stand it.— I own CB gratifies me in a *service* point of view, & it is very useful officially Indeed for clerical correspondence,— but scientifically I rather dislike it.[10]

Private.

P.S. Lyell had a long talk with me about K.t—which he *seems greatly to regret my not accepting*— I could not tell him, how gladly I "not accepted" nor how much I do *not* regret it: nor could I make him understand my feelings in the matter.—that I have no wish **whatever** to be Sir J. at all—but that *if* this must be some day, the least objectionable way is clearly in the Bath, the Bath being a recognition of *public* services &c.[11] I have no *wish* even for that, but I quite feel the force of all my friends wishing it, of my public position suiting it, & of keeping up, (what I & my relations do & my children will take a reasonable pride in) the public recognitions won by so many members of my family *without Court or aristocratic influence.* If I had raised myself wholly unaided, to my scientific position, as Lyell himself, & my father, & Murchison, & Palgrave did, to the positions they attained, then I should have felt that I had earned Knighthood *as they did*, & might have accepted: but my case is wholly different— my Science I owe to my father, ditto my Kew position. My services have been *wholly* under Govt. & if I am entitled to any such recognition as Knighthood at all it is one given for services unmistakeably.[12]

As it is, I am in the horns of a dilemma— I could be Knighted for the saying I wished it tomorrow— the declining is interpreted into despising it, in preference to a riband which I am not offered. Lyell & Murchison say—"take the K.t as a step to K.C.B."— this would be all very well if I really *wanted* the K.C.B.! though even then I do not think I could have *stooped* to any such dodge. Huxley & Lyell are the only persons with whom I have talked over this matter. Huxley quite understands & approves: but then he *despises* K.ts, which I do not.

DAR 103: 39–41

1 See letter to J. D. Hooker, 19 November [1869] and n. 4; *Nature* was published by Alexander Macmillan.

2 The reference is to Kerner von Marilaun 1869; see letter to J. D. Hooker, 19 November [1869] and n. 7. Hooker also refers to Daniel Oliver.

3 J. D. Hooker 1859.

4 *Juglans* is the walnut genus.

5 See the letter to J. D. Hooker, 19 November [1869], for CD's discussion of French translations of *Origin*, including those of Clémence Auguste Royer.

6 Alexander and Anna Russell Agassiz were visiting Britain (see G. R. Agassiz ed. 1913, p. 97). Thomas Henry and Henrietta Anne Huxley hosted the anatomist William Henry Flower, the mineralogist Nevil Maskelyne, and their wives or other family members.

7 For Hooker's interest in Charles Lyell as a possible successor to Edward Sabine as president of the Royal Society, see the letter from J. D. Hooker, 14 November 1869. Hooker also refers to William Robert Grove.

8 William Henslow Hooker was in New Zealand; see letter from J. D. Hooker, 14 November 1869.

9 Hooker had asked Mary Elizabeth Lyell about her husband's physical fitness for the presidency of the Royal Society; see letter from J. D. Hooker, 14 November 1869. The enclosure has not been found; it is not in the Director's Correspondence, Library and Archives, Royal Botanic Gardens, Kew.

10 On Hooker's receiving a CB, see the letter from J. D. Hooker, 14 November 1869.

11 On Hooker's declining the offer of a knighthood, while waiting for a KCB to become vacant, see the letter from J. D. Hooker, 14 November 1869. On the civil KCBs as an award for service in the mid nineteenth century, see Risk 1972, pp. 71–3.

[12] Charles Lyell, William Jackson Hooker, Roderick Impey Murchison, and Francis Palgrave had all been knighted. On the public recognition received by members of Hooker's family, see Allan 1967. For Hooker's father as director of the Royal Botanic Gardens, Kew, and Hooker's replacing him in 1865, see also R. Desmond 1995 and 1999, and Bellon 2001.

From George Henslow 22 November 1869

St John's Parsonage | St John's Wood
Nov. 22 /69

Dear Sir

Mr Henderson Junr. florist of S. Johns wood informed me that *Lapageria rosea* never sets seed unless artificially impregnated.[1] The anthers cluster round the stigma & cover it with pollen; but it appears, from his account, to be powerless on the stigma of the same flower. I should be glad to know if this has been noticed by yrself or others. If not, it would seem to be another case analogous to *Linum Grandifolium Passiflora sp.* & other well known instances.[2]

Yrs truly | G Henslow

DAR 166: 170

[1] The reference is probably to Andrew Henderson of Henderson & Son, nurserymen, 3 Wellington Road NW (*Post Office London directory* 1869, and R. Desmond 1994). Joseph Dalton Hooker had advised CD to obtain *Lapageria rosea*, the Chilean bellflower, from this nursery in his letter of 9 [March] 1864 (*Correspondence* vol. 12).

[2] CD described self-sterility in *Linum* in 'Two forms in species of *Linum*'. For self-sterility in *Passiflora*, see Scott 1864, Munro 1868, and *Correspondence* vols. 10 and 11. *Linum grandifolium* is an error for *Linum grandiflorum*.

From Charles Layton 22 November 1869

16 Little Britain (City)
November 22/69

Sir

Messrs Appleton & Co (Publishers) New York have directed me to write to you and ask if you will supply them with a *Set* of *Stereotype Plates* of your *New book* for the American Market,[1] they will pay the cost of Plates, and give you copyright of 10% on the sale of copies, same as they do with Profr Huxley, Profr Bain[2] and Mr. Herbert Spencer— They would also like to have a *Set of Stereotype Plates* of *New Editn Origin of Species* on same terms[3] an early reply will oblige

Yours Respectfully | Charles Layton | Agent D. Appleton & Co

To | Mr Darwin

Not knowing your address I send this to care of Mr Murray—[4]

DAR 159: 83a

[1] The reference is to *Descent.* On the practice of American publishers of printing from stereotypes, see the letter from Orange Judd & Co., 21 April 1869 and n. 3.
[2] The references are to Thomas Henry Huxley and the philosopher Alexander Bain.
[3] The fifth edition of *Origin* was published in June 1869 (letter from R. F. Cooke, 22 June 1869).
[4] John Murray was CD's publisher.

From George Bentham 23 November 1869

25. Wilton Place. | *S.W.*
Nov 23 /69

My dear Mr Darwin

I understand that you have passed an indulgent judgement on my last Address but at the same time you have met in it with some points on which I am wrong—or what amounts to the same thing—on which you differ from me[1] I do not like encroaching upon your time yet I should feel very much obliged if you could in a few words point out to me what these errors or mistaken views are in order that I may reconsider them and stand corrected. All questions bearing on Geographical distribution have long had a peculiar interest for me and now more particularly since a better acquaintance with the African and Australian floras seem to have placed many things in a different light from that in which they previously appeared, and I am very anxious not to go astray in considering them[2]

What do you think of Galton's Hereditary Genius?[3]

Yours very sincerely | George Bentham

DAR 160: 164

[1] Joseph Dalton Hooker evidently mentioned to Bentham CD's comments on Bentham's anniversary address to the Linnean Society of London (Bentham 1869b); see the letter to J. D. Hooker, 13 November [1869].
[2] Bentham's address was on the subject of 'Geographical Biology' (Bentham 1869b, p. lxv); he considered the Australian and African floras on pp. lxxxvi–lxxxviii; see also pp. lxxiv–lxxv.
[3] The reference is to Francis Galton's *Hereditary genius: an inquiry into its laws and consequences* (Galton 1869).

To J. D. Hooker 24 November [1869][1]

Down. | Beckenham | *Kent. S.E.*
Nov. 24th.

My dear Hooker

I return Lady Lyell's very nice note as you might wish to have it back.[2]

The title of A. Kerner's book is "Die Abhangigkeit der Pflanzen von Klima & Boden: 1869 Innsbruck.— If you like I could lend it you, & finish reading it when returned to me.[3] He does not give the *one* parent form of the genus, but selects two intermediate forms as probably the earliest. The distribution of the sub-species near the margins of the larger areas inhabited by some of the other species is very curious; as is his explanation of this fact.[4]

Please tell Bentham that I will write to him in a few days. Many thanks for your last note—[5]

yours affectly | Ch. Darwin

I may do it inside as I like the looks—

To D^r. Hooker C.B.[6] F.R S. | &c &c &c

LS(A)
DAR 94: 162–3

[1] The year is established by the relationship between this letter and the letter from J. D. Hooker, 21 November 1869.

[2] Hooker had enclosed a letter from Mary Elizabeth Lyell with his letter of 21 November 1869. The letter has not been found.

[3] For CD's earlier comments on Kerner von Marilaun 1869, see the letter to J. D. Hooker, 19 November [1869]. A. Kerner: Anton Kerner von Marilaun.

[4] For Hooker's comment on Kerner von Marilaun's plant genealogy, see his letter of 21 November 1869.

[5] See letter from George Bentham, 23 November 1869. Hooker's most recent extant letter is that of 21 November 1869.

[6] CD may be referring to writing 'C.B.' inside a letter, rather than on the envelope; in his letter of 21 November 1869, Hooker had written: 'Pray do not C.B. your letters to me—'. For more on Hooker's newly acquired CB, see the letter from J. D. Hooker, 14 November 1869.

To Charles Layton 24 November [1869][1]

Down | Beckenham
24 November

[. . .] I have long earnestly wished for a new edition of the Origin in the United States, as it is 92 pages longer than the 2nd. edition, besides endless small though important corrections.[2] I feel sure that the continued large sale of this book in England Germany & France has depended on my keeping up each edition to the existing standard of science.[3] I hope I am right in supposing that Messrs. Appleton are willing to print in some form a new edition, for though unwilling to act in a disobliging manner towards them I had resolved soon to write to Professor Asa Gray to ask him to find some publisher who would print the new edition of the Origin, on condition of my supplying him with the sheets of my new book as they were printed & which book will probably have a large sale[4] . . . I should like in case the answer is favorable to send in M.S. half a dozen small corrections for the Origin[5] [. . .]

Incomplete[6]
Parke-Benet (31 March 1964), Sotheby's, London (21 May 1968)

[1] The year is established by the relationship between this letter and the letter from Charles Layton, 22 November 1869.

[2] See letter from Charles Layton, 22 November 1869. The first US edition of *Origin* was published from stereotypes of the second English edition by D. Appleton & Co. in 1860.

[3] On foreign language translations of *Origin*, see Freeman 1977.

4 CD had asked Orange Judd & Co, the American publishers of *Variation*, about the possibility of publishing a new American edition of *Origin*; see the letter from Orange Judd & Co., 21 April 1869. CD did not favour use of stereotype plates as suggested in the letter from Charles Layton, 22 November 1869; see letter to Asa Gray, 1 June [1869] and n. 9.

5 CD had just sent a few corrections to the fifth English edition of *Origin* (1869) for its publication in French and German (see letter to J. V. Carus, 20 November 1869, and letter to J. J. Moulinié, 20 November 1869).

6 The original letter is complete and is described in the sale catalogues as being four pages long.

From A. B. Meyer 24 November 1869

Zimmerstrasse 77 | Berlin
24. Nov 1869

Dear Sir

I have just received your kind letter of Nov. 20. with a copy of the list of your works and papers and the sketch of your life and beg to offer you my thanks for the same.[1]

The editor of the papers in question wishes to insert as frontispice your photograph as well as Mr. Wallace's, a wish that I approve highly.[2] If you are inclined to consent to this—and I hope you do—I should think it best that the photographs should be taken in England from the original negatives and care being taken the size of each be the same. But perhaps you will not object to have your photograph taken with Mr Wallace in *one* picture, and if this idea meets your approval you might perhaps carry it out when you go to London at Christmas time or earlier if *possible*. At all events I shall be very glad if you will arrange about the matter with Mr. W., to whom I also prefer the same request today.

500 copies will be required and I shall of course bear the expense, although I do not know to what it will amount.

Mr. W. also sent me a list of his papers and a sketch of his life.

I willingly accept your kind offer to send a copy of your and Mr W's paper, which I am about to translate, for the copy I am now using does not belong to me.[3]

I hope the translation of Mr. W's work is come to hand.[4] I took the liberty of enclosing a new paper of mine, concerning the influence of the par vagum on the heart, which you will be kind enough to accept.[5]

Before leaving Europe I shall certainly follow your advice and examine a few caves, for instance in Belgium and those near the Rhine and hope when in the tropics to be able to make use of the knowledge I shall there obtain.[6]

On Pag. 50 of your paper: "on the variation etc" you quote the "dancing rock-thrush of Guiana". I do not know if that is another title for the "cock of the rock", the Rupicola aurantia (crocea?), which I believe is meant by it.[7] Please let me know.

Awaiting a favourable answer in respect of the photographies | I am | Yours very faithfully | D[r] Adolf Bernhard Meyer

Charles Darwin Esq

DAR 171: 167

CD ANNOTATIONS
6.1 I took ... heart, 6.3] *triple scored blue crayon*
8.3 the Rupicola aurantia (crocea?),] *double scored blue crayon*
8.3 Rupicola aurantia] *underl blue crayon*

1 CD's letter to Meyer has not been found. For the information that Meyer requested, see the letter from A. B. Meyer, 16 November 1869.

2 No photograph of CD or Alfred Russel Wallace is included in Meyer 1870; there is no frontispiece.

3 The reference is to C. Darwin and Wallace 1858.

4 Meyer had sent CD a copy of Wallace 1869c; see the letter from A. B. Meyer, 16 November 1869.

5 Meyer refers to his short book *Das Hemmungsnervensystem des Herzens, Kritisches und Experimentelles* (Meyer 1869b). The *par vagum* is now called the pneumogastric nerve.

6 The letter with CD's advice has not been found; however, CD had advised another traveller to look for fossil bones in caves in the Malayan Archipelago; see *Correspondence* vol. 14, letter to Cuthbert Collingwood, 16 February [1866]. Meyer was soon leaving for south-east Asia (see letter from A. B. Meyer, 16 November 1869).

7 Meyer refers to C. Darwin and Wallace 1858. *Rupicolia aurantia* (now *R. rupicola*) is known as the Guianan cock-of-the-rock; see also *Correspondence* vol. 15, letter from J. V. Carus, 18 January 1867. In *Descent* 2: 87, CD referred to this species by the name *R. crocea*.

To George Bentham 25 November [1869][1]

Down. | Beckenham | *Kent. S.E.*
Nov. 25

My dear Mr Bentham

I was greatly interested by your address, which I have now read thrice, & which I believe will have much influence on all who read it. But you are mistaken in thinking that I ever said you were wrong on any point. All that I meant was that on certain points, & these very doubtful points, I was inclined and had to differ from you.[2] And now, on further considering the point, on which even 2 or 3 months ago I felt most inclined to differ, viz on isolation,[3] I find I differ very little. What I have to say is *really* not worth saying, but as I sh^d^ be very sorry not to do whatever you asked, I will scribble down the slightly dissentient thoughts which have occurred to me. It w^d^ be an endless job to specify the points in which you have interested me; but I may just mention the relation of the extreme W. flora of Europe (some such very vague thoughts have crossed my mind, relating to the Glacial period) with S. Africa[4]—& your remarks on the contrast of passive & active distribution.—[5]

p. Lxx. I think the contingency of a rising isl^d^, not as yet fully stocked with plants, ought always to be kept in mind, when speaking of colonisation.—[6]

p. Lxxiv. I have met with nothing which makes me in the least doubt, that large genera present a greater number of vars. relatively to their size, than do small genera.— Hooker was formerly convinced by my data, never as yet published in full, only abstracted in the Origin.—[7]

p. Lxxviii.— I dispute whether a new race or species is necessarily, or even generally, descended from a single or pair of parents: the whole body of individuals, I believe, becomes altered together—like our race-horses, and like all domestic breeds which are changed through "unconscious selection" by man.[8]

p. d°. When such great lengths of time are considered, as are necessary to change a specific form, I greatly doubt whether more or less rapid powers of multiplication have more than the most insignificant weight.[9] These powers, I think, are chiefly related to greater or less destruction in early life.

p. Lxxix. I still think you rather underrate the importance of isolation. I have come to think it very important from various grounds; the anomalous and quasi-extinct forms on islands &c. &c. &c.[10] With respect to areas with numerous "individually durable" forms, can it be said that they generally present a "broken" surface with "impassible barriers"? This no doubt is true in certain cases, as Teneriffe.[11] But does this hold with S.W. Australia and the Cape? I much doubt. I have been accustomed to look at the cause of so many forms, as being partly an arid or dry climate (as De Candolle insists) which indirectly leads to diversified conditions, and, secondly to isolation from the rest of the world during a very long period, so that other more dominant forms have not entered, and there has been ample time for much specification and adaptation of character.[12]

Lxxx.— I suppose you think that the Restiaceae, Proteaceae, etc. etc. once extended over the world, leaving fragments in the South.[13]

You in several places speak of distribution of plants, as if exclusively governed by soil and climate: I know that you do not mean this, but I regret whenever a chance is omitted of pointing out that the struggle with other plants (and hostile animals) is far more important.[14] I told you that I had nothing worth saying, but I have given you my *thoughts*.

How detestable are the Roman numerals. why should not the President's addresses, which are often, & I am sure in this case, worth more than all the rest of the Number, be paged with Christian figures?—

My dear Mr. Bentham | Your's very sincerely | Ch. Darwin

Royal Botanic Gardens, Kew (GEB/1/3)

1 The year is established by the relationship between this letter and the letter from George Bentham, 23 November 1869.

2 For CD's comments on Bentham 1869b, see the letters to J. D. Hooker, 22 June [1869] and 13 November [1869].

3 See letter to J. D. Hooker, 22 June [1869].

4 See Bentham 1869b, p. lxxxvii.

5 See Bentham 1869b, pp. xcv–xcvi.

6 See *Origin* 5th ed., pp. 466–7 and 472.

7 CD discussed the greater number of varieties in large genera (genera containing many species) in *Origin*, pp. 55–6. In Bentham 1869b, p. lxxiv, Bentham suggested that Joseph Dalton Hooker's similar opinion expressed in *On the flora of Australia, its origin, affinities, and distribution; being an introductory essay to the flora of Tasmania* (J. D. Hooker 1859) had probably since changed.

8 See *Origin*, pp. 355–6. CD discussed 'unconscious selection' in *Origin*, pp. 34–40.

9 See, for example, *Origin*, pp. 302–3.

10 CD discussed the role of isolation, including isolation on islands, in forming new species in *Origin*, pp. 104–7. CD made some additions to this section in the fifth edition of *Origin*; see Peckham ed. 1959, p. 196.

[11] For contemporary knowledge of the botany of Tenerife, one of the Canary Islands, see C. J. F. Bunbury 1855; see also *Correspondence* vol. 15.

[12] CD refers to south-western Australia and the Cape of Good Hope in South Africa. Alphonse de Candolle referred to isolation in dry climates several times in *Géographie botanique raisonnée* (A. de Candolle 1855); see, for example, vol. 1, p. 562.

[13] Bentham mentioned these and additional names of southern hemisphere families in Bentham 1869b, p. lxxiii, in reference to possible ancient connections between their representatives in South Africa and Australia. Restiaceae is roughly equivalent to the modern family Restionaceae.

[14] CD mentions the importance of interaction with other plants and animals in determining the range of a species in, for example, *Origin*, p. 175.

From Charles Boner 25 November [1869][1]

5 Louisen S⟨trasse⟩ | Muni⟨ch⟩
Nov 25. 18⟨69⟩

Dear Sir,

I take the liberty of sending you an account of a family in which blindness is inherited. To an investigator like yourself I think every additional fact that can be obtained must be important and interesting.[2]

Count Tannenburg, in Tyrol, when young had the full use of his eyes. Later in life he became blind from amaurosis. While yet in possession of his sight he begot 1 son, Rudolph, born blind, and ⟨3 dau⟩ghters, of whom 2 were ⟨bli⟩nd and the third not so.[3] This last daughter is still alive, married to Count Enzenberg,[4] and though now an old lady enjoys the full use of her eyes. She has 5 sons and one daughter, none of whom are blind. One daughter is married and has several children, none of them blind. But 2 of the sons—24 and 27 years of age—have the infirmity of being unable to distinguish red and blue colors.

I cannot say for certain if the Count Tannenburg mentioned above—the father of Rudolf—was the first of the family afflicted with blindness, but I think not.

You will, I am sure, excuse me if I presume to make a remark on an assertion of yours in your work on "Animals & Plants under Domestication". At page 32 Vol II. you say "Sheep never become feral".[5] My experience is at variance with this assertion; and in the mountains when a sheep strays from the flock it becomes as wild as a chamois and is shot like any wild animal. Such sheep never return and can never be recovered, for they flee at the mere approach of any human being. On this subject which is extremely curious, you will some facts in a note at Page 92 of the 2nd edition of my "Chamois Hunting in the mountains of Bavaria"[6] At p 103 are also some remarks about the goat.

In mountanious districts—I speak of Bavaria & the Tyrol— it is of common occurrence that one or two sheep run wild when on the mountain pastures.

The same thing has been observed in Scotland, as it would seem from a line I found in some scott⟨ish⟩ ⟨au⟩thor but whether in ⟨ ⟩ or the Ettrick shepherd ⟨was⟩ unable to say.

"Even sheep ran wod (mad) and furious":

The circumstance was accounted for as being the work of a witch Marjore Nutchie.[7] However it is, as it seems to me, interesting as affirming the fact that sheep do occasionally return to a wild state.

A note at P. 117 of another book of mine "Forest Creatures"[8] might perhaps interest you, as showing how varieties of the same race may live in a wild state in juxtaposition and still keep distinct.

Among wild boars I ⟨have⟩ seen occasionally one ⟨ ⟩ litter of a white color like the domesticated pig. And it is impossible that this could have arisen from a chance cross with the common pig, as such forests where the wild boars were, had strong fences round them, so that no animal enter could enter.

Such an animal varying from the black wild-boar type, is always shot at once, to prevent any propagation of this deviation from the original color.

Your very Obedient Servant | Charles Boner

DAR 160: 238

CD ANNOTATIONS
0.1 5 Louisen ... colors. 2.8] *crossed pencil*
4.8 "Chamois ... Bavaria"] *underl pencil*
6.1 I found ... state. 8.3] *crossed pencil*
9.1 "Forest Creatures"] *underl pencil*
Top of letter: '2. Books creation of Wild Vars — | Feral sheep | Wild Boar occasionally producing white Pigs' *pencil*

1 The year is established by the relationship between this letter and the letter from Charles Boner, 8 January 1870 (*Correspondence* vol. 18).
2 CD discussed inherited eye problems, including blindness, in *Variation* 2: 8–10.
3 Boner refers to Ignaz and Rudolph von Tannenberg. The two blind daughters have not been identified.
4 The reference is to Ottilie and Franz von Enzenberg.
5 CD wrote: 'Sheep have never become feral, and would be destroyed by almost every beast of prey' (*Variation* 2: 32).
6 Boner refers to Boner 1860. After further discussion with Boner, CD added Boner's altered account to *Variation* 2d ed., 2: 5; see letter from Charles Boner, [December 1869 – early January 1870].
7 The Scottish poet and novelist, James Hogg, was sometimes referred to as the Ettrick shepherd (*ODNB*); the phrase and the witch have not been identified in his writings.
8 Boner 1861.

From Charles Layton 25 November 1869

16 Little Britain (City)
Nov[r] 25/69

Charles Darwin Esq
Dear Sir

Your favour of yesterday is received—[1] If the *type* is still standing of new Editn *Origin Species*, Mess[rs] Appletons would be glad if you would supply them with a set of Stereotype Plates, there would be such a great saving in cost, as *composition* is the heaviest item in making books, and but for this your later Editions would have been republished in America—[2]

The demand in America for books of this class is not nearly so great as in England and the Continent of Europe, but Mess[rs] Appletons wish now to reproduce your book, and pay copyright on all copies sold and of course would pay for Stereotype Plates, but if it is *impossible* to have plates I have no doubt they will incur the extra expense and have it set up in New York, only in that case perhaps you could let them have *a copy in advance to print from* as I was told yesterday in Paternoster Row the new Editn was not yet ready—[3]

With regard to the new book whenever it is ready I suppose you would be able to let them have Stereo Plates, as Casts could be taken as the book is composed in London, and I shall be glad, if there is yet time to have the same done with New Editn Origin of Species—[4]

Besides not having the same large number of buyers for good Scientific Books in America as here, the New York Publishers cannot get the same high prices as M[r] Murray and Mess[rs] Longmans in London

Yours Respectfully | Charles Layton | Agent D. Appleton & Co

DAR 159: 84

1 Letter to Charles Layton, 24 November [1869].
2 John Murray did not use stereotype plates in publishing any editions of *Origin* (see Freeman 1977). On the typesetting of the first five English editions of *Origin*, see Shillingsburg 2007.
3 *Origin* 5th ed. was published in the week of 22 June 1869; see letter from R. F. Cooke, 22 June 1869. Layton refers to Paternoster Row, Cheapside, London, where several publishers and booksellers (but not Murray, who was in Albemarle Street) had premises.
4 Layton refers to *Descent* and to the fifth edition of *Origin*.

From George Bentham 26 November 1869

25. Wilton Place. | S.W.
Nov 26 /69

My dear M[r] Darwin

I really am very much obliged to you for having given your time to the observations on my address which I received this morning[1]—which are just what I hoped to obtain from you and which will give me much to reconsider as to some of my views.

I do not wish to be supposed to underrate the importance of isolation but all I meant is that I do not think with M. Wagner that it is *absolutely indispensable* for the formation and maintenance of races which may in time become species[2]—a process evidently influenced by a great variety of causes many of which it is impossible to appreciate ⟨al⟩thou⟨gh⟩ much more study and data ⟨ ⟩ ⟨have⟩ at command—but the study of which will I trust be much facilitated by the publication of the second portion of your great work.[3]

I cannot think that an arid and dry climate has much effect in the concentration of a large number of species in a small space as at the Cape and S.W. Australia without the very varied protection to individuals given by a broken surface.[4] A large proportion of the desert flora of Australia extends in identical species from the Murray and

Darling northward to the Brigalow scrub of Queensland, to the N.W. Coast and to the Murchison and Sharks Bay on the W. Coast[5] and as far as we now know is not in any part so rich as the true S.W. flora which does not extend much to the eastward of King Georges Sound nor much to the north of Avon River—[6] but if I live to complete the Australian flora,[7] with the additi⟨onal⟩ materials that are pouring in ⟨ ⟩ ⟨ ⟩ the interior and N. West I hope we may be able to obtain some more accurate generalisations as to that most interesting vegetation.

The flora of the arid regions of N.E Africa and Arabia is I think a widely spread and not a particularly rich one.

One effect upon species of a hot dry arid climate must be the result of propagation under those circumstances being chiefly by seed, and long preservation of individuals by underground thick rhizomes or woody bases of stems without division whilst in wet climates propagation by division—rooting of stems, suckers and runners etc assures larger proportions renovating and multiplying individuals almost without changing them—but still I do not think that extensive uniform districts have a more varied Flora when accompanied by a dry than by a moist climate. The ⟨ ⟩ in our herbaria from which any tolerably well conclusions or ⟨ ⟩ ⟨ ⟩ can be drawn are ⟨ ⟩ ⟨ ⟩ worked up.[8]

With regard to Restiaceæ Proteaceæ etc. I did not intend to intimate that I thought they once extended over the world leaving fragments in the South but that those of S. Africa and Australia must have had a common origin of very great antiquity with a sufficiently varied flora to have produced at least as many genera as there are now tribes or Orders common to the two but that the isolation has been so long and so great as to have left no species common to both countries and comparatively few groups of a higher order—[9] What was the extent of the area of the original parents I have not presumed to conjecture. These original parents and the land they inhabited may all have perished, leaving descendants in two different countries which have diverged in different directions to the total extinction of the parent form in both countries. There are however a few plants common to Australia and South Africa which have diverged in different directions in the two countries apparently with the preservation of the common type in both—such as Pelargonium grossularioides or australe which has innumerable varieties in both countries the Australian ones varying in one set of characters the African in another,[10] but one form (and that a rare one) identical in the two. Such plants as these I take to be of less antical[11] transportation from the one to the other country or from a common source— probably from Africa to Australia as Pelargonium is an African not an Australian type. We have no evidence of Restiaceæ Proteaceæ or any Australian African type properly so called having found its way into the northern hemisphere although Australian ones not in Africa or African ones not in Australia may have done so in very ancient times.

I had worked up some instances of evidences of common origin in widely dispersed disjointed species or rather groups of species apropos of Cassia in a paper I read last winter at the Linnean Society but which I deferred printing till I had verified some of De Candolle's species which I did this summer—[12] and now there are other

Transaction papers printing which I suppose will put off mine till the spring
With many excuses for taking up your time with my

Incomplete
DAR 160: 165

CD ANNOTATIONS
1.1 I really . . . work. 2.7] *crossed pencil*
6.11 Australia . . . source— 6.17] *scored pencil*
Top of letter: 'Relations of distant *[*species*]* | G Bentham' *pencil*
End of letter: 'G. Bentham' *pencil*

1 See the letter to George Bentham, 25 November [1869], regarding Bentham's anniversary address to the Linnean Society of London (Bentham 1869b).
2 See Bentham 1869b, p. lxxix, and letter to George Bentham, 25 November [1869] and nn. 3 and 10. CD had added a reference to Moritz Wagner's essay in *Origin* 5th ed., p. 120; he disagreed with Wagner over whether migration and isolation were necessary for species formation; see Wagner 1868b, and *Correspondence* vol. 16, letter to Moritz Wagner, [April–June 1868].
3 Bentham refers to CD's original plan to publish a book to follow *Variation* on the 'variability of organic beings in a state of nature' (see *Variation* 1: 4).
4 See letter to George Bentham, 25 November [1869] and n. 12.
5 Bentham describes an area extending from the Murray and Darling Rivers in south-eastern Australia, north to Brigalow (acacia) scrubland of New South Wales and Queensland, to the north-western coast, and then south to the Murchison River and Shark Bay of western Australia.
6 Bentham describes the south-western corner of Western Australia, west of King George's Sound on the southern coast, and south of the Avon River, which flows into the Swan River near Perth.
7 Bentham refers to his seven-volume work, *Flora australiensis: a description of the plants of the Australian territory* (Bentham 1863–78).
8 Bentham refers to the herbarium at the Royal Botanic Gardens, Kew.
9 See letter to George Bentham, 25 November [1869] and n. 13.
10 *Pelargonium grossularioides* is the gooseberry-leaved, gooseberry, or coconut pelargonium; *P. australe* is sometimes called the southern storksbill.
11 Antical: 'fronting external objects, and thus remote from the axis' (botanical term; *OED*).
12 Bentham refers to 'Revision of the genus *Cassia*' (Bentham 1869a). On Bentham's verification of some species in a volume of Augustine Pyramus de Candolle and Alphonse de Candolle's *Prodromus* (A. P. Candolle and Candolle 1824–73), see Bentham 1869a, pp. 503 n., and 504.

From E. A. Darwin 26 [November 1869 or later][1]

26th

Dear Charles

I was talking to Sylvester yesterday and he asked 'has your nephew a great liking for the army, has he a fancy for it, ah well or it is quite a pity that he does not leave it & go to Cambridge he'd be certain to become a fellow—[2] I'nt that a compliment considering that the summum bonum[3] of this life might be obtained—

His opponent he said (Cardew?) was a regular genius & might become anything—a nephew of Ld Westbury—[4] he'll have to go into F Galton's list as well as Lenny who I see has already a niche in the temple of Fame[5]

Yours aff | E D

DAR 105: B66

[1] The date range is established by the reference to Galton 1869, published in mid-November 1869 (*Publisher's Circular* (1869): 791); see n. 5, below.
[2] James Joseph Sylvester was a professor at the Royal Military Academy, Woolwich, where Leonard Darwin was a student. Erasmus Darwin also refers to the University of Cambridge.
[3] *Summum bonum*: the highest good.
[4] Philip Cardew passed first into the Royal Military Academy, Woolwich, 1868 (*ODNB*); Leonard Darwin passed second (see *Correspondence* vol. 16). Cardew's mother was the second daughter of Richard Bethell, first Baron Westbury; Lord Westbury was lord chancellor from 1861 to 1865 (*ODNB*).
[5] In Francis Galton's *Hereditary genius: an inquiry into its laws and consequences*, Leonard was mentioned as coming second in the 1868 examination for the Royal Military Academy, Woolwich (Galton 1869, p. 209).

To T. H. Farrer [27 November 1869][1]

Down, | Beckenham, Kent.
Sat.

My dear Mr Farrer

I send the following extract from a letter recd yesterday from F. Müller in S. Brazil, as a caution about Passiflora in contrast with Tacsonia.[2]

Do not trouble yourself to acknowledge this

Yours very sincerely | Ch Darwin

"In one of your letters you called my attention on perennial *Passiflora*. I have in consequence planted several species into my garden; only 2 have as yet flowered, & these are self-sterile. In one species the first flowers, which appeared after transplantion, had contabescent anthers.

As to the fertilization of Passiflora, *Delpino* thinks in his interesting "Ulteriori osservazioni sulla Dicogamia", which I recd a few days ago) that in the larger species this is effected by humming birds. I had come to the same conclusion by observing how frequently these flowers are visited by humming-birds, while I did not yet see neither humble-bees, nor butterflies, nor any other large insects able to fertilize them."

L(S)
Linnean Society of London (Farrer: 12)

[1] The date is established by the relationship between this letter and the letter from T. H. Farrer, 28 November 1869; in 1869, the Saturday preceding 28 November was 27 November.
[2] The extract is from the letter from Fritz Müller, 18 October 1869. Farrer had recently suggested to CD that *Tacsonia* could be pollinated by hummingbirds or large moths (see letter from T. H. Farrer, 13 October 1869). The genus *Tacsonia* has since been subsumed into *Passiflora*.

To A. B. Meyer 27 November [1869][1]

Down. | Beckenham Kent S.E.
Novr. 27th.

Dear Sir.

I despatched the pamphlet yesterday—[2] You are quite at liberty to give a Photograph of me; but it seems useless, & I feel sure that the English Photographers will charge you a very large price. But I am not willing to sit on purpose; it is what I hate doing & wastes a whole day owing to my weak health; and to sit with another person would cause still more trouble & delay—[3] Photographs can be procured at Messrs Elliot & Fry, or Messrs Edwards & Bult both residing in Baker S^t. London.

The Rock thrush is the Rupicola, but I forget whether that of Guiana is the crocea or aurantia.[4]

Your translation of Wallace has not yet arrived but books are often long delayed in coming from Germany[5]

Pray believe me | Dear Sir | Yours very faithfully | Ch. Darwin

P.S | I am very sorry to be disobliging about the Photographs, but I cannot endure the thought of sitting again, and I have refused 3 or 4 Photographers lately.

Messrs Elliot & Fry are the last who have taken me, & they came down here on purpose—[6]

Copy
DAR 146: 362

[1] The year is established by the relationship between this letter and the letter from A. B. Meyer, 24 November 1869.
[2] The reference is to C. Darwin and Wallace 1858; see the letter from A. B. Meyer, 24 November 1869.
[3] Meyer had asked whether CD would sit with Alfred Russel Wallace for a frontispiece to Meyer 1870, a translation of C. Darwin and Wallace 1858.
[4] See letter from A. B. Meyer, 24 November 1869 and n. 7.
[5] CD refers to Wallace 1869c; see letter from A. B. Meyer, 16 November 1869.
[6] CD entered a payment to Elliott & Fry on 25 July 1869 in his banking account (CD's Account books–banking account (Down House MS)).

From T. H. Farrer 28 November 1869

Abinger Hall, | Reigate. (Post Town) | Gomshall (Station) S.E.R.
28 Nov /69

My dear Mr Darwin

I am extremely obliged for the note on Passiflora.[1]

It quite satisfies me that it wise to delay.[2] An ounce of observation is worth a pound of conjecture. And I am in a great puzzle too about names & species, for Hooker tells me that scarcely any of them are true— there has been so much hybridization.[3]

It would be very interesting to know from your correspondent—[4]

1. Whether in the species of Passiflora on which he has seen humming birds there are any stiff projections or corners which would make it difficult for a straight stiff object to reach the nectary without boring holes. There are certainly such in those which I have seen.

2. What is the relative position of the humming bird to the anthers or stigmas at the time it is sucking? In none of those which I have seen is there space for a hovering humming bird between the anthers and the entrance to the nectary. This is just one of those things which actual observation alone can tell.

I am sure about the Bee on P. Coerulea, for I have seen it.[5]

Sincerely yours | T H Farrer
C. Darwin Esqr FRS

[Enclosure]

I open my note to express a doubt whether one is right in assuming that no bees now visit any flowers. Last Sunday I was marking some plants of Ulex Europæus[6] in flower. And I was surprized to find—that whilst in most of those in blossom, not a single flower was deflowered—(ie had an open keel and projecting stigma) there were one or two plants on which almost every flower was thus deflowered— I cut some branches from these two plants as well as from other plants; took off all the flowers then open and placed them in water in a South window (closed). After a week I find all these branches in full blossom—with something like 150 flowers on all—and amongst them only three deflowered. This looks as if some insect still visited—though sparsely and capriciously—some of the flowering plants out of doors. If so the winter flowering plants are more easily explained.

The stout hairy calyx of the Ulex Europæus, is a capital protection to the winter flower, as compared with the thin and comparatively naked calyx of the September Ulex Manus.[7]

I am delighted with old Sprengel: and his quaint positiveness.[8] He is sometimes over-ingenious, it seems to me—but there is a world of suggestion

DAR 164: 61

1 Letter to T. H. Farrer, [27 November 1869].
2 Farrer is evidently referring to CD's 'caution' about contrasting the pollination of *Passiflora* to that of *Tacsonia*; see letter to T. H. Farrer, [27 November 1869]. For Farrer's thoughts on *Passiflora* and *Tacsonia*, see the letters from T. H. Farrer, 13 October 1869 and 27 October 1869.
3 For Joseph Dalton Hooker and George Bentham's list of species in the 'Order Passifloreæ', see Bentham and Hooker 1862–83, 1: 807–16. *Tacsonia* has since been subsumed into *Passiflora*.
4 In his letter to Farrer of [27 November 1869], CD enclosed an extract from a letter from Fritz Müller.
5 Farrer refers to *Passiflora caerulea*, sometimes called the blue passionflower. See enclosure to letter to T. H. Farrer, [27 November 1869].
6 *Ulex europaeus*, the common gorse, or furze, is winter flowering.
7 *Ulex manus* is now *U. minor* (dwarf gorse or dwarf furze).
8 Farrer also commented on Sprengel 1793 in his letters of 13 October 1869 and 17 October 1869.

From C. S. Bate 29 November 1869

8, Mulgrave Place, | Plymouth.
Nov 29— 69

My dear Sir

Probably the enclosed may not be new information to you but I am loth to run the risk of not conveying to you that which may be of interest.

Walking in the nursery garden this afternoon I observed a smooth leaved Holly *one* branch of which was of the ordinary prickly kind. The Gardener assures me that there has been no graft, nor is it probable considering the position of the branch.[1]

He says that the branch had been cut back for the purpose of not allowing it to interfere with the pathway & that the young branch came forth in this altered variety:

Should you feel interest in the circumstance I dare say that I could obtain further information or probably supply you with the branch.

I am my dear Sir | Yours very sincerely | C. Spence Bate

DAR 160: 55

[1] CD discussed the effects of 'bud-variation' or sports in plants that appeared to have been grafted in *Variation* 1: 387–94.

From Charles Boner [December 1869 – early January 1870][1]

My dear Sir,

On ⟨*3 or 4 words missing*⟩ for which I beg to thank ⟨you⟩ ⟨ ⟩ though⟨t⟩ over what you said about the she⟨ep⟩ *breeding* in their changed wild state.[2] It occurred to me that the opportunity for their doing so was hardly given, as they were shot at ⟨the⟩ end of the season, the Jäger[3] being only too glad to make use of their right to take them after a certain day, as the booty afforded them a feast of which they all partook. I was unable from my own experience to say whether any escaped being shot, & thus had an opportunity of testing their power of living through the Winter. On several points I was desirous of hearing the opinions of others, for it was a matter of conscience not to mislead you, and I also wished to furnish you with whatever information I could gather on the subject. I spoke to friends here; men who, like myself, were chamois hunters and were well-acquainted with mountain life; and afterwards wrote to a Jäger in the bavarian ⟨*4 or 5 words missing*⟩ opinions, at ⟨*3 or 4 words missing*⟩ ⟨ ⟩ing to him cer⟨tain⟩ questi⟨ons⟩ which I begged him ⟨to a⟩nswer ⟨ ⟩ and decidedly. I ⟨ ⟩ asked him to speak the matter over with his comrades, so to obtain their experience as well. His letter you will fi⟨nd⟩ interesting, I think; and I translate it and send it herewith. You will see that what he writes rather tends to disprove what I said about the sheep *remaining* wild, which was the result of *my* experience.[4] Truth however is, or ought to be, the end & aim of all investigation, and I give you the whole of his letter though it tells somewhat against me.

"I now proceed to answer your questions as well as I can.

Yes; every year nearly some sheep stay away from the herd and grow wild, however not as wild as a chamois or a red-deer—so I think at least.*

It is true one must stalk them very ca⟨refully⟩ ⟨*5 or 6 words missing*⟩ best wind in ⟨*4 or 5 words missing*⟩ ⟨other⟩wise—a whistle ⟨*4 or 5 words missing*⟩ have vanished. But ⟨still I should⟩ look on such sheep as p⟨erfectl⟩y wild because when a small herd of 3 or ⟨4⟩ with their sheep-bell are driven into their neighbourhood, the wild sheep soon associate with the flock, and thus they are driven home later with the rest and the wild sheep grow again as tame as the others & remain so. To recover a sheep in this way the terrain must however of course be favorable. When wild-sheep are not caught or shot there is no thought of them propagating themselves. The wild sheep can to be sure, subsist easily and find food; that is to say when there is no snow or only a very little.

Such sheep keep in Winter on the southern slopes of the mountains in hollows where grass of the year before is always to be found. But if such spots be covered by snow-drifts then the sheep must wander further to look for food; and then the animal is lost; for then the snow hangs about the long wool to such a degree that the creature gets exhausted and thus perishes.

An acquaintance asserts that the wool of sheep which are not shorn in Spring & Autumn gets too long and the animals begin to grow bare-naked—and then they die a miserable death.

⟨*4 or 5 words missing*⟩ sheep living in ⟨*4 or 5 words missing*⟩ ⟨kn⟩ow how to ma⟨ ⟩ ⟨*4 or 5 words missing*⟩ and frozen. Once ⟨*2 or 3 words missing*⟩ ⟨in⟩ November on the steep ⟨ ⟩ of the ⟨mou⟩ntain and 5 days later went out ⟨to⟩ fetch him. (To shoot him is what he means.) However the night before was a hard frost, and on awaking the ram was precipitated down into a terrible abyss: a proof more that under certain circumstances it is not always so easy for them to subsist as one supposes.

My own experience has shown me that the ram is less likely to run wild than the ewe. This arises probably from the circumstance that the ewe likes to remain away from the herd some days before she drops her lamb, and afterwards wants to remain with her young one quite separated from the others. As long as the young one can be caught the mother naturally runs about with it: if however some weeks have passed then both are wild. I am persuaded that in some instances a lamb is to be seen with the ewe grown wild.

A neighbour tells me that once two ewes subsisted through the Winter, each of which had a lamb. Three died, & the old sheep that survived came in Spring to the flock and was ⟨*3 or 4 words missing*⟩ I ought howev⟨er,⟩ ⟨*3 or 4 words missing*⟩ the ewe's ⟨*3 or 4 words missing*⟩ as if she still has ⟨*2 or 3 words missing*⟩ one, nor did it ever ⟨ ⟩ ⟨re⟩turn ⟨to⟩ the natural size. This ⟨ ⟩ prob⟨ably⟩ from the cold & deficient food."[5]

Such, dear Sir, is the res⟨ult⟩ of my questioning. The informat⟨ion⟩ is not without intere⟨st⟩ at all I think it right to give it you bearing on the subject of the fir⟨st⟩ letter I had the honor to send you.

Is Beckenham your temporary address, or will a letter find you there **always**? The work in which you specially devote your investigations to the human race I look forward to with impatient interest.[6] Is it soon to appear?

Yours, my dear Sir | Most obediently | Charles Boner

* In my letter to him in asking about the sheep & their growing wild, I had said "as wild as a chamois or red-deer."

To Charles Darwin, Esq[q] | &c—&c—&c

DAR 160: 237

CD ANNOTATIONS

1.1 On … least. 3.2] *crossed blue crayon*

5.3 and then ... perishes. 5.5] *double scored blue crayon*
6.1 An ... death. 6.3] *double scored blue crayon*
9.2 lamb. . . . died,] *underl blue crayon*

CD notes:

Boner ⟨ ⟩ does ⟨*2 or three words missing*⟩ good. ⟨*2 or three words missing*⟩ apt to ⟨*2 or three words missing*⟩ hungry ⟨*2 or three words missing*⟩ have lost skill ⟨ ⟩ pass over steep */rocky/* slopes— In one case 2 ewes passed the ⟨winter⟩ free, but three lambs perished.—[7]

⟨*4 or 5 words missing*⟩ Hunting) in the ⟨*4 or 5 words missing*⟩ 1860 p. 92) that ⟨*3 or 4 words missing*⟩ w⟨il⟩d, (but goats never) ⟨*2 or 3 words missing*⟩ mountains & have to be shot;[8] but ⟨*2 or 3 words missing*⟩ were left & found capable [*interl*] */reportedly/* breeding, they ought hardly to be called feral.—

(Feral sheep) [*square brackets in MS*]

Vol. 2. p. 32 I say never feral[9] [*added pencil*]

1 The date range is established by the relationship between this letter, the letter from Charles Boner, 25 November [1869] (see n. 2, below), and the letter to Charles Boner, [before 8 January 1870] (*Correspondence* vol. 18).

2 In a letter that has not been found, CD evidently asked whether the feral sheep described by Boner had bred successfully in their wild state; see letter from Charles Boner, 25 November [1869].

3 Jäger: 'hunter' (German).

4 See letter from Charles Boner, 25 November [1869].

5 In *Variation* 2d ed., 2: 5 n. 10, CD cited Boner 1860, p. 92, adding that Boner later found that strayed sheep were not able to establish themselves: 'they generally perish from the frozen snow clinging to their wool, and they have lost the skill necessary to pass over steep icy slopes. On one occasion two ewes survived the winter, but their lambs perished.'

6 The reference is to *Descent*.

7 See n. 5, above.

8 See letter from Charles Boner, 25 November [1869] and n. 6.

9 See letter from Charles Boner, 25 November [1869] and n. 5. CD refers to *Variation* 2: 32.

To Fritz Müller 1 December [1869][1]

Down Beckenham | Kent
Dec 1st.

My dear Sir

I am much obliged for your letter of oct 18th, with the curious account of Abutilon & for the seeds.[2] A friend of mine, Mr Farrer has lately been studying the fertilisation of Passiflora, & concluded from the curiously crooked passage into the nectary that it could not be fertilised by humming-birds; but that Tacsonia was thus fertilised.[3] Therefore I sent him the passage from your letter, & I enclose a copy of his answer.[4] If you are inclined to gratify him by making a few observations on this subject, I shall be much obliged & will send them on to him.

I enclose a copy of my rough notes, on your Escholtzias, as you might like to see them.

Somebody has sent me from Germany two papers by you, one with a most curious account of Alisma,[5] & the other on Crustaceans. Your observations on the Bronchiæ & heart have interested me extremely.[6]

Alex. Agassiz has just paid me a visit with his wife. He has been in England two or three months, & is now going to tour over the continent to see all the zoologists.[7]

We liked him very much. He is a great admirer of yours, & he tells me that your correspondence & book first made him believe in Evolution.[8] This must have been a great blow to his father, who, as he tells me, is very well & so vigorous that he can work twice as long as he, the son, can.[9] I have been very sorry to hear from him that Dana is quite broken down in health.[10] As I was sure that you would wish it, I gave him a copy of the English translation of your book.[11] By the way Mr Murray takes stock of his books in November, & informs me that of the thousand published of your book, 537 are unsold; & this I think is a very fair sale for a purely scientific work.[12]

Dr Meyer has sent me his translation of Wallace's Malay Archipelago, which is a valuable work;[13] & as I have no use for the translation, I will this day forward it to you by post, but to save postage viâ England. With every good wish believe me, my dear Sir, | Yours very sincerely | Ch. Darwin

[Enclosure 1]

Abinger Hall—
28. Nov—

My dear Mr Darwin

I am extremely obliged for the notes on Passiflora. It quite satisfies me that it is wise to delay. An ounce of observation is worth a pound of conjecture. And I am in a great puzzle too about names & species, for Hooker tells me that scarcely any of them are true, there has been so much hybridization.

It w^d be very interesting to know from your correspondent—

1 Whether in the species of Passiflora on which he has seen humming birds there are any stiff projections or corners which w^d make it difficult for a straight stiff object to reach the nectary without boring holes. There are certainly such projections in those species which I have seen.

2 What is the relative position of the humming bird to the anthers or stigmas at the time it is sucking?

3. Are the flowers upright or dependent?

In none of those which I have seen is there space for a hovering humming bird between the anthers & the entrance to the nectary. This is just one of those things which actual observation alone can tell.

I am sure about the bee in P Cœrulea for I have seen it.

Sincerely yours | T H Farrer

[Enclosure 2]

Escholtzia Californica 1869.[14]

Many plants were raised from crossed seed from self-sterile plants, from S. Brazil, sent by F. Müller.[15] Two plants were covered with nets; & 8 flowers on the two were crossed with pollen from distinct plants, & all produced very fine pods; a medium one contained 80 seed & none contained much fewer seed.

8 flowers on the 2 plants were fertilized with pollen from same flower & produced 7 pods; the finest of these contained 25 seed, the next finest 16 seed, & several others from 4 to 7 or 8 seed; average about 12 seed.

Later in the season, though the uncovered plants still produced pods, 12 flowers were self-fertilized, & they produced only 2 pods, containing 3 & 6 seed; so the colder temperature checked self-fertilization.

There was this difference between the 2 covered plants, that one *spontaneously* produced only *1* pod with no seed, yet it produced some when artificially self-fertilized; whilst the other plant spontaneously produced 8 pods, the finest of which contained 30 seed, the next finest 12, & several others from 3 to 6 seed.— Hence these 2 plants differed a little in their self-sterility. It was most curious to observe the more sterile plant of the two after it had been uncovered for about a week & insects had access, how it became completely covered with young pods;—wonderfully good evidence of benefit of a cross. These Brazilian plants appear under our climate much *more* self-fertile than their parents in Brazil, & very much *less* self-fertile than our English plants— Effect partly inherited, partly the result of external conditions.[16]

British Library (Loan 10: 31) (by kind permission of English Heritage)

[1] The year is established by the relationship between this letter and the letter from Fritz Müller, 18 October 1869.

[2] Müller evidently enclosed seeds with his letter of 18 October 1869 to replace those that had been damaged earlier (see letter to Fritz Müller, 18 July [1869] and n. 5).

[3] For Thomas Henry Farrer's observations, see Appendix IV.

[4] CD sent a copy of the extract from Müller's letter of 18 October 1869 with his letter to T. H. Farrer, [27 November 1869]. For Farrer's original reply, see the letter from T. H. Farrer, 28 November 1869. Both the enclosures to the letter to Müller are copies made by Emma Darwin.

[5] The reference is evidently to a pre-publication copy of 'Die Bewegung des Blüthenstieles von Alisma' (The movement of the pedicle in *Alisma*; F. Müller 1870a); CD's heavily annotated copy is in the Darwin Pamphlet Collection–CUL. See also *Correspondence* vol. 16, letter to Fritz Müller, 3 April [1868] and n. 3. The person in Germany who sent the papers has not been identified.

[6] The second article sent was 'Bemerkungen über Cypridina' (F. Müller 1870c); an annotated copy is in the Darwin Pamphlet Collection–CUL. Müller observed that two species of *Cypridina* with branchiae had hearts, while a third species lacked both (F. Müller 1870d, p. 269).

[7] The reference is to Alexander and Anna Russell Agassiz; for more on their European tour, see G. R. Agassiz ed. 1913, pp. 97–114.

[8] CD refers to *Für Darwin* (F. Müller 1864a). For more on Agassiz's correspondence with Fritz Müller, see Dobbs 2005, pp. 100–1; some of Agassiz's letters to Müller are in G. R. Agassiz ed. 1913, pp. 48–52, 91–4.

[9] Louis Agassiz had been an outspoken critic of CD's theories; for his views on evolution in the late 1860s and early 1870s, see Lurie 1988, pp. 372–7, and Winsor 1991, pp. 150–1. See also letter from Louis Agassiz, 6 July 1869.

[10] On the health problems of James Dwight Dana, see *ANB*.

[11] The reference is to the translation of F. Müller 1864a (Dallas trans. 1869).

[12] See letter from John Murray, 17 November 1869. Murray was the publisher of Dallas trans. 1869.

[13] Adolf Bernhard Meyer had sent CD his German translation (Wallace 1869c) of Alfred Russel Wallace's *The Malay Archipelago: the land of the orang-utan, and the bird of paradise* (Wallace 1869a). See letter from A. B. Meyer, 16 November 1869.

14 The enclosure, written in Emma Darwin's hand, is a copy of notes written by CD that are now in DAR 76: B31–2.

15 Müller had sent CD seeds of *Eschscholzia californica* (the California poppy) earlier in the year, and CD had reported on their growth; see letter from Fritz Müller, 12 January 1869, and letter to Fritz Müller, 18 July [1869] and n. 6. CD included the information in *Cross and self fertilisation*, pp. 111–12.

16 For CD's published discussion on the self-sterility of *Eschscholzia californica* in Brazil, and the effect of changed conditions on the plant's reproductive system, see *Cross and self fertilisation*, pp. 343, 358, 444, and 449.

From G. H. Darwin [2 December 1869][1]

14 Arlington St. | Piccadilly
Thursday

My dear Father,

After grinding away for some time at the calculation of the red & white animals it suddenly occurred to me that I could find without any bother at all the ratio of the reds to the browns at any period, that is to say the percentage of the whole n° of animals *existing at that period* which are brown. In fact the death rate has *no effect* whatever on this percentage, for this is the very meaning of the death rate being the same for browns & reds. If this is all you want I can do it in five minutes, but if you want the actual numbers of reds and browns at each period—I can easily calculate it, I had in fact actually done it for the first period when I perceived this thing wh. had been staring me in the face all the time.—[2] Will you write me a line to Cambridge & let me know what you want. I have just been dining here with a man that I have known for a good time at Cambridge called Hammond, & he tells me that his Uncle was a lieutenant on the Beagle.[3] As I thought you might like to know what had become of him I found out that he is banking down in Norfolk or Yorkshire I forget which.[4]

I also find that he knows Sullivan a little.[5] Again I fd out today that Whiteford (a man at Tathum's) old Col. H Smith of Plymouth's grandson knows Mrs Clem. Wedgwood very well;[6] how marvellously small the world is. I am going to take a holiday on Saturday & Monday & I am going down to Camb. tomorrow night.

Balfour & Strutt I there & it is the last of Balfour that I shall see at Camb. as he has just been in for his degree & goes down next term.[7] I went to Nilsson's concert last night & thought it extremely good.—[8]

Your affectionate Son | G. H. Darwin

DAR 210.2: 17

1 The date is established by the reference to Arthur James Balfour's taking his degree (see n. 7, below) and by the reference to Christine Nilsson's evening concert (see n. 8, below). In 1869, 2 December was a Thursday.

2 George was making calculation about a hypothetical population of red and brown animals. A statement of the problem that George was working on is in George's hand in DAR 85: B17:

> Let there be a number of young animals all born brown & let them turn red (the operation being supposed instantaneous for each animal) at such a uniform rate that (if none died)

half of the whole no: wd be red at the end of 8 quarter years; also let them perish irrespective of colour at such an uniformly diminishing rate that $\frac{2}{3}$ the original no: are left after 4 quarter years & $\frac{1}{2}$ orig: no: are left after 8 quarter years— It is then required to find the nos: of red & browns at any period.

George's calculations are in DAR 85: B1b–17.

3 George refers to Thomas Astley Horace Hamond. His uncle was Robert Nicholas Hamond.

4 Robert Nicholas Hamond appears at Tun Street, Fakenham, Norfolk, in the *Post Office directory of Cambridge, Norfolk, and Suffolk* 1869.

5 George refers to Bartholomew James Sulivan. See *Correspondence* vol. 18, letter from B. J. Sulivan, 27 June 1870.

6 George refers to Ferdinand Mauger Whiteford, Charles Meaburn Tatham, Charles Hamilton Smith, and Emily Wedgwood, wife of Clement Francis Wedgwood.

7 George refers to Arthur James Balfour and John William Strutt. The examination for the moral sciences tripos took place on 29 November 1869 (*Cambridge University calendar* 1869).

8 Nilsson appeared in a series of evening concerts at Exeter Hall in London in 1869; the last two were on 1 and 15 December (*The Times*, 10 September 1870, p. 1).

From A. R. Wallace 4 December [1869][1]

9, St. Mark's Crescent N.W.
December 4th

Dear Darwin

Dr. Adolf Bernhard Meyer who translated my book into German has written to me for permission to translate my original paper in the Linnæan Proc. with yours,—and wants to put my photograph & yours in it.[2] If you have given him permission to translate the papers (which I suppose he can do without permission if he pleases) I write to ask which of your photographs you would wish to represent you in Germany,—the last, or the previous one by Ernest Edwards, which I think much the best, as if you like I will undertake to order them and save you any more trouble about it.[3] It is of course out of the question our meeting to be photographed together, as Mr. Meyer coolly proposes.

Hoping you are well, believe me yours very faithfully | Alfred R. Wallace

P.S. I have written a paper on *Geological Time*, which will appear in "Nature",—& I *think* I have hit upon a solution of our greatest difficulties in that matter.[4]

A. R. W.

DAR 106: B88–9

1 The year is established by the relationship between this letter and the letter from A. B. Meyer, 16 November1869 (see n. 2, below).

2 Meyer translated Wallace's *Malay Archipelago* (Wallace 1869a; the translation is Wallace 1869c). Meyer had also asked CD for permission to translate CD's and Wallace's papers in the 1858 *Proceedings of the Linnean Society* (C. Darwin and Wallace 1858; see letter from A. B. Meyer, 16 November 1869). Meyer requested CD's photograph in his letter of 24 November 1869.

3 CD agreed to Meyer's translation of C. Darwin and Wallace 1858 in a letter that has not been found; see letter from A. B. Meyer, 24 November 1869. The most recent photograph of CD by Ernest Edwards was taken in 1866; see *Correspondence* vol. 14, plate facing page 88. The previous photograph by Ernest

Edwards was taken in 1865; see *Correspondence* vol. 13, frontispiece, and letter from E. A. Darwin to Emma Darwin, 25 [November 1865] and nn. 2 and 3.

4 'The measurement of geological time' was published in the 17 February and 3 March 1870 issues of *Nature* (Wallace 1870). A copy is in the Darwin Pamphlet Collection–CUL.

To A. R. Wallace 5 December [1869][1]

Down. | Beckenham | *Kent. S.E.*
Dec. 5

My dear Wallace

I wrote to Dr. Meyer that the Photograph in England wd. cost much, & that they did not seem to me worth the cost to him, but that I of course had no sort of objection.—[2] I shd. be greatly obliged if you wd kindly take the trouble to order any one which you think best:[3] possibly it would be best to wait, unless you feel sure, till you hear again from Dr. M.

I sent him a copy of our joint paper.—[4] He has kindly sent me the translation of your book, which is splendidly got up & which I thought I could not better use than by sending it to Fritz Müller in Brazil, who will appreciate it.—[5]

I liked your Reviews on Mr Murphy very much; they are capitally written like every thing which is turned out of your workshop.— I was specially glad about the Eye.—[6] If you agree with me take some opportunity of bringing forward case of perfected greyhound or race-horse, as proofs of possibility of the selection of many correlated variations. I have remarks on this head in my last book.[7] If you throw light on the want of geological time, may honour, eternal glory & blessings crowd thick on your head.[8]

Yours most sincerely | Ch. Darwin

I forgot to say that I wrote to Dr M. to say that I shd. not soon be in London, & that of all things in the world I hate most, the bother of sitting for photograph, so I declined with many apologies.—[9] I have recently refused several applications.—

Dear Mr Wallace

I enclose a diminished photo, from Mrs Camerons but it looks heavy & unclear.[10] I like best the profile of Ernest Edwards, or a $\frac{3}{4}$ face vignette which Messrs Elliot & Fry (Baker St) & which is a strong likeness & pleasing are now making some copies of for us,[11] but I cannot send you one, or one of Mr Edwards'.

British Library Board (Add 46434 f. 194–5)

1 The year is established by the relationship between this letter and the letter from A. R. Wallace, 4 December [1869].

2 CD refers to Adolf Bernhard Meyer. See letter to A. B. Meyer, 27 November [1869].

3 For the photographs Wallace suggested, see the letter from A. R. Wallace, 4 December [1869] and n. 3.

4 See letter to A. B. Meyer, 27 November [1869] and n. 2. CD refers to C. Darwin and Wallace 1858.

5 See letter to Fritz Müller, 1 December [1869]. CD refers to Meyer's German translation of Wallace's *Malay Archipelago* (Wallace 1869a; the translation is Wallace 1869c).

6 Wallace reviewed Joseph John Murphy's *Habit and intelligence, in their connection with the laws of matter and force* in *Nature*, 25 November and 2 December 1869 (Wallace 1869d, Murphy 1869); see also letter from A. R. Wallace, 20 October 1869. Wallace argued against Murphy that spontaneous variations could have produced all the 'delicate adjustments' of the eye (Wallace 1869d, pp. 106–7). CD quoted part of Wallace's discussion of the eye in *Origin* 6th ed., p. 145.

7 In *Variation* 2: 221–2, CD had offered the example of the development of the greyhound breed through the gradual alteration of separate characteristics to support his contention, against Murphy, that the eye could have developed through natural selection. See Murphy 1869, 1: 326, 345–7, and Wallace 1869d, p. 106. On correlated variation, see *Variation* 2: 319–38.

8 See letter from A. R. Wallace, 4 December [1869] and n. 4.

9 See letter to A. B. Meyer, 27 November [1869].

10 Emma Darwin's note refers to one of Julia Margaret Cameron's 1868 photographs of CD. See *Correspondence* vol. 16, letter from J. D. Hooker, 30 August 1868 and n. 12, and plate facing p. 630.

11 See *Correspondence* vol. 14, plate facing p. 88, for the profile by Ernest Edwards, probably taken in 1866. On the photograph by Elliott & Fry, see the letter to A. B. Meyer, 27 November [1869] and n. 6. Meyer's translation of C. Darwin and Wallace 1858 (in Meyer 1870) did not include photographs.

From E. A. Darwin 7 December [1869][1]

Dec 7

Dear Charles

I enclose you E Hollands[2] letter. I have written to him to say so & I should think you need not take any further notice

yours aff | E D

[Enclosure]

London
Dec 6th 1869

My dear Erasmus

Where is Charles? I am not sure of his Surrey address and therefore through you beg to communicate with him—[3] Today I came up from Portsmouth where I have been since Tuesday—and where I return on Thursday to bid my son Swinton[4] farewell before he leaves for China. He is appointed to the Sylvia (now in Chinese waters) as her first Lieutenant. The Sylvia is a surveying ship and will have to deal with *deep* waters in *inner* seas— Swinton is a naturalist by taste and habit—and if Charles Darwin has any particular instructions to give him regarding any objects in which he is particularly interested, I think that he would find Swinton carrying out those instructions con amore—[5]

He sails in the Donegal—probably on Friday next—but if a letter were addressed forthwith to

Lieut S. C. Holland RN
H.M.S. Donegal—Portsmouth

it would reach him before the ship leaves—and it is through an oversight of mine that I have delayed thus announcing to Charles, the departure for China of a ready worker in Nat: History.

Hoping that this will find you well believe me sincerely ys | Ed Holland

DAR 105: B67, DAR 166: 234

1 The year is established by the date of the enclosure.
2 Edward Holland.
3 CD lived in Kent, but had relatives in Surrey.
4 Swinton Colthurst Holland.
5 Swinton Holland later presented some of his anthropological observations to the Anthropological Institute of Great Britain and Ireland, and contributed photographs and implements of the Ainu people of Japan to the Christy collection of the British Museum (see Holland 1873). No correspondence between CD and Swinton Holland has been found.

From John Lubbock 8 December [1869][1]

15, Lombard Street, E.C.
8 Dec.

Dear Mr. Darwin

I am very glad you like my new Edition.[2]

I have just returned from Scotland where I have been lecturing.

Enclosed is a cheque for my £2.10.[3]

Believe me | Yours most sincerely | John Lubbock

DAR 170: 75

1 The year is established by the reference to Lubbock 1869a; see n. 2, below.
2 Lubbock refers to the second edition of *Pre-historic times, as illustrated by ancient remains, and the manners and customs of modern savages* (Lubbock 1869a). No letter from CD regarding the new edition has been found.
3 CD recorded a contribution of £2 from Lubbock in 1869 in his Down Coal and Clothing Club account book (Down House MS). CD was treasurer of the club, which supplied parishioners with coal and clothes in exchange for regular savings.

From C. S. Bate 11 December 1869

8, Mulgrave Place, | Plymouth.
Decr. 11. 69

My dear Sir

The Ilex, the Gardener says on referring to the name is Ilex Scotica.[1] It is a tree about 18 feet high and smooth leafed all over except this branch, on some of the other branches that, like this one, have been cut back for the purpose of keeping the walk clear, over which they are likely to encroach, the leaves without being spinous shew a tendency to that character, as you may see by the enclosed leaf:

I can send you the branch if you would like to have it, or if you think it more desirable that it should be retained on the tree: I will watch it & report any feature that may appear worth recording.

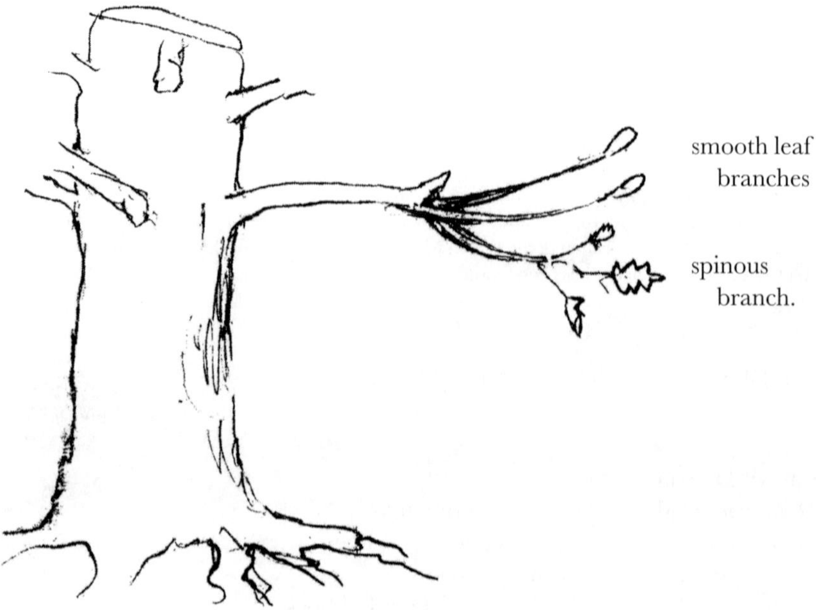

The gardener is positive that there has been no artificial budding or grafting but the variation is the result of the branch having been cut back: *He* considers I. Scotica as being a *variety* of the common Species & not as I thought a distinct species:[2]

Believe me | Yours sincerely | C. Spence Bate

DAR 160: 56

[1] CD's letter, a reply to the letter from C. S. Bate, 29 November 1869, has not been found. He evidently asked questions about variation in a smooth-leaved holly.

[2] *Ilex aquifolium*, the common holly, is prickly; it has a smooth-leaved variety now known as *Ilex aquifolium* 'Scotica'. See also *Variation* 1: 360–1.

From Francis Galton 11 December 1869

42 Rutland Gate. S.W.
Dec 11. 69

My dear Darwin

I wonder if you could help me. I want to make some peculiar experiments that have occurred to me, in breeding animals and want to procure a few couples of rabbits of marked and assured breeds—viz:

Lop-ears with as little tendency to Albinism as possible.
Common rabbits ditto
Angora albinos

and I find myself wholly unable to get them, though I have asked many people. Do you know anybody who has such things.?[1]

I write without your book in reach, but you there especially mention a breeder of Angoras.[2] Also you quote with approbation from Delamer's little book.[3] Are either or both of those men accessible & likely to help.?

Pray excuse my troubling you— the interest of the proposed experiment—for it is really a curious one,—must be my justification

Very sincerely yr.s | Francis Galton

DAR 105: 1–2

1 On Galton's published experiments on heredity in rabbits, see Galton 1871; see page 400 for the breeds of rabbits he worked with. For accounts of the experiments, see also Pearson 1914–30, 2: 156–66, and Bulmer 2003, pp. 116–18.

2 CD mentioned Robert Heron as a breeder of angora rabbits in *Variation* 2: 92; the source of his information was probably an extract included in the letter from William Yarrell, [*c.* 17 December 1838] (*Correspondence* vol. 2).

3 CD quoted Eugene Sebastian Delamer's *Pigeons and rabbits, in their wild, domestic, and captive states* (Delamer 1854) in *Variation* 1: 107. There is an annotated copy of Delamer 1854 in the Darwin Library–CUL (see *Marginalia* 1: 190). Eugene Sebastian Delamer was a pseudonym of Edmund Saul Dixon. See also *Correspondence* vol. 5, letter from Edward Blyth, 4 August 1855 and n. 7.

From J. T. Moggridge 12 December 1869

Maison Gastaldy | Mentone | (*Alpes Maritimes*)
Dec. 12 1869.

Dear M.r Darwin

I have obained a few seeds of *Lathyrus Clymenum* L. & *Lath*: *Ochrus* DC. to send you, & these I enclose.[1]

I find that my botanical friends generally have never thought of collecting seed of such common plants as these for growing, so that I fear I shall not be likely to obtain more for you before next year.

I hope however that these may germinate. Some weeks ago I collected the small quantity of seed of *Lath: Clymenum* L. which I send, by looking under the low bushes which were still covered with the dead haulm & gaping pods of this plant. I was much struck by the scarcity of the seeds when compared with the abundance of the pods, &, as the few seeds which I secured were obtained by searching in the thickest & thorniest places, I feel sure that they have been eaten for the most part by birds & mice.—

This fact serves to point out an additional advantage which climbing plants gain, beside support & abundance of light, namely the opportunity of shedding their seeds, or some of them, in places secure from the depredation of animals.

I now remember that I obtained last year a good supply of fallen seed from the cruelly-spiny *Astragalus Tragacantha* L. by searching underneath the matted branches, which act as protection from all ordinary intruders . . . .

I have now, in my little plot of garden here, two seedling plants of *Ophrys aranifera* Huds., which sprang the self-sown seed of capsules produced by the artificial fertilization of a plant which I had moved from the wild ground last spring.[2]

I am much surprised to find that the seedlings are nearly as large as the thriving parent plant, thus indicating an extremely rapid growth— —.

I find some difficulty in reconciling M. Delphino's statement as to the sterility of *Oph*: *aranifera* Huds. with the fact that it is abundant in all this region, the plant being one which can only be propagated by seed—[3] Moreover, whenever I have attempted the fertilization of this subspecies, fine capsules have been developed.

There are, no doubt, a great many obstacles to the *germination* of the seed; & this the example of my solitary pair of seedlings tends to shew, for I fertilized a considerable number of plants in the same plot of garden-ground; & the two fine capsules of the successful plant must have furnished seed enough to sow the whole garden, yet only the two seedlings appeared—

My tiny plot of garden was left absolutely untouched during the months from May to October, & I returned to find it a jungle of weeds. This may have injured the chances of the *Ophrys* seeds; but, on the other hand, I find that very few plants have been materially injured, & some have certainly gained by the shade which was afforded by the tall, much-branched intruders.

A plant of *Convolvulus sabatius* Viv, a perennial of somewhat woody growth, & which does not twine in its native habitat, was curiously affected, the branches being greatly lengthened & '*drawn*', so that in several instances they twined round one another, one pair of stems making 8 complete turns—!

I have been able to carry on & to extend my observations on *Arbutus*, & hope to get some good results from them. . .[4]

I have satisfied myself that the offspring do in some cases closely reproduce the details of the parents, but I still think that close resemblance is rare. I have obtained a curious record of the difference between the individuals in respect of their season of flowering, differences which, reckoned as they are by months, must expose the plants to very different conditions.

That these & similar differences of condition are either secondary in their influence, or without influence, I have now gathered some evidence to shew.

I am now collecting the details of variation in seedling plants, & in luxuriant shoots springing from trees which have suffered from cutting . . . . My general observations lead to the conclusion that there is in *Arbutus Unedo* L. a great amount of variability, but that among these variable forms a peculiar type of fruit—the globose-depressed—prevails— But as on the same tree the fruits do not always present the same shape, three forms being sometimes found together, I cannot avoid speculating whether there may not have been, or may not be, a selective process at work among the individual fruits, as there has been or may be among the individual trees; for these polymorphic individuals always have one predominant type of fruit, the three forms not being found together in equal or nearly equal numbers.

These differences among the fruits of one branch, are certainly not due to any circumstance such as position, crowding or the like, & there is great reason to suppose that they are due to independant variation—

I must, however, apologise for the liberty I take in troubling you with details which are, I fear, too incomplete to be of interest I hope, if my health continues to improve, to get my notes into order for publication before the summer; trusting that the Linnean Society will not reject them on account of the minute, &, to some, perhaps tedious, details, & the cost of reproducing a large series of illustrations.[5]

Believe me | yrs. very sincerely | J. Traherne Moggridge.

I hope that you will not scruple to give me any commissions which you think I can do for you— I esteem all such as favours recieved.

DAR 171: 214

CD ANNOTATIONS
1.1 I have ... intruders . . . . 5.3] *crossed blue crayon*
13.1 I have ... conditions. 13.5] *scored blue crayon*

1 See letter from J. T. Moggridge, 18 September 1869. *Lathyrus ochrus* is the cypress vetch; *L. clymenum* the Spanish vetchling.

2 For Moggridge's earlier work on *Ophrys aranifera* (the spider orchid), and CD's interest in it, see *Correspondence* vols. 13 and 14.

3 Moggridge evidently refers to Federico Delpino's statement in Delpino 1868–75, part 1: 177, that in Liguria hardly one flower of *Ophrys aranifera* out of 3000 set a capsule; see *Orchids* 2d ed., pp. 50–1. See also Delpino 1867b, pp. 19–20, and 'Fertilization of orchids', p. 144 (*Collected papers* 2: 141).

4 On Moggridge's earlier observations of an *Arbutus* (madrone) species, see the letter to J. T. Moggridge, 7 June 1869, and letter from J. T. Moggridge, 16 September [1869].

5 The Linnean Society did not publish a paper by Moggridge on *Arbutus*.

From Camille Dareste[1] 13 December [1869][2]

Lille
13 décembre

Monsieur

Je vous suis bien reconnaissant de l'éclatant témoinage que vous avez bien voulu m'adresser sur mes travaux.[3] Il a produit un grand effet, et vivement impressionné tout les personnes qui avaient à voter sur ma candidature:

Je regrette d'avoir à vous apprendre que ma candidature, bien qu'appuyée de votre nom, a echoué.[4] Je suis arrivé aussi près que possible du bas; car j'ai eu, un très grand nombre de voix parmi les membres du Conseil acadèmique et de la faculté des Sciences qui dressaient leurs listes de présentation. Mais j'ai rencontré devant moi une hostilité qui s'est manifestée de la manière la plus enérgique; celle de Claude Bernard qui est aujourd'hui tout puissant même en dehors de la Science. Il n'admet pas que l'on puisse traiter la physiologie autrement qu'il ne la fait lui-même; c'est à dire en ne suivant pas la méthode des vivisections.[5] Il m'a donc combattu en criant très haut et partout que je n'étais pas physiologiste. Sans lui, j'aurais très certainement réussi.

Je suis donc retourné à Lille, très froissé, mais non découragé; et j'ai repris la serié interrompue de mes travaux.[6] J'ai tout lieu de croire, par les resultats de mes expériences de l'année, que j'arriverai prochainement au terme d'une série de recherches qui m'occupent depuis près de dix-huit ans. Peut-être, lorsque je publierai mon travail d'ensemble, pourrai-je trouver dans mon pays un accueil plus favourable que celui que j'ai rencontré jusqu'à présent.[7] Du reste, si j'étais d'une façon quelconque, abattu par mon insuccès actuel, j'aurai, pour me relever, l'opinion d'un Savant comme vous; elle ne prouve que je n'ai point fait facille route; et qu'en cherchant à étudier par la méthode expérimentale, la production des monstrosités, je suis entré dans une carrière où je pourrai rendre de véritables services à la science de la vie, quoiqu'en dire Claude Bernard et son école si exclusive.

Veuillez recevoir, Monsieur, l'assurance de mon profond respect et de ma vive reconnaissance. | Camille Dareste

DAR 162: 46

1 For a translation of this letter, see Appendix I.
2 The year is established by the relationship between this letter and the letter to Camille Dareste, 20 November 1869.
3 See letter to Camille Dareste, 20 November 1869.
4 On the position Dareste had been seeking, see the letter from Camille Dareste, 18 November [1869].
5 For Claude Bernard's position in French science and philosophy, and for his promotion of vivisection, see *DSB* and Holmes 1974.
6 Dareste was a professor of zoology at the University of Lille (*DBF*).
7 For Dareste's recent work on animal monstrosities, see Dareste 1862, 1863, and 1867. Dareste later published *Recherches sur la production artificielle des monstruosités, ou essai de tératogénie expérimentale* (Dareste 1877).

From Fritz Müller 18 December 1869

Itajahy, S[a] Catharina, Brazil
Decbr. 18. 1869

My dear Sir

I have to thank you for having kindly favoured me with a copy of your "Notes on the fertilization of Orchids" which contain a large number of interesting facts quite new to me.[1]

A few days ago I returned from an excursion, on which I made a rich harvest of dimorphic and trimorphic plants. There were about half a dozen dimorphic Rubiaceæ, and among them the beautiful *Rudgea eriantha*.[2]

But the most interesting plant was a *Pontederia*, which I had long ago suspected to be trimorphic and which now indeed I found to be so.[3]

⟨*equivalent of 5 or 6 lines excised*⟩

This is, I think, the first case of trimorphism or dimorphism among monocotyledons and in irregular flowers. The three longer stamens do not alternate with the shorter ones, nor belong to the same whorl; one of the longer stamens is inserted to the anterior sepal, two to the lateral petals; one of the shorter ones to the posterior petal, which is marked with a yellow spot, and two to the lateral sepals. (In the allied monomorphic genus *Heteranthera*, at least in *H. reniformis*, the three shorter stamens of Pontederia are wanting.)[4] I was much surprised, that among hundreds of plants I could not find a single mid-styled one, and I have have hardly any doubt, that this form does not exist in the *Itajahy-mirim*,[5] where this species grows abundantly.— The long-styled and short-styled plants could be distinguished even from a distance, the flowers of the former being paler, than those of the latter; in the dried flowers this difference has disappeared.— In a former letter I have sent you a mid-styled flower of a second species of *Pontederia*, which has been introduced here some years ago from Rio de Janeiro.—[6]

As to dimorphic Rubiaceæ, Professor Hildebrand told me, that Bernouilli (in Guatemala) had observed *Coffea arabica* to be dimorphic, producing hermaphrodite and small female flowers; but this is an error.[7] The small flowers are rudimentary and sterile ones which occasionally appear among the normal flowers. I have lately seen these aborted flowers in unusual number on coffee-trees, which had been damaged by hail some weeks before.

Of the *Begonia* with monstrous male flowers I have raised four lots of seedlings and it is curious, in how odd a manner the monstrosity has been transmitted.[8] The female flowers, from which the seeds were obtained, had all been fertilized with pollen of monstrous male flowers.

⟨ ⟩ seeds of a normal female flower. ⟨ ⟩ ⟨o⟩f which have normal and 2 monstrous ⟨ ⟩ of ⟨ ⟩ male ⟨ ⟩

3d lot. Seeds of a female flower, which had 6 petals instead of 5.— 20 plants; 17 have flowered, all having normal male flowers.

4th lot. Seeds of two female flowers of a second plant, which also had 6 petals.— 22 plants, of which 7 are flowering, only one of them producing monstrous male flowers.— Some of the 9 plants of the second lot, which had monstrous male flowers, showed this monstrosity only for a short time; the *first* male flowers were monstrous, but afterwards they produced and are now producing normal male flowers. *Gærtner* has already observed, that the first flowers ("Erstlingsblumen") are often distinguished from the later ones by a greater number of stamens or pistils, by greater fertility etc. and I have myself lately observed some curious instances of this.[9] In a plant of *Eschscholtzia* the first six or eight flowers had from 5 to 7 petals; afterwards not a single flower with more than 4 petals has appeared on this plant.

On a plant of *Siphocampylus* the first flowers of several branches had six petals instead of 5; the first of all flowers which this plant produced, had even 7 petals. A few sixpetaled flowers have also appeared among the other flowers of this plant; but these 6 petaled flowers are at least 100 times more numerous among the first flowers of the branches, than among the rest.—

On an umbel of *Agapanthus* the first flower had 5 segments instead of six.— On a *Jussieua*,[10] which is extremely common on the banks of the Itajahy, and the flowers of which have normally 4 petals, the first flowers of the branches have not unfrequently 5 petals, and in this case there are generally also 5 sepals, 10 stamens and a 5 locular germen; but in one single flower there were 6 sepals, 5 petals, 10 stamens and a 4 locular germen.—

—To the list of self-impotent plants I may now add *Tabernæmontana echinata*, on which I lately tried some experiments. A *Calonyction* with very large white flowers appears to be self-impotent also;[11] at least a plant in my garden is quite sterile with its own pollen; but I have no second plant and am therefore unable to say, whether it would be fertile with pollen of a distinct plant.—

A very strange case of graft-hybrids is recorded in one of the last numbers of the "Auxiliador da Industria nacional" (April 1869 pg 158). A "fazendeiro" of the province of Rio de Janeiro, Señr J. F. Silveira da Motta, says that he has obtained a green and red striped variety of sugar-cane by planting close together plants of the green and of the red sugar-cane; as well as yellow and red striped sugar-cane by planting in the same hole cuttings of yellow and of red sugar-cane.[12] I confess that I have some doubts, whether these statements may be trusted; but I hope that I shall soon be able to repeat the experiments of Señr Silveira da Motta.

Permit me to repeat my thanks for all your kindness and believe me, dear Sir, with profound respect

very truly yours | Fritz Müller.

DAR 109: B125–6

CD ANNOTATIONS

2.2 There were ... *eriantha.* 2.3] *scored red crayon*
10.6 *Gærtner* ... this. 10.8] *double scored blue crayon*
10.8 In a ... germen.— 12.6] *scored blue crayon*
14.1 A very ... sugar-cane. 14.6] *crossed blue crayon*
Top of letter: 'Law of Variation Keep' *blue crayon*
End of letter: 'J Jenner Weirs case'[13] *pencil*; 'In a second letter he mentions case of a Cassia,— in which the first flowers on several racemes produced 2 carpels.— Hooker, mentions genus in which 2–5 carpels are normally produced.—'[14] *ink*

1 Müller refers to 'Fertilization of orchids'; see letter to Fritz Müller, 8 September [1869] and n. 10.

2 A specimen of *Rudgea eriantha* (now *R. coriacea*), long styled, is stuck to the letter at this point.

3 Specimens of *Pontederia* species were evidently attached to an excised portion of the letter beneath the *Rudgea* specimen. There is an empty envelope in DAR 109: B124 on which CD had written, 'Pontederia from F. Müller introduced from Rio Janeiro & new seeds. suppose trimorphic Measure Pollen—'. On Müller's earlier observations of *Pontederia*, see *Correspondence* vol. 16, letter to Fritz Müller, 3 April [1868] and n. 5. Müller published a description of three forms of *Pontederia* in F. Müller 1871. CD described heterostyly in *Pontederia* (now *Eichhornia*, the water hyacinth), including Müller's specimens, in *Forms of flowers*, pp. 183–7.

4 *Heteranthera* (mudplantains) and *Pontederia* (now *Eichhornia*) are genera in family Pontederiaceae. *Heteranthera reniformis* is the kidneyleaf mudplantain.

5 The reference is to a tributary of the Itajahy river (see West 2003, map, p. 70); the main river is now called Itajaí Açu.

6 Müller may have sent a specimen of *Pontederia aquatica* (now *Eichhornia azurea*) with a now incomplete letter in which he evidently discussed the plant; see *Correspondence* vol. 14, letter from Fritz Müller, [2 November 1866] and n. 16.

7 The references are to Friedrich Hildebrand and to Carl Gustav Bernoulli, a German coffee planter in Guatemala. Müller and CD had been discussing dimorphic flowers in the family Rubiaceae since 1866; see especially *Correspondence* vol. 16, letter from Fritz Müller, 31 October 1868.

8 See letter from Fritz Müller, 18 October 1869 and n. 9.

9 For examples of Karl Friedrich von Gärtner's observations of differences in early flowers, see Gärtner 1844, pp. 68, 100, and 220. There is a heavily annotated copy of Gärtner 1844 in the Darwin Library–CUL (*Marginalia* 1: 248–98).

10 *Jussieua* was a common misspelling of *Jussiaea*, a synonym of *Ludwigia*.

11 Müller and CD had been exchanging information on self-sterile plants since 1866 (see *Correspondence* vols. 14–16). CD included Müller's reference to *Tabernaemontana echinata* and the *Calonyction* in his notes (DAR 76: 154b), and referred to the self-sterility of the former as observed by Müller in *Cross and self fertilisation*, p. 331. The name *Tabernaemontana echinata* has since been rejected (see *Taxon* 50 (2001): 1180); Müller probably referred to what is now called *T. siphilitica. Calonyction* is a synonym for *Ipomoea.*

12 The article by Ignácio Francisco Silveira da Mota was published in the Brazilian journal *O Auxiliador da Industria Nacional. Fazendeiro*: plantation-owner, rancher (Portuguese). For CD's interest in graft-hybrids, see *Variation* 1: 390–7.

13 See letter to Fritz Müller, 12 May 1870 (*Correspondence* vol. 18). John Jenner Weir reported to CD the case of a horse born to a mare that had previously had a foal by a quagga; the horse had quagga-like characteristics (*ibid.*, letter from J. J. Weir, 17 March 1870). In his reply to Weir, CD quoted a similar case reported to him by Müller in a letter that has not been found (*ibid.*, letter to J. J. Weir, 17 March [1870]).

14 Müller mentioned *Cassia* in his letter of 29 March 1870 (*Correspondence* vol. 18). CD also refers to Joseph Dalton Hooker.

From John Scott 21 December 1869

Royal Botanic Gardens | Calcutta
Decr 21st /69

Dear Sir

I often think on subjects, material I may have collected information likely to interest you, but I get together even in India so slowly little scraps that I am long in getting sufficient to fill a sheet or so. I was indeed very much pleased to find that my observations on expressions were valued of you.[1] I do not think I have anything noteworthy to add to them … I have particularly observed, since I heard from you, the colour of hair and beard, and invariably seen that when they do differ the former is of the darker shade.[2] (*beard always lighter coloured than the hair*) I have indeed seen it somewhat marked in the thinly bearded Lepchas and Mechis of Sikkim: the *Bhotea* also show it as do many Burmese and Chinamen—all of which have very little hair in general on the face.[3]

I have frequently thought—though you are in all probability already familiar—that the native mode of swimming might interest you: the action is so known, by that of the lower animals. It is different of course with some of the sea-faring races as the Burmese and Malays who have been much in contact with Europeans. I refer only to those races who have come little in contact with Europeans. The Santhals, or Dhangars of Nagpore illustrate it well,[4] as do also the generality of the country nurtured Bengalee and various other Indian races. These in the act of swimming pose the body at a high angle (*nearly* **45°**) in the water, throw the head slightly backward so that the water level is in a line with the middle of the chin base of the ears and nape of neck; thus strikingly different from the Europeans. Again the progressive action is simply "pawing", the manner of short, quick, alternate, almost vertical strokes, with slightly bent or curved hands, the motion of feet spurring, the right limb being drawn forward and struck out again with the stroke of the left hand and so on alternately in a *somewhat devious course* with opposing limbs and arms. I was oft struck with their truly animal like mode of progression and frequently thought of mentioning it to you. Lately however, I had the opporunity of seeing one of our big Hanuman monkeys[5] fall accidentally from a tree into one of our garden tanks in making a long spring. After composing himself somewhat his movements resembled very strikingly indeed those which I have above described: this has induced me to bring it to your notice.

I have often thought also of monkey like mode natives here generally adopt in carrying their children. It is striking to see the youngest infants tied astride the mothers hip with its slender arms athwart her body. Truly they need to have but a slight elongation of the latter so that they might clasp the body to complete the comparison.

I shall now add a few notes on a different subject which I have thought might interest you. *Polydactylism*[6] This occurs in several native families, from whom I have had the following [records]. In the first case is a family of two sons and a daughter, of whom the latter and one of the former had six digits; the supernumerary being a finger attached to the metacarpal bone on the outer margin of the hand. The feet had the ordinary number of digits in the feet. In this case the grandfathers brother had six digits on one hand. Both brothers and sister were married but none of their children have inherited this peculiarity.

Another case is of a man with six digits on one foot (supernumerary on outer margin) the left, while this man's *grandfather* had a sixth on the right foot. It has appeared in none of the other members of their families. A third had six fingers (very perfectly developed) on both hands, as had also a brother and uncle—but none of their descendants, which are as yet however confined to a single generation. A fourth family had a supernumary thumb (perfectly developed) supported on the metacarpal bone, and this *man's grandson* (an infant of 10 months) has also six digits but in this case a finger on the outer margin of the left hand.— I have also to note a Bengale⟨e⟩ family which for three generations have had all its members strongly prognathic. The present family consists of three male and two female children all of whom present the peculiarity and which has been inherited from the father and grandfather. I have seen all the present male members of the family and found in each with the upper maxillary bone projecting considerably more than lower with exceedingly large though equably developed teeth: the canine are same plane and not longer than the others; *previously* to *seeing* the *family I had been told that they projected* and *were much longer.*

I do not think that I have anything further worth communicating (if indeed what I have already written is) I have been much bothered with fever and ague since the close of the rains and have really been little able for work of any sort. I am now however I think fairly clear of them—I have just sent to Dr. Hooker a paper on the tree-ferns of Sikkim in which I treat of their distribution, economic uses, structures of caudex, and relations with the higher plants and a few other peculiarities with descriptions and figures of the species—also a number of plates illustrations of their structure. Dr. Hooker promises to look it over for me and if it proves worthy he will communicate it to the Linnean.[7] I am just commencing a book on Indian Horticulture, in which I devote a chapter to acclimatization.[8]

I am Dear Sir | Yours respectfully | John Scott.

DAR 85: A106–6a

CD ANNOTATIONS
1.1 I often ... them ... 1.5] *scored blue crayon*
1.5 I have] *opening square bracket added, blue crayon*
1.7 (*beard* ... *hair*) 1.7] *two crosses in margin, pencil*
1.7 (*beard* ... face. 1.10] *double scored blue crayon*
2.1 I ... comparison. 3.4] *crossed blue crayon*
4.1 I shall ... peculiarity. 4.8] 'Supernumerary Digit' *margin, blue crayon*
5.8 I have ... acclimatization 6.10] *crossed blue crayon*
Top of letter: 'J. Scott' *blue crayon*

1 For previous communication between Scott and CD on human expression, see *Correspondence* vol. 16, letter from John Scott, 4 May 1868, and letter to John Scott, 3 June 1868. Scott sent further observations in his letter of 2 July 1869.

2 See letter from John Scott, 2 July 1869 and n. 11.

3 CD included Scott's information on the beard and hair colour of 'two races in Sikhim, the Bhoteas, Hindoos, Burmese, and Chinese' in *Descent* 2: 319. On the Mechi, or Mech, and Lepcha people, see the letter from John Scott, 2 July 1869 and nn. 4 and 6. The Bhutia are a Himalayan people who migrated southward from Tibet, settling in Bhutan and India, particularly in Sikkim (*EB*).

4 The Santals, or Santhals, were a large group of people in eastern India; they now constitute one of the largest tribal communities in the country. The Dhangar people are based in Maharashtra state, which now includes the city of Nagpur in central India; in 1861 the Nagpur province of British India was merged into the Central Provinces. (*EB*.)

5 Scott refers to the Hanuman langur (*Semnopithecus entellus*).

6 CD discussed polydactylism in *Variation* 2: 12–17.

7 The reference is to Joseph Dalton Hooker. 'Notes on the tree ferns of British Sikkim, with descriptions of three new species, and a few supplemental remarks on their relations to palms and cycads' (Scott 1870) was read at the Linnean Society of London on 17 February 1870 by John Anderson.

8 CD and Scott previously corresponded on the acclimatisation of plants; see *Correspondence* vols. 15 and 16, and Scott 1868. Scott never completed the book.

From James Leask Sinclair to John Murray 22 December 1869

Auckland New Zealand
22 Decr. 1869.

To Mr Murray Esq
Sir,

I beg to enclose some memoranda containing scientific theories that I wish brought before the notice of D^r. Darwin, and, if considered of any value, reviewed in an early number of the "Academy".—[1]

Oblige me by sending a number of this periodical by post to my address, and say by letter if you desire to have any colonial contributors. I should also like to have your Catalogue, as most of my information concerning new books is taken from the advertisements in the "Saturday Review," and "Notes on Books" that I regularly receive from the Messrs. Longman.[2]

In literature we have done nothing, in fact, the colony is too young to do much in this way, but we certainly excel in science, for you have nothing to compare with our earthquakes. On the strong and healthy their effect seems to be exhilarating, on the weak and sickly most depressing, but here in the South we have so much more sunshine, that it would be very difficult to dishearten us for any length of time.—

I am Sir | Your obedient Servant | J. L. Sinclair.

Memoranda from James L. Sinclair to D^{r}. Darwin.

The following are extracts from some letters of my wife[3] to a literary friend in the Scottish Metropolis and are little else than an amplification of some notes sent through Professor Fotheringham of Paris to the French Institute in 1866.—[4] She considers the theories extremely problematical and the practical illustrations as ridiculously common yet I consider they may be at least suggestive to you and Professor Huxley.—[5] The dates of her letters I cannot give but I think the earliest is July 1869.—

'xxxx When his apprenticeship expired in 1847 he went to Glasgow and armed with a letter of introduction attempted to secure a place in harmony with his tastes by getting into the St. Rollox Works.[6] But M^{r}. Sinclair was of a shy retiring disposition and M^{r}. Gibbs being of opinion he wished a place in the counting-room— he was told that M^{r}. Lennart filled all these places with his own friends.[7] Thus disappointed in pursuing his scientific researches he thought of entering the artillery but not receiving permission and some of his friends being decidedly opposed to it he took various places as a clerk and went to America in 1848.—

Returning in the winter of 1849 he went to Kirkwall and resided with his father[8] and here early in 1850 he resumed his scientific researches. In the spring of 1850 he came to the conclusion that the story of The Fall in Genesis was merely a veiled description of some of the effects of the law of reproduction on the human race. Soon after Hugh Miller's Footprints of the Creator[9] fell in his way and he testifies his admiration in his usual way by copious notes. Lying before me are 15 folio pages of extracts closely written but M^{r}. Sinclair maintained that Hugh's Theory of Degradation was altogether wrong and Sceptic in Religion though he happened to be at the time he clung to Genesis against the Great Geologist. God he says looked upon every thing he had made and he thought them '*very good*' Hugh Miller looks upon every thing and finds them very bad and this peculiar view M^{r}. Sinclair was disposed to attribute to M^{r}. Miller's theological education.[10] And he could never be induced to believe that the flat fish he had speared so often in Stromness Harbour and Kirkwall basin were degraded fish. Taking Birds his favourites among the animal creation and considered by him the most symetrical and spending much of his leisure in shooting he carefully dissected them small and large in the endeavour to find what this so called degradation meant. In 1854 he arrived at the following conclusion: That in every healthy organism there is a perfect physiological equilibrium maintained under the law of Growth till the Law of Reproduction begins to act when the physiological equilibrium is destroyed though M^r. Sinclair believed that this organic disturbance like the perturbations of the planets in Astronomy had certain limits. Among his papers it appears as Lex Lapsus or Law of the Fall Lex Avium or Law of Birds The Fall of Man (Genesis) Animal Degradation (Hugh Miller) Lex Animalis or Law of the Animal Lex Vitalis or Law of the Living Lex Vitalitatis or Law of Life and finally in 1866 as Lex Animalium or Law of Animals though properly speaking it is merely a Theory of Organic Disturbance.

He then began a series of researches as to whether the disturbance was heightened or lessened by particular parts of the body the unequal size of the lobes of the lungs the different motions of the two sides of the body the expansion of the right lung at the right side and the expansion of the left lung and pulsation of the heart at the left side the greater weight of the left side of the brain— and he arrived at the conclusion that malformation of the brain exercised a prodigious influence over the physiological disturbance caused by the Law of Reproduction and that a parent by carelessly or thoughtlessly keeping a child sucking at one breast or even laying it down to sleep always on the same side might seriously affect its future happiness or misery.— Professor Owen[11] believes there is an innate tendency to change M^r. Sinclair maintains that under the Law of Growth there is no tendency to change till the Law of Reproduction begins to operate and the physiological disturbance caused by this law is heightened or lessened by the malformation or symmetry of other parts of the body and especially of the brain".—

Extract from Second Letter.—

"xx The Chevalier de Lamarck[12] was the first I believe who promulgated the idea that changes in the organic world are as much the result of law as those that take place among inorganic substances.— But M^r. Sinclair went a step farther for he suggests that combinations in the organic world follow Dalton's Famous Law of definite proportion[13] and that sex is merely the union of the male and female elements or germs in different combining proportions.—

During his researches on this important subject he wrote to Professor Fotheringham to ascertain if there were any well authenticated instances in the records of the Jardin Des Plantes[14] of the Continuance of a race by means of hybrids but receiving no reply he was obliged to prosecute his enquiries by means of Books and personal observation. As the result of his investigations I gather from his memoranda that in the case of the human family he considers the offspring to be affected by the physical social intellectual & moral condition of the parents and also by the season of the year at which the birth may take place. Not only he says are general and individual characteristics transmitted from parent to progeny but temporary tastes are liable to be inherited. He has also some remarks on what he calls mental atavism or atavism as increased or heightened by hearing or reading of particular personal traits of one's ancestry. What he says is much wanted among civilized communities is a Register of the average vitality of individuals showing what description of constitution they entered the world with the conditions of life to which they were subject and the result of the struggle for existence. He admits however that the data would be very complicated and materially affected by the influence of Temperance and Christianity. In the very remarkable case of the Longevity of the Jews he seems to be of opinion that it may be owing to the greater care taken of the Jewish children by their parents and not from any superior potency of race as many have supposed. In the case of the Inferior animals his opinion is altogether unfavourable to the transmission of a race by means of hybrids. Here however his views are not so definite. For instance the asinine appearance of the mule and hinny he first ascribes to the greater plasticity of the horse but after he tries to make out that it is owing to the greater combining proportion of the element of the ass. He seems to doubt also that domestication can eliminate the sterility of hybrids.— All he says that domestication can do in these cases is to modify and make more uniform the conditions of life. But these he continues cannot cause great or sudden alterations in the animal economy." M.M.[15] 30 September 1869.

DAR 177: 172

[1] Murray recently started publishing the monthly *Academy*; see letter from John Murray, 18 September [1869].
[2] Sinclair refers to the *Saturday Review of Politics, Literature, Science and Art* (North 1997, 6: 425), and to the quarterly *Notes on Books*, published by Longman (North 1997, 5: 3565).
[3] Mary Sinclair.
[4] Professor Fotheringham and the literary friend have not been further identified.
[5] Thomas Henry Huxley.
[6] Sinclair was apprenticed as a clerk to a tea merchant in Leith (Saint-Clair 1898). Mary Sinclair also refers either to the St Rollox chemical works or to the St Rollox railway works in Springburn, north-east Glasgow.
[7] Mr Gibbs and Mr Lennart have not been identified.
[8] William Sinclair resided on Broad Street, Kirkwall, Orkney (Census returns 1851 (General Register Office for Scotland, Kirkwall and St Ola 4/5)).
[9] H. Miller 1849.
[10] For more on Miller and his work, see Shortland ed. 1996.
[11] Richard Owen.
[12] Jean Baptiste Lamarck.
[13] The reference is to John Dalton and his law of constant composition, which was one of the strongest evidences for Dalton's atomic theory.
[14] The Jardin des Plantes was part of the Muséum d'Histoire Naturelle in Paris.
[15] Mary Sinclair's surname was Mowat before she married (Saint-Clair 1898).

To Francis Galton 23 December [1869][1]

Down. | Beckenham | *Kent. S.E.*
Dec. 23[d]

My dear Galton

I have only read about 50 pages of your Book (to the Judges) but I must exhale myself, else something will go wrong in my inside.[2] I do not think I ever in all my life read anything more interesting & original. And how well & clearly you put every point! George,[3] who has finished the Book, & who expressed himself just in the same terms, tells me the earlier chapters are nothing in interest to the latter ones! It will take me some time to get to these latter chapters,, as it is read aloud to me by my wife, who is also much interested.— You have made a convert of an opponent in one sense, for I have always maintained that, excepting fools, men did not differ much in intellect, only in zeal & hard work; & I still think there is an *eminently* important difference.

I congratulate you on producing what I am convinced will prove a memorable work.—

I look forward with intense interest to each reading, but it sets me thinking so much that I find it very hard work; but that is wholly the fault of my brain & not of your beautifully clear style.—

Yours most sincerely | Ch. Darwin

University College London, Library Services, Special Collections, The Galton Papers (39E)

1 The year is established by the reference to Galton's *Hereditary genius* (Galton 1869).
2 CD was reading *Hereditary genius: an inquiry into its laws and consequences* (Galton 1869).
3 George Howard Darwin.

From Charles Layton 23 December 1869

16 Little Britain (City)
Dec[br] 23/69

Charles Darwin Esq
Dear Sir

I have just received letter from Mess[rs] Appletons, in which they say, your proposal is accepted by them, and they will reprint from last Editn Origin of Species, making the corrections you sent,[1] and receiving from you when ready early sheets of *New Work on Man*, with Electros of Illustrs and paying on that work 10% copyright—[2]

Mess[rs] Appletons also mention that they had already an agreement with Prof[r] Gray to remit you 5% on previous Sales of Origin of Species, and next Month you will receive a Small remittance on that a/c—[3]

Yours Respectfully | Charles Layton | Agent D. Appleton & Co

DAR 159: 85

1 Layton evidently refers to CD's proposal in his letter to Layton of 24 November [1869]; Layton was the London agent for D. Appleton & Co. in New York City. D. Appleton & Co. published the fifth English edition of *Origin* 'with additions and corrections' in 1870 (*Origin* 2d US ed.).
2 See letter to Charles Layton, 24 November [1869]. In 1871, D. Appleton & Co. issued three printings of *Descent* (see Freeman 1977).
3 On the agreement with Asa Gray regarding the remittance and previous sales of *Origin*, see *Correspondence* vol. 14, letter from Asa Gray, 7 May 1866 and n. 3. For D. Appleton & Co.'s statement on sales of

Origin by February 1870, see *Correspondence* vol. 18, letter from D. Appleton & Co. to Asa Gray, 1 February 1870. On 12 October 1870, CD recorded receiving a cheque from Layton for the 'sale of Origin in the United States 1866–71'; the amount was £62 5*s.* 2*d.* (CD's Account books–banking account (Down House MS)).

From Francis Galton 24 December 1869

42 Rutland Gate S.w.
Dec 24— 1869

My dear Darwin,

It would be idle to speak of the delight your letter has given me,[1] for there is no one in the world whose approbation in these matters can have the same weight as yours. Neither is there any one whose approbation I prize more highly, on purely personal grounds, because I always think of you in the same way as converts from barbarism think of the teacher who first relieved them from the intollerable burden of their superstition. I used to be wretched under the weight of the old fashioned 'arguments from design',[2] of which I felt though I was unable to prove to myself, the worthlessness. Consequently the appearance of your 'Origin of Species' formed a real crisis in my life; your book drove away the constraint of my old superstition as if it had been a nightmare and was the first to give me freedom of thought.

I now look forward anxiously to your final opinion after you have quite gone through the book.—[3] Wishing you and yours a very happy Christmas

Believe me very sincerely yours | Francis Galton

DAR 105: 3–4

[1] Letter to Francis Galton, 23 December [1869].
[2] Galton refers to works such as William Paley's *Natural theology; or, evidences of the existence and attributes of the Deity, collected from the appearances of nature* (Paley 1802).
[3] CD was reading *Hereditary genius* (Galton 1869).

To Anton Dohrn 25 December 1869

Down. | Beckenham | *Kent. S.E.*
Dec 25 1869

Dear Sir

I thank you sincerely for your very kind present received yesterday, & I congratulate you on the production of your work which must have cost you so much labour, & which I trust will prove a memorable event in our science—[1]

The plates are magnificent & extremely clear. I almost daily regret that I am not a better German scholar but I must endeavour to understand your views.

With my best thanks I beg leave to remain | My dear Sir | yours very faithfully | Ch. Darwin

LS
Bayerische Staatsbibliothek (Ana 525, Ba696)

[1] Dohrn sent the first volume of *Untersuchungen über Bau und Entwicklung der Arthropoden* (Investigations on the structure and development of arthropods; Dohrn 1870) to CD on 11 November 1869 (Groeben 1982, p. 90 n. 11). A copy of this volume, inscribed to CD, is at Down House.

From W. W. Reade 26 December 1869

C/o Hon. C. Heddle[1] | Sierra Leone
Dec.r 26. 1869

My dear Sir

I received and read your letter of July 27 with very great pleasure.[2] I wish I had brought from my journey to Bouré any information worth giving you.[3] I fear I have not.

Respecting expression I have seen a curious method of expressing surprise—of clapping the hand to the mouth.[4] It does express it admirably, I suppose by suggesting the idea that the person is going to cry out. I saw it in one man only—the King of Falaba[5]—when I returned. Major Laing observed it also in the King's father.[6]

I have not been able to discover anything of importance bearing on sexual selection but the following may interest you. The Jollofs a tribe of Senegambia are remarkable for their uniformly fine appearance. How is it said my informant to a Jollof that every one I meet is so fine looking not only your men but your women? Very easily explained said the Jollof. It has always been our custom to pick out our worse looking slaves and to sell them".[7] This was not done of course to improve the race: but the negro's knowledge of the power of selection is very suggestive.

I have seen a negress with blue eyes. I forget if I mentioned this to you in my previous letters.[8] No possibility of white parentage.

There is a race of dogs on the banks of the Niger which has pendent ears: & I have seen dogs with ears apparently in a state of transition—the points of the ears only, drooping.[9]

I gave a Mohammedan friend of mine here instructions to make inquiries about the Guinea rams. The breeder told him the horns appeared at no particular time varying from two months to three months.[10] Not a very satisfactory answer.

I see that your new book is about to appear,[11] and no doubt all the world of science is on tip toe and the world of theology also.

I fear it will have a very serious effect upon Agassiz's health, but it will *I know* be as warmly welcomed by Wyman and by all the rising Naturalists of America, as it will be by Huxley & the naturalists of England.[12] Darwinism is a *fait accompli* in the States: everyone when I was there in '67 accepted your hypothesis, or at least "was investigating in that direction". Asa Gray's defence on religious grounds was not necessary[13]—except for a very few—the New Englanders who with rare exceptions are the working men in science have been cured of all nonsense in religion by Theodore Parker, Emerson[14] &c.

In all probability I shall remain out here another twelvemonth for the purpose of going up the Niger next July. Be sure my dear Sir that I will do my best to obtain any information you may require for your second edition or for future works, &c

Yours very truly | Winwood Reade

DAR 176: 35

1 Reade stayed with Charles Heddle when in Sierra Leone (see letter from W. W. Reade, 28 June [1869]).
2 No letter from CD to Reade of 27 July 1869 has been found.
3 On Reade's travels through the gold region of Bouré on the upper Niger River, in what is now north-eastern Guinea and south-western Mali, see Hargreaves 1957, p. 310.
4 In *Expression*, p. 289, CD included Reade's account of the gesture of surprise with further information from Reade's letter of 4 June 1870 (*Correspondence* vol. 18).
5 Manga Sewa, the king of Falaba, ruled Solimina State; Falaba was its capital (*Dictionary of African biography*).
6 Alexander Gordon Laing mentioned the king of Falaba, 'Assana Yeera' (Asan Yira), in his account of his travels in Africa (see Laing 1825, pp. 223, 228; for his description of the gesture, see p. 249). Reade had recently read Laing 1825 (Hargreaves 1957, p. 309).
7 CD quoted much of Reade's account of the 'Jollofs' in *Descent* 2: 357, writing also: 'It need hardly be added that with all savages female slaves serve as concubines.' The Wolof people (also spelled Ouolof) now live primarily in Senegal, Gambia, and Mauritania (Appiah and Gates eds. 2005, 5: 430). On Senegambia, see the letter from W. W. Reade, 17 January 1869 and n. 4.
8 Reade mentioned this in his letter of 28 June [1869].
9 CD had discussed pendant ears in dogs in *Variation* 2: 301.
10 On the Guinea rams, see the letter from W. W. Reade, 28 June [1869] and n. 2. CD evidently asked for more information on the breed in his missing letter to Reade; see n. 2, above.
11 The publication of *Descent* had recently been advertised in the *Academy* (see letter from John Murray, 18 September [1869] and n. 3); it was published in 1871.
12 The references are to Louis Agassiz, Jeffries Wyman, and Thomas Henry Huxley.
13 Reade probably refers to *Natural selection not inconsistent with natural theology. A free examination of Darwin's treatise on the origin of species, and of its American reviewers* (A. Gray 1861); Gray also addressed theological questions related to CD's theory in A. Gray 1860a, 1860b, and 1860c.
14 Ralph Waldo Emerson.

From L. E. Becker 29 December 1869

28 Jackson's Row | Albert Square | Manchester
Dec 29. 1869

My dear Sir

I have thought of sending my paper on the Lychnis to some magazine or Journal of Botany[1] There are in it verbatim extracts from letters of yours to me about these flowers and I write to ask if you will give me permission to print these extracts.[2] If you desire it, I will send the paper to you that you may see what the extracts are—

Do you think that Seeman's Journal of Botany would be a good periodical to send it to or that it would be better to have it inserted in the Transactions of the Manchester Literary & Philosophical Society.[3] I wish that the matter could be brought under the notice of botanists so that attention may be directed to the appearances during the ensuing spring.

Believe me to be | yours sincerely | Lydia E Becker
Chas Darwin Esq

DAR 160: 118

1 Becker read a paper on *Lychnis diurna* earlier in the year; she sent a copy of the abstract in the *Report of the thirty-ninth meeting of the British Association for the Advancement of Science held at Exeter* to CD (Becker 1869a; see letter from L. E. Becker, 14 October 1869).

2 A summary of Becker's work on *Lychnis*, as reported at the Exeter meeting of the British Association, appeared in the 1869 volume of *Journal of Botany* (Becker 1869b); a quotation is included from a letter of CD's to Becker that must have been written on 26 or 27 May 1863 (*Calendar* no. 4185g, to be published in next supplement; see also *Correspondence* vol. 11, letter from L. E. Becker, 28 May [1863]). No other part of the letter of [26 or 27 May 1863] has been found. No further correspondence between CD and Becker has been found.

3 The *Journal of Botany* was edited by Berthold Carl Seemann. Becker also refers to the *Memoirs of the Manchester Literary and Philosophical Society*.

From Anton Dohrn 30 December 1869

My dear Sir!

It is long, since I did not send you anything of my zoological working, to show you, that I made good progress in applying your theory for elucidating the Morphology of the Arthropoda.[1]

But I daresay, I have not been idle; though you know best, how difficult these questions are, how one may drop easily into errors and mistakes.

Thus I must beforehand openly confess, that my former views on these questions cannot be holden any longer: they are wrong without the least doubt.[2] But I did not arrive at that conclusion, without bringing out something better.

So I have gone through almost the whole Embryology of Crustacea, looking everywhere for rudimentary organs, which should lead towards a true genealogical System, and I may say, I have surpassed all my hopes. I have not only arrived at a pretty well established Genealogy of the whole Class, but I have even tried to write the History of the whole tribe, as far as by observing their way of life and their functions one could form an abstraction, how these functions were developed, and how they developed their organs.

The end is, that I am now scarcely able to adopt any of the old Separations of Orders in the Class, finding, that you cannot trace the limits between them with the least distinctness. The truth seems to me that the Crustacea have begun with *Nauplius*, as stated by that genial little work of Fritz Müller, but that *all* have moved to *Zoëa*, as well Malacostraca as Entomostraca, which I hope to prove by rudimentary organs.[3] Then follows a long pause, of which we have scarcely any knowledge, and after that the *Phyllopoda*. And it is from these, that all the other Orders are taking their origin, Malacostraka, Copepoda, Cirripeds, Daphnidae, and Ostracoda.[4] This result I think I may prove with the fullest certainty.

You observe, how it differs from Fritz Müllers and Haeckels opinions who both think to adopt different lines from Nauplius for the different Orders.[5] There is a good quantity of general conclusions, which I hope to put forth regarding the method and the task of genealogical investigation, and I shall be in the utmost glad and proud, if I should find your consent. I hope I may be able to send you the whole next summer, in Manuscript it is ready and worked over again and again.[6]

I have too tried to find the union between Crustacea and Insects. But though I formerly thought, I could show such link,[7] I think now quite the reverse. The better and deeper I have studied the Embryology of both the Classes, the less I think of any direct union for them, and I am inclined to adopt a different origin for them within the Worms. But this may be decided after a longer course of investigations in the Embryology of Insects, Myriapoda and Spiders on the one side and Annelids on the other.

There are now two other points, for which I should dare to apply to you. The first is, that I have not studied the Embryology of Limulus, and should be extremely glad to do so, as just Limulus would tell us perhaps by its Embryology, how to understand its own Morphology and that of the Trilobites.[8] I could perhaps go to North-America for money-sake, my father would willingly pay the expense;[9] but I have not so much time, the Embryology of Insects and some important questions about the true nature of the so-called nutritory-Vitellus ("Nahrungsdotter") keeping me tight to my working-table in Jena. Would it be possible by any means to get Limulus ♀ with eggs, living in the Aquarium of the London or the Liverpool Garden? Could I perhaps make an application to the Directory of one of these Institutions and rely in the mean time upon your name? All the expenses, I am sure, would be paid by my father, since I myself I have no money of my own. Perhaps Limulus does not carry the eggs? This I never could bring out, since I don't possess but two male specimens.

The second point is still more important to me.

Having stayed now several times on the seashore for zoological studies, I have found how difficult it is to study Embryology without an Aquarium. This want has suggested me the idea of founding not only Aquariums, but Zoological Stations or Laboratories on different points of our European coast. Such a Station should consist of a little house of perhaps four rooms, an Aquarium connected with the sea and the house,—the Aquarium of perhaps 60 feet in Cubus, + where one might have streaming water,— a boat for dredging work, dredges, nets, ropes,—in short, all that is necessary for a marine Zoologist. Besides glasses larger tumlers, bottles; Acids and other chemical objects, and lastly a library. All this might be had at a not to high prize at such a place as Messina, where I thought of founding the first Station.[10] For the library, I intend to apply to all living Zoologists, to send each one copy of their works and articles to the Station. For the House, boat, Aquarium and other apparatus I will apply to some of my richer friends and to my father, and I hoped to get so much money during this winter and the next summer, that I might build next winter in Messina the little house and arrange it so as to have it ready in winter 1871.

I have spoken about this already to several Zoologists, such as Professor v. Siebold,[11] Carl Vogt, Haeckel etc. All are quite satisfied with my plan and have offered their help. But we think altogether, if *you* would give me your consent and approval in some lines, which I might show, whereever I thought it necessary to prove, that my intention is worth to be followed up, I might risk no difficulty in raising as much money as might be necessary to construct and furnish the first Station. Already has the Meeting of the Russian Association for Science voted 1500 Rubels for three Stations to be built in the Black Sea, and Carl Ernst v. Baer has applauded greatly to my plan,[12]—but I think it best, before applying to the general public, I might have constructed one of the Stations completely, to prove that there is no humbug whatever in it.

I would now beseech you, to send me a few lines, wherein you tell me, I might rely upon your consent,—with that I am sure I might prove to everybody, that my intentions are good and useful, and I might get the means necessary for the building.

If such a Station is ready, every Zoologist might go there, have all instruments at hand, even more, than he would ever have afforded by his own means. He would now be obliged to add all his experiences about the Fauna, the Habitat etc. etc. into a large Diary, the "Diary of the Zoological Station of Messina" that every one, who comes after him, might enjoy at once the experiences, the other had made, so as to spare to himself all the unnecessary trouble with looking for the best dredging ground, and discovering anew things, which all his predecessors have discovered also. So may by and by arise a thorough good knowledge of the Fauna, and if this Diary is published,—certainly after careful examination—every Zoologist may know at once, where to find certain animals, which he wants to study, with indications of their spawning time etc.

I know, that the same fisherman, who has attended Johannes Müller in his researches on the Larvae of the Echinoderms, has afterwards attended also Kölliker, Geganbaur, Haeckel, Claus, and all following Zoologists in Messina, and I know him myself as a good and honest man, so as to rely upon him for everything concerning the future Station.[13] Besides some of my friends in Messina are taking a great interest in my idea, and one of them the Swedish Consul, a young german merchant,[14] has promised to serve as Superintendent of the whole Station,—so that no dammage might be done neither to the house nor to the different apparatus and instruments. And if every Zoologist, before using the Instruments of the Station, does oblige himself to repair whatever he might spoil,—the Station and its property might be kept up pretty well.

I think this plan has nothing fantastical or utopian round it,—so I may not risk your displeasure in communicating it to you; in the contrary I hope to meet your consent, which is the chief thing I could now strive at.

Pardon this long letter,—but I found no possibility of being shorter, which was certainly my will. I hope for a short answer.

Should you add a few words, that the State of your health is well, you would afford the greatest joy

to yours ever sincerely | devoted | Anton Dohrn

Stettin. 30th Decemb. 1869.

My address after the 8th of January will be again | Dr. A. D. Jena.

In this very moment, I meet the great joy of your kind letter.[15] My best thanks for it. It shall always be a day of festival for me whenever I shall be allowed to write to you or to read a letter from you. I am once for ever devoted to the course of thoughts, that bear your name, and I shall be proud, if you judge me capable of doing something good and useful.

My sincerest heart-wishes for your health are joined to this letter.

Yours— | Anton Dohrn

DAR 162: 204

CD ANNOTATIONS
8.8 Would … name? 8.11] *scored pencil*
End of letter: 'Dana — U. States.' *ink*

CD note:[16]

I enclose separate letter, but I am sure you think much too much on my judgment ['opin' *del*] that of one, who /studied the sea/, w^d be worth far more— I fear your plan will cost you endless time in writing letters, & arrangements but then you would be the best judge.

Lastly let me thank you, for your too kind & too strong expressions towards myself & your interesting letter, ['with it in' *del*] & ['Health' *blue crayon del*] for your information about your present views.— ['Let me' *del*] Forgive me for suggesting one caution, as Demosth said— action action action, was the soul,[17] so is caution almost the soul of science— Bear in mind, that if an author is once put down as not trusting it takes long years ['to' *del*] before others will trust his observations.— Gärtner made a great blunder when he began his experiments in Hyb, & even to this day his error is sometimes remembered & his admirable works forgotten.[18]

But forgive me, ['& beli' *del*]

You ask about my [*interl*] Health.— I cannot say much for it, if I live the life of a Hermit I can work /some/ hours daily.

With the most sincere good wishes.

I would suggest delaying the prop of /starting/ a Library till you have shown that you can carry out the new /Foundation/ part of your plan.

1 Dohrn had sent CD a paper on the morphology of the Arthropoda in 1867 (Dohrn 1867; see *Correspondence* vol. 15, letter to Anton Dohrn, 26 November [1867]). See also letter to Anton Dohrn, 25 December 1869.

2 Dohrn refers to his earlier work on insect–crustacean homologies; see *Correspondence* vol. 15, letter to Anton Dohrn, 26 November [1867], and letter from Anton Dohrn, 30 November 1867 and n. 5. See also Heuss 1991, pp. 67–8, 70–3. CD had been sceptical of conclusions on this subject published in Dohrn 1867 (see *Correspondence* vol. 16, letter to Ernst Haeckel, 6 February [1868]).

3 Fritz Müller argued that if all Crustacea shared a common progenitor they ought to pass through similar nauplius-like conditions during their development (Dallas trans. 1869, p. 13; Müller 1864a, p. 8); a nauplius is a free-swimming early larval stage. He further argued that since the segmentation in crabs was similar to that in the Amphipoda and Isopoda, these should pass through a similar zoea stage, and a later larval stage (*ibid.*, pp. 14–15; Müller 1864a, pp. 9–11). Entomostraca is a former division that included most lower forms of Crustacea not in Malacostraca (Leftwich 1973). A zoea stage is not characteristic in the Entomostraca.

4 Phyllopoda was an order (or suborder) of lower Crustacea (Entomostraca). In modern classification, Phyllopoda is a subclass of the class Branchiopoda in the subphylum Crustacea. Malacostraca and Ostracoda are classes; Cirripedia and Copepoda are subclasses of the class Maxillopoda; Daphnioidea is a superfamily in the order Cladocera (class Branchiopoda). For more on classification within the Crustacea, see Bowman and Abele 1982.

[5] For Müller's views on evolution in Crustacea, see F. Müller 1864a, pp. 81–91 (Dallas trans. 1869, pp. 122–41). For Ernst Haeckel's phylogenetic classification of Crustacea, see Haeckel 1866, 2: LXXXVII–XCIII, and plate ('Taf.') V.

[6] Dohrn published his investigations in papers that were reprinted in volume 2 of Dohrn 1870.

[7] Dohrn had attempted to homologise larval insects and crustaceans (Dohrn 1867, p. 86); see n. 2, above.

[8] *Limulus* (the horseshoe crab of North and Central America) and the extinct trilobite are both marine arthropods. Dohrn later published on the embryology and morphology of *Limulus* (Dohrn 1871).

[9] Carl August Dohrn was a wealthy businessman and entomologist (Heuss 1991).

[10] Dohrn had worked in Messina from October 1868 until the spring of 1869 (Heuss 1991, pp. 74–9). Messina is on the north-east coast of the island of Sicily. Soon after writing this letter, Dohrn decided instead on Naples as the site for his zoological station (see Groeben 1982, p. 91 n. 17).

[11] Karl Theodor Ernst von Siebold.

[12] Dohrn refers to the second congress of Russian naturalists and physicians held in Moscow in August 1869 and its decision to establish a network of Russian marine biological stations (see Pavlov and Shishkin 2003, p. 422). See also Groeben 1982, pp. 91–2 n. 20. Karl Ernst von Baer had been a professor of zoology in St Petersburg until his retirement in 1867, and was a friend of Carl August Dohrn.

[13] The references are to Johannes Peter Müller, Rudolf Albert von Kölliker, Carl Gegenbaur, and Carl Friedrich Claus. The assisting fisherman has not been identified.

[14] Dohrn refers to Julius Alfred Klostermann

[15] Letter to Anton Dohrn, 25 December 1869.

[16] CD's note is a draft of his letter to Anton Dohrn, 4 January 1870 (*Correspondence* vol. 18).

[17] CD refers to the Greek orator Demosthenes and the saying attributed to him that action was the soul of eloquence. The anecdote about Demosthenes is related in Pseudo-Plutarch, *Lives of the ten orators* (Demosthenes). 'Action' is sometimes rendered 'delivery'.

[18] CD refers to Karl Friedrich von Gärtner and his experiments on hybridity. On Gärtner's error, see *Correspondence* vol. 11, letter to the *Journal of Horticulture and Cottage Gardener*, [before 3 February 1863] and nn. 6 and 7.

From J. L. Sinclair 31 December 1869

Auckland New Zealand
31 Dec.r 1869.

Dear Sir,

By the last English Mail, I forwarded some memoranda to M.r Murray, which I requested him to show you, in order that they may be noticed in "*The Academy*".—[1] I am extremely anxious to have your opinion on my "*Theory of organic Disturbance*", and also on my Hypothetical "*Law of Organic Combination*", and if you should not have leisure to write an article on them for the "*Academy*", would you favour me with a letter on the subject?—[2]

I enclose a copy of a letter to Sir Henry Holland. which contains the leading points of my psychological researches.— This letter, & the one I sent to M.r Murray, contain the results of my scientific labours, which you will think are most meagre, but I have had such a struggle for life that I shall never be able to do more.— Here I have been (like Kepler)[3] calculating Almanacks for a livelihood, and now the season is over, I am out of employment, & have a wife & 4 children depending on my exertions![4] This would be nothing, if my wife did not happen to be a person of a nervous habit, &

subject to frequent epileptic attacks, which hinders & harasses me very much. I write this hurriedly, as I do not know when I may be able again to have the postage.

If you should think my psychological investigations worth reviewing in the "*Academy*", I shall be glad to see them noticed.—

By the way, I observe Mr. Murray has issued a cheap edition of my favourite Hymn writer—Heber:[5] please tell him to send me a copy by post, & I shall contrive to pay him somehow.—

I am Dear Sir | very truly yours | James Leask Sinclair | Auckland. | N.Z.

To Dr. Darwin | Author of the "*Origin of Species*".

[Enclosure]

Copy of a Letter from James Leask Sinclair to Sir Henry Holland London 1869.

"Sometime in 1844, my attention was drawn to the subject of Psychology, by the study, or rather the perusal, of a little book of Forbes Winslow's that I picked up in an Edinburgh Library.[6]

This thin volume was a great treat to me at the time, as it related chiefly to nervous disorders, and some passages in it on Dreams made a very serious impression on me.

Afterwards I read something on the same subject in Brougham's "Introductory Discourse to Paley's Natural Theology,"[7] and this I made the basis of my own researches on the subject. From 1850 to 1862, I had leisure for such investigations, and I now beg to send you a brief summary of the results.—

Taking the impressions made upon the sense of sight during a day, I endeavoured to ascertain whether their reproduction in Dreams followed one or more laws, and I found, that, excluding such dreams as might be considered to arise from the state of the secretions, or from a temporary derangement of the digestive powers, they could be reduced to two, which I named "*Lex Fortuitæ Impressionis*", or The Law of Oblique Impression (in Dreams), and "*Lex Intensæ Impressionis*", or The Law of Intense Impression (in Dreams): in plain language, "The Law of Casual Impression," and "The Law of Strong Impression." In the former, the Impression enters, as it were, obliquely, or, at an angle; it is entirely unexpected, you were not seeking for it, and yet it is as liable to be reproduced as those deeper impressions which Shakespeare has so admirably depicted in his Romeo and Juliet

In 1860, & 1861, I studied the subject of Mesmerism, and arrived at another law, which I termed "*Lex Habitus*", or Law of Habit. This Law is merely "That, in abnormal states of the body, the impressions made upon it, are in exact accordance with its habits in the normal state". For instance, in the Mesmeric coma, a man speaks when his eyes are touched, not because the organ of language has been stirred, but because, from his earliest years, whenever his eyes were covered by a companion coming behind him, his first impulse has been invariably to speak. When touched on the side of the head, he may assume an offensive or defensive attitude, not because the organ of combativeness has been awakened, but because, from his earliest youth, he has been accustomed to be cuffed on the side of the head. When touched on the top of the head, again, he may appear to be elevated, not because the organ of self-esteem has been excited, but because he has been accustomed to be patted on the head in his youthful years, and the erect posture of the body that is necessary to sustain the head when touched in this way, is the ordinary attitude of people in whom sanity predominates.—"

DAR 177: 173

1 See letter from J. L. Sinclair to John Murray, 22 December 1869.
2 For Sinclair's theories, see the enclosures to the letter from J. L. Sinclair to John Murray, 22 December 1869. No letter of reply from CD has been found.
3 Johann Kepler.
4 Sinclair worked as a journalist and teacher while in New Zealand. His wife was Mary Sinclair.
5 The references are to Heber 1870 and Reginald Heber.
6 Forbes Benignus Winslow wrote several short books before 1844; Sinclair probably refers to *On the preservation of the health of body and mind* (Winslow 1842).
7 Sinclair refers to Henry Peter Brougham's contribution to *Paley's Natural theology* (Paley 1845).

APPENDIX I
Translations of letters

From C. F. Claus[1] 24 January 1869

Marburg
24/1 1869

Dear Sir!

While I venture to send you, most respectfully, the enclosed pamphlet on the reproduction of *Leptodera*,[2] I would also like to take the liberty of addressing a few lines to you and thereby to fulfil a wish that I have long held, that is, to express my sincere respect and admiration for you. You will no doubt know, most esteemed Sir, that among German zoologists there are numerous admirers and adherents of your theory of selection who are busy searching for facts that will provide further evidence for the transmutation of species, without which the relations of the phenomena of organic life will remain a mystery to us. I also count myself among them, though without losing sight of the necessity of producing convincing proof, unlike *Haeckel*, whose works you will surely agree are, in spite of numerous excellent and stimulating ideas, a *not mature*, too enthusiastic attempt at natural philosophy.[3] Just this winter, I have been particularly occupied with this topic, and as I am giving a course of public lectures to a sizeable audience of students ("*Explanation and critique of the Darwinian theory*"), I've been forced to work through the material as carefully as possible and examine its persuasive force. While doing this it has unfortunately become clear to me that H. fails to analyse with sufficient rigour, and always treats his expression of belief as proof. I am unable to grant Wagner's law of migration, which indeed is already contained in full in your work, the exclusive importance that its author claims for it;[4] clearly he has not adequately understood your theory, and accordingly has failed to appreciate the influence of *indirect* potential adaptation within natural selection. It is possible that I might publish the contents of my lectures in a small brochure, as an attempt at an evaluation of the evidence for your theory. However, I want to be cautious in this, to present the subject in lectures for a number of times yet, and preferably to leave it to others, for the time being, to refute Haeckel's self-deceptions.[5] For the present I am occupied with a publication on another subject, "The development of the Lepadidae", about which perhaps my observations might hold some interest for you.[6] Here too it is your work that forms the foundation of the subject, but as there is no topic that is not taken up again after a number of years, where further advancement—even after a most excellent earlier treatment—would not be possible, so it is the case here. To be sure, *Pagenstecher* has already made such an attempt, but unfortunately not with the care and thoroughness that is necessary to further your work.[7]

Now, I would not be able to finish my little investigation unless I was sure that I could not add to the material at my disposal. If you, dear sir, should be able and willing to send me rather soon some specimens of pupae,[8] and young Lepadidae still in their pupal stage or of *Balanidae*, I would be deeply grateful to you. It would be of special interest for me to be able to study the pupae of *Proteolepas*, *Ibla Cryptophialus Alcippe*, at the very least these peculiar forms and their complemental males.[9] Hoping you will not deem my request too immoderate, but will kindly excuse it as the desire of a naturae curiosi,[10] I remain | respectfully yours | C. Claus | professor of zoology.

DAR 161: 176

1 For a transcription of this letter in its original German, see pp. 40–1.

2 Claus enclosed an offprint of an article on the organisation and reproduction of the nematode *Leptodera appendiculata* (Claus 1868a). CD's copy is in the Darwin Pamphlet Collection–CUL.

3 Claus refers to Ernst Haeckel. For more on Claus's objections to Haeckel's version of Darwinism, see Nyhart 1995, pp. 186–93.

4 Claus refers to Moritz Wagner and the 'law of migration' elucidated in Wagner 1868a and 1868b. In a letter to August Weismann of 22 October 1868 (*Correspondence* vol. 16), CD had remarked that Wagner greatly overrated the necessity for emigration and isolation. In *Origin* 5th ed., p. 120, CD referred to Wagner's essay, noting that while geographical isolation was important for preventing crosses between new varieties, it was not necessary for the formation of new species.

5 Claus never published the lectures on Darwinian theory; but in later work he was highly critical of many aspects of Haeckel's work, most notably the latter's gastraea theory (see Claus 1874 and Claus 1876; see also Di Gregorio 2005, pp. 297–301).

6 Claus enclosed an offprint of an article on the cypris-like larvae of cirripedes and their transformation into sedentary adults (Claus 1868b). The offprint is in the Darwin Pamphlet Collection–CUL, and is lightly annotated.

7 Claus refers to Heinrich Alexander Pagenstecher and Pagenstecher 1863.

8 Claus followed CD's usage in referring to the last larval stage in barnacles as 'pupal' (see *Living Cirripedia* (1854), p. 13); this stage is now more commonly referred to as the cypris stage (see *Living Cirripedia* (1851), p. 14).

9 The genus *Alcippe* is now *Trypetesa*. *Proteolepas* is no longer classed as a cirripede but instead is classed as a parasitic isopod of the family Cabiropidae (see *Living Cirripedia* (1854), p. 588). Claus refers to CD's discovery of 'complemental males' in several species of Cirripedia. CD had observed minute males attached to the bodies of hermaphrodites, and differing greatly from them in size and structure (see *Living Cirripedia* (1851), pp. 231–2 and 281–93, *Living Cirripedia* (1854), pp. 23–30, and Newman 1993, pp. 377–81). In his letter to J. D. Hooker, 10 May 1848 (*Correspondence* vol. 4), CD remarked that the discovery was suggestive of the development of separate sexes, and thus of significance for his theory of species (see also *Correspondence* vol. 4, Appendix II, pp. 399–400). CD later suggested that complemental males might have developed in cirripedes to enable intercrossing between two hermaphrodites (see 'On the males and complemental males of certain cirripedes, and on rudimentary structures', *Nature* 8 (1873): 431–2; *Collected papers* 2: 177–82).

10 *Naturae curiosi*: inquirer into nature (Latin).

From David Forbes 30 January 1869

[Enclosure][1]

Extract from a letter from Professor Zirkel[2] of Kiel to DF.

What do people in England think of Jenzsch's new discovery: a microscopic fossil flora and fauna in the constituents of volcanic rocks?[3] Everyone with whom I've discussed this in conversation or by letter thinks it is humbug— Until I'm better advised, it is also my view. A fine new scene: a region of quartz in melaphyre,[4] in the foreground an idyllic grazing 'infusorium' which happily gnaws at a layer of green algal cells with its outstretched snout, next to it a nascent rotifer, in the background a few fluid-cavities with moving bubbles. A picture worthy of the brush of Sir Edwin Landseer![5] That such a discovery had to escape *us*! I have written to Jenzsch and asked him to send a few specimens for a day and to give me some pieces of melaphyre so I could make my own specimens. By way of an answer, he said he needs his specimens *himself* and his melaphyre samples are so tightly packed that he himself could not get one. "Sapienti sat."[6]

If you do not yet have this wonderful piece of writing which of course marvellously impresses all ultratransformists and dilettantes, I can send it to you, I have three copies.

DAR 164: 143–143/2

1 For a transcription of this enclosure in its original German, and the letter with which it was enclosed, see pp. 49–50.
2 Ferdinand Zirkel.
3 Zirkel refers to Gustav Jenzsch and Jenzsch 1868.
4 Melaphyre: 'any of various dark-coloured porphyritic (esp. basaltic) rocks. Later: (specifically) an altered basalt, esp. of Carboniferous or Permian age' (*OED*).
5 Edwin Henry Landseer.
6 *Sapienti sat.*: abbreviation of '*Verbum sapienti satis est*', a word to the wise is sufficient. (Latin.)

From F. A. von Hartsen[1] 5 February 1869

Cannes
5 Febr—1869

Mr Charles Darwin Esq
Most esteemed Sir

I would not have dared to write again to your honour if it were not for my heartfelt interest in your research and to science in general.[2]

A number of objections have been raised against your theories. To me, however, it looks as though most opponents of these theories expected them to meet far too difficult requirements. In my opinion, therefore, it is vital for the defence of these theories to admit in advance the following: that in most cases it is quite impossible *to prove* that all the individuals of a group have had a common ancestor. In most cases, I think, one can't do more than show that this would not constitute an *inconsistency*.

I maintain, it is rarely possible to prove that the individuals of a natural group have a common ancestor. What I mean by this is the following. If it is possible for a natural process, through *natural selection*, to produce a group of higher organisms from a simple organism, then there is no reason why this process should occur only at *a single* place on earth. For example, I can imagine that in East India, *palm trees* gradually developed from a very simple plant, and that quite independently in South America palms *also* originated from a very simple plant. In short, I can imagine that at 2, 3 or more places on earth *the same* genus, perhaps even *the same* species have emerged from the realm of inorganic matter.

Hence, according to your theory we can in my view very well surmise that some group or other of organisms did not have *one* but rather *several* centres of origin. The theory therefore would remain valid even if the individuals of a group, for geographical reasons, could not possibly have a common origin. I can imagine anyway that inorganic mass has produced living creatures in *a number of* places on earth and that these creatures continued to evolve *along similar lines* in these various places, so that e.g. the snails of England have a different origin from the snails of America.

To me this observation seemed not unsuited for the refutation of geographical arguments against your doctrine.[3] This is why I take the liberty of communicating it.—

"Carl Vogt" has been giving lectures in Holland, and in them he has interwoven your theories with his own materialism.[4] As a result, some people over here have been led to assume that materialism and atheism are the ultimate conclusion of Darwinism. I have now completed a popular account of your doctrines, in which I have shown that they actually do not decide the cardinal questions of religion.[5] It seems unlikely, however, that Your Honour reads Dutch, hence Dutch works must be of little interest to you.—

Prof. Fechner writes me that a Dr. Pfaff has criticised your views.[6] I hope to read Pfaff's publication, although I do not place much stock on objections of German provenance, since these are mostly derived from prejudice.[7] I regret only that we do not know how organic life develops on *other planets*. A comparative "organismogeny" would be most valuable! Perhaps one day we will acquire one!

Mr Traherne Moggridge is fortunately doing better.[8] He is a very enthusiastic disciple of yours!

I also thank you most humbly for your kind letter with which you have honoured me upon receiving my essay.[9] Had I known that you intended to reply in such detail, I should perhaps not have had the courage to send you my book. With heartfelt wishes for your health I remain | highly honoured Sir | your devoted servant | F. A. v. Hartsen

DAR 166: 114

[1] For a transcription of this letter in its original German, see pp. 65–6.

[2] See *Correspondence* vol. 16, letter from F. A. von Hartsen, 5 January 1868. The abbreviation 'Eu Hw.' is not standard, but Hartsen evidently intended 'Eure Hochwohlgeboren'.

[3] In *Origin*, p. 351, CD had argued that species of the same genus, although widely separated geographically, must originally have proceeded from the same source, as they had descended from the same progenitor. He then discussed geographical distribution and the means of dispersal of species to new locations (see *Origin*, pp. 356–410). In subsequent editions of *Origin*, CD added material to counter objections to his theory based on difficulties associated with dispersal.

[4] For more on Vogt's popular lectures of 1868 and 1869, see Kockerbeck ed. 1999, pp. 23–4.

[5] Hartsen refers to his book, *Darwin en de Godsdienst* (Darwin and religion; Hartsen 1869b).

[6] Hartsen refers to Gustav Theodor Fechner and Friedrich Pfaff.

[7] In his book, *Die neuesten Forschungen und Theorien auf dem Gebiete der Schöpfungsgeschichte* (Latest research and theories on the history of creation; Pfaff 1868), Pfaff argued that CD's theory was untenable in the light of geological findings.

[8] Hartsen refers to John Traherne Moggridge. In his letter of 5 January 1868 (*Correspondence* vol. 16), Hartsen had mentioned that Moggridge's health was bad.

[9] CD's reply to Hartsen's letter of 5 January 1868 (*Correspondence* vol. 16) has not been found. Hartsen had sent CD a copy of his *Grundlegung von Aesthetik, Moral und Erziehung* (Foundation of aesthetics, morals and education; Hartsen 1869a).

From C. F. Claus[1] 6 February 1869

Marburg
6/II 1869

Most esteemed Sir!

With your kind lines you have given me as much pleasure as honour, and indeed I scarcely know how I can show my gratitude for the research material you promised me.[2] I hope, through the most careful and conscientious examination possible of these very interesting and important subjects, to prove myself in some way worthy of your great kindness and of your trust. I will accept the specimens referred to with my deepest gratitude. As I have hitherto been able to analyse only the larvae and pupae of Lepadidae, I would also be particularly obliged to you if you could possibly provide me with an opportunity of examining the pupae of Balanus; I should think that the postal service would easily accept and deliver such a little box which I would ask you to address to the Königliche Zoologische Institut—

In a small note that I have permitted myself to send to you under wrapper, along with a number of other works, I have already in 1862 made an attempt to trace back the morphology of the *cirripede body* to that of *copepods*.[3] My more recent studies of the larvae, pupae and young cirripedes that are still enclosed in their pupae have, so I hope, largely borne out this earlier analysis.[4] The organs in the horns of the larva are in any case sensory organs and equipped with hairs comparable to the sensory hairs (Leydig organs) of lower crustaceans; they are obviously identical with what you presented as the acoustic organs of the pupa, whose *sternal edges* correspond to the sternal horns of the larva.[5] Unfortunately, a finer determination of these organs can only be carried out on fresh pupae, and regrettably I was only able to examine spirit specimens.

Regarding Wagner's pamphlet,[6] surely it should also be emphasised that gradual tellurian and climatic changes have been greatly underestimated as a cause of change in the conditions of existence and the interactions among organisms. Migration may be of importance especially in accounting for geographical varieties that develop side by side, the latter, on the other hand, in accounting for *chronologically* successive variations. In the latter case migration is secondary and an effect, not, as Wagner believes, a primary cause. Furthermore you have, I should have thought, with your account of the 2 imaginary examples, particularly that of the long-legged, faster, and shorter-legged wolves, made it very probable that even without such strict isolation as is produced by mountains, streams and oceans, variations can develop side by side in the same area. This Wagner has not refuted, but merely contrasted it with his own assertion—.

I feel it a special honour, esteemed Sir, to be able to enclose my photograph.[7] I would consider myself fortunate to own one of you, signed by your own hand, and to be able to insert it in my zoological album at the summit of naturalists who through their work have had a powerfully radical, transforming and constructive impact on the development of science.

I expect you have not yet heard that *R Leuckart*[8] has been offered and accepted a position at the University of Leipzig.

With the greatest respect and veneration I remain your most devoted | C. Claus

P.S. In a pamphlet "A Darwinian account of the copepod fauna of Nice" I questioned the value of Fr. Müller's remarks on the two male forms of Tanais d. (smellers and graspers).[9] Surely I may be allowed not to withhold from you my doubts about the *nauplius larvae* of the prawn. Fr. Müller has neither proven that the so-called *nauplius larvae* develop from the eggs of *Peneus* nor that the zoea stage develops from nauplius larvae after the skin has been shed. The simultaneous occurrence of the forms of nauplius and the stages of zoea does not warrant the conclusion that the former are the earlier stages of the latter, *which still remains to be shown.*[10]

It may be perfectly true, but the *certainty* is missing. And my doubts are now aroused by the *size* of the eggs of a species of Peneus as yet not precisely identified. From an egg of this size one may infer a fuller development of both embryo and hatching larva—

In about *four* weeks the holidays will begin and I shall have *six* weeks to apply myself exclusively to the study of cirripedes. In the meantime, should you, esteemed Sir, not have the time to send me the material until then, I can well wait until summer, for I should not at any price want to take up your time at the wrong moment. Yours | C. Claus.

DAR 161: 177

1 For a transcription of this letter in its original German, see pp. 67–9.

2 CD had offered to send some specimens of barnacles. See letter to C. F. Claus, 28 January 1869.

3 Claus refers to his paper on the morphological relations between copepods and other crustacean groups, including cirripedes (Claus 1862; for the section on cirripedes, see pp. 161–2). CD's copy of the paper has not been found.

4 With his letter of 24 January 1869, Claus had enclosed an offprint of his article on the cypris-like larvae of cirripedes and their transformation into sedentary adults (Claus 1868b).

[5] Claus refers to CD's description of acoustic organs in the final pre-adult stage (cypris) of *Lepas australis* (*Living Cirripedia* (1851), p. 15). For more on CD's view of the homologies of the naupliar, cyprid, and adult horns and antennae, see Newman 1993, pp. 372–4. In modern terminology, the 'Leydig organ' is the organ found near the oesophagus of sharks and other members of the subclass Elasmobranchii, and is associated with the immune system. Claus probably refers to an organ described by Franz von Leydig in a monograph on *Daphnia* as an analogue of the green gland in decapods (Leydig 1860, pp. 25–6).

[6] Claus refers to Moritz Wagner and Wagner 1868a or 1868b. See letter from C. F. Claus, 24 January 1869 and n. 4.

[7] The photograph has not been found.

[8] Rudolf Leuckart.

[9] Claus refers to his study, *Die Copepoden-Fauna von Nizza. Ein Beitrag zur Charakteristik der Formen und deren Abänderungen 'im Sinne Darwin's'* (The copepod fauna of Nice. A contribution on the diagnosis of its form and its modification 'in the Darwinian sense'; Claus 1866). In *Für Darwin* (F. Müller 1864a, pp. 13–16; Dallas trans. 1869, pp. 20–4), Fritz Müller had described two forms of male in *Tanais dubius* (now *Leptochelia dubia*). Claus questioned whether one might be the sexually mature form of the other, as was the case with the amphipod *Phronima* (see Claus 1866, pp. 1 and 2). In *Origin* 4th ed., p. 50, CD had noted Müller's discovery of the two forms of male; the passage remained unchanged in subsequent editions. In *Descent* 1: 328–9, CD cited Müller at greater length on *Tanais*.

[10] Müller had suggested that *Penaeus* emerged from the egg in its nauplius state in a paper, 'Die Verwandlung der Garneelen' (F. Müller 1863, p. 8), that appeared in English as 'On the metamorphoses of the prawns' (F. Müller 1864b, p. 104). Müller's observation was only confirmed much later (Kishinouye 1900), but Claus evidently accepted Müller's account when he made his own observations on development in the genus (see Claus 1876, pp. 10–13).

From Georg Recht[1] 9 February 1869

Munich Schommergasse N.2/3
9 February 1869

Your Honour!

In sending you my pamphlet, "die Weltgesetze",[2] I would like to show that I, too, respect you as the master of natural sciences. It would be an infinite honour for me if you would communicate this text to the scientific society in London, and if you would give me your opinion of its worth or lack thereof; perhaps you would even convey to me the opinion of that society, for my circumstances hardly allow me to dispatch a separate copy with a letter to the latter.[3] My family and I have only one gulden for our daily requirements and lack the very essentials of life.

This is also the reason that I was able to send an abstract of the above-named work, through the goodness of a friend, and had to leave out the more complete exposition as well as a comparison of my view with the earlier theories on the topic.

I would be glad if circumstances would permit me to continue to work on a task that to me seems Very promising, but that rests in higher hands.

With my greatest respect | Your Honour | most devotedly D^r. Recht

DAR 176: 74

[1] For a transcription of this letter in its original German, see pp. 73–4.

[2] Recht refers to his monograph, *Die Entwicklung der Weltgesetze* (Development of the laws of the universe; Recht 1869).

[3] CD evidently sent the copy of Recht 1869 to the Royal Society of London; it is still in their collection (Royal Society Library, Tract 525/5). An inscription at the front of the book reads, 'Herrn Charles Darwin | Esq. M. A | der Verfasser' (to Charles Darwin Esq. M.A., from the author).

From Ernest Faivre[1] 18 February 1869

Lyon
18 February 1869

Dear Sir,

One of My students has translated most of your interesting work on the *Fertilisation of Orchids* into French;[2] Before pursuing his project any further, he asks you to do him the kindness of giving him some information which I shall venture to request from you.

Has your book on the fertilisation of the orchids been translated into French? Would you be disposed to authorise a translation, and could it be produced by taking the volume published in London by John Murray in 1862 as a guide?

As science has made some progress since then, perhaps you are currently preparing a new edition? If not, it would no doubt suit you to revise the edition that has appeared and to add notes; would it be agreeable to you if the translation were submitted to you in advance and in stages?

I take the liberty, dear Sir, to ask your advice on these questions that greatly preoccupy my young disciple, and I dare to hope that you will be so good as to let us know your feelings.

Perhaps you may find difficulties that we have not foreseen in carrying out the work? I beg you not to dissimulate them to us.

In any case, dear Sir, rest assured that nothing will be done without your express consent.

I deeply regretted not being able to converse with you when I came to London last August on the occasion of the Norwich Congress, to which I was invited.[3]

Knowing your state of health, I feared to importune you and disturb your much-needed rest.

Allow me to thank you for sending your second volume on the variation of animals and plants, I am reading it, or rather studying it attentively.[4]

I dare not abuse your precious moments any further and I beg you, Sir, to rest assured of my admiration and deep respect.

Ernest Faivre | Professor at the Faculty of Sciences at Lyon | Curator of the Botanical Garden etc.

DAR 164: 4

[1] For a transcription of this letter in its original French, see pp. 85–6.

[2] Faivre refers to Louis Rérolle and to *Orchids.*

[3] Faivre refers to the meeting of the British Association for the Advancement of Science held in Norwich from 19 to 26 August 1868 (*Report of the 38th meeting of the British Association for the Advancement of Science*, p. lvii).

[4] Faivre refers to the second volume of the French translation of *Variation* (Moulinié trans. 1868).

From Georg Recht[1] 1 March 1869

Munich Schommergasse N. 2/III
1st March 1869

Your Honour!
Noble patron & benefactor!

Yesterday I received payment of your transfer, and I will employ the first of the money in conveying my thanks to you for your assistance, which will relieve me of all worries for a few months, and which will enable me to continue to send out my work.[2] My most cordial thanks for this unexpected support. Now I am able now send you your copy of my pamphlet as a small token of my appreciation.[3]

For the time being I cannot say more about its content, but permit me to remark that d'Alembert's principle is its main support. This principle states: in every state of motion, force and resistance are in balance.[4]

In this simple form, the aforementioned principle appears to be self-evident and of no further significance beyond supplying necessary equations. However, understood in conjunction with the idea that in any movement, obstacles are not a reciprocal but merely a constraining force, the result is uniform motion and its speed as a measure of the same, while a reciprocal force would result in a state of rest. In regular or uniform motion the momentum of the resistance is equal to the momentum of force or the momentum of released heat.

Everywhere in nature we see uniform motion under the influence of two forces or two velocities, namely from the basic elements of the universe—space & time—right down to the organs of the lowest animals. Space is the given stationary or inhibiting element, time the moving one, and thus every organ is composed of two such elements. Every movable bone, for example, is faced by a stationary or restraining one. The application of the d'Alembertian principle to nature, especially to chemistry, promises, in my opinion, to yield amazing results and makes this science accessible to calculus. Right now I would like to write a large work consisting only of inferences whose truth is more or less confirmed by experiment, and the truth of these inferences allows me to conclude that I have set down the correct principles in my treatise.[5]

Please excuse the didactic seeming style of my writing, as I await a later opportunity to communicate further results to you. Allow me to add just one further remark; namely: human skin, with its coiled helical glands, consists of inner and outer layers. The outer layer nourishes man with light through the eyes, with electricity through the ears and heat through the sense of touch; the inner, through scent & air, through taste and through digestion.

By this I want to say that I distinguish between six instead of five senses, and that each sense is a nutritive organ.

With the most cordial request to accept the enclosed copy as a token of my respect & veneration, I thank you again.[6]

Your Honour | most devotedly | D Recht

DAR 176: 75

[1] For a transcription of this letter in its original German, see pp. 103–4.
[2] See letter from Georg Recht, 9 February 1869. A payment of £3 to Recht on 13 February 1869 is recorded in CD's Account Books–banking account (Down House MS).
[3] See letter from Georg Recht, 9 February 1869 and nn. 2 and 3.
[4] Recht refers to Jean le Rond d'Alembert and to the principle of classical mechanics first elucidated in his *Traité de dynamique* (Alembert 1743).
[5] Recht's proposed larger work was apparently never completed. Recht also refers to Recht 1869 (see n. 6, below).
[6] Recht enclosed a copy of his work on the development of world laws (Recht 1869; see also n. 3, above).

From Ernest Faivre[1] 15 March 1869

15 March 1869

Dear Sir,

I want to make a brief response to your letter of 21 February last, in order to leave my student, Mr Rérolle, the task of finishing it and thus entering into direct communication with you[2]

The translation of your important work on the orchids is agreed in principle: it will certainly have significance, thanks to the notes that you have been so kind as to promise to send.

The question now is to find a publisher, and I hope that this will be resolved by around the fifteenth of next month.

We shall have the honour of informing you directly the agreement has been made and settled; and at that point it will be appropriate to make contact with your London publisher on the matter of stereotypes and to discuss the notes you so kindly promised.[3]

I look forward in advance to being able to contribute, even in an indirect way, to making one of your most original and interesting works known in France, | and I beg you to accept, Sir, my respectful admiration | Ernest Faivre

DAR 176: 129

[1] For a transcription of this letter in its original French, see p. 134.
[2] CD's letter has not been found, but see the letter to R. F. Cooke, [22 February 1869]. Faivre had asked whether CD would agree to a translation of *Orchids* by Louis Rérolle (see letter from Ernest Faivre, 18 February 1869). See also the letter from Louis Rérolle, 17 March 1869, which was written on the second leaf of the same folio as this letter.
[3] CD had already written to his publisher, John Murray, about providing stereotypes (see letter to R. F. Cooke, [22 February 1869]). For more on CD's notes to the French translation of *Orchids*, see letter to J. D. Hooker, 8 March [1869] and n. 5.

From C. F. Claus[1] 19 March 1869

Marburg
19.III.69.

Most esteemed Sir!

In your kind letter of 10 February this year, you informed me that you had instructed Messrs *Williams* and *Norgate* to dispatch, via booksellers, the cirripede material that you intended for me.[2] As 5 weeks have since passed and the little parcel has not yet arrived, and as an inquiry through Elwert's bookshop[3] to Leipzig has also failed to yield any results, I am beginning to fear that possibly the little parcel may have been delivered to another address or to another town, or that some other mishap has befallen it, and I presume to notify you of this with the request that you inform Messrs Williams and Norgate, as this seems the best way to clear up a potential misunderstanding. I am very sorry to have to abuse your kindness with this perhaps somewhat presumptuous yet surely justified request, and I hope that you will kindly forgive my frankness.

With my highest respect | Your devoted | C Claus

DAR 161: 178

[1] For a transcription of this letter in its original German, see p. 138.
[2] CD's letter of 10 February 1868 has not been found, but see the letter from C. F. Claus, 6 February 1869.
[3] Claus refers to the Marburg publisher and bookseller, N. G. Elwertsche Verlagsbuchhandlung (Würffel 2000).

From Karl von Seidlitz[1] 22 March 1869

Livonia | Dorpat
[10]/22 March 1869[2]

Dear Sir!

I see from your letter of 31 January 1869 that my bookseller has *not* dispatched to you the book that I would now like to present you with as evidence of the recognition that your pathbreaking ideas have won among us for quite some years.[3]

That *inheritance* is a powerful factor in the history of the development of organic beings has long been acknowledged. But I should like to go further and to suggest that the *power of transmission* be viewed as an immanent property of all organised matter—from the first amorphous slime to the most developed life forms. Only by virtue of this basic property can the development of all organic nature on our Earth have taken place and continue to advance also in future. What gravity is in the inorganic world,—the power of transmission is in the organic world. I have expressed my views on this accordingly on pag. 52–54.

Since my brochure, which I have enclosed under wrapper, will hardly have reached the booktrade, perhaps an announcement in some English journal would help make known how your hypothesis has been interpreted in the so-called land of the polar bears.

With my greatest respect | Your most devoted | Dr. C. Seidlitz

DAR 177: 132

[1] For a transcription of this letter in its original German, see pp. 148–9.
[2] Dorpat: now Tartu, in Estonia. Seidlitz gives both the Julian (10 March) and Gregorian (22 March) calendar dates.
[3] Seidlitz sent his monograph *Ueber die Vererbung der Lebensformen, Eigenschaften und Fähigkeiten organischer Wesen auf ihre Nachkommen in Bezug auf Physiologie und praktische Heilkunst* (On the transmission of life structures, properties and vitality of organic beings to their descendants in relation to physiology and practical medicine; Seidlitz 1865). CD's copy is in the Darwin Pamphlet Collection–CUL.

From Martin Wilckens[1] 25 March 1869

Pogarth | *near Prieborn in Silesia* | Prussia.
25 March 1869.

Mr Charles Darwin | in | Down Bromley Kent
Most esteemed Sir

I venture to enclose two little pieces of research of mine, an earlier one, which discusses your "on the origin of species etc.", as well as an annual report on the scientific advances in animal husbandry, which critically reviews your work translated by Carus from the perspective of animal husbandry.[2] I ask you to confirm its arrival and hope that an opposing view of this subject will not be unwelcome to you.

I remain with the greatest respect | Your most devoted | Dr. M. Wilckens. | esquire.

DAR 96: 67v

1 For a transcription of this letter in its original German, see p. 156.
2 Wilckens enclosed 'Darwin's Theorie in Beziehung zur landwirtschaftl. Viehzucht' (Darwin's theory in relation to cattle breeding; Wilckens 1866) and 'Bericht über die wissenschaftlichen Fortschritte der Thierzucht im Jahre 1867/68' (Report on scientific advances in animal husbandry; Wilckens 1869). CD's annotated copies are in the Darwin Pamphlet Collection–CUL. Wilckens also refers to the German editions of *Origin* (Bronn trans. 1862) and *Variation* (Carus trans. 1868) and to Julius Victor Carus.

From Armand de Quatrefages[1] 29 March 1869

Paris
29 March 69

My dear Colleague

I wanted to put off replying to your letter until after I had had the pleasure of seeing your son.[2] But it seems that fate does not want me to have that pleasure. He came twice to my house and I went twice to his, without our ever meeting each other. And yet for three months in a row I do not go out except to visit the Institute in the evenings.[3] I hope that I shall end by being more fortunate and that I shall have the pleasure of making his acquaintance one of these days.

I must thank you for the way in which you have appreciated my studies concerning your theory. It had become impossible for me not to undertake this discussion. In the name of transformism in general, in the name of your ideas in particular, I saw what I regarded as the very foundations of the natural sciences attacked every day and represented as outworn ideas. I have often personally been pointed out as a man behind the times, wilfully shutting his eyes to scientific progress.[4] So I was obliged to defend myself. In my turn, I had to discuss the theories with which I was being opposed. You are incontestably the leader of all transformist theories, the only person to have proposed a theory founded solely on scientific considerations and embracing all the problems posed by living nature. So I had to combat you above all. I did what I could to show that scientific reason is on my side. But I hope that I have never failed to recognise the greatness of your work, and on completing this lengthy study with one final article, which is due to appear on the 1st of April, I was happy to express publicly all my esteem for you as a man and a man of science— As soon as I have received my *offprints* the first copy shall be sent to you.[5]

I shall take the opportunity to express my regrets about an inaccuracy that has crept into my third article.— In examining what you say concerning the *titmouse* and the *nuthatch* in your book on species, I had added a note in which I drew attention to the fact that you were not precisely speaking of a genealogy; that you were not affirming anything positively. This note vanished in the final printing along with many others. Probably some typographical arrangement was the cause of its disappearance. It will be restored in the book edition in which these articles are due to appear But I am anxious to let you know at this point that I am not guilty of the matter.[6]

I dearly hope, dear Sir and colleague, that schisms of opinion will never affect our relationship. It is in that hope that I beg you to accept my most cordial and devoted regards | De Quatrefages

DAR 175: 5

1 For a transcription of this letter in its original French, see pp. 162–3.
2 CD's letter has not been found. George Howard Darwin had probably travelled to Paris on 5 March 1869 (see letter from G. H. Darwin, [23 February 1869] and n. 7).
3 Quatrefages refers to the Institut de France. The institute was the home of five learned societies, including the Académie des Sciences, of which Quatrefages was a member (*DSB*).
4 For more on the French debates over transformism, see Conry 1974, pp. 359–92. See also *Correspondence* vol. 16, letter from Camille Dareste, 3 April 1868.
5 Quatrefages later sent CD an offprint of a series of five articles he had written for the *Revue des Deux Mondes* between December 1868 and April 1869, 'Origines des espèces animales et végétales' (Quatrefages 1868–9). CD's annotated copy of the offprint is in the Darwin Pamphlet Collection–CUL.
6 In Quatrefages 1868–9, p. 63, Quatrefages had questioned whether CD had demonstrated that the nuthatch was the 'grandson' of the titmouse. When Quatrefages later published a book based on these articles, a note was added giving a reference to the section of *Origin* where CD had commented on the titmouse and nuthatch (see *Origin* 4th ed., p. 281), but there was no mention in the note that CD had not, in fact, suggested a genealogical connection between these species (Quatrefages 1870, p. 156).

From Georg Recht[1] 21 April 1869

Munich
21st. April 1869

Your Honour!

A meeting today with Baron von Liebig on the enclosed article in the Allgemeine Zeitung has left me with the depressing certainty that my poor and wretched little fatherland regards me as a fantasist and is abandoning me or rather leaving me to starve to death.[2] Wretched I call Bavaria, for not a single scholar who thinks can be found on her soil, while thinking, especially practical thinking is completely unknown to our statesmen and administrators, of whom we possess a whole host, who draw on the marrow of the people even more than our absurd military, and moreover they are imbued with a prodigious sense of self-importance. Poor, however, I call my fatherland because general impoverishment is spreading in a horrifying manner as a result of a lack of trade and industry and because our unnatural soil conditions have allowed all capital to be pulled out of land and converted into foreign securities. Shall labour not despair and must the soil not become desolate, if they are deserted by capital and intelligence! if the taxable capacity flows into the coffers of the state, and not into pensions! if the people live from the soil itself and no longer from its yield! From such a people learning and thinking flee, for they feel no cultivation.

In this despair I dare once again address your Honour with the question whether you could not find me an English sponsor. In vain I am working in Bavaria on several treatises in support of the introduction of English conditions, for which I live and die.[3]

For the present I beg your pardon for writing in German and for sending you my essay in the German language. While I am diligently practising my English, I have as yet not progressed sufficiently to move easily in your tongue, and I lack the funds to pay for a translation. I presume that you received my last letter of thanks for your support and the copy of my pamphlet which I intended for you.[4]

In the enclosed essay I tried to show that all branches of natural science find their common ground in mechanics, but that analytical mechanics is mistaken insofar as its fundamental principles are not given in nature, which is also why no conclusions can be drawn from it on the essence of nature, its bodies & its phenomena.

Its results are only correct as a result of two errors that compensate one another, for it considers pressure or constant force as a continuous impact, despite the fact that the nature of the latter is motion, while the nature of the former is rest; furthermore, it considers obstacles as reciprocal forces, when actually they must be regarded as a constantly renewed restraint, or what we call mass, and which can be designated as a tendency towards rest resulting from continual renewal, which the operation of the fundamental law of mechanics, namely the law of gravity, excludes, while force constitutes a tendency towards motion. Matter in constant renewal is tendency towards rest, elasticity in constant renewal is tendency towards motion, and both of these elements reside in every body. Rotation and transposition together constitute development, and an effect of two bodies on one another is possible only if they are first united into one single body, when one of them constitutes the mobile, and the other the impeding element. Any uniform linear motion consists in a *stable* equilibrium between force and the obstacle which has transformed itself into a force through motion.

Mechanics must adopt these principles and the natural sciences must consider the whole of nature as a single organic body, and if so, then not a single problem will any longer resist a correct solution.

Noble benefactor & patron, accept these principles & me, for mankind will be carried to hitherto unimagined heights by them, and current and future generations will be grateful to you.

I beg you cordially for a speedy reply and grant you *unlimited* use of this letter.
Your honour | most devoted | D Recht

DAR 176: 76

[1] For a transcription of this letter in its original German, see pp. 188–90.
[2] Recht refers to Justus von Liebig, president of the Bavarian academy of sciences. Recht had published an article, 'Die Mechanik in ihrer Anwendung auf die Natur' (Mechanics and its application to nature), in the *Allgemeine Zeitung* of 3 April 1869. It has not been found in the Darwin Archive–CUL.
[3] The treatises have not been identified.
[4] See letter from Georg Recht, 1 March 1869. Recht refers to Recht 1869.

From Spiridion Brusina[1] 29 April 1869

Agram[2]
29 April 1869.

Dear Sir!

With the help of some colleagues, I have founded a society of natural history. This will have for its object the study of the fauna, flora, and terra of the triple kingdom of Dalmatia, Croatia and Slavonia, and of other Slavic countries—southern Carniola, Serbia, Bosnia, Herzegovina, Montenegro etc.; then it aims to spread among us the knowledge of those natural sciences whose importance has been so much increased by the classic works that you have published.

I would like the diplomas that are to be distributed among the members to be adorned with four pictures of the most distinguished naturalists and as it was decided that you ought to be among the four, unless it is forbidden to have it, I am asking you to have the goodness to send me a photograph of you to be reproduced.

Requesting a thousand pardons for such an intrusion and offering in advance many thanks, I have the honour to pronounce myself your most obedient and devoted servant | Spiridion Brusina

Custodian of the National Museum of Natural History

DAR 160: 354

CD ANNOTATIONS
Top of letter: 'Austria | **via Belgium** | 6^{d} for 1/2 oz | 1^{s} for 1 oz' *pencil*

[1] For a transcription of this letter in its original Italian, see p. 196.
[2] Agram is the German name for Zagreb (Croatia).

From W. C. Tait[1] 10 May 1869

[Enclosure]

Description of Drosophyllum Lusitanicum from Brotero's Flora Lusitanica page 215.[2]

"4 Spergula droseroides.

S with a stemmed root: having multi-branched stems above; with subulate, keeled leaves, twisted in spirals from the top, pilose, with dewy, glandular hairs: with seeds, with pedicels

Portugal Herva Pinheira orvalhada

Drosera Lusitanica Linn.

Chamaeleontioides. Grisl Vir Lusit[3] N1351

(Grisleyus in Virid Lusit sometimes describes the same single plant under a second name, thus Apium hortense is called Seleri, N 122, or cultivated Apium palustre N 1309.

By both of which he means to indicate, Apium graveolens Linn var β)

It dwells in dry sandy areas across the Tagus around Seixal and Arrentella, around Torres Vedras, Montejunto, Chaõ de Maçãas, Redinha, and also in dry hills from Olisipone to Aveiro, up to five leagues from the shores of the Ocean It flowers in the summer, perennially, as a half-shrub.

Perennial root, with branches about the thickness of a goose feather, which gradually raises itself above the earth every year, and having a stem two to four inches long, wreathed at the top with dead leaves: there new leaves germinate as in Palms, as in the first year there is no generation in the earth except from seeds. It puts forth a single stem when the plant is young, and later two, three, or four from the top of the root, a half foot tall or a little higher, terete, with a few leaves, upright, branched above, with alternate erect single-flowered branches, alternating higher and lower, all haired as the stem, the hairs at the top with glandular heads that discharge a viscous juice. Radical leaves, and those which are at the top of the root, collected into a sphere, slightly more upright, a little shorter than the stem, subulate, carinate, fairly flat at the top, twisted back in a spiral from the top, glands especially hairy at the edges: similarly cauline, another above gradually shorter upwards from the top of the stem. Four to seven flowers, at the ends of the stem and branches. Five-leafed calyx, with ovate, lanceolated folioles that are taper-pointed, glabrous inside, with glandular hairs outside to the edges; the corolla has shorter folioles.

The corolla has yellow obovate petals.

There are ten uniform stamens. There are five pistils The capsula is about twice as long as the calyx, glabrous, glabrous, having one loculus, five-valved, with ovato-lanceolated valves. Conical seeds one line long, black, wide and concave at the base, navel-like there in the middle, ten to twelve, several abortions, all at the bottom of the capsula, each attached to a small pedicel. Two menbranaceous coverings.

Albumen the same shape as the seed, dingy, slightly white, cartiliginous, with the apex where the embryo is, hollowed out in a curve, the Embryo turned about, very small, white, conical, on a narrow apex and sitting in the narrow, hollowed-out apex of the albumen— Cotyledons obovate, thick, concave from here to the plumula, thence convex. Conical plumula, white, twice-divided at the apex, shorter by half than the cotyledons: radical very short, slightly acute.

The plant, considered carpologically, is unique; for the fruit is neither Drosera's nor Spergula's. Nonetheless, from its sandy, dry place of habitation, its branching, leafy stem, and taking into account the number of its stamens, I believed that there was a stronger case, especially for the convenience of beginners, for it to be grouped with the Spergulas rather than with the Droseras. The juice of the glands tints paper with a rich purple hue.'

DAR 58.1: 19–20, DAR 178: 47

1 For a transcription of this enclosure in its original Latin, and the letter that accompanied it, see pp. 226–9.
2 Tait refers to Félix de Avellar Brotero and Brotero 1804, vol. 2.
3 Grisley [1661].

From Max Schmidt[1] 14 June 1869

Frankfurt a/M.
14 June 1869.

Dear Sir!

A number of work-related travels and a heavy workload have prevented me from replying to your honoured letter of 29 April until now, and I beg you to kindly forgive me.[2] Unfortunately I am even now not in a position to supply the information that you requested, for we do not have two adult mandrills in our garden, but only a male drill (Mormon leucophaeus),[3] which has been with us for eleven years now and as he was quite young when he arrived, he must be around twelve years old now. If a more detailed description of this animal would be of use to you, I am happy to put myself at your disposal and compose one at once, since at present I have somewhat more free time again.

Whether the zoological garden at Antwerp has *a pair* of mandrills I do not know, but it could possibly be the only one among such institutions that would be likely to. In any case I would advise you to ask director Vekemans there and I am convinced that for him, as well, your name will be the best recommendation.[4] Should you also wish to give my name as a reference, you may feel free to do so.

If I can be of service to you in any other matter, I am ready to do so with pleasure and beg you to consider me completely at your disposal

With my greatest respects | your most devoted | Dr. Max Schmidt | Director of the Zoological Garden.

DAR 177: 57

1 For a transcription of this letter in its original German, see pp. 269–70.
2 See letter to Max Schmidt, [29 April 1869].
3 The drill is now *Mandrillus leucophaeus*.
4 No correspondence between CD and Jacobus Vekemans has been found.

From Gerolamo Boccardo[1] 15 June 1869

Genoa
15 June | 1869

Illustrious Sir

Accept, I beg you, with your usual kindness, my new book, titled "Sismopirologia–Terremoti, Vulcani e lente Oscillazioni del suolo", whose only merit is perhaps that of paying to you at every stroke the profound respect I bear towards you.[2]

I am not sending to England any other copy besides this, since I am afraid that what happened already to my "Fisica del Globo", of which I made a present to Sir Charles Lyell and several scientific societies, without receiving so much as a sign to assure me that my gift had arrived, will happen to the "Sismopirologia".[3]

Grant me, illustrious Sir, your precious forbearance, and believe me full of respect | your most devoted | G. Boccardo

DAR 160: 234

[1] For a transcription of this letter in its original Italian, see p. 270.
[2] Boccardo refers to Boccardo 1869 (in this version, 'Sismopirologia: earthquakes, volcanoes, and slow oscillations of the ground': the word 'slow' (lente) is apparently not in the printed title of the book). No copy has been found in the Darwin Libraries at CUL or Down.
[3] See *Correspondence* vol. 16, letter from Gerolamo Boccardo, 23 April 1868. Boccardo refers to Boccardo 1868 (Physics of the world).

From Eduard Koch[1] 17 June 1869

Stuttgart
17 June 1869.

Most esteemed Sir!

You will have heard already through Prof. Carus—Leipzig that the fourth ed. of "Entstehung der Arten" will soon be begun. Mr Carus tells me that you were so kind as to send him the proof-sheets for the fifth English edition of the work.[2]

I am very content with the sales of "Variiren" and can tell you that in about two years a second edition will be needed.[3]

Professor Dub—Berlin has been in touch with me regarding the publication of an exposition of your theory that would be easily comprehensible even for the general public, and for this purpose we will embellish it with a number of illustrations, which will make certain points clearer to the lay person.[4]

My best thanks for your support on this occasion of my rights to the German translation and I shall regard the best possible preparation of your books as a duty of gratitude towards you.

Recently, a brochure by Prof. Jaeger on "the position of the Darwinian theory with regard to *morals* and *religion*" has appeared here.[5] Should you not have received it as yet I can send it to you, and remain

Respectfully and most devotedly yours | E. Schweizerbart'sche Verlagshandlung Eduard Koch.

DAR 169: 44

[1] For a transcription of this letter in its original German, see p. 273.
[2] Julius Victor Carus had written to CD about making a fourth German edition of *Origin* using the fifth English edition (see letter from J. V. Carus, 3 June 1869 and n. 2).
[3] Koch refers to the German translation of *Variation* (Carus trans. 1868). A second edition was published in 1873 (Carus trans. 1873).

[4] Julius Dub; his popular exposition of Darwinism was Dub 1870. There is a copy in the Darwin Library–Down.

[5] Koch refers to Gustav Jäger and to his *Die Darwin'sche Theorie und ihre Stellung zur Moral und Religion* (Jäger [1869]). There is an annotated copy in the Darwin Library–CUL (see *Marginalia* 1: 429).

From Ernst Haeckel[1] 2 July 1869

Jena
2. July 69.

Esteemed dear friend!

A few days ago I received the fifth edition of your "Origin of Species" and I thank you cordially for this new evidence of your kindness.[2] I am very glad to see that your epoch-making work is still finding so many new readers every year, even though it is already spread all over the world in so many thousand copies and so many translations. I am especially pleased that you should live to experience the great satisfaction of witnessing with your own eyes the immense effect of the colossal advance which your work has created within biology, and to judge from this just how great its fame will be in the more distant future.

I am now preparing the second edition of my "Natürlichen Schöpfungsge schichte", which will appear next year.[3] The number of works that develop the various aspects of your theory increases by the week, and so it is with the number of your followers. Among German naturalists of the *younger* generation there is now no important opponent of your theory. I am especially pleased, however, that even most eminent older naturalists are converting to the theory of descent, such as for example *P. Harting* in Utrecht and *M. Sars* in Christiania, both of whom have written to me lately and expressed their full agreement.[4]

In a few weeks' time you will receive my "Entwickelungsgeschichte der Siphonophoren", which has now finally come out and which is accompanied by fourteen tables.[5] At present I have been working diligently for almost a year on the *calcareous sponges* (Calcispongiae),[6] a most interesting class of animals which is also of very great importance for your theory. The organic "species" can be observed in some Grantia quasi "*in statu nascenti*". In order to compile the most complete natural history possible of these animals, I am going for a few weeks at the end of July to *Norway* (Christiania, Bergen, Drontjhem)[7], where calcareous sponges can apparently be found in large numbers. I shall send you my preliminary results in a few weeks. I'm planning to publish a comprehensive "*Monograph on calcareous sponges*" next year.[8] So that the systematic part of the work will be quite complete, I am now requesting material from all sides. Should you or one of your friends be able to send me calcareous sponges (even if only the *very common* species of the British coast, Grantia ciliata,[9] Leucosolenia botryoides etc), I would be very grateful indeed. Due to *the extraordinary variability of calcareous sponges* I can *not* compare *enough individuals* from different locations. In particular, from the British coast, I have as yet received only very little.—

I hope you and your dear family are fine. I am quite well. My boy is now 9 months old and he is already beginning to lose much of the pithecoid, and even to philosophise already![10]

With the most cordial greetings | Your loyally devoted | Haeckel

DAR 166: 52

[1] For a transcription of this letter in its original German, see pp. 297–8.
[2] *Origin* 5th ed. was published in June 1869 (letter from R. F. Cooke, 22 June 1869).
[3] The second edition of Haeckel's *Natural history of creation* appeared in 1870 (Haeckel 1870a). CD's annotated copy of Haeckel 1868 (the first edition) is in the Darwin Library–CUL (see *Marginalia* 1: 358–60). His copy of Haeckel 1870a is in the Darwin Library–Down.
[4] Haeckel refers to Pieter Harting and Michael Sars.
[5] Haeckel refers to his study of the development of siphonophores (Haeckel 1869a). CD's annotated copy is in the Darwin Library–CUL (see *Marginalia* 1: 355). Siphonophores are hydroids of the order Siphonophorae.
[6] The current name for the class of calcareous sponges is Calcarea.
[7] Christiania is now Oslo; Haeckel also refers to Trondheim.
[8] *Die Kalkschwämme* (Calcareous sponges) was published in 1872 (Haeckel 1872). CD's annotated copy is in the Darwin Library–CUL (see *Marginalia* 1: 357–8).
[9] *Grantia ciliata* is now *Sycon ciliatum*.
[10] Haeckel refers to his son, Walter Haeckel. In an earlier letter, Haeckel had described his son's 'atavistic movements' (see *Correspondence* vol. 16, letter from Ernst Haeckel, 9 November 1868).

From Federico Delpino[1] 22 August 1869

Florence
22. August 1869

Most famous man!

Encouraged by the courtesy with which you were kind enough to appreciate two works of mine published in 1867, I take the liberty to send you three later works of mine.[2] Please accept the homage of this humble offering.

The doctrines that are fortunate enough to be propounded by the highest intellect of our century can count me amongst their supporters, albeit one poor in intellectual power but rich in good will. I have tried to make contributions to as great an extent as I could to the illustrations of the doctrine of dichogamy and the variability of species.

On the other hand I regret not to be able to consent to some general ideas. Perhaps because of intellectual error but certainly with profound conviction, I am a teleologist. I believe that the rational principle by which man is conscious of being animated rules all other beings as well, plants included. And I believe that the first cause of all variations, whether for better or worse, progressive or regressive, lies exactly in this inner principle, inborn in all beings.[3]

Be that as it may, from my botanical researches I reach the conclusion that the variability of species cannot be called a theory: all aspects of science concur, at least according to me, to show that it is clearly an absolutely true fact.

Praise be the intellect that was able to give so much impetus to human thought. Praise be to the foremost naturalist of the century, of which I have the honour of remaining | most humble disciple | Federico Delpino

Assistant in botany at the Museo di Storia Naturale, Florence

DAR 162: 143

1 For a transcription of this letter in its original Italian, see p. 368.
2 Delpino refers to 'Pensieri sulla biologia vegetale, sulla tassonomia, sul valore tassonomico dei caratteri biologici, e proposta di un genere nuovo della famiglia delle labiate' (Thoughts on plant biology, taxonomy, the taxonomic value of biological characters, and the proposal of a new genus of the labiate family; Delpino 1867c), the first part of *Ulteriori osservazioni sulla dicogamia nel regno vegetale* (Further observations on dichogamy in the vegetable kingdom; Delpino 1868–75), and 'Sulla darwiniana teoria della pangenesi' (On the Darwinian theory of pangenesis; Delpino 1869a). Delpino had sent CD two papers on dichogamy in 1867 (Delpino 1867a and 1867b; see *Correspondence* vol. 15, letter from Federico Delpino, 5 September 1867 and n. 2).
3 For more on Delpino's teleological approach to CD's theory, see Pancaldi 1991, pp. 107–36.

From Alphonse de Candolle[1] 28 September 1869

Geneva
28 Sept. 1869.

My dear Sir

I am sending you some seeds that have a genuine interest from the point of view with which we are both concerned and I want to suggest that you sow them in order to pursue an experiment that is already underway and to confirm it for yourself.

In 1868 I had seeds from some *spontaneous* species sent to me from four or five locations that are very far apart in Europe: *Palermo* and *Montpellier*, *Kharkov*[2] and *Moscow*, *Edinburgh*, in order to sow them in Geneva and to see whether the same species had become more precocious or more retarded as a result of having had ancestors that had been exposed to very different climates for centuries.[3] I chose species that are easily recognised and that do not have any very distinct varieties. Unfortunately, not all my correspondents collected all the species I asked for and I have had to limit the experiment to a few species. Also unfortunately, the gardener on whom I had thought I could rely for certain details did not raise all the plants that had been sowed suitably. Nevertheless, from my sowings of 1869, it transpired that in three species that can readily be compared, the *Senecio vulgaris* and *Sisymbrium officinale* coming from the north were quicker to develop and flower. Conversely in *Trifolium repens*, the plants from the north were slower, but they were also weaker, smaller in all their parts, and one might say that they had the appearance of a *minor* variety.[4]

In 1869 I sowed *Senecio vulgaris* from all 4 sources at Geneva, and in this second generation, the precocity of the northern plants persisted.

How long will it last? That's a subject not lacking in interest.

The rigour of our climate and the lack of greenhouses and gardeners available to me do not allow me to make a new sowing this season. Perhaps you may be disposed to make one at your house, or if you prefer, you might communicate with Dr Hooker, for instance, in order to sow these so that they would be carefully observed.[5]

I am sending you:

Sisymbrium officinale of 1868, collected at	Edinburgh Moscow Kharkov Palermo
Trifolium repens, of 1868, collected at	Moscow Palermo
Senecio vulgaris, of 1868, collected at	Moscow Palermo
Id. of 1869, collected at Geneva, originating at	Edinburgh Moscow Montpellier Palermo

These last are the most interesting.

You will know that several indicators suggest the same species feels the effects of temperature differently according to climate. Thus our European trees come into leaf in Madeira in temperatures analogous to those of an English summer, and remain leafless in winter temperatures similar to those of our springs. I believed I had found the cause in the fact that in cold regions the sap is more completely dormant, a dormancy that prepares for vigorous vegetative growth in the springtime in northern countries; a dormancy that is imperfect or almost completely absent in Madeira, Naples, etc. Mr Lindseer, a young astronomer of St Petersburg, has since published a memoir in the Mémoires de l'Académie de St Petersbg on the times of foliation, flowering, etc, of various species in various points of Europe, and he reaches the conclusion that the same species has a different propensity for feeling the heat in the north and the south, with species being more hasty in the north, and he believes that this relates to heredity.[6]

Probably the twin causes of heredity and the condition of plants in winter act together, at least in perennial woody species, but the experiments I have just made on annual species show well the effects of heredity alone. *Trifolium repens* was different— Does this come from the condition of the seeds or from varietal difference? I don't know.

I should have liked to have had species that were more reliably indigenous, since Senecio, Sisymbrium and Trifolium might well derive from seeds imported along with merchandise, seed, etc. However, I took care to recommend that they should be collected outside gardens, far from dwellings, and that has been done. These species were chosen because there are really very few species common to European countries that are so far apart, yet easy to find and without different varieties.

I have sent my Senecio from the 1869 Genevan harvest, to *Florence*, where I hope to find them sown when I go to that town in the autumn. My plan is actually to go and spend 2 months in the north of Italy, from the middle of next October and I shall remain at least a fortnight in Florence—completely at your service should you wish for any information.

Please accept, my dear Sir, the assurance of my most devoted sentiments | Alph. de Candolle.

DAR 261.11: 32.i

[1] For a transcription of this letter in its original French, see pp. 410–12.

[2] Kharkov (now Kharkiv) is a city in eastern Ukraine, in the Donets valley (*Columbia gazetteer of the world*).

[3] CD had written extensively on the effect of changes in climate and other conditions of life on various domestic plants and animals (see *Variation* 1: 271–92).

[4] *Senecio vulgaris* is common groundsel, a winter annual. *Sisymbrium officinale* is hedge-mustard. *Trifolium repens* is white clover.

[5] CD forwarded Candolle's letter and most of the seed to Joseph Dalton Hooker (see letter to J. D. Hooker, 2 October [1869]).

[6] Candolle refers to Carl Linsser and his monograph 'Die periodischen Ersheinungen des Pflanzenlebens in ihrem Verhältniss zu den Wärmeerscheinungen' (The periodic appearance of plant life in relation to heat-phenomena; Linsser 1867). For a discussion of Linsser's work on the effects of temperature on plant development, see Thornthwaite 1943.

From Ernst Haeckel[1] 29 September 1869

Jena
29. Sept. 69

Most esteemed, dear friend!

Just last night I returned from a two-month trip to Norway, and found your kind letter of 5 August, which unfortunately arrived only after I had already departed.[2] I am extremely grateful for your great kindness and for all the trouble you have gone to on account of my Calcispongiae; but I almost regret having caused you so much trouble and loss of time with my request.[3] Hopefully the results of my work will be so satisfactory, and so useful especially for "Darwinism", that at least all your efforts will not have been in vain.— The kind offer of Mr. Norman to send me the English calcareous sponges I accept very gratefully, as I have so far been able to get hold of only very few English calcareous sponges.[4] Until now only Mr. Allman was so kind as to send me sponges from the Edinburgh collection.[5] On the other hand, Mr. *Bowerbank*, to whom I have written on this matter *twice* already this summer, has answered neither of my letters.[6] He probably believes that he *alone* has the monopoly on understanding sponges and regards other works as superfluous.

I am enclosing the preliminary report on the results to date of my researches on sponges; the passages that might interest you I have marked in *red* (p. 218, 220, 222, 233) 243).[7]

I also enclose another essay that includes a description of a new form of medusa (Crambessa) and a number of fossil medusae.[8] It is a purely descriptive work, without any special ideas. Furthermore, may I kindly ask you to send the enclosed *second copy* of both works (sponges and medusae) to Mr. Norman, whose exact address I do not know for sure. Since you also so kindly offered to forward a letter to Mr. Norman, I enclose this as well.

In the meantime I hope you will have received my Entwickelungsgeschichte der Siphonophoren, which I already despatched to you through the publisher at the end of July.[9] The following sections may interest you: IV (p. 38), VII (p. 73), VIII (p. 80) and X (p. 92, 97, 100, 103).

My journey to Norway was interesting, though not very comfortable. In Christiania[10] I was very pleased to meet an enthusiastic *Darwinist* in the excellent zoologist *Sars*,[11] who also specifically approved of my hypothesis on the origins of echinoderms.[12] He is already quite old, but mentally still very vigorous. From Christiania I travelled through the interior of Norway (*Valders*) to the *Sogne-Fjord. I saw as many as 12 or 16 ponies with excellent zebra-stripes* on legs and shoulders. They are supposed to be *very common* there.[13] (N. B. here in Jena there also is an *ass* with such stripes on its legs running around (but not a professor, as you may well be thinking!). I fished for the longest time in *Bergen*, where I also met *Mr* Gwyn-Jeffreys, and where I found an admirer of your theory in the excellent naturalist *Danielson.*[14] Likewise I found at Hamburg almost everything *Darwinistic.* Your portrait is on display in the famous aquarium.—[15]

Again many, many thanks, dear friend! I hope you are quite well. Ever devotedly your Haeckel

N. B. The calcareous sponges observed in Norway are very important for your theory

DAR 166: 53

1 For a transcription of this letter in its original German, see pp. 413–14.

2 See letter to Ernst Haeckel, 3 August [1869].

3 Haeckel had asked CD to help him obtain specimens of calcareous sponges. See letter from Ernst Haeckel, 2 July 1869, and letter to Ernst Haeckel, 3 August [1869].

4 Haeckel refers to Alfred Merle Norman. See letter from A. M. Norman, 2 August 1869.

5 George James Allman was professor of natural history and keeper of the natural history museum at Edinburgh University (*ODNB*). See also letter from A. M. Norman, 2 August 1869.

6 James Scott Bowerbank had a large private collection and was working on a multi-volume monograph on British sponges (Bowerbank 1864–82; *ODNB*).

7 Haeckel refers to his paper, 'Ueber den Organismus der Schwämme und ihre Verwandtschaft mit den Corallen' (Haeckel 1869b; later translated as 'On the organisation of sponges, and their relationship to the corals', Haeckel 1870b). CD's lightly annotated copy is in the Darwin Pamphlet Collection–CUL.

8 Haeckel refers to 'Ueber die Crambessiden, eine neue Medusen-Familie aus der Rhizostomeengruppe' (On the Crambessidae, a new medusa family of the rhizostomean group; Haeckel 1869c). CD's copy is in the Darwin Pamphlet Collection–CUL. The genus *Crambessa* and the family Crambessidae are now considered invalid; the family corresponds roughly to the modern family Catostylidae in the order Rhizostomeae.

9 Haeckel refers to *Zur Entwickelungsgeschichte der Siphonophoren* (On the developmental history of the siphonophores; Haeckel 1869a). See letter from Ernst Haeckel, 2 July 1869 and n. 5.

10 Christiania is now Oslo, the capital of Norway.

11 Michael Sars was professor of zoology at Christiania University.

12 For Haeckel's view on the descent of echinoderms, see Haeckel 1868c, pp. 420–1 and tables I and III.

13 CD had discussed stripes in the Norwegian horse in *Variation* 1: 58.

14 Haeckel refers to John Gwyn Jeffreys and Daniel Cornelius Danielssen.

15 The Hamburg aquarium opened in 1864 and was regarded as one of the finest in Europe (Taylor 1876, p. 17).

From Victor Masson[1] 29 September 1869

Paris, Place de l'Ecole-de-Médecine,
29 Sept. 1869.

Dear Sir

We were absolutely unaware that a fifth edition of your *Origin of Species* had been published.[2]

It goes without saying that our interest and that of Madame Royer are the same as yours in this affair, and that we had no reason to do anything which might be disagreeable to you, while at the same time diminishing the value of the work we were publishing.[3]

Madame Royer, who had not had the pleasure of receiving any news from you for a long time, bought your recent work as a source of inspiration in her work.[4] Had she known of its existence, she had no motive not to buy the fifth edition of *the origin of Species* as well, and to compare its text with that of her translation.

We are writing by this post to our London correspondent to send us a copy, which we shall give to Madame Royer on her return from Italy, where she is at present.

If there should be in your new edition any point which you would particularly regret to see omitted from the French edition, for our part we should have no objection to covering the cost of an appendix of several pages, and we beg you to indicate this to us.

Never having had the honour of being in direct communication with you until now, we deeply regret having, most unwittingly, displeased you to such an extent that your first letter contains reproaches that are most out of keeping with our character, our practices, and, we dare to claim, our reputation among the scientific public.

Please accept, Sir, the assurance of our most distinguished regards | Victor Masson junior

Mr Charles Darwin | Down | Beckenham | Kent. (S.E.)

DAR 171: 64

1 For a transcription of this letter in its original French, see pp. 415–16.

2 The fifth edition of *Origin* was published in June (see letter from R. F. Cooke, 22 June 1869). CD had written to Masson to complain that a third French edition of *Origin* had been prepared without reference to the fifth English edition (see letter to Victor Masson, [after 29 September 1869]); however, no letter to Masson on this subject has been found.

3 Clémence Auguste Royer was the French translator of *Origin*. Two editions had been published (Royer trans. 1862 and 1866). A third edition dated 1870 evidently appeared in 1869 (see letter to J. J. Moulinié, 23 October [1869]).

4 Masson refers to *Variation*, which Royer discussed in her preface to Royer trans. 1870 (see letter to J. J. Moulinié, 23 October [1869] and n. 5).

From Federico Delpino[1] 9 October 1869

Florence
9. October 1869

Most illustrious man!

I received the works "On the specific difference between Primula veris etc," "Offspring from the illegitimate unions of dimorphic and trimorphic plants etc." notes on the fertilization of orchids.[2] I thank you wholeheartedly for them. I am studying them with the utmost care, as one must do with anything coming from the pen and the mind of the greatest naturalist of this century.

I venture a few respectful observations. An infinite number of cases convince me that honeysucking birds (Trochilus, Ornismya, Nectariania, Melitreptus and similar),[3] must take a very great and important part in the fecundation of many tropical flowers (and probably of most orchids). Either I am gravely mistaken or I cannot understand how you did not take such pollinators into special consideration.

Thus, while contemplating the structure of the flowers of Corianthes and noticing the enormous nectary cups, and the enormous secretion of sweetish liquid, I said to myself: this flower must be fertilised by a very special trochilus. Now, while skimming through Gould's work on trochilidae, I happened to see a figure showing a singular trochilus with a curiously bent beak visiting the flowers of corianthes and sucking the abundant liquid from the nectar cup.[4] Dr Crüger's explanation of this seems to me hardly reliable. Nature does not produce honey glands or perfectly designed nectar cups simply to moisten the wings of an Euglossa.[5] I would entrust myself to your judgment on this point.

It was extremely informative for me to learn that some insects visit the flowers of orchids not so much to suck honey as to nibble some special outgrowths.[6] This idea had been going around in my head for quite some time already. I was pleased that facts correspond to this idea. No doubt the same thing must happen amongst our local orchids, in the genus Serapias (S. lingua, cordigera, neglecta etc.).[7] Inside the flower, there is no honey; instead, one finds a large *purple outgrowth*, the function of which is precisely to provide food for the pollinator.

I have received an issue of the journal in which a part of my work on pangenesis has appeared.[8] Pangenesis can be considered from two points of view, that of monism and that of dualism.

From the monistic point of view, it seems to me that, unless I am mistaken, pangenesis is inconceivable within the laws of chemistry.

From a dualistic point of view, pangenesis is a wonderful and great theory that groups together successfully, in my opinion, all phenomena of the genesis of organisms. By admitting the existence of gemmules of non-corporeal substance, we avoid the objections coming from chemistry, we explain metagenesis, the alternation of generations, the dislocation, multiplication, and diminution of organs, the phenomena of atavism etc. etc. I am full of admiration for you for formulating such a great theory, and I will be pleased if I can soon publish a work that is not unworthy of this eminent topic, entitled "*Apologia della pangenesi sotto il punto di vista dualistico*".[9]

Please accept the humble gift of two pamphlets of mine in which I endeavoured to use the notions of dichogamy for botanic geography and to extend the genealogical tree of Marantaceae whose style I recognised as homologous with the rostellum (!) of orchids.[10]

I have the honour of declaring myself Your Honour's | Most humble disciple | Federico Delpino

DAR 162: 144

1 For a transcription of this letter in its original Italian, see pp. 421–3.
2 Delpino refers to 'Specific difference in *Primula*', 'Illegitimate offspring of dimorphic and trimorphic plants', and 'Fertilization of orchids'.
3 Delpino refers to the genus *Trochilus* and the now invalid genus *Ornismya* in the family Trochilidae (hummingbirds), the genus *Nectarinia* in the family Nectariniidae (sunbirds), and the genus *Melithreptus* in the family Meliphagidae (honeyeaters).
4 Delpino refers to John Gould and Gould 1861, 1: plate 3, which depicts *Eutoxeres aquila* (the white-tipped sicklebill) at a flower of *Coryanthes speciosa*, a species of bucket orchid.
5 In 'Fertilization of orchids', p. 157, CD had described Hermann Crüger's observations on the visits of bees of the genus *Euglossa* to *Coryanthes*. CD wrote, 'The fluid in the bucket formed by the basal part of the labellum is not nectar and does not attract insects, but serves, by wetting their wings, to prevent them from crawling out except through the small passages close to the anther and stigma.' For more on pollination of orchids by *Euglossa*, see Dressler 1968.
6 See 'Fertilization of orchids', p. 157.
7 *Serapias lingua* is the tongue orchid; *S. cordigera* is the heart-flowered serapias; *S. neglecta* is the scarce tongue orchid. For more on pollination in this genus, see Jersáková *et al.* 2006, p. 222.
8 Delpino refers to the first part of the English translation of his article 'Sulla darwiniana teoria della pangenesi' (On the Darwinian theory of pangenesis; Delpino 1869a), published in the 29 September 1869 issue of *Scientific Opinion* (Delpino 1869b).
9 The proposed work (Defence of pangenesis from a dualistic point of view) was never published.
10 The articles referred to are 'Alcuni appunti di geografia botanica a proposito delle tabelle fitografiche del Prof Hoffmann' (Some remarks on geographical botany with reference to Prof Hoffmann's phytographical tables; Delpino 1869c) and 'Breve cenno sulle relazioni biologiche e genealogiche delle Marantacee' (Brief remarks on the biology and genealogy of the Marantaceae; Delpino 1869d). CD's annotated copies of these papers are in the Darwin Pamphlet Collection–CUL.

From Emile Alglave[1] 22 October 1869

Paris
22 October *1869*

Dear Sir

I have learnt that you are going to publish a new work in a few months comprising three parts, I, Descent of Man.— II, On Sexual Selection, III Expression of the emotions.[2]

I should very much like to have part of this work translated for the *Revue des cours scientifiques*.[3] By the same circumstance the publisher of the Revue des cours would be very happy to publish a translation of the complete work in his publishing house.[4] This combination would facilitate both matters and would immediately assure your work extensive publicity in France, since the Revue des cours scientifiques *on its own* prints nearly as many copies as all the other French scientific periodicals put together. The French translation could be made using the sheets of the English work progressively as they are printed so as to appear almost at the same time, and even earlier for some sections, in the Revue des cours scientifiques.— In any case, if you do not accept this combination for the French publication of the complete work, there is nothing to prevent the translation of some chapters from the book for the Revue des cours scientifiques anyway

Please accept, dear Sir, my most distinguished regards | Em Alglave | director of the Revue des cours scientifiques

DAR 159: A35

1 For a transcription of this letter in its original French, see p. 449.

2 Alglave had evidently seen the notice that appeared in *Academy* 1 (1869–70): 15–16 of *Descent* and *Expression.*

3 *Revue des cours scientifiques de la France et de l'étranger* was a weekly journal that reported on the lectures of prominent European scientists. Excerpts from *Descent* eventually appeared in the journal in 1871.

4 The publisher of *Revue des cours scientifiques* was J. B. Baillière et fils.

From Carl Vogt[1] 26 October 1869

Geneva
26 Oct 1869

Dear Sir and Master!

Monsieur Moulinié has already written to you, I believe, on the subject of the French translation of your book on man, the announcement of which we have read in "the Academy".[2] So I do not need to go over all that again—I believe you were content with the translation he made before and he will take as much care over this book as over the one on domestication.[3]

Today I received a letter from my publisher, Mr. J. Ricker in Giessen,[4] asking if I would like to undertake the German translation of your forthcoming work on man—he promises to take every care to produce an edition worthy of you, and I think that in this respect Mr. Ricker has proved himself—just as I would do in producing a translation. In spite of my many occupations, I would take all care over this more important task.

You replied, when I wrote to you concerning the translation of your book on domestication, that you had committed yourself to Messrs. Càrus and Schweizerbart.[5] I do not know whether this commitment extends to your new book and though I am not afraid to say that Mr Ricker will do better, far better than Schweizerbart, I should not want to compete with Mr. Càrus in any way, who is a worthy and brave young man who had the great misfortune to be turned down for the chair of zoology at Leipzig, certainly in large part because he professed your doctrines out loud instead of toeing the official line like his successful rival Mr. Leukart.[6] Last year in Norwich,[7] Mr. Carus told me that he was no longer doing translations—I think it was in the hope of being appointed to the professorship— I do not know whether in the current circumstances, which are so trying for him, he might not have changed his resolve. But I believe in any case that it is my duty to abstain in the event that he has manifested his desire to translate your new work and should that be the case, let it be as if I had not spoken.

I was very jealous, last year, of Mr. Moulinié, who was able to make your acquaintance in person, while circumstances outside my control prevented me from doing so. But I hope that the German proverb "Aufgeschoben ist nicht Aufgehoben" will find its meaning here[8] and in that hope I remain

Your devoted | C Vogt

DAR 180: 13

[1] For a transcription of this letter in its original French, see pp. 452–3.
[2] See letter from J. J. Moulinié, 20 October 1869 and n. 1. The reference is to *Descent*.
[3] Vogt refers to *Variation* and to Moulinié trans. 1868.
[4] Vogt refers to the firm J. Ricker'sche Buchhandlung.
[5] See *Correspondence* vol. 15, letter to Carl Vogt, 12 April [1867]. The references are to Julius Victor Carus and the firm E. Schweizerbart'sche Verlagsbuchhandlung.
[6] For more on the appointment committee's comments on the three leading candidates for the chair of zoology at Leipzig, see Nyhart 1995, p. 173; on Rudolf Leuckart's appointment to the post and his approach to zoology, see *ibid.*, pp. 310–11.
[7] Vogt evidently met Carus at the thirty-eighth meeting of the British Association for the Advancement of Science, held at Norwich in August 1868.
[8] According to Emma Darwin's diary (DAR 242), Moulinié visited CD on 31 August 1868. *Aufgeschoben ist nicht Aufgehoben*: put off is not called off.

From Eduard Koch[1] 30 October 1869

Stuttgart
30 October 1869

Highly honoured Sir!

As I write, I venture to send you a copy of the just released

Dub, Kurze Darstellung[2]

and I believe that the work, both in its content and presentation will meet with your approval; in case you wish to receive further copies please let me know.

The German press is spreading the news from the English journal "Academy" that a new work by you will appear in the course of the coming year, the subject of which is man.[3]

Professor Carus, to whom I have therefore already written about it, will probably have already communicated this development to you.[4]

In the agreeable hope that this new work, too, like earlier ones will be adapted into German by Mr Carus for my press, please kindly notify me whether this new work will appear and what its content will be.

Having read all of your works with great attention, and having noted that until now you have drawn no conclusion concerning man, I can only guess that the latest work of the duke of Argyll on prehistoric man has induced you to address this problem.[5]

At this moment we are in the process of printing the fourth edition of the "Origin of species", and I shall not fail to send you, as soon as it appears, a number of copies which you will yet need to specify.[6]

Taking this opportunity to assure *[you]* of my highest respect | I remain your | wholly devoted | Eduard Koch
Firm: E. Schweizerbart'sche Verl.

DAR 169: 45

[1] For a transcription of this letter in its original German, see pp. 455–6.
[2] The reference is to *Kurze Darstellung der Lehre Darwin's über die Entstehung der Arten der Organismen: mit erläuternden Bemerkungen* (Brief description of Darwin's theory on the origin of species, with explanatory remarks; Dub 1870); see also letter to Julius Dub, 20 March 1869.
[3] Koch refers to *Descent* and to the notice in *Academy* 1 (1869–70): 15–16.
[4] The reference is to Julius Victor Carus. See letter from J. V. Carus, 20 October 1869.
[5] Koch refers to George Douglas Campbell and to *Primeval man* (G. D. Campbell 1869).
[6] The reference is to the fourth German edition of *Origin* (Bronn and Carus trans. 1870).

From Federico Delpino[1] 1 November 1869

Florence
1. November 1869

Most famous man!

I thank you both for your tolerant and gracious letter recently sent to me and for the kind present of Dr Crüger's interesting memoir, which is written with such clarity and acuity that it greatly undermined my doubt about the reality of the fertilisation of *Coryanthes* through the mediation of *Euglossae* and the very strange process of *cold bath* there described.[2] This notwithstanding, I still have to deepen my understanding of the rationality of the floral forms of *Coryanthes* and *Gongora*. I would admit straightaway that it puzzles me that the liquid collected in the labellum by distillation, *is produced by two peculiar nectar-producing glands.* Nature is very economical, and I am not happy that a liquid having simply the function indicated by Dr. Crüger, should be produced by a secretory and glandular apparatus.[3]

I am especially pleased that you mentioned that, according to your discoveries at the Cape of Good Hope, *Strelitzia* is fertilised by nectarinae.[4] This confirms my views. Prof. Hildebrand thought that the fertilisation of this flower was effected by insects; but that did not seem very likely to me, for insects, even of large size, would not have sufficient strength to open wide the lobes of the box containing the anthers.[5] I set great store by this point of yours, and, if you agree, I will use it on a suitable occasion.

I am especially pleased with your suggestion that I should study the fertilisation of wheat,[6] especially considering that I worked on that last Summer. I think I dispelled any doubt, 1st that every floret of wheat stays open for about a quarter of an hour, a period of time in which stigmatic fecundation has not yet taken place; 2nd that consequently dichogamy in wheat is very possible; 3rd that, however, the flower of wheat can be fertilised by its own pollen. I found that in rye dichogamic fertilisation is by far the easiest. I found out that in *Hordeum vulgare* one third of the flowers do not open at all (and therefore they cannot be fertilised dichogamically), whereas two thirds of them open and can be fertilised dichogamically.[7] In *Hordeum distichum*[8] (a strange exception) homogamic (and not dichogamic) fertilisation is really necessary, for it takes place when the ears are not yet out of their foliaceous sheaths. But then, as a wonderful alternative, this species of barley has two kinds of flowers: some (very few) that open and are predestined to dichogamy; and others (very many) that do not open, and are predestined to homogamy. So that in this plant we see the repetition of the well known phenomenon that takes place in the flowers of *Viola*, *Vicia amphicarpa* etc. etc.[9] Thus, even in *Hordeum distichum* the great law remains intact "no hermaphrodite fertilises itself for a perpetuity of generations."[10] At present I am writing a short paper about this, and I will send it to you as soon as it is printed.[11]

I received only the first issue of "Scientific Opinion", that is N°. 48. I came to know by chance that you inserted a reply to my work in N° 51.[12] I humbly beg you to be kind enough to send me a copy, and if I have your permission, I will translate *in deference to you* your reply to appear in the *Revista contemporanea*, which today changes its name to *Revista Europea*.[13]

I received *Axell*'s Swedish book.[14] It seems to me very good and written with unusual care. I am really pleased that Axell praises very greatly indeed the works of the greatest naturalist of our century, of whom I have the honour of confirming myself

a very respectful disciple | Federico Delpino

DAR 162: 145

1 For a transcription of this letter in its original Italian, see pp. 458–9.

2 See letter to Federico Delpino, 14 October 1869. Hermann Crüger's paper described how the pollen-mass of *Coryanthes* was secured onto the back of the bee (*Euglossa*) as it tried to force its way out through a passage after falling into a 'bucket' of nectar; when the bee next fell into the bucket of the same or another flower, the pollen was deposited on the stigma as it tried to leave (Crüger 1864, p. 130). See *Origin* 4th ed., pp. 229–30, 'Fertilization of orchids', p. 157 (*Collected papers* 2: 153–4), *Orchids* 2d ed., pp. 173–6, and *Correspondence* vol. 12.

3 See also the letter from Federico Delpino, 9 October 1869. Visits of bees to flowers of the orchid genus *Gongora* were also described in Crüger 1864.

4 For CD's remark on the Nectarinae (now the family Nectariniidae, sunbirds) fertilising *Strelitzia*, see the letter to Federico Delpino, 14 October 1869.

5 Friedrich Hildebrand recorded his opinion on *Strelitzia* pollination in Hildebrand 1869, pp. 508–9.

6 CD had suggested that Delpino study the fertilisation of 'Gramineæ', the grass family; see the letter to Federico Delpino, 14 October 1869.

7 On dichogamy, see the letter to W. C. Tait, 17 July [1869]. *Hordeum vulgare* is common barley.

8 *Hordeum distichum* is an alternative spelling of *H. distichon*, now considered to be a variety of *H. vulgare vulgare*. It was known as two-rowed barley.

9 Delpino refers to violets and to the vetch *Vicia amphicarpa* (now *Vicia sativa amphicarpa*); see also letters from T. H. Farrer, 8 August 1869 and 12 August 1869.

10 Delpino refers to CD's conclusion in *Orchids*, p. 359: 'Nature thus tells us, in the most emphatic manner, that she abhors perpetual self-fertilisation.'

11 Delpino probably refers to his paper on dichogamy in cereals (Delpino 1871); an inscribed copy is in the Darwin Pamphlet Collection–CUL.

[12] Issue 48 (29 September 1869) of *Scientific Opinion* included the first part of Delpino 1869, the translation of Delpino's review of pangenesis. In his letter to Delpino of 14 October 1869, CD had offered to send the remaining parts. CD replied to Delpino's criticism in the letter to *Scientific Opinion*, [before 20 October 1869]; this was published in issue 51 (20 October 1869).

[13] Delpino's translation of CD's letter to *Scientific Opinion*, [before 20 October 1869], appeared in the first volume of *Rivista Europea* (1869): 118–24.

[14] The references are to Johan Severin Axell and Axell 1869.

From F. C. Donders[1] 12 November 1869

Utrecht
12 November | 1869

Please do not believe, My dear and most honoured Sir, that I have forgotten to reply to the questions which you did me the honour of addressing to me.[2] The fact is that I have found the answer more difficult than I at first thought. This is why I have begun, touching the first question, a series of experiments, which, as usual, have led me further and taken me longer than I had expected. But seeing with some alarm that you have been entitled to expect my reply for nearly two months, I felt the need to banish any suspicion of a negligence on my part that would accord badly with the respect that I have for yourself and your works.

Be so kind as to let me know, therefore, my dear Sir, if you are need some solution to your questions quickly; in that case I will immediately give you an account of the results obtained so far. If, on the contrary, you are not in a hurry, I hope that in six weeks my reply may be less imperfect. In the meantime, I owe you thanks for having directed my attention towards a question which is truly important from various perspectives.

I was very happy, Mr Darwin, to have had the advantage of making your personal acquaintance as well as that of Mrs Darwin, and most grateful for the welcome that your kindness had prepared for me. I shall never forget my few hours spent at Down.—[3] | Allow me to offer you the assurance of my affectionate regards and my high esteem | F. C. Donders

DAR 162: 223

[1] For a transcription of this letter in its original French, see p. 475.

[2] Memorandum to F. C. Donders, 6 September 1869.

[3] Donders and William Bowman visited Down, probably on 7 September 1869; see the letter from William Bowman, 3 September [1869], and the letter to J. D. Hooker, 10 September [1869].

From A. B. Meyer[1] 16 November 1869

Zimmerstrasse 77. 1 Treppe | Berlin,
Nov. 16. 1869.

Highly honoured Sir

I have yet to thank you for your kind letter of 22 September last year.[2] Today, I have taken the liberty of dispatching, by post, a translation of Mr Wallace's work that I am presenting to the German public,[3] so that you, dear Sir, may see again how much we Germans appreciate the excellent work of English naturalists and how we hasten to introduce it. Since first of all I am planning to visit the Malayan archipelago,[4] this inspired work captured my interest most significantly, and while studying it I decided to make it more accessible to my fellow countrymen in a German edition.

This, however, is not the reason for my writing today, but rather the following:

I was asked by a publisher to translate and present to the German public the first announcement of the new theory of the "Origin of species", which you and Mr Wallace published in 1858, in the "Journal of the Proceedings of the Linnean Society", for these letters and treatises are of such significant—not only historical, interest.[5] They would appear in the form of a small pamphlet, which I should like to preface with a brief summary of your and Mr Wallace's biographies.[6]

My request to you, dear Sir, is thus the following

1) to authorise me to publish this treatise in German and

2) to supply me with a few data about your life, which the German audience is most interested in, for due to the difference in the conditions of life in England and Germany, they are uninformed about many things. Above all I am asking for a *complete* list of your scientific *treatises* and works.

The preparations for my tropical journey are now almost completed, so that I now am just waiting to settle a few family matters before I can leave. I hope to visit England in spring. I was delighted to hear that your health has improved again, which joyous news was received with great satisfaction over here.— A small treatise by myself on the poison apparatus of snakes, which I sent you, has, I hope, been received.[7]

Asking you, highly esteemed Sir, most kindly to speedily send me a favourable reply, and requesting your forgiveness and consideration for my desire

I sign | most respectfully | Dr Adolf Bernhard Meyer.

DAR 171: 166

1 For a transcription of this letter in its original German, see pp. 481–2.

2 CD's letter to Meyer of 22 September 1868 has not been found.

3 The reference is to Wallace 1869c, a translation of Alfred Russel Wallace's *The Malay Archipelago: the land of the orang-utan, and the bird of paradise* (Wallace 1869a).

4 Meyer spent at least three years (1870 to 1873) travelling and collecting in the Celebes, visiting New Guinea and the Philippines (*Auk* 28: 519).

5 Meyer refers to the joint publication of CD and Wallace's papers on natural selection. 'On the tendency of species to form varieties; and on the perpetuation of varieties and species by natural means of selection' was published in the *Journal of the Proceedings of the Linnean Society* (*Zoology*), and included the two essays, an abstract of a letter from CD to Asa Gray, and an introductory explanation by Charles Lyell and Joseph Dalton Hooker (see C. Darwin and Wallace 1858).

6 Meyer published *Charles Darwin und Alfred Russel Wallace. Ihre Ersten Publicationen über die 'Entstehung der Arten' nebst einer Skizze Ihres Lebens und einem Verzeichniss Ihrer Schriften* in 1870; a copy is in the Darwin Pamphlet Collection–CUL.

7 There is a lightly annotated copy of Meyer 1869a in the Darwin Pamphlet Collection–CUL.

From Camille Dareste[1] 18 November [1869][2]

Paris
18 November

Dear Sir

Allow me to appeal to your obligingness in a circumstance that is most important to me.

I am currently a candidate for a vacant chair of Comparative Physiology at the Faculté des Sciences at Paris. I have found many supporters, but also opponents who, with little understanding of the new scientific route to which I have committed myself, do not want to allow that my work on the artificial production of monstrosities belongs within the scope of physiology.[3]

You, Sir, are the man who has best understood the importance of the studies to which I have dedicated my life; and the manner in which you cited me in your last work proves to me that you hold them in high esteem.[4]

Would it be an indiscretion on my part, to ask you to send me a testimonial indicating what you think of my work. Your name, and the great position you have won in European science would give such a testimonial great value, and might ensure the success of my candidacy.

Please excuse me, Sir, for asking this service of you; but the question is very important for me, for our university organisation is so imperfect, that in the provinces we can have resources of all kinds that men of science can procure for themselves in Paris.[5]

Please accept, dear Sir, the assurance of my profound respect | Camille Dareste

Paris: rue de fleurus 37 bis

DAR 162: 45

1 For a transcription of this letter in the original French, see pp. 485.
2 The year is established by the relationship between this letter and the letter to Camille Dareste, 20 November 1869.
3 For Dareste's research on animal monstrosities, see Dareste 1862, 1863, and 1867.
4 See *Variation* 1: 262, and 2: 289, 331, 340.
5 Dareste apparently forgot to include the 'ne . . . pas' in the sentence that would have indicated the lack of resources in the provinces. Dareste was employed at the University of Lille (*DBF*).

From Camille Dareste[1] 13 December [1869][2]

Lille
13 December

Dear Sir,

I am very grateful to you for the brilliant recommendation that you kindly sent to me about my work.[3] It produced a fine effect, and keenly impressed all those who voted for my candidature:

I regret to have to tell you that my candidature, although well-supported by your name, has failed.[4] I came as close as possible to the bottom; for I had a great many votes among the members of the Conseil Académique and the Faculté des Sciences who drew up their presentation lists. But I faced a hostility that was manifested in the most energetic manner; that of Claude Bernard, who is nowadays all-powerful even outside science. He does not allow that physiology can be treated in any other way than he does himself, that is to say following the method of vivisections.[5] Therefore he fought me by shouting very loudly and everywhere that I was not a physiologist. Without him, I would have very certainly succeeded.

I have since returned to Lille, very bruised but not discouraged; and I have resumed the interrupted course of my work.[6] I have every reason to believe, from the results of my experiments this year, that I will soon reach the end of a course of research that has occupied me for nearly eighteen years. Perhaps, when I have published my work as a whole, I might find a more favourable reception in my country than that which I have encountered up to now.[7] Besides, if I have been to some degree disheartened by my current lack of success, I have the opinion of a scholar like you to lift my spirits; it just proves that I did not take the easy road; and that in seeking to study the production of monstrosities by the experimental method, I entered a career where I could render real services to the life sciences, whatever Claude Bernard and his very exclusive school may say about it.

Please accept, dear Sir, the assurance of my profound respect and continuing regard. | Camille Dareste

DAR 162: 46

1 For a transcription of this letter in its original French, see pp. 521–2.

2 The year is established by the relationship between this letter and the letter to Camille Dareste, 20 November 1869.

3 See letter to Camille Dareste, 20 November 1869.

4 On the position Dareste had been seeking, see the letter from Camille Dareste, 18 November [1869].

5 For Claude Bernard's position in French science and philosophy, and for his promotion of vivisection, see *DSB* and Holmes 1974.

6 Dareste was a professor of zoology at the University of Lille (*DBF*).

7 For Dareste's recent work on animal monstrosities, see Dareste 1862, 1863, and 1867. Dareste later published *Recherches sur la production artificielle des monstruosités, ou essai de tératogénie expérimentale* (Dareste 1877).

APPENDIX II
Chronology 1869

This appendix contains a transcription of Darwin's 'Journal' for the year 1869. Darwin commenced his 'Journal' in August 1838 and continued to maintain it until December 1881. In this small notebook, measuring 3 inches by $4\frac{1}{2}$ inches, Darwin recorded the periods he was away from home, the progress and publication of his work, and important events in his family life.

The version published by Sir Gavin de Beer as 'Darwin's Journal' (de Beer ed. 1959) was edited before the original 'Journal' had been found and relied upon a transcription made by an unknown copyist. The original, now in the Darwin Archive in Cambridge University Library (DAR 158), reveals that the copyist did not clearly distinguish between the various types of entries it contains and that the transcription made was incomplete.

From 1845 onward, Darwin recorded all that pertained to his work (including his illnesses, since these accounted for time lost from work) on the left-hand pages of the 'Journal', while the periods he was away from home, and family events, were noted on the right-hand pages. In order to show clearly Darwin's deliberate separation of the types of entries he made in his 'Journal', the transcription has the left- and right-hand pages labelled.

All alterations, interlineations, additions, and the use of a different ink or pencil have been noted. In addition, the editors have inserted additional information relevant to Darwin's correspondence throughout this transcription of the 'Journal' for 1869. These interpolations are enclosed in square brackets to distinguish them from Darwin's own entries, the source of the information being given in the footnotes.

[**Left**]

1869

Feb. 10th Finished 5th Edit of Origin:[1] has taken me 46 days—-

Feb. 11th Sexual Selection of Mammals—& Man & Preliminary Chapt on Sexual selection[2] (with 10 days for notes on Orchids)[3] to June 10th when I went to N. Wales—[4] On Aug. 4th recommenced going over all Chapts on Sexual Selection

[Week beginning 22 June. *Origin* 5th ed. published.][5]

[Right]

1869

[29 January. Mary Elizabeth Lyell, Charles Lyell, and Hyacinth Symonds visit.][6]

Feb. 16–24th to Erasmus[7]

[8 or 9 April. CD falls off his horse, Tommy.][8]

[25 April. Joseph Dalton Hooker and John Tyndall visit.][9]

Jun 10th Started for Caerdeon, Barmouth, sleeping at Shrewsbury— Returned July 31, having slept at Stafford.— Weak & Unwell—

[25 May – 4 June. Henrietta Anne Huxley, Thomas Henry Huxley, and their children visit.][10]

[28 August. Asa Gray and Jane Loring Gray visit.][11]

[25–6 September. Joseph Dalton Hooker visits.][12]

[30 September – 1 October. Vladimir Onufrievich Kovalevsky and Sofia Vasilyevna Kovalevskaya visit.][13]

[8 October. Thomas Henry Farrer visits.][14]

Nov. 1st to 9th Erasmus.—

[1 December or shortly before. Alexander Agassiz and Anna Russell Agassiz visit.][15]

1 CD refers to *Origin* 5th ed.
2 CD's work on sexual selection was in preparation for writing *Descent*.
3 CD prepared additional notes on orchids for the French edition of *Orchids* (Rérolle trans. 1870). He published these as 'Fertilization of orchids'.
4 CD went to Caerdeon, Barmouth, North Wales from 10 June to 31 July.
5 See letter from R. F. Cooke, 22 June 1869.
6 Emma Darwin's diary (DAR 242).
7 CD refers to his brother, Erasmus Alvey Darwin, who lived at 6 Queen Anne Street, London.
8 See letter from Henrietta Emma Darwin to George Howard Darwin, 10 April [1869] (DAR 245: 291) and Emma Darwin's diary (DAR 242).
9 Emma Darwin's diary (DAR 242).
10 Emma Darwin's diary (DAR 242).
11 Emma Darwin's diary (DAR 242).
12 See letter from J. D. Hooker, 24 September 1869, and letter to E. F. Lubbock, [26 September 1869].
13 Emma Darwin's diary (DAR 242).
14 See lettter from T. H. Farrer, 9 October 1869.
15 See letter to Fritz Müller, 1 December [1869].

APPENDIX III
A diploma presented to Charles Darwin

In 1869, Darwin received the following diploma. Although not letters in the conventional sense, diplomas represent significant communication between Darwin and scientific organisations, and the citations in such diplomas often provide valuable indications of those aspects of Darwin's work that were considered worthy of honour. In view of this, it has been included here.

From the American Philosophical Society 15 October 1869

To all persons to whom these presents shall come,
GREETING.

The American Philosophical Society held at Philadelphia for promoting useful Knowledge, desirous of advancing the Interest of the Society by associating to themselves Men of distinguished Eminence, and of conferring Marks of their Esteem upon Persons of literary Merit, have Elected

Charles Darwin of England.

Member of the said Philosophical Society, hereby granting unto him all the Rights of Fellowship, with all the Liberties and Privileges thereunto belonging.

In Testimony whereof the said Society have caused the Seal of their Corporation to be annexed to this Certificate, and the same to be attested by the Names of the proper Officers this fifteenth *Day of* October *in the Year of our Lord One thousand* Eight Hundred and Sixty Nine.

Attested

Chas. B. Trego[1] E. Otis Kendall[2] J. P. Lesley[3]	*Secretaries*
Geo. B. Wood[4]	*President*
[John] C. Cresson[5] Isaac Lea Fredk Fraley[6]	*Vice Presidents*

DAR 229: 22

1 Charles B. Trego.
2 Ezra Otis Kendall.
3 J. Peter Lesley.
4 George Bacon Wood.
5 John Chapman Cresson.
6 Frederick Fraley.

APPENDIX IV
Notes on Passiflora and Tacsonia

Thomas Henry Farrer enclosed notes on *Passiflora* and *Tacsonia* with his letter to Darwin of 17 October 1869. Darwin read the notes and commented on them in a letter to Farrer of 20 October [1869]. He advised Farrer to examine more species, and Farrer reported on 27 October 1869 that he had received more specimens from Joseph Dalton Hooker at the Royal Botanic Gardens, Kew. Farrer evidently soon had a fair copy of his notes, together with some introductory material on Christian Konrad Sprengel's observations, drawn up. The notes are in the hand of an amanuensis with some corrections by Farrer. The diagrams were drawn and annotated by Farrer and he added notations to indicate where they should be placed in the manuscript. In the present transcription, the diagrams have been integrated into the text. The last page of the fair copy was dated 13 November 1869 by Farrer. In addition, two further manuscript pages in Farrer's own hand and dated November 1869 were added later. These pages refer to *P. princeps*, a species that Hooker had not been able to supply Farrer with in October 1869 (letter from J. D. Hooker to T. H. Farrer, 23 October 1869; Linnean Society of London, MSS Case 6B, MS505).

Although it is not possible to reproduce the original enclosure to Farrer's letter of 17 October 1869, the following transcription of the later version is probably very close to what Darwin would have seen. Sections that were clearly later additions, written on separate sheets in Farrer's hand with instructions as to where they were to be added to the paper, and dated November 1869, have not been transcribed here, but Farrer's remarks on Sprengel, which were probably not in the notes that Darwin saw, have been included here because they are an integral part of the paper.

Farrer's paper was never read at a meeting or published by the Linnean Society of London.

Notes on the fertilization of certain Passifloras and Tacsonias.

C. K. Sprengel in his 'Entdeckte Geheimniss der Natur' gives an excellent account of Passiflora Cærulea (p.p. 160. 166) with careful drawings Table V. 1.8.11.[1] He describes the nectary at the base of the flower (shewn opposite in the diagram No 1) the process marked (a) in the same diagram proceeding from the column above the nectary; the inner corona, marked b2. b.3: the intermediate corona (marked c.) with its short upright rays; & the third and largest corona (marked d.) with its long horizontal rays.[2]

Diagram of perpendicular section.

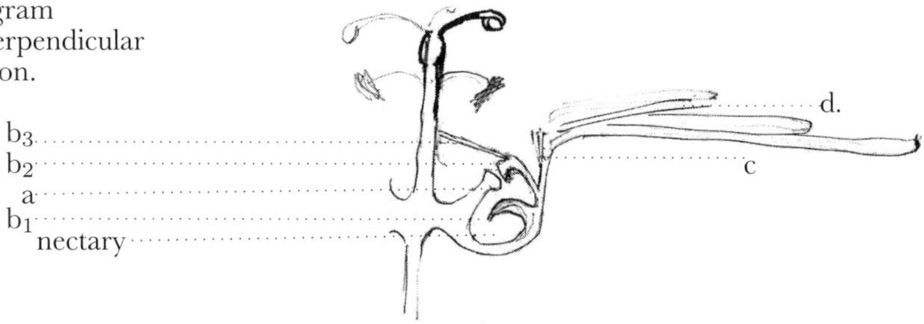

He describes the colours of the coronas disposed in rings, so as to attract an insect to the annular nectary in its centre, and he points out that the mechanism of these coronas and processes protect the nectary from rain. After thus describing what we may call the insects part of the flower, he proceeds to describe the peculiar stamens and styles with their remarkable successive movements, the filaments bending down in the first instance so as to bring the pendent anthers with their opened cells downwards in contact with an insect settling on the coronas: and the styles bending down after the anthers have shed their pollen so as in their turn to bring the reversed stigmas in contact with another insect similarly placed.

He then quotes "Medicus" & Linnæus[3] who had observed the depression of the stigmas and had supposed it to be a means of self fertilization, and proceeds to combat this opinion in words which are sufficiently quaint and characteristic to be worth quoting He says "If the fertilization happens in the mechanical way assumed by Linnæus and Medicus, I maintain that the beautiful Passion flower, which is the admiration of the learned and the wonder of the ignorant, is a wretched bungle of nature. For in that case, the beautiful arrangement of parts which, as I have shewn above, stands in immediate relation to the insect, is in no way connected with any purpose proper to the flower itself. In that case the sexual parts of the flower alone are there for the sake of the flower; and the remaining and far larger part is there, not for the sake of the flower, but of the insect. In that case, the flower is not one single beautiful whole: it consists of two wholes, which regarded with a view to their purpose, have not the least resemblance to each other, and are stuck together in the oddest way so as make a sham unnatural whole (Schein-ganzen) In that case the flower is no better than the Centaurs, the Sphynx, the Hippogriff, Pegasus, and other abortions of the imagination. In that case nature would have done much better if she had separated these two distinct and inconsistent wholes from one another, so that the plant might bring forth on the one hand flowers without calyx, corona, nectary, cover to the nectary, and guide to the nectary; and on the other, flowers without sexual organs. For if she had done this one would at least know, what to make of such flowers. The former would be devoted to the production of fruit and would be fertilized precisely as supposed by Linnæus and Medicus; whilst the latter would serve for the nourishment of insects. The order in which Linnæus would have arranged these flowers might have been called 'Gynandrea (vere) Frustranea'.[4] So we are reduced to one of two alternatives. Either nature has gone to sleep when producing these flowers or Linnæus and Medicus have been dreaming when explaining their fertilization. The former is impossible: the latter not only possible but certain."

"In truth the flower is a Dicbiogamist,[5] with the male organs preceding the female." I have quoted Sprengels words, irreverent as they are towards a great name, not only for their own quaint truth; but because, as we shall see, there is still scope for his sarcasm even amongst the most intelligent pupils of his own school— Sprengel goes on to describe at length how an insect attracted by the corona to the centre of the flower in its earlier stage, must with its back sweep the anthers and carry away pollen, and how in its later stage, when the stigmas are bowed down between and even below the stamens, another insect, similarly attracted and bearing pollen from some other flower in a earlier stage, must sweep the stigmas and deposit that pollen upon them And he concludes. "Nature then has not gone to sleep whilst producing this flower: she has not worked up two dissimilar wholes into an inconsistent anomalous sham whole; but has produced a beautiful whole, in which nothing is deficient, nothing is superfluous; in which each part is in exact accordance with every other, and everything has relation to her one great object, the fertilization of the ovary: in short, a whole which does her honour."

He then says that the flower is intended for the day, since it has colours leading to the nectary; that it blossoms one day as a male; shuts up at night, and blossoms the next day as a female.— He has never seen an insect upon it; does not know what insect fertilizes it or whether it is fertilized at all by insects in our climate; but is satisfied that if it is not, the only reason is that the traveller who brought over the plant failed to bring the appropriate insect with it. Had he seen this Passiflora abundant in the open air, as it is with us now, he would certainly have been satisfied that bees can and do perform this office for it.

He then observes that the anthers are so suspended on their filaments as to swing round easily and remarks that this little point probably gives help in some way towards securing fertilization Had he been able to watch the bees upon the flowers he would have had no difficulty in seeing what this help is. The flower is large compared to the size of the bee, and she walks round the column, thrusting her proboscis as she goes at various points of the circle into the annular nectary. Her back, or, most likely, the back—of her head, touches one end of the long, suspended anther; and turns it, so that it becomes parallel to the back of the insect and is swept by it from end to end as she moves.[6] Just as the hair-clad style in Lobelia forces the pollen out of the staminal tube: just as a similar service is performed by the brush on the style or by the dilated filaments in many of the Papilionaceæ:[7] just as the openings at the base of the two foremost of the five united anthers serve the same purpose in Viola: just as the hairs on the edges of the anther cells shake the pollen from them in many of the Labratæ: just as the tails of the anthers in the Ericacæ,[8] jarring against the bell of the corolla, shake the dusty pollen out of the narrow downward pointing pores of the anther cells: just as, most wonderful of all, the various pollinia of the Orchids attach themselves to the heads and trunks of bees and moths:— so here is the pollen so placed as to be swept away and carried off with the greatest economy and in the most complete way by the nectar seeking bee.—

But this is not the point on which I am anxious to dwell— Sprengel, it would seem, had scarcely seen the flower in growth and action. He had certainly never seen a bee upon it: and he had never seen the beautiful Tacsonias, which I venture to think may throw a new light on parts of the structure, and especially upon that very remarkable part, the corona.

Mr Darwin suggested to me the question. What is the use of the corona? and observed that he had a pendent Tacsonia (Van Volxemus) without one.[9]

Now Sprengel says that the object of the coronas of Passiflora Cærulea is two fold First that by their rings of colour, disposed, not in lines towards a single opening, as in many flowers, but in concentric zones parallel with and above the annular nectary, it attracts insects to the latter.

This seems very probable:— But secondly he says that they operate as so many successive coverings to the nectary for the purpose of keeping out rain. This seems to me a very inadequate explanation. He thinks that the short upright corona marked (c.) in the diagram would keep drops from rolling down the rays of the outer corona (d) upon the inner corona (b.) and that if they get through this they would be stopped by the process marked (a). Whether these processes would actually have this effect in Passiflora cærulea: whether, if this were their object, some of them would not be superfluous and useless , even in Passiflora cærulea, and still more so in other species, may well be doubted. And it is also worth consideration whether, having regard to the short time the flower is in blossom and the climate of its native country, it needs this protection.

Signor Delpino in his extremely interesting observations on Dichogamy in the Vegetable Kingdom published in Milan 1868.9. page 172. has another explanation of some of these structures[10] He gives an account of Passiflora Princeps and of the three spaces or chambers formed by the various processes which inclose the nectary, and then observes. "Here it is clear that Nature with the object of preserving the honey for special fertilizing creatures has conceived and executed one of the strangest imaginable traps ('trappole'). The creatures that want to suck the honey from this flower must not only have a long proboscis but must be singularly intelligent"

And he then suggests a sort of competitive examination between three different Lepidoptera of different degrees of intelligence and perseverance, of which the first only gets as far as the first chamber; the second, turns back on finding no honey in the second chamber, whilst the third insect, who has the sense to persevere till he reaches the third chamber, is rewarded by a copious draught of nectar. It is not a little strange that with Sprengel's words before him, so able a man as Delpino should have put forward such a suggestion. If the old German Botanist were alive again and heard that the Passion flower possesses peculiar organs for the purpose of testing developing or rewarding the intelligence of moths and butterflies, he would certainly repeat his sarcasms on the blunders Nature must be supposed to have committed in producing such an inconsistent abortion. It seems to me still more strange that such an admirable observer of Nature, and so thorough a student of Darwin as Delpino, should be led by his Teleology into assuming that the existence of organs in one living creature is to be explained by supposing them to have been created for the benefit of another.

Let us see of what several organs or parts this curious corona consists, in 'Passiflora Cærulea' and in a few other Passifloras and Tacsonias

First in Passiflora Cærulea

See Diagram opposite page I[11]

In this as in all other species I have examined, the corona in the early bud is developed much later than the calyx corolla stamens or pistil.

And it is also as might be expected much more variable. It seems in fact to be that part which most readily changes so as to suit changed position or changed external circumstances Schleiden observes this late development; and argues from it that the corona must be developed from an axial, not from a foliar organ, a distinction which I leave to those who are more competent to deal with it than I am.[12]

In Passiflora Cærulea the corona is I believe fourfold. There is the outer corona with long spreading horizontal rays, marked (d.) in the diagram. There is the next corona marked (c.) consisting of short rays, having little knobs at the top, standing upright at right angles to the rays of the outer corona, being in in length not above an eighth of the length of those rays: and inserted immediately within those rays. Next there is the corona marked (b.) The upper part of this (marked (b.3.)) is obviously a corona or part of one. It consists of rays separated from each other for nearly their whole length and inclined so that their upper ends rest against the column, forming a circular blunt pyramid of open grating round the column and above the nectary. These rays are inserted into a base (b.2.) which thickens and bends over so as nearly to meet the interior process (a) and then is prolonged till it joins the calyx or disk. Below its insertion there is a further process marked (b 1) projecting into the interior of the flower: stiff, solid fleshy, and curving over and downwards at the interior End or rim, so as to form an annular nectary at the base of the flower. I have marked it (b.1.) and thus treated it as the base of the corona (b.) rather than as a separate corona because I find it later in development than the other coronas, sometimes, as in Passiflora princeps, very nearly disappearing; and never in any flower I have seen developing into a separate corona with fringed and coloured edges.

The nectary thus formed is a large annular chamber and abounds with nectar.

Still proceeding inwards there is a fourth process marked (a) thick, solid, and having a dilated head or rim, which projects from the column so that its head or rim is above (b1) and almost meets (b.2.) I have called it a separate corona because it appears to be developed as soon as the other coronas and in some species or varieties e.g. Passiflora cærulea—racemosa,[13] hybrida, floribunda and Imperatrice Eugenie is coloured and fringed at the top like a true corona.

Whether these names are morphologically correct or not may be doubtful; but they are at any rate convenient in comparing different flowers. I shall accordingly, proceeding from the centre of the flower call process (a) the first corona, process

b. including b1, b2, b3, the second; (c.) the third; & (d.) the fourth.

Before quitting Passiflora Cærulea let us observe.

1. That the flower is erect.

2. That it has no tube,

3. That the third corona (c) at right angles to the fourth is short, and that correlatively the rays of the second corona (b.3.) are inclined almost horizontally towards the column. The anthers and stigmas when depressed hang over (c.) so that an insect having lighted on the calyx, corolla, or outer corona (d), would in getting to and round the nectary have to clamber over and amongst the rays of (c.) so as to bring his back in contact with the open anther cells or stigmas. I have been much interested in watching a bee thus struggling over the unequal footing and projections of the two outer coronas.

4. That the grating formed by the rays of the second corona (b.1.) is open from the base of the rays. i.e. for a considerable space from the column upwards to their extremity, so that any thin object, stiff or not, could, *so far as this grating is concerned,* easily reach the nectary

5. That there are three stiff projections overlapping one another above the nectary— viz! the base of the second corona (b.2) & the lower process of the same corona (b.1.) which project from the calyx into the cavity; and between them the solid first corona (a), with its thickened rim; so that the three together completely close the path into the nectary to any object, which is not, like an insects proboscis, long, thin, flexible, and thus capable of turning corners.

Passiflora Kermesina — Diagram 2.

The flower is comparatively small: the distance between the anthers or stigmas and nectaries is also small: so that a bee would easily reach the nectary whilst touching them. The flower is erect or sub-erect and has no tube. The corolla and calyx are a bright purplish red or crimson, and the outer corona marked (d.) is brilliantly coloured, the outer ends being white, and the inner circle purple.

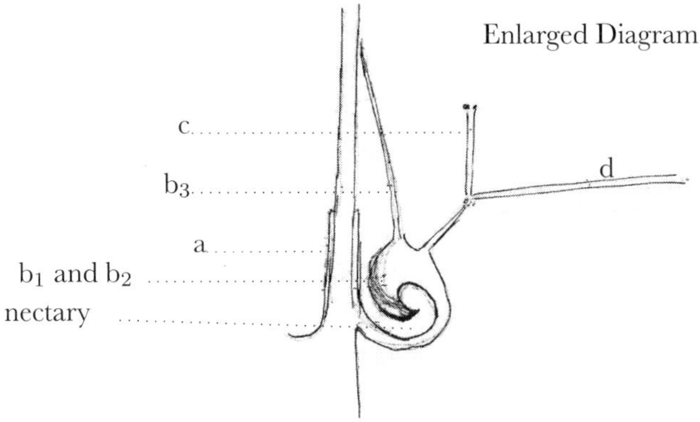

As regards the interior coronas.

The first or inner corona (a) appears as soon and as completely as the others in the bud, but is scarcely developed further at a later stage, being then a mere protuberance on the base of the column.

In the second corona the process b1. at its base includes the annular nectary as in Passiflora Cærulea, but it is united with (b2), the centre part of this corona, and forms with it a stiff solid fleshy ring leaving a narrow winding passage between it and the column through which a flexible object can pass to the nectary, but a stiff one cannot. The upper part of the corona (b.3) is not divided into separate rays until near the top and the short rays are there appressed against the column. An entering insect must therefore insert its proboscis close to the column and thrust it perpendicularly down the side of the column. When it is in that position, it cannot possibly reach the nectary unless it is flexible and can turn corners.

The third corona (c) is upright and so placed and of such a length as to make an insect mount to the point in the second corona (b) where the separating rays (b3) give access to the interior of the flower.

Passiflora Princeps. Diagrams 3. & 3a

I speak with some hesitation about this species since the only specimens I have seen are some late flowers not fully developed, or opened, and without nectar in the nectary. And Dr. Hooker tells me that it is nearly hybridored out of cultivation.[14] Certainly the plant in the Botanical Gardens of the Regents Park does not agree in several respects with the description given by S Delpino in the work quoted above.

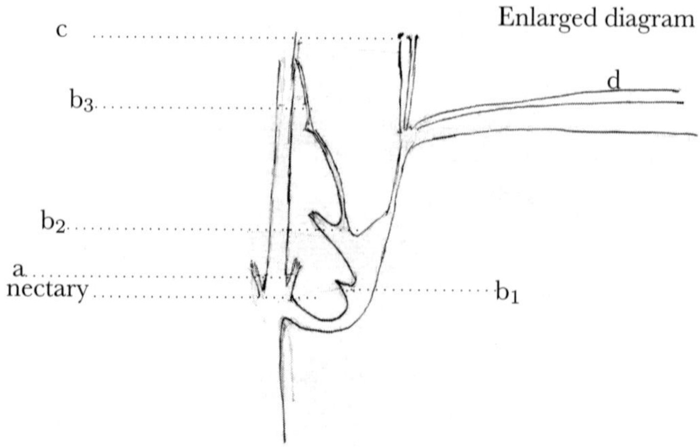

The flower is erect, and the buds are erect also. The colour of the calyx and corolla is bright red. The positions of the pendent anthers or stigmas, of the open grating at the top of the corona (b.) and of the nectary, are such that a large insect placed on the top of the corona might reach the nectar with his proboscis, and at the same time sweep the sexual organs with his back or wings. If however, as M. Delpino suggests the distance between the parts is ever too great for insects, and the work is done by humming birds,[15] it would be curious to observe whether, correlatively, the large openings in Passiflora Princeps leading towards the nectary which are noticed below, do really give a more direct access to it than in the other Passifloras If so, Passiflora princeps will be an intermediate stage between Passiflora Cærulea and Tacsonia. But in that case what is the use of the coronas?

As regards the coronas—

The first or inner corona (a) is less developed than in Passiflora Cærulea, more so than in Passiflora Kermesina. It is a mere short and thin process projecting upwards from the swelling base of the column so as to form an acute angle with the column.

Diagram of transverse section above nectary

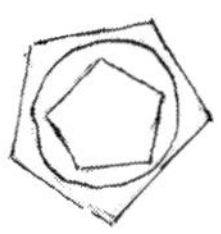

Inner pentagon—section of column
Intermediate circle—summit of process (a) or innermost corona
Outer pentagon—summit of process b2.

The process (b1) or base of second corona is very little developed, but on the other hand the process (b2) or middle part of the second corona is largely developed into a thin elastic rim, so as almost to intimate the possibility of its being developed into another corona. It does not fit close round the column, nor is it applied to process (a) so as to close the nectary but if looked on from above (see diagram) it is pentagonal, with its angles opposite to the sides of the column thus leaving large apertures leading to the nectary at each angle. The upper part of the second corona (b.3) is connate till near the top, and the short rays are there applied close against the column so as to force an entering object, as in Passiflora Kermesina, to enter perpendicularly down the side of the column.

The height of the 3rd. corona (c) is such as to force an insect to climb upwards to the point in the second corona (b3.) where it can obtain an entrance

There appears at first sight to be a straight passage to the nectary; but on thrusting a pin or other stiff object between the short rays of the second corona and the column, it will be found that if it goes perpendicularly down it is caught by the process (a) and stopped at the angle which that process makes with the column, whilst if it is thrust down obliquely it is stopped by the projection of b.2.

Passiflora Cæruleo Racemosa[16]
Diagram 4

This being a garden Hybrid cannot be appealed to as shewing an organization which has enabled it to prosper in the battle of life, but there are some interesting points.

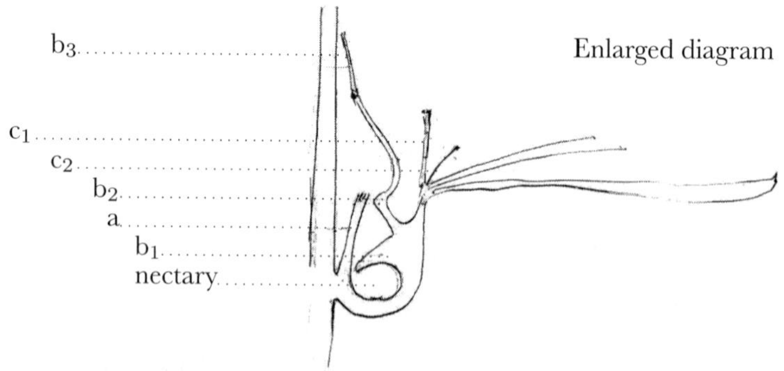

The inner process or corona (a) instead of being a thick fleshy process as in Passiflora Cærulea: or abortive as in Passiflora Kermesina, or being short, as in Passiflora Princeps, is considerably developed and divided into a fringe at the top. It is here obviously an inner corona.

The nectary is thoroughly closed and protected by the thick fleshy process b.1. The process (b.2.) is much less pronounced than in Passiflora Cærulea and is apparently only a fold or bend of the second corona and this fold or bend is applied generally but with some intervals to the inner corona (a). A straight stiff object directed downwards between the short open rays of the second corona (b.3) would be stopped, as in Passiflora princeps, by the first corona (a) or if it passed that guard would not get round the fleshy process (b.1.)

A flexible proboscis of the length of half or $\frac{3}{4}$ of an inch would reach the nectary.

There is a fourth corona ((c.2) in diagram) between the third corona (c.1) and the long outer or fifth corona. This corona is inclined at an angle of about 45^0 and would help an insect to mount to top of (c.1.)

I found that a pin inserted between the rays of the second corona (b.3) would not reach the nectary. It is stopped either by the process (b.2.) or by the inner process (a). On the contrary a horse hair thrust into the same place and wriggled about returns covered with nectar.

The flower is erect or sub erect.

Passiflora Hybrida Floribunda[17]
another Hybrid. Diagram 5

The flowers are erect or sub-erect

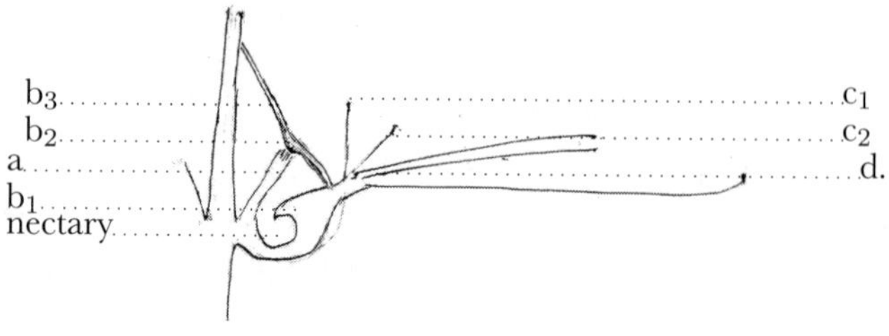

Process (a) is here again developed and divided at the top like a true corona.

Process (b.1) is thick and fleshy as in Passiflora Cærulea. The rays of the second corona (b3) are divided about half way down its length at the point where they unite process (b.2.) thickens and the lower surface is closely applied to the summit of process (a)

Some of the rays of the third corona (c.) are long and upright so as to carry an insect up to the openings in b.3: others are shorter

No straight stiff object can enter the nectary.

Passiflora Imperatrice Eugènie[18]
another garden variety

Diagram 6.
This flower is erect or sub erect.

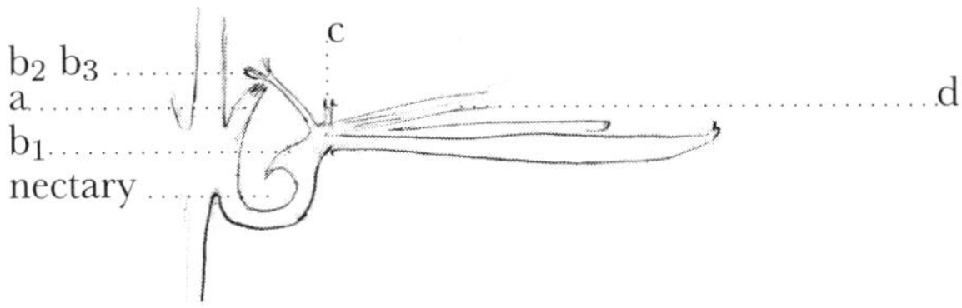

The process (a) or inner corona is much developed and the upper edge is divided or fringed at the top and coloured a deep purple, shewing that it is a true corona

Process (b.1.) is thick and fleshy as in Passiflora Cærulea.

Process (b2. and b3) are much shortened and brought together. Instead of the rays reaching up to the central column they scarcely reach above the summit of (a) to which they are closely applied. The third corona (c) is correlatively short also

It is obvious that here again a stiff straight object could not reach the nectary

It will be observed that all these Passifloras are alike in the following respects

1. Though the flowering branch is pendent the flowers are erect or sub erect.

2. The nectary is inclosed by a variety of processes, seemingly parts of different coronas, but all having one apparent function. viz[t] to keep a straight stiff object from reaching the nectary, whilst they allow a flexible object to do so. This is effected sometimes by making the passage wind round corners through stiff, thick, fleshy walls: sometimes by a narrow grating above and a circular screen below, so contrived that a straight object entering the former would be caught by the latter.

3. The third corona (c) is of such a height as to lead an insect to the openings in the grating of the screened corona (b.3) which furnish the way to the nectary.

4. In the flowers which I have seen the interval interposed by the length of the column between the pendent anthers or stigmas, and the coronas is such that a large insect might sweep the former whilst standing on the latter. But having regard to S. Delpino's remarks on Passiflora princeps this requires further observation

Let us now compare two Tacsonias viz[t] Tacsonia mollissima and Tacsonia Van Volxemii.

First Tacsonia mollissima. Diagram 7.

The flower is pendent

The length of the tube and column, and consequently the distance between the nectary and the anthers and stigmas is very great; 4 or 5 inches.

There is no true corona but merely a rudimentary trace at the bottom of the limb of the corolla.

The limb of the corolla is very handsome; bright pink and bell shaped.

The nectary is at the base of the tube i.e., the flower being pendent, at the higher end of the tube. It contains an immense quantity of nectar, which sometimes escapes and runs down the column.

The various processes or coronas which we have marked as a. and b. in Passiflora Cærulea and in the other Passifloras are here replaced as follows

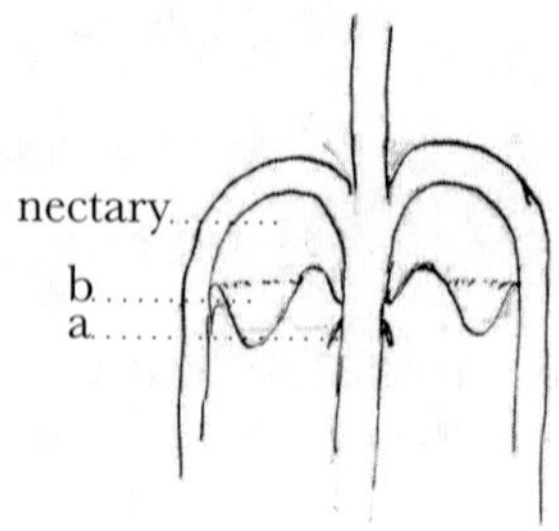

The process (a) by one process, almost rudimentary, thin and soft, growing from the column.

The process (b.) by a second process, growing inwards from the calyx half an inch from its base, and then turning inwards and upwards till it reaches the column, when it folds over so that the soft folded end fits close round the column. This process is soft, thin, leathery and quite elastic. It gives way to any stiff object thrust up the tube and returns to its place. It is in fact an elastic circular valve forming a chamber at the base or higher end of the tube and thus retaining the copious nectar found there.

Tacsonia Van Volxemii
Diagram 8.

The flower is pendent.

There is a mere vestage of a true corona.

The corolla and inner side of sepals are a brilliant red.

The tube of the corolla is from an inch to an inch and half long, but the column is prolonged so that the anthers and stigmas hang three inches lower; or four or five inches below the nectary

The nectary as before is at the base or higher end of the tube.

Diagram of section of nectary

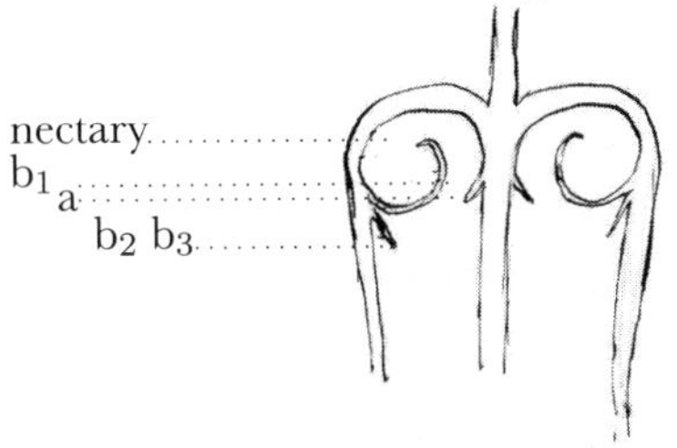

The Processes described in Passiflora Cærulea i.e. the first corona (a) and the upper part of the second corona (b2, b3) are represented in T. V. Volxemii by two small short elastic rudimentary processes, the one growing from the column, the other from the calyx. These processes scarcely meet each other and would not stop or turn any entering object.

The third process, (b1) in Passiflora Cærulea, is represented by a third process growing from the calyx nearer to the base of the flower than the two former It curls inwards towards the column so as to form an annular cavity round the internal base of the flower and thus prevents the nectar, which is abundant, from running down the tube. This process is elastic: it yields to a stiff object and returns to its place.

Now admitting Sprengels explanation of the value of the corona in Passiflora Cærulea in attracting bees, and of the inner corona and other processes to keep out rain, this does not explain.

1. The comparative length of tube and column in the two Tacsonias as compared with Passiflora.

2. The difference in the processes (a) (b) and (c).

But let us suppose that Passiflora Cærulea and the other Passifloras are fertilized by bees or other insects, as no doubt some of them are in this country, and that Tacsonia mollissima and Tacsonia Van Volxemii are fertilized by Humming Birds. Then the structure becomes intelligible.

The outer coronas in Passiflora Cærulea may and probably do attract insects and guide them to the centre of the flower, but they probably also raise them & force them to struggle so that their backs touch rub & sweep, in an earlier flower, the suspended and opened anthers, in a later flower, the lowered and mature stigmas. It is, as I have said, interesting to watch the bees struggling upon the rays of the corona. The inner coronas and the processes belonging to them admit the flexible proboscis of the bee or other insect to the nectary below, but would not admit the stiff beak of the humming bird.

In Passiflora Cærulea, the distance from the corona, where the insect is, to the nectary at the base of the flower is, compared to that in P. princeps, small; and the column is short; so that the insect easily gets his nectar and at the same time touches the fertilizing apparatus.

In Tacsonia the flower is pendent and can be sucked more easily by a humming bird which does not need an alighting place than by an insect, like a bee, that does.

The tube in Tacsonia Mollissima is far too long for an insects proboscis but might be reached by a humming birds beak.

The processes which in Passiflora would keep out a stiff object are in Tacsonia Mollissima, so reduced, thin and elastic, that they, or rather the third of them, would whilst keeping the nectar in its place, easily admit a birds stiff beak, and when it is withdrawn, would resume their place so as to keep in the nectar.

In Tacsonia Van Volxemii the structure is intermediate, but the function is the same as in Tacsonia mollissima. The tube is not so long as in Tacsonia mollissima but too long for a bee, possibly it is sucked by a humming bird with a comparatively short beak and long tail, so as to reach the far projecting anthers and stigmas. The processes inclosing the nectary are structurally more like those of Passiflora than are those of Tacsonia mollissima, but in function they agree with those of Tacsonia mollissima for they would admit a humming birds beak and then return to their place.

So far therefore it would seem that in these flowers.

An upright flower, a short column, largely developed coronas with processes inclosing the nectary and shutting it against a humming birds beak, are correlated so as to enable insects to get at and fertilize the flower. And that

A pendent flower, no visible corona; a long column; and thin elastic processes are correlated, so as to enable humming birds to do the same.

I am aware that the above speculations require confirmation by the examination of further species, and above all by observations of the habits of the plants where they grow naturally and where they are adapted to the creatures which fertilize them. D^r^. Hooker has kindly written to a friend in S. America to learn whether Tacsonia is visited by humming birds. But I have written this paper with the desire of calling attention to the subject. M^r^. Darwin in his work on orchids, and following him S. Delpino and Herr Hildebrand of Bonn have shewn in how many of our common flowers the structure can be explained by reference to the visits of bees and other insects known to us in this country[19] It seems to be an interesting subject for inquiry whether the structure and functions of flowers in countries where humming birds abound is modified so as to suit these differently organised visitors.

In the case of Fuchsia for instance there seems strong reasons to suspect this to be the case.

T H Farrer | 13 November 1869

Linnean Society of London, MSS Case 6B, MS 506

1 Farrer refers to Sprengel 1793, pp. 160–6 (see letter from T. H. Farrer, 13 October 1869 and n. 4).

2 Each diagram is accompanied by a faint preliminary pencil sketch which it was not possible to reproduce here. The diagrams are not represented in their original size.

3 The references are to Friedrich Casimir Medicus and Carl von Linné.

4 'Gynandrea (vere) Frustranea': Gynandria (a Linnaean class referring to flowers in which the stamens and pistils are united in a column) (truly) in vain (a botanical joke derived from Latin and Greek).

5 The copyist evidently mistranscribed 'dichogamist', a flower in which the male and female sexual organs mature at different times.

6 The rest of this paragraph is crossed through in ink. 'Scarcely analogous' is written in pencil in the margin in an unknown hand.

7 The family Papilionaceae is now Fabaceae.

8 The copyist evidently mistranscribed the plant family names Labiatae (now Lamiaceae) and Ericaceae.

9 CD had evidently discussed the function of the corona when Farrer visited in October. In his letter to CD of 9 October 1869, Farrer hypothesised that in *Passiflora* it might serve to elevate the body of a bee so that it would make more efficient contact with the anthers and style. Farrer refers to *Tacsonia van volxemii* (now *Passiflora antioquiensis*, the red banana passionflower). *Passiflora antioquiensis* does have a corona, though it is much smaller than that of *P. caerulea* (see Ulmer *et al.* 2004, p. 336).

10 Farrer refers to Federico Delpino and Delpino 1868–75, 1: 172.

11 The reference is to the original manuscript pagination.

12 Matthias Jacob Schleiden had observed the development of the corona in *Passiflora* and argued that it could not be a foliar organ (see Schleiden 1845, 2: 580–1 and Plate III, figs. 7 and 10).

13 *Passiflora caeruleo-racemosa* is now *P.* x *violacea.*

14 Farrer refers to Joseph Dalton Hooker. The copyist mistranscribed the word 'hybridized'.

15 Farrer refers to Delpino 1868–75, 1: 172.

16 Farrer has deleted *Hybrid Floribunda* and written *Cæruleo Racemosa* above. He wrote in the margin, 'I am not quite sure about the names of this and the following variety'.

17 The name *Passiflora hybrida floribunda* no longer exists.

18 *Passiflora* 'Impératrice Eugénie' is a cultivar of *P.* x *belotii*, which is a hybrid of *P. alata* and *P. caerulea.*

19 Farrer refers to CD's *Orchids* and Friedrich Hildebrand (Hildebrand 1867).

APPENDIX V
Portrait of Charles Darwin by Laura Russell

The painting

The portrait of Darwin reproduced in the frontispiece is an oil painting on a canvas measuring 7 x 10 inches. There is a painted inscription on the back: '23d. Augt. 1869 / Down / from / nature by Laura Russell'. The painting is in private possession and has been since it was painted. The heavy, loose, dark jacket Darwin is shown wearing corresponds to ones in photographs of him in the 1870s. His son Francis noted in *LL* 1:112: 'He wore dark clothes, of a loose and easy fit.'

Laura Russell and her husband

Laura Russell was daughter of Jules, vicomte de Peyronnet, translator of Thomas Babington Macaulay's *History of England* into French. The Peyronnets were known in Paris in the 1850s for their writings and independence of mind. Laura was an accomplished amateur painter. In 1865, she married Arthur Russell MP, son of Lord William Russell and Elizabeth Ann Rawdon. He had become MP for Tavistock in 1857 and in the following years became known in London as a savant and a philosopher. When his elder brother Hastings took over the Woburn estate in 1861 before becoming ninth Duke of Bedford in 1872, he followed his grandfather the sixth Duke (d. 1839) and his uncle the seventh Duke (d. 1861) in supporting research for agricultural improvements. When together at Woburn, Hastings, Arthur and their brother Odo, a diplomat, would recapture their childhood passion for natural history. 'The numerous ponds at Woburn were fished and dragged by the hour with boyish enthusiasm; frogs, toads and water-tortoises in variety were procured from zoological shops and from abroad, to be released in suitable areas, and their survival or increase was noted by the brothers with scientific interest.' (Blakiston 1980, p. 214.) In the 1860s Arthur Russell was a member of the Athenaeum and Brooks's, and his diaries which survive in his family show that he met Joseph Dalton Hooker, William Benjamin Carpenter, Roderick Impey Murchison, John Lubbock, James Anthony Froude, John Tyndall, Charles Eliot Norton of Boston, and Robert Browning among many others in London scientific and literary circles. He was one of the first members of the Metaphysical Society, founded in April 1869 by James Thomas Knowles and Alfred Tennyson for free discussion of current questions about the existence of God across the intellectual and denominational boundaries of the time. In 1872, when Hastings succeeded to the dukedom, Arthur was raised to the rank of a duke's son. In 1873, he gave a paper to the Metaphysical Society on 'Darwinians and idealists'. In the paper, he remarked on the idealistic implications of Darwin's comment on the preference for beauty among female animals and concluded that 'the phenomena of physics are practically interpretable only by the methods and formulae of mind'. The paper was privately printed in *Papers read at the meetings of the Metaphysical Society by Lord Arthur Russell* (1896). See A. W. Brown 1947, pp. 150 and 324. Laura died in 1910.

The meetings and sitting in August 1869

Arthur and Laura Russell were friends of Edward Stanley, fifteenth earl of Derby, and in August 1869 while he was living at Holwood House just north of Down village, they visited him for a few weeks. At the time Lord Derby was close to the dowager marchioness of Salisbury, Mary Catherine Gascoyne-Cecil, who was a cousin of Emma Darwin's mother, and the Darwins were also frequent visitors to Holwood House. Lady Salisbury married the Earl of Derby in 1870 and they maintained their close friendship with the Darwins until Darwin's death. His relationship with Lady Derby is illustrated by a comment of his recorded in 1880 (Grant Duff, 1898, 2: 283). Lady Derby reported that 'she had lately explained to Darwin the state of her sight, which is very peculiar. Darwin had said: "Ah! Lady Derby … how I should like to dissect you!"')

At the time of the Russells' visit to Holwood, Laura was eight months pregnant with her second child. She gave birth to her daughter Flora on 28 September. Lord Arthur wrote in his diary for 5 August: 'We left for Bromley at 5 p.m. Lady Salisbury at the Station. Holwood. What a charming place!' On 6 August, he wrote: 'How pleasant to walk in a garden before breakfast.' During the following days, the Russells visited a number of friends in the neighbourhood. On 10 August: 'L. & I with Mary ate luncheon with Darwin at Down. Mrs Darwin grand dr of Wedgwood 2 d. 3 sons at home.' On 23 August 1869, Arthur Russell wrote: 'We went to Down at 4 $\frac{1}{2}$ and Darwin gave L. a sitting for his picture.' On 29 August: 'Clouds hid the Sun at 8 a.m. Sudden change of temperature after the great heat. We walked to Down to take leave of Darwin. He was not well. We saw Mrs. Darwin.' At Down House, Emma noted in her diary: 'Lady Salisbury & Russells called [-] v. cold' (DAR 242). The next day, Arthur wrote, 'Our last day at Holwood. We have been very happy here!'

Later contacts between the Russells and the Darwins

In later years Lord Arthur was in occasional touch with Darwin about natural history specimens. In August 1878, Lord Arthur took a house near Leith Hill Place, Surrey, where Josiah Wedgwood III and his wife Caroline, Darwin's sister, lived. Charles and Emma were staying there at the time. Lord Arthur wrote in his diary for 9 August: 'We had lunch at Leith Hill Place to meet C. Darwin—was moved to see that great man again—he has grown much whiter.'

On 20 April 1882, the day after Darwin's death, Lord Arthur noted in his diary: 'Fine afternoon. I took 4 children to Crystal Palace driving by the road. They had not been there for years. . . . Darwin died at 4 yesterday afternoon.' Lord Arthur's daughter Diana, who lived until 1971, wrote at the bottom of the page: 'I remember seeing the news of D.'s death on the posters as we drove home from the C. P.' In the following days, Lord Arthur was among the signatories of Sir John Lubbock's petition for Darwin to be buried in Westminster Abbey (A. Desmond and Moore 1991, p. 667).

Position among portraits of Darwin

Laura Russell's portrait of Darwin at Down House is one of a small group with two others made by lady artists who met the Darwins at their home. It compares closely with Louisa Nash's ink sketch between 1873 and 1880 (in possession of a Nash descendant), and Marian Huxley's drawing in 1878 at Down House (National Portrait Gallery). In each case the picture was an informal likeness taken in the strictly private setting in which Darwin chose to spend most of his time.

Randal Keynes
15 December 2008

APPENDIX VI
Darwin's *Queries about expression*

In 1868, Darwin sent out a number of printed copies of a list of queries on human expression (see, for example, enclosure to letter to G. H. K. Thwaites, 31 January [1868]). In the Darwin Archive–CUL, three copies of these queries have been found (DAR 53: B2, DAR 181: 41, DAR 186: 1). There is also what is apparently a proof copy of this printed version (DAR 96: 1) with handwritten corrections on it, some by Darwin, that are incorporated into the three other copies. It is likely that this printed version was not made until late 1867 or early 1868 (see Freeman and Gautrey 1972 and 1975), and that it represents a near final version of the queries; in particular, questions 2, 5, 6, 10, and 13 contain additions that rarely appear in extant early versions of the queries, but that do appear in the version that Darwin reproduced in the introduction to *Expression*. The final version of the printed queries in the Darwin Archive–CUL is transcribed below for its own intrinsic interest and as a convenience to readers wishing to see the printed questions when answers to a now unavailable list of queries appear in the *Correspondence*. At least two additional copies of Darwin's printed list of queries survive in other archives (see Freeman and Gautrey 1975).

QUERIES ABOUT EXPRESSION

(1.) Is astonishment expressed by the eyes and mouth being opened wide, and by the eyebrows being raised?

(2.) Does shame excite a blush when the colour of the skin allows it to be visible? and especially how low down the body does the blush extend?

(3.) When a man is indignant or defiant does he frown, hold his body and head erect, square his shoulders and clench his fists?

(4.) When considering deeply on any subject, or trying to understand any puzzle, does he frown, or wrinkle the skin beneath the lower eyelids.

(5.) When in low spirits, are the corners of the mouth depressed, and the inner corner of the eyebrows raised by that muscle which the French call the "Grief muscle?" The eyebrow in this state becomes slightly oblique, with a little swelling at the inner end; and the forehead is transversely wrinkled in the middle part; but not across the whole breadth, as when the eyebrows are raised in surprise.

(6.) When in good spirits do the eyes sparkle, with the skin a little wrinkled round and under them, and with the mouth a little drawn back at the corners?

(7.) When a man sneers or snarls at another, is the corner of the upper lip over the canine or eye tooth raised on the side facing the man whom he addresses?

(8.) Can a dogged or obstinate expression be recognized, which is chiefly shewn by the mouth being firmly closed, a lowering brow and a slight frown?

(9.) Is contempt expressed by a slight protrusion of the lips and by turning up the nose, with a slight expiration?

(10.) Is disgust shewn by the lower lip being turned down, the upper lip slightly raised, with a sudden expiration, something like incipient vomiting, or like something spat out of the mouth?

(11.) Is extreme fear expressed in the same general manner as with Europeans?

(12.) Is laughter ever carried to such an extreme as to bring tears into the eyes?

(13.) When a man wishes to shew that he cannot prevent something being done, does he shrug his shoulders, turn inwards his elbows, extend outward his hands and open the palms; with the eyebrows raised?

(14.) Do children when sulky, pout or greatly protrude the lips?

(15.) Can guilty, or sly, or jealous expressions be recognized? though I know not how these can be defined.

(16.) As a sign to keep silent, is a gentle hiss uttered?

(17.) Is the head nodded vertically in affirmation, and shaken laterally in negation?

Observations on natives who have had little communication with Europeans would be of course most valuable, though those made on any natives would be of much interest to me.

General remarks on expression are of comparatively little value; and memory is so deceptive that I earnestly beg it may not be trusted.

A definite description of the countenance under any emotion or frame of mind, with a statement of the circumstances under which it occurred, would possess much value. An answer within six or eight months, or even a year, to any *single* one of the foregoing questions would be gratefully accepted. In sending answers, the questions need not be copied, but reference may be made to the numbers of each query.

CHARLES DARWIN,

DOWN, BROMLEY, KENT,
1867.

MANUSCRIPT ALTERATIONS AND COMMENTS

The alteration notes and comments are keyed to the letter texts by paragraph and line numbers. The precise section of the letter text to which the note applies precedes the square bracket. The changes recorded are those made to the manuscript by CD; changes of hand in letters written partly by CD and partly by amanuenses are also recorded. Readers should consult the Note on editorial policy in the front matter for details of editorial practice and intent. The following terms are used in the notes as here defined:

del	deleted
illeg	illegible
interl	interlined, i.e., inserted between existing text lines
omitted	omitted by the editors to clarify the transcription
over	written over, i.e., superimposed

To Julius Victor Carus 5 January 1869
0.2 5.] *inserted in CD's hand*
2.5 in which . . . horned] *interl in CD's hand*

To Joseph Dalton Hooker 13 January 1869
4.1 The . . . improvement.—] *in CD's hand*

To J. D. Hooker 16 January [1869]
1.3 after . . . you] *interl in CD's hand*
1.5 from Calcutta] *interl in CD's hand*
2.9 If you have] 'have' *interl in CD's hand*
5.1 It is . . . Review. 5.5] *in CD's hand*

To J. D. Hooker [22 January 1869]
1.2 &c.] *interl*
2.3 that we] 'that' *interl*
2.4 generation are] 'are' *interl*
2.4 for classification,] *interl*
2.4 the manner 2.5] *above del* 'meani'
2.5 for same purpose of] *interl*
2.7 was able] 'was' *interl*
3.2 you] *interl*

To A. R. Wallace 22 January [1869]
3.2 improved] *interl in CD's hand*

To C. F. Claus 28 January 1869
6.1 that] *interl in CD's hand*
6.1 only] *interl in CD's hand*
7.1 formerly] *interl in CD's hand*
11.1 With . . . naturalists.—12.2] *in CD's hand*

To James Croll 31 January [1869]
2.1 especially] *interl in CD's hand*
5.8 force] *in CD's hand above del* 'power'
8.1 I wish . . . discovered 8.4] *in CD's hand*

To John Murray [1–3 February 1869]
1.2 I] *over illeg*

To W. C. Tait 2 February [1869]
4.1 P.S. . . . information.— 4.2] *in CD's hand*

To A. R. Wallace 2 February [1869]
1.6 in my former note] *interl*

To James Croll 6 February [1869]
2.1 or 4] *interl in CD's hand*

To David Forbes 7 February [1869]
1.2 me] *interl in CD's hand*
2.2 again] *interl in CD's hand*
2.6 on the Indians] *interl in CD's hand*
3.1 P.S. . . . yesterday. 3.2] *in CD's hand*

To John Lubbock [before 13 February 1869]
0.2 L.] *after del* 'Sir'
1.5 agree] *above del* 'not object'
1.6 my grandfather matern] *interl*
1.9 valuable.] *altered from* 'valuable to'
1.11 support. She] *altered from* 'support,—she'
1.11 on the whole] *interl*
1.12 list] *after del* 'civil'

To R. F. Cooke 23 [February 1869]
3.1 (7)] *interl*
3.1 (12)] *interl*
3.2 by letter] *interl*
6.2 (Kindly ... Rail.)] *square brackets in MS*

To James Croll 24 February [1869]
1.3 & great] *interl*
2.1 In Haste] *added*

To W. C. Tait 24 February 1869
0.1 *Kent.*] *before del 'S.E.'*
1.1 If the] 'the' *over* 'it'
1.1 Box] *after del illeg*
2.1 at head of this note] *above del* 'above'
2.2 outside] *interl*
4.5 or at 3] *interl*
4.7 rams] *interl*
4.7 much] *interl*

To W. B. Tegetmeier 25 February [1869]
2.6 in colour] *interl*
3.4 doubts] *after del* 'I'
4.3 any] *interl*
5.1 breed of] *interl*

To Thomas Roscoe Rede Stebbing 3 March 1869
0.1 *Kent.*] *before del 'S.E.' in CD's hand*
1.3 spreading] *altered from* 'spread' *in CD's hand*

To J. D. Hooker 8 March [1869]
6.1 Though . . . note. 6.3] *in CD's hand*

To R. F. Cooke 9 March [1869]
1.6 **nine**] *interl*

To Vernon Lushington [12 March 1869]
1.1 *[3 words illeg]*] *interl pencil*
1.1 for him 1.2] *interl over illeg*
1.2 He] *after del* 'I am sorry to say that'
1.2 have at] *above del* 'have an'
1.4 imply] *before del* '*[*express*]*'
1.4 him] *after del* '*[*firstly*]*'
1.4 imply . . . firstly] *interl*
1.8 prevented] *after del* 'destroy'
2.5 for] *after del illeg*
2.7 I remember] *after del* 'I used'
2.7 in consequence] *interl*
2.7 doubting,] *after del illeg*
2.8 however able] *interl*
2.8 This will appear] *after del* 'Here'

To W. C. Tait 12 and 16 March 1869
5.8 my] *interl in CD's hand*
5.9 what suits best.—] *interl in CD's hand*
5.10 shallow] *interl in CD's hand*
5.11 on a few,] *after del comma in CD's hand*
5.12 2 or] *interl in CD's hand*
5.13 revive] *after del* 'or 4' *in CD's hand*

To Fritz Müller 14 March 1869
1.4 the busiest] 'the' *interl*
2.5 the presentation] 'the' *added*
2.6 they are] *over* 'it is'
2.7 of course] *interl*
2.8 such] *above del illeg*
2.8 as] *above del* 'which'
3.2 There . . . printed.] *added*
4.4 by experiment] *interl*
4.8 rather] *interl*
7.5 for our purpose] *interl*
7.6 I . . . autumn] *interl*
7.7 or 3] *interl*
7.7 without aid,—] *interl*
10.4 petals.—] *after del* 'flowers'

To J. D. Hooker 17 March [1869]
2.1 now] *interl*
2.2 has a] *above del* 'is'
4.6 at Oporto] *above del* 'there'

To Julius Dub 20 March 1869
1.1 be] *after del* 'naturally'
1.3 literally . . . Origin.] *interl*
1.3 You must] *after del* 'in literal translation.'
1.5 believe] *above del* 'I am'
1.6 of] *before del* 'opinion'
1.6 sell.] *above del* 'be sold'
1.7 popular] *after del* 'series of'

1.8 your] *after del* 'the'
1.9 give] *after del* 'properly'
1.9 receiving the distinct] *interl above del* 'the'
1.10 A new] *after del* 'I am *[now]*'
1.10 English] *interl*
1.10 is] *after del* 'in English'
1.11 The Edt] *after del* '& a'
1.12 additions] *below del* 'corr'
1.13 given] *above del* '*[give]*'
1.14 if . . . do so,] *interl after del illeg*
1.14 gladly] *interl*
1.16 point] *interl*
1.17 become] *above del* 'been'
1.17 slight . . . & that 1.18] *interl*
1.18 have largely] *after del illeg*
1.18 largely] *interl*
2.1 you] *transposed from before* 'on reflection'
2.2 I . . . this case] *interl*
2.3 will] *above del* 'shall be happy to'
3.1 Believe . . . Sir] *below del illeg*

To Charles Lyell 20 March [1869]
1.7 assumes,] *altered from* 'assume, was'

To A. R. Wallace 22 March [1869]
9.5 up] *interl in CD's hand*
15.1 Believe . . . Papuans.— 16.5] *in CD's hand*

To A. R. Wallace 27 March [1869]
1.1 note] *interl*
2.2 had] *interl*
5.2 of Origin] *interl*
5.6 is] *above del* 'was'
7.4 in tint] *after del* 'which'

To J. D. Hooker 30 March [1869]
1.6 tuff.] *point over dash before del illeg*
2.1 collected.] *after del* '& Bird'
2.3 is] *interl*
4.5 here] *interl*

To C. F. Claus 31 March 1869
1.5 in Leipzig.—] *interl*
1.6 to you.] *interl*

To Benjamin Dann Walsh 3 April [1869]
0.1 Beckenham] *after del* '*Bromley.*'

To W. C. Tait 18 April [1869]
1.3 for a week] *interl in CD's hand*
2.1 Book] *interl in CD's hand*

To W. B. Tegetmeier 18 April [1869]
5.1 of the Fowls] *interl in CD's hand*
5.2 telling . . . have them] *interl in CD's hand*
7.1 P.S. . . . me.— 7.3] *in CD's hand*

To George Charles Wallich 18 April [1869]
0.1 Beckenham.] *before del* '*Bromley*'
2.4 also] *interl*
2.5 dark] *interl*

To J. Noordhoek Hegt 19 April 1869
2.6 of Nat. Hist] *interl in CD's hand*

To J. N. Hegt 27 April [1869]
0.1 Beckenham] *interl before del* '*Bromley*'
1.5 rate of] *interl*

To George Maw 27 April [1869]
0.1 Beckenham] *before del* '*Bromley.*'

To John Murray 29 April [1869]
0.1 Beckenham] *before del* '*Bromley.*'
1.2 Wallace] 'W' *over* 'L'

To James Paget 29 April [1869]
0.1 Beckenham] *before del* '*Bromley.*'
5.1 the] *after del illeg*
5.2 naked] *interl*

To Max Schmidt [29 April 1869]
1.2 that] *after del* 'but whether correctly I know not,'
1.2 sexes.—] *before del* '& perhaps the young'
1.3 whether this] 'the' *above del illeg*
1.3 species] *after del* 'colour'
1.4 *almost or quite*] *interl*
1.4 Is not] *below del* 'I believe'
1.5 *[much]*] *after del* 'in'
1.5 a larger] *after del* 'whether'
1.6 Some . . . the ♂. 1.7] *interl*
1.7 The description] *below del* 'I find that a'
1.7 of the colour] *after del* 'in *[work]* of but that'
1.8 Mandrills . . . be] *interl*
1.9 in . . . sexes 1.10] *interl above del* 'are'
1.10 with the] *after del* 'in ♂ & ♀,'
1.11 asking . . . favour,] *above del* 'troubling you'
1.11 vainly] *after del* 'been'
1.12 If you . . . ZG. 1.13] *interl*
1.14 *[Cologne]*] *after del* '*[Female]* of'
2.2 if you are] *after del* 'could you *[illeg]*'

To Charles Lyell 4 May [1869]
0.1 Beckenham.] *before del* '*Bromley.*'
1.1 (carte de visite)] *interl*
1.2 of Hon. members] *interl*

2.1 satisfied] *second* 's' *over* 'f'

To T. H. Farrer 6 May [1869]
0.1 Beckenham] *before del* '*Bromley.*'
2.1 & still dislike] *interl*
2.2 V. canina] *after del* 'the other species'

To J. D. Hooker [before 7 May 1869]
1.1 as Dutch] *interl*

To George Bentham 10 May [1869]
0.1 Beckenham] *before del* '*Bromley.*'
2.8 attached] *interl*

To Linnean Society, President and Council [10 May 1869]
3.1 To ... Soc.] *in CD's hand*

To Charles Lyell 20 May 1869
2.4 chief] *interl in CD's hand*
6.1 My . . . affectionately] *in CD's hand*

To George Maw 22 May [1869]
0.1 Beckenham] *before del* '*Bromley.*'

To Asa Gray 1 June [1869]
0.1 Beckenham] *before del* '*Bromley.*'
4.4 references] *before del comma*
5.3 soul] *before del comma*
7.3 Beard & Hair] *interl after del* 'the'

To Stephen Paul Engleheart 2 June [1869]
1.2 especially . . . labour] *interl*
1.3 round] *interl above del* 'of'
1.5 from suffering 1.6] *interl*
1.7 the nose 1.8] *interl above del* 'it'
1.8 very] *added*

To J. V. Carus 21 June [1869]
1.4 by] *after del* 'on'
3.2 which] *over* '—'

To J. D. Hooker 22 June [1869]
1.2 , as] *after del* 'that'
4.4 from him] *interl*
6.2 eyes] *after del illeg*

To A. R. Wallace 25 June [1869]
2.4 very] *interl*
2.4 Expression.—] 'E' *over* 'e'
2.5 months] *after del* 'tim'
4.1 to Origin] *interl*
8.1 your] *after del* 'the'

To Edward A. Smith [before 30 June 1869]
5.1 XI] *in CD's hand above del* '11'
5.2 3 figures 5.3] *interl in CD's hand*
5.3 close together] *interl in CD's hand*
5.3 , &] *after del* 'with fig 3a'
6.1 Four . . . block 6.2] *in CD's hand*
8.1 on same block] *interl in CD's hand*
8.2 runs] *after* 'from r' *interl and del in CD's hand*
8.2 from r] *interl in CD's hand*
9.1 Cut off top of.] *in CD's hand, pencil*
13.1 Down . . . called for 14.1] *in CD's hand*; 'The *Plates* . . . called for' *in pencil*

To Edward A. Smith 30 June [1869]
5.1 to] *after del* 'as'
6.2 lately] *first* 'l' *over* 'b'

To W. B. Tegetmeier 2 July [1869]
1.1 tell] *over illeg*
2.1 home,] *comma over illeg*

To W. E. Darwin 7 [July 1869]
3.2 injuriously] *interl*
4.2 crawled] *below del* 'crawed'
4.2 & flexible] *interl*
4.3 but] *interl above del comma*

To William Ogle 7 July [1869]
1.2 out,] *interl*
5.1 private] *interl*

To T. H. Huxley 9 July [1869]
2.1 to] *after del* 'for' *and over* 'N'
2.3 with . . . success] *interl*

To G. R. Gray 12 July 1869
1.1 in the . . . manner 1.2] *interl*

To Albany Hancock 17 July [1869]
1.7 last] *after del* 'on'

To Fritz Müller 18 July [1869]
1.5 all] *interl*
1.10 Some . . . appeared.] *added*
2.3 some] *above del* 'any'
2.5 to which place] *above del* 'where'
3.7 & . . . will] *above del* 'when I'
5.1 the beginning of] *interl*

To W. B. Tegetmeier 21 July [1869]
5.1 soon] *interl*
5.2 any Cuckoo] 'any' *interl*
5.2 any Spangled] 'any' *interl*
6.1 let me hear &] *interl above del* 'you'
6.3 formerly] *interl*
Enclosure

5.2 crescent] *after del illeg*

To J. D. Hooker 24 July [1869]
2.7 cured] *above del* 'cured'
3.1 North] *after del* 'B'
3.9 a] *interl*
3.11 faced] *above del* 'faced'
3.11 the] *interl*
4.1 still] *interl*
5.2 ever] *interl*
6.1 does] *after del* 'is'

To Walter Elliot 3 August [1869]
0.1 Beckenham] *before del* '*Bromley.*'

To John Lubbock 3 August 1869
2.1 (& . . . woodcut)] *interl*
2.2 ♂] *interl*
4.2 do you know?] *interl*
4.3 or forceps] *interl*
5.1 the sentence] 'the' *interl*

To M. C. Lloyd 4 August [1869]
0.1 Beckenham.] *before del* '*Bromley.*'

To J. D. Hooker 7 August [1869]
0.1 Beckenham] *before del* '*Bromley.*'
1.1 which] *interl*
1.2 at] *interl*
1.2 of wheat] *interl*
1.3 quickly] *above del* 'quickly'
2.2 formerly] *interl*
3.1 &] *over del* 'w'
3.4 the author] *above del* 'he'
6.2 of strength.] *interl*
7.3 I] *after del* 'if'
8.1 to read] *interl above del* 'for'
9.2 stupidly] *interl*
9.3 (*but...otherwise*)] *parentheses over commas*
10.1 the recent] *interl*
10.1 would . . . it? 10.2] *interl*

To T. H. Farrer 10 August [1869]
0.1 Beckenham] *before del* '*Bromley.*'
1.3 free] *above del* 'free'
2.1 before Linnean Soc. 2.2] *interl*
4.5 apparently] *interl*
5.1 of not 5.2] 'of' *over illeg*
5.2 use] *over* 'as'
5.2 —a mere . . . outside,—] *interl*
5.5 some Frenchman says] *above del* 'I suspect'
6.2 —] *after del illeg*

To Federico Delpino 24 and 25 August [1869]
0.1 Beckenham] *before del* '*Bromley.*'
1.3 for] *over* 'but'
1.4 I am] 'I' *after del* 'my'
3.2 honouring] 'ho' *over* 'co'

To Fritz Müller 8 September [1869]
0.1 Beckenham] *before del* '*Bromley.*'
2.1 dimorphic 2.2] *interl*
3.1 (N.B. . . . Californica) 3.2] *square brackets in MS*
3.3 produced] *interl*
3.3 number)] 'm' *added*
6.2 yours] *above del* 'yours'

To J. D. Hooker 10 September [1869]
0.1 Beckenham] *before del* '*Bromley.*'
1.4 N. B] *interl*

To Edward Blyth 14 September [1869]
0.1 Beckenham] *before del* '*Bromley.*'
2.4 have] *interl*

To Samuel Hubbard Scudder 15 September [1869]
0.1 Beckenham] *before del* '*Bromley.*'

To J. D. Hooker 18 [September 1869]
0.1 Beckenham] *before del* '*Bromley.*'
2.3 the surrounding] 'the' *interl*
2.4 this] *interl*

To John Murray [after 18 September 1869]
0.1 Beckenham] *before del* '*Bromley.*'
4.1 later in the] *above del* 'in'

To Albert Günther 21 September 1869
Enclosure
0.1 for Dr Günther] *in CD's hand*
2.2 the female,] 'the' *interl in CD's hand*
3.4 must] *in CD's hand above del* 'may'
5.2 males] 'ales' *in CD's hand*
5.2 female;] 'emale' *in CD's hand*
11.1 (9) . . . females?] *after and below del* 'Ch. Darwin.'; *in CD's hand*

To T. H. Huxley 21 September [1869]
0.1 Beckenham] *before del* '*Bromley.*'
1.1 found] *after del* 'travelled &'
1.3 from U. States] *interl*
1.4 learn] *above del* 'know'
1.6 in Nat. Hist 1.7] *interl*

To [?] 21 September [1869]
0.1 Beckenham] *before del* '*Bromley.*'

To Albert Günther 25 September 1869
0.1 Beckenham] *before del* '*Bromley.*'

2.2 can ever] 'can' *interl*
3.1 the] *interl*

To Ellen Frances Lubbock [26 September 1869?]
0.1 *Kent.*] *after del* '*Bromley.*'

To Albert Günther 27 September [1869]
0.1 Beckenham] *before del* '*Bromley.*'

To W. B. Tegetmeier 28 September [1869]
0.1 Beckenham] *before del* '*Bromley.*'
1.2 **very much**] *interl*
1.4 stations] *interl*

To Victor Masson [after 29 September 1869]
1.2 5th] *above del* 'last'
1.2 I] *over del* '8'
1.3 I] *after del* 'I'
1.4 ought] *above del* 'were bound'
1.5 all] *interl*
2.2 to print ... pages] *interl*
2.3 made] *after del illeg*
2.3 some of extended length] *interl*
2.4 that ... you] *added*
2.4 on this,] *before del* 'w^d be useless'
2.5 the] *over del* '& I'

To T. H. Huxley 1 October [1869]
0.1 Beckenham] *before del* '*Bromley*'
2.1 your] *after del illeg*
5.1 He] *after del illeg*

To W. B. Tegetmeier 1 October [1869]
0.1 Beckenham] *before del* '*Bromley.*'

To J. D. Hooker 2 October [1869]
2.3 of lots] 'of' *interl*

To Ernst Haeckel 3 October [1869]
0.1 Beckenham] *before del* '*Bromley.*'
2.2 stands as] 'as' *over del* 'at' *interl*
2.3 next to be read.] *interl*

To James Orton 7 October [1869]
0.1 Beckenham] *before del* '*Bromley.*'
2.2 The] *before del illeg*

To T. Thompson 9 October 1869
1.1 County] *interl*
1.2 means] *after del* 'plan'
1.2 payment] *after del* 'amount'
1.2 which ... is] *interl pencil*
1.2 is] *after* 'are' *pencil del pencil*
2.1 condition] *pencil above del pencil* 'state'
2.1 is unable] *above del* 'cannot'
2.2 to address] 'to' *over illeg*
2.3 I] *after del* '&'
2.4 payment,] *after del* 'their allowance'
2.4 correct] *interl pencil after del pencil* 'the case' *interl*
2.5 any one member] *above del* 'all'
2.5 may] *interl*
2.5 being refused.] *above del* 'not equally receiving their right.—'
3.1 W Reeves'] *above del* 'his demand'
3.1 claim.] *under del* 'his demand'

To T. H. Farrer 10 October [1869]
0.1 Beckenham] *before del* '*Bromley.*'
1.1 been] *interl*
2.1 a] *after del* 'sum'
4.3 collecting] *interl*
5.2 & ... winter.— 5.3] *added*

To T. H. Huxley 14 October 1869
1.2 (*requiring no answer*)] *parentheses over commas in CD's hand*
6.1 Good ... you.—] *in CD's hand*

To J. B. Innes 18 October 1869
0.1 Beckenham] *before del* '*Bromley.*'
1.8 I] *after del* 'Expect'
3.3 occasionally] *interl*
4.3 though] *after del* 'so'

To T. H. Farrer 20 October [1869]
0.1 Beckenham] *before del* '*Bromley.*'
1.3 , a splendid species,] *interl*
1.5 or holes] *interl*
3.2 an] *interl*

To A. R. Wallace 21 October 1869
1.11 not yet read] *interl in CD's hand*

To J. V. Carus 22 October [1869]
0.1 Beckenham] *before del* '*Bromley.*'
2.7 in Germany] *interl*
2.8 to me] *interl*

To J. J. Moulinié 23 October [1869]
0.1 Beckenham] *before del* '*Bromley.*'
3.7 & additions] *interl*
3.12 French] *interl*

To ? 30 October [1869 or 1870]
0.2 Oct. 30th.] *in CD's hand*
3.2 had reason to] *interl in CD's hand*
3.2 suspect] *altered from* 'suspected' *in CD's hand*
3.3 the reversed] 'the' *interl in CD's hand*

To Charles Lyell [3 November 1869]
2.1 long endurance] *above del* 'permanence'
2.2 Origin—] 'O' *over* 'o'
2.7 carrying] *above del* 'denuding'

To Antoinette Louisa Brown Blackwell 8 November [1869]
1.4 to public] *after blotted illeg*

To John Murray 8 November [1869]
0.1 Down Beckenham Kent] *above del* '6 Queen Anne S^{t}'
4.1 you] *over* 'I'
4.1 of translation] *interl*

To J. V. Carus 9 November [1869]
0.1 Beckenham] *before del* '*Bromley.*'

To Frederick Smith 9 November [1869]
0.1 Beckenham.] *before del* '*Bromley.*'

To J. D. Hooker 13 November [1869]
0.1 Beckenham] *before del* '*Bromley.*'
3.1 sometime to hear] *interl*
10.1)] *over* '.—'

To J. J. Moulinié 15 November [1869]
0.1 Beckenham] *before del* '*Bromley.*'

To Christian Gottfried Ehrenberg 18 November [1869]
0.1 Beckenham] *before del* '*Bromley.*'

To J. D. Hooker 19 November [1869]
0.1 Beckenham] *before del* '*Bromley.*'
1.5 Murchison] *above del* 'he'
2.1 I] *after del* 'do'
3.2 by] *after del* 'on'
3.6 the formation] 'the' *over* 'their'
3.6 of a species.] *interl*
3.6 almost] *interl*
3.7 just] *after del illeg*
3.9 newer] *interl*
3.9 inhabiting rather] *interl above del* 'into'
4.2 whose] 'se' *over* 'm'
4.2 2^{d}. Edit] *interl*
4.7 her] *interl*
4.8 at once] *interl*
4.9 Competition] 'C' *over* 'c'

To George Cupples 20 November [1869]
0.1 Beckenham] *before del* '*Bromley.*'

To J. D. Hooker 24 November [1869]
0.1 Beckenham] *after del* '*Bromley.*'
2.1 A.] *interl*
2.1 Die . . . Innsbruck.— 2.2] *inserted*
2.4 probably] *interl*

To George Bentham 25 November [1869]
0.1 Beckenham] *before del* '*Bromley.*'
1.3 that] *interl*
1.5 further] *interl*
1.8 slightly dissentient] *interl*
1.10 (some . . . period) 1.11] *interl*
2.1 fully] *interl*
5.4 greater or less] *interl*
6.5 This is... But 6.6] *interl*
6.6 does] 'd' *over* 'D'
6.7 partly] *over del* 'partly'
6.7 arid or dry] *over del illeg*

To Fritz Müller 1 December [1869]
1.3 into the nectary] *interl*
3.1 from Germany] *interl*
4.10 537] *after del* '33'
Enclosure 1
3.3 projections] *interl in CD's hand*
3.4 species] *interl in CD's hand*
5.1 3. Are ... dependent?] *added in CD's hand*
Enclosure 2
2.2 Two plants] 'plants' *interl in CD's hand*
2.4 seed.] *square bracket added in CD's hand*
3.1 8] *square bracket added in CD's hand*
5.3 plant] *interl in CD's hand*

To A. R. Wallace 5 December [1869]
0.1 Beckenham] *before del* '*Bromley.*'
2.1 of our] 'of' *interl*
3.3 with me] *interl*
3.5 you] *interl*
5.2 most,] *added*

To Francis Galton 23 December [1869]
0.1 Beckenham] *before del* '*Bromley.*'
1.6 these] *over* 'them'
1.6 latter chapters,] *interl*
3.2 much] *interl*

BIOGRAPHICAL REGISTER

This list includes all correspondents and all persons mentioned in the letters and notes that the editors have been able to identify. Dates of letters to and from correspondents are given in chronological order. Letters to correspondents are listed in roman type; letters from correspondents in italic type; third-party letters are listed in italic type with the name of the recipient given in parentheses.

Abdülaziz (1830–76). Sultan of Turkey, 1861–76. (*EB.*)

Acland, Thomas Dyke (1787–1871). Politician and philanthropist. Resided at Killerton, Broad Clyst, near Exeter, in Devon. Succeeded as tenth baronet, 1794. Tory MP for Devon, 1812–18, 1820–30; North Devon, 1837–57. Supported Catholic emancipation. (*ODNB.*)

Agassiz, Alexander (1835–1910). Swiss-born zoologist, oceanographer, and mining engineer. Son of Louis Agassiz. Emigrated to the United States in 1849. Joined the US survey of Washington Territory boundaries in 1859. Superintendent, Calumet copper mine, Michigan, 1867; later president of the corporation. Curator of the Harvard Museum of Comparative Zoology, 1874. (*DAB.*)

Agassiz, Anna Russell (1839/40–73). Daughter of George Russell, a Boston textile merchant. Married Alexander Agassiz in 1860. Gave birth to three sons in 1862, 1866, and 1871. (Dobbs 2005.)

Agassiz, Jean Louis Rodolphe (Louis) (1807–73). Swiss-born zoologist and geologist. Professor of natural history, Neuchâtel, 1832–46. Emigrated to the United States in 1846. Professor of zoology and geology, Harvard University, 1847–73. Established the Museum of Comparative Zoology at Harvard in 1859. Foreign member, Royal Society of London, 1838. (*ANB*, *DAB*, *DSB*, *Record of the Royal Society of London.*)

6 July 1869

Aitchison, William (1800/1–73). Farmer. Of Linhope and Brieryhill, near Hawick in the Scottish borders. Involved with the Hawick Farmers' Club and West Teviotdale Agricultural Society. First president of the Teviotdale Farmers' Club, 1859–73. (Douglas Scott, *A Hawick word book*, 9 October 2007, http://www.astro.ubc.ca/people/scott/book.pdf, p. 17.)

16 June 1869 (to George Cupples), *25 June 1869* (to George Cupples), *30 June 1869* (to George Cupples)

Alembert, Jean le Rond d' (1717–83). French mathematician, physicist, and philosopher. (*DSB.*)

Alexander II (1818–81). Tsar of Russia, 1855–81. (*EB.*)

Alglave, Emile (1842–1928). French economist and editor. Studied palaeography and law at the Ecole des Chartes, Paris, and received a doctorate in law, 1868. Professor, faculty of law, Douai, 1870; taught political economy, Lille, 1873; professor, scientific finance, faculty of law, Paris, 1878. Co-founder and director of the journals *Revue des cours scientifiques* and *Revue des cours littéraires*, 1864–80. Worked mainly in the area of tax reform. (*DBF.*)

22 October 1869

Allman, George James (1812–98). Botanist and zoologist. Professor of botany, Dublin University, 1844; regius professor of natural history, Edinburgh University, 1855–70. FRS 1854. (*ODNB.*)

Anderson, John (1833–1900). Zoologist and ethnologist. Younger brother of Thomas Anderson (1832–70). Studied medicine and zoology at Edinburgh University (MD 1862). In charge of the Indian Museum at Calcutta, 1865–86. Attached as naturalist to the Yunnan (China) expeditions, 1867–8 and 1875–6, and an expedition to the Mergui archipelago, Burma, 1881–2. Returned to England in 1886, wintering in Egypt from 1891, and studying Egyptian zoology. FRS 1879. (*ODNB.*)

Anderson-Henry, Isaac (1800–84). Scottish lawyer and horticulturalist. Practised as a solicitor in Edinburgh from 1834 or earlier. Retired from practice in 1861, when his wife succeeded to the estates of Woodend, Perthshire, and, as a condition of the entail, took the additional name Henry. Established a garden in Edinburgh known for its many rare plants. Collected plants from the Andes, north-western Himalayas, and New Zealand. Interested in plant hybridisation and acclimatisation. Member of various scientific societies in London and Edinburgh. President of the Botanical Society of Edinburgh, 1867–8. (*County families* 1878; *Curtis's botanical magazine dedications*, pp. 175–6; *Gardeners' Chronicle*, 22 March 1873, p. 399; *Post Office Edinburgh directory* 1834–84 (1861–84 s.v. Henry, Isaac Anderson); *Proceedings of the Linnean Society of London* (1886–7): 42–4.)

Andersson, Nils Johan (1821–80). Swedish botanist. Explored the Americas on the *Eugenie* expedition, 1851–3. Lectured on botany at Lund University, 1855. (*SBL.*)

Angelico, Fra (1387–1455). Italian painter and Dominican friar. (*EB.*)

Annals and Magazine of Natural History

23 July 1869

Anon.

6 April [1869–71], 23 June 1869, 21 September [1869], 30 October [1869 or 1870]

Appleton, Charles Edward Cutts Birchall (1841–79). Philosopher and journal editor. Lecturer in philosophy at St John's College, Oxford, from 1867. Founded the *Academy: a Monthly Record of Literature, Learning, Science, and Art*; editor, 1869–79. (*ODNB.*)

D. Appleton & Co. New York publishing house. Founded by Daniel Appleton (1785–1849) in 1831. His son William Henry Appleton (1814–99) was taken into

D. Appleton & Co, cont.
partnership in 1838. American publishers of works by CD and Herbert Spencer. (*ANB.*)
1 February 1869 (to Asa Gray)

Athenæum
19 June 1869, 7 [July] 1869

Atkinson, Edward (1819–1915). Scholar and university administrator. BA, Cambridge, and fellow, Clare College, 1842; tutor, 1850; master 1856–1915. Ordained priest, 1846. Lecturer in classics. Vice-chancellor, 1862, 1868, 1869, 1876, 1877. (*Alum. Cantab.*)

Audubon, John James La Forest (John James) (1785–1851). American ornithologist and illustrator. His published illustrations, such as those in *The birds of America* (1827–38), introduced him to the scientific community in Europe. (*DAB, DSB.*)

Axell, Johan Severin (1843–92). Swedish botanist. Studied at Uppsala University, 1861–9; DPhil., 1869. Lecturer in botany, 1869–70. Wholesale dealer, 1870–91. (*Taxonomic literature. Supplement.*)

Ayres, Philip Burnard (1813–63). Surgeon and botanist. Pupil of John Lindley; studied cryptogams. Practised medicine in Thame, Oxfordshire, for ten years. Lecturer on chemistry at Charing Cross Hospital; physician to the Islington dispensary, 1851. Edited the *Pharmaceutical Times.* Superintendent of quarantine on Mauritius, 1856–63. (R. Desmond 1994, *Modern English biography.*)

Azara, Félix d' (1746–1811). Spanish explorer and army officer. Sent to South America to survey Spanish and Portuguese territories. Published works on the fauna of Paraguay. (*NBU.*)

Babbage, Charles (1791–1871). Mathematician and pioneer in the design of mechanical computers. FRS 1816. (*DSB, ODNB.*)

Babington, Charles Cardale (1808–95). Botanist, entomologist, and archaeologist. Involved in natural history activities at Cambridge for more than forty years; an expert on plant taxonomy. A founding member of the Cambridge Entomological Society and the Cambridge Antiquarian Society. Editor of *Annals and Magazine of Natural History* from 1842. Chairman, Cambrian Archaeological Association, 1855–85. Professor of botany, Cambridge University, 1861–95. FRS 1851. (*DNB, DSB.*)

Babington, Churchill (1821–89). Clergyman and scholar. BA, Cambridge, 1843; MA, 1846. Perpetual curate, Horningsea, Cambridgeshire, 1848–61; rector, Cockfield in Suffolk, 1866. Disney Professor of archaeology, Cambridge, 1865–80. (*ODNB.*)

Baer, Karl Ernst von (1792–1876). Estonian zoologist and embryologist. Professor of anatomy at Königsberg University, 1819; professor of zoology, 1826–34. Professor of zoology at the Academy of Sciences, St Petersburg, 1834–67. Demonstrated the existence of the mammalian egg, 1826. Propounded the influential view that embryological development proceeds from the general to the specific. Foreign member, Royal Society of London, 1854. (*DSB*, *NDB*, *Record of the Royal Society of London*.)

Bain, Alexander (1818–1903). Scottish philosopher and psychologist. Graduated from Marischal College, Aberdeen, 1840; stayed as assistant and teacher in philosophy. Assistant secretary, metropolitan sanitary commission of the Board of Health, London, 1846. Lecturer at Bedford College, 1851, with other posts following. Professor of logic, University of Aberdeen, 1860. Founded the journal *Mind* in 1876; editor until 1892. Considered a founder of modern psychology. (*ODNB*.)

Baker, Charles Newcomb. Bird and live-animal dealer with premises at 3 Half-moon Passage, Gracechurch Street, London. (*Post Office London directory* 1869.)

Baker, John Gilbert (1834–1920). Botanist. Draper in Thirsk, Yorkshire, 1847–64. Active in the Thirsk Natural Historical Society. Assistant in the herbarium at the Royal Botanic Gardens, Kew, 1866–90; keeper of the herbarium and library, 1890–9. Lectured on botany at the London Hospital Medical School, 1869–81, and at the Chelsea Physic Garden, 1882–96. Contributed to a wide range of colonial floras. Authority on ferns. FRS 1878. (*DSB*, *ODNB*.)

Balfour, Arthur James, first earl of Balfour (1848–1930). Prime minister and philosopher. At Trinity College, Cambridge, 1866–9. Prime minister, 1902–5. FRS 1888. (*ODNB*.)

Barber, Mary Elizabeth (1818–99). British-born naturalist, artist, and writer in South Africa. Sister of James Henry Bowker. Emigrated to South Africa with her family in 1820. Married Frederick William Barber, a chemist, in 1845. Studied birds, moths, reptiles, and plants, and corresponded with a number of leading scientists, providing them with specimens and drawings. Published a number of scientific papers. (*DSAB*, *ODNB*.)

Barkly, Henry (1815–98). Colonial administrator. Served as governor of British Guiana, 1849–53; Jamaica, 1853–6; Victoria, 1856–63; Mauritius, 1863–70; and Cape Colony, 1870–77. Sent plants to the Royal Botanic Gardens, Kew. Investigated the ferns of Jamaica and Mauritius and its dependencies. Knighted, 1853. FRS 1864. (Gunn and Codd 1981, *ODNB*, *Proceedings of the Royal Society of London* 75 (1905): 23–5.)

Barkly, Anne Maria (1837–1932). Botanical illustrator. Daughter of Thomas Pratt. Second wife of Henry Barkly, whom she married in 1860. Made drawings of plants collected by her husband in South Africa, 1870–7; some of these are at the Royal Botanic Gardens, Kew. Published on the ferns of South Africa. (Gunn and Codd 1981, p. 88, *ODNB* s.v. Barkly, Henry.)

Barr, J. G. R. Dog breeder. Of Longsight House, Manchester. Correspondent of George Cupples. Left Manchester for the continent due to ill health *circa* 1869. (*Correspondence* vol. 16, letter from J. G. R. Barr to George Cupples, [after 11 May 1868?]; this volume, letter from George Cupples, 11 March 1869.)

Barron, Archibald F. (1835–1903). Gardener. Superintendent of the Royal Horticultural Society gardens at South Kensington and Chiswick. (R. Desmond 1994.)

Bartlett, Abraham Dee (1812–97). Taxidermist and zoo superintendent. Taxidermist, *circa* 1834–52. Superintendent of the natural history department, Crystal Palace, 1852–9; of the Zoological Society's gardens, Regent's Park, 1859–97. (*Modern English biography*, *ODNB*.)

Bate, Charles Spence (1819–89). Dentist and scientific writer. Practised dentistry in Swansea, 1841–51, then in Plymouth. Secretary of the Plymouth Institution, 1854–60; president, 1861–2 and 1869–70. Authority on the Crustacea. FRS 1861. (*ODNB*.)

1 February 1869, 1 March 1869, 31 July 1869, 29 November 1869, 11 December 1869

Bates, Henry Walter (1825–92). Entomologist. Undertook a joint expedition to the Amazon with Alfred Russel Wallace, 1848–9; continued to explore the area, after Wallace returned to England, until 1859. Provided the first comprehensive scientific explanation of the phenomenon subsequently known as Batesian mimicry. Published an account of his travels, *The naturalist on the River Amazons*, in 1863. Assistant secretary, Royal Geographical Society of London, 1864–92. President, Entomological Society of London, 1868, 1869, and 1878. FRS 1881. (*DSB*, *ODNB*.)

8 June 1869

Beck, John (1801–73). Banker. Baptised by the Reverend George Augustus Case at the Unitarian Church, High Street, Shrewsbury, Shropshire, 20 July 1801. Attended Case's day-school in Shrewsbury with Erasmus Alvey Darwin, then Shrewsbury School, leaving in 1816. Married Susanna Margaret Badger in the parish church of Ellesmere, Shropshire, in 1824; their three children were also baptised in the Church of England. A partner in the Shrewsbury and Welshpool Bank, Shrewsbury; retired between 1829 and 1831. By 1841 he was living apart from his family. Left an estate valued at under £100. (Parish records, Shropshire Records and Research Centre (Shrewsbury High St. baptism register, 20 July 1801; Ellesmere marriage register, 30 March 1824; Shrewsbury St Julian baptism register, 12 March 1825; Meole Brace baptism register, 30 September 1829, 1 April 1831; Will of John Beck, 1873/257); Death certificate, district of St Chad, Shrewsbury, 15 February 1873; Census returns 1841 and 1851 (Public Record Office HO 107/925/65 and HO 107/1992/749); letter from John Beck, 6 October 1864; *Commercial directory for Shropshire*; *Shrewsbury School register*.)

15 May 1869

Becker, Lydia Ernestine (1827–90). Leading member of the women's suffrage movement, botanist, and astronomer. Published *Botany for novices* (1864); awarded a Horticultural Society Gold Medal, 1865. Founder and president of the Manchester Ladies' Literary Society, 1867. Secretary to the Manchester National Society for Women's Suffrage from 1867. Member of the Manchester School Board, 1870. Editor of and regular contributor to the *Women's Suffrage Journal* from 1870. Secretary to the London central committee of the National Society for Women's Suffrage from 1881. (Blackburn 1902, *Journal of Botany* 3 (1865): 164, *Macmillan dictionary of women's biography*, *ODNB*.)

13 January 1869, 14 October 1869, 29 December 1869

Beddoe, John (1826–1911). Physician and anthropologist. Began medical practice in Bristol in 1857. Physician, Bristol Royal Infirmary, 1862–73; consulting physician, Bristol Children's Hospital, 1866–1911. Began active researches on ethnology during an extended European tour prior to setting up practice. An authority on the physical characteristics of living European races. Founder member of the Ethnological Society. President of the Anthropological Society, 1869–70. FRS 1873. (*DNB*.)

21 August 1869, [22 August – 12 September 1869], 13 September 1869

Bell, Charles (1774–1842). Anatomist and surgeon. Best known for his investigations of the nervous system and the expression of emotions in humans. Illustrated his own works. Co-owner of and principal lecturer at the Great Windmill Street School of Anatomy, London, 1814–25. Surgeon at the Middlesex Hospital, 1812–36. Professor of surgery, Edinburgh University, 1836. Knighted, 1831. FRS 1826. (*DSB*, *ODNB*, *Record of the Royal Society of London*.)

Bennet, Charles Augustus, 6th earl of Tankerville (1810–99). Styled Lord Ossulston, 1822–59. MP for North Northumberland, 1832–59. Became sixth earl in 1859. Queen Victoria's lord steward of the household, 1867–8. (*Burke's peerage* 1999, *Modern English biography*.)

Bennett, Alfred William (1833–1902). Botanist, bookseller, and publisher. Proprietor and editor of the *Friend*, the monthly journal of the Society of Friends. Lecturer in botany, Bedford College, London, 1868; Westminster Hospital, London, 1869–73. Author of several papers on pollination, 1871–3, and other botanical works. Biological sub-editor for *Nature*, 1871–4. Vice-president of the Microscopical Society, 1892, 1899, and 1900; editor of the society's journal, 1897–1902. Vice-president of the Linnean Society of London, 1891–2. (R. Desmond 1994; *Journal of the Royal Microscopical Society* 1879–1902; *List of the Linnean Society of London*, 1891–2; *Medical directory* 1869–73; *ODNB*.)

22 April 1869, 3 May 1869, 8 November 1869 (to *Nature*)

Bentham, George (1800–84). Botanist. Moved his botanical library and collections to the Royal Botanic Gardens, Kew, in 1854, and was provided with facilities there for his research from 1861. President of the Linnean Society of London, 1861–74. Published *Genera plantarum* (1862–83) with Joseph Dalton Hooker. FRS 1862. (*DSB*, *ODNB*.)

7 May 1869, 10 May [1869], *10 May 1869*, *23 November 1869*, 25 November [1869], *26 November 1869*

Bernard, Claude (1813–78). French physiologist. Professor of general physiology, Faculty of Sciences, Paris, 1854; Muséum d'Histoire Naturelle, 1868. Professor of medicine, Collège de France, 1855. Made significant discoveries in digestion, neurology, and other aspects of animal physiology. Contributed to the philosophy of experimental methodology. (*DSB*.)

Bernoulli, Carl Gustav (1834–78). Swiss physician, naturalist, and coffee planter. Studied medicine at the University of Basel and elsewhere; practised in Basel

Bernoulli, Carl Gustav, cont.
until 1858. Resided in Guatemala, where he owned a coffee plantation, 1858–78. Collected primarily botanical specimens. (Cambranes 1985, pp. 121, 131, *Taxonomic literature. Supplement.*)

Bethell, Richard, 1st Baron Westbury (1800–73). Legal reformer and lord chancellor. Graduated in classics and mathematics, Oxford University, 1819. Called to the bar as a member of the Middle Temple, 1820. Practised in the equity courts from 1823. MP for Aylesbury, 1851. Appointed vice-chancellor of the duchy of Lancaster, 1851. Became solicitor-general, 1852; attorney-general, 1856; lord chancellor, 1861–5. Took the title Baron Westbury of Westbury in 1861. Knighted, 1852. (*ODNB.*)

Bidard, Monsieur (*fl.* 1860–80). French botanist. Director of the Laboratory of Agricultural Chemistry of Seine Inferieure. (*Catalogue de la Bibliothèque Nationale.*)

Birkbeck, William Lloyd (1806–88). Scholar. BA, Cambridge, and fellow, Trinity College, 1830. Called to the bar, 1833. Downing Professor of laws, 1860–85. (*Alum. Cantab.*)

Bishop, Henry Rowley (1786–1855). Composer. Wrote opera and music for the stage, notably for plays of Shakespeare. (*ODNB.*)

Blackwall, John (1790–1881). Zoologist. Specialised in the study of spiders. Author of the *History of the spiders of Great Britain and Ireland* (1861–4). Fellow of the Linnean Society, 1827. (*ODNB.*)
10 August 1869, 8 September 1869

Blackwell, Antoinette Louisa Brown (1825–1921). American reformer, cleric, and author. Attended Oberlin College, 1846–50. Became an independent lecturer, speaking for women's rights, temperance, and the abolition of slavery. Ordained minister, Congregational Church of South Butler, N.Y., 1853. Left the pastorate in 1854, owing to religious doubts. Co-founded the American Woman Suffrage Association, 1869, and other reforming organisations. Author of at least eight books on science, religion, and philosophy, 1869–1902. Minister, American Unitarian Association, from 1878. Minister emeritus, All Souls Church, Elizabeth, N.J., 1908. Honorary AM, Oberlin College, 1878; DD, 1908. (*ANB.*)
8 November [1869]

Blyth, Edward (1810–73). Zoologist. Druggist in Tooting, London, *circa* 1832–7. Wrote and edited zoological works under the pseudonym Zoophilus. Curator of the museum of the Asiatic Society of Bengal, Calcutta, India, 1841–62. Provided CD with information on the plants and animals of India in correspondence between 1855 and 1858. Returned to Britain in 1863, and continued to write on zoology and on the question of the origin of species. (*Correspondence* vols. 5–7, *DSB*, *ODNB.*)
1 March 1869, 17 April 1869, 26 April 1869, 13 September 1869, 14 September [1869]

Boardman, Alexander F. (1819–76). American dry goods merchant. Born in Haiti. Educated in New Hampshire schools; attended Bowdoin College, but left because of weak eyes. Established a large and successful dry goods business in

Brunswick, Maine, in 1840. (Wheeler and Wheeler 1878.)
23 June 1869

Boccardo, Gerolamo (1829–1904). Italian economist, lawyer, and statesman. Professor of political economy, faculty of law, Genoa, 1860; senator, 1877; state council member, 1888. Applied a 'Darwinian' model to economics. (*DBI*, Tort 1996.)
15 June 1896

Bodinus, Karl August Heinrich (Heinrich) (1818–84). German zoologist. Studied medicine, natural science, and zoology at the University of Greifswald, 1833–6. Physician in Bergen (Rügen) and Greifswald. Appointed to construct the Cologne Zoological Garden in 1859; senior director of Berlin Zoological Garden, 1869. (*DBE*.)

Bojer, Wenzel (1797–1856). Bohemian botanist and explorer. Explored islands off the east coast of Africa for the Austrian government; returned with a large herbarium in 1821. Returned to Mauritius after receiving a state pension; founder and later vice-president of the Mauritius natural history society. Published a major work on the flora of Mauritius (1837). (*OBL*.)

Boner, Charles (1815–70). Journalist, translator, and poet. Tutored the two eldest sons of John Constable, 1831–7. Employed by the family of Prince Thurn und Taxis in Regensburg, Germany, 1840–60. Translated several German works, wrote verse, articles, and books; travelled as a naturalist and huntsman. Settled in Munich in 1860. (*ODNB*.)
25 November [1869], [December 1869 – early January 1870]

Bonney, Thomas George (1833–1923). Geologist. BA, Cambridge (St John's College), 1856. Ordained priest, 1858. Fellow of St John's, 1859; tutor, 1868. College lecturer in geology, 1869–81. Yates-Goldsmid Professor of geology, University College, London, 1877–1901. FRS 1878. (*ODNB*.)

Bory de Saint-Vincent, Jean Baptiste Georges Marie (1778–1846). French army officer and naturalist. Leader of several botanical collecting expeditions. Zoogeographer with a particular interest in the faunas of islands. Editor of the *Dictionnaire classique d'histoire naturelle*, 1822–31. (*DBF*, *DSB*.)

Bouché, Carl David (1809–81). German botanist. Inspector (technical director), Royal Botanic Garden, Berlin, 1843–81. (Leuenberger 2004, p. 310.)

Bowerbank, James Scott (1797–1877). London distiller and geologist. Had a special interest in London Clay fossils; devoted his later career to the study of sponges. Founder of the Palaeontographical Society, 1847. FRS 1842. (*ODNB*, *Record of the Royal Society of London*, Sarjeant 1980–96.)

Bowman, William (1816–92). Ophthalmic surgeon. Author of works on physiology. Assistant surgeon, King's College Hospital, London, 1839–56; surgeon, 1856–62; elected joint professor of physiology and of general and morbid anatomy, King's College, London, 1848; assistant surgeon at the Royal London Ophthalmic Hospital, Moorfields, 1846–51; surgeon, 1851–76. Created baronet, 1884. FRS 1841. (*DSB*, *ODNB*.)
3 September [1869]

Bradshaw, Henry (1831–86). Librarian and scholar. BA (King's College),

Bradshaw, Henry, cont.
Cambridge, 1854; fellow, 1853. Librarian, University Library, Cambridge, 1867–86. (*ODNB*.)

Brandt, Johann Friedrich von (1802–79). German zoologist and palaeontologist. Studied medicine at the University of Berlin, 1821–6. Taught medical botany and pharmacology there, 1829–31. Appointed to the Academy of Sciences, St Petersburg, 1831; director of the Zoological Museum until 1879. Professor of zoology and comparative anatomy at the Medical Surgical Academy, St Petersburg, 1851–69. (*DBE*.)

Bree, Charles Robert (1811–86). Medical practitioner and zoologist. Medical officer in the Polish army, 1831; qualified to practise medicine in Britain, 1833. MD, Edinburgh, 1859. General practitioner, Colchester, Essex, and physician, Essex and Colchester Hospital, 1860–82. Entomological editor of the *Naturalist*, 1858. Published articles critical of CD's theory of the origin of species. (*CDEL*, Freeman 1978, *Medical directory* 1847–87.)

Brewer, Thomas Mayo (1814–80). American ornithologist, journalist, and publisher. Practised medicine briefly after graduating from Harvard Medical School in 1838. Editor of the Boston *Atlas* and Washington correspondent for the paper during the 1840s and 1850s. Became a partner in the publishing firm Swan & Tileston, 1857; later head of Brewer & Tileston, a position he held until his retirement in 1875. Worked on birds' eggs and contributed to bird guides and lists. (*ANB*.)

Bridgman, William Kencely (1812–84). Dentist and naturalist. Practised as a surgeon-dentist in Norwich. (*NUC*, *Post Office directory of Cambridge, Norfolk, and Suffolk* 1865.)

[before 21 January 1869] (to J. G. Jeffreys), 21 January [1869], *23 January 1869*

Bright, John (1811–89). Manufacturer, statesman, and orator. Joined his father's cotton business in the late 1820s. A leader, with Richard Cobden, of the Anti-Corn-Law League. MP for Durham, 1843; Manchester, 1847–56. Denounced the Crimean war, 1854. Elected MP for Birmingham, 1857. A supporter of parliamentary reform. President of the Board of Trade, 1868–70. Chancellor of the duchy of Lancaster, 1873–4, 1880–2; retired from the cabinet, 1882. (*ODNB*.)

Bristowe, Henry Fox (1824–93). Barrister. Called to the bar, 1847; incorporated at Lincoln's Inn, 1854. Nephew of William Darwin Fox. Married Selina Bridgeman in 1850. Knighted, 1887. (*Alum. Cantab.*, *Darwin pedigree*.)

Brockhaus, Heinrich (1804–74). German publisher. Son of Friedrich Arnold Brockhaus, the founder of the Leipzig publishing firm F. A. Brockhaus Verlag, which he took over with his brother Friedrich Brockhaus after their father's death in 1823. After his brother's death in 1864, became head of the firm. (Würffel 2000.)

Brodie, Benjamin Collins (1817–80). Chemist. Awarded a doctorate at Giessen, Germany, in 1850 for his analysis of beeswax. Waynflete Professor of chemistry, Oxford, 1865–72. Devised an alternative approach to chemistry from that based on atomic theory, and developed a calculus of chemical operations. Created baronet, 1862. FRS 1849. (*Alum. Oxon.*, *DNB*, *DSB*.)

Bronn, Heinrich Georg (1800–62). German palaeontologist. Professor of natural science at Heidelberg University, 1833. Translated and superintended the first German editions of *Origin* (1860) and *Orchids* (1862). (*DSB*, *NDB*.)

Brooke, James (1803–68). Soldier and colonial governor. East India Company military service, 1819–30; served in the Burma War. At the invitation of Muda Hassim, uncle of the Malay sultan of Brunei and nominal ruler of Borneo, became raja of Sarawak, 1841. Commissioner and consul-general to the sultan and the independent chiefs in Borneo, 1847–55, and governor of the island of Labuan and its dependencies, 1847–52. Knighted, 1848. (*Foreign Office list* 1856, *ODNB*.)

Brotero, Félix de Avellar (1744–1828). Portuguese botanist. Chair of botany and agriculture, Coimbra, 1791; director of the botanic gardens and royal museum, 1800. Director, botanic gardens, Ajuda, 1811. First botanist to describe native Portuguese plants. (*Grande encyclopédie*.)

Brougham, Henry Peter, 1st Baron Brougham and Vaux (1778–1868). Scottish-born statesman, lawyer, reformer, and educationalist. A founder of the *Edinburgh Review*, 1802, the Society for the Diffusion of Useful Knowledge, 1826, and London University, 1826. Prolific author. Lord chancellor, 1830–4. A founder of the National Association for the Promotion of Social Science, 1857. Elected chancellor of Edinburgh University, 1859. Created Baron Brougham and Vaux, 1830. FRS 1803. (*ODNB*, *Record of the Royal Society of London*.)

Browning, Oscar (1837–1923). Teacher and historian. BA (King's College), Cambridge, 1860; MA, 1863; fellow 1859–1923. Assistant master, Eton, 1860–75. Lecturer, King's College, Cambridge, 1880; university lecturer, 1884. (*ODNB*.)

Browning, Robert (1812–89). Poet. (*ODNB*.)

Brusina, Spiridion (1845–1908). Croatian conchologist, stratigrapher, and invertebrate palaeontologist. Director of the National Museum, Zagreb. (Sarjeant 1980–96.)

29 April 1869, 8 May [1869]

Buckle, George (1819/20–1900). Clergyman. BA (Corpus Christi College), Oxford, 1842; MA (Oriel College), 1845. Deacon, 1846; priest, 1848. Vicar of Twerton-on-Avon, Somerset, 1852–76. Rector, Weston-super-Mare, 1876–88. Prebendary of Wells, 1868; canon, 1887; precentor, 1888. (*Alum. Oxon.*, *The Times* 4 January 1900, p. 8.)

Buckler, William (1814–84). Artist and entomologist. Studied at the Royal Academy; became a portrait artist. Illustrated entomological works, devoting most of the last twenty-five years of his life to *Larvae of the British butterflies and moths*, which was published posthumously in nine volumes between 1886 and 1901. (*ODNB*, Salmon 2000, pp. 157–9.)

[after 8 June 1869]

Bunbury, Charles James Fox, 8th baronet (1809–86). Botanist. Collected plants in South America, 1833–4; in South Africa, 1838–9. Accompanied Charles

Bunbury, Charles James Fox, 8th baronet, cont.
Lyell to Madeira in 1853. Justice of the peace and deputy lieutenant for Suffolk; high sheriff, 1868. Succeeded to the baronetcy, 1860. FRS 1851. (*County families* 1875, R. Desmond 1994, Sarjeant 1980–96.)

Bunge, Alexander von (1803–90). Russian-born German botanist. MD, Dorpat (Tartu), 1825. Collected plants in North China and Altai, Russia. Professor extraordinarius of botany, University of Kasan (Uzbekistan), 1833. Professor of botany and director of the botanic gardens, Dorpat, 1836–67. (*ADB*.)

Burmeister, Karl Hermann Konrad (1807–92). German zoologist, ethnographer, and geologist. Published extensively on entomology. Professor of zoology, University of Halle, 1837–61. Travelled in Brazil, 1850–2; Argentina, 1857–60. Director of the Museo Nacional in Buenos Aires, 1861–80. (Berg 1895, *NBU*, Sarjeant 1980–96.)

Burn, Robert (1829–1904). Clergyman, scholar, and archaeologist. Published on the archaeology, literature, and art of ancient Rome. Fellow of Trinity College, Cambridge, 1854–1904; tutor, 1856–72; praelector in Roman archaeology, 1873–85. Ordained priest, 1862. (*Alum. Cantab*, *ODNB*.)

Busk, George (1807–86). Russian-born naval surgeon and naturalist. Served on the hospital ship at Greenwich, 1832–55. Retired from medical practice in 1855. Member of several scientific societies. President of the Microscopical Society, 1848–9; of the Anthropological Institute, 1873–4. Zoological secretary of the Linnean Society of London, 1857–68. Hunterian Professor of comparative anatomy, Royal College of Surgeons of England, 1856–9; council member, 1863; member of board of examiners, 1868; president, 1871. Specialised in palaeontology and in the study of Bryozoa. FRS 1850. (*DNB*, *DSB*, Plarr 1930.)
7 November 1869

Buxton, Charles (1822–71). Politician. BA, Cambridge, 1845; MA, 1850. MP for Newport, 1857; Maidstone, 1859; East Surrey, 1865–71. (*ODNB*.)

Buxton, Hannah (1783–1872). Daughter of John Gurney of Earlham Hall and wife of Thomas Fowell Buxton, whose country seat was Northrepps Hall, Norfolk. Mother of Charles Buxton. (*ODNB* s.v. Buxton, Charles, and Buxton, Thomas Fowell.)

Cambridge, Octavius Pickard (1828–1917). Anglican clergyman and zoologist. Ordained 1858; succeeded his father as rector of Bloxworth, Dorset, 1868. An expert on spiders, on which he published numerous papers and monographs. FRS 1887. (*Proceedings of the Royal Society* series B, 91 (1920): xlix–liii.)

Cameron, Julia Margaret (1815–79). Photographer. Born in India, but spent much of her childhood in England and France. Became a friend of John Herschel's in South Africa, where she went in 1835 to convalesce from an illness. Married Charles Hay Cameron in Calcutta in 1838, and acted as a social organiser and hostess for the governor-general of India, Lord Henry Hardinge, in the 1840s. The family returned to England in 1848, and moved to Freshwater, Isle of Wight, in 1860. Cameron took up photography in 1863; was elected to the

Photographic Society of London, 1864, and organised her first solo exhibition in 1865. She exhibited regularly in Europe. Emigrated to Ceylon (Sri Lanka) with her husband in 1875. (*ODNB*.)

Campbell, Archibald (1805–74). Surgeon and administrator. Assistant surgeon, East India Company, 1827. Superintendent of the Darjeeling station, where he implemented social and economic improvements and developed profitable tea plantations, 1840–62; responsible for political relations between the British and the Sikkim government. Wrote many papers on Himalayan geography. Travelled with Joseph Dalton Hooker in Sikkim in 1849, and was imprisoned with Hooker by the Sikkim rajah. (Balfour 1885, R. Desmond 1994, J. D. Hooker 1854, L. Huxley ed. 1918, W. B. Jackson 1854.)

Campbell, Dudley (1833–1900). Barrister. MA, Cambridge, 1855. Student at Lincoln's Inn, 1853; called to the bar, 1859. Member of the south-eastern circuit. (*Alum. Cantab.*)
29 July [1869]

Campbell, George Douglas, 8th duke of Argyll (1823–1900). Scottish statesman and author of works on science, religion, and politics. A defender of the concept of design in nature. Chancellor of St Andrews University, 1851–1900. President of the Royal Society of Edinburgh, 1860–4. Privy seal, 1852–5, 1859–60, 1860–6, and 1880–1; postmaster-general, 1855–8 and 1860; secretary of state for India, 1868–74. Succeeded to the dukedom in 1847. FRS 1851. (*ODNB*.)

Candolle, Alphonse de (1806–93). Swiss botanist, lawyer, and politician. Active in the administration of the city of Geneva until 1860. Responsible for the introduction of postage stamps to Switzerland. Professor of botany and director of the botanic gardens, Geneva, from 1835. Concentrated on his own research after 1850. Foreign member, Royal Society of London, 1869. (*DSB*, *Record of the Royal Society of London*.)
28 September 1869, 11 October 1869

Candolle, Augustin Pyramus de (1778–1841). Swiss botanist. In Paris, 1796–1808; professor of botany, Ecole de Médecine and Faculté des Sciences, Montpellier, 1808–16; professor of natural history, Academy of Geneva, 1816–35. Foreign member, Royal Society of London, 1822. (*DSB*, *Record of the Royal Society of London*.)

Canestrini, Giovanni (1835–1900). Italian zoologist. Studied philosophy and natural sciences at Vienna, receiving his degree in 1861. Professor of natural history, Modena, 1862–9. Professor of zoology and comparative anatomy and physiology, Padua, 1869–1900. Translated the first Italian edition of *Origin* (1864) in association with Leonardo Salimbeni. Worked mainly in ichthyology and later on human origins. (*DBI*, Pancaldi 1991.)

Cardew, Philip (1851–1910). Army officer and electrical engineer. Passed first into the Royal Military Academy, Woolwich, 1868. Commissioned lieutenant in the Royal Engineers, 1871. Invented the hot-wire galvanometer (voltmeter), and the vibratory transmitter for telegraphy. Captain, 1883; major, 1889; from 1882, instructor in electricity, Chatham (School of Military Engineering). Appointed

Cardew, Philip, cont.

the first electrical adviser to the Board of Trade, 1889. Upon retirement, entered into partnership with Sir William Preece & Sons, consulting engineers. Served on the council of the Institution of Electrical Engineers. (*ODNB*.)

Carpenter, William Benjamin (1813–85). Naturalist. Fullerian Professor of physiology at the Royal Institution of Great Britain, 1844–8; physiology lecturer, London Hospital, 1845–56; professor of forensic medicine, University College, London, 1849–59. Registrar of the University of London, 1856–79. President of the British Association for the Advancement of Science, 1872. Founding member of the Marine Biological Association. FRS 1844. (*DNB*, *DSB*, *Modern English biography*, Royal Institution of Great Britain.)

Carrière, Elie Abel (1818–96). French botanist and horticulturalist. Started work at the Muséum d'Histoire Naturelle at the age of 14 and rose to the rank of head perennial grower. Head of the botanic garden, Zaragossa, Spain, in the early 1860s; head of the nursery, Muséum d'Histoire Naturelle, 1862–9. Editor of the *Revue Horticole* from 1862; editor-in-chief, 1869–96. Worked on peach trees, dimorphism, and hybridity. (Barnhart comp. 1965, *Revue Horticole* (1896): 389–97, Tort 1996.)

Carus, Julius Victor (1823–1903). German comparative anatomist. Conservator of the Museum of Comparative Anatomy, Oxford University, 1849–51. Professor extraordinarius of comparative anatomy and director of the zoological museum, University of Leipzig, 1853. Translated the third German edition of *Origin* (1867) and, subsequently, twelve other works by CD. (*DSB*, *NDB*.)

5 January 1869, *22 January 1869*, *2 February 1869*, 5 February 1869, *29 April 1869*, 4 May 1869, *3 June 1869*, 21 June [1869], *20 October 1869*, 22 October [1869], [30 October 1869], 30 October 1869, *6 November 1869*, 9 November [1869], 20 November 1869

Caton, Charles Edward (b. 1843). Son of John Dean Caton. (*IGI* (accessed 12 March 2008).)

Caton, John Dean (1812–95). American jurist, author, and naturalist. Opened Chicago's first law office, 1833. Admitted to the Illinois bar. Appointed associate justice of the supreme court of Illinois, 1842, eventually serving as chief justice in 1855, 1857–64. Owned a large property in Ottawa, Illinois. Travelled widely in the United States, Europe, and Asia. Wrote several books, including *The antelope and deer of America* (1877). (*DAB*.)

20 March 1869, *5 May 1869*, 24 May 1869

Cavendish, William, 7th duke of Devonshire and 2d earl of Burlington (1808–91). Statesman and benefactor of scientific and industrial enterprises. Whig MP for Cambridge University, 1829–31; for Malton, Yorkshire, 1831; for Derbyshire, 1831; for North Derbyshire, 1832–4. Succeeded to the earldom, 1834; to the dukedom, 1858. Chancellor of London University, 1836–56; of Cambridge University, 1861–91. President of the Royal Agricultural Society, 1870. Privy councillor, 1876. Knighted, 1858. FRS 1829. (*DNB*, *Record of the Royal Society of London*.)

Cecil, James Brownlow William Gascoyne-, 2d marquess of Salisbury (1791–1868). Nobleman. MA (Christ Church), Oxford, 1813; DCL, 1834. Lord president of the Council, 1858–9. (*Burke's peerage*, *ODNB* s.v. Stanley, Mary Catherine.)

Cecil, Mary Catherine Gascoyne-, countess of Derby (1824–1900). Political hostess. Daughter of George Sackville-West, fifth Earl De La Warr (1791–1869). Married James Brownlow William Gascoyne-Cecil in 1847. After his death, married Edward Henry Stanley in 1870. Deeply involved in conservative politics. (*ODNB* s.v. Stanley, Mary Catherine, countess of Derby.)

Challis, James (1803–82). Astronomer and physicist. Ordained, 1830. Plumian professor of astronomy and experimental philosophy, Cambridge University, 1836–82. FRS 1848. (*ODNB*.)

Chambers, Robert (1802–71). Publisher, writer, and geologist. Partner, with his brother William Chambers (*ODNB*), in the Edinburgh publishing company W. & R. Chambers. Joint editor of *Chambers's Journal* from 1832. Anonymous author of *Vestiges of the natural history of creation* (1844). (*DSB*, *ODNB*.)

Chapman, John (1822–94). Physician, author, and publisher. Studied medicine in Paris, 1842. Publisher and bookseller at 142 Strand until 1860. Editor and proprietor of the *Westminster Review*, 1851–94. MD, St Andrews, 1857; practised as a physician. Advocated the application of an ice-bag to the spine as a remedy particularly for diarrhoea and cholera. (*ODNB*.)

Chapman, Thomas Algernon (1842–1921). Scottish physician and entomologist. MD, Glasgow University. Resident physician and surgeon, Glasgow Royal Infirmary. Assistant Medical Officer of the Joint Counties Asylum, Abergavenny. Medical superintendent of the City and County Asylum, Hereford. Retired in 1896. Continued entomological pursuits, travelling extensively in Europe. Specialised in life histories of the Lycaenidae. FRS 1918. (*Proceedings of the Royal Society of London* Series B 94 (1923): xxx–xxxi.)

Christison, Robert (1797–1882). Scottish physician. Professor of medical jurisprudence, Edinburgh University, 1822–32; of clinical medicine, 1832–55; of materia medica and therapeutics, 1832–77. President of the Royal College of Physicians of Edinburgh, 1839 and 1848. President of the Royal Society of Edinburgh, 1868–73. President of the British Medical Association, 1875. Physician-in-ordinary to the queen in Scotland, 1848. Created baronet, 1871. (*DNB*.)

Church, Richard William (1815–90). Author and clergyman. BA (Wadham College), Oxford, 1836; MA, 1839. Fellow of Oriel College, 1838–54. Rector of Whatley, Somerset, 1852–71. Dean of St Paul's, 1871–90. Created DCL, 1875. (*Alum. Oxon.*, *ODNB*.)

Claparède, Jean Louis René Antoine Edouard (Edouard) (1832–71). Swiss naturalist and invertebrate zoologist. Professor of zoology and comparative

Claparède, Jean Louis René Antoine Edouard (Edouard), cont.
anatomy, Academy of Geneva, 1862. Specialised in invertebrate anatomy, histology, embryology, and evolution. One of the first Swiss naturalists to endorse CD's theory. (*Dictionnaire historique & biographique de la Suisse*, Gilbert 1977.)

Clark, John Willis (1833–1910). University administrator. BA (Trinity College), Cambridge, 1856; elected fellow, 1858. Superintendent of the Zoological Museum, Cambridge University, 1866–91. (*ODNB.*)

Clarke, Alexander Ross (1828–1914). Geodesist. Graduated from Royal Military Academy, Woolwich, and commissioned as a second lieutenant in the Royal Engineers, 1847. Posted to the Ordnance Survey in Southampton, 1850; chief mathematical and scientific adviser, 1856–81. In 1860 began a comparison of different national standards of length. FRS 1862. (*ODNB.*)

Claus, Carl Friedrich (1835–99). German zoologist. Studied medicine, mathematics, and zoology at Marburg and Giessen, 1854–7. Professor of zoology, Marburg, 1863; Göttingen, 1870. Professor of zoology and head of the institute of zoological and comparative anatomy at Vienna, 1873. Founder and first director of the zoological research station at Trieste, 1873. Did major research in environmental influences on variability, especially in Crustacea. A strong supporter of CD in both his writing and lecturing. His zoology textbook was a standard work in the last three decades of the nineteenth century. (*DBE*, *NDB*, *OBL.*)
24 January 1869, 28 January 1869, *6 February 1869*, *19 March 1869*, 31 March 1869

Clifford, William Kingdon (1845–79). Mathematician and philosopher of science. BA (Trinity College), Cambridge, 1867; fellow, 1868; MA, 1870. Professor, applied mathematics, University College, London, 1871. Gave many public lectures on scientific topics. FRS 1874. (*ODNB.*)

William Clowes & Sons. Printers. William Clowes (1807–83), eldest son of William Clowes (1779–1847), joined his father's printing business in 1823; the name of the firm was changed to William Clowes & Sons in 1839. Printed the official catalogue of the Great Exhibition of 1851. Introduced improvements in music printing. Printers to John Murray. (*ODNB.*)

Cobbe, Frances Power (1822–1904). Writer and philanthropist. Wrote extensively on religious and ethical subjects. Leading campaigner for women's rights and against animal vivisection. (*ODNB.*)

Cobbold, Thomas Spencer (1828–86). Physician and zoologist. Studied medicine at Edinburgh University, graduating in 1851; appointed curator of the university anatomical museum. Moved to London in 1857; appointed botany lecturer at St Mary's Hospital. Obtained a similar position at the Middlesex Hospital in 1861, lecturing on zoology and comparative anatomy. Established a medical practice in London as a consultant in cases where internal parasites were suspected. Swiney Lecturer in geology at the British Museum, 1868–73. From 1873, professor of botany at the Royal Veterinary College, which shortly afterwards instituted a helminthology professorship for him. FRS 1864. (*ODNB.*)
9 August [1869], *1 September [1869]*

Colenso, William (1811–99). English-born New Zealand botanist and missionary. Worked with the London firm of Richard Watts, printer to the Church Missionary Society, 1833. Left to run a printing press for the CMS in Paihia in the Bay of Islands, New Zealand, in 1834. Published Christian religious works in the Maori language and official publications in English and Maori. Ordained deacon, 1844, and went with his family to a new mission station in Hawke's Bay. Had a son by a Maori woman in 1851, and was suspended as deacon and dismissed from the mission, 1852. Elected to the Provincial Council, 1859, and the General Assembly, 1861–6. Published scientific papers and was commissioned by the General Assembly to produce a Maori dictionary, which he worked on until his death. FRS 1886. (*DNZB*, *Record of the Royal Society of London*.)

Colvin, Walter Mytton (1847–1908). Lawyer. LLB, Cambridge, 1870. Admitted to the Middle Temple, 1866; called to the bar, 1871. Practised criminal law in Allahabad; member of the government of India police commission of 1902–3. Knighted, 1904. (*Alum. Cantab.*, *ODNB* s.v. Colvin, Sir Auckland.)

Comte, Isidore Auguste Marie François Xavier (Auguste) (1798–1857). French philosopher. Private secretary to Claude Henri de Rouvroy, comte de Saint-Simon, 1817–23. Founded the Association Polytechnique, a group devoted to the education of the working classes, in 1830; the Société Positiviste, devoted to the promulgation of the 'Cult of Humanity', in 1848. Adopted the term 'positivism' for his philosophy. (*DSB*.)

Congreve, Richard (1818–99). Philosopher. Ordained 1843. Fellow of Wadham College, Oxford, 1844–54. The leading proponent in England of the positivist philosophy of Auguste Comte. Supported legalisation of trades unions, Irish independence, and parliamentary reform. Member, Royal College of Physicians, 1866. (*ODNB*.)

Cooke, Robert Francis (1816–91). Publisher. Partner in John Murray's publishing company, which published many of CD's books. Cousin of John Murray. (*Correspondence* vol. 13, letter from John Murray, 1 April 1865, *Modern English biography*.)

[22 February 1869], *22 February 1869*, *23 February 1869*, 23 [February 1869], 9 March [1869], *10 March 1869*, *12 March 1869*, *29 April 1869*, *22 June 1869*, 23 June [1869], 2 July [1869], *2 November 1869*

Corrie, George Elwes (1793–1885). College head. BA, Cambridge, and fellow, St Catharine's College, 1817. Tutor until 1849. Master of Jesus College, 1849–85. (*Alum. Cantab.*, *ODNB*.)

Craven, William Augustus Frederick (Viscount Uffington) (1838–65). Captain, Grenadier Guards, 1862. (*Complete peerage*.)

Cresson, John Chapman (1806–77). American agriculturalist. Worked as a farmer until 1834, then moved to Philadelphia. Member of the Franklin Institute from 1831; professor of mechanics and natural philosophy from 1837; president from 1855. Superintendent of the gas works, 1836–64. Vice-president, American Philosophical Society, 1857–77. (*Proceedings of the American Philosophical Society* 17 (1877–8): 149–62.)

Crichton-Browne, James (1840–1938). Scottish physician, psychiatrist, and

Crichton-Browne, James, cont.
author. Graduated from the College of Surgeons, Edinburgh, in 1862. Medical director, West Riding Asylum, Wakefield, 1866–75. Lord chancellor's visitor in lunacy, 1875–1922. Pioneering researcher into the causes and management of mental illness. Editor, *West Riding Lunatic Asylum medical reports*, 1871–5. One of the founders of *Brain*, 1878; co-editor, 1878–85. Author of a number of popular works on mental health and social policy. Knighted, 1886. FRS 1883. (*DNB*.)

22 May 1869, *1 June 1869*, 8 June 1869

Croll, James (1821–90). Scottish geologist. Keeper, Andersonian Museum, Glasgow, 1859. In the 1860s began publishing papers on physical geology. Secretary to the Geological Survey of Scotland, 1867; retired after suffering a mild stroke in 1880. Until his death, wrote papers and books on cosmology, on oceanic circulation patterns, and on climate change and the causes of the glacial epoch. FRS 1876. (*ODNB*.)

31 January [1869], *4 February 1869*, 6 February [1869], *15 February 1869*, 24 February [1869], *23 June 1869*

Crüger, Hermann (1818–64). German pharmacist and botanist. Apothecary in Trinidad in the West Indies from 1841; government botanist and director of the botanic garden, Trinidad, from 1857. Collected plants in Jamaica, Trinidad, and Venezuela. (R. Desmond 1994, S[chlechtenda]l 1864.)

Cruickshank, Anthony (b. 1812). Breeder of shorthorn cattle at Sittyton farm, Aberdeenshire. With his brother Amos, developed a type of cattle that dominated both British and North American herds by the latter half of the century. (Derry 2003, pp. 28–30; *IGI* (accessed 12 March 2008).)

Cunningham, Robert Oliver (1841–1918). Naturalist. MD Edinburgh. Collected plants in South America. Naturalist to HMS *Nassau* on an expedition to the Strait of Magellan and the west coast of Patagonia, 1866–9. Professor of natural history, Belfast, 1871–1902. (Cunningham 1871, R. Desmond 1994, Newmann comp. 1993.)

Cupples, Anne Jane (1839–98). Scottish author. Second daughter of Archibald Douglas. Married George Cupples in 1858. Wrote children's books. Lived in New Zealand from 1891. (*Modern English biography*.)

Cupples, George (1822–91). Scottish writer and dog breeder. Served as an apprentice on an eighteen-month voyage to India and back on the *Patriot King*, *circa* 1838; had his indentures cancelled on his return. Studied arts and theology at Edinburgh University for eight years. Published a number of novels and other books, and wrote many articles and stories for journals. Bred Scottish deer-hounds. (*Correspondence* vol. 16, letter from George Cupples, 1 May 1868, Cupples 1894, *Modern English biography*.)

4 January 1869, [6–9? January 1869], *11 January 1869*, *15 January 1869*, *21 January 1869*, *11 March 1869*, *22 March 1869*, *1 June 1869*, *20 June 1869*, *13 September 1869*, *[4–10 November 1869]*, 20 November [1869]

Cuvier, Clémentine (1804/5–27). Daughter of Georges Cuvier. (*DSB* s.v. Cuvier, Georges.)

Cuvier, Jean Léopold Nicolas Frédéric (Georges) (1769–1832). French systematist, comparative anatomist, palaeontologist, and administrator. Professor of natural history, Collège de France, 1800–32; professor of comparative anatomy, Muséum d'Histoire Naturelle, 1802–32. Permanent secretary to the Académie des Sciences from 1803. Foreign member, Royal Society of London, 1806. (*DBF*, *DSB*.)

Dallas, William Sweetland (1824–90). Entomologist, author, and translator. Prepared lists of insects for the British Museum, 1847–58. Curator of the Yorkshire Philosophical Society's museum, 1858–68. Assistant secretary to the Geological Society of London, 1868–90. Translated Fritz Müller, *Für Darwin* (1869); prepared the index for *Variation* and the glossary for *Origin* 6th ed. Editor, *Annals and Magazine of Natural History*, 1868–90, *Popular Science Review*, 1877–80. (Freeman 1978, *Geological Magazine* n.s. decade 3, vol. 7 (1890): 333–6, *Modern English biography*, Sarjeant 1980–96.)

Dalton, John (1766–1844). Chemist and natural philosopher. Originator of the chemical atomic theory. (*DSB*, *ODNB*.)

Dana, James Dwight (1813–95). American geologist and zoologist. Geologist and mineralogist with Charles Wilkes's expedition to the South Seas, 1838–42; wrote reports on the geology, zoophytes, and Crustacea. An associate editor of the *American Journal of Science and Arts* from 1846. Professor of natural history, Yale University, 1855–64; professor of geology and mineralogy, 1864–90. Foreign member, Royal Society of London, 1884. (*ANB*, *DSB*, *Record of the Royal Society of London*.)

Danielssen, Daniel Cornelius (1815–94). Norwegian physician and zoologist. Specialist in skin diseases, especially leprosy. Director of the Bergen Museum. (*NBL*.)

Dareste, Gabriel-Madeleine-Camille (Camille) (1822–99). French zoologist. A specialist in experimental embryology. Doctor of medicine, 1847. Doctor of science, 1851. Taught natural history at various provincial institutions. Professor of zoology, University of Lille, 1864–72. Professor of ichthyology and herpetology, Muséum d'Histoire Naturelle, Paris, 1872. Director of the laboratory of teratology, later attached to the Ecole des Hautes-Etudes, 1875. Awarded the grand prize in physiology by the Académie des Sciences for *Recherches sur la production artificielle de monstruosités* (1877). (*DBF*, *Dictionnaire universel des contemporains*.)
18 November [1869], 20 November 1869, *13 December [1869]*

Darwin, Edward (1782–1829). Army officer. Eldest son of Erasmus and Elizabeth Collier Darwin. (*Darwin pedigree*.)

Darwin, Elizabeth (Bessy/Lizzy) (1847–1926). CD's daughter. (*Darwin pedigree*, Freeman 1978.)

Darwin, Emma (1808–96). Youngest daughter of Josiah Wedgwood II. Married CD, her cousin, in 1839. (*Emma Darwin* (1904) and (1915).)

Darwin, Erasmus (1731–1802). CD's grandfather. Physician, botanist, and poet. Advanced a theory of transmutation similar to that subsequently propounded by Jean Baptiste de Lamarck. FRS 1761. (*DSB*, King-Hele 1999, *ODNB*.)

Darwin, Erasmus Alvey (1804–81). CD's brother. Attended Shrewsbury School, 1815–22. Matriculated at Christ's College, Cambridge, 1822; Edinburgh University, 1825–6. Qualified in medicine but never practised. Lived in London from 1829. (*Alum. Cantab.*, Freeman 1978.)

[after 21 April 1869], 11 November [1869], 26 [November 1869 or later], 7 December [1869]

Darwin, Francis (1848–1925). CD's son. Botanist. BA, Trinity College, Cambridge, 1870. Qualified as a physician but did not practise. Collaborated with CD on several botanical projects, 1875–82. Lecturer in botany, Cambridge University, 1884; reader, 1888–1904. Published *LL* and *ML*. President of the British Association for the Advancement of Science, 1908. Knighted, 1913. FRS 1882. (*DSB, ODNB.*)

Darwin, George Howard (1845–1912). CD's son. Mathematician. BA (Trinity College), Cambridge, 1868; fellow, 1868–78. Studied law in London, 1869–72; called to the bar in 1872 but did not practise. Plumian Professor of astronomy and experimental philosophy, Cambridge University, 1883–1912. President of the British Association for the Advancement of Science, 1905. Knighted, 1905. FRS 1879. (*DSB, Men-at-the-bar, ODNB.*)

6 February [1869], *6 February 1869, 14 February 1869, [23 February 1869], [18 October 1869] , [2 December 1869]*

Darwin, Henrietta Emma (1843–1927). CD's daughter. Married Richard Buckley Litchfield (*Alum. Cantab.*) in 1871. Assisted CD with some of his work. Edited *Emma Darwin* (1904) and (1915). (*Burke's landed gentry* 1952, *Correspondence* vol. 11, Freeman 1978.)

Darwin, Leonard (1850–1943). CD's son. Military engineer. Attended the Royal Military Academy, Woolwich. Commissioned in the Royal Engineers, 1871; major, 1889. Served on several scientific expeditions, including those for the observation of the transit of Venus in 1874 and 1882. Instructor in chemistry and photography, School of Military Engineering, Chatham, 1877–82. Intelligence service, War Office, 1885–90. Liberal Unionist MP, Lichfield division of Staffordshire, 1892–5. President, Royal Geographical Society of London, 1908–11; Eugenics Education Society, 1911–28. Chairman, Bedford College, London University, 1913–20. (Margaret Keynes 1943, RMA Gentleman Cadet Register (RMAS archives, Sandhurst collection), Sarjeant 1980–96, *WWW.*)

Darwin, Robert Waring (1766–1848). CD's father. Physician. Had a large practice in Shrewsbury and resided at The Mount. Son of Erasmus Darwin (*ODNB*) and his first wife, Mary Howard. Married Susannah, daughter of Josiah Wedgwood I (*ODNB*), in 1796. FRS 1788. (Freeman 1978.)

Darwin, William Erasmus (1839–1914). CD's eldest son. Banker. BA, Christ's College, Cambridge, 1862. Partner in the Southampton and Hampshire Bank, Southampton, 1861. Chairman of the Southampton Water Company. Amateur photographer. (*Alum. Cantab.*, F. Darwin 1914.)

[January 1869], 5 July [1869], 7 [July 1869], *12 July 1869*

Dawkins, William Boyd (1837–1929). Geologist and palaeontologist. Member of the Geological Survey of Great Britain, 1861–9. Curator of natural history, Manchester Museum, 1869. Professor of geology, Owens College, Manchester, 1874–1908. Specialised in fossil mammals. FRS 1867. (*ODNB*.)
17 July 1869, 19 July [1869], *19 October 1869*

Decaisne, Joseph (1807–82). Belgian botanist. Gardener at the Jardin des Plantes, Paris, 1824. Professor of agricultural statistics, Collège de France, 1848. Professor of plant cultivation, Muséum d'Histoire Naturelle, 1850. Foreign member, Royal Society of London, 1877. (*DBF*, *NBU*.)

Delamer, Eugene Sebastian. *See* Dixon, Edmund Saul.

De la Rue, Warren (1815–89). Printer, astronomer, and chemist. Educated in Paris. Entered his father's printing firm, De la Rue & Sons. Published researches on practical chemistry and astronomy that he conducted at his private laboratory and observatory. Noted for his work on celestial photography. President of the Royal Astronomical Society, 1864–6; of the Chemical Society, 1867–9 and 1879–80. FRS 1850. (*DSB*, *ODNB*.)

Delaunay, Charles Eugène (1816–72). French astronomer and mathematician. Developed the most comprehensive lunar theory of his time. (*DSB*.)

Delpino, Federico (1833–1905). Italian botanist. Travelled extensively for botanical purposes as a youth and in 1873. Civil servant, ministry of finances, Turin, 1852–6; assistant in the botanic garden and museum, Florence, 1867; lecturer, Vallombrosa school of forestry, 1871; professor of botany and director of the botanic garden, Genoa, 1875–84; professor, University of Bologna, 1884; professor of botany and head of the botanic garden, Naples, 1894. (*DBI*, Mayerhöfer 1959–70, Penzig 1905.)
22 August 1869, 24 and 25 August [1869], *9 October 1869*, 14 October 1869, *1 November 1869*

Demosthenes (384–322 BCE). Athenian orator. (*Oxford classical dictionary*.)

Disraeli, Benjamin (1804–81). Author and Conservative politician. Prime minister, 1868, 1874–6. Created earl of Beaconsfield, 1876. (*ODNB*.)

Dixon, Edmund Saul (1809–93). Clergyman and poultry-fancier. Rector of Intwood with Keswick, Norfolk, 1842–93. Author of books on the history and management of poultry. Also published under the pseudonym Eugene Sebastian Delamer. (*Modern English biography*.)

Dohrn, Carl August (1806–92). German entomologist and businessman. Travelled to Algeria and Brazil to study language, literature, and zoology. Returned to Stettin, 1837. Headed the editorial staff of *Entomologische Zeitung*, 1843–87. Published sixteen volumes of *Linnaea entomologica*, 1846–66. Became director of a sugar refining plant in Stettin when his father died in 1852. Father of Anton Dohrn. (*DBE*.)

Dohrn, Felix Anton (Anton) (1840–1909). German zoologist. Studied medicine and zoology at various German universities. PhD, Breslau, 1865. Studied with Ernst Haeckel and became Haeckel's first assistant at Jena, where he habilitated

Dohrn, Felix Anton (Anton), cont.

in 1868. Founded the Zoological Station at Naples, built between 1872 and 1874. The station was the first marine laboratory, and served as a model for other similar institutions throughout the world. (*DBE*, *DSB*, Heuss 1991.)

25 December 1869, *30 December 1869*

Domeyko, Ignacy (1802–89). Lithuanian-born Polish geologist, mineralogist, and ethnographer. Graduated from Vilnius University in 1822. Student in Paris, 1832–6; graduated in mining engineering, Ecole des Mines, in 1837. Professor of mineralogy and chemistry, La Serena College, Coquimbo, Chile, 1838–46. Rector, University of Chile, Santiago, four times between 1867 and 1883. (Grigelas 2005.)

Donders, Frans Cornelis (Franciscus Cornelius) (1818–89). Physiologist and eye specialist. MD, Leiden University, 1840. Taught at the University of Utrecht; professor, 1852. Established a teaching and research institute in ophthalmology in 1858. Professor of physiology, 1862. (*DSB*.)

6 September 1869, *12 November 1869*, 17 November [1869]

Doran, John (1807–78). Writer and literary editor. Tutor to the sons of the nobility. Literary editor of the *Church and State Gazette* from 1841. A regular contributor to the *Athenæum*. Editor of *Notes and Queries* from 1872. A prolific author on historical and literary subjects. (*ODNB*.)

Doubleday, Edward (1811–49). Entomologist. Brother of Henry Doubleday. (*ODNB*.)

Dub, Christoph Julius (Julius) (1817–73). German teacher and physicist. Professor of physics at the Gymnasium zum Grauen Kloster, Berlin. Wrote mostly on electromagnetism. (*BHGW*.)

20 March 1869

Duberry, Amy (b. 1834/5). Dressmaker and Sunday School teacher in Down, Kent. In 1871, was head of a household consisting of her three younger brothers, Albert, Mark, and George, and her sister Elizabeth. (Census returns 1871 (Public Record Office RG/10/875); *Correspondence* vol. 16, letter from J. B. Innes, 18 June [1868]; this volume, letter to J. B. Innes, 18 October 1869.)

Duchenne, Guillaume Benjamin Amand (1806–75). French physician. Studied medicine at Paris, then practised in his native town of Boulogne-sur-Mer. Returned to Paris in 1855 and carried out experiments on the therapeutic use of electricity. One of the founders of neurology. (*DBF*.)

Duncan, J. W. (*fl.* 1860s–90s). Son of James Duncan, curator of the Botanic Garden, Mauritius, 1849–65. Presented drawings of Mauritius plants to the Royal Botanic Gardens, Kew, in 1894. (R. Desmond 1994, s.v. Duncan, James.)

Duncan, James (1802–76). Botanist. Curator, Botanic Garden, Mauritius, 1849–65. (R. Desmond 1994.)

Duncker, Alexander Friedrich Wilhelm (Alexander) (1813–97). German publisher. Took over the retail side of the publishing firm Duncker & Humblot in 1837 and established his own publishing firm, Alexander Duncker Verlag, which he directed until his death. (Würffel 2000.)

Duvaucel, Antoinette Sophie Laure (Sophie) (1789–1867). French aide-naturaliste. Stepdaughter of Georges Cuvier. Acted as secretary to Cuvier and worked unofficially as aide-naturaliste at the Jardin des Plantes. Married Ducrest de Villeneuve, Alexandre-Louis. (*DBF* s.v. Ducrest de Villeneuve, Alexandre-Louis; Orr 2007; Jacques Seynaeve, *Morts pour la France et militaires divers sépultures communales individuelles*, http://pagesperso-orange.fr/seynaeve/page_d.htm (accessed 9 December 2008).)

Eastlake, Elizabeth (1809–93). Journalist and writer on art. Daughter of Dr Edward Rigby. Studied literature and art in the British Museum and the National Gallery, 1832. Travelled widely and wrote travel books from the mid-1830s to the 1840s. Married Charles Lock Eastlake in 1849; became Lady Eastlake in 1850. From the 1860s, wrote mostly on art history. (*ODNB*.)

Edwards, Ernest (1837–1903). Photographer. Ran a photographic studio in London, specialising in portraits, 1864–9. Moved to Boston, and later New York, where he opened commercial firms in landscape photography and colour printing. (W. S. Johnson 1990, *Post Office London directory*, Pritchard 1994.)

Egerton, Philip de Malpas Grey-, 10th baronet (1806–81). Palaeontologist. Of Oulton Park, Cheshire. Tory MP for South Cheshire, 1835–68; for West Cheshire, 1868–81. Specialised in fossil fish. FRS 1831. (*ODNB*, Sarjeant 1980–96.)

18 February 1869

Ehrenberg, Christian Gottfried (1795–1876). German zoologist, comparative anatomist, and microscopist. Wrote extensively on infusoria and the development of coral reefs. Professor of medicine at Berlin University, 1839. Foreign member, Royal Society of London, 1837. (*DSB*, *NDB*.)

18 November [1869]

Ellice, Russell (1799–1873). Businessman. Harrow school, 1811–15. BA (Brasenose College), Oxford, 1821; MA, 1824. Lincoln's Inn, 1822. Director, East India Company, 1832–58. Partner in the banking firm Robarts, Lubbock & Co. (*Alum. Oxon.*, Census returns 1841 (Public Record Office HO107/680/24), *Harrow School register*, *Post Office London directory* 1868.)

Elliot, Robert. Scottish sheep-farmer. Farmed in Roxburgh, Selkirk, Dumfries, and Perthshire for more than forty years. (Letter from Robert Elliot to George Cupples, 4 June 1869.)

4 June 1869 (to George Cupples), *21 June 1869* (to George Cupples), *25 June 1869* (to George Cupples)

Elliot, Walter (1803–87). East India Company servant and archaeologist. Commissioner for the administration of the Northern Circars, 1845–54. Member of the council of the governor of Madras, 1854–60. Wrote articles on Indian natural history and culture. Knighted, 1866. FRS 1878. (*ODNB*, *Record of the Royal Society of London*.)

3 August [1869], *7 August 1869*, *7 September 1869*

Ellman, John (1753–1832). Agriculturalist. Tenant of Place Farm, Glynde, 1780–1829. Developed the Southdown breed of sheep; co-founder of the Lewes wool fair. Wrote on a variety of agricultural subjects. (*ODNB*.)

Ellman, John (1787–1867). Farmer. Son of John Ellman (1753–1832). Tenant of Place Farm, Glynde, from 1829. (Farrant 1978, *IGI*.)

Ellman, Robert Harvey (1816/17–93). Farmer. Son of John Ellman (1787–1867). Farmed at Landport Farm, Lewes, Sussex, after his father's death. (*BMD* (*Death index*), Census returns 1871 (Public Record Office RG10/1067/120/9).)

Emerson, Ralph Waldo (1803–82). American clergyman, essayist, and poet. (*ANB*.)

Endlicher, Stephan Ladislaus (1804–49). German botanist. (*NDB*.)

Engleheart, Stephen Paul (1831/2–85). Surgeon. Member of the Royal College of Surgeons of England, 1859; licentiate of the Royal College of Surgeons of Edinburgh, 1860. Surgeon in Down, Kent, 1861–70. Medical officer, Second District, Bromley Union, 1863–70; divisional surgeon of police, 1863–70. Resident in Shelton, Norfolk, 1870–81; in Old Calabar, Nigeria, 1882–5. (*Medical directory* 1861–86, *Post Office directory of the six home counties* 1862.)

2 June [1869]

Enzenberg, Franz von. Son-in-law of Ignaz von Tannenberg. (*BLKO* s.v. Tannenberg, Ignaz von.)

Enzenberg, Ottilia von (b. 1801). Daughter of Ignaz von Tannenberg. Married Franz von Enzenberg in 1831. (*BLKO* s.v. Tannenberg, Ignaz von.)

Eugénie (Marie Eugénie Ignace Augustine de Montijo) (1826–1920). Empress of the French. Married Napoleon III in 1853. Lived in exile in England, from 1870, at Chislehurst and Farnborough. (*DBF*, *EB* (1970).)

Evans, Margaret (b. 1831/2). Cook. Cook at Down House until about 1890. (Atkins 1974, pp. 74–5; Census returns 1861 (Public Record Office RG9/462).)

Ewbank, Lucas (1835–1916). Scholar and university administrator. BA (Clare College), Cambridge, 1857; fellow, 1858–1916. Bursar and mathematical lecturer, 1863–95. (*Alum. Cantab.*)

Eyton, Thomas Campbell (1809–80). Shropshire naturalist. Friend and Cambridge contemporary of CD. Author of several works on natural history. On coming into possession of the family estate at Eyton, Shropshire, in 1855, he built a museum for which he formed a collection of skins and skeletons of European birds. An active magistrate throughout his adult life. (*ODNB*.)

[before 23 January 1869], 24 January 1869, *26 January [1869]*, *4 March 1869*, 6 March [1869]

Eyton, William Campbell (1849–79). Son of Thomas Campbell Eyton. Matriculated, Trinity College, Cambridge, Michelmas 1868. (*Alum. Cantab.*)

Faivre, Jean-Joseph-August-Ernest (Ernest) (1827–79). French physiologist and botanist. MD, Paris, 1853; PhD, 1858. Lecturer in botany, Lille, 1858; Lyons, 1859; professor of botany, Lyons, 1860. Director, botanic gardens, Lyons, 1860. Worked on both plant and animal physiology. (*DBF*, Tort 1996.)

18 February 1869, *15 March 1869*

Falconer, Hugh (1808–65). Palaeontologist and botanist. Superintendent of the botanic garden, Saharanpur, India, 1832–42. Superintended the arrangement of Indian fossils for the British Museum in 1844. Superintendent of the Calcutta botanic garden and professor of botany, Calcutta Medical College, 1848–55. Retired owing to ill health and returned to Britain in 1855; pursued palaeontological research while travelling in southern Europe. Vice-president of the Royal Society of London and foreign secretary of the Geological Society of London, 1865. FRS 1845. (*DSB*, *ODNB*.)

Faraday, Michael (1791–1867). Natural philosopher. Apprentice to a bookbinder, 1805. Appointed chemical assistant at the Royal Institution of Great Britain, 1813; director of the laboratory, 1825; Fullerian Professor of chemistry, 1833. Noted for his popular lectures and for his extensive researches in electrochemistry, magnetism, and electricity. FRS 1824. (*DSB*, *ODNB*.)

Farrer, Frances (1833–70). Daughter of Maitland and William Erskine. Her mother, Maitland, a daughter of Sir James Mackintosh, was a half-sister of Frances Emma Elizabeth Wedgwood (1800–89). Married Thomas Henry Farrer in 1854. (*ODNB* s.v. Farrer, Thomas Henry; Wedgwood and Wedgwood 1980 (family tree).)

Farrer, Thomas Henry, first Baron Farrer (1819–99). Civil servant. BA, Oxford (Balliol College), 1840. Called to the bar, 1844; ceased to practise in 1848. Secretary of the marine department, Board of Trade, 1850, rising to sole permanent secretary of the Board of Trade, 1867–86. In 1854, married Frances Erskine; in 1873, married Katherine Euphemia Wedgwood, daughter of Frances and Hensleigh Wedgwood. Made a baronet, 1883; created Baron Farrer of Abinger, 1893. (*ODNB*, Wedgwood and Wedgwood 1980.)

15 January 1869, *5 May 1869*, 6 May [1869], *9 May 1869*, *8 August 1869*, 10 August [1869], *12 August 1869*, *18 September 1869*, *9 October 1869*, 10 October [1869], *13 October 1869*, *17 October 1869*, 20 October [1869], *27 October 1869*, *5 November 1869*, [27 November 1869], *28 November 1869*

Fechner, Gustav Theodor (1801–87). German physicist, philosopher, and psychophysicist. Studied medicine at Leipzig but never practised; MD, 1822. Professor of physics, Leipzig, 1834–9. Suffered a breakdown in 1839, resuming academic life in 1843, when he began to lecture on philosophy, ethics, and the mind–body problem. Founded the discipline of psychophysics, which he described as the scientific study of the functional relations between body and mind. Introduced statistical methods of psychophysical measurement. (*DBE*, *DSB*.)

Fenn, Robert (1817–1912). Steward at Woodstock Rectory, Oxfordshire. Later smallholder at Sulhamstead. Experimented in cross-breeding potatoes. (R. Desmond 1994.)

Fergusson, William, 1st baronet (1808–77). Scottish surgeon. Professor of surgery, King's College, London, 1840–70. Surgeon-in-ordinary to the prince consort, 1849; surgeon extraordinary to Queen Victoria, 1855. Created baronet, 1866. FRS 1848. (*ODNB*.)

Fish, David Taylor (1824–1901). Scottish-born gardener and horticultural

Fish, David Taylor, cont.
journalist. Head gardener, Broke Hall, Suffolk; Hardwick House, Bury St Edmunds, Suffolk. Later became county council lecturer for Cambridgeshire. Returned to Scotland in 1896 and continued to lecture on horticulture, especially in Edinburgh and Dundee. Edited Cassell's *Dictionary of popular gardening.* (R. Desmond 1994, *Gardeners' Chronicle* (1901) (1): 288–9)
8 May 1869 (to *Gardeners' Chronicle*)

Fitzmaurice, Edmond George Petty- (1846–1935). Politician and historian. BA (Trinity College), Cambridge, 1868; MA, 1872. Entered Parliament, 1868. Parliamentary under-secretary at the Foreign Office, 1883. Retired, 1885–98, due to ill health. Returned to Parliament, 1898–1909. (*Alum. Cantab.*, *ODNB.*)

FitzRoy, Robert (1805–65). Naval officer, hydrographer, and meteorologist. Commander of HMS *Beagle*, 1828–36. Tory MP for Durham, 1841–3. Governor of New Zealand, 1843–5. Superintendent of the dockyard at Woolwich, 1848–50. Chief of the meteorological department at the Board of Trade, 1854; chief of the Meteorological Office from 1855. Rear-admiral, 1857; vice-admiral, 1863. FRS 1851. (*DSB*, *ODNB.*)

Fitzwilliam, William Wentworth, Viscount Milton (1839–77). Politician. Entered Trinity College, Cambridge, in 1859. MP for South West Yorkshire, 1865–72. (*Alum. Cantab.*, *Complete peerage* vol. 5, *Modern English biography.*)

Fleury, Jules Joseph (*fl.* 1860s). French consul-general in London, appointed in 1860. (*The Times*, 5 September 1860, p. 5.)
23 July 1869

Flower, William Henry (1831–99). Anatomist and zoologist. Curator of the Hunterian Museum, Royal College of Surgeons of England, 1861–84; Hunterian Professor of comparative anatomy, Royal College of Surgeons, 1870–84. Director of the Natural History Museum, London, 1884–98. President of the Zoological Society of London, 1879–99. President of the British Association for the Advancement of Science, 1889. Knighted, 1892. FRS 1864. (*DNB.*)

Forbes, David (1828–76). Geologist and mining engineer. Brother of Edward Forbes. Supervised mining and metallurgical works in Norway, 1848–57. Travelled in South America, 1857–60, in search of ores of nickel and cobalt for the firm of Evans & Askins, nickel–cobalt refiners of Birmingham. After more years of travelling, he settled in England and became foreign secretary to the Iron and Steel Institute in 1869. FRS 1858. (*ODNB.*)
30 January 1869, 7 February [1869]

Forbes, Edward (1815–54). Zoologist, botanist, and palaeontologist. Naturalist on board HMS *Beacon*, 1841–2. Appointed professor of botany, King's College, London, and curator of the museum of the Geological Society of London, 1842. Palaeontologist with the Geological Survey of Great Britain, 1844–54. Professor of natural history, Edinburgh University, 1854. FRS 1845. (*DSB*, *ODNB.*)

Ford, George Henry (1809–76). South African artist. Assistant to Andrew Smith, 1821–4. Artist, Cape Town Museum, 1825; expedition for exploring Central

Africa, 1834–6. Moved to England in 1837. Artist, British Museum, 1837 until *circa* 1875. Friend of Albert Günther and illustrator of several of his works; provided illustrations for the second volume of *Descent.* (Gunther 1972, Gunther 1975.)

Fortune, Robert (1813–80). Traveller and botanist. Collected plants in China for the Horticultural Society, 1842–6. Curator of the Chelsea botanic garden, 1846–8. Collected plants and seeds of the tea-shrub in China for the East India Company from 1848 and wrote several books on the subject. (*DNB.*)

John Fox & Sons. Brewers at Farnborough, Beckenham, Kent. (*Post Office directory of the six home counties* 1870.)
12 October 1869

Fraley, Frederick (1802/3–1901). American businessman. Partner in a mercantile firm, secretary of the American Fire Insurance Company, president of the Schuylkill Navigation Company and the Western Saving Fund Society. One of the founders of the Franklin Institute, 1824. President of the National Board of Trade, 1868–1901. Secretary and vice-president of the American Philosophical Society; president, 1880–1901. (*Proceedings of the American Philosophical Society* 40 (1901): i–ix.)

Francis, William (1817–1904). Editor and publisher. Studied science at Berlin and Giessen, 1839–42. PhD in chemistry, Giessen, 1842. Editor of *Philosophical Magazine*, 1851–1904; *Annals and Magazine of Natural History*, 1859–97. Partner in the publishing firm Taylor & Francis, 1852; head of the firm, 1858. (*Annals and Magazine of Natural History*, 1859–97; Brock and Meadows 1998; *ODNB* s.v. Taylor, Richard.)

Frederick of the Netherlands (1797–1881). Prince of Orange-Nassau. Second son of King William I of the Netherlands. (*Europäischen Stammtafeln.*)

Freeman, Philip (1818–75). Anglican clergyman. Vicar of Thorverton, Devon, from 1858; archdeacon of Exeter from 1865. (*ODNB.*)

Froude, James Anthony (1818–94). Historian and writer. Disciple of Thomas Carlyle; published *History of England* (1856–70), *The English in Ireland in the eighteenth century* (1872–4), and *The English in the West Indies* (1888). Editor of *Fraser's Magazine* 1860–74. Regius professor of modern history, Oxford University, 1892–4. (*DNB.*)

Gall, Franz Joseph (1758–1828). Physician, neuroanatomist, and psychologist. His theories were applied in the form of phrenology. (*DBF*, *DSB.*)

Galton, Douglas Strutt (1822–99). Military engineer and civil servant. Officer in the Royal Engineers. Joined the Ordnance Survey in 1846. Secretary, Railway Commission, 1847; Railway Department, Board of Trade, 1854. Assistant permanent under-secretary of state for war, 1862. Director of public works and buildings, Office of Works, 1869–75. Author of numerous government reports on sanitation, telegraphy, and railways. Member, British Association for the Advancement of Science, 1860; general secretary, 1871–95; president, 1895. Knighted, 1887. FRS 1859. (*ODNB.*)

Galton, Francis (1822–1911). Traveller, statistician, and scientific writer. CD's

Galton, Francis, cont.

cousin. Explored in south-western Africa, 1850–2. Carried out various researches on heredity. Founder of the eugenics movement. FRS 1860. (*DSB*, *ODNB*.)

11 December 1869, 23 December [1869], *24 December 1869*

Galton, Isabella. Eldest daughter of Joseph Strutt of Derby, wife of John Howard Galton. (*ODNB* s.v. Galton, Douglas Strutt.)

Garbett, Edward Lacy (d. 1900). Architect and religious writer. Son of the architect Edward Garbett. Author of *Treatise on the principles of design in architecture* (1850), and of numerous religious tracts. (Colvin 1995, s.v. Garbett, William.)

29 June 1869 (to the *Athenæum*), *19 July 1869* (to the *Athenæum*)

Gardeners' Chronicle.

9 May [1869]

Gärtner, Karl Friedrich von (1772–1850). German physician and botanist. Practised medicine in Calw, Germany, from 1796, but left medical practice in 1800 to pursue a career in botany. Travelled in England and Holland in 1802. Studied plant hybridisation from *circa* 1824. Elected a member of the Deutsche Akademie der Naturforscher Leopoldina, 1826. Ennobled, 1846. (*ADB*, *DBE*, *DSB*.)

Gaudry, Albert-Jean (Albert) (1827–1908). French palaeontologist. Assistant to his brother-in-law, Alcide d'Orbigny, professor of palaeontology at the Muséum d'Histoire Naturelle. Carried out excavations at Pikermi, Attica, in 1855 and 1860, and published *Animaux fossiles et géologie de l'Attique* (1862–7). Studied the fossils of small reptiles and batrachians, 1866–92. Taught a course in palaeontology at the Sorbonne, 1868–71; appointed professor of palaeontology at the Muséum, 1872. (*DBF*, *DSB*.)

Geach, Frederick F. Mining engineer in Malaya. Worked for the Portuguese government in Timor, where he met Alfred Russel Wallace in 1861, and was working in Malacca (Melaka) in 1867. (*Correspondence* vol. 15, letter from A. R. Wallace, 2 March [1867]; Marchant ed. 1916.)

Geach, Robert Edgar (1827/8–92). Solicitor. Of Woodbridge Road, Guildford. Brother of Frederick F. Geach. (*BMD* (*Death index*), letter from A. R. Wallace, 23 June [1869], *Post Office directory of the six home counties* 1870.)

Gegenbaur, Carl (or Karl) (1826–1903). German anatomist and zoologist. A supporter of CD; emphasised the importance of comparative anatomy in evolutionary reconstruction. Professor extraordinarius of zoology, Jena, 1855–8; professor of anatomy and zoology, 1858–62; of anatomy, 1862–73. Professor of anatomy and comparative anatomy, Heidelberg, 1873–1901. Elected to the Deutsche Akademie der Naturforscher Leopoldina, 1857. (*DBE*, *DSB*, *NDB*.)

Geikie, Archibald (1835–1924). Scottish geologist. Appointed member of the Scottish branch of the Geological Survey in 1855. Director of the Geological Survey of Scotland, 1867–82. Director-general of the Geological Survey of Great Britain, 1882–1901. Murchison Professor of geology and mineralogy, Edinburgh University, 1871–81. Knighted, 1891. FRS 1865. (*DNB*, *DSB*.)

Geoffroy Saint-Hilaire, Isidore (1805–61). French zoologist. Succeeded his father, Etienne Geoffroy Saint-Hilaire, as professor at the Muséum d'Histoire Naturelle in 1841. Continued his father's work in teratology. Became professor of zoology at the Sorbonne in 1850. (*DBF*, *DSB*.)

Gilbert, Joseph Henry (1817–1901). Agricultural chemist. Collaborated with John Bennet Lawes at the Rothamsted Agricultural Station, 1843–1900. Sibthorpian Professor of rural economy, Oxford University, 1884–90. Knighted, 1893. FRS 1860. (*ODNB*.)
3 June [1869]

Gladstone, William Ewart (1809–98). Statesman and author. Chancellor of the Exchequer, 1852–5 and 1859–66. Prime minister, 1868–74. FRS 1881. (*ODNB*.)

Glenie, Samuel Owen (1811–75). Clergyman. BA, Peterhouse, Cambridge, 1829. MA, Lambeth, 1849. Appointed colonial chaplain, Colombo, Ceylon (Sri Lanka), 1834. Also appointed chaplain at Trincomalee, and at Kandy, at various times from 1840; returned to Trincomalee in 1851. Retired on pension, 1871. Fellow of the Linnean Society of London, 1863–73. (*Alum. Cantab.*, *Colonial Office list*, R. Desmond 1994.)
20 March 1869

Goethe, Johann Wolfgang von (1749–1832). German poet and naturalist. (*DSB*, *NDB*.)

Goeze, Edmund (1838–1929). German botanist. Worked in the botanic garden at Greifswald before becoming director of the botanic garden at Coimbra, Portugal. Director of the Lisbon botanic garden, 1873–6. Later head of the botanic gardens, Greifswald, Prussia. (*GEPB*, *Proceedings of the Royal Society of London* series B 98 (1925): ii.)

Gosse, Philip Henry (1810–88). Zoologist, traveller, and writer. Collected zoological specimens in the West Indies for the British Museum, 1844–6. Made his living as a writer from 1847. Lived at St Marychurch, near Torquay, Devonshire, from 1857. Cultivated orchids. Studied marine invertebrates. FRS 1856. (*DNB*.)

Gould, John (1804–81). Ornithologist and artist. Taxidermist to the Zoological Society of London, 1828–81. Described the birds collected by CD on the *Beagle* expedition. FRS 1843. (*ODNB*.)

Graham, George Augustus (1833–1909). Military officer and dog-breeder. Served in the 28th Regiment Native Infantry, 1851–64, leaving as captain. Lived after retirement at Rednock, Dursley, Gloucestershire. Revived the Irish wolfhound breed and founded the Irish wolfhound Club in 1885. (*India list*, *NUC*, *The Times*, 26 October 1909, p. 11.)

Graham, John. Scottish sheep-farmer; farmed at Lockerbie. (Letter from John Graham to George Cupples, 6 July 1869.)
6 July 1869 (to George Cupples)

Grant, Robert Edmond (1793–1874). Scottish physician and zoologist. An early supporter of the theory of the transmutation of species. Befriended CD in Edinburgh. Professor of comparative anatomy and zoology, University College, London, 1827–74. FRS 1836. (*ODNB*.)

Gratiolet, Louis Pierre (1815–65). French anatomist and anthropologist. Laboratory assistant, Muséum d'Histoire Naturelle, Paris, 1842–53; lectured on anatomy, 1844–50; director of anatomical studies, 1853–62. Deputy to the professor of zoology, Faculty of Science, Paris, 1862–3; professor, 1863–5. (*DSB*.)

Gray, Asa (1810–88). American botanist. Fisher Professor of natural history, Harvard University, 1842–88. Wrote numerous botanical textbooks and works on North American flora. President of the American Academy of Arts and Sciences, 1863–73; of the American Association for the Advancement of Science, 1872; a regent of the Smithsonian Institution, 1874–88. Foreign member, Royal Society of London, 1873. (*DAB*, *DSB*, *Record of the Royal Society of London*.)
8 and 9 May 1869 (and Jane Loring Gray), 1 June [1869], *23 August 1869*, *20 October 1869*, *26 and 29 October [1869]*

Gray, George Robert (1808–72). Zoologist. An expert on insects and birds. Assistant in the zoological department of the British Museum, 1831–72. Brother of John Edward Gray. FRS 1865. (*ODNB*.)
9 July 1869, 12 July 1869

Gray, Jane Loring (1821–1909). Daughter of Charles Greely Loring, Boston lawyer and politician, and Anna Pierce Brace. Married Asa Gray in 1848. Edited the *Letters of Asa Gray* (1893). (Barnhart comp. 1965; Dupree 1959, pp. 177–84.)
8 and 9 May 1869 (and Asa Gray)

Gray, John Edward (1800–75). Botanist and zoologist. Assistant keeper of the zoological collections at the British Museum, 1824; keeper, 1840–74. President, Botanical Society of London, 1836–57. FRS 1832. (R. Desmond 1994, *ODNB*.)

Greg, William Rathbone (1809–81). Essayist. Mill-owner, 1832–50. Author of the *Creed of Christendom* (1851). Wrote articles for the leading quarterlies, and books, mostly on politics and economics. Comptroller of the Stationery Office, 1864–77. (*ODNB*, Tort 1996.)

Grey, George (1812–98). Army officer, explorer, and colonial governor. Governor of South Australia, 1841–5; of New Zealand, 1845–53 and 1861–8; of Cape Colony, 1854–61. Settled in New Zealand, 1870–94. Prime minister of New Zealand, 1877–9. Knighted, 1848. (*Aust. dict. biog.*, *ODNB*.)

Grote, George (1794–1871). Historian. One of the founders of University College, London; vice-chancellor of the University of London, 1862–71; FRS 1857. (*DNB*.)

Grove, William Robert (1811–96). Lawyer, judge, and natural philosopher. Professor of experimental philosophy, London Institution, 1841. An active member of the Royal Society of London; founder member of the Philosophical Club, 1847; Royal Medallist, 1847. Appointed judge, Court of Common Pleas, 1871. Knighted, 1872. FRS 1840. (*DSB*, *ODNB*.)

Günther, Albrecht Carl Ludwig Gotthilf (Albert) (1830–1914). German-born zoologist. Began his association with the British Museum in 1857; made catalogues of the museum's specimens of amphibia, reptiles, and fish; officially joined the staff in 1862. Assistant keeper of the zoological department, 1872–5; keeper,

1875–95. Edited the *Record of Zoological Literature*, 1864–9. FRS 1867. (*NDB*, *ODNB*.)

21 September 1869, *23 September 1869*, 25 September 1869, *[before 27 September 1869]*, 27 September [1869]

Gunther, Robert William Theodore (1869–1940). Historian of science. Founded the Museum of the History of Science, Oxford, in 1924. Son of Albert and Roberta Günther. (*ODNB*.)

Günther, Roberta (1843–69). Natural history artist. Daughter of the baillie John M'Intosh of St Andrews and Elizabeth Mitchell. Met Albert Günther at the British Association for the Advancement of Science meeting in Dundee in 1867, and married him in 1868. Died after giving birth to her son, Robert William Theodore Gunther. (*IGI* (accessed 27 August 2008); *ODNB* s.v. Günther, Albert Charles Lewis Gotthilf, and Gunther, Robert William Theodore.)

Habel, Simon (1820/1–79). Austrian physician and naturalist. Graduated from Vienna University, 1846. Resident physician of hospital. Emigrated to New York City, 1847. Travelled to Central and South America, including the Galápagos Islands. Returned to New York in 1877. The Smithsonian Institution published his archaeological research. (*Annual Report of the Board of Regents of the Smithsonian Institution* 1879, pp. 15–16; *New York, 1820–1850 Passenger and Immigration Lists*, microfilm roll M237_69 (http://www.ancestry.com).)

Haeckel, Agnes (1842–1915). Daughter of Emil Huschke. Second wife of Ernst Haeckel, whom she married in 1867. (Krauße 1987.)

Haeckel, Ernst Philipp August (Ernst) (1834–1919). German zoologist. MD, Berlin, 1857. Lecturer in comparative anatomy, University of Jena, 1861–2; professor extraordinarius of zoology, 1862–5; professor of zoology and director of the Zoological Institute, 1865–1909. Specialist in marine invertebrates. Leading populariser of evolutionary theory. His *Generelle Morphologie der Organismen* (1866) linked morphology to the study of the phylogenetic evolution of organisms. (*DSB*, *NDB*.)

2 July 1869, 3 August [1869], *29 September 1869*, 3 October [1869]

Haeckel, Walter (1868–1939). Artist. Son of Ernst Haeckel and his second wife, Agnes Huschke. (Krauße 1987.)

Haliburton, Robert Grant (1831–1901). Canadian lawyer and ethnologist. Active in many Nova Scotian societies; founder (1862) and vice-president of the Nova Scotia Institute of the Natural Sciences. Secretary to the Nova Scotia commissioners for the International Exhibition in London, 1862. Advocated protectionism; a founding member of the Canada First Movement. Published anthropological and ethnological studies. (*DCB*.)

Hallett, Frederic Francis (*fl.* 1860s). Brewer and agriculturalist. Partner in the brewing firm Hallett & Abbey, Hove, Sussex. Farmed at Black Rock, Brighton, Sussex. Developed pedigree strains of wheat. Fellow of the Linnean Society, 1862. (*British Farmer's Magazine* n.s. 44 (1863): 207–9; *Gentleman's Magazine* n.s. 12

Hallett, Frederic Francis, cont.
(1862): 655; *List of the Linnean Society of London* 1862; *Post Office Directory of the six home counties* 1869.)

Hamilton, Alfred Douglas (*fl.* 1860s). Barrister, Church Porch Chambers, Temple, London. Home circuit, Hertford, Colchester & Essex sessions. (*Men-at-the-bar*, *Post Office London directory* 1868.)

Hamond, Robert Nicholas (1809–83). Naval officer. Lieutenant, 1827. (*Burke's landed gentry* 1952, O'Byrne 1849.)

Hamond, Thomas Astley Horace (1844/5–1917). Solicitor. Studied at Magdalene College, Cambridge; BA 1868, MA 1871. Practised as a solicitor in Lincoln's Inn Fields. Lord of the Manor of Swaffham, Norfolk. (*Alum. Cantab.*)

Hancock, Albany (1806–73). Zoologist and palaeontologist. Collaborated with Joshua Alder on the *Monograph of British nudibranchiate Mollusca* (1845–55). Contributed several papers on the boring apparatus of sponges, molluscs, and cirripedes to the Tyneside Naturalists' Field Club, which he had helped to found in 1846. (*ODNB*.)

17 July [1869], *25 July 1869*

Hardy, Thomas Duffus (1804–78). Historian and archivist. Junior clerk, record office, Tower of London, 1819–56. Senior assistant keeper, Public Record Office, 1840–61; deputy keeper, 1861–78. (*ODNB*.)

Harris, Thaddeus William (1795–1856). American entomologist and librarian. Librarian of Harvard College, 1831–56. Gave instruction in natural history, 1837–42. Contributed articles on entomology and horticulture to scientific journals. Author of section 8, 'Insects', of the *Catalogue of the animals and plants of Massachusetts* (1833), also published as part 4 of Edward Hitchcock's *Report on the geology, mineralogy, botany, and zoology of Massachusetts* (1833), and of the *Report on the insects of Massachusetts injurious to vegetation* (1841). (*ANB*.)

Harting, Pieter (1812–85). Dutch zoologist and microscopist. Doctorate in medicine, Utrecht, 1835; obstetrics, 1837. Professor of pharmacy, Athenaeum, Franeker, 1841; of zoology, Utrecht, 1846–82. (*BHGW*, *DSB*.)

Hartsen, Frederik Anthony (1838–77). Dutch botanist and philosopher. Studied philosophy at Leiden, and medicine and philosophy at Utrecht, taking a PhD at Utrecht in 1861. Moved to Cannes in 1863 after suffering from tuberculosis. Published many works of philosophy and psychology in German and Dutch. (*NNBW*.)

5 February 1869

Haultain, Theodore Minet (1817–1902). English-born New Zealand soldier and politician. Attended Sandhurst; served in India for ten years. Emigrated to New Zealand in 1849. Resigned from the army in 1857. Member of the New Zealand House of Representives, 1858–60. Organised a defence force after the outbreak of fighting at Taranaki in 1860, and served in the Waikato Militia, 1863–4. Organised the settlement of the militia on confiscated land. Resigned his commission in 1865. Minister of defence, 1865–9. (*DNZB*, Yaldwyn and Hobbs eds. 1998, p. 249.)

Haussoullier, Charles Phipps (b. 1814). English-born French translator. (*Catalogue général de la librairie française.*)
31 January 1869

Heber, Reginald (1783–1826). Clergyman, hymn writer, and poet. Bishop of Calcutta, 1823–6. (*ODNB.*)

Hector, James (1834–1907). Scottish geologist. Surgeon and geologist on the government expedition to the western parts of British North America, 1857–60. Geologist to the provincial government of Otago, New Zealand, 1861–5. Director of the Geological Survey of New Zealand, 1865. Director of the meteorological department of the New Zealand Institute, of the Colonial Museum, and of the botanical garden, Wellington, 1866–1903. Knighted, 1887. FRS 1866. (*DNZB.*)

Heddle, Charles William Maxwell (Charles) (1811/12–99). Senegal-born merchant. Born in Senegal of a Scottish father and Senegalese mother. Educated in Scotland. Worked for the London firm Forster and Smith, importers and exporters for west Africa. Returned to west Africa in 1834; set up as an import and export trader in Freetown, Sierra Leone, in 1838. Appointed member of the governor's council (later the legislative council), 1845. Retired to France in 1871. One of the richest traders in west Africa. (*ODNB.*)

Hegel, Georg Wilhelm Friedrich (1770–1831). German philosopher. Professor of philosophy, University of Berlin, from 1818. (*EB.*)

Hegt, J. Noordhoek (1835–1930). Dutch merchant and zookeeper. Director of a trading company based in the Dutch East Indies and later assistant director of the Amsterdam zoo, 'Natura Artis Magistra'. (Archief van de familie Westerman en aanverwante families (Archive of the Westerman family and their related families), No. 399. Gemeentearchief Amsterdam.)
19 April 1869, *[23 April 1869]*, 27 April [1869]

Hellins, John (1829–87). Clergyman and entomologist. BA, Oxford, 1851; MA, 1857. Deacon, 1852; priest, 1854. Chaplain of Devon County Prison, Exeter, 1859–79. As an entomologist, best known for his work on the larvae of Lepidoptera, which was undertaken to assist William Buckler. (*Alum. Oxon.*, *Entomologist's Monthly Magazine* 24 (1887–8): 20.)
10 May 1869

Helmholtz, Hermann von (1821–94). German physiologist and physicist. Professor of anatomy and physiology at Bonn from 1855, and at Heidelberg from 1858. A close friend of William Thomson from 1853. Appointed professor of physics, Berlin, 1871. Foreign member, Royal Society of London, 1860. Copley Medal, 1873. (*DSB*, *Record of the Royal Society.*)

Henderson, Andrew (b. *c.* 1823 d. 1906). Nurseryman. Proprietor of the family business, Wellington Road Nurseries, at 3 Wellington Road, St John's Wood, London. Took charge of the nursery at his father's death, with Pineapple

Henderson, Andrew, cont.
Nursery, Maida Vale. Later emigrated to Florida. (R. Desmond 1994, *Post Office London directory* 1869.)

Henslow, George (1835–1925). Clergyman, teacher, and botanist. BA, Christ's College, Cambridge, 1858. Curate of Steyning, 1859–61; of St John's Wood Chapel, 1868–70; of St James's Marylebone, 1870–87. Headmaster at Hampton Lucy Grammar School, Warwick, 1861–4; at the Grammar School, Store Street, London, 1865–72. Lecturer in botany at St Bartholomew's Hospital, 1866–80. Honorary professor of botany at the Royal Horticultural Society, 1880–1918. Author of a number of religious books, including *Plants of the Bible* (1907), and of children's books on natural history. Younger son of John Stevens Henslow. (*Alum. Cantab.*, *Crockford's clerical directory*, R. Desmond 1994, Lightman ed. 2004, *ODNB* s.v. Henslow, John Stevens.)
22 February 1869, [after 22 February 1869], 25 September 1869, 22 November 1869

Henslow, John Stevens (1796–1861). Clergyman, botanist, and mineralogist. CD's teacher and friend. Professor of mineralogy, Cambridge University, 1822–7; professor of botany, 1825–61. Extended and remodelled the Cambridge botanic garden. Curate of Little St Mary's Church, Cambridge, 1824–32; vicar of Cholsey-cum-Moulsford, Berkshire, 1832–7; rector of Hitcham, Suffolk, 1837–61. (*DSB*, *Historical register of the University of Cambridge*, *ODNB*.)

Heron, Robert, 2d baronet (1765–1854). Whig politician. MP for Grimsby, 1812–18; Peterborough, 1819–47. Succeeded to a baronetcy upon his father's death in 1805. (*ODNB*.)

Herschel, John Frederick William, 1st baronet (1792–1871). Astronomer, mathematician, chemist, and philosopher. Member of many learned societies. Carried out astronomical observations at the Cape of Good Hope, 1834–8. Master of the Royal Mint, 1850–5. Created baronet, 1838. FRS 1813. (*DSB*, *ODNB*.)

Hewitt, Edward. Judge of poultry exhibitions. Resided in Sparkbrook, Birmingham. (Wingfield and Johnson 1856–7, p. 89.)

Higginson, James Macaulay (1805–85). Irish-born colonial administrator. Governor of Mauritius and the Seychelles, 1850–7. Knighted, 1857. (*DCB*.)

Higginson, Mary (*fl.* 1850s–90s). Daughter of James Macaulay Higginson. Married Charles Henry Higginson of Springmount. (Public Record Office of Northern Ireland, *Introduction: Higginson papers*, November 2007, http://www.proni.gov.uk/ (accessed 6 May 2008).)

Hildebrand, Friedrich Hermann Gustav (Friedrich) (1835–1915). German botanist. After studying mineralogy, geology, and agriculture at Berlin, he took up botany, studying at Bonn, then from 1855 to 1858 at Berlin, where he received his doctorate. Habilitated at Bonn, becoming privat-dozent there, in 1859. Professor of botany, Freiburg im Breisgau, 1868–1907. Worked mainly on hybridity, dimorphism, and generation. (Correns 1916, Junker 1989, Tort 1996.)

Hoffmann, Carl (1802–83). German publisher and bookseller. Acquired a bookshop with a lending library in Stuttgart, 1826. After 1833 devoted himself exclusively to an added publishing house. Publications included popular scientific works and botanical works. Later included a printing press and lithographic office. (*DBE.*)

Hogarth, William (1697–1764). Painter and engraver. (*ODNB.*)

Hogg, James (b. *c.* 1770 d. 1835). Scottish poet and novelist. Son of a shepherd of Ettrick Hall in Ettrick Forest. Worked as a shepherd in Ettrick Forest, then southwestern Scotland, 1790–1810. Published his first book in 1801, and then regularly submitted to magazines. In 1810, moved to Edinburgh, where he established himself as a leading literary figure. (*ODNB.*)

Hogg, Robert (1818–97). Scottish horticulturist and pomologist. Gained practical horticultural training with Peter Lawson in Edinburgh and Hugh Ronalds in Brentford, after which he studied in Paris, Belgium, and Germany. Partner in a nursery in Brompton Park, London, 1842–51. Co-editor of the *Cottage Gardener* (from 1861, *Journal of Horticulture*), 1855–95. (R. Desmond 1994.)

Holdsworth, Edmund William Hunt (1829–1915). Naturalist. Secretary to the Commision of Inquiry into the sea fisheries of the United Kingdom, 1863–5. Investigated the state of the pearl fishery in Ceylon (Sri Lanka), 1864–9. (Collinge comp. 1984, no. 107; Fergusson and Fergusson 1888, pp. 383; *NUC.*)

Hollams, John (b. 1820/1). Solicitor with the firm of Thomas & Hollams of Mincing Lane, London. (Census returns 1861 (Public Record Office RG9/413/23), *Post Office London directory* 1868.)

Holland, Edward (1806–75). CD's second cousin. BA (Trinity College), Cambridge, 1829. Liberal MP for East Worcestershire, 1835–7; Evesham, 1855–68. President of the Royal Agricultural Society. (*Alum. Cantab.*, Stenton 1976.)

Holland, Elizabeth. Of Knutsford, Cheshire. (*Post Office directory of Cheshire* 1857.)

Holland, Hannah. Of Knutsford, Cheshire. (*Post Office directory of Cheshire* 1857.)

Holland, Henry, 1st baronet (1788–1873). Physician. Related to Josiah Wedgwood I. Physician-in-ordinary to Prince Albert, 1840; to Queen Victoria, 1852. President of the Royal Institution of Great Britain, 1865–73. Created baronet, 1853. FRS 1815. (Caroe 1985, *DNB*, *Emma Darwin* (1904), *Record of the Royal Society of London.*)

24 February [1869] (to E. A. Darwin)

Holland, Lucy. Of Knutsford, Cheshire. (*Post Office directory of Cheshire* 1857.)

Holland, Mary. Of Knutsford, Cheshire. (*Post Office directory of Cheshire* 1857.)

Holland, Swinton Colthurst (1844–1922). Naval officer. Entered the Royal Navy, 1857. Lieutenant, 1865; commander, 1878; captain, 1884; rear-admiral, 1899; vice-admiral, 1903; admiral, 1907. Commodore in charge, Hong Kong, 1896. Admiral superintendent of Chatham Dockyard, 1899–1902. Retired in 1908. Fellow, Royal Geographical Society. (H. B. Mason ed. 1908, *WWW* 1916–28.)

Hooker, Brian Harvey Hodgson (1860–1932). Fifth child of Frances Harriet and Joseph Dalton Hooker. (Allan 1967 s.v. 'Hooker pedigree'.)

Hooker, Charles Paget (1855–1933). Physician and surgeon. Third child of Frances Harriet and Joseph Dalton Hooker. Trained at St Bartholomew's Hospital, London; made a licentiate of the Royal College of Physicians of London and the Royal College of Surgeons, Edinburgh, 1879, before being appointed to the staff of the Hertfordshire General Infirmary. Cottishall Cottage Hospital, Norfolk, 1880–5; Cirencester Cottage Hospital, Gloucestershire, 1885–1912. (Allan 1967, *Medical directory* 1881–1933, *Medical who's who* 1914.)

Hooker, Frances Harriet (1825–74). Daughter of John Stevens Henslow. Married Joseph Dalton Hooker in 1851. Assisted her husband significantly in his published work. Translated *A general system of botany, descriptive and analytical*, by Emmanuel Le Maout and Joseph Decaisne (1873). (Allan 1967, Lightman ed. 2004.)

Hooker, Joseph Dalton (1817–1911). Botanist. Worked chiefly on taxonomy and plant geography. Son of William Jackson Hooker. Friend and confidant of CD. Accompanied James Clark Ross on his Antarctic expedition, 1839–43, and published the botanical results of the voyage. Appointed palaeobotanist to the Geological Survey of Great Britain, 1846. Travelled in the Himalayas, 1847–9. Assistant director, Royal Botanic Gardens, Kew, 1855–65; director, 1865–85. Knighted, 1877. FRS 1847. (*DSB*, *ODNB*.)

13 January 1869, *14 [January] 1869*, *15 January 1869*, 16 January [1869], *18 January 1869*, [22 January 1869], *[25 January 1869]*, 8 March [1869], *11 March 1869*, 17 March [1869], *29 March 1869* (to Emma Darwin), 30 March [1869], [before 7 May 1869], *6 June 1869*, 22 June [1869], *24 June 1869*, 8 July [1869], *17 July 1869*, 24 July [1869], *5 August 1869*, 7 August [1869], *13 August 1869*, 14 August [1869], *7 September 1869*, 10 September [1869], *17 September 1869*, 18 [September 1869], *24 September 1869*, 2 October [1869], 11 October 1869, 13 November [1869], *14 November 1869*, 19 November [1869], *21 November 1869*, 24 November [1869]

Hooker, Maria (1797–1872). Eldest daughter of Dawson Turner. Married William Jackson Hooker in 1815; acted as her husband's secretary. Mother of Joseph Dalton Hooker. (Allan 1967 s.v. 'Turner pedigree', R. Desmond 1994.)

Hooker, William Henslow (1853–1942). Eldest child of Frances Harriet and Joseph Dalton Hooker. Civil servant, India Office, 1877–1904. Encouraged imperial ties between metropolitan institutions (particularly the Royal Botanic Gardens, Kew) and British East Africa, *circa* 1896–1906. (Allan 1967; *India list* 1904–5; *Zanzibar Gazette*, 5 February 1896, p. 6, and 28 November 1900, p. 5.)

Hooker, William Jackson (1785–1865). Botanist. Father of Joseph Dalton Hooker. Regius professor of botany, Glasgow University, 1820–41. Appointed first director of the Royal Botanic Gardens, Kew, 1841. Knighted, 1836. FRS 1812. (*DSB*, *ODNB*.)

Horne, John (1835–1905). Plant collector and botanist. Worked at Royal Botanic Gardens, Kew, 1859–60; Botanic Garden, Mauritius, 1861–91 (director from 1877). Collected plants in Mauritius, the Seychelles, and Fiji. (R. Desmond 1994.)

Horner, Anne Susan (Susan) (1816–1900). Author and translator. Daughter of Leonard Horner. Author of works on Italian history, literature, and other subjects. Lived in Florence from soon after 1864 until 1890. ('Susan Horner Collection', *The British Institute of Florence Library and Archive*, http://Britishinstitute.it/en/library/archive_horner.html (accessed 9 December 2008); Alyson Pryce, British Institute of Florence (personal communication).)

Horner, Joanna Baillie (1823–1916). Author and translator. Daughter of Leonard Horner. Author, with Susan Horner, of *Walks in Florence and its environs* (1873) and translator, with Leonora Pertz, of German works. Lived in Florence from soon after 1864 until 1890. ('Susan Horner Collection', *The British Institute of Florence Library and Archive*, http://Britishinstitute.it/en/library/archive_horner.html (accessed 9 December 2008); Alyson Pryce, British Institute of Florence (personal communication).)

Humboldt, Friedrich Wilhelm Heinrich Alexander (Alexander) von (1769–1859). Prussian naturalist, geographer, and traveller. Official in the Prussian mining service, 1792–6. Explored northern South America, Cuba, Mexico, and the United States, 1799–1804. Travelled in Siberia in 1829. Foreign member, Royal Society of London, 1815. (*DBE, DSB, NDB.*)

Hutton, Richard Holt (1826–97). Journalist and theologian. Studied philosophy and mathematics at University College, London, 1841–5; Bonn 1845–6; Heidelberg 1846–7. Prepared for a Unitarian ministry at Manchester New College, 1847–8. Vice-principal and chaplain of University Hall, Gordon Square, London, 1850. Professor of mathematics, Bedford College, London, 1856–7. Entered the Church of England in 1862. Proprietor and joint editor of the *Spectator*, 1861–97. (*ODNB.*)

Huxley, Henrietta Anne (1825–1915). Born Henrietta Anne Heathorn. Emigrated to Australia in 1843. Met Thomas Henry Huxley in Sydney, Australia, in 1847, and married him in 1855. (A. Desmond 1994–7, Freeman 1978.)

Huxley, Marian (1859–87). Artist. Daughter of Henrietta Anne and Thomas Henry Huxley. Studied art at the Slade School, London. Exhibited at the Royal Academy, 1880–4. Married artist–author John Collier, 1879. Her sketch of CD, made when she was 18, hangs in the National Portrait Gallery. (Bibby 1959; *Bryan's dictionary of painters and engravers*; Clark 1968, p. 97 and *passim*; A. Desmond 1994–7; Petteys 1985.)

Huxley, Thomas Henry (1825–95). Zoologist. Assistant-surgeon on HMS *Rattlesnake*, 1846–50, during which time he investigated Hydrozoa and other marine invertebrates. Lecturer in natural history, Royal School of Mines, 1854; professor, 1857. Appointed naturalist to the Geological Survey of Great Britain, 1854. Hunterian Professor, Royal College of Surgeons of England, 1862–9. Fullerian Professor of physiology, Royal Institution of Great Britain, 1855–8, 1866–9. President of the Royal Society of London, 1883–5. FRS 1851. (Clark 1968, A. Desmond 1994–7, *DSB, ODNB.*)

10 March 1869, *11 March 1869*, 12 March [1869], *17 March 1869*, 19 March

Huxley, Thomas Henry, cont.
[1869], *7 May 1869*, 8 May [1869], 9 July [1869], *16 July 1869*, 21 September [1869], *28 September 1869*, 1 October [1869], 14 October 1869

Hywel Dda (Hywel Dda ap Cadell) (d. 949/50). Welsh king. Said to have first codified and promulgated Welsh customary law. (*ODNB*.)

Innes, Eliza Mary Brodie. Daughter of John Laidlaw. Married John Innes (later John Brodie Innes) in 1847. (*Burke's landed gentry* 1879 s.v. Brodie-Innes, John.)

Innes, John Brodie (1817–94). Clergyman. Perpetual curate of Down, 1846–68; vicar, 1868–9. Left Down in 1862 after inheriting an entailed estate at Milton Brodie, near Forres, Scotland; changed his name to Brodie Innes in 1861 as required by the entail. Priest in charge of Milton Brodie Mission and general licentiate of the diocese of Moray, 1861. Chaplain to the Bishop of Moray, 1861–80 and 1886–94. (*Clergy list*, *County families* 1864, *Crockford's clerical directory*, Freeman 1978, J. R. Moore 1985.)
15 October 1869, 18 October 1869, *20 October 1869*

Innes, John William Brodie (1848–1923). Barrister and novelist. Son of John Brodie Innes. Called to the bar, 1876. Advocate at the Scottish bar, 1888. Interested in antiquarian research, romance, demonology, witchcraft, and criminology. (*Alum. Cantab.*, Freeman 1978.)

Jacob-Makoy, Lambert (1790–1873). Belgian nurseryman and horticulturalist. Florist's apprentice, 1806. Gardener, 1807–10. From 1810, headed the Jacob-Makoy nursery in Liege. Founding member of the horticultural society of Liege, 1830. Developed several new varieties of garden plants, especially orchids and peonies. (*La Belgique Horticole: Annales d'Horticulture Belge et Etrangère* (1874): 5–18.)

Jäger, Gustav (1832–1917). German physician and zoologist. Co-founder and director of the Vienna zoological garden, 1858–66. Professor of zoology and anthropology, Hohenheim Academy, from 1867; Stuttgart Polytechnic, from 1870. Taught physiology and histology at the Veterinary School, Stuttgart, from 1874. Co-editor of *Kosmos*, 1877–9. Returned to medical practice in 1884. (*DBE*, Freeman 1978, *NDB*, Weinreich 1993.)
9 September 1869

Janson, Edward Westey (1822–91). Natural history dealer. Educated for the medical profession. Assisted his father as a city merchant and later was secretary of the Dutch Rhenish railway. On the death of his father, in 1867, went into business as a bookseller and natural history dealer. Specialised in Coleoptera and Elateridae, of which he amassed a large collection. Curator of the collections of the Entomological Society of London, 1850–63; librarian until 1874; secretary, 1857–61. (Gilbert 1977, *Proceedings of the Entomological Society of London* (1891): xlix–l.)

Jeffreys, John Gwyn (1809–85). Conchologist, zoologist, and lawyer. Solicitor in Swansea until 1856, when he was called to the bar and moved to London. Justice of the peace for Glamorgan, for Brecon, and for Hertfordshire. Deputy lieutenant, Hertfordshire; high sheriff, 1877. Treasurer of the Linnean Society of London and the Geological Society of London for many years. Conducted extensive deep-sea dredging researches. FRS 1840. (*DSB*, *ODNB*.)

Jenkin, Henry Charles Fleeming (1833–85). Engineer and university teacher. Studied natural philosophy at Genoa. Apprenticed at Fairbairn's works, Manchester, 1851. Worked as a draughtsman and marine telegraph engineer in London from 1855. Carried out important experiments on the resistance and insulation of electrical cables with William Thomson (later Lord Kelvin). Largely occupied in the fitting out of submarine telegraph cables, 1858–73. Appointed professor of civil engineering, University College, London, 1866; professor of engineering, Edinburgh University, 1868. Promoted the formation of a sanitary association, Edinburgh, 1877–8. Wrote miscellaneous papers on literature, science, and political economy. FRS 1865. (*ODNB.*)

Jenzsch, Gustav Julius Siegmund (Gustav) (1830–77). German mineralogist. (*BHGW.*)

Johnson, Samuel (1709–84). Lexicographer and man of letters. (*ODNB.*)

Johnstone, James (*fl.* 1860s). Lieutenant, Indian army. In charge of the training of elephants for the native infantry in Bengal. (*India list*, July 1868.)

Jones, Albert Hugh (1840/1–1924). Bank clerk and entomologist. By 1891, principal clerk in Drummonds Bank. (Census returns 1871, 1881 and 1891 (Public Record Office RG10/90: 40, RG11/72: 12, RG12/104: 30); Gilbert 1977; letter from A. H. Jones, 17 November 1869.)

17 November 1869

Jones, Edward. Sheep farmer at Hendrewallog, Merionethshire. (Letter from Edward Jones to M. A. Ruck, 28 April 1869.)

28 April 1869 (to M. A. Ruck)

Jones, Henry Bence (1814–73). Physician and chemist. Physician to St George's Hospital, 1846–62. Secretary of the Royal Institution, 1860–72. In his chemical studies, devoted himself especially to the application of chemistry to pathology and medicine. FRS 1846. (*ODNB.*)

Jones, John Winter (1805–81). Librarian. Appointed assistant, library of the British Museum, 1837; assistant keeper of printed books, 1850; keeper, 1856; principal librarian, 1866. (*ODNB.*)

Orange Judd & Co. American publishers.

21 April 1869

Jukes, Joseph Beete (1811–69). Geologist. Geological surveyor of Newfoundland, 1839–40. Naturalist aboard HMS *Fly* in the survey of the north-east coast of Australia, 1842–6. Geologist with the Geological Survey of Great Britain working in North Wales and the English Midlands, 1846–50. Local director of the Geological Survey of Ireland, 1850–67; director, 1867–9. Lecturer on geology at the Royal College of Science, Dublin, from 1854. President of the Geological Society of Dublin, 1853–4. FRS 1853. (*ODNB.*)

Kaup, Johann Jakob (1803–73). German naturalist and palaeontologist. (*NDB*, Sarjeant 1980–96.)

Kelaart, Edward Frederick (*c.* 1818–60). Ceylonese-born physician and botanist. Served in the Ceylon Medical Service. In Gibraltar, 1843–5. An active member of the Ceylon branch of the Royal Asiatic Society. (R. Desmond 1994.)

Kendall, Ezra Otis (1816/17–99). American mathematician and astronomer. Professor of mathematics and astronomy at the University of Pennsylvania. Secretary of the American Philosophical Society, 1849–76; vice-president, 1877–8. (*New York Times*, 6 January 1899, p. 7; *Proceedings of the American Philosophical Society* 38 (1899): 3, 73 (1934): 371–97.)

Kepler, Johann (1571–1630). German astronomer. Discovered that the earth and other planets travel around the sun in elliptical orbits. (*EB*.)

Kerner von Marilaun, Anton (1831–98). Austrian botanist. Studied medicine in Vienna, 1848–53. MD 1854. Teacher of natural history, Oberrealshule, Ofen, 1855; professor of natural history, Josefs-Polytechnikum, 1858–60. Professor of natural history and director of the botanic gardens and museum of natural history, University of Innsbruck, 1860–78. Professor of systematic botany and director of the botanic gardens, University of Vienna, 1878–98. Established an experimental alpine garden near the top of Mt Blaser, above Trins in the Gschnitztal (Tirol). Studied the effects of climate on the morphology of plants. (*NDB*, *OBL*.)

King, Philip Gidley (1817–1904). Australian farmer and mining company manager. Son of Phillip Parker King. Midshipman in HMS *Beagle*, 1831–6. Settled in Australia in 1836. Entered the service of the Australian Agricultural Company in 1842; superintendent of stock, 1851. New South Wales manager for the Peel River Land and Mineral Company, 1852–81. (*Aust. dict. biog.*)
25 February 1869, 24 April 1869

Kingsley, Charles (1819–75). Author and clergyman. Appointed professor of English, Queen's College for Women, London, 1848. Regius professor of modern history, Cambridge University, 1860–9. Rector of Eversley, Hampshire, 1844–75. Chaplain to the queen, 1859–75. (*ODNB*.)

Kippist, Richard (1812–82). Botanist. Librarian of the Linnean Society of London, 1842–80. Specialist in Australian plants. (R. Desmond 1994, *ODNB*.)
31 January [1869?]

Kirby, William (1759–1850). Clergyman and entomologist. One of the authors of the Bridgewater Treatises (1833–6) and, in collaboration with William Spence, co-author of *Introduction to entomology* (1815–26). Vicar of Barham, Suffolk, 1797–1850. One of the original fellows of the Linnean Society, 1788. Honorary president of the Entomological Society, 1837. FRS 1818. (*Alum. Cantab.*, *DNB*.)

Klostermann, Julius Alfred (1826–1900). German merchant and consul-general. Consul-general of Sweden and Norway in Messina, Sicily, 1851–82. Merchant in Berlin until 1897. Honoured in the flatworm name *Pseudostomum klostermanni* (von Graf, 1874), or *Turbella klostermanni* (von Graff, 1874). (Hans G. Hansson, *Database of biographical etymology of marine organism names* (BEMON), April 2008, http://www.tmbl.gu.se/libdb/libdb.html (accessed 9 December 2008).)

Knowles, James Thomas (1831–1908). Journal editor and architect. Worked as an architect in London, 1854–83. Editor of the *Contemporary Review*, 1870–7; of *Nineteenth Century*, 1877–1908. (*ODNB*.)

Koch, Eduard Friedrich (Eduard) (1838–97). German publisher. Took over

E. Schweizerbart'sche Verlagsbuchhandlung in 1867, after which the firm published mostly scientific titles. Published a multi-volume edition of CD's works, translated by Julius Victor Carus. (*Biographisches Jahrbuch und deutscher Nekrolog* 2 (1898): 227.)

17 June 1869, 22 June [1869], *30 October 1869*, 4 November 1869

Koch, Karl Heinrich Emil (1809–79). German botanist, dendrologist, and horticulturalist. MD, University of Würzburg, 1833. DPhil, Jena, 1834, privat-dozent, 1834, professor extraordinarius, 1836. A member of two botanical expeditions to Asia, between 1838 and 1844. From 1847 *Adjunkt* at the Royal Botanic Garden, Berlin, where he was responsible for expanding the tree collection. Appointed director of the state nursery in Berlin, 1851. Founded a society for pomology and fruit cultivation in 1853. Published a series on Asian flora, and a major work on dendrology. (*ADB*, *Taxonomic literature*.)

Kölliker, Rudolf Albert von (1817–1905). Swiss anatomist and physiologist. Professor of physiology and comparative anatomy, University of Würzburg, 1847–64; professor of anatomy, 1849–97. Foreign member, Royal Society of London, 1860. (*DSB*.)

Kovalevskaya, Sofia Vasilyevna (1850–91). Russian mathematician and author. Wife of Vladimir Onufrievich Kovalevsky. Studied mathematics with private tutors, in Heidelberg from 1869, then in Berlin under Karl Weierstrass from 1871. Doctorate in absentia from Göttingen, 1874. Lecturer in mathematics, University of Stockholm, 1883; professor, 1884. Life professorship, 1889. Wrote mathematical papers and autobiographical novels. (*BDWS* s.v. Kovalevskaia, Sofia Vasilyevna.)

Kovalevsky, Alexander Onufrievich (1840–1901). Russian embryologist. Brother of Vladimir Onufrievich Kovalevsky. Held academic posts at various Russian universities; professor of histology, St Petersburg, 1890–4. His studies of ascidian embryology revealed that tunicates were chordates and gave strong support to Darwinian transmutation theory. Foreign member of the Royal Society of London, 1885. (*DSB*, *GSE* s.v. Tunicata.)

Kovalevsky, Vladimir Onufrievich (1842–83). Russian palaeontologist. Graduated from the School of Jurisprudence in 1861. Thereafter published, translated, and edited works by CD, Charles Lyell, Louis Agassiz, and others. Studied natural science and palaeontology, travelling throughout Europe, 1869–74. Submitted doctoral thesis on the palaeontology of horses at the university of Jena in 1872. Associate professor, Moscow University, 1880–3. (*DSB*.)

[January–March 1869], *13 September 1869*

Lacroix, Sylvestre François (1765–1843). French mathematician. Wrote on calculus, analytic geometry, and history of mathematics. (*DSB*.)

Lagrange, Joseph Louis (Lagrangia, Giuseppe Lodovico) (1736–1813). Italian mathematician and astronomer. Studied mathematics in his native Turin. Appointed director of the mathematics section of the Berlin Academy, 1766. Moved to the Academy of Sciences in Paris in 1787. Made significant discoveries

Lagrange, Joseph Louis (Lagrangia, Giuseppe Lodovico), cont.
in the fields of calculus, number theory, planetary movement, and fluid dynamics. (*DSB.*)

Laing, Alexander Gordon (1794–1826). Scottish army officer and explorer. Educated at Edinburgh University. Held various military posts in Barbados, West Indies, Jamaica, Honduras, and Sierra Leone. Captain to the Royal African Colonial Corps, 1822; posted to Sierra Leone. Charted the Rokel and Niger rivers, 1822–3. Took part in the Ashanti War of 1823–4. Investigated the hydrography of the Niger basin. First European to reach Timbuktu, 1826. (*ODNB.*)

Lamarck, Jean Baptiste Pierre Antoine de Monet (Jean Baptiste) de (1744–1829). French naturalist. Held various botanical positions at the Jardin du Roi, 1788–93. Appointed professor of zoology, Muséum d'Histoire Naturelle, 1793. Believed in spontaneous generation and the progressive development of animal types; propounded a theory of transmutation. (*DSB.*)

Landor, Henry (1815–77). Welsh-born physician. Superintendent of Malden Lunatic Asylum, Amherstburg, Ontario, Canada; thereafter of the Asylum for the Insane, London, Ontario. (*IGI* (accessed 1 March 2008); Linda G. Bunn, ed., *Amherstburg Echo: index to births, marriages, deaths*, Canada (1995); *Registration of deaths, 1869–1932*, Toronto, Archives of Ontario, roll: MS935_16, http://www.ancestry.com (accessed 4 March 2008).)
10 March 1869

Landseer, Edwin Henry (1802–73). Animal painter. Pupil of Benjamin Robert Haydon, 1813–16; studied at the Royal Academy Schools from 1816. Made several expeditions to Scotland between 1824 and 1835. Known for animal paintings especially in Scottish settings. (*ODNB.*)

Langstaff, Charles (1832/3–84). Surgeon. MRCS, England, 1856. MD, Edinburgh, 1858. President of the Southampton Medical Society. Medical Officer at the Female Orphan Asylum, Southampton. Medical Officer in charge of the Royal Engineers and to the Ordnance Survey. Surgeon at the Royal South Hampshire Infirmary. (*Medical directory.*)

Lankester, Edwin Ray (1847–1929). Zoologist. Studied natural sciences at Oxford under George Rolleston, 1866–8; physiology at Leipzig and Vienna; morphology under Ernst Haeckel at Jena; marine zoology with Anton Dohrn in Naples, 1871–2. Fellow and tutor, Exeter College, Oxford, 1872–5; professor of zoology, University College, London, 1875–91; Linacre Professor of comparative anatomy, Oxford, 1891–8; director of the natural history departments and keeper of zoology, British Museum, 1898–1907. Knighted, 1907. FRS 1875. (*ODNB.*)
25 June 1869

Laplace, Pierre Simon, marquis de (1749–1827). French mathematician, physicist, and cosmologist. (*DSB.*)

Latey, John Lash (1808–91). Poet and editor. Editor of the *Illustrated London News*, 1858–90. (*Modern English biography.*)

Layard, Edgar Leopold (1824–1900). Civil servant and naturalist. Served in the

Ceylon civil service, 1846–54; in the Cape of Good Hope civil service, 1854–70. Studied the birds and shells of Ceylon (Sri Lanka). Founded the South African museum; curator, 1855–72. Published on the birds of South Africa (1867). (*DSAB*, *Modern English biography*.)

Layton, Charles (b. 1826/7). Publisher's agent. London agent for D. Appleton & Co., New York City. (Census returns 1881 (Public Record Office RG11/68: 13); letter from Charles Layton, 22 November 1869.)

22 November 1869, 24 November [1869], *25 November 1869*, *23 December 1869*

Lea, Isaac (1792–1886). American naturalist and publisher. President, Academy of Natural Sciences of Philadelphia, 1858–63; American Association for the Advancement of Science, 1860. Did extensive work on freshwater molluscs, publishing a series of thirteen volumes on the genus *Unio*. (*ANB*, *DAB*.)

Le Couteur, John (1794–1875). Soldier and agriculturalist. Son of John Le Couteur (1760–1835). Commander of the Jersey Militia. Founder and president, Royal Jersey Agricultural Society. Author of a well-known work on wheat (1836). Displayed 104 varieties of wheat at the Great Exhibition of 1851. FRS 1843. (R. Desmond 1994, *ODNB* s.v. Le Couteur, John (1760–1835).)

Lecouteux, Edouard (1819–93). French agriculturalist, writer, and editor. Director of cultivation, Institut agronomique de Versailles. Editor of *Journal d'agriculture pratique*. Founder and later vice-president of the Société des agriculteurs de France, 1868. Professor of rural economy, Institut agronomique de Paris, 1876. Landowner in Cerçay. (*Catalogue général de la librairie française*, *La Nature* 21 (1893): 366.)

Leech, Jane (b. 1808/9). Housewife. Wife of John Leech of 14 Commercial Street, Wakefield, Yorkshire. Admitted to Wakefield Pauper Lunatic Asylum on two occasions, the first in about 1863 when she was a patient for two and a half years, and the second on 16 March 1869; discharged 5 November 1874. (Wakefield Record Office: C85/3/6/22/p573-576/4.)

Leidy, Joseph (1823–91). American anatomist, microscopist, and palaeontologist. Appointed dissecting assistant to William Horner, professor of anatomy at the University of Pennsylvania, 1844. Professor of anatomy, University of Pennsylvania, 1853–91. President of the Academy of Natural Sciences of Philadelphia, 1881–91. Published extensively on North American fossil vertebrates, and also studied freshwater Protozoa. (*ANB*, *DSB*.)

Leifchild, John R. (b. 1815). Author and reviewer. Son of John Leifchild, a leading evangelical minister (*ODNB*), and Elizabeth Stormonth. Visiting government commissioner to coal mines in Newcastle and Durham. Author of works on mining and coal. Reviewed *Origin* in the *Athenaeum* in 1859 and *Orchids* in 1862. (*BLC*, *NUC*, *Wellesley index*.)

Leigh, Augustus Austen (1840–1905). College head. BA (King's College), Cambridge, 1863; fellow, 1862; MA, 1866; tutor, 1868–81; dean, 1871–3, 1882–5; vice-provost, 1877–89; provost, from 1889. Member of the Council of the Senate of Cambridge University, 1876–80, 1886–90; vice-chancellor, 1893–5. (*ODNB*.)

Lenné, Peter Joseph (1789–1866). German gardener and landscape architect. Assistant gardener, Sanssouci, 1816; director, 1822. Founded an academy for gardeners in Schöneberg and Potsdam, 1823. Prussian garden director-general, 1854. From 1840, in charge of urban planning in Berlin. (*ADB*, *DBE*, *NDB*.)

Lesley, J. Peter (1819–1903). American geologist. Pastor, Congregational church, Milton, Massachusetts, 1847–52. Employed by the Pennsylvania geological survey, 1852. Professor of mining, University of Pennsylvania, 1859; dean of the science department, 1872. Visited Europe in 1863 to study rail manufacture and steel processing on behalf of the Pennsylvania Railroad. Secretary, American Iron Association, 1856–64. State geologist of Pennsylvania, 1873–87. Librarian, American Philosophical Society, 1859–85; secretary, 1859–87; vice-president, 1887. (*DAB*.)

Lettington, Henry (b. 1822/3). Gardener in the village of Down, Kent. Worked as a gardener at Down House, 1854–79, and occasionally in following years. Assisted CD with botanical experiments. Son-in-law of William Brooks, who was also employed by the Darwins. (CD's Classed account books (Down House MS); Census returns 1861 (Public Record Office RG9/462: 70); F. Darwin 1920, pp. 56–7; letter from Emma Darwin to George Darwin (DAR 210.3: 110); Recollections of CD by Francis Darwin (DAR 140.3: 90a).)

Leuckart, Karl Georg Friedrich Rudolf (Rudolf) (1822–98). German zoologist. Lecturer in zoology, University of Göttingen, 1847–50. Professor extraordinarius of zoology, University of Giessen, 1850–5; professor, 1855–69. Appointed professor of zoology in Leipzig, 1869; dean of the philosophical faculty, 1873; rector, 1877–8. Carried out important work in morphology and parasitology. (*DBE*, *DSB*, *NDB*.)

Lewes, George Henry (1817–78). Writer. Author of a biography of Goethe (1855). Contributed articles on literary and philosophical subjects to numerous journals. Editor, *Fortnightly Review*, 1865–6. Published on physiology and on the nervous system in the 1860s and 1870s. Lived with Mary Ann Evans (George Eliot) from 1854. (Ashton 1991, *ODNB*.)

Leydig, Franz von (1821–1908). German zoologist. Lecturer in physiology, University of Würzburg, 1849–55; professor extraordinarius, 1855–7. Professor of zoology, Tübingen, 1857–75; of comparative anatomy, Bonn, 1875–87. Foreign member, Royal Society of London, 1901. (*DSB*, *NDB*.)

Liebig, Justus von (1803–73). German organic chemist. Professor of chemistry, University of Giessen, 1825–51; University of Munich, 1852–73. Played a major role in the development of organic chemistry. Developed artificial agricultural fertilisers. President of the Bavarian Academy of Science, 1859–73. Foreign member, Royal Society of London, 1840. (*ADB*, *DBE*, *DSB*, *NDB*.)

Lightfoot, Joseph Barber (1828–89). Clergyman and biblical scholar. BA (Trinity College), Cambridge, 1851; fellow, 1852. Ordained priest, 1858. Tutor, 1857. Appointed Hulsean Professor of divinity, 1861. Member of the Council of the Senate, 1860–78. Bishop of Durham, 1879–89. (*ODNB*.)

Lindley, John (1799–1865). Botanist and horticulturalist. Assistant in Joseph Banks's library and herbarium, 1819. Garden assistant secretary, Horticultural Society of London, 1822–6; assistant secretary, 1826–41; vice-secretary, 1841–58; honorary secretary, 1858–62. Lecturer on botany, Apothecaries' Company, from 1836. Professor of botany, London University (later University College, London), 1829–60. Horticultural editor of the *Gardeners' Chronicle* from 1841. FRS 1828. (R. Desmond 1994, *DSB*, *ODNB*.)

Linnaeus, Carolus. *See* Linné, Carl von.

Linné, Carl von (Carolus Linnaeus) (1707–78). Swedish botanist and zoologist. Professor of practical medicine, University of Uppsala, 1741; professor of botany, diatetics, and materia medica, 1742; court physician, 1747. Proposed a system for the classification of the natural world, and reformed scientific nomenclature. FRS 1753. (*DSB*, *Record of the Royal Society of London*.)

Linnean Society, President and Council of the
10 May 1869, [10 May 1869]

Linsser, Carl (1837–69). German astronomer. Studied astronomy in Berlin, 1859. From 1860 worked as an arithmetician at the astronomical observatory at Pulkowa, near St Petersburg. (*BHGW* vol. 3.)

Lloyd, Gertrude Jane Mary (b. *c.* 1815). Daughter of Philip Lake Godsal. Married John Lloyd in 1847. Sister-in-law of Mary Charlotte Lloyd. Superintended the collection of fossil remains found in caves on the estate at Rhagatt near Corwen, Wales. (*IGI*, s.v. Godsal, Gertrude Jane Mary; Nicholas 1872, 2: 705.)

Lloyd, Mary Charlotte (1819–96). Welsh landowner and artist. Studied in Italy with the sculptor John Gibson. Met Frances Power Cobbe there and became her lifelong companion. Neighbour of the Darwin family during their stay at Caerdeon, Barmouth, Wales, in summer 1869. (Lucas 2007.)
[24 July 1869], 4 August [1869]

Lockwood, Ruth (1833/4–70). Factory worker. Lived in Kirkgate, Wakefield, Yorkshire. Admitted to Wakefield Pauper Lunatic Asylum, 28 July 1862; discharged, 16 September 1862. Admitted for a second time, 23 May 1868, described as an unmarried mill-hand; died in the asylum on 4 August 1870. (Wakefield Record Office: C85/3/6/17/p167/1, C85/3/6/21/p499A-500A/1.)

Lockyer, Joseph Norman (1836–1920). Astronomer. Civil servant in the War Office, 1857–75; published papers on solar physics. Secretary to the Royal Commission on Scientific Instruction and the Advancement of Science, 1870–5; employee of the Science and Art Department at South Kensington from 1875; first director of the Solar Physics Observatory, and professor of astronomical physics, Royal College of Science, South Kensington, 1890–1911. Established the journal *Nature* in 1869. Knighted, 1897. FRS 1869. (*DNB*, *DSB*.)

Locock, Charles, 1st baronet (1799–1875). Obstetric physician. MD Edinburgh, 1821. Had a large midwifery practice in London. First physician accoucheur to Queen Victoria, 1840. Retired and was created baronet, 1857. FRS 1864. (*ODNB*.)

Longley, Charles Thomas (1794–1868). Clergyman. Attended Christ Church, Oxford; BA, 1815; MA, 1818. Ordained priest, 1819. Appointed headmaster of Harrow, 1829. Appointed bishop of Ripon, 1836; of Durham, 1856; archbishop of York, 1860; of Canterbury, 1862. (*ODNB*.)

Lord, John Keast (1818–72). Naturalist and traveller. Served as veterinary surgeon in the Crimea, 1855–6. Naturalist to the Boundary Commission sent to British Columbia, 1858. Joined the staff of the journal *Land and Water* in 1866 before being engaged by the viceroy in archaeological and scientific research in Egypt. Appointed manager of the newly established Brighton Aquarium four months before his death. (*ODNB*.)

Loring, Charles Greely (1828–1902). American soldier, traveller, and curator. Graduated from Harvard, 1848; AM degree, 1851. Joined Louis Agassiz's Lake Superior expedition, and attended Harvard Scientific School. Travelled in Europe and the Middle East with his father for several years. Achieved the rank of major-general of volunteers in the Civil War. Visited Egypt and studied Egyptology. Installed the Way collection of Egyptian antiquities in Museum of Fine Arts, Boston; became trustee, and then curator in 1876. (*National cyclopædia of American biography* 25 (1936): 301–2.)

Lovegrove, Henrietta (b. 1832/3). Wife of Charles Lovegrove. (Census returns 1861 (Public Record Office RG9/462: 73).)

Lowe, Robert, 1st Viscount Sherbrooke (1811–92). Politician, lawyer, and colonialist. Younger brother of Henry Porter Lowe. BA (University College), Oxford, 1833. Lived in Australia, 1842–50. Writer of lead articles for *The Times*, 1850–68. MP, 1850–80. FRS 1871. (*DNB*.)

Lubbock, Ellen Frances (1834/5–79). Daughter of Peter Hordern, clergyman, of Chorlton-cum-Hardy, Lancashire. Married John Lubbock in 1856. (*Burke's peerage* 1970, Census returns 1861 (Public Record Office RG9/462: 75).)

[26 September 1869?]

Lubbock, Frances Mary (1845–1922). Daughter of the Reverend Henry Turton and Amelia St George. Married Henry James Lubbock in 1866. (*Burke's peerage* 1929 s.v. Avebury, *IGI* s.v. Turton, Frances Mary.)

Lubbock, Henry James (1838–1910). Banker. Second son of John William Lubbock. High sheriff for the county of London, 1897. (*WWW*.)

Lubbock, John, 4th baronet and 1st Baron Avebury (1834–1913). Banker, politician, and naturalist. Son of John William Lubbock and a neighbour of CD's in Down. Studied entomology and anthropology. Worked at the family bank from 1849; head of the bank from 1865. Liberal MP for Maidstone, Kent, 1870–80; for London University, 1880–1900. Succeeded to the baronetcy in 1865. Created Baron Avebury, 1900. FRS 1858. (*DSB*, Hutchinson 1914, *ODNB*, *Record of the Royal Society of London*.)

8 February 1869, [before 13 February 1869], *13 February [1869]*, *2 July [1869]*, *20 July [1869]*, 3 August 1869, *5 August [1869]*, *[after 5 August 1869]*, *8 December [1869]*

Lushington, Godfrey (1832–1907). Civil servant. BA (Balliol College), Oxford, 1854; fellow, All Souls College, 1854. Called to the bar at the Inner Temple, 1858. Legal adviser at the Home Office, 1869; legal assistant under-secretary, 1876; permanent under-secretary, 1885–95. (*ODNB*.)

Lushington, Vernon (1832–1912). Lawyer. Attended Trinity College, Cambridge; BA, 1855; LLB, 1859; LLM, 1885. Called to the bar, 1857. Deputy judge-advocate-general, 1864–9; became secretary to the Admiralty in 1869. (*ODNB*.)
[12 March 1869]

Lütke, Feodor Petrovich (1797–1882). Russian navigator, geographer, and arctic explorer. Admiral, 1855; count, 1866. Founding member, Russian Geographical Society, 1845; director, 1845–50, 1857–72. President, St Petersburg Academy of Sciences, 1864–82. (*GSE* s.v. Litke, Fedor Petrovich.)

Lyell, Charles, 1st baronet (1797–1875). Scottish geologist. Uniformitarian geologist whose *Principles of geology* (1830–3), *Elements of geology* (1838), and *Antiquity of man* (1863) appeared in many editions. Professor of geology, King's College, London, 1831. President of the Geological Society of London, 1835–7 and 1849–51; of the British Association for the Advancement of Science, 1864. Travelled widely and published accounts of his trips to the United States. CD's scientific mentor and friend. Knighted, 1848; created baronet, 1864. FRS 1826. (*DSB*, *ODNB*.)
20 March [1869], 4 May [1869], *5 May 1869*, 20 May 1869, *2 November 1869*, [3 November 1869]

Lyell, Leonard, 1st baron (1850–1926). Eldest son of Henry Lyell, Charles Lyell's brother, and Katharine Murray Lyell. MP for Orkney and Shetland, 1885–1900. Created baronet, 1894; peer, 1914. (*Burke's peerage* 1980.)

Lyell, Mary Elizabeth (1808–73). Eldest child of Leonard Horner. Married Charles Lyell in 1832. (Freeman 1978.)

Macaulay, Thomas Babington (1800–59). Historian, writer, and politician. Took the title Baron Macaulay of Rothley in 1857. (*ODNB*.)

McCann, James (b. 1829/30). Clergyman. MA, University of Giessen, 1863; DD, 1868. Deacon, 1863; priest, 1864. Curate of Holmfirth, 1863; of St Paul's, Huddersfield, 1864–8. Incumbent of St Paul's Episcopal, Glasgow, 1868–78. (*Crockford's clerical directory* 1870, 1879; *Scotland census* 1871 Glasgow Barony 45/26.)

McCombie, William (1805–80). Scottish farmer and politician. Farmed at Tillyfour near Aberdeen. Liberal MP for West Aberdeenshire, 1868–76; president of the Scottish chamber of agriculture. (*ODNB*; *The Times*, 3 February 1880, p. 5.)

M'Intosh, Roberta. *See* Günther, Roberta.

McLennan, John Ferguson (1827–81). Scottish lawyer and social anthropologist. MA, King's College, Aberdeen, 1849. Studied mathematics at Trinity College, Cambridge. Began practising law in Edinburgh, 1857. Published on kinship, marriage, and the law. Moved to London, 1870. Appointed parliamentary draughtsman for Scotland, 1871. Regarded as one of the founders of modern British social anthropology. (*ODNB*.)

Macmillan, Alexander (1818–96). Publisher. Co-founder with his brother Daniel

Macmillan, Alexander, cont.
of the firm Macmillan & Co. in 1843. After moving the firm's headquarters to London in the early 1860s, Macmillan began commissioning textbooks on botany, chemistry, and geology for children in elementary schools. (*ODNB.*)

McNeill, Archibald (1803–70). Scottish lawyer and dog breeder. Brother of Duncan McNeill, Baron Colonsay and Oronsay, and Sir John McNeill. Writer to the signet, 1829; director and principal clerk at Chancery Office, Edinburgh, 1843–58; one of the principal clerks of session, 1858–70. (*Modern English biography* s.v. M'Neill, Archibald.)
24 January 1869, 20 February 1869

McNeill, Duncan, Baron Colonsay and Oronsay (1793–1874). Lawyer and politician. Became a member of the Scottish bar in 1816. Solicitor-general for Scotland, 1834–5, 1841–2. Appointed lord advocate, 1842. Conservative MP for Argyll, 1843–6. Appointed judge of the Court of Session, 1851, taking the title Lord Colonsay and Oronsay; lord justice-general and lord president of the Court of Session, 1852; appeals judge, 1867, taking the title Baron Colonsay and Oronsay. (*ODNB.*)

Malven, Ferdinand Maria (*fl.* 1860s). Librarian to the Austrian Imperial and Royal Board of Trade. (Letter from F. M. Malven, 12 February [1869].)
12 February [1869]

Maskelyne, Mervyn Herbert Nevil Story- (Nevil) (1823–1911). Mineralogist. Lectured on mineralogy, Oxford University, 1850–7; professor of mineralogy, 1856–95. Keeper of the mineral department, British Museum, 1857–80. MP for Cricklade, 1880–5; North Wiltshire, 1885–92. FRS 1870. (*Alum. Oxon.*, Morton 1987, *ODNB.*)

Masson, Victor (1807–79). French publisher. Joined the publishing firm of Crochard in 1838, becoming the sole proprietor in 1846. Published elegant editions of scientific works. (*Dictionnaire universel des contemporains.*)
29 September 1869, [after 29 September 1869]

Masters, Maxwell Tylden (1833–1907). Botanist, journal editor, and general medical practitioner. Sub-curator, Fielding Herbarium, University of Oxford, *circa* 1853–7. GP at Peckham from 1856. Lecturer on botany at St George's Hospital medical school, 1855–68. Editor of the *Gardeners' Chronicle*, 1865–1907. Active in the Royal Horticultural Society, succeeding Joseph Dalton Hooker as the chairman of the scientific committee; secretary of the International Horticultural Congress, 1866. FRS 1870. (Clokie 1964, pp. 106, 208; *ODNB.*)
21 January 1869, 30 April 1869, 21 September 1869

Maudsley, Henry (1835–1918). Medical doctor, psychologist, psychiatrist, and author. Studied medicine at University College Hospital, London. MD 1857. Assistant medical officer, Wakefield Asylum, 1857–8; medical superintendent, Manchester Royal Lunatic Asylum, 1859–62. Appointed physician, West London Hospital, 1864; lecturer on insanity, St Mary's Hospital, London, 1868–81; professor,

medical jurisprudence, University College, London, 1869–79. Joint editor, *Journal of Mental Science*, 1863–78. Developed a lucrative private practice. Researched heredity and degeneration, and promoted belief in the physical causes of mental illness. Wrote extensively on medical jurisprudence. Elected president, Medico-Psychological Association, 1870. (*ODNB*.)
20 May 1869

Maw, George (1832–1912). Tile manufacturer, geologist, botanist, and antiquarian. Partner with his younger brother Arthur in the encaustic tile company, Maw & Co., of Brosley, Shropshire. Established a well-known garden at his residence at Benthall Hall, Shropshire; an expert on crocuses. Wrote on the geology of western England and North Wales. Travelled to Morocco and Algeria with Joseph Dalton Hooker in 1871 and independently in 1873, writing on the geology of these countries. (Benthall 1980; R. Desmond 1994; *Gardeners' Chronicle*, 12 February 1881, pp. 205–6, 208, 209; Sarjeant 1980–96.)
9 January 1869, 13 January 1869, *14 January 1869*, 17 January [1869], *18 April 1869*, 27 April [1869], *17 May 1869*, 22 May [1869]

Max Müller, Friedrich (1823–1900). German-born orientalist and philologist. Published an edition of the 'Rigveda', the most important of the sacred books of the Brahmans, 1849–73. Moved to Paris in 1845; settled in Oxford in 1848 after fleeing the revolution in France. Deputy Taylorian Professor of modern European languages, Oxford University, 1850–4; professor, 1854–68; professor of comparative philology, 1868–75. Curator of the Bodleian Library, Oxford, 1856–63 and 1881–94. (*DNB*.)

Mayne, Richard Charles (1835–92). Naval officer. Joined the navy in 1847. Served in the Crimean War, 1854–6; on Vancouver Island and in British Columbia, 1857–61; in New Zealand, 1863–4. Commanded the survey expedition to the Strait of Magellan, 1866–9. (*ODNB* s.v. Mayne, Sir Richard.)

Medici, Lorenzo di Pierfrancesca de' (1449–92). Italian statesman and patron of the arts. (*EB*.)

Medicus, Friedrich Casimir (1736–1808). German botanist and physician. Studied medicine in Tübingen, Strasbourg, and Heidelberg; received his doctorate in 1758. Court physician to Christian IV of Pfalz-Zweibrücken. Established a botanic garden at Mannheim and promoted the introduction of American trees to German forests. (*DBE*.)

Meehan, Thomas (1826–1901). English-born botanist, horticulturalist, and author. Gardener at the Royal Botanic Gardens, Kew, 1845–8. In 1848, emigrated to the United States, where he worked as a gardener. Established a nursery in Germantown, Pennsylvania, *circa* 1853. Editor, *Gardener's Monthly*, 1859–87; *Meehan's Monthly*, 1891–1901. Botanist on the Philadelphia state board of agriculture, 1877–1901. Elected to the Academy of Natural Sciences of Philadelphia, 1860; to the American Philosophical Society, 1871. (Baker 1965, *DAB*.)

Meller, Charles James (b. *c.* 1835 d. 1869). Surgeon and botanist. Surgeon

Meller, Charles James, cont.
and naturalist on David Livingstone's African expedition, 1860–3. Superintendent, Botanic Garden, Mauritius, 1865. Collected plants on the Zambesi, and in Madagascar and Mozambique. (R. Desmond 1994.)

Menzies, Fletcher Norton (b. 1818/19). Scottish landowner. Secretary to the Highland Society, Edinburgh. Brother of Sir Robert Menzies, baronet. (*Alum. Oxon.*, letter from R. W. Elliot to George Cupples, 4 June 1869.)

Merivale, Herman (1806–74). Civil servant and economist. Professor of political economy, Oxford University, 1837–42. Assistant under-secretary of state for the colonies, 1847; permanent under-secretary, 1848. Permanent under-secretary, India Office, 1860. (*ODNB.*)

Merrell, S. A. (b. 1828). American homeopathic doctor. Graduated from Illinois College, Jacksonville, 1853; Eclectic Medical School of Cincinnati. Worked at the United States Military Hospital, Kansas City, Missouri, 1863–5. Moved to Council Bluffs, Iowa, in 1865. (Cleave 1873.)
1 February 1869

Meteyard, Eliza (1816–79). Author. Born in Liverpool; lived in Shrewsbury, 1818–29; in Thorpe, near Norwich, 1829–42; thereafter in London. Contributed fiction and social articles to numerous periodicals under the pen-name Silverpen. Her novels include *Struggles for fame* (1845). An active member of the Whittington Club from 1846. Published a number of works about the Wedgwoods including a two-volume life of Josiah Wedgwood I (1865–6), and *The Wedgwood handbook* (1875). (*ODNB.*)
[18 February 1869], *19 February 1869*

Meyer, Adolf Bernhard (1840–1911). German zoologist, anthropologist, and museum director. Trained as a physician in Berlin. Travelled extensively in the Eastern Archipelago, publishing on the birds of the Celebes. Director, Royal Museum of Zoology, Anthropology, and Ethnography in Dresden, from 1874. (*Auk* 28 (1911): 519.)
16 November 1869, *24 November 1869*, 27 November [1869]

Michelangelo (di Lodovico Buonarroti Simoni) (1475–1564). Italian artist. (*EB.*)

Millais, John Everett (1829–96). Painter. An originator of the Pre-Raphaelite movement. Knighted, 1885. (*DNB.*)

Miller, Hugh (1802–56). Scottish quarryman, poet, ecclesiastical journalist, and geologist. Examined the stratigraphy of parts of Great Britain, in particular the Edinburgh district and the Hebrides. Author of popular works on geology. (*Modern English biography*, Sarjeant 1980–96.)

Milman, Archibald John Scott (1834–1902). Civil servant. BA (Trinity College), Cambridge, 1856; MA, 1859. Entered service in the House of Commons, 1857; second clerk assistant, 1870; clerk assistant, 1886–1900; clerk of the House of Commons, 1900–2. (*Alum. Cantab.*)

Milne-Edwards, Alphonse (1835–85). French naturalist. Trained as a physician. Assistant to his father, the naturalist Henri Milne Edwards. Professor, school of pharmacy, Paris, 1865. Professor of zoology, Muséum d'Histoire Naturelle, Paris, 1876, in succession to his father; director, 1891. (*Dictionnaire universel des contemporains.*)

Miquel, Friedrich Anton Wilhelm (1811–71). Dutch botanist. Described the flora of the Dutch East Indies. Director, Rotterdam botanic garden, 1835–46. Professor of botany, Amsterdam, 1846–59; Utrecht, 1859–71. (*DSB.*)

Moggridge, John Traherne (1842–74). Entomologist and botanist. Wintered in Mentone, France, and studied the flora of the area. (R. Desmond 1994, *Gardeners' Chronicle* n.s. 2 (1874): 723.)

7 June 1869, *16 September [1869]*, *18 September 1869*, *12 December 1869*

Molendo, Ludwig (1833–1902). German bryologist and journalist. Editor of the *Bayreuther Zeitung*, 1866–8; the *Passauer Zeitung*, 1868–75; the *Oberpfälzer Kurier* (Regensburg), 1875–7; various journals in Bayreuth, 1877–8, and Munich. (*Taxonomic literature.*)

Moore, Charles (1820–1905). Botanist. Gardener at the Trinity College Botanic Garden, Dublin, 1835–8. Gardener at the Royal Botanic Gardens, Kew, 1847. Emigrated to Sydney, Australia, in 1848. Toured the South Pacific on HMS *Havannah* in 1850. Director of the Botanic Gardens, Sydney, 1848–96. (R. Desmond 1994.)

Moore, Thomas (1821–87). Gardener. Gardener, Botanic Garden at Regent's Park, London, 1844–7. Curator, Physic Garden, Chelsea, London, 1848–87. Published major works on ferns. Assistant editor, *Gardeners' Chronicle*, 1841–66; co-editor, 1866–81. Co-editor, *Gardener's Magazine of Botany*, 1850–1; *Garden Companion and Florist's Guide*, 1852; *Orchid Album*, 1881–7. Editor, *Florist and Pomologist*, 1868–74. (*Annals of Botany* (1888): 409–10, R. Desmond 1994, *ODNB*, *Proceedings of the Linnean Society of London* (1886–7): 41–2.)

More, Alexander Goodman (1830–95). Botanist and natural historian. Matriculated at Trinity College, Cambridge, 1850. Suffered ill health for many years and later settled in Ireland. Published his chief work, *Cybele Hibernica*, with David Moore, in 1866. Assistant, Natural History Museum, Dublin, 1867; curator, 1881–7. (*Alum. Cantab.*, R. Desmond 1994, *Journal of Botany* 33 (1895): 225–7.)

24 June 1869

Morris, Francis Orpen (1810–93). Clergyman, naturalist, and writer. While at Worcester College, Oxford, arranged the insect collection at the Ashmolean museum. BA, 1833. Ordained deacon, 1834; priest 1835. Held several clerical posts until settling as rector of Nunburnholme, Yorkshire, in 1854. Publications included works on birds, bird protection, butterflies and moths, and opposition to Darwinism. (*DNB*, *ODNB*.)

Mortlock, Edmund John (1833–1902). Cambridge banker. BA, Cambridge, 1854; MA, 1857. (*Alum. Cantab.*, *County families.*)

Morton, Thomas (1764?–1838). Dramatist. Author of *Speed the plough* (1798). (*ODNB.*)

Moseley, Henry (1801–72). Mathematician and writer on mechanics. BA,

Moseley, Henry, cont.
Cambridge, 1826; MA, 1836. Ordained deacon, 1827; priest, 1828. Professor of natural and experimental philosophy and astronomy, King's College, London, 1831–44. Inspector of normal schools, 1844. FRS 1839. (*ODNB*.)

Mota, Ignácio Francisco Silveira da (*fl.* 1849–69). Brazilian governor and sugar-cane planter. Born in Goiás. Took his degree at the University of São Paulo. Married into an established family of sugar-cane planters. Appointed president of several provinces: Piauí, 1849, then Ceará, and later Rio de Janeiro. Published articles on plants and agriculture. (Needell 2006, pp. 125–6, 210.)

Moulinié, Jean Jacques (1830–73). Swiss zoologist and militia inspector. Translated *Variation*, *Descent*, and the sixth edition of *Origin* into French. (Tort 1996.)
15 February 1869, 22 February [1869], *20 October 1869*, 23 October [1869], *5 November 1869*, 15 November [1869], 20 November 1869

Mozley, James Bowling (1813–78). Theologian and journalist. BA, Oxford, 1834; MA, 1838; fellow, Magdalen College, 1840; BD, 1846. Canon of Worcester, 1869. Regius professor of divinity at Oxford, 1871. (*ODNB*.)

Mueller, Ferdinand Jakob Heinrich (Ferdinand) von (1825–96). German-born explorer and botanist. Emigrated to Australia in 1847. Government botanist, Victoria, 1852. Botanist to the North West Australia Expedition, 1855–7. Director of the Botanical Gardens, Melbourne, 1857–73. President of the Australian Association for the Advancement of Science, 1890. FRS 1861. (*Aust. dict. biog.*, R. Desmond 1994.)

Müller, Heinrich Ludwig Hermann (Hermann) (1829–83). German botanist and entomologist. Brother of Johann Friedrich Theodor (Fritz) Müller. Schoolteacher in Schwerin, 1854–5. Studied blind cave insects in Krain, 1855. Teacher of natural sciences at the Realschule in Lippstadt, 1855–83; became director of the school. After settling in Lippstadt, studied the local flora, in particular the mosses. CD's *Orchids* directed Müller's attention to the pollination and fertilisation of flowers, on which he published several papers and books. (Gilbert 1977, Krause 1883, *Science* 2 (1883): 487–8.)
22 September 1869, *12 November 1869*

Müller, Johann Friedrich Theodor (Fritz) (1822–97). German naturalist. Emigrated to the German colony in Blumenau, Brazil, in 1852. Taught mathematics at the Lyceum in Destêrro (now Florianópolis), 1856–67. *Naturalista Viajante* of the National Museum, Rio de Janeiro, 1876–92. His anatomical studies on invertebrates and work on mimicry provided important support for CD's theories. (*ADB*, *DBE*, Möller ed. 1915–21, *NDB*, West 2003.)
12 January 1869, 14 March 1869, *14 March 1869*, 18 March 1869, *15 June 1869*, 18 July [1869], 8 September [1869], *18 October 1869*, 1 December [1869], *18 December 1869*

Müller, Johannes Peter (1801–58). German comparative anatomist, physiologist, and zoologist. Became professor of anatomy and physiology at Berlin University in 1833. Foreign member, Royal Society, 1840. (*DSB*, *Record of the Royal Society of London*.)

Munby, Giles (1813–76). Studied medicine and botany in Edinburgh and Paris. Collected plants in Algeria, 1839–59. (R. Desmond 1994, *DNB*.)

Murchison, Roderick Impey, 1st baronet (1792–1871). Geologist and army officer. Served in the British army, 1807–15. Noted for his work on the Silurian system. President of the Geological Society of London, 1831–3 and 1841–3; of the British Association for the Advancement of Science, 1846; of the Royal Geographical Society of London, 1843–4, 1851, 1857–8, 1862–70. Director-general of the Geological Survey of Great Britain, 1855. Knighted, 1863; created baronet, 1866. FRS 1826. (*DSB*, *Journal of the Royal Geographical Society*, *ODNB*.)

Murphy, Joseph John (1827–94). Irish mill-owner, merchant, and writer. Published poetry and works on natural theology. Originally a Quaker, but became a member of Church of Ireland. Resided in Belfast. (*Modern English biography*.)

Murray, Andrew Dickson (Andrew) (1812–78). Lawyer, entomologist, and botanist. Practised law in Edinburgh; later moved to London. Assistant secretary to the Royal Horticultural Society, 1861; member of the council from 1865; member of the scientific committee from 1868. An expert on insects harmful to crops. In entomology, specialised in the Coleoptera; in botany, in the Coniferae. (*ODNB*.)

Murray, John (1808–92). Publisher, and author of guide-books. CD's publisher from 1845. (Freeman 1978, *ODNB* s.v. Murray family, publishers.)

[1–3 February 1869], *3 February [1869]*, *28 April [1869]*, 29 April [1869], *18 September [1869]*, [after 18 September 1869], 8 November [1869], *8 November [1869]*, *17 November 1869*

Nägeli, Carl Wilhelm von (1817–91). Swiss botanist. Maintained a teleological view of evolution. Originally studied medicine, but transferred to botany under Alphonse de Candolle at Geneva. Worked for eighteen months with Matthais Jacob Schleiden at the University of Jena, then worked in Zurich, where he collaborated with Carl Cramer, 1845–52. Professor of botany, University of Freiburg, 1852; professor of botany, University of Munich, 1857. (*DSB* s.v. Naegeli, Carl Wilhelm von.)

Nash, Henry Allden (b. 1838). Clergyman. Minister of the Independent Chapel at South Bar, Banbury, Oxfordshire, 1869. Ordained, 1870. Curate of St Thomas's, Lambeth, 1869–70; of St Margaret's, Birmingham, 1870–5; vicar of St Margaret's, 1875–90; of Ogley-Hay, 1890–3; of St Mark's, Liverpool, 1893–1909. (*Crockford's clerical directory* 1915; Foster 1893; *Post Office directory of Northamptonshire etc.* 1869.)

Nash, Louisa Ann. Wife of Wallis Nash (1873–1926); they lived in Down during some part of the 1870s, and later emigrated to the US. Painted a portrait of CD in india ink, 1873. (Freeman 1978.)

Nature

13 November [1869]

Newton, Alfred (1829–1907). Zoologist and ornithologist. Travelled extensively throughout northern Europe and North America on ornithological expeditions, 1854–63. Editor of *Ibis*, the journal of the British Ornithologists' Union, 1865–70. Professor of zoology and comparative anatomy, Cambridge University, 1866–1907. FRS 1870. (*DNB*.)

9 April 1869

Newton, Edward (1832–97). Ornithologist and colonial official. Held a succession of appointments in the colonial service, including colonial secretary of Mauritius, 1868–78, and lieutenant-governor of Jamaica, 1878–83. Co-authored papers with his elder brother, Alfred Newton. Knighted, 1887. (*Modern English biography*, A. F. R. Wollaston 1921.)

Nilsson, Christine (1843–1921). Swedish soprano. Studied in Stockholm and Paris. Made her London début in 1864, sang at Covent Garden in 1869 and Drury Lane in 1870. Embarked on a tour of the US in 1870. Later visited St Petersburg, Moscow, Brussels, Vienna, and Munich. (Grove 1980.)

Nilsson, Sven (1787–1883). Swedish zoologist and archaeologist. Professor of natural history, Lund, 1831. (*NBU*, Nilsson 1868, *SMK*.)

5 January 1869

Noble, Charles (b. 1817/18). Nurseryman. Partner in Sunningdale Nursery at Bagshot, Surrey, 1846. Retired, 1898. (Census returns 1861 (Public Record Office RG9/423/122/24), Leeds 2000, R. Desmond 1994.)

Norman, Alfred Merle (1831–1918). Clergyman and invertebrate zoologist. BA, Oxford, 1852; deacon 1856; priest, 1857. Incumbent of Burnmoor, 1866–95. Honorary canon of Durham Cathedral, 1885. Rector, Houghton-le-Spring, 1895–8. Spent all his summer vacations dredging around the British Isles, Norway, and Madeira and built up a large collection of marine invertebrates from these regions; the collection was later catalogued and transferred to the British Museum. (*Proceedings of the Malacological Society of London* 13 (1919): 116–17.)

29 July [1869], *2 August 1869*

Norris, Richard (b. 1831/2, d. *c.* 1917). Physiologist. MD, St Andrews, 1862. Member of the Royal College of Surgeons of England, 1860. Professor of physiology, Queen's College, Birmingham. (Census returns 1881 (Public Record Office RG11/3040/44/1), *Medical directory*.)

Norton, Charles Eliot (1827–1908). American editor, literary critic, and professor of art history. Graduated from Harvard College in 1846. Apprenticed himself in the East India trade, travelling widely in India and Europe. Gradually shifted to a literary career; wrote, translated, and edited books; contributed to the *Atlantic Monthly*; co-edited the *North American Review*, 1863–68; and co-founded and wrote for the *Nation*. Travelled and lived in England and continental Europe, 1868–73. Taught history of art and literature at Harvard, 1874–98. (*ANB*.)

Norton, Susan Ridley (1838–72). Daughter of Sara Ashburner and the American legal theorist Theodore Sedgwick. Grew up in New York and Massachusetts. Married Charles Eliot Norton in 1862. Died after giving birth to her sixth child in Dresden, Germany. (Turner 1999.)

Odling, William (1829–1921). Chemist. Student, Guy's Hospital, London; demonstrator in chemistry, 1850; MB, 1851. Appointed lecturer, St Bartholomew's Hospital, 1863. Fullerian Professor of chemistry, Royal Institution, 1868. Waynflete Professor of chemistry, Oxford, 1872–1912. Fellow, Royal College of Physicians, 1859. Recognised for his research in both medical and theoretical chemistry. FRS 1859. (*ODNB*.)

Oersted, Hans Christian (1777–1851). Danish physicist and chemist. Professor of natural philosophy, University of Copenhagen. (*DSB*.)

Ogle, William (1827–1912). Physician and naturalist. MD, 1861. Lecturer on physiology at the medical school, and assistant physician, St George's Hospital, 1869–72. Superintendent of statistics, General Register Office, 1880–1903. Translated Aristotle's *De partibus animalium* into English in 1882. Published on flower structure and mechanisms for pollination. (*Alum. Oxon.*, *The Times*, 15 April 1912, p. 9.)

7 July [1869], *[after 7 July 1869]*

Oliver, Daniel (1830–1916). Botanist. Assistant in the herbarium of the Royal Botanic Gardens, Kew, 1858; librarian, 1860–90; keeper, 1864–90. Professor of botany, University College, London, 1861–88. FRS 1863. (R. Desmond 1994, *List of the Linnean Society of London*, 1859–91.)

Oliver, James (b. 1801). Scottish farmer. Tenant at Borthwickbrae Burnfoot; later farmed at Howpasley, near Hawick, in the Scottish borders. (Douglas Scott, *A Hawick word book*, 9 October 2007, http://www.astro.ubc.ca/people/scott/book.pdf (accessed 10 October 2007), p. 713.)

3 July 1869 (to George Cupples)

Orbigny, Alcide Charles Victor Dessalines d' (1802–57). French palaeontologist. Travelled widely in South America, 1826–34, on a commission for the Muséum d'Histoire Naturelle. Professor of palaeontology, Muséum d'Histoire Naturelle, 1853. (*DSB*.)

Ormerod, George Wareing (1810–91). Geologist. BA, Oxford, 1833; MA, 1836. Practised as a solicitor in Manchester until 1855, after which he moved to Devon. Published on the geology of Devon. Compiled an index to the publications of the Geological Society, 1858; second edition, 1870. (*ODNB*.)

Orton, James (1830–77). American natural historian, explorer, and educator. AB, Williams College, 1855; graduated from Andover Theological Seminary, 1858. Ordained in the Presbyterian ministry, 1860. Instructor in natural history, University of Rochester, 1866. Leader, Andes and Amazon expedition, 1867–9, 1873, 1876. Professor of natural history, Vassar College, 1869–77. Wrote a textbook on comparative zoology and also published on women's education. (*ANB*, Elliott 1979.)

4 January 1869, 23 January [1869], *8 September 1869*, 7 October [1869]

Owen, Richard (1804–92). Comparative anatomist. Assistant conservator of the Hunterian Museum, Royal College of Surgeons of England, 1827; Hunterian Professor of comparative anatomy and physiology, 1836–56. Superintendent of the natural history departments, British Museum, 1856–84; prime mover in establishing the Natural History Museum, South Kensington, 1881. President of

Owen, Richard, cont.
the British Association for the Advancement of Science, 1858. Described the *Beagle* fossil mammal specimens. Knighted, 1884. FRS 1834. (*DSB*, *ODNB*.)

Pagan, Allan C. Scottish sheep-farmer. Farmed at Crieff. (Letter from A. C. Pagan to George Cupples, 16 July 1869.)
16 July 1869 (to George Cupples)

Pagenstecher, Heinrich Alexander (1825–89). German zoologist and surgeon. Professor of zoology and palaeontology, Heidelberg University, from 1863. Director of the Zoological Institute, Heidelberg. Director of the Natural History Museum, Hamburg, 1882. Author of *Allgemeine Zoologie* (1875–81). (*NDB*.)

Paget, James, 1st baronet (1814–99). Surgeon. Assistant surgeon at St Bartholomew's Hospital, London, 1847; surgeon, 1861–71. Arris and Gale Professor of anatomy and surgery at the Royal College of Surgeons of England, 1847–52. Lectured on physiology in the medical school, St Bartholomew's, 1859–61; on surgery, 1865–9. Appointed surgeon-extraordinary to Queen Victoria, 1858; serjeant-surgeon, 1877. Created baronet, 1871. FRS 1851. (*ODNB*.)
[1869], 29 April [1869]

Paley, William (1743–1805). Anglican clergyman and philosopher. Propounded a popular system of natural theology. (*DSB*, *ODNB*.)

Palgrave, Francis (1788–1861). Historian. Author of several works on the history of England and on antiquarian topics. Deputy keeper of the queen's records, 1838–61. Married Joseph Dalton Hooker's aunt, Elizabeth Turner, in 1823. Knighted, 1832. FRS 1821. (*DNB*.)

Palgrave, Francis Turner (1824–97). Poet, critic, and educationalist. Assistant private secretary to William Ewart Gladstone, 1846. Joined the department of education in 1848; vice-principal, Kneller Hall, 1850–5; examiner and assistant secretary, 1855–84. Art critic for the *Saturday Review*. Professor of poetry, Oxford University, 1885–95. (*DNB*.)

Pallas, Pyotr Simon (1741–1811). German naturalist and geographer. Travelled widely in the Russian empire. Foreign member, Royal Society of London, 1764. (*ADB*, *DSB*, *Record of the Royal Society of London*.)

Parker, Francis (Frank) (1829–71). Solicitor in Chester, Cheshire. CD's nephew. Third son of Henry and Marianne Parker. (*Darwin pedigree*.)

Parker, Theodore (1810–60). American Unitarian clergyman. Campaigned against slavery and was actively involved in attempts to protect fugitive slaves. (*ANB*.)

Paul, John (1806–72). American physician. Educated in Vermont. Practised medicine in Canada, Ohio, and Illlinois. Moved to Ottawa, Illinois, in 1868. (*Ottawa Free Trader*, 28 September 1872.)

Payer, Jean Baptiste (1818–60). French botanist. Professor of geology and mineralogy at Rennes, 1840. Professor of botany at the Ecole Normale, Paris, 1841. Professor of plant organography, Faculté des Sciences, Paris, 1852. Member of the Académie des Sciences, Paris, 1854. (*IBF*.)

Pertz, Georg Heinrich (1795–1876). German historian. Historian of medieval Germany; published important source material in *Monumenta Germaniae historica*. Archives secretary, royal library, Hanover, 1823; principal keeper, 1827. Called to Berlin as chief librarian, 1842. Married Leonora Horner in 1853. (*ADB*, *EB*.)

Pertz, Leonora (b. 1818). Daughter of Leonard Horner. Translator, with her sister Joanna Baillie Horner, of German works. Married Georg Heinrich Pertz in 1853. (*ADB* s.v. Pertz, Georg Heinrich, *NUC*.)

Peters, Wilhelm Karl Hartwig (Wilhelm) (1815–83). German zoologist and explorer. Professor of zoology, Berlin, 1858. (*DBE*.)

Peyronnet, Jules, vicomte de (1804–1872). French landowner and translator. Owned land in the West Indies. Married Georgina Frances Whitfield (1815–95), an Englishwoman. Moved to St Vincent in 1840 and returned to France in 1844. Translated into English the *History of England* by Thomas Babington Macaulay (1853). (*ODNB* s.v. Peyronnet, (Georgina) Frances de.)

Pfaff, Alexius Burkhard Immanuel Friedrich (Friedrich) (1825–86). German mineralogist and geologist. Studied mathematics and mineralogy at Erlangen, where he habilitated in 1853. Professor of mineralogy, Erlangen, 1859–86. Attempted to reconcile the results of geological research with biblical accounts and opposed CD's theory. (*ADB*.)

Phillips, John (1800–74). Geologist. Keeper of the Yorkshire Museum, 1825–40. Assistant secretary, British Association for the Advancement of Science, 1832–62. Professor of geology, King's College, London, 1834–9. Palaeontologist to the Geological Survey of Great Britain, 1841–4. Deputy reader in geology, Oxford University, 1853; professor, 1860–74. FRS 1834. (*DSB*, *ODNB*.)

27 January 1869

Pike, Nicholas (1818–1905). American naturalist and diplomat. Director, Brooklyn Institute of Sciences, 1850. Consul general to Portugal, 1852; consul to Mauritius, 1866. (*Condor* 30 (1928): 291–2.)

Playfair, John (1748–1819). Mathematician and geologist. Professor of mathematics, Edinburgh University, 1785–1805; professor of natural philosophy, 1805. Expounded the geological theories of James Hutton in a short and clear form, 1802. FRS 1807. (*DNB*, *DSB*.)

Pollock, Frederick (1845–1937). Jurist. BA (Trinity College), Cambridge, 1867; fellow, 1868. Called to the bar, 1871. Corpus Professor of jurisprudence, Oxford, 1883–1903. Founding editor of the *Law Quarterly Review*, 1885. Wrote extensively on legal topics. (*ODNB*.)

Ponderer. Pseudonymous letter-writer to the *Athenæum*.

[before 5 June 1869] (to the *Athenæum*)

Powell, Henry (1839/40–92). Clergyman. BA, Cambridge, 1861; MA, 1865. Deacon, 1862; priest, 1864. Vicar at Down, 1869–71. (*Alum. Cantab.*)

Preyer, William Thierry (1841–97). English-born German physiologist. Studied medicine and natural science at Bonn, Heidelberg, Berlin, and Vienna. PhD,

Preyer, William Thierry, cont.

Heidelberg, 1862. Habilitated at Bonn, 1865. MD, 1866, Jena. Professor of physiology, Jena 1869; Berlin, 1888–94. Worked mainly on developmental physiology. A proponent of science teaching in schools. (*DBE.*)

[before 21 March 1869], *21 March 1869*, 29 March 1869

Price, John (1803–87). Scholar, schoolteacher, and naturalist. Educated at Shrewsbury School, 1818–22, and St John's College, Cambridge. Assistant master, Shrewsbury, 1826–7. Headmaster of the junior department at Bristol College, then classics principal at the Liverpool High School, before settling in Chester. A founding member of the Chester Natural Science Society. (*Alum. Cantab.*, *Eagle* (St John's College, Cambridge) 15 (1888): 169–72, *Modern English biography.*)

24 April 1869

Pritchard, Charles (1808–93). Clergyman and astronomer. Headmaster, 1834–62, of Clapham Grammar School, where he established an observatory. Hulsean Lecturer, Cambridge University, 1867. Savilian Professor of astronomy, Oxford University, 1870–93. President of the Royal Astronomical Society, 1866–8. FRS 1840. (*Alum. Cantab.*, *DSB*, *ODNB.*)

Pryor, Marlborough Robert (1848–1920). Businessman. BA, Cambridge, 1870; MA, 1873; fellow of Trinity College, 1870. Chairman, Sun life Insurance Co. (*Alum. Cantab.*)

Pulvermacher, Isaac Louis (1814/15–84). Austrian-born electrical engineer. Proprietor of Isaac Louis Pulvermacher & Co, galvanic establishment, 200 Regent's Street, London. Published pamphlets on his 'volta-electric chain bands'. (*BMD* (*Death index*); *Chemist* n.s. 3 (1851–2): 569; *Post Office London directory* 1869.)

Quatrefages de Bréau, Jean Louis Armand de (Armand de Quatrefages) (1810–92). French zoologist and anthropologist. Doctorate in the physical sciences, University of Strasbourg, 1830; doctorate in medicine, 1832. Founded *Journal de médicine et de chirurgie de Toulouse*, 1836. Moved to Paris and took a doctorate in the natural sciences, 1840. Professor of natural history at the Lycée Henri IV, Paris, 1850; professor of anthropology, Muséum d'Histoire Naturelle, 1855. Foreign member, Royal Society of London, 1879. (*DSB*, *Record of the Royal Society of London.*)

29 March 1869

Quetelet, Lambert Adolphe Jacques (1796–1874). Belgian statistician. Astronomer, Brussels Royal Observatory, 1828–74. Secretary, Académie Royale des Sciences et Belles-Lettres, Brussels, 1834–74. Foreign member, Royal Society, 1839. (*BNB*, *DSB.*)

Radclyffe, William Frederick (1805/6–80). Clergyman and horticulturalist. BA, Oxford, 1829; MA, 1837. Deacon, 1829; priest, 1830. Rector, Tarrant Rushton and Tarrant Rawston, Dorset, 1852–80. Wrote horticultural articles. (*Alum. Oxon.*, *Crockford's clerical directory*, R. Desmond 1994.)

Rait, James. Forester at Castle Forbes, Aberdeenshire. (J. Brown 1861, p. 630.)
20 July 1869 (to George Cupples)

Ramsay, Andrew Crombie (1814–91). Geologist. Appointed to the Geological Survey of Great Britain, 1841; senior director for England and Wales, 1862; director-general, 1871–81. Professor of geology, University College, London, 1847–52; lecturer on geology at the Royal School of Mines, 1852–71. President of the Geological Society of London, 1862–4. Knighted, 1881. FRS 1862. (*DNB*, *DSB*.)
2 February 1869, 3 February [1869]

Ramsey, George M'Ilvaine (1820–1902). American physician, writer, and farmer. After working as a carpenter and schoolteacher, graduated from Jefferson Medical College, Philadelphia, in 1852. Practised in New York City, 1853–61, 1865–76. Surgeon, Ninety-fifth Regiment of New York, Army of the Potomac, 1861–5. Returned to his family farm in Washington county, Pennsylvania, in 1876. Published several books linking astronomy, meteorology, geology, and theology. (Crumrine 1882; personal communication from the Washington County Historical Society (Pennsylvania) (Cemetery listings).)
5 April 1869

Rawdon, Elizabeth Anne (1793–1874). Daughter of John Theophilus Rawdon and Frances Stevenson. Married George William Russell in 1817. (*ODNB* s.v. Russell, Lord George William.)

Rawlins, Francis Hay (1850–1920). Teacher and classicist. BA (King's College), Cambridge, 1874; MA, 1877; fellow, 1875–1920. Assistant master, Eton, 1875–1905; lower master, 1905–16; vice-provost, 1916–20. (*Alum. Cantab.*)

Reade, William Winwood (1838–75). Traveller, novelist, and journalist. Travelled in West Africa, 1861–3. Studied at St Mary's Hospital, London, in 1865; worked at a cholera hospital in Southampton, 1866. Travelled in West Africa, 1868–70, and again in 1873 as correspondent for *The Times*. Wrote novels as well as travel observations. (Letter from W. W. Reade, 28 June [1869], *Correspondence* vol. 18, letter from W. W. Reade, 4 June 1870, *ODNB*.)
17 January 1869, 28 June [1869], 26 December 1869

Recht, Georg (b. *c.* 1816 d. 73). German mathematician. Studied mathematics at Munich; PhD, 1840; habilitated, 1842; professor extraordinarius, 1849–67. Wrote on mathematics and philosophy. (Bosl ed. 1983.)
9 February 1869, 1 March 1869, 21 April 1869

Reeves, William (b. 1814/15). Gamekeeper, employed by John Lubbock. (Census returns 1871 (Public Record Office RG10/875: 2 s.v. Willm Reaves); DAR 84.2: 202.)

Reeves, William (b. 1834/5). Blacksmith and fly proprietor at Down. (Census returns 1871 (Public Record Office RG10/875 p. 6 s.v. Will Reaves); Census returns 1881 (Public Record Office RG11/855 p. 11).)

Reinwald, Charles-Ferdinand (1812–91). German-born bookseller and editor.

Reinwald, Charles-Ferdinand, cont.

Founded a business exporting French books in Paris, 1849. Editor in particular of foreign scientific works, and of the *Dictionnaire universel de la langue française*, by M. P. Poitevin. Published the *Catalogue annuel de la librairie française*. (*Dictionnaire universel des contemporains*.)

13 November 1869

Rérolle, Louis (1849–1928). French naturalist and museum director. Completed a degree in natural science, Lyon, 1876. Travelled in South America for the museum of Lyon, 1877–8. Teacher, agricultural school, Ecully, 1880–2; Société d'ensignement professionnel du Rhône,1881–5. Director, natural history museum of Grenoble, 1887–1919. (Personal communication from the Museum d'Histoire Naturelle, Grenoble.)

17 March 1869, 10 May 1869, 30 July 1869

Reuter, Adolf (1825–1901). German horticulturalist. From 1843, studied horticulture in the Prussian royal botanic gardens at Sanssouci and Berlin, and from 1850 to 1854 in Liege, Ghent, Paris, and Kew. Planteur in the royal nursery, Sanssouci, 1854; inspector, 1865, head gardener, 1866. In 1894 received the title of head court gardener. (*Gartenflora* 51 (1902): 64–8.)

30 May 1869, 2 June 1869, *18 July 1869*, *23 September 1869*, 27 September 1869

Richardson, H. D. Scottish author and dog-breeder. Born in Scotland, resident in Dublin some time before his death. Wrote on deer, dogs, bees, pigs, and horses. (*CDEL*.)

Richardson, John (1787–1865). Arctic explorer and naturalist. After serving as assistant surgeon and surgeon on various ships, acted as surgeon and naturalist on John Franklin's polar expeditions, 1819–22 and 1825–7. Surgeon to the Chatham division of marines, 1824–38. Appointed physician to the Royal Hospital, Haslar, 1838. Conducted a search expedition for John Franklin, 1847–9. Retired to Grasmere, where he accomplished much literary work. Knighted, 1846. FRS 1825. (*DNB*; R. Johnson 1976.)

Ricker, Franz Anton (1816–92). German bookseller and publisher. Took over his brother Joseph's publishing business, J. Ricker'sche Buchhandlung, in Giessen, Hesse, in 1835. (*Hessische Biographien* 3: 301.)

Ridpeth, Thomas Hardwick (b. 1835). Resident of Poplar House, Wilmslow Road, Manchester. (*IGI* (accessed 10 March 2008); *Post Office directory of Lancashire, Liverpool and Manchester* 1867.)

Rigby, Anne. *See* Wahl, Anne de.

Rigby, Gertrude. *See* Rosen, Gertrude de.

Rigby, Maria Justine. *See* Rosen, Maria Justine de.

Riley, Charles Valentine (1843–95). Entomologist. Emigrated to the United States *circa* 1859. Worked as a livestock farmer in Illinois. Wrote and illustrated for the *Prairie Farmer* in Chicago, 1863–8; served in a volunteer regiment in 1864. State entomologist of Missouri, 1868–76. Became a US citizen in 1869. Chief of the Department of the Interior's US Entomological Commission, 1877–82. Entomologist with the Division of Entomology of the Department of Agriculture, 1878, 1881–94. Appointed honorary curator of the Department of Insects of the Smithsonian Institution, 1881. Organised the American Association of Economic Entomologists, 1889. (*ANB*.)

Robertson, Peter. Scottish forester. Head-forester of Breadalbane forest in Scotland. (*Correspondence* vol. 16, letter from George Cupples, 13 July 1868; this volume, letter from George Cupples, 11 January 1869.)

Robinson, John Warburton (b. 1837/8). Clergyman. BA, Trinity College, Dublin, 1859; MA, 1863. Admitted *comitatis causa*, Oxford, 1864. Ordained deacon, 1864; priest, 1860. Appointed curate of St Mary's, Haggerston, London, 1864. Curate of Down, 30 August 1868 to 4 February 1869. Appointed curate of Brasted, Kent, 1869. Curate of Lynsted near Sittingbourne, Kent, 1872–4. Appointed curate of Blisworth, Northants, 1876. Emigrated to Australia *circa* 1882. (*Alum. Dublin.*, J. R. Moore 1985, p. 477, Census returns 1871 (Public Record Office RG10/924/6/4).)

1 February 1869

Rolfe, Robert Monsey, 1st Baron Cranworth of Cranworth (1790–1868). Statesman and jurist. Whig MP for Penryn and Falmouth, 1832–9. Solicitor-general, 1834 and 1835–9. Created Baron Cranworth of Cranworth, 1850. Lord justice of appeal, 1851–2. Lord chancellor, 1852–8 and 1865–6. Holwood Park, his country residence, was a mile and a half north of Down House, the Darwins' residence. (*Dod's parliamentary companion*, Freeman 1978, *ODNB*.)

Rolle, Friedrich (1827–87). German geologist, palaeontologist, and natural history dealer. Assistant at the Kaiserlich-königliche Hofmineralien-Cabinett, Vienna, 1857–9; associate, 1859–62. Returned to Bad Homburg in 1862. Author of *Ch. Darwin's Lehre von der Entstehung der Arten und ihre Anwendung auf die Schöpfungsgeschichte* (1863). (*ADB*, *BLKO*, Martin and Uschmann 1969, Sarjeant 1980–96.)

Rosen, Gertrude de (1813–59). Daughter of Anne Palgrave and Edward Rigby. Cousin of Maria Turner, J. D. Hooker's mother. Married Theophile de Rosen in Estonia. (L. Huxley 1918, 1: 17–8.)

Rosen, Maria Justina de (1808–89). Daughter of Anne Palgrave and Edward Rigby. Cousin of Maria Turner, J. D. Hooker's mother. Married Baron Robert de Rosen in Estonia. (L. Huxley 1918, 1: 17–18.)

Ross, James Clark (1800–62). Naval officer and polar explorer. Joined the navy in 1812. Discovered the northern magnetic pole in 1831. Employed on the magnetic survey of the United Kingdom, 1835–8. Commander of an expedition to the Antarctic, 1839–43; and of a search expedition for John Franklin (*ODNB*), 1848–9. Knighted, 1843. FRS 1828. (*ODNB*.)

Royer, Clémence Auguste (1830–1902). French author and economist. Studied natural science and philosophy in Switzerland. In Lausanne in 1859, founded a

Royer, Clémence Auguste, cont.
course on logic for women. Translated *Origin* into French in 1862. (*Dictionnaire universel des contemporains*, Harvey 1997.)

Ruck, Mary Anne (1820/1–1905). Née Matthews; married Lawrence Ruck in 1859. Lived at Pantlludw, near Machynlleth, Wales. Mother of Amy Richenda Ruck, who married Francis Darwin in 1874. (*BMD* (*Death index*, *Marriage index*), *Darwin pedigree*, *ODNB* s.v. Ruck, Amy Roberta.)
[29 April – 27 May 1869] (to Francis Darwin)

Ruck, Richard Matthews (1851–1935). Army officer. Brother of Amy Ruck, the first wife of Francis Darwin. Attended the Royal Military Academy, Woolwich. Entered the Royal Engineers, 1871; captain, 1883, major, 1889; lieutenant-colonel, 1896; colonel, 1904; major-general, 1908. Knighted, 1920. (Ruck 1935, *WWW* 1929–40.)

Russell, Arthur John Edward (Arthur) (1825–92). Member of Parliament. Brother of the ninth duke of Bedford, Hastings Russell. Private secretary to Lord John Russell, 1849–54. Liberal MP for Tavistock, 1857–85. Member of the Royal Geographical, Linnean, and Zoological Societies of London, and the Anthropological Institute. (Grant Duff 1903, 2: 112–28; Stenton 1976.)

Russell, Caroline Diana Rosalind (Diana) (1874–1971). Daughter of Laura and Arthur Russell. (*Burke's peerage* 2003.)

Russell, Flora Magdalen Isabel (1869–1967). Daughter of Laura and Arthur Russell. (*Burke's peerage* 2003.)

Russell, Frances Anna Maria (Fanny), Countess Russell (formerly Lady John Russell) (1815–98). Scottish-born political hostess. Daughter of Gilbert Elliot Murray Kynynmound, second earl of Minto. Whig in politics, and Presbyterian, later Unitarian, in religion. Married Lord John Russell in 1841. Resided at Pembroke Lodge in Richmond Park. (*ODNB*.)

Russell, Francis Charles Hastings (Hastings), 9th duke of Bedford (1819–91). Agriculturalist. MP for Bedfordshire 1847–72. Succeeded to the dukedom in 1872. Carried out agricultural experiments on his estate at Woburn. President of the Royal Agricultural Society from 1879. (*ODNB* s.v. Russell, Lord George William.)

Russell, John. Scottish sheep-farmer. Farmed at Pitbladdo, Cupar, Fife. (Letter from John Russell to George Cupples, 18 June 1869.)
18 June 1869 (to George Cupples)

Russell, John, 1st Earl Russell (1792–1878). Statesman. Liberal prime minister, 1846–52 and 1865–6. Home secretary, 1835–9; secretary of state for foreign affairs, 1852–3; for the colonies, 1839 and 1855. Foreign secretary under Lord Palmerston, 1859–65. Created Earl Russell of Kingston Russell, 1861. In later life, occupied with literary work. President of the Royal Historical Society, 1872. FRS 1847. (*DNB*, *Record of the Royal Society of London*.)

Russell, Laura (d. 1910). Daughter of the vicomte de Peyronnet and Frances de Peyronnet (*ODNB*). Married Arthur Russell in 1865. (*Burke's peerage*.)

Russell, Odo William Leopold (1829–84). Diplomat. Brother of Hastings Russell, ninth duke of Bedford. Worked at the Foreign Office, 1850–2. Sent during the 1850s to embassies in Vienna, Paris, Constantinople, Washington, and Florence. Attached to a special mission to Sicily, 1859–70; to the German army headquarters at Versailles, 1870–1. Ambassador at Berlin, 1871–84. (*ODNB.*)

Rütimeyer, Karl Ludwig (Ludwig) (1825–95). Swiss palaeozoologist and geographer. Professor of zoology and comparative anatomy, University of Basel, 1855; rector, 1865; professor in the medical and philosophical faculties, 1874–93. Made important contributions to the natural history and evolutionary palaeontology of ungulate mammals. (*DSB.*)

Sabine, Edward (1788–1883). Astronomer, geophysicist, and army officer. Entered the army in 1803; major-general, 1856; general, 1870. Astronomer to expeditions in search of a north-west passage, 1818 and 1819–20. Appointed one of three scientific advisers to the Admiralty, 1828. General secretary of the British Association for the Advancement of Science, 1839–52 and 1853–9. Foreign secretary of the Royal Society of London, 1845; treasurer, 1850; president, 1861–71. Knighted, 1869. FRS 1818. (*DNB, DSB.*)

St Clair, George (*fl.* 1860s). Baptist minister in the township of Grimsbury, Banbury, Oxfordshire. (*Post Office directory of Northamptonshire etc.* 1869.)
14 August 1869

Saint-Hilaire, Augustin François César Prouvençal (Auguste) de (1779–1853). French naturalist. Surveyed the flora and fauna of Brazil, 1816–22. Appointed professor at the faculty of sciences, Paris, 1830. (*DSB.*)

Salvin, Francis Henry (1817–1904). Field sport enthusiast. Best known for his work in falconry, but kept several wild animals on his estate at Sutton Place. (*ODNB.*)
31 October 1869

Samuelson, James (1829–1918). Businessman, science teacher, and writer on scientific and social issues. Inherited his father's business of seed crushing and oil extraction. Worked as science teacher in Liverpool. Founder, with Robert Hardwicke, and editor of the *Popular Science Review*, 1862–4. Helped launch the *Quarterly Journal of Science* and occasionally acted as editor until 1870. Called to the bar, 1870, and then campaigned as social reformer and writer. (Lightman ed. 2004.)

Saporta, Louis Charles Joseph Gaston (Gaston), comte de (1823–96). French palaeobotanist. Specialist on the Tertiary and Jurassic flora. Wrote extensively on the relationship between climatic change and palaeobotany. (*DSB.*)

Sars, Michael (1805–69). Norwegian naturalist and cleric. Pastor, Kinn, Norway, 1830; Manger, Hordaland, 1839. Professor of zoology, Christiania, 1854. (*BHGW, DSB.*)

Savery, Roelandt (1576–1639). Dutch painter, draughtsman, and etcher. (*Dictionary of art.*)

Schimper, Wilhelm Philipp (1808–80). Alsatian bryologist and palaeontologist. Assistant in the geological section of the Museum of Natural History, University

Schimper, Wilhelm Philipp, cont.
of Strasbourg, from 1835; director, 1866–80. Professor of geology and palaeontology, University of Strasbourg, 1872–9. Specialist in European mosses and fossil flora. (*Taxonomic literature.*)

Schlegel, Hermann (1804–84). German naturalist. Trained in bronze-casting; assistant, court museum, Vienna, 1824. Taxidermist, Rijksmuseum van Natuurlijke Historie, Leiden, 1825; curator, 1828; director, 1858. (*DBE.*)

Schleiden, Matthias Jacob (1804–81). German botanist, author, and lecturer. Trained and practised as a lawyer but abandoned the profession in 1833 for medical and botanical studies at Göttingen and Berlin. Honorary professor of botany, University of Jena, 1846; professor, 1850–62. Professor of plant chemistry, University of Dorpat, 1863–4. Thereafter a private lecturer, residing in various German cities. (*ADB*, *DBE*, *DSB.*)

Schmidt, Eduard Oskar (Oskar) (1823–86). German zoologist. Professor of zoology and comparative anatomy, Cracow, 1855; Graz, 1857. Professor of zoology and director of the institute of zoology, Strasbourg, 1872. (*DBE.*)

Schmidt, Maximilian (Max) (1834–88). German veterinary surgeon and zoologist. Studied veterinary medicine in Stuttgart and Berlin. Established a veterinary practice in Frankfurt am Main in 1856. Director of the Zoological Gardens from 1859. Director of Berlin Zoological Gardens from 1885. (*DBE.*)
[29 April 1869], *14 June 1869*

Schultze, Max Johann Sigismund (1825–74). German anatomist. MD, Greifswald, 1849; Privat-dozent, 1850–4. Assistant professor of anatomy, Halle, 1854–9. Professor of anatomy and director of the anatomical institute, Bonn, 1859. Founder and editor of *Archiv für mikroskopische Anatomie*, 1865–74. Noted for his work in microscopy, the reform of cell theory, and descriptive and taxonomic studies of rhizopods and sponges. (*DBE*, *DSB.*)

Scientific Opinion
[before 20 October 1869]

Sclater, Philip Lutley (1829–1913). Lawyer and ornithologist. One of the founders of *Ibis*, the journal of the British Ornithologists' Union, 1858; editor, 1858–65 and 1878–1912. Secretary of the Zoological Society of London, 1860–1903. FRS 1861. (*DSB*, Scherren 1905.)

Scott, John (1836–80). Scottish botanist. Gardener at several different country estates, before becoming foreman of the propagating department at the Royal Botanic Garden, Edinburgh, in 1859. Through CD's patronage emigrated to India in 1864, and worked briefly on a *Cinchona* plantation before taking a position as curator of the Calcutta botanic garden in 1865. Seconded to the opium department, 1872–8. Carried out numerous botanical experiments and observations on CD's behalf. Fellow of the Linnean Society of London, 1873. (Lightman ed. 2004, *ODNB.*)
2 July 1869, *21 December 1869*

Scrope, William (1772–1852). Artist and sportsman. Published works on deerstalking and salmon-fishing. (*ODNB.*)

Scudder, Samuel Hubbard (1837–1911). American entomologist. Graduated from the Lawrence Scientific School in 1862; assistant to Louis Agassiz, 1862–4. Librarian and custodian of collections, Boston Society of Natural History, 1864–70; vice-president, 1874–80; president, 1880–7. Palaeontologist, United States Geological Survey, 1886–92. (*ANB.*)
15 September [1869]

Seemann, Berthold Carl (1825–71). German-born traveller and botanist. Studied botany at the Royal Botanic Gardens, Kew, 1844–6. Naturalist to HMS *Herald*, 1847–51. Commissioned in 1860 to report on the Fiji Islands and published a botanical catalogue of the islands. Editor of *Bonplandia*, 1853–62; of the *Journal of Botany, British and Foreign*, 1863–71. Travelled in Venezuela in 1864; in Nicaragua, 1866–7. (R. Desmond 1994, *DNB.*)

Seidlitz, Karl Johann (Karl) von (1798–1885). Baltic German doctor. MD Dorpat (Tartu), 1821. Joined the Russian military and served in the naval hospital at St Petersburg, 1822–6. Held several appointments in military field hospitals. Professor of clinical medicine, medical–surgical academy, St Petersburg, 1837–47. Retired to Dorpat. (*BLA.*)
22 March 1869

Sewa, Manga (d. 1884). Ruler in northern Sierra Leone. Son of Asan Yira (Sori Wuleng) former ruler of Solimana state. Ruler of Solimana state from around 1862. Maintained commercial links with other northern peoples in Sierra Leone and beyond; extended areas of influence to include smaller states in the country. Committed suicide rather than submit during the seige of his capital, Falaba, by Samori Touré. (*Dictionary of African biography* 2: 144.)

Shakespeare, William (1564–1616). Poet and dramatist. (*ODNB.*)

Shepard, Charles Upham (1804–86). American mineralogist. BA, Amherst, 1824. Lecturer in botany, Yale, 1831–2; in natural history, 1833–47. Professor of chemistry, South Carolina Medical College, 1835–61; 1865–9. Professor of chemistry and natural history, Amherst College 1854–77. Built up a large mineral collection, which he bequeathed to Amherst College, and wrote a textbook on mineralogy. (*ANB.*)

Siebold, Karl Theodor Ernst von (1804–85). German zoologist and doctor. District physician, Heilsburg, 1831; Königsberg, 1834. Professor extraordinarius of zoology, comparative anatomy, and veterinary medicine, Erlangen, 1840; Freiburg, 1845. Professor of physiology, Breslau, 1850. Professor of physiology and comparative anatomy, Munich, 1853; of zoology and comparative anatomy, 1855. Curator, Bavarian state zoological collection, 1855. Co-founder and editor of *Zeitschrift für wissenschaftliche Zoologie*, 1848. A supporter of Darwinian transmutation theory; did research on generation, especially parthenogenesis. (*ADB, DBE, DSB.*)

Silliman, Benjamin (1779–1864). American chemist, geologist, and mineralogist. Professor of chemistry and natural history, Yale University, 1802–53. Founder and first editor of the *American Journal of Science and Arts*, 1818. (*DAB, DSB.*)

Sinclair, James Leask (1828–95). Scottish-born journalist. Educated in the

Sinclair, James Leask, cont.

Orkneys. Worked as clerk for a tea merchant, Leith, *circa* 1842–7; then worked for a tea merchant in London. Visited the United States for about two years. Returned to Kirkwall, Orkney, where he helped launch the Orkney newspaper, the *Orcadian*. Moved to London again, returning to Orkney to take over his father's business, and moved later to Stirling. Emigrated to New Zealand, 1865; worked as journalist on the *Southern Cross* newspaper in Auckland, and from 1876 as a teacher in Wellington. Retired in 1884 or 1885 and lived outside Auckland. (Saint-Clair 1898, Tee 1998, pp. 21–2.)

22 December 1869 (to John Murray), *31 December 1869*

Sinclair, Mary (b. 1837/8). Née Mowat. Born in St Andrews, Orkney. Married James Leask Sinclair in 1857. Emigrated to New Zealand, 1865; resided in Auckland and Wellington. Had four children in Britain, and more in New Zealand; a son died on voyage to New Zealand. (Census records 1861 (General Register Office for Scotland, Kirkwall and St Ola 10/16), (http://www.ancestry.com, accessed 25 April 2008); Saint-Clair 1898.)

Sinclair, William (1795–1874). Scottish merchant and harbourmaster. Grocer and harbourmaster in Kirkwall, Orkney. (Census records 1851 (General Register Office for Scotland, Kirkwall and St Ola 4/5), (http://www.ancestry.com, accessed 25 April 2008); Saint-Clair 1898.)

Sison, Robert (*fl.* 1860s). German sheep-breeder. Administrator of the royal merino-breeding farm in Lohmen, Saxony. (Sison 1867.)

Smiles, Samuel (1812–1904). Biographer. Medical diploma, Edinburgh University, 1832. Editor, *Leeds Times*, 1838–42. Assistant secretary, Leeds and Thirsk Railway, 1845–54; secretary, South Eastern Railway, 1854–66. President, National Provident Institution, from 1866. Wrote self-help books and biographies. (*ODNB*.)

Smith, Charles Hamilton (1776–1859). Flemish-born army officer and writer on natural history. Wrote on historical costume and war, and produced many natural history illustrations. Knighted, 1834. FRS 1824. (*ODNB*.)

Smith, Edgar Albert (1847–1916). Zoologist. Son of the entomologist Frederick Smith. Assistant keeper, Zoological department, British Museum, 1867–1913. Published memoirs on mollusca. (Freeman 1978, *Geological Magazine* 53 (1917), pp. 431–2.)

Smith, Edward A. (b. 1842/3). Artist and civil servant. Son of the entomologist Frederick Smith. Provided insect illustrations for CD's *Descent* and other natural history books. Employed by the General Post Office, eventually attaining post of Assistant Controller by 1901. (Census records 1861 (Public Record Office RG9/38/33); Census records 1901 (Public Record Office RG13/435/65/11); letter to E. A. Smith, 30 June [1869]; Peter Dance (personal communication).)

[before 30 June 1869], 30 June [1869]

Smith, Frederick (1805–79). Entomologist. Entomologist in the zoology department of the British Museum from 1849. Specialised in the Hymenoptera.

President of the Entomological Society of London, 1862–3. (*Entomologist* 12 (1879): 89–92, Gilbert 1977.)

26 August 1869, *14 September 1869*, *8 October 1869*, *5 November 1869*, 9 November [1869]

Smith, William (1813–93). Classical and biblical scholar. Taught classics at University College school; professor of classics, Independent college, St John's Wood. Published many classical grammars and text-books. Editor of the *Quarterly Review*, 1867–93. Knighted, 1892. (*ODNB*.)

Smith, William Henry (1825–1891). Newsagent and politician. In 1851, secured a monopoly of bookstalls on the London and North-Western railway and by 1862 had secured exclusive bookstall rights on all the important English railway systems. Member of Parliament, 1868–91, holding cabinet posts in various administrations. (*ODNB*.)

Snow, George (1820/1–85). Coal-dealer, Down, Kent. Operated a weekly carrier service between Down and London. (Census returns 1861 (Public Record Office RG9/462: 72); gravestone inscription, Down churchyard; *Post Office directory of the six home counties* 1862.)

Somerville, Mary (1780–1872). Writer on science and mathematics. Author of a condensed English version of Laplace's *Mécanique céleste* (1831), and of *On the connection of the physical sciences* (1834). Advocate of higher education for women and women's suffrage. (*BDWS*, *DSB*, *ODNB*.)

21 January [1869]

Sorby, Henry Clifton (1826–1908). Geologist. Pioneered microscopic petrology. President of the Geological Society, 1878–80. Established the chair of geology at Sheffield University. FRS 1857. (*DNB*, *DSB*.)

Spencer, Herbert (1820–1903). Philosopher. Apprenticed as a civil engineer on the railways, 1837–41. Became sub-editor of the *Pilot*, a newspaper devoted to the suffrage movement, in 1844. Sub-editor of the *Economist*, 1848–53. From 1852, author of books and papers on transmutation theory, philosophy, and the social sciences. (*DSB*, *ODNB*.)

Sprengel, Christian Konrad (1750–1816). German botanist. Rector of the Great Lutheran Town School, Spandau, where he taught languages and natural science, 1780–94. Moved to Berlin, where he worked as a private tutor. Published his major work on insect-aided fertilisation of flowers in 1793. (*ADB*, *DSB*.)

Spruce, Richard (1817–93). Botanist and schoolteacher. Schoolteacher at St Peter's School, York, 1839–44. Collected plant specimens in the Pyrenees, 1845–6; in South America, 1849–64. PhD, Imperial German Academy, 1864. Retired in poor health to Coneysthorpe, Yorkshire, where he worked on his plant collections. (R. Desmond 1994, *ODNB*.)

[before 1 April 1869], 1 April 1869, *15 April 1869*

Standish, John (1814–75). Nurseryman and gardener. Founded a nursery at Bagshot, Surrey, in the late 1830s. Specialised in plants from the Far East. (R. Desmond 1994.)

Stanley, Edward Henry, 15th earl of Derby (1826–39). Politician and diarist. BA (Trinity College), Cambridge, 1848. MP for King's Lynn, 1848–69. Travelled in the West Indies, 1848–52. First secretary of state for India from 1858. Premier of the Foreign Office from 1868. Colonial secretary from 1882. (*ODNB*.)

Stebbing, Thomas Roscoe Rede (1835–1926). Zoologist and clergyman. BA, King's College, London, 1855. BA, Worcester College, Oxford, 1857; MA, 1859; fellow, 1860–8; tutor 1865–7; vice-provost 1865; dean, 1866. Tutor and schoolmaster, Torquay, 1867–77; in 1877, moved to Tunbridge Wells and devoted his time to natural history. Specialised in amphipod Crustacea. His strong Darwinian views led to his being forbidden a parish. Lectured and wrote essays in favour of Darwinism. FRS 1896. (*ODNB*.)

3 March 1869, *5 March 1869*

Stirling, James Hutchison (1820–1909). Scottish physician and philosopher. Studied at the University of Glasgow, 1833–42. Awarded a diploma by the College of Surgeons, Edinburgh, 1842. Assistant practitioner, Hirwaun ironworks, 1844–6; sole surgeon, 1846–51. Received a large inheritance in 1851. Studied German philosophy in France and Germany, 1851–7. Returned to Britain and wrote works on Hegel from 1857. (*ODNB*.)

Stokes, George Gabriel, 1st baronet (1819–1903). Mathematician and physicist. Lucasian Professor of mathematics, Cambridge University, 1849–1903. Secretary of the Royal Society of London, 1854–85; president, 1885–90. President of the British Association for the Advancement of Science, 1869. Conservative MP for Cambridge University, 1887–91. Master of Pembroke College, Cambridge, 1902–3. Created baronet, 1889. FRS 1851. (*DNB*, *DSB*.)

Strickland, Hugh Edwin (1811–53). Geologist and zoologist. An advocate of reform in zoological nomenclature. FRS 1852. (*ODNB*, Sarjeant 1980–96.)

Strutt, John William, 3d Baron Rayleigh (1842–1919). Experimental and mathematical physicist. BA, Trinity College, Cambridge, 1865; MA, 1868; fellow, 1866–71; Cavendish Professor of experimental physics, 1879–84. Professor, natural philosophy, Royal Institution, 1887–1905. Nobel prize, physics, 1904. Co-discoverer of the inert element argon. FRS 1873. (*Alum. Cantab.*, *ODNB*.)

Strzelecki, Paul Edmund de (1797–1873). Polish-born explorer in Australia. Travelled widely in North and South America, 1831–4. From 1839, explored Australia and produced a large geological map of the area he covered as well as writing a physical description of the country. After his return to England in 1843, became a British subject and did philanthropic work. Knighted, 1869. FRS 1853. (*ODNB*.)

Sulivan, Bartholomew James (1810–90). Naval officer and hydrographer. Lieutenant on HMS *Beagle*, 1831–6. Surveyed the Falkland Islands in HMS *Arrow*, 1838–9. Commander of HMS *Philomel*, 1842–6. Resided in the Falkland Islands, 1848–51. Commanded HMS *Lightning* in the Baltic, 1854–5. Naval officer in the marine department of the Board of Trade, 1856–65. Admiral, 1877. Knighted, 1869. (*ODNB*.)

Sutton, Seth. A keeper at the zoological gardens of London. (Archives of the

Zoological Society of London, 'Daily occurrences', September 1860; *Correspondence* vol. 15, letter from Seth Sutton, 8 August 1867; Freeman 1978.)

Swainson, William (1789–1855). Naturalist and illustrator. Collected in Sicily and in Brazil. Follower of the quinary system of classification. Emigrated to New Zealand in 1841. FRS 1820. (*DSB*, *ODNB*.)

Swaysland, George (1813/14–88). Naturalist. Of 4 Queens Road, Brighton. (*BMD* (*Death index*), Census records 1881 (Public Record Office RG11/1094/59/44).)

12 June 1869

Swinhoe, Robert (1836–77). Diplomat and ornithologist. Attached to the British consulate in Hong Kong, 1854; in Amoy (Xiamen), China, 1855. British vice-consul, Formosa (Taiwan), 1860–5; consul, 1865–73. Acting consul, Amoy, 1865–71; Ning-po (Ningbo), 1871–3. Consul, Ning-po, 1873–5. Collected plants and animals in Eastern Asia. FRS 1876. (*Foreign Office list* 1877, P. B. Hall 1987, *ODNB*.)

16 October 1869

Sylvester, James Joseph (1814–97). Mathematician. Professor of natural philosophy, University College, London, 1837–41. Professor of mathematics, University of Virginia, 1841–5; Royal Military Academy, Woolwich, 1855–70; Johns Hopkins University, 1877–83. Savilian Professor of geometry, University of Oxford, 1883–94. FRS 1839. (*DNB*.)

Symonds, Hyacinth (1842–1921). Daughter of William Samuel Symonds. Married Sir William Jardine of Applegirth, seventh baronet, in 1871; widowed in 1874. Married Joseph Dalton Hooker in 1876. (Allan 1967 s.v. 'Hooker pedigree', *Burke's peerage* 1876.)

Symonds, William Samuel (1818–87). Geologist, author, and clergyman. Curate of Offenham, Worcestershire, 1843–5; rector of Pendock, Worcestershire, from 1845. Wrote papers on local archaeology and geology. Visited the Auvergne and the Ardèche, south-west France, 1874–6, to search for traces of ancient glaciers. (*ODNB*, Sarjeant 1980–96.)

Tait, Peter Guthrie (1831–1901). Scottish mathematician and physicist. Studied mathematics at Edinburgh University and Peterhouse, Cambridge. Fellow at Peterhouse, 1852–4. Professor of mathematics, Queen's College, Belfast, 1854–60; of natural philosophy, Edinburgh, 1860–1901. Published widely on mathematics and physics; published *Natural philosophy* (1867) with William Thomson (later Lord Kelvin). (*ODNB*.)

Tait, William Chester (1844–1928). British-born merchant, landowner, horticulturalist, and sportsman. Resident in Portugal. Head of Tait & Co., merchants and agents for the Royal Marine Steam Packet Co. (*Gardeners' Chronicle*, 5 May 1928, p. 327.)

26 January 1869, 2 February [1869], *19 February 1869*, 24 February 1869, *2 March 1869*, *5 March 1869*, 12 and 16 March 1869, 7 April [1869], *13 April 1869*, 18 April [1869], *10 May 1869*, 1 June [1869], *11 July 1869*, 17 July [1869], *[23 August 1869]*, 27 August [1869], *29 September 1869*

Tannenberg, Ignaz von (1743–1810). Austrian humanist. Born with poor eyesight

Tannenberg, Ignaz von, cont.
due to cataracts, he eventually lost his sight completely. Known for his philanthropic projects. (*BLKO*.)

Tannenberg, Rudolph von (1795–1846). Son of Ignaz von Tannenberg. (*BLKO* s.v. Tannenberg, Ignaz von.)

Tatham, Charles Meaburn (1828–1924). Barrister. BA, Cambridge (St John's), 1851; MA, 1854. Student of Lincoln's Inn, 1851; called to the bar, 1854. Equity draftsman and conveyancer. (*Alum. Cantab.*, *Men-at-the-bar*.)

Teebay, Richard (*fl.* 1850s–80s). Farmer and poultry breeder. Resident at Fulwood, Lancashire. (*Post Office directory of Lancashire, Liverpool and Manchester*; *The Times*, 3 December 1856, p. 12, 13 November 1883, p. 3.)

Tegetmeier, Egbert (b. 1852). Journalist. Son of William Bernhard and Anne Edwards Tegetmeier. Wrote for the *Field* on cycling and natural history. (E. W. Richardson 1916.)

Tegetmeier, William Bernhard (1816–1912). Editor, journalist, lecturer, and naturalist. Pigeon-fancier and expert on poultry. Pigeon and poultry editor of the *Field*, 1864–1907. Secretary of the Apiarian Society of London. (*Field*, 23 November 1912, p. 1070; *ODNB*; E. W. Richardson 1916.)

5 February [1869], 25 February [1869], 17 April 1869, *[before 18 April 1869]*, 18 April [1869], *30 June 1869*, 2 July [1869], *20 July 1869*, 21 July [1869], 25 August 1869, 28 September [1869], 1 October [1869]

Tennyson, Alfred, 1st Baron Tennyson (1809–92). Poet. Poet laureate, 1850. Created Baron Tennyson, 1883. (*ODNB*.)

Thiel, Hugo (1839–1918). German agriculturalist and civil servant. Studied agriculture, Royal Agricultural Academy, Poppelsdorf (Bonn); lecturer, 1865. Professor of agriculture, Darmstadt polytechnic, 1869; Munich polytechnical high school, 1872. General secretary, Prussian council on agriculture, 1873. Trustee of the agricultural high school, Berlin, 1881. Official in the Prussian ministry of agriculture and forests, 1885. (*Kurzbiographien zur Berliner Agrarwissenschaft*, Faculty of agriculture and horticulture, Humboldt University, Berlin, http://www.agrar.hu-berlin.de/profil/geschichte/geschichte/personen_html/ (accessed 10 December 2008).)

18 February 1869, 25 February 1869

Thompson, T. Secretary of the Westerham (Kent) Friendly Club. (Letter to T. Thompson, 9 October 1869.)

9 October 1869

Thompson, William Hepworth (1810–86). College head. BA Cambridge, 1832; fellow, Trinity College, 1834. Ordained priest, 1838. Tutor, 1844–53. Regius professor of Greek, 1853–66. Master of Trinity, 1866–86. (*ODNB*.)

Thomson, Robert (b. 1807/8). Scottish farmer. Of Seggie House, Leuchars, Fifeshire, Scotland. (Census returns 1871 (General Register Office for Scotland, Leuchars 4/3), (http://www.ancestry.com, accessed 28 February 2008).)

Thomson, Thomas (1817–78). Naturalist. MD, Glasgow, 1839. Travelled to India as assistant surgeon to the East India Company. Curator of the Asiatic Society's museum, Calcutta, 1840. Travelled and was taken prisoner in Afghanistan, 1840–2. Accompanied Joseph Dalton Hooker to the Himalayas, 1850–1, and collaborated with him at Kew on various botanical publications. Superintendent of the Calcutta botanic garden and professor of botany at the Calcutta Medical College from 1855 until his return to England in 1861. FRS 1855. (R. Desmond 1994, *ODNB*.)

Thomson, William, Lord Kelvin (1824–1907). Scientist and inventor. Professor of natural philosophy, Glasgow, 1846–99. Formulated laws of equivalence and transformation in thermodynamics and a doctrine of available energy. Pioneered telegraphic systems and assisted in the laying of the first transatlantic cable. Wrote on the age and cooling of the earth. Proposed a hydroelectric scheme for Niagara. Created Baron Kelvin of Largs, 1892. FRS 1851. Awarded the Copley Medal, 1883. (*DSB*, *ODNB*.)

Thwaites, George Henry Kendrick (1811–82). Botanist and entomologist. Superintendent of the Peradeniya botanic gardens, Ceylon (Sri Lanka), 1849; director, 1857–80. FRS 1865. (R. Desmond 1994, *ODNB*.)

Tickel, Thomas Robert Thorne (Thomas) (b. *c.* 1831 d. 1886). British-born trader in Africa. Left England for Cape Colony, 1842, and moved to Lagos and then Badagry in the early 1850s. Became an independent trader in the late 1850s. Appointed vice-consul for Badagry and Porto Novo, 1861; resident agent and collector of customs, Badagry, 1862; acting vice-consul of Okeodan and resident agent of Okeodan, Ado, and Ipokia, 1863–5; special envoy to Dahomey, 1866–7; civil commandant, police magistrate, sub-collector of customs, and justice of the peace for the Western district, including Badagry, 1867–78; political agent and, at intervals, acting district commissioner, 1879–86. (K. Folayan, 'The career of Thomas Tickel', *Journal of the Historical Society of Nigeria* 5 (1969): 27–46.)

Tobler, Johann Georg Christoph (1757–1812). Swiss theologian and translator. (McVaugh 2004, p. 792.)

Trego, Charles B. (1794–1874). American teacher. Taught at a school in Philadelphia, *circa* 1821–35. Member of the House of Representatives, 1835–6, 1841–7. Assistant to the state geologist in the geological survey of Pennsylvania, 1837–41. Professor of geology, University of Pennsylvania. Secretary of the American Philosophical Society, 1848–74; treasurer, 1851–74. Member, City Council of Philadelphia, 1856–62. (*Proceedings of the American Philosophical Society* 14 (1874–5): 356–8.)

Trimen, Roland (1840–1916). Zoologist and entomologist. Emigrated to South Africa in 1858. Arranged the Lepidoptera at the South African Museum. Held civil-service positions in the Commission of Land and Public Works, the governor's office, and the colonial secretary's office. Became part-time curator of the South African Museum in 1873; full-time curator, 1876. FRS 1883. (*DSAB*.)
18 November 1869

Tylor, Edward Burnett (1832–1917). Anthropologist. Educated at the School of the Society of Friends, Grove House, Tottenham. Author of *Primitive culture*

Tylor, Edward Burnett, cont.

(1871). Keeper of Oxford University Museum, 1883; reader in anthropology, Oxford University, 1883; professor, 1896; professor emeritus, 1909. President of the Anthropological Society, 1879–80, 1891–2. Helped to establish anthropology as a legitimate field of scientific enquiry. Knighted, 1912. FRS 1871. (*Men and women of the time* 1899, *ODNB*.)

8 September 1869

Tyndall, John (1820–93). Irish physicist, lecturer, and populariser of science. Studied in Marburg and Berlin, 1848–51. Professor of natural philosophy, Royal Institution of Great Britain, 1853–87; professor of natural philosophy, Royal School of Mines, 1859–68; superintendent of the Royal Institution, 1867–87. Scientific adviser to Trinity House and the Board of Trade, 1866–83. FRS 1852. (*DSB*, *ODNB*.)

Van Houtte, Louis Benoit (1810–76). Belgian horticulturalist and civil servant. Educated at Paris, moved to Brussels in 1828 and became a civil servant in the ministry of finance in the newly independent Belgian government of 1830. Travelled to Brazil, Guatemala, and Honduras collecting plants until 1836, when he returned and was employed by the botanic garden in Brussels. Founded his own nursery in Ghent in 1839. Known for his development of hybrid gladioluses. (*Gartenflora* 25 (1876): 262–6.)

VanSittart, Augustus Arthur (1824–82). Scholar. BA, Cambridge, 1847; fellow, Trinity College, 1848. Called to the bar, 1852. Interested in geology and New Testament scholarship. (*Alum. Cantab.*)

Varley, Cromwell Fleetwood (1828–83). Telegraph engineer. Chief engineer, Electric Telegraph Company, from 1861. Chief electrician, Atlantic Telegraph Company, from 1865. Formed a patent partnership with William Thomson (later Lord Kelvin) and Henry Charles Fleeming Jenkin. In 1874 used a cable galvanometer to investigate the 'materialisations' of the medium Florence Cook. FRS 1871. (*ODNB*.)

Vaucher, Jean-Pierre-Etienne (1763–1841). Swiss botanist and clergyman. Parish priest, Geneva, 1797–1822. Honorary professor of botany, Geneva University, 1798–1807; of ecclesiastical history, 1807–39; rector, 1818–21. Founding member of the Société de Physique et d'Histoire Naturelle de Genève. Wrote extensively on plant physiology. (*DSB*.)

Vekemans, Jacobus-Cornelius (Jacobus) (1815–88). Belgian ornithologist. Director of the Antwerp Zoological Gardens from 1865. (*BNB*.)

Virgil (Publius Vergilius Maro) (70–19 BC). Roman poet. (*Oxford classical dictionary*.)

Vogt, Carl (1817–95). German naturalist. After receiving a doctorate in Giessen, 1839, moved to Switzerland and worked with Louis Agassiz on a treatise on fossil and freshwater fish until 1846. Professor of zoology, Giessen, 1846. Forced to leave the German Federation for political reasons in 1849; settled in Geneva.

Professor of geology, Geneva, 1852; director of the Institute of Zoology, 1872. (*ADB*, *DSB*, Judel 2004.)

26 October 1869, 30 October 1869

Voysey, Charles (1828–1912). Theistic preacher. BA, Oxford, 1851. Incumbent of St Andrew's, Craigton, Jamaica, 1858; curate at St Mark's, Whitechapel, 1861; at St Mark's, Victoria Docks, 1863. Vicar of Healaugh, 1864. Denounced for unorthodoxy, 1869, and expelled from the Church of England, 1871. Founded the Theistic church, which was based in Swallow Street, London from 1885. His reforming sermons were widely published from the late 1860s. (*ODNB*.)

27 July 1869

Wagner, Moritz (1813–87). German zoologist and explorer. Studied at the natural history colleges in Erlangen and Munich, 1834–5. Made a scientific expedition to Algeria, 1836. Editor of the *Augsburger Allgemeine Zeitung*, 1838. Continued his studies in Göttingen, 1840. Made research trips to the Caucasus, Armenia, Persia, and South Asia, 1843–5; to North and Central America, 1852–5; to Panama and Ecuador, 1857–9. Professor at Munich University from 1862. Formulated the migration theory for fauna and flora in 1868. (*DBE*, Tort 1996.)

Wahl, Anne de (1804–69). Daughter of Anne Palgrave and Edward Rigby. Cousin of Maria Hooker, J. D. Hooker's mother. Married George de Wahl in Estonia. (L. Huxley 1918, 1: 17.)

Walford, Henry (*fl.* 1860s). Watchmaker and silversmith in Banbury, Oxfordshire. (*Post Office directory of Northamptonshire etc.* 1869.)

Wallace, Alexander (1829–99). Physician, entomologist, and plantsman. BA, Oxford, 1852; MD, Oxford, 1861. Physician at Colchester from 1859. Tried to introduce silkworm culture to England. Introduced many new varieties of lilies and orchids from Japan. (*Entomologist's Monthly Magazine* 25 (1899): 275–6, *Modern English biography*.)

15 July [1869]

Wallace, Alfred Russel (1823–1913). Naturalist. Collector in the Amazon, 1848–52; in the Malay Archipelago, 1854–62. Independently formulated a theory of evolution by natural selection in 1858. Lecturer and author of works on protective coloration, mimicry, and zoogeography. President of the Land Nationalisation Society, 1881–1913. Wrote on socialism, spiritualism, and vaccination. FRS 1893. (*DSB*, *ODNB*.)

20 January 1869, 22 January [1869], *30 January 1869*, 2 February [1869], 5 March [1869], *10 March 1869*, 22 March [1869], *24 March 1869*, 27 March [1869], 14 April 1869, *18 April [1869]*, *23 June [1869]*, 25 June [1869], *20 October 1869*, 21 October 1869, *4 December [1869]*, 5 December [1869]

Wallace, Violet (b. 1869). Schoolteacher. Daughter of Alfred Russel Wallace. Worked as a kindergarten teacher in Liverpool; later started a school near her parents at Tulgey Wood, Broadstone. (Raby 2001, pp. 199, 262, 282.)

Wallich, George Charles (1815–99). Surgeon and zoologist. Served with the

Wallich, George Charles, cont.
Indian army medical service, 1838–57. Joined HMS *Bulldog* as naturalist to the North Atlantic telegraphic expedition in 1860; discovered evidence of animal life at great oceanic depths and published his findings. Also published on the Protozoa. (*DNB*, *DSB*.)
18 April [1869]

Walsh, Benjamin Dann (1808–69). Entomologist, farmer, and timber merchant. Student at Trinity College, Cambridge, 1827–31; fellow, 1833. Emigrated to the United States, where he farmed in Henry County, Illinois, 1838–51. Lumber merchant, Rock Island, Illinois, 1851–8. Retired from commerce in about 1858 and concentrated on entomology, making contributions to agricultural entomology. Suggested the use of natural enemies to control insect pests. Author of several papers in agricultural journals. Associate editor of the *Practical Entomologist*, 1865. Acting state entomologist, Illinois, 1867. (*Alum. Cantab.*, *DAB*.)
3 April [1869]

Walther, Alexander Wilhelm Hannibal Franz (Alexander) (1813–90). German physician and botanist. Lived in Bayreuth. (*Taxonomic literature.*)

Watson, Frederick (1831–86). Clergyman. BA, Cambridge, 1856. Ordained priest, 1857. Curate of Aslacton, Norfolk, 1856–9. Rector of Salcott-Virley, Essex, 1859–75. Rector of Great and Little Wigborough, 1875–86. (*Alum. Cantab.*)

Weber, Ernst Heinrich (1795–1878). German anatomist and physiologist. Studied medicine in Wittenberg and Leipzig. Obtained a doctorate in comparative anatomy in 1815. Professor of anatomy, University of Leipzig, 1821–71; of physiology, 1840–66. Accomplished significant work in sensory physiology, physiological acoustics, and other areas. (*DBE*.)

Wedgwood, Amy (b. 1835). Daughter of Francis and Frances (née Mosley) Wedgwood. Emma Darwin's niece. (Freeman 1978.)

Wedgwood, Caroline Sarah (1800–88). CD's sister. Married Josiah Wedgwood III, her cousin, in 1837. (*Darwin pedigree.*)

Wedgwood, Clement Francis (1840–89). Second son of Francis Wedgwood and Frances Mosley. Apprenticed in the Wedgwood pottery business, 1859; later a partner in the business and director at the Etruria works in Staffordshire. (Wedgwood and Wedgwood 1980.)

Wedgwood, Emily (1840–1921). Née Rendel. Married Clement Francis Wedgwood in 1866. (Wedgwood and Wedgwood 1980.)

Wedgwood, Frances Emma Elizabeth (Fanny) (1800–89). Second child of James Mackintosh and Catherine Allen. Married Hensleigh Wedgwood in 1832. (Freeman 1978, O'Leary 1989.)

Wedgwood, Hensleigh (1803–91). Philologist. Emma Darwin's brother. Qualified as a barrister in 1828, but never practised. Fellow, Christ's College, Cambridge, 1829–30. Police magistrate at Lambeth, 1831–7; registrar of metropolitan carriages, 1838–49. An original member of the Philological Society, 1842. Published *A dictionary of English etymology* (1859–65). Married Frances Emma Elizabeth Mackintosh in 1832. (Freeman 1978, *ODNB*.)

Wedgwood, Josiah I (1730–95). Master-potter. Founded the Wedgwood pottery works at Etruria, Staffordshire. Grandfather of CD and Emma Darwin. Greatly interested in experimental chemistry. Contributed several papers on the measurement of high temperatures to the Royal Society of London's *Philosophical Transactions*. Actively associated with scientists and scientific societies. FRS 1783. (*DSB*, *ODNB*.)

Wedgwood, Josiah III (1795–1880). Master-potter. Partner in the Wedgwood pottery works at Etruria, Staffordshire, 1841–4; moved to Leith Hill Place, Surrey, in 1844. Emma Darwin's brother. Married CD's sister Caroline, his cousin, in 1837. (Freeman 1978.)

Wedgwood, Margaret Susan. *See* Williams, Margaret Susan Vaughan.

Wedgwood, Sarah Elizabeth (Elizabeth) (1793–1880). Emma Darwin's sister. Resided at Maer Hall, Staffordshire, until 1847, then at The Ridge, Hartfield, Sussex, until 1862. Moved to London before settling in Down in 1868. (*Emma Darwin* (1915), Freeman 1978.)

Weir, Harrison William (1824–1906). Painter and illustrator. Specialised in landscape and natural history subjects. Brother of John Jenner Weir. (*ODNB*.)
23 March 1869, *7 May 1869*

Weir, John Jenner (1822–94). Naturalist and accountant. Worked in HM customs as an accountant, 1839–85. Studied entomology, especially microlepidoptera; conducted experiments on the relations between insects and insectivorous birds and published papers in 1869 and 1870. Member of the Entomological Society of London from 1845, serving many times on the council. Fellow of the Linnean Society of London, 1865; Zoological Society, 1876. (*Science Gossip* n.s. 1 (1894): 49–50.)
[1–13] May 1869, 13 May [1869], 20 May [1869], *26 May 1869*, 27 May [1869], *June 1869*, 1 July [1869]

Weismann, Leopold Friedrich August (August) (1834–1914). German zoologist. Studied medicine at Göttingen, 1852–6; qualified as a doctor, 1858. Physician to Archduke Stephan of Austria, 1861–3. Studied zoology at Gießen in 1861; habilitated at Freiburg (im Breisgau), 1863; privat-dozent in zoology and comparative anatomy, 1863; professor extraordinarius, 1866; full professor and first holder of the chair in zoology, 1874. Best known for his work on heredity, especially the theory of continuity of the germ-plasm. (*DBE*, *DSB*.)

Westring, Niklas (1797–1882). Swedish entomologist. (Gilbert 1977, Sandahl 1882.)

Wheler, Elizabeth Ann (1808–1906). Daughter of Samuel Tertius and Frances Anne Violetta Galton. Sister of Francis Galton. Married Edward Wheler in 1845. (*Darwin pedigree*.)

Wheler, Lucy Elizabeth (b. 1847). Daughter of Edward and Elizabeth Ann Wheler. (*Darwin pedigree*.)

Whitaker, William (1836–1925). Geologist. Employed by the Geological Survey, 1857–96. Considered a pioneer of English hydrogeology. President, Geological Society of London, 1898–1900. FRS 1887. (*ODNB*.)

White, Gilbert (1720–93). Naturalist and clergyman. Author of *The natural history and antiquities of Selborne* (1789). (*ODNB*.)

Whiteford, Ferdinand Mauger (1835–81). Barrister. Son of Charles Whiteford and Eliza Hamilton Whiteford, who was a daughter of Charles Hamilton Smith. Attended Harrow School, leaving in 1852. Solicitor, 1862. Admitted to Lincoln's Inn, 1869; called to the bar 1872. (*Modern English biography*, *Records of Lincoln's Inn*.)

Wicking, Matthew. Brewer of Jenner, Wicking and Jenner, 153 Southwark Bridge Road, London. Breeder of fancy pigeons. Member of the Philoperisteron Society. (*Post Office London directory* 1856.)

Wiggins, Thomas Greene (1849–1908). American musician and composer. Born blind and a slave, he continued in the service of his former owner even after emancipation, after being indentured and then legally declared *non compos* in court proceedings. Performed in concerts in North America and Europe. His compositions were only published in the latter half of the twentieth century. (Southall 1999.)

Wilberforce, Samuel (1805–73). Clergyman. Rector of Brighstone, Isle of Wight, 1830–40; of Alverstoke, Hampshire, 1840–3. Chaplain to Prince Albert, 1841. Dean of Westminster, 1845. Bishop of Oxford, 1845–69. Bishop of Winchester, 1869–73. FRS 1845. (*ODNB*, *Record of the Royal Society of London*.)

Wilckens, Martin (1834–97). German agriculturalist. Studied medicine and natural science at Göttingen, Vienna, and Würzburg. Studied agriculture and economics from 1859; and managed the Pogarth estate in Silesia, 1861–71. Professor of animal physiology and animal husbandry at the Hochschule für Bodenkultur (high school for soil culture/cultivation), Vienna, 1872. (*Der große Brockhaus*, *Meyers enzyklopädisches Lexikon*.)

25 March 1869, [after 25 March 1869]

Wilhelm I (1797–1888). King of Prussia, 1861–88; emperor of Germany, 1871–88. (*EB*.)

Williams, Hervey Wedgwood (1869–1944). Son of Margaret Susan Wedgwood and Arthur Vaughan Williams. (Birth certificate, General Register Office, Dorking; Milo Keynes 1998, p. 255.)

Williams, Margaret Susan Vaughan (1843–1937). Daughter of Caroline Wedgwood and Josiah Wedgwood III. CD's niece. Married Arthur Charles Vaughan Williams in 1869. Mother of Ralph Vaughan Williams (*ODNB*). (*Emma Darwin* (1915), Freeman 1978.)

[after 14 October 1869] (to Henrietta Emma Darwin)

Wilson, Edward (1813–78). Australian newspaper proprietor. Left England for Australia in 1841; proprietor of the *Argus* newspaper from 1848. Founded the Acclimatisation Society of Victoria in 1861. Moved back to England in 1864; moved to Hayes Place, Bromley, Kent, in 1867. Continued to be active in Australian affairs and promoting assisted emigration. A founder of the Colonial Society, 1868, and member of the council from 1868. (*Aust. dict. biog.*, *ODNB*.)

12 February 1869

Winslow, Forbes Benignus (1810–74). Physician and specialist in insanity. MD,

Aberdeen, 1849. Opened two private lunatic asylums in Hammersmith in 1847 and carried on a private psychiatric practice outside the asylum. Founded the *Journal of Psychological Medicine and Mental Pathology* in 1848. Published prolifically on psychiatric subjects. A controversial expert witness in the courts, being one of the most extreme advocates of the non-responsibility of the insane. (*ODNB.*)

Wollaston, Thomas Vernon (1822–78). Entomologist and conchologist. Passed many winters in Madeira, where he collected insects and shells. Wrote a series of works on the Coleoptera. (*ODNB.*)

Wood, George Bacon (1797–1879). American physician, medical author, and teacher. Attending physician, Pennsylvania Institute for the Deaf and Dumb, 1822–44; Pennsylvania Hospital, 1835–59. Professor of chemistry, Philadelphia College of Pharmacy, 1822–31; of materia medica, from 1831. Professor of materia medica and pharmacy, University of Pennsylvania, 1835–50; of the theory and practice of medicine, 1850–60. Helped to revise the *United States pharmacopoeia.* President of the American Philosophical Society, 1859–79. (*ANB.*)

Woodcroft, Bennet (1803–79). Engineer and Patent Office administrator. Partner in his father's textile business, 1828–40. Consulting engineer and patent agent, 1843. Professor of machinery at University College, London, 1847–51. Assistant to the commissioner of patents from 1852. FRS 1859. (*ODNB.*)

Woodward, Samuel Pickworth (1821–65). Naturalist. Sub-curator, Geological Society of London, 1839–45. Professor of geology and natural history at the Royal Agricultural College, Cirencester, 1845. First-class assistant in the department of geology and mineralogy, British Museum, 1848–65. (*DNB*, Sarjeant 1980–96.)

Woolner, Thomas (1825–92). Sculptor and poet. Member of the Pre-Raphaelite brotherhood. Established his reputation in the 1850s with medallion portrait sculptures of Robert Browning, Thomas Carlyle, and William Wordsworth. Went on to make acclaimed busts of CD, Charles Dickens, Thomas Henry Huxley, Adam Sedgwick, and Alfred Tennyson, and life-size studies of Francis Bacon, John Stuart Mill, and William Whewell. (*DNB.*)

Wright, Edward Perceval (1834–1910). Irish naturalist. One of the editors of the *Natural History Review*, 1854–65. Curator of the university museum in Dublin, 1857. Lecturer in zoology, Trinity College, Dublin, 1858–68; professor of botany and keeper of the herbarium, 1869. Resigned the professorship in 1904 owing to ill health, but continued to superintend the herbarium. His principal research was in marine zoology. (*ODNB.*)

Wright, John. Breeder of dogs, horses, and cattle. Of Yeldersley House, Derby. Correspondent of George Cupples. (*Correspondence* vol. 16, letter from George Cupples, 26 May 1868.)

Wyman, Jeffries (1814–74). American comparative anatomist and ethnologist. Curator of the Lowell Institute, Boston, 1839–42. Travelled in Europe, 1841–2. Professor of anatomy and physiology, Hampden-Sydney College, Virginia, 1843–8. Hersey Professor of anatomy, Harvard College, 1847–74. Curator, Peabody Museum of Archaeology and Ethnology, Harvard, 1866–74. (*ANB*, *DSB.*)

Wynn, Anna Williams- (b. before 1840 d. 1926). Daughter of Edward Lloyd. Married Herbert Watkin Williams-Wynn in 1855. (*Burke's peerage* 1927, s.v. Wynn.)

Yira, Asan (*fl.* 1820s). Ruler in northern Sierra Leone. Also called Sori Wuleng. Became ruler of Solimana State, and King of its capital, Falaba. (*Dictionary of African biography* 2: 144, Laing 1825, p. 228.)

Zirkel, Ferdinand (1838–1912). German geologist and mineralogist. PhD, Bonn, 1861. Professor of mineralogy and geology, Lemberg, 1863; full professor, 1865. Professor, Kiel, 1868; Leipzig, 1870. Director of the mineralogical museum, Leipzig, 1870. Did pioneering work in the microscopic examination of rocks. (*DBE*.)

BIBLIOGRAPHY

The following bibliography contains all the books and papers referred to in this volume by author–date reference or by short title. Short titles are used for some standard reference works (e.g., *ODNB*, *OED*), for CD's books and papers, and for editions of his letters and manuscripts (e.g., *Descent*, *LL*, *Notebooks*). Works referred to by short titles are listed in alphabetical order according to the author's surname. Notes on manuscript sources are given at the end of the bibliography.

ADB: *Allgemeine deutsche Biographie.* Under the auspices of the Historical Commission of the Royal Academy of Sciences. 56 vols. Leipzig: Duncker & Humblot. 1875–1912.

Agassiz, George Russell, ed. 1913. *Letters and recollections of Alexander Agassiz, with a sketch of his life and work.* London: Constable & Co.

Agassiz, Jean Louis Rodolphe and Agassiz, Elizabeth. 1868. *A journey in Brazil.* Boston: Ticknor and Fields.

Aharoni, Reuven. 2007. *The pasha's bedouin; tribes and state relations in the Egypt of Mehemet Ali, 1805–1848.* London and New York: Routledge.

Alembert, Jean Le Rond d'. 1743. *Traité de dynamique.* Paris: David l'aîné.

Allan, Mea. 1967. *The Hookers of Kew, 1785–1911.* London: Michael Joseph.

Alum. Cantab.: *Alumni Cantabrigienses. A biographical list of all known students, graduates and holders of office at the University of Cambridge, from the earliest times to 1900.* Compiled by John Venn and J. A. Venn. 10 vols. Cambridge: Cambridge University Press. 1922–54.

Alum. Dublin.: *Alumni Dublinenses. A register of the students, graduates, professors and provosts of Trinity College in the University of Dublin (1593–1860).* New edition with supplement. Edited by George Dames Burtchaell and Thomas Ulick Sadleir. Dublin: Alex. Thom & Co. 1935.

Alum. Oxon.: *Alumni Oxonienses: the members of the University of Oxford, 1500–1886: … with a record of their degrees. Being the matriculation register of the university.* Alphabetically arranged, revised, and annotated by Joseph Foster. 8 vols. London and Oxford: Parker & Co. 1887–91.

ANB: *American national biography.* Edited by John A. Garraty and Mark C. Carnes. 24 vols. and supplement. New York and Oxford: Oxford University Press. 1999–2002.

Anon. 1868. Natural and supernatural selection. *Spectator* 41: 1154–5.

——. 1869a. The alleged failure of natural selection in the case of man. *Quarterly Journal of Science* 21: 152–3.

Anon. 1869b. Review of *Facts and arguments for Darwin* by Fritz Müller, translated by W. S. Dallas. *Scientific Opinion* 1: 473–4.

——. 1869c. Difficulties of the theory of natural selection. *Month* 11: 35–53, 134–53, 274–89.

Appiah, Kwame Anthony and Gates, Henry Louis, Jr., eds. 2005. *Africana: the encyclopedia of the African and African American experience.* 2d edition. 5 vols. Oxford and New York: Oxford University Press.

Ashton, Rosemary. 1991. *G. H. Lewes: a life.* Oxford: Clarendon Press.

Atkins, Hedley. 1974. *Down: the home of the Darwins; the story of a house and the people who lived there.* London: Royal College of Surgeons.

Audubon, John James. 1831–[9]. *Ornithological biography, or an account of the habits of the birds of the United States of America accompanied by descriptions of the objects represented in the work entitled 'The birds of America' and interspersed with delineations of American scenery and manners.* 5 vols. Edinburgh: Adam Black; Adam & Charles Black.

——. 1839. *A synopsis of the birds of North America.* Edinburgh: A. and C. Black.

Aust. dict. biog.: *Australian dictionary of biography.* Edited by Douglas Pike *et al.* 14 vols. [Melbourne]: Melbourne University Press. London and New York: Cambridge University Press. 1966–96.

Autobiography: *The autobiography of Charles Darwin 1809–1882. With original omissions restored.* Edited with appendix and notes by Nora Barlow. London: Collins. 1958.

Axell, Severin. 1869. *Om anordningarna för de fanerogama växternas befruktning.* Stockholm: Iwar Hæggströms Boktryckeri.

Ayres, P. B. 1860. On the geology of Flat and Gabriel Islands. *Transactions of the Royal Society of Arts and Sciences of Mauritius* n. s. 1: 220–32.

Azara, Félix de. 1809. *Voyages dans l'Amérique Méridionale.* Edited by C. A. Walckenaer, with additional notes by G. Cuvier. 4 vols. and atlas. Paris: Dentu.

Babbage, Charles. 1829. A letter to the Right Hon. T. P. Courtenay, on the proportionate number of births of the two sexes under different circumstances. *Edinburgh Journal of Science* n.s. 1: 85–104.

Baker, Herbert G. 1965. Charles Darwin and the perennial flax—a controversy and its implications. *Huntia* 2: 141–61.

Balfour, Edward. 1885. *The cyclopædia of India and of eastern and southern Asia, commercial, industrial, and scientific; products of the mineral, vegetable, and animal kingdoms, useful arts and manufactures.* 3d edition. 3 vols. London: Bernard Quaritch.

Barber, Mary Elizabeth. 1869. On the fertilization and dissemination of *Duvernoia adhatodoides.* [Read 15 April 1869.] *Journal of the Linnean Society (Botany)* 11 (1871): 469–72.

Barbier, Edmond, trans. 1875. *Voyage d'un naturaliste autour du monde: fait à bord du navire le Beagle de 1832 à 1836.* By Charles Darwin. (Translation of *Journal of researches.*) Paris: Reinwald.

Barkly, Henry. 1870. Notes on the flora and fauna of Round Island. In *The Mauritius almanac and colonial register for 1870.* Edited by John B. Kyshe. Mauritius: n.p.

Barnhart, John Hendley, comp. 1965. *Biographical notes upon botanists … maintained in the New York Botanical Garden Library.* 3 vols. Boston, Mass.: G. K. Hall.

Barton, Ruth. 1990. 'An influential set of chaps': the X-Club and Royal Society politics 1864–85. *British Journal for the History of Science* 23: 53–81.

——. 1998. 'Huxley, Lubbock, and Half a Dozen Others': professionals and gentlemen in the formation of the X Club, 1851–1864. *Isis* 89: 410–44.

——. 2004. Scientific authority and scientific controversy in *Nature*: North Britain against the X Club. In *Culture and science in the nineteenth-century media*, edited by Louise Henson *et al.* Aldershot: Ashgate.

Bassett, Thomas J. and Porter, Philip W. 1991. 'From the best authorities': the mountains of Kong in the cartography of West Africa. *Journal of African History* 32: 367–413.

BDWS: *Biographical dictionary of women in science: pioneering lives from ancient times to the mid-20th century.* Edited by Marilyn Ogilvie and Joy Harvey. 2 vols. New York and London: Routledge. 2000.

Becker, Lydia Ernestine. 1869a. On alteration in the structure of *Lychnis diurna*, observed in connexion with the development of a parasitic fungus. [Read August 1869.] *Report of the 39th meeting of the British Association for the Advancement of Science, held at Exeter*, Transactions of the sections, p. 106.

——. 1869b. On an alteration in the structure of *Lychnis dioica* observed in connection with the development of a parasitic fungus. *Journal of Botany* 7: 291–2.

Beddoe, John. 1863. On the supposed increasing prevalence of dark hair in England. *Anthropological Review* 1: 310–12.

——. 1869. On the stature and bulk of man in the British Isles. [Read 15 June 1869.] *Memoirs read before the Anthropological Society of London* 3 (1870): 384–573.

Bell, Charles. 1844. *The anatomy and philosophy of expression as connected with the fine arts.* Preface by George Bell, and an appendix on the nervous system by Alexander Shaw. 3d edition, enlarged. London: John Murray.

Bellon, Richard. 2001. Joseph Hooker's ideals for a professional man of science. *Journal of the History of Biology* 34: 51–82.

Bennett, Alfred William. 1868. Note on the structure and affinities of *Parnassia palustris* L. [Read 19 November 1868.] *Proceedings of the Linnean Society* (*Botany*) 11 (1871): 24–31.

——. 1869. On the fertilisation of winter-flowering plants. *Nature* 1: 11–13.

Benthall, Paul. 1980. George Maw: a versatile Victorian. *National Trust Studies* (1980): 11–20.

Bentham, George. 1863–78. *Flora Australiensis: a description of the plants of the Australian territory.* Assisted by Ferdinand Mueller. 7 vols. London: Lovell Reeve.

——. 1869a. Revision of the genus *Cassia*. [Read 4 March 1869.] *Transactions of the Linnean Society of London* 27 (1871): 503–91.

——. 1869b. Anniversary address. [Read 24 May 1869.] *Proceedings of the Linnean Society of London* (1868–9): lxv–c.

Bentham, George and Hooker, Joseph Dalton. 1862–83. *Genera plantarum. Ad exemplaria imprimis in herbariis Kewensibus servata definita.* 3 vols. in 7. London: A. Black [and others].

Berg, Carlos. 1895. Carlos Germán Conrado Burmeister. Reseña Biográfica. *Anales del Museo Nacional de Buenos Aires* 2d ser. 1: 316–57.

Bhatt, S. C. and Bhargava, Gopal K., eds. 2005. *Land and people of Indian states and union territories.* 36 vols. Delhi: Kalpaz.

BHGW: *Biographisch-literarisches Handwörterbuch zur Geschichte der exacten Wissenschaften enthaltend Nachweisung über Lebensverhältnisse und Leistungen von Mathematikern, Astronomen, Physikern, Chemikern, Mineralogen, Geologen usw.* By Johann Christian Poggendorff. 5 vols. Leipzig: Johann Ambrosius Barth; Verlag Chemie. 1863–1926.

Bibby, Cyril. 1959. *T. H. Huxley. Scientist, humanist and educator.* London: Watts.

Bidard. 1869. On the structure of the flower of the *Gramineæ*, the functions of the organs of which it is composed, and the phenomena which accompany the act of fecundation. *Annals and magazine of natural history* 4th ser. 4: 134–5.

Birds: Pt 3 of *The zoology of the voyage of H.M.S. Beagle.* By John Gould. Edited and superintended by Charles Darwin. London: Smith, Elder and Co. 1838–41.

Birds of the world: *Handbook of the birds of the world.* By Josep del Hoyo *et al.* 13 vols. to date. Barcelona: Lynx editions. 1991–.

Bishop, Henry Rowley. [1864.] *The pilgrim of love: recitative and romance.* London: Robert Cocks & Co.

BLA: *Biographisches Lexikon der hervorragenden Aerzte aller Zeiten und Völker.* Edited by A. Wernich *et al.* 6 vols. Vienna and Leipzig: Urban and Schwarzenberg. 1884–8.

Blackburn, Helen. 1902. *Women's suffrage: a record of the Women's Suffrage Movement in the British Isles with biographical sketches of Miss Becker.* London: Williams & Norgate.

Blackwell, Antoinette Louisa Brown. 1869. *Studies in general science.* New York: G. P. Putnam & Son.

Blakiston, Georgiana. 1980. *Woburn and the Russells.* London: Constable.

BLC: *The British Library general catalogues of printed books to 1975.* 360 vols. and supplement (6 vols.). London: Clive Bingley; K. G. Saur. 1979–88.

BLKO: *Biographisches Lexikon des Kaiserthums Oesterreich, enthaltend die Lebensskizzen der denkwürdigen Personen, welche seit 1750 in den österreichischen Kronländern geboren wurden oder darin gelebt und gewirkt haben.* By Constant von Wurzbach. 60 vols. Vienna: L. C. Zamarski. 1856–91.

BMD: General Register Office, *England and Wales civil registration indexes. England & Wales birth index, 1837–1983. England and Wales marriage index, 1837–1983. England and Wales death index, 1837–1983.* [Online database: The Generations Network. 2006. http://www.ancestry.co.uk.]

BNB: *Biographie nationale publiée par l'Académie Royale des Sciences, des Lettres et des Beaux-arts de Belgique.* 44 vols., including 16 supplements. Brussels: H. Thiry-Van Buggenhoudt [and others]. 1866–1986.

Boccardo, Gerolamo. 1868. *Fisica del globo: spazi, climi e meteore; corso completo di geografia fisica e di meteorologia.* Genoa: R. I. de'Sordo-Muti.

——. 1869. *Sismopirologia: terremoti, vulcani ed oscillazioni del suolo saggio di una teoria di geografia fisica.* Genoa: R. I. de'Sordo-Muti.

Bojer, Wenzel. 1837. *Hortus Mauritianus: ou énumération des plantes, exotiques et indigènes, qui croissent à l'Ile Maurice, disposées d'après la méthode naturelle.* Maurice: Aimé Mamarot et Compagnie.

Boner, Charles. 1860. *Chamois hunting in the mountains of Bavaria and the Tyrol.* New edition. London: Chapman and Hall.

——. 1861. *Forest creatures.* London: Longman, Green, Longman, and Roberts.

Bosl, Karl, ed. 1983. *Bosl's Bayerische Biographie.* Regensburg: Friedrich Pustet.

Bowerbank, James Scott. 1864–82. *A monograph of the British Spongidae.* 4 vols. London: Ray Society.

Bowman, Thomas E. and Abele, Lawrence G. 1982. Classification of the recent Crustacea. In *The biology of Crustacea*, editor-in-chief, Dorothy E. Bliss, vol. 1 of *Systematics, the fossil record, and biogeography*, edited by Lawrence G. Abele. New York and London: Academic Press.

Brandt, Johann Friedrich von. 1866. *Mittheilungen über die Naturgeschichte des Mammuth oder Mamont (Elephas primigenius).* St Petersburg: Kaiserlich Akademie der Wissenschaften.

——. 1869. *De dinotheriorum genere elephantidorum familiae adjungendo nec non de elephantidorum generum craniologia comparata.* St Petersburg: Académie impériale des sciences de St.-Pétersbourg.

Bree, Charles Robert. 1859–63. *A history of the birds of Europe: not observed in the British Isles.* London: Groombridge and sons.

Briggs, Katherine Mary. [1976.] *A dictionary of fairies: hobgoblins, brownies, bogies, and other supernatural creatures.* [London]: Allen Lane.

Brock, William Hodson and Meadows, Arthur Jack. 1998. *The lamp of learning: two centuries of publishing at Taylor & Francis.* 2d edition. London: Taylor & Francis.

Bronn, Heinrich Georg, trans. 1860. *Charles Darwin, über die Entstehung der Arten im Thier- und Pflanzen-Reich durch natürliche Züchtung, oder Erhaltung der vervollkommneten Rassen im Kampfe um's Daseyn.* Stuttgart: E. Schweizerbart.

——, trans. 1862. *Charles Darwin, über die Einrichtungen zur Befruchtung Britischer und ausländischer Orchideen durch Insekten und über die günstigen Erfolge der Wechselbefruchtung.* Stuttgart: E. Schweizerbart.

——, trans. 1863. *Über die Entstehung der Arten im Thier- und Pflanzen-Reich durch natürliche Züchtung; oder, Erhaltung der vervollkommneten Rassen im Kampfe um's Daseyn.* By Charles Darwin. 2d edition. Stuttgart: E. Schweizerbart.

Bronn, Heinrich Georg and Carus, Julius Victor, trans. 1867. *Über die Entstehung der Arten durch natürliche Zuchtwahl oder die Erhaltung der begünstigten Rassen im Kampfe um's Dasein.* 3d edition. Translated by Heinrich Georg Bronn. Revised and corrected from the fourth English edition by Julius Victor Carus. Stuttgart: E. Schweizerbart'sche Verlagshandlung und Druckerei.

Bronn, Heinrich Georg and Carus, Julius Victor, trans. 1870. *Charles Darwin über die Entstehung der Arten durch natürliche Zuchtwahl; oder, Die Erhaltung der begünstigten Rassen im Kampfe um's Dasein.* 4th edition. Stuttgart: E. Schweizerbart.

Brooke, Christopher N. L., ed. 1988–2004. *A history of the University of Cambridge.* 4 vols. Cambridge: Cambridge University Press.

Brotero, Félix de Avellar. 1804. *Flora lusitanica: seu plantarum, quae in Lusitania vel sponte crescunt, vel frequentius coluntur, ex florum praesertim sexubus systematice distributarum, synopsis.* 2 vols. Olissipone: ex typographia regia.

Brown, Alan Willard. 1947. *The Metaphysical Society; Victorian minds in crisis, 1869–1880.* New York: Columbia University Press.

Brown, James. 1861. *The forester: a practical treatise on the planting, rearing and general management of forest trees.* 3d edition. Edinburgh: Blackwood.

Browne, Janet. 1995. *Charles Darwin. Voyaging. Volume I of a biography.* New York: Alfred A. Knopf.

——. 2002. *Charles Darwin. The power of place. Volume II of a biography.* London: Pimlico.

Browning, Oscar. 1910. *Memories of sixty years at Eton, Cambridge and elsewhere.* 2d edition. London and New York: John Lane.

Brush, Stephen G. 1979. Nineteenth-century debates about the inside of the earth: solid, liquid or gas? *Annals of Science* 36: 225–54.

——. 1996. *Transmuted past. The age of the Earth and the evolution of the elements from Lyell to Patterson.* Vol. 2 of *A history of modern planetary physics.* Cambridge: Cambridge University Press.

Buckler, William. 1886–1901. *The larvae of the British butterflies and moths.* 9 vols. London: Ray Society.

Bulmer, Michael. 2003. *Francis Galton: pioneer of heredity and biometry.* Baltimore: The Johns Hopkins University Press.

Bunbury, Charles James Fox. 1855. Remarks on the botany of Madeira and Teneriffe. [Read 6 March and 3 April 1855.] *Journal of the Proceedings of the Linnean Society (Botany)* 1 (1857): 1–35.

Bunbury, Frances Joanna, ed. 1891–3. *Memorials of Sir C. J. F. Bunbury, Bart. Early life* 1 vol.; *Middle life* vols. 1–3; *Later life* vols. 1–5. Mildenhall: S. R. Simpson.

Burchfield, Joe D. 1990. *Lord Kelvin and the age of the earth. With a new afterword.* Chicago and London: University of Chicago Press.

Burke's landed gentry: *A genealogical and heraldic history of the commoners of Great Britain and Ireland enjoying territorial possessions or high official rank but unvisited with heritable honours. Burke's genealogical and heraldic history of the landed gentry.* By John Burke, *et al.* 1st–18th edition. London: Henry Colburn [and others]. 1833–1969.

Burke's peerage: *A genealogical and heraldic dictionary of the peerage and baronetage of the United Kingdom. Burke's peerage and baronetage.* 1st– edition. London: Henry Colburn [& others]. 1826–.

Burmeister, Karl Hermann Konrad. 1854–6. *Systematische Uebersicht der Thiere Brasiliens, welche während einer Reise durch die Provinzen von Rio de Janeiro und Minas geraës gesammelt oder beobachtet wurden.* 3 vols. Berlin: Georg Reimer.

Busk, George. 1852. Account of the Polyzoa, and Sertularian zoophytes, collected in the voyage of the *Rattlesnake*, on the coasts of Australia and the Louisiade Archipelago. In *Narrative of the voyage of H.M.S. Rattlesnake commanded by the late Captain Owen Stanley during the years 1846–1850, including discoveries and surveys in New Guinea, the Louisiade Archipelago, etc.*, by John MacGillivray, vol 1. London: T. & W. Boone.

Buxton, Charles. 1868. Acclimatization of parrots at Northrepps Hall, Norfolk. *Annals and Magazine of Natural History* 4th ser. 2: 381–6.

Calendar: *A calendar of the correspondence of Charles Darwin, 1821–1882. With supplement.* 2d edition. Edited by Frederick Burkhardt *et al.* Cambridge: Cambridge University Press. 1994.

Cambranes, Julio Castellanos. 1985. *Coffee and peasants; the origins of the modern plantation economy in Guatemala, 1853–1897.* English version revised by Carla Clason-Höök. Stockholm, Sweden: Institute of Latin American Studies.

Cambridge, Octavius Pickard. 1869. Descriptions and sketches of two new species of Araneidea, with characters of a new genus. [Read 4 February 1869.] *Journal of the Linnean Society (Zoology)* 10 (1870): 398–405.

Campbell, Archibald. 1869. Notes on the mode of capture of elephants in Assam. [Read 25 February 1869.] *Proceedings of the Zoological Society of London* (1869): 136–40.

Campbell, George Douglas. 1867. *The reign of law.* London: Alexander Strahan.

——. 1869. *Primeval man: an exhibition of some recent speculations.* London: G. Routledge.

Candolle, Alphonse de. 1855. *Géographie botanique raisonnée ou exposition des faits principaux et des lois concernant la distribution géographique des plantes de l'époque actuelle.* 2 vols. Paris: Victor Mason. Geneva: J. Kessmann.

Candolle, Augustin Pyramus de. 1829. *Mémoire sur la famille des ombellifères.* Paris: Treuttel et Würtz.

Candolle, Augustin Pyramus de and Candolle, Alphonse de. 1824–73. *Prodromus systematis naturalis regni vegetabilis, sive enumeratio contracta ordinum generum specierumque plantarum huc usque cognitarum, juxta methodi naturalis normas digesta.* 17 vols. Paris: Treuttel & Würtz [and others].

Candolle, Casimir de. 1865. Théorie de l'angle unique en phyllotaxie. *Bibliothèque Universelle et Revue Suisse. Archives des Sciences Physiques et Naturelles* 23: 199–212.

Caroe, Gwendolen Mary. 1985. *The Royal Institution: an informal history.* With a final chapter by Alban Caroe. London: John Murray.

Carrière, Elie Abel. 1869. *Origine des plantes domestiques démontrée par la culture du radis sauvage.* Paris: the author.

Carus, Julius Victor, trans. 1868. *Das Variiren der Thiere und Pflanzen im Zustande der Domestication.* By Charles Darwin. 2 vols. Stuttgart: E. Schweizerbart'sche Verlagshandlung (E. Koch).

——, trans. 1871. *Die Abstammung des Menschen und die geschlechtliche Zuchtwahl.* By Charles Darwin. 2 vols. Stuttgart: E. Schweizerbart'sche Verlagshandlung.

——. 1872. *Geschichte der Zoologie bis auf Joh. Müller und Charl. Darwin.* Munich: R. Oldenbourg.

Carus, Julius Victor, trans. 1873. *Das Variiren der Thiere und Pflanzen im Zustande der Domestication.* By Charles Darwin. 2d edition. 2 vols. Stuttgart: E. Schweizerbart'sche Verlagshandlung (E. Koch).

Catalogue de la Bibliothèque nationale: *Catalogue général des livres imprimés de la Bibliothèque Nationale.* 231 vols. Paris: Imprimerie nationale. 1897–1981.

Catalogue général de la librairie française: *Catalogue général de la librairie française pendant 25 ans (1840–1865).* Edited by Otto Lorenz. 4 vols. Paris: O. Lorenz. 1868.

Caton, John Dean. 1868. *American Cervus.* Read before the Ottawa Academy of Natural Sciences, 21 May 1868. Ottawa, Illinois: Osman and Hapeman.

——. 1877a. *The antelope and deer of America: a comprehensive scientific treatise upon the natural history, including the characteristics, habits, affinities, and capacity for domestication of the Antilocapra and Cervidae of North America.* New York: Hurd and Houghton. Boston: H. O. Houghton and Company.

——. 1877b. The wild turkey and its domestication. *American Naturalist* 11: 321–30.

——. 1880. *Miscellanies.* Boston: Houghton, Osgood.

CDEL: *A critical dictionary of English literature, and British and American authors, living and deceased, from the earliest accounts to the middle of the nineteenth century … with forty indexes of subjects.* By S. Austin Allibone. 3 vols. London: Trübner. Philadelphia: Childs & Peterson; J. B. Lippincott. 1859–71. *A supplement to Allibone's critical dictionary of English literature and British and American authors. Containing over thirty-seven thousand articles (authors), and enumerating over ninety-three thousand titles.* By John Foster Kirk. 2 vols. Philadelphia and London: J. B. Lippincott. 1891.

Challis, James. 1869. *Notes on the principles of pure and applied calculation: and applications of mathematical principles to theories of the physical forces.* Cambridge: Deighton, Bell and Co.

[Chambers, Robert.] 1844. *Vestiges of the natural history of creation.* London: John Churchill.

Chapman, Thomas Algernon. 1869. Observations on the œconomy of the British species of *Scolytus. Entomologist's Monthly Magazine* 6: 126–31.

Chindahl, George L. 1959. *A history of the circus in America.* Caldwell, Idaho: Caxton Printers.

Church dictionary: *A church dictionary.* Edited by Walter Farquhar Hook and William Richard Wood Stephens. 14th edition. London: Murray. 1887.

Cittadino, Eugene. 1990. *Nature as the laboratory. Darwinian plant ecology in the German Empire, 1880–1900.* Cambridge: Cambridge University Press.

Claparède, Edouard. 1868. Studien an Acariden. *Zeitschrift für wissenschaftliche Zoologie* 18: 445–546.

Clark, Ronald W. 1968. *The Huxleys.* London: Heinemann.

Clarke, Alexander Ross. 1866. Abstract of the results of the comparisons of the standards of length of England, France, Belgium, Prussia, Russia, India, Australia, made at the Ordnance Survey Office, Southampton. [Read 13 December 1866.] *Philosophical Transactions of the Royal Society of London* 157 (1867): 161–80.

Claus, Carl Friedrich. 1862. Ueber die morphologischen Beziehungen der

Copepoden zu den verwandten Crustaceen-gruppen der Malacostraken, Phyllopoden, Cirripedien und Ostracoden. *Würzburger naturwissenschaftliche Zeitschrift* 3: 159–67.

——. 1866. *Die Copepoden-Fauna von Nizza. Ein Beitrag zur Charakteristik der Formen und deren Abänderungen 'im Sinne Darwin's'.* Marburg and Leipzig: N. G. Elwert'sche Universitäts-Buchhandlung. [Reprinted from *Schriften der Gesellschaft zur Befoerderung der gesammten Naturwissenschaften zu Marburg.*]

——. 1868a. Beobachtungen ueber die Organisation und Fortpflanzung von Leptodera appendiculata. *Schriften der Gesellschaft zur Beförderung der gesammten Naturwissenschaften zu Marburg.* Supplement-Heft III. [1869.]

——. 1868b. Die Cypris-aehnliche Larve (Puppe) der Cirripedien und ihre Verwandlung in das festsitzende Their. Ein Beitrag zur Morphologie der Rankenfuessler. *Schriften der Gesellschaft zur Befoerderung der gesammten Naturwissenschaften zu Marburg.* Supplement-Heft V. [1869.]

——. 1874. *Die Typenlehre und E. Haeckels sog. Gastraea-Theorie.* Vienna: Manz.

——. 1876. *Untersuchungen zur Erforschung der genealogischen Grundlage des Crustaceen-Systems. Ein Beitrag zur Descendenzlehre.* Vienna: Carl Gerold's Sohn.

Cleave, Egbert. 1873. *Biographical cyclopaedia of homeopathic physicians and surgeons.* Philadelphia: Galaxy Publishing Co.

Clergy list: *The clergy list … containing an alphabetical list of the clergy.* London: C. Cox [and others]. 1841–89.

Clifford, William Kingdon. 1882. *Mathematical papers.* Edited by Robert Tucker, with an introduction by H. J. Stephen Smith. London: Macmillan and Co.

Clifton, Gloria. 1995. *Directory of British scientific instrument makers 1550–1851.* London: Zwemmer, in association with the National Maritime Museum.

Climbing plants: *On the movements and habits of climbing plants.* By Charles Darwin. London: Longman, Green, Longman, Roberts & Green; Williams & Norgate. 1865.

Clokie, Hermia Newman. 1964. *An account of the herbaria of the department of botany in the University of Oxford.* Oxford: Oxford University Press.

Cobbold, Thomas Spencer. 1885. Notes on parasites collected by the late Charles Darwin, Esq. [Read 3 December 1885.] *Journal of the Linnean Society of London (Zoology)* 19 (1886): 174–8.

Colenso, William. 1868. On the Maori races of New Zealand. *Transactions and Proceedings of the New Zealand Institute* 1: (Essays) 1–76.

Collected papers: *The collected papers of Charles Darwin.* Edited by Paul H. Barrett. 2 vols. Chicago and London: University of Chicago Press. 1977.

Collinge, J. M. comp. 1984. *Officials of Royal Commissions of Inquiry 1815–1870.* Vol. 9 of *Office-holders in modern Britain*, compiled by J. C. Sainty and J. M. Collinge. London: University of London Institute of Historical Research.

Colonial Office list: *The Colonial Office list … or, general register of the colonial dependencies of Great Britain.* London: Edward Stanford; Harrison & Sons. 1862–99.

Columbia gazetteer of the world: *The Columbia gazetteer of the world.* Edited by Saul B. Cohen. 3 vols. New York: Columbia University Press. 1998.

Colvin, Howard. 1995. *A biographical dictionary of British architects, 1600–1840*. 3d edition. New Haven and London: Yale University Press.

Commercial directory for Shropshire: *Pigot & Co.'s national commercial directory; 1828–9; comprising a directory and classification of the merchants, bankers, professional gentlemen, manufacturers and traders, in all the cities, towns, sea-ports, and principal villages in the following counties, viz. Cheshire Derbyshire Nottinghamshire Shropshire.* London and Manchester: J. Pigot & Co. 1828–9.

Complete peerage: *The complete peerage of England, Scotland, Ireland, Great Britain, and the United Kingdom, extant, extinct, or dormant.* By George Edward Cokayne. Revised edition. Edited by Vicary Gibbs, *et al.* 12 vols. London: St Catherine Press. 1910–59.

Conry, Yvette. 1974. *L'introduction du Darwinisme en France au XIXe siècle.* Paris: J. Vrin.

Cope, Edward D. 1868. On the origin of genera. *Proceedings of the Academy of Natural Sciences of Philadelphia* (1868): 242–300.

Coral reefs: *The structure and distribution of coral reefs. Being the first part of the geology of the voyage of the* Beagle, *under the command of Capt. FitzRoy RN, during the years 1832 to 1836.* By Charles Darwin. London: Smith, Elder & Co. 1842.

Correns, C. 1916. Friedrich Hildebrand. *Berichte der deutschen botanischen Gesellschaft* 34 (pt 2): 28–49.

Correspondence: *The correspondence of Charles Darwin.* Edited by Frederick Burkhardt *et al.* 16 vols to date. Cambridge: Cambridge University Press. 1985–.

County families: *The county families of the United Kingdom; or, royal manual of the titled & untitled aristocracy of Great Britain & Ireland.* By Edward Walford. London: Robert Hardwicke; Chatto & Windus. 1860–93. *Walford's county families of the United Kingdom or royal manual of the titled and untitled aristocracy of England, Wales, Scotland, and Ireland.* London: Chatto & Windus; Spottiswoode & Co. 1894–1920.

Crockford's clerical directory: *The clerical directory, a biographical and statistical book of reference for facts relating to the clergy and the church. Crockford's clerical directory etc.* London: John Crockford [and others]. 1858–1900.

Croll, James. 1868. On geological time, and the probable date of the glacial and the upper miocene period. *Philosophical Magazine* 4th ser. 35: 363–84; 36: 141–54, 362–86.

——. 1869. On the physical cause of the motion of glaciers. *Philosophical Magazine* 37: 201–6.

——. 1870. On the cause of the motion of glaciers. *Philosophical Magazine* 40: 153–70.

Cross and self fertilisation: *The effects of cross and self fertilisation in the vegetable kingdom.* By Charles Darwin. London: John Murray. 1876.

Crüger, Hermann. 1864. A few notes on the fecundation of orchids and their morphology. [Read 3 March 1864.] *Journal of the Linnean Society (Botany)* 8 (1865): 127–35.

Crumrine, Boyd. 1882. *History of Washington County, Pennsylvania, with biographical sketches of many of its pioneers and prominent men.* Philadelphia: L. H. Everts & Co.

Cunningham, Robert Oliver. 1871. *Notes on the natural history of the Strait of Magellan and west coast of Patagonia, made during the voyage of H.M.S. 'Nassau' in the years 1866, 67, 68, & 69.* Edinburgh: Edmonston and Douglas.

Cupples, George. 1894. *Scotch deer-hounds and their masters.* With a biographical sketch of the author by James Hutchison Stirling. Edinburgh and London: William Blackwood and Sons.

Curtis's botanical magazine dedications: *Curtis's botanical magazine dedications 1827–1927.* Portraits and biographical notes compiled by Ernest Nelmes and William Cuthbertson. London: Bernard Quaritch for the Royal Horticultural Society. [1931.]

Cuvier, Georges. 1815. *Essay on the theory of the earth.* Translated by Robert Kerr. 2d edition. Edinburgh: William Blackwood.

——. 1817. *Le règne animal distribué d'après son organisation, pour servir de base a l'histoire naturelle des animaux et d'introduction à l'anatomie comparée.* 4 vols. Paris: Deterville.

——. 1827. *Essay on the theory of the earth.* Translated by Robert Kerr. 5th edition. Edinburgh: William Blackwood. London: T. Cadell.

——. 1834–6. *Recherches sur les ossemens fossiles, où l'on rétablit les caractères de plusieurs animaux dont les révolutions du globe ont détruit les espèces.* 4th edition. 10 vols. and 2 atlases. Paris: E. d'Ocagne.

DAB: *Dictionary of American biography.* Under the auspices of the American Council of Learned Societies. 20 vols., index, and 10 supplements. New York: Charles Scribner's Sons; Simon & Shuster Macmillan. London: Oxford University Press; Humphrey Milford. 1928–95.

Dallas, William Sweetland, trans. 1869. *Facts and arguments for Darwin.* By Fritz Müller. London: John Murray.

Dareste, Camille. 1862. Mémoire sur la production artificielle des monstruosités. *Annales des Sciences Naturelles* (*Zoologie*) 4th ser. 18: 243–76.

——. 1863. Recherches sur les conditions de la vie et de la mort chez les monstres ectroméliens, célosomiens et exencéphaliens, produits artificiellement dans l'espèce de la poule. [Read 23 January 1863.] *Mémoires de la Société Impériale des Sciences de l'Agriculture et des Arts de Lille* 10: 39–82. [Reprinted in *Annales des Sciences Naturelles (Zoologie)* 4th ser. 20: 59–99.]

——. 1867. *Rapport sur un veau monstreux.* (Extracted from the *Archives du Comice Agricole de l'Arrondisement Lille.*) Lille: Blocquel-Castiaux.

——. 1877. *Recherches sur la production artificielle des monstruosités, ou essai de tératogénie expérimentale.* Paris: C. Reinwald.

Darwin, Charles and Wallace, Alfred Russel. 1858. On the tendency of species to form varieties; and on the perpetuation of varieties and species by natural means of selection … Communicated by Sir Charles Lyell … and J. D. Hooker. [Read 1 July 1858.] *Journal of the Proceedings of the Linnean Society* (*Zoology*) 3 (1859): 45–62. [*Collected papers* 2: 3–19.]

Darwin, Erasmus. 1794–6. *Zoonomia; or, the laws of organic life.* 2 vols. London: J. Johnson.

Darwin, Francis. 1914. William Erasmus Darwin. *Christ's College Magazine* 29: 16–23.

——. 1920. *Springtime and other essays.* London: John Murray.

Darwin pedigree: *Pedigree of the family of Darwin.* Compiled by H. Farnham Burke. N.p.: privately printed. 1888. [Reprinted in facsimile in *Darwin pedigrees*, by Richard Broke Freeman. London: printed for the author. 1984.]

Davies, Gordon Leslie. [1969.] *The earth in decay: a history of British geomorphology 1578–1878.* London: MacDonald Technical & Scientific.

Dawkins, William Boyd. 1866. On the fossil British oxen. Part I. *Bos urus*, Caesar. [Read 21 March 1866.] *Quarterly Journal of the Geological Society of London* 22: 391–401.

——. 1867. On the dentition of Rhinoceros leptorhinus, Owen. *Quarterly Journal of the Geological Society of London* 23: 213–27.

——. 1868. On the prehistoric Mammalia of Great Britain. *Transactions of the International Congress of Prehistoric Archaeology* 3d session (1868): 269–90.

——. 1870. On the discovery of platycnemic men in Denbighshire. *Journal of the Ethnological Society of London* n.s. 2 (1869–70): 440–68.

Dawkins, William Boyd and Busk, George. 1870. On the discovery of platycnemic men in Denbighshire. [Read September 1870.] *Report of the 40th meeting of the British Association for the Advancement of Science, held at Liverpool*, Transactions of the sections, p. 148.

Dawkins, William Boyd and Sanford, William Ayshford. 1866–72. *British Pleistocene felidae.* Vol. 1 of *A monograph of the British Pleistocene Mammalia.* London: Palaeontographical Society.

DBE: *Deutsche biographische Enzyklopädie.* Edited by Walter Killy *et al.* 12 vols. in 14. Munich: K. G. Saur. 1995–2000.

DBF: *Dictionnaire de biographie Française.* Under the direction of J. Balteau *et al.* 20 vols. (A–Lefèvre) to date. Paris: Librairie Letouzey & Ané. 1933–.

DBI: *Dizionario biografico degli Italiani.* Edited by Alberto M. Ghisalberti *et al.* 69 vols. (A–Marconi) to date. Rome: Istituto della Enciclopedia Italiana. 1960–.

DCB: *Dictionary of Canadian biography.* Edited by George W. Brown *et al.* 13 vols. and index to first 12 vols. Toronto, Ontario: University of Toronto Press. 1966–94.

De Beer, Gavin, ed. 1959. Darwin's journal. *Bulletin of the British Museum (Natural History). Historical Series* 2 (1959–63): 3–21.

Decaisne, Joseph. 1834. Description d'un herbier de l'île de Timor, faisant partie des collections botaniques du Muséum d'Histoire Naturelle. *Nouvelles Annales du Muséum d'Histoire Naturelle* 3: 333–501.

Delamer, Eugene Sebastian [Edmund Saul Dixon]. 1854. *Pigeons and rabbits, in their wild, domestic, and captive states.* London: G. Routledge.

Delaunay, Charles Eugène. 1868a. Sur l'hypothèse de la fluidité intérieure du globe terrestre. *Comptes rendus hebdomadaires des séances de l'Académie des Sciences* 67: 65–70.

——. 1868b. On the hypothesis of the internal fluidity of the terrestrial globe. *Geological Magazine* 5: 507–11.

Delpino, Federico. 1867a. Sull'opera 'La distribuzione dei sessi nelle piante e la legge che osta alla perennità della fecondazione consanguinea' del prof. Hildebrand. Con note critiche. *Atti della Società Italiana delle Scienze Naturali in Milano* 10: 3–34.

——. 1867b. *Sugli apparecchi della fecondazione nelle piante antocarpe (fanerogame): sommario di osservazioni fatte negli anni 1865–1866.* Florence: Cellini.

——. 1867c. Pensieri sulla biologia vegetale, sulla tassonomia, sul valore tassonomico dei caratteri biologici, e proposta di un genere nuovo della famiglia delle labiate. *Il Nuovo Cimento. Giornale di Fisica, Chimica e Storia Naturali* 25: 284–304, 321–98.

——. 1868–75. *Ulteriori osservazioni sulla dicogamia nel regno vegetale.* 2 parts. Milan: Giuseppe Bernardoni.

——. 1869a. Sulla darwiniana teoria della pangenesi. *Rivista Contemporanea Nazionale Italiana* 56: 196–204; 57: 25–38.

——. 1869b. On the Darwinian theory of pangenesis. *Scientific Opinion* 2: 365–7, 391–3, 407–8.

——. 1869c. Alcuni appunti di geografia botanica a proposito delle tabelle fitografiche del Prof Hoffmann. *Bollettino della Società Geografica Italiana* (1869.)

——. 1869d. Breve cenno sulle relazioni biologiche e genealogiche delle Marantacee. *Nuovo Giornale Botanico Italiano* 1: 293–306.

——. 1871. *Sulla dicogamia vegetale e specialmente su quella dei cereali.*

Derry, Margaret Elsinor. 2003. *Bred for perfection. Shorthorn cattle, collies and arabian horses since 1800.* Baltimore and London: Johns Hopkins University Press.

Descent: *The descent of man, and selection in relation to sex.* By Charles Darwin. 2 vols. London: John Murray. 1871.

Desmarest, Anselm-Gaëtan. 1820. *Mammalogie ou description des espèces de mammifères.* Paris: Veuve Agasse.

Desmond, Adrian. 1994–7. *Huxley.* 2 vols. London: Michael Joseph.

Desmond, Adrian and Moore, James. 1991. *Darwin.* London: Michael Joseph.

Desmond, Ray. 1994. *Dictionary of British and Irish botanists and horticulturists including plant collectors, flower painters and garden designers.* New edition, revised with the assistance of Christine Ellwood. London: Taylor & Francis and the Natural History Museum. Bristol, Pa.: Taylor & Francis.

——. 1995. *Kew: the history of the Royal Botanic Gardens.* London: Harvill Press with the Royal Botanic Gardens, Kew.

——. 1999. *Sir Joseph Dalton Hooker, traveller and plant collector.* Woodbridge, Suffolk: Antique Collectors' Club with the Royal Botanic Gardens, Kew.

Dictionary of African biography. Editor-in-chief, L. H. Ofusu-Appiah. 3 vols. to date. New York: Reference Publications. 1977–.

Dictionary of art. Edited by Jane Turner. 34 vols. London: Macmillan Publishers. 1996.

Dictionnaire historique & biographique de la Suisse. Under the direction of Marcel Godet *et al.* 7 vols. and supplement. Neuchâtel: Administration du Dictionnaire Historique & Biographique de la Suisse. 1921–34.

Dictionnaire universel des contemporains: *Dictionnaire universel des contemporains contenant toutes les personnes notables de la France et des pays étrangers … Ouvrage rédigé et continuellement tenu à jour avec le concours d'écrivains et des savants de tous les pays.* Edited by Louis Gustave Vapereau. Paris: Libraire Hachette. 1858. 3d edition, 1865. 4th edition, 1870. 5th edition, 1880. 6th edition, 1893.

Di Gregorio, Mario A. 2005. *From here to eternity: Ernst Haeckel and scientific faith.* Göttingen: Vandenhoeck & Ruprecht.

DNB: *Dictionary of national biography.* Edited by Leslie Stephen and Sidney Lee. 63

vols. and 2 supplements (6 vols.). London: Smith, Elder & Co. 1885–1912. *Dictionary of national biography 1912–90*. Edited by H. W. C. Davis *et al.* 9 vols. London: Oxford University Press. 1927–96.

DNZB: *A dictionary of New Zealand biography.* Edited by G. H. Scholefield. 2 vols. Wellington, New Zealand: Department of Internal Affairs. 1940. *The dictionary of New Zealand biography.* Edited by W. H. Oliver *et al.* 5 vols. Auckland and Wellington, New Zealand: Department of Internal Affairs [and others]. 1990–2000.

Dobbs, David. 2005. *Reef madness: Charles Darwin, Alexander Agassiz, and the meaning of coral.* New York: Pantheon Books.

Dohrn, Anton. 1867. On the morphology of the Arthropoda. [Read before the British Association, 5 September 1867.] *Journal of Anatomy and Physiology* 2 (1868): 80–6.

——. 1870. *Untersuchungen über Bau und Entwicklung der Arthropoden.* 2 vols. Leipzig: W. Engelmann.

——. 1871. Untersuchungen über Bau und Entwickelung der Arthropoden. 12. Zur Embryologie und Morphologie des Limulus Polyphemus. *Jenaische Zeitschrift für Medizin und Naturwissenschaft* 6: 580–640.

Domeyko, Ignacy. 1846. Recherches sur la géologie du Chili. *Annales des Mines: ou recueil de mémoires sur l'exploitation des mines et sur les sciences et les arts qui s'y rapportent* 4th ser. 9: 3–34, 365–540.

Dressler, Robert L. 1968. Pollination by euglossine bees. *Evolution* 22: 202–10.

——. 1981. *The orchids: natural history and classification.* Cambridge, Mass., and London: Harvard University Press.

DSAB: *Dictionary of South African biography.* Edited by W. J. de Kock *et al.* 4 vols. Pretoria and Cape Town: Nasionale Boekhandel Beperk [and others]. 1968–81.

DSB: *Dictionary of scientific biography.* Edited by Charles Coulston Gillispie and Frederic L. Holmes. 18 vols. including index and supplements. New York: Charles Scribner's Sons. 1970–90.

Dub, Julius. 1870. *Kurze Darstellung der Lehre Darwin's über die Entstehung der Arten der Organismen mit erläuternden Bemerkungen.* Stuttgart: E Schweizerbart.

Duchenne, Guillaume Benjamin Amand. 1862. *Mécanisme de la physionomie humaine, ou analyse électro-physiologique de l'expression des passions.* 1 vol. and 'Atlas' of plates. Paris: Ve Jules Renouard, Libraire.

Duncan, James. 1863. *Catalogue of plants in the Royal Botanic Garden, Mauritius.* [Mauritius]: n.p.

Dupree, Anderson Hunter. 1959. *Asa Gray, 1810–1888.* Cambridge, Mass.: Belknap Press of Harvard University.

Earthworms: *The formation of vegetable mould, through the action of worms, with observations on their habits.* By Charles Darwin. London. 1881.

EB: *The Encyclopædia Britannica. A dictionary of arts, sciences, literature and general information.* 11th edition. 29 vols. Cambridge: Cambridge University Press. 1910–11.

EB (1970): *Encyclopaedia Britannica.* 24 vols. Chicago and London: Encyclopaedia

Britannica. 1970.

Ehrenberg, Christian Gottfried. 1869. Uber mächtige Gebirgs-Schichten vorherrschend aus mikroskopischen Bacillarien unter und bei der Stadt Mexiko. [Read 29 April 2869.] *Abhandlungen der königlichen Akademie der Wissenschaften zu Berlin (Physikalische Klasse)* (1869): 1–66.

Ehrlich, Eugene. 1986. *Nil desperandum: a dictionary of Latin tags and useful phrases.* London: Robert Hale.

Elliott, Clark A. 1979. *Biographical dictionary of American science. The seventeenth through nineteenth centuries.* Westport, Connecticut: Greenwood Press.

Emma Darwin (1904): *Emma Darwin, wife of Charles Darwin. A century of family letters.* Edited by Henrietta Litchfield. 2 vols. Cambridge: privately printed by Cambridge University Press. 1904.

Emma Darwin (1915): *Emma Darwin: a century of family letters, 1792–1896.* Edited by Henrietta Litchfield. 2 vols. London: John Murray. 1915.

Endlicher, Stephan Ladislaus. 1836–42. *Genera plantarum secundum ordines naturales disposita.* With 4 supplements; in 2 vols. Vienna: Friedrich Beck.

Europäische Stammtafeln: *Europäische Stammtafeln: Stammtafeln zur Geschichte der europäischen Staaten.* Edited by Wilhelm Karl Prinz zu Issenberg *et al.* 25 vols. to date. Marburg: J. A. Stargardt. 1956–

Expression: *The expression of the emotions in man and animals.* By Charles Darwin. London: John Murray. 1872.

Eyton, Thomas Campbell. 1867–75. *Osteologia avium; or, a sketch of the osteology of birds.* 1 vol. and 2 supplements. Wellington, Salop.: R. Hobson.

Faraday, Michael. 1834. Experimental researches in electricity. [Read 23 January, 6 and 13 February 1834.] *Philosophical Transactions of the Royal Society of London* 7th ser. 124: 77–122.

Farrant, Sue. 1978. John Ellman of Glynde in Sussex. *Agricultural History Review* 26: 77–88.

Farrer, Thomas Henry. 1872. On the fertilisation of a few common papilionaceous flowers. *Nature* 6: 478–80, 498–501.

Fergusson, Alastair Mackenzie and Fergusson, John. 1888. *All about gold, gems, and pearls (also minerals generally) in Ceylon and southern India.* 2d edition. Columbo: A. M. & J. Fergusson. London: John Haddon & Co., [and others].

'Fertilization of orchids': Notes on the fertilization of orchids. By Charles Darwin. *Annals and Magazine of Natural History* 4th ser. 4 (1869): 141–59. [*Collected papers* 2: 138–56.]

Fiala, John L. 1988. *Lilacs: the genus syringa.* London: Christopher Helm.

Fichman, Martin. 2004. *An elusive Victorian; the evolution of Alfred Russel Wallace.* Chicago and London: University of Chicago Press.

Fleming, I. A., and Reynolds, J. D. 2004. Salmonoid breeding systems. In *Evolution illuminated: salmon and their relatives.* Edited by Andrew P. Hendry and Stephen C. Stearns. Oxford: Oxford University Press.

Flower, William Henry. 1873. On a newly discovered extinct ungulate mammal from Patagonia, *Homalodontotherium cunninghami*. [Read 19 June 1873.] *Philosophical Transactions of the Royal Society of London* 164 (1874): 173–82.

Forbes, David. 1860. On the geology of Bolivia and southern Peru. [Read 21 November 1860.] *Quarterly Journal of the Geological Society of London* 17 (1861): 7–62.

——. 1869. The nature of the interior of the earth. *Popular Science Review* 8: 121–30.

——. 1870. On the Aymara Indians of Bolivia and Peru. [Read 21 June 1870.] *Journal of the Ethnological Society of London* n.s. 2 (1869–70): 193–298.

Forbes, Edward. 1846. On the connexion between the distribution of the existing fauna and flora of the British Isles, and the geological changes which have affected their area, especially during the epoch of the Northern Drift. *Memoirs of the Geological Survey of Great Britain, and of the Museum of Economic Geology in London* 1: 336–432.

Foreign Office list: *The Foreign Office list*. London: Harrison & Sons. 1852–1965.

'Formation of mould': On the formation of mould. [Read 1 November 1837.] *Transactions of the Geological Society of London* 2d ser. 5 (1840): 505–9. [*Collected papers* 1: 49–53.]

Forms of flowers: *The different forms of flowers on plants of the same species*. By Charles Darwin. London: John Murray. 1877.

Foster, Joseph. 1893. *Oxford men 1880–1892: with a record of their schools, honours and degrees*. Oxford: J. Parker.

Fotheringham, James Gainsborough. 1884. The Orkneys. *Notes and Queries* 6th ser. 9: 2–4, 63–5, 122–5.

Freeman, Richard Broke. 1977. *The works of Charles Darwin: an annotated bibliographical handlist*. 2d edition. Folkestone, Kent: William Dawson & Sons. Hamden, Conn.: Archon Books, Shoe String Press.

——. 1978. *Charles Darwin: a companion*. Folkestone, Kent: William Dawson & Sons. Hamden, Conn.: Archon Books, Shoe String Press.

Galton, Francis. 1869. *Hereditary genius: an inquiry into its laws and consequences*. London: Macmillan.

——. 1871. Experiments in pangenesis, by breeding from rabbits of a pure variety, into whose circulation blood taken from other varieties had previously been largely transfused. [Read 30 March 1871.] *Proceedings of the Royal Society of London* 19 (1870–1): 393–410.

Gärtner, Karl Friedrich von. 1844. *Versuche und Beobachtungen über die Befruchtungsorgane der vollkommeneren Gewächse und über die natürliche und künstliche Befruchtung durch den eigenen Pollen*. Pt 1 of *Beiträge zur Kenntniss der Befruchtung der vollkommeneren Gewächse*. Stuttgart: E. Schweizerbart.

——. 1849. *Versuche und Beobachtungen über die Bastarderzeugung im Pflanzenreich. Mit Hinweisung auf die ähnlichen Erscheinungen im Thierreiche, ganz umgearbeitete und sehr vermehrte Ausgabe der von der Königlich holländischen Akademie der Wissenschaften*. Stuttgart: E. Schweizerbart.

Gaudry, Albert. 1865. Sur les Hipparions. *Bulletin de la Société Géologique de France* 22: 21–4.

Gayon, Jean. 1998. *Darwinism's struggle for survival: heredity and the hypothesis of natural selection.* Translated by Matthew Cobb. Cambridge: Cambridge University Press.

[Geikie, Archibald.] 1868. Sir Roderick Murchison and modern schools of geology. [Review of Roderick Murchison's *Siluria: a history of the oldest rocks in the British Isles and other countries.* 4th edition.] *Quarterly Review* 125: 188–217.

Geoffroy Saint-Hilaire, Isidore. 1832–7. *Histoire générale et particulière des anomalies de l'organisation chez l'homme et les animaux, ouvrage comprenant des recherches sur les charactères, la classification, l'influence physiologique et pathologique, les rapports généraux, les lois et les causes des monstruosites, des variétés et des vices de conformation, ou traité de tératologie.* 3 vols. and atlas. Paris: J. B. Baillière.

GEPB: *Grande enciclopédia portuguesa e brasileira.* Lisbon: Editorial Enciclopédia. [1940–60.]

Gilbert, Pamela. 1977. *A compendium of the biographical literature on deceased entomologists.* London: British Museum (Natural History).

Giles, Barbara E. *et al.* 2006. Natural selection on floral traits of female *Silene dioica* by a sexually transmitted disease. *New Phytologist* 169: 729–39.

Gosse, Philip Henry. 1847. *The birds of Jamaica.* London: John Van Voorst.

Gould, John. 1861. *A monograph of the Trochilidæ, or family of humming-birds.* 5 vols. London: John Gould.

Grande encyclopédie: *La grande encyclopédie inventaire raisonné des sciences, des lettres et des arts.* Edited by F. Camille Dreyfus *et al.* 31 vols. Paris: H. Lamirault; Société Anonyme de la Grande Encyclopédie. [1886–1902.]

Grant Duff, Mountstuart E. 1897–1905. *Notes from a diary.* 14 vols. London: John Murray.

Grant Duff, Mountstuart E. 1903. *Out of the past: some biographical essays.* 2 vols. London: John Murray.

Gratiolet, Pierre. [1865.] *De la physionomie et des mouvements d'expression.* Paris: J. Hetzel.

Gray, Asa. 1856. Note on *Obolaria virginica*, L. [Read 15 April 1856.] *Journal of the Proceedings of the Linnean Society* Botanical Papers 1: 129–30.

——. 1860a. Discussion between two readers of Darwin's treatise on the origin of species, upon its natural theology. *American Journal of Science and Arts* 2d ser. 30: 226–39.

——. 1860b. Review of Darwin's theory on the origin of species by means of natural selection. *American Journal of Science and Arts* 2d ser. 29: 153–84.

[——.] 1860c. Darwin on the origin of species. *Atlantic Monthly* 6: 109–16, 229–39; Darwin and his reviewers. *Atlantic Monthly* 6: 406–25.

——. 1861. *Natural selection not inconsistent with natural theology. A free examination of Darwin's treatise on the origin of species, and of its American reviewers.* London: Trübner. Boston, Mass.: Ticknor & Fields.

Gray, Jane Loring, ed. 1893. *Letters of Asa Gray.* 2 vols. London: Macmillan and Co.

Gray, John Edward. 1852. *Catalogue of the specimens of Mammalia in the collection of the British Museum. Part III. Ungulata Furcipeda.* London: British Museum.

——. 1868. Notes on the skulls of the species of dogs, wolves, and foxes (*Canidæ*) in the collection of the British Museum. [Read 25 June 1868.] *Proceedings of the Zoological Society of London* (1868): 492–525.

[Greg, William Rathbone.] 1868. On the failure of 'natural selection' in the case of man. *Fraser's Magazine for Town and Country* 78: 353–62.

Grigelas, Algimantas. 2005. Ignacy Domeyko—an early investigator of Andean geology. *Episodes* 28: 279–85.

Grisley, Gabriel. [1661]. *Viridarium Lusitanum: in quo arborum fruticum & herbarum differentiæ onomasti insertæ, quas Ager Vlyssiponensis vltrà citràque Tagum ad trigesimum usque lapidem profert.* Lisbon: Antonius Craesbeeck.

Groeben, Christiane. 1982. *Charles Darwin 1809–1882, Anton Dohrn 1840–1909: correspondence.* Naples: Macchiaroli.

Der große Brockhaus: *Der große Brockhaus. Handbuch des Wissens in zwanzig Bänden.* 15th edition. 20 vols. and supplement. Leipzig: F. A. Brockhaus. 1928–35.

Grote, George. 1872. *Aristotle.* 2 vols. Edited by Alexander Bain and G. Croom Robertson. London: John Murray.

Grove, George. 1980. *The new Grove dictionary of music and musicians.* Edited by Stanley Sadie. 20 vols. London: Macmillan.

GSE: *Great Soviet encyclopedia.* Edited by Jean Paradise *et al.* 31 vols. New York: Macmillan. London: Collier Macmillan. 1973–83. [Translation of the 3d edition of *Bol'shaia Sovetskaia entsiklopediia*, edited by A. M. Prokhorov.]

Gunn, Mary and Codd, L. E. 1981. *Botanical exploration of Southern Africa.* Cape Town: A. A. Balkema.

Günther, Albert Charles Lewis Gotthilf. 1864. *The reptiles of British India.* London: Ray Society.

——. 1864–6. An account of the fishes of the states of Central America, based on collections made by Capt. J. M. Dow, F. Godman, Esq., and O. Salvin, Esq. [Read 22 March 1864 and 13 December 1866.] *Transactions of the Zoological Society of London* 6 (1863–7): 377–494.

Gunther, Albert E. 1972. The original drawings of George Henry Ford. *Journal of the Society for the Bibliography of Natural History* 6 (1971–4): 139–42.

——. 1975. *A century of zoology at the British Museum through the lives of two keepers, 1815–1914.* London: Dawsons of Pall Mall.

Haeckel, Ernst. 1866. *Generelle Morphologie der Organismen. Allgemeine Grundzüge der organischen Formen-Wissenschaft, mechanisch begründet durch die von Charles Darwin reformirte Descendenz-Theorie.* 2 vols. Berlin: Georg Reimer.

——. 1868. *Natürliche Schöpfungsgeschichte. Gemeinverständliche wissenschaftliche Vorträge über die Entwickelungslehre im Allgemeinen und diejenige von Darwin, Goethe und Lamarck im Besonderen, über die Anwendung derselben auf den Ursprung des Menschen und andere damit zusammenhängende Grundfragen der Naturwissenschaft.* Berlin: Georg Reimer.

——. 1869a. *Zur Entwickelungsgeschichte der Siphonophoren.* Utrecht: C. van der Post, Jr.

——. 1869b. Ueber den Organismus der Schwämme und ihre Verwandtschaft mit den Corallen. *Jenaische Zeitschrift für Medicin und Naturwissenschaft* 5 (1869–70): 207–35.

——. 1869c. Ueber die Crambessiden, eine neue Medusen-Familie aus der Rhizostomeengruppe. *Zeitschrift für wissenschaftliche Zoologie* 19: 509–37.

——. 1870a. *Natürliche Schöpfungsgeschichte. Gemeinverständliche wissenschaftliche Vorträge über die Entwickelungslehre im allgemeinen und diejenige von Darwin, Goethe und Lamarck im Besonderen, über die Anwendung derselben auf den Ursprung des Menschen und andere damit zusammenhängende Grundfragen der Naturwissenschaft.* 2d edition. Berlin: Georg Reimer.

——. 1870b. On the organisation of sponges, and their relationship to the corals. *Annals and Magazine of Natural History* 4th ser., 5: 1–13, 107–120.

——. 1872. *Die Kalkschwämme: eine Monographie.* 3 vols. Berlin: G. Reimer.

Haliburton, Robert Grant. 1863. *New materials for the history of man, derived from a comparison of the customs and superstitions of nations.* Halifax, Nova Scotia: n.p.

Hall, Alfred Rupert. 1969. *The Cambridge Philosophical Society: a history, 1819–1969.* Cambridge: Cambridge Philosophical Society.

Hall, Philip B. 1987. Robert Swinhoe (1836–1877), FRS, FZS, FRGS: a Victorian naturalist in Treaty Port China. *Geographical Journal* 153: 37–47.

Hallett, Frederic F. 1861. On 'pedigree' in wheat as a means of increasing the crop. *Journal of the Royal Agricultural Society of England* 22: 371–81.

Hancock, Albany. 1849. On the excavating powers of certain sponges belonging to the genus *Cliona*; with descriptions of several new species, and an allied generic form. *Annals and Magazine of Natural History* 2d ser. 3: 321–48.

Hargreaves, J. D. 1957. Winwood Reade and the discovery of Africa. *African Affairs* 56: 306–16.

Harris, Thaddeus William. 1842. *A treatise on some of the insects of New England, which are injurious to vegetation.* Cambridge, Mass.: John Owen.

——. 1862. *A treatise on some of the insects injurious to vegetation.* New edition. Edited by Charles L. Flint. Boston: Crosby and Nichols.

——. 1869. *Entomological correspondence of Thaddeus William Harris.* Edited by Samuel H. Scudder. Boston: Boston Society of Natural History.

Harrison, Charles, *et al.*, eds. 2000. *Art in theory, 1648–1815: an anthology of changing ideas.* Oxford: Blackwell.

Harrow School register: *The Harrow School register, 1800–1911.* 3d edition. Edited by M. G. Dauglish and P. K. Stephenson. London: Longmans, Green, and Co. 1911.

Hartsen, Frederik Anthony von. 1869a. *Grundlegung von Aesthetik, Moral und Erziehung, vom empirischen Standpunkt.* Halle: C. E. M. Pfeffer.

——. 1869b. *Darwin en de Godsdienst: eene populaire uiteenzetting van het Darwinisme, en beschouwingen over dit stelsel.* Leyden: De Breuk & Smits.

Harvey, Joy. 1997. *'Almost a man of genius': Clémence Royer, feminism, and nineteenth-century science.* New Brunswick, N.J., and London: Rutgers University Press.

Heber, Reginald. 1870. *The poetical works of Reginald Heber.* New edition. London: John Murray.

Helmholtz, Hermann von. 1856. On the interaction of natural forces. *Philosophical Magazine* 4th ser. 11: 489–518.

Henslow, George. 1868. On the variations of the angular divergencies of the leaves of *Helianthus tuberosus.* [Read 16 April 1868.] *Transactions of the Linnean Society of London* 26 (1868–70): 647–59.

Herbert, Sandra. 2005. *Charles Darwin, geologist.* Ithaca and London: Cornell University Press.

[Herschel, John Frederick William.] 1850. Review of *Lettres à S.A.R. le duc régnant de Saxe-Coburg et Gotha: sur la théorie des probabilités, appliquée aux sciences morales et politiques*, by Adolphe Quetelet. *Edinburgh Review* 92: 1–57.

Hertz, Anselm. 1981. *Fra Angelico.* Frieburg: Herder.

Hessische Biographien. In association with Karl Esselborn and Georg Lehnert; edited by Herman Haupt. 3 vols. Darmstadt: Hessischer Staatsverlag. 1918–34.

Heuss, Theodor. 1991. *Anton Dohrn: a life for science.* Translated from the German by Liselotte Dieckmann. Berlin: Springer Verlag.

Hildebrand, Friedrich. 1866. Ueber die Befruchtung der Salviaarten mit Hülfe von Insekten. *Jahrbücher für wissenschaftliche Botanik* 4 (1865–6): 451–78.

——. 1867. *Die Geschlechter-Vertheilung bei den Pflanzen und das Gesetz der vermiedenen und unvortheilhaften stetigen Selbstbefruchtung.* Leipzig: Wilhelm Engelmann.

——. 1869. Weitere Beobachtungen über die Bestäubungsverhältnisse an Blüthen. *Botanische Zeitung* 27: 473–81, 489–95, 505–12.

——. 1870. Ueber die Bestäubungsvorrichtungen bei den Fumariaceen. *Jahrbücher für wissenschaftliche Botanik* 7 (1869–70): 423–71.

Historical register of the University of Cambridge: *The historical register of the University of Cambridge, being a supplement to the* Calendar *with a record of university offices, honours, and distinctions to the year 1910.* Edited by J. R. Tanner. Cambridge: Cambridge University Press. 1917.

Hogarth, William. 1753. *The analysis of beauty: written with a view of fixing the fluctuating ideas of taste.* London: the author.

Holland, Swinton Colthurst. 1873. On the Ainos. [Read 17 June 1873.] *Journal of the Anthropological Institute of Great Britain and Ireland* 3 (1874): 233–44.

Holmes, Frederic Lawrence. 1974. *Claude Bernard and animal chemistry; the emergence of a scientist.* Cambridge: Harvard University Press.

Hooker, Joseph Dalton. 1854. *Himalayan journals; or, notes of a naturalist in Bengal, the Sikkim and Nepal Himalayas, the Khasia Mountains, &c.* 2 vols. London: John Murray.

——. 1859. *On the flora of Australia, its origin, affinities, and distribution; being an introductory essay to the flora of Tasmania.* London: Lovell Reeve.

——. 1860. Outlines of the distribution of Arctic plants. [Read 21 June 1860.] *Transactions of the Linnean Society of London* 23 (1862): 251–348.

——. 1866. Insular floras. [Read 27 August 1866.] *Gardeners' Chronicle* (1867): 6–7, 27,

50–1, 75–6.
——. 1868. Address of the president. *Report of the thirty-eighth meeting of the British Association for the Advancement of Science, held at Norwich*, pp. lviii–lxxv.
——. 1869. Vegetable palæontology. *Nature* 1: 48–50.
——. 1870. *The student's flora of the British Islands.* London: Macmillan.
——. 1875–97. *The flora of British India.* Assisted by various botanists. 7 vols. London: L. Reeve & Co.
Hooker, William Jackson. 1846–64. *Species filicum: being descriptions of the known ferns, particularly of such as exist in the author's herbarium, or are with sufficient accuracy described in works to which he has had access; accompanied with numerous figures.* 5 vols. London: W. Pamplin.
Hooker, William Jackson and Baker, John Gilbert. 1868. *Synopsis filicum; or, a synopsis of all known ferns including the Osmundaceæ, Schizæaceæ, Marattiaceæ, and Ophioglossaceæ (chiefly derived from the Kew Herbarium).* London: Robert Hardwicke.
Huet, Marie-Hélène. 1993. *Monstrous imagination.* Cambridge, Mass., and London: Harvard University Press.
Hutchinson, Horace Gordon. 1914. *Life of Sir John Lubbock, Lord Avebury.* 2 vols. London: Macmillan.
Hutton, Richard Holt. 1869. [Paper in session on phases of unbelief; read 5 October 1869.] *Authorized Report of the Church Congress, held at Liverpool*, pp. 98–106. [Also printed in the *Guardian*, 13 October 1869, pp. 1134–5.]
Huxley, Leonard, ed. 1900. *Life and letters of Thomas Henry Huxley.* 2 vols. London: Macmillan.
——, ed. 1918. *Life and letters of Sir Joseph Dalton Hooker, OM, GCSI.* Based on materials collected and arranged by Lady Hooker. 2 vols. London: John Murray.
Huxley, Thomas Henry. 1859. On the anatomy and development of *Pyrosoma.* [Read 1 December 1859.] *Transactions of the Linnean Society of London* 23: 193–250.
[——.] 1860. Darwin on the origin of species. *Westminster Review* n.s. 17: 541–70.
——. 1862. Notes upon human remains from the Valley of the Trent, and from the Heathery Burn Cave, Durham. *Geologist* 5: 201–4.
[——.] 1864. Criticisms on 'The origin of species'. *Natural History Review* n.s. 4: 566–80.
——. 1869a. On the physical basis of life. *Fortnightly Review* n.s. 5: 129–45.
——. 1869b. The scientific aspects of positivism. *Fortnightly Review* n.s. 5: 653–70.
——. 1869c. Anniversary address of the president. [Read 19 February 1869.] *Quarterly Journal of the Geological Society of London* 25: xxviii–liii.
——. 1869d. Nature: aphorisms by Goethe. *Nature* 1: 9–11.
——. 1870. *Lay sermons, addresses, and reviews.* London.
IBF: *Index biographique français.* Edited by Helen and Barry Dwyer. 4 vols. London: K. G. Saur. 1993.
IGI: *International genealogical index.* Church of Jesus Christ of Latter-day Saints. 2008. http://www.familysearch.org.
'Illegitimate offspring of dimorphic and trimorphic plants': On the character and

hybrid-like nature of the offspring from the illegitimate unions of dimorphic and trimorphic plants. By Charles Darwin. [Read 20 February 1868.] *Journal of the Linnean Society of London (Botany)* 10 (1869): 393–437.

India list: *The East-India register and directory.* 1803–44. *The East-India register and army list.* 1845–60. *The Indian Army and civil service list.* 1861–76. *The India list, civil and military.* 1877–95. *The India list and India Office list.* 1896–1917. London: Wm. H. Allen [and others].

Insectivorous plants. By Charles Darwin. London: John Murray. 1875.

Jackson, Benjamin Daydon. 1900. *A glossary of botanic terms: with their description and accent.* London: Duckworth & Co. Philadelphia: J. B. Lippencott Company.

Jackson, Welby B. 1854. *Selections from the records of the Bengal Government. No. XVII. Report on Darjeeling.* Calcutta: Thos. Jones.

Jäger, Gustav. [1869.] *Die Darwin'sche Theorie und ihre Stellung zur Moral und Religion.* Stuttgart: Julius Hoffman.

Janzen, Daniel H. 1968. Reproductive behavior in the Passifloraceae and some of its pollinators in Central America. *Behaviour* 32: 33–48.

[Jenkin, Henry Charles Fleeming.] 1867. The origin of species. *North British Review* 46: 277–318.

Jenzsch, Gustav. 1868. *Ueber eine mikroskopische Flora und Fauna krystallinischer Massengesteine (Eruptivgesteine).* Leipzig: W. Engelmann.

Jersáková, Jana, *et al.* 2006. Mechanisms and evolution of deceptive pollination in orchids. *Biological Reviews of the Cambridge Philosophical Society* 81: 219–35.

Johnson, Robert. 1976. *Sir John Richardson: arctic explorer, natural historian, naturalist.* London: Taylor & Francis.

Johnson, Samuel. 1755. *A dictionary of the English language: in which the words are deduced from their originals, and illustrated in their different significations by examples from the best writers.* 2 vols. London: J. & P. Knapton [and others].

Johnson, William S. 1990. *Nineteenth-century photography: an annotated bibliography, 1839–1879.* Boston: G. K. Hall & Co.

Johnstone, J. 1868. Notes on elephants. [Read 6 May 1868.] *Proceedings of the Asiatic Society of Bengal* (1868): 127–9.

Journal of researches: *Journal of researches into the geology and natural history of the various countries visited by HMS Beagle, under the command of Captain FitzRoy, RN, from 1832 to 1836.* By Charles Darwin. London: Henry Colburn. 1839.

Journal of researches 2d ed.: *Journal of researches into the natural history and geology of the countries visited during the voyage of HMS Beagle round the world, under the command of Capt. FitzRoy RN.* 2d edition, corrected, with additions. By Charles Darwin. London: John Murray. 1845.

Journal of researches (1860): *Journal of researches into the natural history and geology of the countries visited during the voyage of HMS Beagle around the world, under the command of Capt. FitzRoy RN.* By Charles Darwin. Reprint edition. London: John Murray. 1860.

Judel, Claus Günther. 2004. Der Liebigschüler Carl Vogt als Wissenschaftlicher, Philosoph und Politiker. *Giessener Universitätsblätter* 37: 51–6.

Junker, Thomas. 1989. *Darwinismus und Botanik. Rezeption, Kritik und theoretische Alternativen im Deutschland des 19. Jahrhunderts.* Stuttgart: Deutscher Apotheker Verlag.

Kelaart, Edward Frederick. 1846. *Flora Calpensis. Contributions to the botany and topography of Gibraltar and its neighbourhood: with plan and views of the rock.* London: John Van Voorst.

Kent, Christopher. 1978. *Brains and numbers: elitism, Comtism, and democracy in mid-Victorian England.* Toronto: University of Toronto Press.

Kerner von Marilaun, Anton. 1869. *Die Abhängigkeit der Pflanzengestalt von Klima und Boden.* Innsbruck: Wagner'schen Universitæts-Buchhandlung.

Keynes, Margaret. 1943. *Leonard Darwin, 1850–1943.* Cambridge: privately printed at Cambridge University Press.

Keynes, Milo. 1998. The Portland vase: Sir William Hamilton, Josiah Wedgwood and the Darwins. *Notes and Records of the Royal Society* 52: 237–59.

Keynes, Richard Darwin, ed. 2000. *Charles Darwin's zoology notes & specimen lists from H.M.S. Beagle.* Cambridge: Cambridge University Press.

King-Hele, Desmond. 1999. *Erasmus Darwin. A life of unequalled achievement.* London: Giles de la Mare Publishers.

Kirby, William. 1802. *Monographia apum Angliæ: or, an attempt to divide into their natural genera and families, such species of the Linnean genus Apis as have been discovered in England; with descriptions and observations: to which are prefixed some introductory remarks upon the class Hymenoptera, and a synoptical table of the external parts of these insects.* Ipswich: Printed for the author by J. Raw.

Kirby, William and Spence, William. 1815–26. *An introduction to entomology: or elements of the natural history of insects.* 4 vols. London: Longman, Hurst, Rees, Orme, and Brown.

Kirchner, Wilhelm. 1904. *Das Landwirtschaftliche Institut der Universität Leipzig in seiner neuen Gestalt, seine Einrichtungen und seine Tätigkeit. Vortrag, gehalten am 8. Januar 1904 in der Ökonomischen Gesellschaft im Königreiche Sachsen zu Dresden.* Leipzig: Reichenbach'sche Verlagsbuchhandlung.

Kishinouye, K. 1900. On the nauplius stage of Penaeus. *Zoologischer Anzeiger* 23: 73–4.

Koblitz, Ann Hibner. 1983. *A convergence of lives: Sofia Kovalevskaia, scientist, writer, revolutionary.* Boston: Birkhäuser.

Koch, Karl Heinrich Emil. 1869–73. *Dendrologie. Bäume Sträucher und Halbsträucher, welche in Mittel- und Nord-Europa im Freien kultivirt werden.* 2 vols. Erlangen: F. Enke.

Kockerbeck, Christoph, ed. 1999. *Carl Vogt, Jacob Moleschott, Ludwig Büchner, Ernst Haeckel: Briefwechsel.* Vol. 4 of *Acta biohistorica.* Marburg: Basilisken-Presse.

Kottler, Malcolm Jay. 1974. Alfred Russel Wallace, the origin of man, and spiritualism. *Isis* 65: 145–92.

Kovalevsky, Alexander Onufrievich. 1866. Entwickelungsgeschichte der einfachen Ascidien. [Read 1 November 1866.] *Mémoires de L'Académie Impériale des Sciences de St.-Pétersbourg* 7th ser. 10.15: 1–19.

——. 1875. Ueber die Entwickelungsgeschichte der Pyrosoma. *Archiv für mikroskopische Anatomie* 11: 597–635.

Kovalevsky, Vladimir Onufrievich, trans. 1868–9. *Proiskhozhdenie vidov. Otdel 1. Izmeneniya zhivotnyk i rastenii vsledstvie priruchenіya. Priruchenniya zhivotniya i vozdelanniya rasteniya.* By Charles Darwin. (Russian translation of *Variation.*) Edited by Ivan Mikhailovich Sechenov, the botanical parts by Alexander Heard. 2 vols. St Petersburg: F. S. Suchchinskii.

Krause, Ernst. 1883. Prof. Dr. Hermann Müller von Lippstadt. Ein Gedenkblatt. *Kosmos* 13: 393–401.

Kraußе, Erika. 1987. *Ernst Haeckel.* 2d edition. Leipzig: B. G. Teubner.

Lacroix, Sylvestre François. 1816. *Traité élémentaire du calcul des probabilités.* Paris: Courcier.

Laing, Alexander Gordon. 1825. *Travels in the Timannee, Kooranko, and Soolima countries, in western Africa.* London: John Murray.

Landois, Hermann. 1867. Die Ton- und Stimmapparate der Insecten in anatomisch-physiologisher und akusticher Beziehung. *Zeitschrift für wissenschaftliche Zoologie* 17: 105–84.

Laplace, Pierre Simon. 1814. *Essai philosophique sur les probabilités.* Paris: Courcier.

Le Couteur, John. 1836. *On the varieties, properties and classification of wheat.* London: H. Wright.

Leeds, Everett. 2000. Charles Noble of Sunningdale. *Clematis* (Winter 2000): 45–8.

Leftwich, A. W. 1973. *A dictionary of zoology.* 3d edition. London: Constable.

[Leifchild, John R.] 1869. [Review of *Facts and arguments for Darwin* by Fritz Müller. Translated by W. S. Dallas.] *Athenæum* (Jan–Jun 1869): 431.

Le Maout, Emmanuel and Decaisne, Joseph. 1868. *Traité générale de botanique descriptive et analytique.* Paris: Didot frères, fils et cie.

LepIndex: The Global Lepidoptera Names Index. Edited by G. W. Beccaloni *et al.* 2003. http://www.nhm.ac.uk/entomology/lepindex.

Lesley, J. Peter. 1862. Section of coal-measures on the Cape Breton coast. *Proceedings of the American Philosophical Society* 9 (1862–3): 93–109.

——. 1866. Notes on a map intended to illustrate five types of earth-surface in the United States, between Cincinnati and the Atlantic seaboard. [Read 4 May 1866.] *Transactions of the American Philosophical Society* n.s. 13 (1865–9): 305–12.

Lester, Joseph. 1995. *E. Ray Lankester and the making of modern British biology.* Edited and with additional material by Peter J. Bowler. Oxford: British Society for the History of Science.

Leuenberger, Beat Ernst. 2004. The *Cactaceae* of the Willdenow herbarium, and of Willdenow (1813). *Willdenowia* 34: 309–22.

Lewes, George Henry. 1868. Mr. Darwin's hypotheses. *Fortnightly Review* n.s. 3: 353–73, 611–28; 4: 61–80, 492–509.

Leydig, Franz. 1860. *Naturgeschichte der Daphniden (Crustacea cladocera).* Tübingen: H. Laupp'sche Buchhandlung, Laupp & Siebeck.

Lichtenstein, Hinrich. 1811–12. *Reisen im südlichen Africa: in den jahren 1803, 1804, 1805 und 1806.* Berlin: C. Salfeld.

Lightman, Bernard, ed. 2004. *Dictionary of nineteenth-century British scientists.* 4 vols. Bristol: Thoemmes Press.

Lindley, John. 1830–40. *The genera and species of orchidaceous plants.* London: Ridgways.

Linnaeus, Carolus (Carl von Linné). 1751. *Philosophia botanica in qua explicantur fundamenta botanica cum definitionibus partium, exemplis terminorum, observationibus rariorum, adjectis figuris æneis.* Stockholm: Godofr. Kiesewetter.

Linsser, Carl. 1867. Die periodischen Erscheinungen des Pflanzenlebens in ihrem Verhältniss zu den Wärmëerscheinungen. *Mémoires Présentés à l'Académie Impériale des Sciences de St Pétersbourg par divers savants et lus dans ses Assemblées*, ser. 7, vol. 11, number 7.

List of the Linnean Society of London. London: [Linnean Society of London]. 1805–1939.

Living Cirripedia (1851): *A monograph of the sub-class Cirripedia, with figures of all the species. The Lepadidæ; or, pedunculated cirripedes.* By Charles Darwin. London: Ray Society. 1851.

Living Cirripedia (1854): *A monograph of the sub-class Cirripedia, with figures of all the species. The Balanidæ (or sessile cirripedes); the Verrucidæ, etc.* By Charles Darwin. London: Ray Society. 1854.

LL: *The life and letters of Charles Darwin, including an autobiographical chapter.* Edited by Francis Darwin. 3 vols. London: John Murray. 1887–8.

Lloyd, Christopher. 1992. *Fra Angelico.* London: Phaidon.

Lockwood, Jeffrey Alan. 2004. *Locust: the devastating rise and mysterious disappearance of the insect that shaped the American frontier.* New York: Basic Books.

Lubbock, John. 1853. Description of a new genus of *Calanidæ. Annals and Magazine of Natural History* 2d ser. 11: 25–9.

——. 1867. On the origin of civilisation and the primitive condition of man. [Read 26 November 1867.] *Transactions of the Ethnological Society of London* n.s. 6: 328–41.

——. 1869a. *Pre-historic times, as illustrated by ancient remains, and the manners and customs of modern savages.* 2d edition. London and Edinburgh: Williams & Norgate.

——. 1869b. On the origin of civilization and the primitive condition of man. Part II. *Report of the thirty-ninth meeting of the British Association for the Advancement of Science; held at Exeter in August 1869*, Transactions of the sections, pp. 137–151.

——. 1870. Description of the Park Cwm tumulus. *Journal of the Ethnological Society of London* n.s. 2 (1869–70): 416–9.

Lucas, Peter. 2002. 'A most glorious country': Charles Darwin and North Wales, especially his 1831 geological tour. *Archives of Natural History* 29: 1–26.

——. 2007. Charles Darwin, 'little Dawkins' and the platycnemic Yale men: introducing a bioarchaeological tale of the descent of man. *Archives of Natural History* 34: 318–45.

Lurie, Edward. 1988. *Louis Agassiz: a life in science.* Baltimore and London: The Johns Hopkins University Press.

Lyell, Charles. 1830–3. *Principles of geology, being an attempt to explain the former changes of the earth's surface, by reference to causes now in operation.* 3 vols. London: John Murray.

Lyell, Charles. 1865. *Elements of geology, or the ancient changes of the earth and its inhabitants as illustrated by geological monuments.* 6th edition, revised. London: John Murray.

——. 1867–8. *Principles of geology or the modern changes of the earth and its inhabitants considered as illustrative of geology.* 10th edition. 2 vols. London: John Murray.

——. 1872. *Principles of geology or the modern changes of the earth and its inhabitants considered as illustrative of geology.* 11th edition. 2 vols. London: John Murray.

Lyell, Katherine Murray, ed. 1881. *Life, letters and journals of Sir Charles Lyell, Bart.* 2 vols. London: John Murray.

Lyte, H. C. Maxwell. 1911. *A history of Eton College (1440–1910).* 4th edition. London: Macmillan.

Mabberley, David J. 1997. *The plant-book. A portable dictionary of the vascular plants.* 2d edition. Cambridge: Cambridge University Press.

McCann, James. 1869. *Anti-Darwinism.* Glasgow: David Bryce & Co.

MacKillop, James. 1998. *Dictionary of Celtic mythology.* Oxford, New York: Oxford University Press.

McLennan, John Ferguson. 1865. *Primitive marriage: an inquiry into the origin of the form of capture in marriage ceremonies.* Edinburgh: Adam and Charles Black.

[——.] 1869. The early history of man. *North British Review* n.s. 11: 516–49.

Macmillan dictionary of women's biography: *The Macmillan dictionary of women's biography.* 2d edition. Compiled and edited by Jennifer Uglow, assisted (1st edition) by Frances Hinton. London and Basingstoke, Hampshire: Macmillan Press. 1989.

McVaugh, Robert. 2004. Nature. In *Encyclopedia of the romantic era, 1760–1850*, edited by John Christopher Murray. New York: Fitzroy Dearborn.

Marchant, James, ed. 1916. *Alfred Russel Wallace. Letters and reminiscences.* 2 vols. London: Cassell and Company.

Marginalia: *Charles Darwin's marginalia.* Edited by Mario A. Di Gregorio with the assistance of Nicholas W. Gill. Vol. 1. New York and London: Garland Publishing. 1990.

Martin, Gerald P. R. and Uschmann, Georg. 1969. *Friedrich Rolle 1827–1887, ein Vorkämpfer neuen biologischen Denkens in Deutschland.* Leipzig: Johann Ambrosius Barth.

Mason, Herbert B., ed. 1908. *Encyclopædia of ships and shipping.* London: Shipping Encyclopædia.

Mason, Ian Lauder. 1951. *The classification of West African livestock.* Farnham Royal, Slough: Commonwealth Agricultural Bureaux.

Masters, Maxwell Tylden. 1856. Note on a monstrosity of the flowers of *Saponaria officinalis*, L. [Read 18 November 1856.] *Journal of the Proceedings of the Linnean Society* Botanical Papers. 1: 159–62.

Maw, George. 1868. On the disposition of iron in variegated strata. [Read 22 April 1868.] *Quarterly Journal of the Geological Society of London* 24: 351–400.

Max Müller, Friedrich. 1861. *Lectures on the science of language delivered at the Royal Institution of Great Britain in April, May, and June, 1861.* London: Longman, Green, Longman & Roberts.

Max Müller, Friedrich. 1864. *Lectures on the science of language.* 2d series. London:

Longmans.

Mayerhöfer, Josef. 1959–70. *Lexikon der Geschichte der Naturwissenschaften: Biographien, Sachwörter und Bibliographien.* Vol. 1. Vienna: Verlag Brüder Hollinek.

Medical directory: *The London medical directory … every physician, surgeon, and general practitioner resident in London.* London: C. Mitchell. 1845. *The London and provincial medical directory.* London: John Churchill. 1848–60. *The London & provincial medical directory, inclusive of the medical directory for Scotland, and the medical directory for Ireland, and general medical register.* London: John Churchill. 1861–9. *The medical directory … including the London and provincial medical directory, the medical directory for Scotland, the medical directory for Ireland.* London: J. & A. Churchill. 1870–1905.

Medical who's who: *The medical who's who.* London: London & Counties Press Association [and others]. 1913–27.

Meehan, Thomas. 1868a. Variations in *Epigæa repens. Proceedings of the Academy of Natural Sciences of Philadelphia* (1868): 153–6.

——. 1868b. *Mitchella repens*, L., a diœcious plant. *Proceedings of the Academy of Natural Sciences of Philadelphia* (1868): 183–4.

Men-at-the-bar: *Men-at-the-bar: a biographical hand-list of the members of the various inns of court, including Her Majesty's judges, etc.* By Joseph Foster. London: Reeves & Turner. 1885.

Menière, Prosper. 1855. Note sur la sécrétion d'un liquide abondant par l'organe glanduleux des *Coryanthes.* [Read 11 May 1855.] *Bulletin de la Société Botanique de France* 2: 351–5.

Meteyard, Eliza. 1865–6. *The life of Josiah Wedgwood from his private correspondence and family papers … with an introductory sketch of the art of pottery in England.* 2 vols. London: Hurst & Blackett.

——. 1970. *The life of Josiah Wedgwood from his private correspondence and family papers … with an introductory sketch of the art of pottery in England.* Reprint edition, with an introduction by R. W. Lightbown. 2 vols. London: Cornmarket Press.

Meyer, Adolf Bernhard. 1869a. Über den Giftapparat der Schlangen, insbesondere über den der Gattung Callophis (Gray). *Monatsbericht der Königl. Academie der Wissenschaften* (March 1869): 193–214.

——. 1869b. *Das Hemmungsnervensystem des Herzens, Kritisches und Experimentelles; eine vergleichend physiologische Studie.* Berlin: Hirschwald.

——. 1870. *Charles Darwin und Alfred Russel Wallace. Ihre Ersten Publicationen Aüber die 'Entstehung der Arten' nebst einer Skizze Ihres Lebens und einem Verzeichniss Ihrer Schriften.* Erlangen: Eduard Besold.

Meyers enzyklopädisches Lexikon. Ninth edition. 25 vols. with supplement (2 editions), atlas and index of names. Mannheim: Bibliographisches Institut, Lexikonverlag. 1971–84.

Miller, Douglas, ed. and trans. 1988. *Johann Wolfgang von Goethe: scientific studies.* New York: Suhrkamp Publishers.

Miller, Hugh. 1849. *Footprints of the Creator: or, the Asterolepis of Stromness.* London: Johnstone and Hunter.

Mills, Eric L. 1980. One 'different kind of gentleman': Alfred Merle Norman

(1831–1918), invertebrate zoologist. *Zoological Journal of the Linnean Society* 68 (1980): 69–98.

Milne-Edwards, Alphonse. 1867–71. *Recherches anatomiques et paléontologiques pour servir à l'histoire des oiseaux fossiles de la France.* 2 vols. 2 atlases. Paris: Victor Masson.

Milne-Edwards, Henri. 1834–40. *Histoire naturelle des Crustacés, comprenant l'anatomie, la physiologie et la classification de ces animaux.* 4 vols. Paris: Librairie encyclopédique de Roret.

Milton, William Fitzwilliam and Cheadle, Walter Butler. [1865.] *The North-West passage by land: being the narrative of an expedition from the Atlantic to the Pacific.* London: Cassell, Petter, and Galpin.

Mitchell, Sally, *et al.* eds. 1988. *Victorian Britain: an encyclopedia.* New York, London: Garland Publishing

Moço, Gisela, *et al.* 2006. The ibex *Capra pyrenaica* returns to its former Portuguese range. *Oryx* 40: 351–4.

Modern English biography: *Modern English biography, containing many thousand concise memoirs of persons who have died since the year 1850.* By Frederick Boase. 3 vols. and supplement (3 vols.). Truro, Cornwall: printed for the author. 1892–1921.

Moggridge, John Traherne. 1864. Observations on some orchids of the south of France. [Read 3 November 1864.] *Journal of the Linnean Society* (*Botany*) 8 (1865): 256–8.

Mohapatra, Hemanta Kumar. 2005. Oriya speaking feudatory states. *Orissa Review* (August 2005): 17–26.

Möller, Alfred, ed. 1915–21. *Fritz Müller. Werke, Briefe und Leben.* 3 vols in 5. Jena: Gustav Fischer.

Montgolfier, Adelaide de and Belloc, Louise Swanton, trans. 1860. Voyages d'un naturaliste: l'Archipel Galapagos et les attolls de coraux. [Translation of part of *Journal of researches.*] *Le Tour du Monde* 2: 139–59.

Moore, Charles. 1869. Lord Howe's Island. *Gardeners' Chronicle* (1869): 968–9.

——. 1870. Sketch of the vegetation of Lord Howe Island. In *Lord Howe Island: official visit by the water police magistrate and the director of the Botanic Gardens, Sydney; together with a description of the island*, by Edward S. Hill. Sydney: Thomas Richards.

Moore, James Richard. 1985. Darwin of Down: the evolutionist as squarson-naturalist. In *The Darwinian heritage*, edited by David Kohn. Princeton, N.J.: Princeton University Press in association with Nova Pacifica.

Morgan, Lewis Henry. 1868. *The American beaver and his works.* Philadelphia: J. B. Lippincott & Co.

Morton, Vanda. 1987. *Oxford rebels. The life and friends of Nevil Story Maskelyne, 1823–1911, pioneer Oxford scientist, photographer and politician.* Gloucester: Alan Sutton.

Moseley, Henry. 1869a. On the mechanical possibility of the descent of glaciers by their weight only. [Read 7 January 1869.] *Proceedings of the Royal Society of London* 17 (1868–9): 202–8.

——. 1869b. On the mechanical possibility of the descent of glaciers by their weight only. *Scientific Opinion* 1: 191–2.

Moulinié, Jean-Jacques, trans. 1868. *De la variation des animaux et des plantes sous l'action de la domestication.* By Charles Darwin. Preface by Carl Vogt. Paris: C. Reinwald.

——, trans. 1872. *La descendance de l'homme et la sélection sexuelle.* By Charles Darwin. Preface by Carl Vogt. Paris: C. Reinwald & Cie.

——, trans. 1873. *L'origine des espèces au moyen de la sélection naturelle, ou, la lutte pour l'existence dans la nature.* By Charles Darwin. Paris: C. Reinwald.

Movement in plants: *The power of movement in plants.* By Charles Darwin. London: John Murray. 1880.

[Mozley, James Bowling.] 1869. The argument of design. *Quarterly Review* 127: 134–76.

Mueller, Ferdinand. 1858–82. *Fragmenta phytographiæ australiæ.* 12 vols. Melbourne: Joannes Ferres.

Müller, Fritz. 1863. Die Verwandlung der Garneelen. *Archiv für Naturgeschichte* 29: 8–23.

——. 1864a. *Für Darwin.* Leipzig: Wilhelm Engelmann.

——. 1864b. On the metamorphoses of the prawns. *Annals and Magazine of Natural History* 3d ser. 14: 104–15.

——. 1866. Ueber die Befruchtung der *Martha (Posoqueria?) fragrans. Botanische Zeitung* 24: 129–33.

——. 1870a. Die Bewegung des Blüthenstieles von Alisma. *Jenaische Zeitschrift für Medizin und Naturwissenschaft* 5: 133–7.

——. 1870b. Umwandlung von Staubgefässen in Stempel bei *Begonia.* Uebergang von Zwitterblüthigkeit in Getrenntblüthigkeit bei *Chamissoa.* Triandrische Varietät eines monandrischen Epidendrum. Aus einem Briefe an H. Müller. *Botanische Zeitung* 28: 149–53.

——. 1870c. Bemerkungen über Cypridina. *Jenaische Zeitschrift für Medizin und Naturwissenschaft* 5: 255–76.

——. 1871. Ueber den Trimorphismus der Pontederien. *Jenaische Zeitschrift für Medicin und Naturwissenschaft* 6: 74–8.

——. 1872–3. Bestaubungsversuche an Abutilon-Arten. *Jenaische Zeitschrift für Medicin und Naturwissenschaft* 7: 22–45, 441–50.

——. 1883. Biologische Beobachtungen an Blumen Südbrasiliens. *Berichte der Deutschen Botanischen Gesellschaft* 1: 165–9.

——. 1885. Einige Nachträge zu Hildebrand's Buche: die Verbreitungsmittel der Pflanzen. *Kosmos* 17: 438-42.

Müller, Hermann. 1868. Beobachtungen an westfälischen Orchideeen. *Verhandlungen des naturhistorischen Vereines der preussischen Rheinlande und Westphalens* (*Botanik*) 25: 1–62.

——. 1869. Die Anwendung der Darwin'schen Lehre auf Blumen und blumenbesuchende Insekten. [Read 18 May 1869.] *Verhandlungen des naturhistorischen Vereines der preussischen Rheinlande und Westphalens* (Botanik, Correspondenzblatt) 26: 43–66.

Munby, Giles. 1847. *Flore d'Algèrie: ou catalogue des plantes indigènes du royaume d'Alger: accompagné des descriptions de quelques espèces nouvelles ou peu connues.* Paris: J.B. Baillière.

Munro, Robertson. 1868. On reproduction and cross-fertilisation of Passifloras. [Read 11 June 1868.] *Transactions of the Botanical Society of Edinburgh* 9: 299–402.

Murchison, Roderick Impey. 1867. *Siluria: a history of the oldest rocks in the British Isles and other countries; with sketches of the origin and distribution of native gold, the general succession of geological formations, and changes of the earth's surface.* 4th edition including 'The Silurian system'. London: John Murray.

Murphy, Joseph John. 1869. *Habit and intelligence in their connexion with the laws of matter and force: a series of scientific essays.* 2 vols. London: Macmillan and Co.

Nägeli, Carl Wilhelm von. 1865. *Entstehung und Begriff der naturhistorischen Art.* Munich: Verlag der königl. Akademie.

——. 1867. Die Piloselliformia. [Read 4 May 1867.] *Sitzungsberichte der königlichen Bayerischen Akademie der Wissenschaften zu München* 1: 450–78.

Nairn, Ian and Pevsner, Nikolaus. 1971. *The buildings of England: Surrey.* 2d edition revised by Bridget Cherry. Harmondsworth: Penguin.

National cyclopædia of American biography. 63 vols. New York: James T. White & Co. 1892–1984.

Natural selection: *Charles Darwin's Natural selection: being the second part of his big species book written from 1856 to 1858.* Edited by R. C. Stauffer. Cambridge: Cambridge University Press. 1975.

NBL: *Norsk biografisk leksikon.* Edited by Edvard Bull *et al.* 19 vols. Oslo: H. Aschehoug & Co. 1923–83.

NBU: *Nouvelle biographie universelle depuis les temps les plus reculés jusqu'à nos jours, avec les renseignements bibliographiques et l'indication des sources a consulter.* Edited by Jean Chrétien Ferdinand Hoefer. 46 vols. in 23. Paris: Firmin Didot Frères. 1852–66.

NDB: *Neue deutsche Biographie.* Under the auspices of the Historical Commission of the Bavarian Academy of Sciences. 21 vols. (A–Rohlfs) to date. Berlin: Duncker & Humblot. 1953–.

Needell, Jeffrey D. 2006. *The party of order: the conservatives, the state, and slavery in the Brazilian monarchy, 1831–1871.* Stanford, California: Stanford University Press.

Newman, William. 1993. Darwin and cirripedology. *History of Carcinology. Crustacean Issues* 8: 349–434.

Newmann, Kate, comp. 1993. *Dictionary of Ulster biography.* Belfast: Institute of Irish Studies, Queen's University of Belfast.

Nicholas, Thomas. 1872. *Annals and antiquities of the counties and county families of Wales.* 2 vols. London: Longmans, Green, Reader, & Co.

Nilsson, Sven. 1868. *The primitive inhabitants of Scandinavia. An essay on comparative ethnography, and a contribution to the history of the development of mankind.* Edited by John Lubbock. 3d edition. London: Longmans, Green & Co.

NNBW: *Nieuw Nederlandsch biografisch Woordenboek.* Edited by P. C. Molhuysen *et al.* 10 vols. Leiden: A. W. Sijthoff's Uitgevers-Maatschappij. 1911–37.

Norman, Alfred Merle. 1868. Shetland final dredging report. Part 2. On the Crustacea, Tunicata, Polyzoa, Echinodermata, Actinozoa, Hydrozoa, and Porifera. *Report of the thirty-eighth meeting of the British Association for the Advancement of Science; held at Norwich in August 1868*, pp. 247–336, 341–2.

North, John S. 1989. *The Waterloo directory of Scottish newspapers and periodicals, 1800–1900.* 2 vols. Waterloo, Ontario: North Waterloo Academic Press.

——. 1997. *The Waterloo directory of English newspapers and periodicals, 1800–1900.* 10 vols. Waterloo, Ontario: North Waterloo Academic Press.

Notebooks: *Charles Darwin's notebooks, 1836–1844. Geology, transmutation of species, metaphysical enquiries.* Transcribed and edited by Paul H. Barrett *et al.* Cambridge: Cambridge University Press for the British Museum (Natural History). 1987.

Nowell-Smith, Simon. 1968. *International copyright law and the publisher in the reign of Queen Victoria.* Oxford: Clarendon Press.

NUC: *The national union catalog. Pre-1956 imprints.* 685 vols. and supplement (69 vols.). London and Chicago: Mansell. 1968–81.

Nyhart, Lynn K. 1995. *Biology takes form. Animal morphology and the German universities, 1800–1900.* Chicago and London: University of Chicago Press.

OBL: *Österreichisches biographisches Lexikon 1815–1950.* Edited by Leo Santifaller *et al.* 11 vols. and 3 fascicles of vol. 12 (A–Slavik Ernst) to date. Vienna: Osterreichischen Akademie der Wissenschaften. 1957–.

O'Byrne, William R. 1849. *A naval biographical dictionary: comprising the life and services of every living officer in Her Majesty's Navy, from the rank of admiral of the fleet to that of lieutenant, inclusive.* London: John Murray.

ODNB: *Oxford dictionary of national biography: from the earliest times to the year 2000.* [Revised edition.] Edited by H. C. G. Matthew and Brian Harrison. 60 vols. and index. Oxford: Oxford University Press. 2004.

OED: *The Oxford English dictionary. Being a corrected re-issue with an introduction, supplement and bibliography of a new English dictionary.* Edited by James A. H. Murray, *et al.* 12 vols. and supplement. Oxford: Clarendon Press. 1970. *A supplement to the Oxford English dictionary.* 4 vols. Edited by R. W. Burchfield. Oxford: Clarendon Press. 1972–86. *The Oxford English dictionary.* 2d edition. 20 vols. Prepared by J. A. Simpson and E. S. C. Weiner. Oxford: Clarendon Press. 1989. *Oxford English dictionary additional series.* 3 vols. Edited by John Simpson *et al.* Oxford: Clarendon Press. 1993–7.

Ogle, William. 1869. The fertilisation of Salvia and of some other flowers. *Popular Science Review* 8: 261–73.

Oliver, Daniel. 1855. Observations on the growth in diameter of dicotyledonous (exogenous) stems. [Read 15 November 1855.] *Transactions of the Tyneside Naturalists' Field Club* 3 (1854–8): 64–8.

Oliver, Daniel. 1868–77. *Flora of tropical Africa.* 3 vols. London: L. Reeve and Co.

Orchids: *On the various contrivances by which British and foreign orchids are fertilised by insects, and on the good effects of intercrossing.* By Charles Darwin. London: John Murray. 1862.

Orchids 2d ed.: *The various contrivances by which British and foreign orchids are fertilised by insects, and the good effects of intercrossing.* By Charles Darwin. Revised. London: John Murray. 1877.

Origin: *On the origin of species by means of natural selection, or the preservation of favoured races in the struggle for life.* By Charles Darwin. London: John Murray. 1859.

Origin 2d ed.: *On the origin of species by means of natural selection, or the preservation of favoured races in the struggle for life.* By Charles Darwin. London: John Murray. 1860.

Origin 3d ed.: *On the origin of species by means of natural selection, or the preservation of favoured races in the struggle for life.* With additions and corrections. By Charles Darwin. London: John Murray. 1861.

Origin 4th ed.: *On the origin of species by means of natural selection, or the preservation of favoured races in the struggle for life.* With additions and corrections. By Charles Darwin. London: John Murray. 1866.

Origin 5th ed.: *On the origin of species by means of natural selection, or the preservation of favoured races in the struggle for life.* With additions and corrections. By Charles Darwin. London: John Murray. 1869.

Origin 6th ed.: *The origin of species by means of natural selection, or the preservation of favoured races in the struggle for life.* With additions and corrections. By Charles Darwin. London: John Murray. 1872.

Origin US ed.: *On the origin of species by means of natural selection, or the preservation of favoured races in the struggle for life.* A new edition, revised and augmented by the author. By Charles Darwin. New York: D. Appleton. 1860.

Origin 2d US ed.: *On the origin of species by means of natural selection, or the preservation of favoured races in the struggle for life.* By Charles Darwin. From the fifth London edition, with additions and corrections. New York: D. Appleton. 1870.

Orr, M. 2007. Keeping it in the family: the extraordinary case of Cuvier's daughters. In *The role of women in the history of geology*, edited by Cynthia V. Burek and Bettie Higgs. London: Geological Society.

Orton, James. 1868. Physical observations on the Andes and the Amazons. *American Journal of Science and Arts* 2d ser. 46: 203–13.

——. 1870. *The Andes and the Amazon; or, across the continent of South America.* New York: Harper.

Owen, Richard. 1869. On fossil remains of equines from Central and South America referable to *Equus conversidens, Ow.*, *Equus tau, Ow.*, and *Equus arcidens, Ow.* [Read 4 February 1869.] *Philosophical Transactions of the Royal Society of London* 159: 559–73.

Oxford classical dictionary. 3d edition, revised. Edited by Simon Hornblower and Antony Spawforth. Oxford: Oxford University Press. 2003.

Oxford companion to gardens: *The Oxford companion to gardens.* Consultant editors Sir Geoffrey Jellicoe and Susan Jellicoe; executive editors Patrick Goode and Michael Lancaster. Oxford: Oxford University Press. 1986.

Pagenstecher, Heinrich Alexander. 1863. Untersuchungen über niedere Seethiere an Cette. II Abtheilung. Beitrag zur Anatomie und Entwickelungsgeschichte von Lepas pectinata. *Zeitschrift für wissenschaftliche Zoologie* 13: 86–106.

Paget, Stephen, ed. 1901. *Memoirs and letters of Sir James Paget.* London: Longmans, Green, and Co.

Paley, William. 1802. *Natural theology; or, evidences of the existence and attributes of the Deity, collected from the appearances of nature.* London: R. Faulder.

——. 1845. *Paley's Natural theology, with illustrative notes by Henry, Lord Brougham and Sir C. Bell; and an introductory discourse of natural theology by Lord Brougham; to which are added supplementary dissertations and a treatise on animal mechanics by Sir Charles Bell.* 4 vols. London: C. Knight.

Pancaldi, Giuliano. 1991. *Darwin in Italy. Science across cultural frontiers.* Translated by Ruey Brodine Morelli. Updated and expanded edition. Bloomington and Indianapolis, Ind.: Indiana University Press.

Pauly, Daniel. 2004. *Darwin's fishes. An encyclopedia of ichthyology, ecology, and evolution.* Cambridge: Cambridge University Press.

Pavlov, D. S. and Shishkin, V. S. 2003. Significance of biological stations for progress of science and education. *Biology Bulletin of the Russian Academy of Sciences* 30: 422–4.

Payer, Jean-Baptiste. 1857. *Traité d'organogénie comparée de la fleur.* 1 vol. and atlas. Paris: Masson.

Pearson, Karl. 1914–30. *The life, letters and labours of Francis Galton.* 3 vols. in 4. Cambridge: Cambridge University Press.

Peckham, Morse, ed. 1959. *The Origin of Species by Charles Darwin: a variorum text.* Philadelphia: University of Pennsylvania Press.

Penzig, Ottone. 1905. Commemorazione di Federico Delpino. *Malpighia* 19: 294–310.

Petteys, Chris. 1985. *Dictionary of women artists. An international dictionary of women artists born before 1900.* Boston, Mass.: G. K. Hall & Co.

Pfaff, Friedrich. 1868. *Die neuesten Forschungen und Theorieen auf dem Gebiete der Schöpfungsgeschichte.* Frankfurt am Main: Heyder & Zimmer.

Phillips, John. 1869. *Vesuvius.* Oxford: The Clarendon Press.

Plarr, Victor Gustave. 1930. *Plarr's lives of the fellows of the Royal College of Surgeons of England.* Revised by Sir D'Arcy Power. 2 vols. London: Simpkin Marshall.

Playfair, Robert Lambert and Günther, Albert Charles Lewis Gotthilf. 1866. *The fishes of Zanzibar.* London: John Van Voorst.

Ponting, Kenneth. 1980. *Sheep of the world.* Poole: Blandford Press.

Post Office directory of Cambridge, Norfolk, and Suffolk: *Post Office directory of Cambridge, Norfolk, and Suffolk. Kelly's directory of Cambridgeshire, Norfolk and Suffolk.* London: W. Kelly; Kelly & Co. 1853–1900.

Post Office directory of Cheshire: *Post Office directory of Cheshire. Kelly's directory of Cheshire.* London: Kelly & Co. 1857–1902.

Post Office directory of Lancashire, Liverpool and Manchester: *Post Office directory of Lancashire. Post Office directory of Lancashire, Liverpool and Manchester.* London: Kelly & Co. 1858–92.

Post Office directory of Northamptonshire etc.: *Post Office directory of Berkshire, Northamptonshire, Oxfordshire, with Bedfordshire, Buckinghamshire, and Huntingdonshire. The Post Office directory of Northamptonshire, Huntingdonshire, Bedfordshire, Buckinghamshire, Berkshire, and Oxfordshire. The Post Office directory of the counties of Bedford, Huntingdon, Northampton, Berks, Buckingham and Oxford.* London: Kelly and Co. [1847]–77.

Post Office directory of the six home counties: *Post Office directory of the six home counties, viz., Essex, Herts, Kent, Middlesex, Surrey and Sussex.* London: W. Kelly & Co. 1845–78.

Post Office Edinburgh directory: *Post-Office annual directory and calendar. Post-Office Edinburgh and Leith directory.* Edinburgh: Ballantyne & Hughes [and others]. 1845–1908.

Post Office London directory: *Post-Office annual directory. . . . A list of the principal merchants, traders of eminence, &c. in the cities of London and Westminster, the borough of Southwark, and parts adjacent . . . general and special information relating to the Post Office. Post Office London directory.* London: His Majesty's Postmaster-General [and others]. 1802–1967.

Preyer, William Thierry. 1869. *Der Kampf um das Dasein. Ein populärer Vortrag.* Bonn: Eduard Weber Buchhandlung.

——. 1896. *Darwin. Sein leben und wirken.* Berlin: E. Hofmann.

Pritchard, Michael. 1994. *A directory of London photographers 1841–1908.* Revised and expanded edition. Watford, Hertfordshire: PhotoResearch.

Quatrefages de Bréau, Jean Louis Armand de. 1868–9. Origines des espèces animales et végétales. *Revue des Deux Mondes* 78 (1868): 832–60; 79 (1869): 208–40; 80 (1869): 64–95, 397–432, 638–72.

——. 1870. *Charles Darwin et ses précurseurs français. Etude sur le transformisme.* Paris: Germer Baillière.

Quetelet, Adolphe. 1835. *Sur l'homme et le développement de ses facultés, ou, Essai de physique sociale.* 2 vols. Paris: Bachelier.

——. 1846. *Lettres à S.A.R. le duc régnant de Saxe-Coburg et Gotha: sur la théorie des probabilités, appliquée aux sciences morales et politiques.* Brussels: M. Hayez.

——. 1849. *Letters addressed to H.R.H. the Grand Duke of Saxe Coburg and Gotha, on the theory of probabilities, as applied to the moral and political sciences.* Translated by Olinthus Gregory Downes. London: C. & E. Layton.

Rabson, Stephen and O'Donoghue, Kevin. 1988. *P&O: a fleet history.* Kendal: World Ship Society.

Raby, Peter. 2001. *Alfred Russel Wallace: a life.* London: Chatto & Windus.

Rachinskii, Sergei A., trans. 1864. *Proiskhozhdenie vidov putem estestvennogo podbora.* [*Origin.*] By Charles Darwin. St Petersburg: A. I. Glazunov.

——, trans. 1865. *Proiskhozhdenie vidov putem estestvennogo podbora.* [*Origin.*] By Charles Darwin. 2d edition, revised. St Petersburg: A. I. Glazunov.

Ramsay, Andrew Crombie. 1862. On the glacial origin of certain lakes in Switzerland, the Black Forest, Great Britain, Sweden, North America, and elsewhere. [Read 5 March 1862.] *Quarterly Journal of the Geological Society of London* 18: 185–204.

——. 1863. *The physical geology and geography of Great Britain: a course of six lectures delivered to working men in the Museum of Practical Geology, Jermyn Street, in January and February 1863.* London: Edward Stanford.

——. 1864. *The physical geology and geography of Great Britain. Six lectures to working men delivered in the Royal School of Mines in 1863.* 2d edition. London: Edward Stanford.

Ramsey, George M'Ilvaine. 1870. *Cosmology.* Boston: W. White and Co.

——. 1877. *Overturning the world; or, cause and effects of atmospheric motion.* New York: P. F. McBreen.

——. 1897. *Philosophy of phenomena.* Boston: Banner of Light Publishing.

Randau, R. 1869. Un naturaliste dans l'Archipel Malais. [Review of A. R. Wallace *The Malay Archipelago.*] *Revue des Deux Mondes* 2d ser. 83: 675–706.

Reade, William Winwood. 1873. *The African sketch-book.* 2 vols. London: Smith, Elder, and Co.

Recht, Georg. 1869. *Die Entwicklung der Weltgesetze.* Munich: Schurich.

Record of the Royal Society of London: *The record of the Royal Society of London for the promotion of natural knowledge.* 4th edition. London: Royal Society. 1940.

Records of Lincoln's Inn: *The records of the honorable society of Lincoln's Inn: admissions.* Edited by William Paley Baildon. 2 vols. London: Lincoln's Inn. 1896.

Rérolle, Louis, trans. 1870. *De la fécondation des orchidées par les insectes et des bons résultats du croisement.* By Charles Darwin. Paris: C. Reinwald.

Richardson, Edmund William. 1916. *A veteran naturalist; being the life and work of W. B. Tegetmeier.* London: Witherby & Co.

Richardson, H. D. 1847. *Dogs: their origin and varieties, directions as to their general management, and simple instructions as to their treatment under disease.* Dublin: James McGlashan. London: W. S. Orr and Co. Edinburgh: Fraser and Co.

Richardson, John. 1829–37. *Fauna Boreali-Americana; or, the zoology of the northern parts of British America.* Assisted by William Swainson and William Kirby. 4 vols. London and Norwich: John Murray; Richard Bentley; J. Fletcher.

Risk, James C. 1972. *The history of the Order of the Bath and its insignia.* London: Spink & Son.

Risley, Herbert Hope. 1891. *The tribes and castes of Bengal: ethnographic glossary.* 2 vols. Calcutta: Bengal Secretariat Press.

Ritvo, Harriet. 1992. Race, breed, and myths of origin: Chillingham cattle as ancient Britons. *Representations* 39: 1–22.

Rolle, Friedrich. 1863. *Chs. Darwin's Lehre von der Entstehung der Arten im Pflanzen- und Thierreich in ihrer Anwendung auf die Schöpfungsgeschichte.* Frankfurt: J. C. Hermann.

Rose, Lionel. 1988. *'Rogues and vagabonds': vagrant underworld in Britain 1815–1985.* London: Routledge.

Royer, Clémence Auguste, trans. 1862. *De l'origine des espèces ou des lois du progrès chez les êtres organisés.* By Charles Darwin. With preface and notes by the translator. [Translated from the 3d English edition.] Paris: Guillaumin & Cie; Victor Masson.

——, trans. 1866. *De l'origine des espèces par sélection naturelle ou des lois de transformation des êtres organisés.* By Charles Darwin. With preface and notes by the translator. 2d edition. Augmented in accordance with notes by the author. Paris: Guillaumin et Cie; Victor Masson et fils.

——, trans. 1870. *De l'origine des espèces par sélection naturelle ou des lois de transformation des êtres organisés.* By Charles Darwin. With preface and notes by the translator. 3d edition. Paris: V. Masson et fils.

Ruck, Berta. 1935. *A story-teller tells the truth: reminiscences & notes.* London: Hutchinson & Co.

Rudall, Paula J. and Bateman, Richard M. 2004. Evolution of zygomorphy in monocot flowers: iterative patterns and developmental constraints. *New Phytologist* 162: 25–44.

Rudall, Paula J. *et al.* 2005. Evolution of reproductive structures in grasses (Poaceae) inferred by sister-group comparison with their putative closest living relatives, Ecdeiocoleaceae. *American Journal of Botany* 92: 1432–43.

Rupke, Nicolaas A. 1994. *Richard Owen, Victorian naturalist.* New Haven, Conn., and London: Yale University Press.

Saint-Clair, Roland William. 1898. *The Saint-Clairs of the Isles.* Auckland: H. Brett.

Saint-Hilaire, Auguste de. 1841. *Leçons de botanique: comprenant principalement la morphologie végétale, la terminologie, la botanique comparée, l'examen de la valeur des charactères dans les diverses familles naturelles, etc.* Paris: P. J. Loss.

Salmon, Michael A. 2000. *The Aurelian legacy: British butterflies and their collectors.* With additional material by Peter Marren and Basil Harley. Colchester: Harley Books.

Salvin, Osbert. 1875. On the avifauna of the Galapagos Archipelago. [Read 6 April 1875.] *Transactions of the Zoological Society of London* 9 (1877): 447–510.

Sandahl, Oskar Th. 1882. Nekrolog. Niklas Westring. *Entomologisk Tidskrift* 2: 9–12.

Saporta, Louis Charles Joseph Gaston de. 1869. L'école transformiste et ses derniérs travaux. *Revue des Deux Mondes* 2d ser. 83: 635–74.

Sarjeant, William A. S. 1980–96. *Geologists and the history of geology: an international bibliography.* 10 vols. including supplements. London: Macmillan. Malabar, Fla.: Robert E. Krieger Publishing.

Saunders, William Wilson. 1860. On the genus *Erateina*, Doubl.; with descriptions of some new species. [Read 7 May 1860.] *Transactions of the Entomological Society of London* n.s. 5 (1858–61): 261–7.

SBL: *Svenskt biografiskt lexikon.* Edited by Bertil Boëthius *et al.* 29 vols. and 4 fascicles of vol. 30 (A–Rudbeck) to date. Stockholm: Albert Bonnier and P. A. Norstedt. 1918–.

Scherren, Henry. 1905. *The Zoological Society of London: a sketch of its foundation and development and the story of its farm, museum, gardens, menagerie and library.* London: Cassell.

Schimper, Wilhelm Philipp. 1869–74. *Traité de paléontologie végétale: ou, la flore du monde primitif dans ses rapports avec les formations géologiques et la flore du monde actuel.* 3 vols. and atlas. Paris: J. B. Baillière et fils.

S[chlechtenda]l, [D. F. L.]. 1864. Personal-Nachrichten [Hermann Crüger]. *Botanische Zeitung* 22: 119–20.

Schleiden, Matthias Jacob. 1845. *Die Botanik als inductive Wissenschaft.* 2 vols. 2d edition. Leipzig: Engelmann.

Scott, John. 1864. Notes on the sterility and hybridization of certain species of *Passiflora, Disemma,* and *Tacsonia.* [Read 16 June 1864.] *Journal of the Linnean Society (Botany)* 8 (1865): 197–206.

——. 1868. Report on trial sowings of flower-seeds from Messrs. James Carter & Co. of London and Messrs. Vilmorin Andrieux & Co. of Paris. [Extracts read 21

January 1868.] *Journal of the Agricultural and Horticultural Society of India* n.s. 1 (1869): 191–9.

——. 1869. A list of the higher cryptogams cultivated in the Royal Botanical Gardens, Calcutta. *Journal of the Agricultural and Horticultural Society of India* n.s. 1 (1869): 200–64.

——. 1870. Notes on the tree ferns of British Sikkim, with descriptions of three new species, and a few supplemental remarks on their relations to palms and cycads. [Read 17 February 1870.] *Transactions of the Linnean Society of London* 30 (1875): 1–44.

Scrope, William. 1838. *The art of deer-stalking; illustrated by a narrative of a few days' sport in the forest of Atholl, with some account of the nature and habits of red deer, and a short description of the Scotch forests; legends; superstitions; stories of poachers and freebooters, &c. &c.* London: John Murray.

Scudder, Samuel Hubbard. 1861. On the genus Raphidophora, Serville; with descriptions of four species from the caves of Kentucky, and from the Pacific coast. *Proceedings of the Boston Society of Natural History* 8 (1861–2): 6–14.

——. 1868. Supplement to a list of the butterflies of New England. *Proceedings of the Boston Society of Natural History* 11: 375–84.

——. 1869a. *Revision of the large, stylated, fossorial crickets.* Salem, Massachusetts: Peabody Academy of Science.

——. 1869b. A preliminary list of the butterflies of Iowa. *Transactions of the Chicago Academy of Sciences* 1: 326–37.

——. 1869c. Report upon a collection of diurnal Lepidoptera made in Alaska by the scientific corps of the Russo-American telegraph expedition under the direction of Lieut. W. H. Dall. *Proceedings of the Boston Society of Natural History* 12 (1868–9): 404–8.

Seidlitz, Karl von. 1865. *Ueber die Vererbung der Lebensformen, Eigenschaften und Fähigkeiten organischer Wesen auf ihre Nachkommen in Bezug auf Physiologie und praktische Heilkunst.* St Petersburg: H. Schmißdorff.

Seilacher, Adolf. 2005. Whale barnacles: exaptational access to a forbidden paradise. *Paleobiology* 31 (2): 27–35.

Seville, Catherine. 2006. *The internationalisation of copyright law: books, buccaneers, and the black flag in the nineteenth century.* Cambridge: Cambridge University Press.

Shephard, Sue. 2003. *Seeds of fortune: a gardening dynasty.* London: Bloomsbury.

Shillingsburg, Peter L. 2007. The first five English editions of Charles Darwin's *On the origin of species. Variants* 5: 221–48.

Shortland, Michael, ed. 1996. *Hugh Miller and the controversies of Victorian science.* Oxford: Clarendon.

Shrewsbury School register: *Shrewsbury School register. 1734–1908.* Edited by J. E. Auden. Oswestry: Woodall, Minshall, Thomas and Co., Caxton Press. 1909.

Sison, Robert. 1867. Lohmen. Königliche Merino Stammschäferei Sachsens v. 1765. *Agronomische Zeitung* 22: 209–12.

Smith, Frederick. 1853–9. *Catalogue of the hymenopterous insects in the collection of the British Museum.* 7 parts. London: printed by order of the Trustees.

SMK: *Svenska män och kvinnor biografisk uppslagsbok*. Edited by Nils Bohman *et al.* 8 vols. Stockholm: Albert Bonniers. 1942–55.

Somerset Fry, Peter and Somerset Fry, Fiona. 1988. *A history of Ireland*. London and New York: Routledge.

Somerville, Mary. 1869. *On molecular and microscopic science*. 2 vols. London: John Murray.

South America: *Geological observations on South America. Being the third part of the geology of the voyage of the Beagle, under the command of Capt. FitzRoy RN, during the years 1832 to 1836*. By Charles Darwin. London: Smith, Elder & Co. 1846.

Southall, Geneva Handy. 1999. *Blind Tom, the black pianist-composer (1849–1908): continually enslaved*. Lanham, Md., and London: Scarecrow.

'Specific difference in *Primula*': On the specific difference between *Primula veris*, Brit. Fl. (var. *officinalis* of Linn.), *P. vulgaris*, Brit. Fl. (var. *acaulis*, Linn.), and *P elatior*, Jacq.; and on the hybrid nature of the common oxlip. With supplementary remarks on naturally produced hybrids in the genus *Verbascum*. By Charles Darwin. [Read 19 March 1868.] *Journal of the Linnean Society* (*Botany*) 10 (1869): 437–54.

Spencer, Herbert. 1860–2. *First principles*. London: George Manwaring; Williams & Norgate.

——. 1864–7. *The principles of biology*. 2 vols. London: Williams & Norgate.

Sprengel, Christian Konrad. 1793. *Das entdeckte Geheimniss der Natur im Bau und in der Befruchtung der Blumen*. Berlin: Friedrich Vieweg.

Spruce, Richard. 1861. *Report on the expedition to procure seeds and plants of the* Cinchona succirubra *or red bark tree*. London: Her Majesty's Stationery Office.

——. 1869. *Palmae Amazonicae*, sive enumeratio palmarum in itinere suo per regiones Americae aequatoriales lectarum. [Read 21 January 1869.] *Journal of the Linnean Society* (*Botany*) 11 (1871): 65–183.

——. 1908. *Notes of a botanist on the Amazon & Andes, being records of travel … during the years 1849–1864*. Edited by Alfred Russel Wallace. 2 vols. London: Macmillan and Co.

Starkey, Paul and Janet, eds. 2001. *Interpreting the Orient: travellers in Egypt and the Near East*. Reading: Ithaca Press.

Stebbing, Thomas Roscoe Rede. 1871. *Essays on Darwinism*. London: Longmans, Green, and Co.

Stenton, Michael. 1976. *Who's who of British members of Parliament, 1832–1885*. Brighton: Hanover Press.

Stirling, James Hutchinson. 1865. *The secret of Hegel; being the Hegelian system in origin, principle, form and matter*. 2 vols. London: Longman, Green, Longman, Roberts & Green.

——. 1869. *As regards protoplasm, in relation to Professor Huxley's essay, On the physical basis of life*. Edinburgh and London: William Blackwood and sons.

Stokes, George Gabriel. 1869. Presidential address. *Report of the thirty-ninth meeting of the British Association for the Advancement of Science; held at Exeter in August 1869*, lxxxix–cv.

Strzelecki, Paul Edmund de. 1845. *Physical description of New South Wales and Van*

Diemen's Land: accompanied by a geological map, sections and diagrams, and figures of the organic remains. London: Longman, Brown, Green and Longmans.

[Tait, Peter Guthrie.] 1869. Geological time. *North British Review* n.s. 11: 406–39.

Taxonomic literature: *Taxonomic literature. A selective guide to botanical publications and collections with dates, commentaries and types.* By Frans A. Stafleu and Richard S. Cowan. 2d edition. 7 vols. Utrecht, Netherlands: Bohn, Scheltema & Holkema. The Hague, Netherlands: W. Junk. 1976–88.

Taxonomic literature. Supplement: *Taxonomic literaure: a selective guide to botanical publications and collections with dates, commentaries and types.* Supplement. By Frans A. Stafleu and Erik A. Mennega. 7 vols. to date. Königstein: Koeltz Sceintific Books. 1992–.

Taylor, John Ellor. 1876. *The aquarium: its inhabitants, structure & management.* London: Hardwicke.

Tee, Garry J. 1998. Professor and Mrs Aldis: mathematics, feminism and astronomy in Victorian Auckland. *Southern Stars: Journal of the Royal Astronomical Society of New Zealand* 38 (1998): 18–27.

Tegetmeier, William Bernhard. 1867. *The poultry book: comprising the breeding and management of profitable and ornamental poultry, their qualities and characteristics; to which is added 'The standard of excellence in exhibition birds', authorized by the Poultry Club.* London and New York: George Routledge & Sons.

Thiel, Hugo. [1868?] Ueber einige Formen der landwirtschaftlichen Genossenschaften. [Offprint from unknown journal, pp. 132–45]. Darwin Pamphlet Collection–CUL.

Thomson, William. 1862a. On the age of the sun's heat. *Macmillan's Magazine* 5: 388–93.

——. 1862b. On the secular cooling of the earth. [Read 28 April 1862.] *Transactions of the Royal Society of Edinburgh* 23 (1864): 157–70.

——. 1865. The 'Doctrine of uniformity' in geology briefly refuted. [Read 18 December 1865.] *Proceedings of the Royal Society of Edinburgh* 5 (1866): 512–13.

Thomson, William and Tait, Peter Guthrie. 1867. *Treatise on natural philosophy.* Vol. 1 (no other volumes published). Oxford: Clarendon Press.

Thornthwaite, Charles Warren. 1943. Problems in the classification of climates. *Geographical Review* 33: 233–55.

'Three forms of *Lythrum salicaria*': On the sexual relations of the three forms of *Lythrum salicaria.* By Charles Darwin. [Read 16 June 1864.] *Journal of the Linnean Society (Botany)* 8 (1865): 169–96. [*Collected papers* 2: 106–31.]

Todd, Dennis. 1995. *Imagining monsters. Miscreations of the self in eighteenth-century England.* Chicago and London: University of Chicago Press.

Todd, Robert Bentley, ed. 1835–59. *The cyclopaedia of anatomy and physiology.* 5 vols. London: Sherwood, Gilbert, and Piper.

Tort, Patrick. 1996. *Dictionnaire du Darwinisme et de l'evolution.* 3 vols. Paris: Presses Universitaires de France.

Turner, James. 1999. *The liberal education of Charles Eliot Norton.* Baltimore and London: The Johns Hopkins University Press.

Tweddle, Dennis, *et al.* 1998. Cichlid spawning structures–bowers or nests? *Environmental Biology of Fishes* 51: 107–9

'Two forms in species of *Linum*': On the existence of two forms, and on their reciprocal sexual relation, in several species of the genus *Linum.* By Charles Darwin. [Read 5 February 1863.] *Journal of the Proceedings of the Linnean Society (Botany)* 7 (1864): 69–83. [*Collected papers* 2: 93–105.]

Tylor, Edward Burnett. 1869. On the survival of savage thought in modern civilization. *Proceedings of the Royal Institution of Great Britain* 5 (1866–9): 522–35.

——. 1871. *Primitive culture: researches into the development of mythology, philosophy, religion, art, and custom.* 2 vols. London: John Murray.

Tyndall, John. 1860. *The glaciers of the Alps. Being a narrative of excursions and ascents, an account of the origin and phenomena of glaciers, and an exposition of the physical principles to which they are related.* London: J. Murray.

Ulmer, Torsten, *et al.* 2004. *Passiflora: passionflowers of the World.* Portland and Cambridge: Timber Press.

Variation: *The variation of animals and plants under domestication.* By Charles Darwin. 2 vols. London: John Murray. 1868.

Variation 2d ed.: *The variation of animals and plants under domestication.* By Charles Darwin. 2 vols. London: John Murray. 1875.

Variation US ed.: *The variation of animals and plants under domestication.* By Charles Darwin. 2 vols. New York: Orange Judd & Co. [1868.]

Vaucher, Jean Pierre Etienne. 1841. *Histoire physiologique des plantes d'Europe ou exposition des phénomènes qu'elles présentent dans les diverses périodes de leur développement.* 4 vols. Paris: Marc Aurel Frères.

Volcanic islands: *Geological observations on the volcanic islands, visited during the voyage of HMS Beagle, together with some brief notices on the geology of Australia and the Cape of Good Hope. Being the second part of the geology of the voyage of the Beagle, under the command of Capt. FitzRoy RN, during the years 1832 to 1836.* By Charles Darwin. London: Smith, Elder & Co. 1844.

'Volcanic phenomena and the formation of mountain chains': On the connexion of certain volcanic phenomena in South America; and on the formation of mountain chains and volcanos, as the effect of the same power by which continents are elevated. By Charles Darwin. [Read 7 March 1838.] *Transactions of the Geological Society of London* 2d ser. 5 (1840): 601–31. [*Collected papers* 1: 53–86.]

Wagner, Moritz. 1868a. Ueber die Darwin'sche Theorie in Bezug auf das geographischen Verbreitung der Organismen. [Read 7 March 1868.] *Sitzungberichte der königlichen bayerischen Akademie der Wissenschaften zu München* (1868) pt 1: 359–95.

——. 1868b. *Die Darwin'sche Theorie und das Migrationsgesetz der Organismen.* Leipzig: Duncker & Humblot.

Walckenaer, C. A. and Gervais, P. 1837–47. *Histoire naturelle des insectes. Aptères. (Suites à Buffon).* 4 vols. and atlas. Paris: Librarie encyclopédique de Roret.

Wallace, Alfred Russel. 1867. Creation by law. *Quarterly Journal of Science* 4: 471–88.

——. 1869a. *The Malay Archipelago: the land of the orang-utan, and the bird of paradise. A narrative of travel, with studies of man and nature.* 2 vols. London: Macmillan and Co.

[——.] 1869b. Geological climates and the origin of species. [Review of Charles Lyell's *Principles of geology*, 10th edition, and *Elements of geology*, 6th edition.] *Quarterly Review* 126: 359–94.

——. 1869c. *Der Malayische Archipel: die Heimath des Orang-Utan und des Paradiesvogels: Reiseerlebnisse und Studien über Land und Leute.* Translated by Adolf Bernhard Meyer. 2 vols. Brunswick: Westermann.

——. 1869d. The origin of species controversy. [Review of *Habit and intelligence, in their connection with the laws of matter and force*, by Joseph John Murphy.] *Nature* 1: 105–7, 132–3.

——. 1870. The measurement of geological time. *Nature* 1: 399–401, 452–55.

——. [1870?] *L'archipel malaisien: patrie de l'orang-outang et de l'oiseau de paradis: récits de voyage et étude de l'homme et de la nature.* 2 vols. Paris: Librarie Hachette.

——. 1870–1. *Insulinde: het land van den orang-oetan en den paradijsvogel.* Translated by P. J. Veth. 2 vols. Amsterdam: P. N. van Kampen.

——. 1876. *The geographical distribution of animals, with a study of the relations of living and extinct faunas as elucidating the past changes of the earth's surface.* 2 vols. London: Macmillan.

Wallich, George Charles. 1870. *Eminent men of the day.* Scientific series. London: John van Voorst.

Walmsley, Hugh Mulleneux. 1869. *The ruined cities of Zulu land.* 2 vols. London: Chapman & Hall.

Walsh, Benjamin Dann. 1864. On dimorphism in the hymenopterous genus *Cynips*, with an appendix, containing hints for a new classification of Cynipidæ and a list of Cynipidæ, including descriptions of several new species, inhabiting the oak-galls of Illinois. *Proceedings of the Entomological Society of Philadelphia* 3: 443–500.

——. 1866. On the insects, coleopterous, hymenopterous and dipterous, inhabiting the galls of certain species of willow.— Part 2d and last. *Proceedings of the Entomological Society of Philadelphia* 6: 223–88.

Walsh, Benjamin Dann and Riley, Charles Valentine. 1868a. The periodical cicada. *American Entomologist* 1: 63–72.

[——.] 1868b. The hateful or Colorado grasshopper. *American Entomologist* 1: 73–6.

[——.] 1869a. Galls and their architects. *American Entomologist* 1: 101–10.

[——.] 1869b. Wasps and their habits. *American Entomologist* 1: 122–43.

Walsh, John Henry ('Stonehenge'). 1875. *The greyhound: a treatise on the art of breeding, rearing and training greyhounds for public running.* New edition. London: Longmans, Green, and Co.

Walters, Stuart Max and Stow, E. Anne. 2001. *Darwin's mentor. John Stevens Henslow 1796–1861.* Cambridge: Cambridge University Press.

Walther, Alexander and Molendo, Ludwig. 1868. *Die Laubmoose Oberfrankens. Beiträge zur Pflanzengeographie und Systematik und zur Theorie vom Ursprunge der Arten.* Leipzig: Wilhelm Engelmann.

Wandruszka, Adam. 1958. *Geschichte einer Zeitung: Das Schicksal der 'Presse' und der 'Neuen*

Freien Presse' von 1848 zur Zweiten Republik. Vienna: Neue Wiener Presse Druck- und Verlagsgesellschaft.

Wedgwood, Barbara and Wedgwood, Hensleigh. 1980. *The Wedgwood circle, 1730–1897: four generations of a family and their friends.* London: Studio Vista.

Weinreich, Heinrich. 1993. *Duftstofftheorie: Gustav Jaeger (1832–1917); vom Biologen zum 'Seelenriecher'.* Stuttgart: Wissenschaftliche Verlagsgesellschaft.

Weir, John Jenner. 1869. On insects and insectivorous birds; and especially on the relation between the colour and the edibility of Lepidoptera and their larvae. [Read 1 March 1869.] *Transactions of the Entomological Society of London* (1869): 21–6.

Weismann, August. 1868. *Über die Berechtigung der Darwin'schen Theorie: ein akademischer Vortrag gehalten am 8 Juli 1868 in der Aula der Universität zu Freiburg im Breisgau.* Leipzig: W. Engelmann.

Wellesley index: *The Wellesley index to Victorian periodicals 1824–1900.* Edited by Walter E. Houghton *et al.* 5 vols. Toronto: University of Toronto Press. London: Routledge & Kegan Paul. 1966–89.

West, David A. 2003. *Fritz Müller. A naturalist in Brazil.* Blacksburg, Va.: Pocahontas Press.

Westring, Niklas. 1861. *Araneae svecicae: descriptae.* Gothenburg: D. F. Bonnier.

Wheeler, George Augustus and Wheeler, Henry Warren. 1878. *History of Brunswick, Topsham, and Harpswell, Maine, including the ancient territory known as Pejepscot.* Boston: Alfred Mudge and Son.

Whitaker, William. 1867. On subaërial denudation, and on cliffs and escarpments of the chalk and lower tertiary beds. *Geological Magazine* 4: 447–54, 483–93.

White, Gilbert. 1789. *The natural history and antiquities of Selborne, in the county of Southampton.* 2 vols. London.

——. 1825. *The natural history of Selborne.* New edition. 2 vols. London.

Whiting, Sydney. 1855. *Heliondé; or, adventures in the sun.* London: Chapman and Hall.

Wilckens, Martin. 1866. Darwin's Theorie in Beziehung zur landwirthschaftl. Thierzucht. *Jahrbuch der deutschen Thierzucht* (1866): 67–93.

——. 1869. Bericht über die wissenschaftlichen Fortschritte der Thierzucht im Jahre 1867/68. *Jahrbuch der Landwirthschaft* (1869): 299–360.

Willis, John Christopher. 1973. *A dictionary of the flowering plants and ferns.* 8th edition. Revised by H. K. Airy Shaw. Cambridge: Cambridge University Press.

Wilson, Dudley. 1993. *Signs and portents. Monstrous births from the Middle Ages to the Enlightenment.* London and New York: Routledge.

Wingfield, William and Johnson, George William. 1856–7. *The poultry book; comprising the characteristics, management, breeding, and medical treatment of poultry; being the results of personal observation and the practice of the best breeders, including Captain W. W. Hornby, R N.; Edward Bond, Esq.; Thomas Sturgeon, Esq.; Charles Punchard, Esq.; Edward Hewitt, Esq.; and others.* Another edition. Rearranged and edited by W. B. Tegetmeier. London: Wm. S. Orr and Co.

Winslow, Forbes. 1842. *On the preservation of the health of body and mind.* London: Renshaw.

Winsor, Mary Pickard. 1991. *Reading the shape of nature. Comparative zoology at the Agassiz museum.* Chicago and London: University of Chicago Press.

Wollaston, Alexander Frederick Richmond. 1921. *Life of Alfred Newton, professor of comparative anatomy, Cambridge University, 1866–1907.* With a preface by Sir Archibald Geikie. London: John Murray.

Wollaston, Thomas Vernon. 1860. On certain musical *Curculionidæ*; with descriptions of two new *Plinthi*. *Annals and Magazine of Natural History* 3d ser. 6: 14–19.

Wood, Roger J. and Orel, Vítězslav. 2001. *Genetic prehistory in selective breeding: a prelude to Mendel.* Oxford: Oxford University Press.

Woodward, Horace B. 1907. *The history of the Geological Society of London.* London: Geological Society.

Wright, Edward Perceval. 1868. Contributions towards a knowledge of the flora of the Seychelles. [Read 14 December 1868.] *Transactions of the Royal Irish Academy* 24 (1871): 571–8.

Würffel, Rheinhard. 2000. *Lexikon deutscher Verlage von A–Z: 1071 Verlage und 2800 Verlagssignete vom Anfang der Buchdruckerkunst bis 1945; Adressen, Daten, Fakten, Namen.* Berlin: Verlag Grotesk.

WWW: *Who was who: a companion to Who's who, containing the biographies of those who died during the period* [*1897–1995*]. 9 vols. and cumulated index (1897–1990). London: Adam & Charles Black. 1920–96.

Yack, Jayne E. *et al.* 2000. Sound production and hearing in the blue cracker butterfly *Hamadryas feronia* (Lepidoptera, Nymphalidae) from Venezuela. *Journal of Experimental Biology* 203: 3689–3702.

Yaldwyn, John and Hobbs, Juliet, eds. 1998. *My dear Hector. Letters from Joseph Dalton Hooker to James Hector, 1862–1893.* Museum of New Zealand Te Papa Tongarewa Technical Report 31.

Zoology: *The zoology of the voyage of HMS Beagle, under the command of Captain FitzRoy RN, during the years 1832 to 1836.* Edited and superintended by Charles Darwin. 5 pts. London: Smith, Elder & Co. 1838–43.

NOTES ON MANUSCRIPT SOURCES

The majority of the manuscript sources cited in the footnotes to the letters are either in the Darwin Archive, Cambridge University Library, or at Down House, Downe, Kent. Further details about the Darwin Archive are available in the *Handlist of Darwin papers at the University Library Cambridge* (Cambridge: Cambridge University Press, 1960) and the unpublished supplementary handlist available at the library; a new catalogue of the papers is currently being prepared. Further details about the manuscripts at Down House are available in Philip Titheradge, ed. *The Charles Darwin Memorial at Down House, Downe, Kent*, revised ed. ([Downe: Down House Museum], 1981) and from the curator (The Curator, Down House, Downe, Kent, BR6 7JT). In addition, there are a number of named sources that are commonly used in the footnotes: for each of these, the editors have provided brief descriptive notes.

CD's Account books (Down House MS). This series of seventeen account books begins on 12 February 1839, a fortnight after CD and Emma's marriage, and ends with CD's death. The books contain two sets of accounts. From the start, CD recorded his *cash account* according to a system of double-entry book-keeping. On each left-hand page he recorded credits (i.e., withdrawals from the bank, either in the form of cash paid to himself or cheques drawn for others), and on each right-hand page he recorded debits (i.e., cash or cheques paid to others). CD also recorded details of his *banking account* from the start, but only noted them down in a single column at the bottom of the left-hand page of his cash account. In August 1848, however, he began a system of detailing his banking account according to double-entry book-keeping, in a separate chronological section at the back of each account book. On the left, he recorded credits to the account in the form of income (i.e., investments, rent, book sales, etc.). On the right, he recorded debits to the account (i.e., cash or cheque withdrawals).

CD's Classed account books (Down House MS). This series of four account books, covering the years 1839–81, runs parallel to CD's Account books. For each year, September–August (after 1867, January–December), CD divided his expenditure into different classes; in addition, he made a tally for the year of his income, expenditure, cash in hand, and money in the bank. From 1843, CD also compiled at the back of each book a separate account of the total expenditure under the various headings in each year, and from 1844 he added a full account of his income in each year, and of capital invested and 'paid' up.

CD's Experimental notebook (DAR 157a). This notebook contains notes on some of the experiments carried out between 13 November 1855 (with some back references) and 20 May 1868; the majority of the notes date from before

1863. Often only the details of the experiment attempted are given, usually with cross-references to results recorded in CD's portfolios of notes. The notebook also contains a number of letters to CD.

CD's Investment book (Down House MS). This book records for each of CD's investments the income received during the period 1846–81.

CD's 'Journal'. *See* Appendix II.

CD's Library catalogue (DAR 240). This manuscript catalogue of CD's scientific library was compiled by Thomas W. Newton, assistant librarian of the Museum of Practical Geology, London, in August 1875. Additions to the catalogue were later made by Francis Darwin (who inherited most of his father's scientific library) and by H. W. Rutherford, who apparently used this catalogue as a basis for compiling his *Catalogue of the library of Charles Darwin now in the Botany School, Cambridge* (Cambridge: Cambridge University Press, 1908). However, there are items listed in this manuscript catalogue that do not appear in Rutherford's published catalogue, and which must have been dispersed after being listed.

Down Coal and Clothing Club account book (Down House MS). CD was for some years treasurer of this charitable organisation. The account book records subscriptions made by honorary subscribers between 1841 and 1876; between 1848 and 1869 the entries are in CD's handwriting. For the years 1841–8 and 1868–76, there is also a statement of expenditures, though not in CD's handwriting.

Emma Darwin's diary (DAR 242). This is a series of small pocket diaries, in which Emma recorded details of the health of family members, trips made by herself, CD, and their children, school holidays, and visits to Down by others. The collection at CUL comprises diaries for the years 1824, 1833–4, 1839–45, and 1848–96.

H. E. Litchfield's autobiography (DAR 246). This unfinished autobiography, written in 1926 on forty-two loose leaves, and chiefly concerning Henrietta Emma Darwin's childhood, has never been published.

List of pamphlets (DAR 252.4). This is a catalogue of CD's pamphlet collection prepared by CD and Francis Darwin in 1878 (see the letter from Emma Darwin to Henrietta Emma Litchfield, [June 1878] (DAR 219.9: 175)). From about 1878 CD began to arrange the articles, papers, and reprints he received into a numbered collection. CD maintained this reprint collection until his death, when it was taken over by Francis Darwin. Francis continued the collection, adding new items, the numbers running consecutively from those of his father. Evidently, until this catalogue was prepared, CD used a working index similar to that of his 'List of reviews'. The catalogue is in two sections, a list of the quarto collection and one of the general collection. Both sections are alphabetically arranged with the entries pasted on sheets in a loose-leaf folder.

List of reviews (DAR 262.8: 9–18 (English Heritage MS: 88206151–60)). This manuscript, headed 'List Reviews of Origin of Sp & of C. Darwins Books', was CD's working index to his collection of reviews of his own books. It corresponds approximately to the review collection in the Darwin Pamphlet Collection–CUL, but includes some items that were dispersed after being listed.

Reading notebooks. *See Correspondence* vol. 4, Appendix IV. These notebooks are divided into sections entitled 'Books Read' and 'Books to be Read'. CD's entries in 'Books Read' often include a brief opinion of the work.

Scrapbook of reviews (DAR 226.1 and 226.2). Many of the reviews contained in these two volumes bear CD's annotations and thus were evidently collected by CD. However, the scrapbook seems to have been assembled by Francis Darwin: the tables of contents are in the handwriting of H. W. Rutherford, an assistant at Cambridge University Library who acted as a copyist for Francis on several occasions (see *ML*, 1: x, and Francis Darwin, ed. *The foundations of the Origin of Species. Two essays written in 1842 and 1844 by Charles Darwin* (Cambridge: Cambridge University Press, 1909)). In addition, the scrapbook is identified as Francis's in a note (DAR 226.1:132a) made in 1935 by Arthur Keith, whose appeal led to the purchase of Down House as a Darwin memorial (see Arthur Keith, *An autobiography* (London: Watts & Co., 1950)). DAR 226.1 bears the inscription 'Reviews of C. Darwin's works' on the spine, and contains, among others, reviews of *Origin* and *Orchids*; DAR 226.2 is inscribed: 'Reviews. Descent. Expression. Insect. Pl. Eras. D.'

W. E. Darwin's botanical notebook (DAR 117). This notebook contains observational and experimental notes on plants made by William, often in consultation with CD. The first observation bears the date 13 July 1862, and, although the date of the last observation is 26 June 1870, most of the notes were made between 1862 and 1864. The notebook originally contained letters from CD, but these were later removed. William entered notes made from botanical textbooks in a separate notebook (DAR 234).

W. E. Darwin's botanical sketchbook (DAR 186: 43). This sketchbook, which contains entries dated 1862–72, was evidently begun in parallel to William's botanical notebook. It contains ink drawings of various parts of plants, and of sections, together with descriptions, which are sometimes very extensive.

Index

The dates of letters to and from Darwin's correspondents are listed in the biographical register and are not repeated here. Darwin's works are indexed under the short titles used throughout this volume and listed in the bibliography.

Table of

- Robert Darwin 1682–1754 = Elizabeth Hill 1702–97
  - William Alvey Darwin 1726–83 = Jane Brown 1746–1835
    - Samuel Fox 1765–1851 = Ann 1771–1859
      - Mary Ann 1800–29 = Samuel Ellis Bristowe 1800–55
      - Eliza 1801–86
      - Emma 1803–85
      - William Darwin 1805–80 =(1) Harriet Fletcher 1799–1842; =(2) Ellen Sophia Woodd 1820–87
      - Frances Jane b.1806 = John Hughes 1794–1873
      - Julia b.1809
  - Erasmus Darwin 1731–1802 =(1) Mary Howard 1740–70
    - Charles 1758–78
    - Erasmus 1759–99
    - Robert Waring 1766–1848 =
      - Marianne 1798–1858 = Henry Parker 1788–1856
      - Susan Elizabeth 1803–66
      - Erasmus Alvey 1804–81
      - Emily Catherine 1810–66
  - Erasmus Darwin =(2) Elizabeth Collier Pole 1747–1832
    - Edward 1782–1829
    - Emma 1784–1818
    - Francis Sacheverel 1786–1859 = Jane Harriett Ryle 1794–1866
    - John 1787–1818
    - Harriot 1790–1825 = Thomas James Maling 1778–1849
    - Frances Anne Violetta 1783–1874 = Samuel Tertius Galton 1783–1844
      - Elizabeth Ann (Bessy) 1808–1906
      - Lucy Harriot 1809–48
      - Milicent Adèle 1810–83
      - Emma Sophia 1811–1904
      - Darwin 1814–1903
      - Erasmus 1815–1909
      - Francis 1822–1911 = Louisa Jane Butler d.1897